Quantum Science and Technology

Aims and Scope

The book series Quantum Science and Technology is dedicated to one of today's most active and rapidly expanding fields of research and development. In particular, the series will be a showcase for the growing number of experimental implementations and practical applications of quantum systems. These will include, but are not restricted to: quantum information processing, quantum computing, and quantum simulation; quantum communication and quantum cryptography; entanglement and other quantum resources; quantum interfaces and hybrid quantum systems; quantum memories and quantum repeaters; measurement-based quantum control and quantum feedback; quantum nanomechanics, quantum optomechanics and quantum transducers; quantum sensing and quantum metrology; as well as quantum effects in biology. Last but not least, the series will include books on the theoretical and mathematical questions relevant to designing and understanding these systems and devices, as well as foundational issues concerning the quantum phenomena themselves. Written and edited by leading experts, the treatments will be designed for graduate students and other researchers already working in, or intending to enter the field of quantum science and technology.

More information about this series at http://www.springer.com/series/10039

Felipe Fernandes Fanchini
Diogo de Oliveira Soares Pinto
Gerardo Adesso
Editors

Lectures on General Quantum Correlations and their Applications

Springer

Editors
Felipe Fernandes Fanchini
São Paulo State University
São Paulo
Brazil

Diogo de Oliveira Soares Pinto
São Carlos Institute of Physics
University of São Paulo
São Paulo
Brazil

Gerardo Adesso
School of Mathematical Sciences
University of Nottingham
Nottingham
UK

ISSN 2364-9054 ISSN 2364-9062 (electronic)
Quantum Science and Technology
ISBN 978-3-319-53410-7 ISBN 978-3-319-53412-1 (eBook)
DOI 10.1007/978-3-319-53412-1

Library of Congress Control Number: 2017931538

Printed on acid-free paper

This Springer imprint is published by Springer Nature
The registered company is Springer International Publishing AG
The registered company address is: Gewerbestrasse 11, 6330 Cham, Switzerland

Foreword

It is by now widely appreciated that quantum physics enables tasks that are much more difficult or even impossible to accomplish by relying on purely classical phenomena. The applications that benefit from "quantumness" include information processing (e.g., computing), high-precision measurements, and communications. The strategies that lead to these benefits vary, and their practical implementations often have to contend with limitations that are also quantum-specific (e.g., prohibition on cloning or decoherence) and can be "friend or foe", depending on the attempted implementation.

The obvious question that arose with the inception of the field quantum information concerns the "magic ingredient", the precise aspect of quantum theory that is responsible for this "quantum advantage". Quantum entanglement was the early suspect, and it remains in a lineup of possible culprits. By now, it is however abundantly clear that the list of possible suspects should be extended to include other, less flagrantly quantum correlations (e.g, these with non-zero discord). For example, while there are quantum algorithms that depend on entanglement in at least part of their execution, to date there is no general proof that entanglement is indispensable in every quantum computation. Moreover, in a number of relevant settings including quantum estimation and communication, one can get an advantage by suitably exploiting discord even in absence of entanglement.

This volume presents a collection of chapters authored by the leading contributors to the study of general quantum correlations and their applications and is organized in four parts. The first part of the book discusses the foundations of quantum correlations beyond entanglement and how to characterize them. Various forms of quantumness of states are exemplified and several quantitative methods are introduced, considering both bipartite and multipartite cases. In the second part of the book, operational interpretations and applications of quantum correlations come into play. Approaches from broadcasting and distribution of correlations to quantum metrology and cryptography are revisited, elucidating in particular the importance of such correlations for quantum information processing. In the third part of the book, the role of quantum correlations in the dynamics of open systems is explored. Sudden-change phenomena, robustness to decoherence and revivals are among the topics discussed, as well as quantumness indicators in interference and synchronization effects. Finally, the last part of the book revolves around physical realizations and experimental demonstrations. Investigations of quantum correlations in different physical systems like nuclear magnetic resonance, solid-state spin systems, and optical systems are reported.

The whole book samples the diversity of approaches found in the literature along the past years of research and overall constitutes a comprehensive source to guide and inspire both experienced researchers and beginners in the fascinating field of quantum correlations.

Felipe Fernandes Fanchini
Diogo de Oliveira Soares Pinto
Gerardo Adesso
Wojciech H. Zurek

Contents

Part I
Measures of Quantum Correlations Beyond Entanglement

Foundations of Quantum Discord

Vlatko Vedral

Abstract This paper summarizes the basics of the notion of quantum discord and how it relates to other types of correlations in quantum physics. We take the fundamental information theoretic approach and illustrate our exposition with a number of simple examples.

1 Introduction

In order to understand quantum discord we need to first understand the difference between classical and quantum correlations as quantified by mutual information. Mutual information is originally a classical measure of correlations. It is defined as

$$I(A, B) = H(A) + H(B) - H(A, B) \tag{1}$$

where $S = -\sum_n p_n \log p_n$ is the Shannon entropy [1] and A and B are two random variables whose probability distribution is given. There is a different way of writing this quantity using the conditional entropy, $H(A/B) = H(A, B) - H(B)$, namely,

$$I(A, B) = H(A) - H(A/B) \tag{2}$$

V. Vedral (✉)
Clarendon Laboratory, University of Oxford, Parks Road, Oxford OX1 3PU, UK
e-mail: vlatko.vedral@qubit.org

V. Vedral
Centre for Quantum Technologies, National University of Singapore,
3 Science Drive 2, Singapore 117543, Singapore

V. Vedral
Department of Physics, National University of Singapore,
2 Science Drive 3, Singapore 117542, Singapore

V. Vedral
Center for Quantum Information, Institute for Interdisciplinary Information Sciences,
Tsinghua University, Beijing 100084, China

F.F. Fanchini et al. (eds.), *Lectures on General Quantum Correlations and their Applications*, Quantum Science and Technology,
DOI 10.1007/978-3-319-53412-1_1

Therefore, there are two equivalent ways of thinking about classical correlations (there are many more, but they are not necesarily relevant to our topic). One is that they are measured by the difference in the sum of local entropies and the total entropy and the other one is that they tell us by how much we can reduce the entropy of one random variable by measuring the other.

When generalizing the concept of mutual information to quantum physics, it is straightforward to do it using the first expression. All we need do is use the von Neumann instead of the Shannon entropy. The quantum mutual information is defined as

$$I_Q = S(A) + S(B) - S(A, B) \tag{3}$$

where $S(A) = -tr\rho_A \log \rho_A$ and the subscript Q just indicates that this is a quantum measure. The second expression for the mutual information, involving the conditional entropy, is, however, harder to upgrade to quantum physics. The reason is that if we do the same substitution of the Shannon with the von Neumann entropy, the resulting entity

$$S(A, B) - S(B) \tag{4}$$

can actually be negative for bipartite quantum systems. Take any pure entangled state of two subsystems (such as $a|00\rangle + b|11\rangle$) and its total entropy vanishes, while the reduced entropy is non-zero. It is therefore hard to call this quantum conditional entropy since the quantity can be negative (and given that entropy quantifies disorder it is hard to see how disorder can be smaller than zero).

A way out of this is to define the quantum conditional entropy S(A/B) as the average entropy $\sum_n p_n S(\rho_{A,n})$ of states of A after a measurement is made on B. There are infinitely many measurements we can perform on B, so we will choose the one that makes $S(A/B)$ minimum (we want to learn as much about A by measuring B). It is clear that the upper bound on this quantum conditional entropy is $S(A)$, while the lower bound is zero.

So $S(A/B)$ is always positive when defined as the average entropy reduction, but now we have another problem. The quantum mutual information defined above is not equal to $S(A) - S(A/B)$! Unlike in classical information, the two ways of expressing quantum mutual information are actually different.

This is because the quantum mutual information can actually reach the value of $2S(A)$, while the quantity to $S(A) - S(A/B)$ can at most be $S(A)$. What does the difference between two quantum quantities (I_Q and $S(A) - S(A/B)$) signify, if anything?

This question was first asked by Lindblad [2] (he phrased is slightly differently but that was the spirit). His answer was that the difference is actually due to quantum entanglement (more precisely, he says: "This extra correlation is of course the cause of the Einstein-Podolsky-Roscn "paradox" and is thus a typical quantum effect."). And for pure states he was perfectly correct (as we will see below), but at that time there was virtually no work on mixed state entanglement (which only properly took off in the mid-nineties) and so it was difficult for him to anticipate many subtleties involved.

2 Discord: What It Is and How to Quantify It

I came upon this difference of mutual informations after my initial work on entanglement because I was asking if the quantum mutual information, which quantifies all correlations, can actually be written as a sum of entanglement and classical correlations. I defined classical correlations as $C(A, B) = S(A/B)$ and I quantified entanglement using the relative entropy of entanglement $E(A, B)$.

My then student, Leah Henderson, and I discovered that the sum $C + E$ is mostly smaller than I_Q for mixed states [3]. In other words, there is more to quantum correlations than just entanglement when it comes to mixed states. For pure states, entanglement and classical correlations are equal to one another and the sum is then exactly equal to the quantum mutual information which explains why quantum mutual information is twice as big as the classical mutual information (as anticipated by Lindblad).

A few months later, Ollivier and Zurek [4] wrote a paper where they named this difference between the two ways of defining quantum mutual information quantum discord. They defined it slightly differently as they had the open system setting in mind, but I do not wish to enter any subtleties in this introductory article (an interested reader is encouraged to consult the review in [5]). Physically, quantum discord, according to Zurek [6], represents the difference between the efficiency of classical and quantum Maxwell's demons, while in other interpretations it has also been linked to the fidelity of remote state preparation as well as to the difference in information extraction by local and global means (mathematically, at least, a protocol that is somewhat related to the Maxwell's demon interpretation).

So discord seems to measure quantum correlations that go beyond just entanglement. Disentangled states can actually possess non-zero quantum discord. But is discord really a form of correlation? To answer that, we need to discuss an important property of any measure of correlation.

One of the features of correlations is that they cannot increase by local operations (LO). If we do something to A alone, and, independently, to B, we should not be able to correlate them to a higher degree than we started with. The intuition behind this is clear: we cannot correlate things more unless we are allowed to act on them jointly. Any separate action can only degrade the initial correlation (or, at best, preserve it).

Both mutual information and entanglement are decreasing under LO (entanglement, in fact, under an even more general class, but this need not concern us here [7]). This is actually straightforward to prove if we express both in terms of the quantum relative entropy $S(\sigma||\rho) = tr(\sigma \log \sigma - \sigma \log \rho)$. Entanglement is then the relative entropy to the closest disentangled (separable) state [7, 8] (whose form is $\sum_n p_n \rho_A^n \otimes \rho_B^n$, where p_n is any probability distribution), while the quantum mutual information is the relative entropy to the closest product states (which happens to be the product of the reduced states $\sigma_A \otimes \sigma_B$). Since quantum relative entropy is monotonic under completely positive maps and as the sets of product states and separable states are invariant under LO (which is a special set of completely positive maps), it follows that the quantum mutual information and entanglement are monotones (non-increasing) under LO (all the relevant proofs can be found in [9]).

However, monotonicity under LO is not true for discord! We can start with a state with no discord and actually create some by LO. A simple example is a stating state which is an equal mixture of $|00\rangle$ and $|11\rangle$, which can be converted by LO into a mixture of $|00\rangle$ and $|1+\rangle$. It is therefore hard to think of discord as a form of correlations. Also, given that it can be created by local means, it is questionable if we can think of (all) discordant states as useful for quantum information processing. Having said this, there are examples of protocols where discord has an operational meaning [10, 11]. It is also still an open question if universal quantum computation can be done without entanglmenet in the general case of mixed states. Maybe not all, but certainly some kind of discord could be of importance.

This does not mean that we cannot quantify discord using the quantum relative entropy. We can take the relative entropy from a given state to the closest classically correlated state. This set, however, is not invariant under LOs which is why this measure fails to be a monotone (as examplified in the previous paragraph).

3 Outlook

One should emphasise that though this article has dealt with bipartite systems only (for clarity, as well as for historical reasons), correlation measures can be generalized to many partite systems (see e.g. [12] for entanglement in many-body systems and [5] for discord and related measures). A way to do that is using the same relative entropy based logic outlines above (see also [13] for a unified view of all correlations based of the quantum relative entropy).

Also, we did not discuss how we can tell if a given state has discord. The method is simple and it boils down to showing that correlations are non-vanishing in more than one basis [14]. Classical correlations, according to this logic, are the ones that exist only in one basis (though this basis could be different for different subsystems, depending on how they couple to their environments, for instance).

In conclusion, discord without entanglement can be seen as a form of classical correlation aided with quantum coherence (superpositions) at the level of individual subsystems. This is why the research on discord has naturally led to the research on quantifying quantum coherence.

Acknowledgements The author acknowledges funding from the John Templeton Foundation, the National Research Foundation (Singapore), the Ministry of Education (Singapore), the Engineering and Physical Sciences Research Council (UK), the Leverhulme Trust, the Oxford Martin School, and Wolfson College, University of Oxford. This research is also supported by the National Research Foundation, Prime Ministers Office, Singapore under its Competitive Research Programme (CRP Award No. NRF- CRP14-2014-02) and administered by Centre for Quantum Technologies, National University of Singapore.

References

1. C. Shannon, Bell Syst. Tech. J. **27**(379–423), 623–656 (1948)
2. G. Lindbad, Commun. Math. Phys. **33**, 305 (1973)
3. L. Henderson, V. Vedral, J. Phys. A: Math. Gen. **34**, 6899 (2001)
4. H. Ollivier, W.H. Zurek, Phys. Rev. Lett. **88**, 017901 (2001)
5. K. Modi, A. Brodutch, H. Cable, T. Paterek, V. Vedral, Rev. Mod. Phys. **84**, 1655 (2012)
6. W.H. Zurek, Phys. Rev. A **67**, 012320 (2003)
7. V. Vedral, M.B. Plenio, M. Rippin, P.L. Knight, Phys. Rev. Lett. **78**, 2275 (1997)
8. V. Vedral, M.B. Plenio, Phys. Rev. A **57**, 1619 (1998)
9. V. Vedral, Rev. Mod. Phys. **74**, 197 (2002)
10. B. Dakic et al., Nat. Phys. **8**, 666 (2012)
11. M. Gu et al., Nat. Phys. **8**, 671 (2012)
12. L. Amico, R. Fazio, A. Osterloh, V. Vedral, Rev. Mod. Phys. **80**, 1 (2008)
13. K. Modi, T. Paterek, W. Son, V. Vedral, M. Williamson, Phys. Rev. Lett. **104**, 080501 (2010)
14. B. Dakic, V. Vedral, C. Brukner, Phys. Rev. Lett. **105**, 190502 (2010)

From Discord to Entanglement

Shunlong Luo

Abstract Two prominent and widely studied notions of quantum correlations are discord and entanglement, with the latter occupying a central place in quantum information theory, while the former being regarded of marginal significance and even being criticized by some researchers, although the deep relations between them have been revealed in recent years. Discord and entanglement, being indistinguishable for pure states, only differ for mixed states. The aim of this work is to subsume entanglement under discord by identifying entanglement as the minimal shadow of discord over extended systems. For this purpose, we first present a brief and concise review of some historical aspects of discord and entanglement, emphasizing the ideas leading to them and the intimate relations between them. Then by exploiting an intrinsic connection between classicality and separability of correlations, we derive entanglement from discord in terms of state extensions, and put discord in a more primitive place than entanglement in this context. We comment that the entanglement of pure states studied by EPR and Schrödinger can actually also be well understood as discord, only with the emergence of nonlocality characterized by the Bell inequalities involving mixed states rather than pure states, the LOCC paradigm for mixed-state entanglement becomes significant and attracts great interests. Discord and entanglement are different manifestations of the same global quantum substrate, with discord conceptually more ubiquitous in quantum information and more deeply rooted in quantum measurements.

1 Introduction

Correlations permeate our interpretation and understanding of the physical world. To extract correlation information from physical systems, whether classical or quantum,

S. Luo (✉)
Academy of Mathematics and Systems Science, Chinese Academy of Sciences,
Beijing 100190, People's Republic of China
e-mail: luosl@amt.ac.cn

S. Luo
School of Mathematical Sciences, University of Chinese Academy of Sciences,
Beijing 100490, People's Republic of China

F.F. Fanchini et al. (eds.), *Lectures on General Quantum Correlations and their Applications*, Quantum Science and Technology,
DOI 10.1007/978-3-319-53412-1_2

correlations in the bivariate probability distribution $p^{ab} = \{p_{ij}^{ab} = p_i^a \delta_{ij}\}$, or in its quantum formalism, $\rho^{ab} = \sum_i p_i^a |i\rangle_a\langle i|_a \otimes |i\rangle_b\langle i|$ with $\{|i\rangle_a\}$ and $\{|i\rangle_b\}$ orthonormal bases for parties a and b, respectively. The amount of total correlations coincides with the marginal entropy, i.e.,

$$I(\rho^{ab}) = I(p^{ab}) = H(p^a) = S(\rho^a).$$

This is also the amount of classical correlations, and there are no quantum correlations here. In contrast, for the quantum case, consider the quantum pure state $\sigma^{ab} = |\Psi^{ab}\rangle\langle\Psi^{ab}|$ with the Schmidt decomposition $|\Psi^{ab}\rangle = \sum_i \sqrt{p_i^a}|i\rangle_a \otimes |i\rangle_b$ and the marginal $\sigma^a = \mathrm{tr}_b|\Psi^{ab}\rangle\langle\Psi^{ab}| = \rho^a$, the amount of total correlations is

$$I(\sigma^{ab}) = 2S(\sigma^a) = 2H(p^a) = 2S(\rho^a).$$

The extra amount of correlations in the quantum case, $I(\sigma^{ab}) - I(\rho^{ab}) = H(p^a)$, is the root lurking in the EPR argument and the state steering [2, 5–7].

The total correlations in a classically correlated state can be fully extracted by certain measurements, but this is not true for genuinely quantum correlated states. To see this and to facilitate the comparison between the classical and the quantum, we cast the classical bivariate probability distribution $p^{ab} = \{p_{ij}^{ab}\}$ in the quantum formalism as

$$\rho^{ab} = \sum_{ij} p_{ij}^{ab}|i\rangle_a\langle i| \otimes |j\rangle_b\langle j|$$

where $\{|i\rangle_a\}$ and $\{|j\rangle_b\}$ are orthonormal bases for parties a and b, respectively. The amount of total correlations in this state, as quantified by the quantum mutual information $I(\rho^{ab})$, coincides with the Shannon mutual information $I(p^{ab})$ in the bivariate probability distribution p^{ab}, i.e., $I(\rho^{ab}) = I(p^{ab})$. This can be interpreted as that all correlations in $\rho^{ab} = \sum_{ij} p_{ij}^{ab}|i\rangle_a\langle i| \otimes |j\rangle_b\langle j|$ are classical, and there are no quantum correlations in this state. Indeed, the state ρ^{ab} is left undisturbed after the local von Neumann measurements $\Pi^a = \{\Pi_i^a := |i\rangle_a\langle i|\}$ and $\Pi^b = \{\Pi_j^b := |j\rangle_b\langle j|\}$ by parties a and b, respectively, in the sense that $\rho^{ab} = \Pi(\rho^{ab})$, where

$$\Pi(\rho^{ab}) := \sum_{ij}(\Pi_i^a \otimes \Pi_j^b)\rho^{ab}(\Pi_i^a \otimes \Pi_j^b)$$

is the post-measurement state. All the correlations in this state are extracted by these measurements.

A characteristic feature of classicality is the invariance under certain quantum measurements. In contrast, disturbance under quantum measurements signifies quantumness. In the context of correlations, one may define a bipartite state to be classically correlated if it is left undisturbed under certain von Neumann measurement [70]. More precisely, for a bipartite state σ^{ab}, if there exist local von Neumann measurements $\{\Pi_i^a\}$ and $\{\Pi_j^b\}$ such that $\sigma^{ab} = \Pi(\sigma^{ab}) := \sum_{ij}(\Pi_i^a \otimes \Pi_j^b)\sigma^{ab}(\Pi_i^a \otimes \Pi_j^b)$, then σ^{ab} can be considered as a classically correlated state, and the correlations therein can be fully extracted without loss. In this case, σ^{ab} can be identified with the classical bivariate probability distribution $p^{ab} = \{p^{ab}_{ij} := \mathrm{tr}(\Pi^a_i \otimes \Pi^b_j)\sigma^{ab}\}$.

We have the following equivalent characterizations for the classically correlated states, which justify the notion of classicality of correlations [70–72]:

(1) σ^{ab} is classically correlated.
(2) σ^{ab} can be represented as $\sigma^{ab} = \sum_{ij} p^{ab}_{ij} \Pi^a_i \otimes \Pi^b_j$, where $p^{ab} = \{p^{ab}_{ij}\}$ is a bivariate probability distribution, Π^a_i and Π^b_j are orthogonal projections for parties a and b, respectively [70].
(3) σ^{ab} commutes with each $\Pi^a_i \otimes \Pi^b_j$, where Π^a_i and Π^b_j are the spectral projections of the reduced states $\sigma^a = \mathrm{tr}_b \sigma^{ab}$ and $\sigma^b = \mathrm{tr}_a \sigma^{ab}$, respectively [70].
(4) The correlations in σ^{ab} can be locally broadcast [71].
(5) Both parties a and b can establish perfect correlations with other systems [72].

Although a state σ^{ab} may not be classically correlated, the post-measurement state $\Pi(\sigma^{ab}) := \sum_{ij} (\Pi^a_i \otimes \Pi^b_j)\sigma^{ab}(\Pi^a_i \otimes \Pi^b_j)$ is always a classical state after any local von Neumann measurement $\Pi = \{\Pi^a_i \otimes \Pi^b_j\}$. By the monotonicity of quantum relative entropy [65],

$$I(\Pi(\sigma^{ab})) \leq I(\sigma^{ab}),$$

and the difference $I(\sigma^{ab}) - I(\Pi(\sigma^{ab}))$ signifies the loss caused by the measurements and captures quantumness of correlations.

Similarly, one may also define classicality of correlations with respect to one party. More precisely, one defines σ^{ab} to be classical-quantum if there exists a local von Neumann measurement $\Pi^a = \{\Pi^a_i\}$ for party a which leaves the state undisturbed in the sense that $\sigma^{ab} = \Pi^a(\sigma^{ab})$, where

$$\Pi^a(\sigma^{ab}) := \sum_i (\Pi^a_i \otimes \mathbf{1}^b)\sigma^{ab}(\Pi^a_i \otimes \mathbf{1}^b)$$

is the post-measurement state after party a performs the quantum measurement Π^a. Analogously, the following characterizations of classical-quantum states are equivalent [70, 73]:

(1) σ^{ab} is classical quantum.
(2) σ^{ab} can be represented as $\sigma^{ab} = \sum_i p_i \Pi^a_i \otimes \sigma^b_i$, where $\{p_i\}$ is a probability distribution, Π^a_i are orthogonal projections for party a, and σ^b_i are local states for party b.
(3) σ^{ab} commutes with each $\Pi^a_i \otimes \mathbf{1}^b$, where Π^a_i are the spectral projections of $\sigma^a := \mathrm{tr}_b \sigma^{ab}$.
(4) The correlations in σ^{ab} can be locally broadcast by party a [73].

In general, a classical-quantum state may not be classically correlated due to the non-commutativity of σ^b_i for party b, and it is impossible to identify such a state with a classical bivariate probability distribution in general.

$S(\rho^{ab}|\rho^b) := S(\rho^{ab}) - S(\rho^b)$ is the quantum conditional entropy, $\mathcal{C}^b(\rho^{ab})$ and $\mathcal{Q}^b(\rho^{ab})$, similar to $C^b(\rho^{ab})$ and $Q^b(\rho^{ab}) = I(\rho^{ab}) - C^b(\rho^{ab})$, are the measures of classical correlations and discord defined in terms of general POVMs rather than von Neumann measurements [12].

3 Entanglement as Discord

A remarkable relation between the two classification schemes for correlations, classical/quantum [70, 71] and separable/entanglement [52], is that on one hand, a classically correlated state is always separable, on the other hand, any separable state can be imbedded into a classically correlated state in the sense that for any separable state ρ^{ab}, there is a classically correlated state $\rho^{a'a:bb'}$ shared between aa' and bb' such that

$$\rho^{ab} = \mathrm{tr}_{a'b'}\rho^{a'a:bb'},$$

where a' and b' are two ancillary systems [82]. Any entangled state does not admit such an extension. Phrased alternatively, a bipartite state is separable if and only if it admits an extension which is classically correlated with the natural bipartition, i.e., with a and b in different parties. This identifies entanglement as truly nonlocal quantum correlations, and has some interesting consequences [83–85]. Here we will exploit it to define entanglement in terms of discord. More precisely, for any reasonable measure of discord $\mathcal{Q}(\cdot)$, not necessary defined in terms of the quantum mutual information as the original one, we define

$$\mathcal{E}(\rho^{ab}) := \min_{\mathrm{tr}_{a'b'}\rho^{a'a:bb'}=\rho^{ab}} \mathcal{Q}(\rho^{a'a:bb'}),$$

where the minimization is over all state extensions $\rho^{a'a:bb'}$ of ρ^{ab} (i.e., $\rho^{ab} = \mathrm{tr}_{a'b'}\rho^{a'a:bb'}$), including the cases when a' or b' is trivial (one dimensional), and the discord $\mathcal{Q}(\rho^{a'a:bb'})$ is taken with respect to the bipartition $a'a : bb'$. This renders entanglement to a kind of shadow of discord, i.e., the minimal discord over state extensions.

Clearly, $\mathcal{E}(\rho^{ab}) = 0$ for separable ρ^{ab}. This follows from the theorem in Li and Luo [82] concerning the relation between separable states and classical states: A bipartite state ρ^{ab} is separable if and only if it can be extended to a certain classical state $\rho^{a'a:bb'}$ (with respect to the bipartition $a'a : bb'$).

The entanglement measure $\mathcal{E}(\cdot)$ has the nice property that it is automatically dominated by the discord in the sense that

$$\mathcal{E}(\rho^{ab}) \leq \mathcal{Q}(\rho^{ab})$$

since $\rho^{a'a:bb'} = \rho^{ab}$ with the a' and b' being trivial (one-dimensional) can be regarded as a state extension of ρ^{ab} itself.

With the above property, we may decompose the total correlations, as quantified by the quantum mutual information $I(\rho^{ab})$, into classical correlations $C(\rho^{ab})$ plus dissonance $\mathcal{D}(\rho^{ab})$ plus entanglement $\mathcal{E}(\rho^{ab})$:

$$I(\rho^{ab}) = C(\rho^{ab}) + \mathcal{D}(\rho^{ab}) + \mathcal{E}(\rho^{ab}),$$

where the difference

$$\mathcal{D}(\rho^{ab}) := \mathcal{Q}(\rho^{ab}) - \mathcal{E}(\rho^{ab})$$

is interpreted as a measure of dissonance as termed by Kavan et al. [23].

$\mathcal{E}(\cdot)$ is locally unitary invariant in the sense that

$$\mathcal{E}((U^a \otimes U^b)\rho^{ab}(U^a \otimes U^b)^\dagger) = \mathcal{E}(\rho^{ab})$$

for any unitary operators U^a and U^b on parties a and b, respectively, as long as the discord is invariant under local unitary transformations.

Since any pure state $\rho^{ab} = |\Psi^{ab}\rangle\langle\Psi^{ab}|$ has only trivial extensions of the form $\rho^{a'b'} \otimes |\Psi^{ab}\rangle\langle\Psi^{ab}|$, it follows that the entanglement $\mathcal{E}(\rho^{ab})$ coincides with the discord $\mathcal{Q}(\rho^{ab})$, i.e., $\mathcal{E}(\rho^{ab}) = \mathcal{Q}(\rho^{ab})$, for any pure state ρ^{ab}, as long as the discord has the decreasing property $\mathcal{Q}(\rho^{a'b'} \otimes |\Psi^{ab}\rangle\langle\Psi^{ab}|) \geq \mathcal{Q}(|\Psi^{ab}\rangle\langle\Psi^{ab}|)$.

Since any state extension $\rho^{a''a'a:bb'b''}$ of $\rho^{a'a:bb'}$ is necessarily a state extension of the reduced state $\rho^{ab} = \mathrm{tr}_{a'b'}\rho^{a'a:bb'}$, it follows from

$$\begin{aligned}\mathcal{E}(\rho^{ab}) &\leq \min_{\mathrm{tr}_{a''a'b'b''}\rho^{a''a'a:bb'b''}=\rho^{ab}} \mathcal{Q}(\rho^{a''a'a:bb'b''}) \\ &\leq \min_{\mathrm{tr}_{a''b''}\rho^{a''a'a:bb'b''}=\rho^{a'a:bb'}} \mathcal{Q}(\rho^{a''a'a:bb'b''}) \\ &= \mathcal{E}(\rho^{a'a:bb'})\end{aligned}$$

that $\mathcal{E}(\cdot)$ is non-increasing under local partial trace (state reduction) in the sense that

$$\mathcal{E}(\rho^{ab}) \leq \mathcal{E}(\rho^{a'a:bb'})$$

for any state extension $\rho^{a'a:bb'}$ of ρ^{ab}.

We list some important and interesting problems requiring further investigations:

(1) Classify the discord measures such that the induced entanglement measures are convex in the sense that

$$\mathcal{E}(\sum_i p_i\rho_i^{ab}) \leq \sum_i p_i\mathcal{E}(\rho_i^{ab}),$$

where ρ_i^{ab} are bipartite states shared by parties a and b, and $p_i \geq 0$, $\sum_i p_i = 1$. We remark that this may be related to the direct sum property of the discord measures.

(2) More generally, classify the discord measures such that the induced entanglement measures are entanglement monotones.

(3) How to evaluate the entanglement measures? On may try to find some analytical formulas for some highly symmetric states, and establish some bounds for general cases.

(4) What are the relations between the relative entropy of entanglement and the relative entropy of quantumness? If one defines an entanglement measure induced by the relative entropy of quantumness $Q_{\rm rel}(\cdot)$ as

$$\mathcal{E}_{\rm rel}(\rho^{ab}) := \min_{{\rm tr}_{a'b'}\rho^{a'a:bb'}=\rho^{ab}} Q_{\rm rel}(\rho^{a'a:bb'}),$$

where the minimization is over all state extensions $\rho^{a'a:bb'}$ of ρ^{ab} (i.e., $\rho^{ab} = {\rm tr}_{a'b'}\rho^{a'a:bb'}$), then an interesting question arises as the relation between this induced entanglement measure $\mathcal{E}_{\rm rel}(\cdot)$ and the original relative entropy of entanglement $E_{\rm rel}(\cdot)$: Does it hold that

$$E_{\rm rel}(\rho^{ab}) = \mathcal{E}_{\rm rel}(\rho^{ab})?$$

Since

$$E_{\rm rel}(\rho^{ab}) \le E_{\rm rel}(\rho^{a'a:bb'}) \le Q(\rho^{a'a:bb'}),$$

we have

$$E_{\rm rel}(\rho^{ab}) \le \mathcal{E}_{\rm rel}(\rho^{ab}),$$

thus it remains to establish the reversed inequality.

4 Discussions

Discord stems directly from the pivotal and ubiquitous notion of quantum measurements, while entanglement is widely regarded as a key feature of quantum information. We have reviewed briefly several aspects of discord and entanglement with emphasis on their intertwining, and have illustrated that discord is not only a kind of quantum correlations beyond entanglement, but also that quantum discord contracts to entanglement, i.e., entanglement can be interpreted as the irreducible residue, as the minimal shadow, of discord over all state extensions. This puts discord, conceptually, in a more primitive place than entanglement, sheds lights on the fundamental importance of quantumness in characterizing quantum correlations, and highlights the significance of the interplay between quantum measurements and state extensions in quantum information science. We have outlined some problems for further investigations.

Acknowledgements This work was supported by the National Natural Science Foundation of China, Grant Nos. 11375259, 61134008, the National Center for Mathematics and Interdisciplinary Sciences, CAS, Grant No. Y029152K51.

References

1. J.A. Wheeler, W.H. Zurek, *Quantum Theory and Measurement* (Princeton University Press, Princeton, 1983)
2. A. Einstein, B. Podolsky, N. Rosen, Can quantum-mechanical description of physical reality be considered complete? Phys. Rev. **47**, 77 (1935)
3. N. Bohr, Can quantum-mechanical description of physical reality be considered complete? Phys. Rev. **48**, 696 (1935)
4. H.M. Wiseman, Quantum discord is Bohr's notion of non-mechanical disturbance introduced to counter the Einstein-Podolsky-Rosen argument. Ann. Phys. **338**, 361 (2013)
5. E. Schrödinger, Discussion of probability relations between separated systems. Math. Proc. Camb. Philo. Soc. **31**, 555 (1935)
6. E. Schrödinger, Probability relations between separated systems. Math. Proc. Camb. Philo. Soc. **32**, 446 (1936)
7. E. Schrödinger, Die gegenwärtige situation in der quantenmechanik (The present situation in quantum mechanics), Naturwissenschaften **23**, 807, 823, 844 (1935)
8. W.H. Zurek, Decoherence, einselection, and the quantum origins of the classical. Rev. Mod. Phys. **75**, 715 (2003)
9. E. Joos, H.D. Zeh, C. Kiefer, D.J.W. Giulini, J. Kupsch, I.-O. Stamatescu, *Decoherence and the Appearance of a Classical World in Quantum Theory*, 2nd edn. (Springer, Berlin, 2003)
10. M.A. Schlosshauer, Decoherence, the measurement problem, and interpretations of quantum mechanics. Rev. Mod. Phys. **76**, 1267 (2005)
11. H. Ollivier, W.H. Zurek, Quantum discord: a measure of the quantumness of correlations. Phys. Rev. Lett. **88**, 017901 (2001)
12. L. Henderson, V. Vedral, Classical, quantum and total correlations. J. Phys. A **34**, 6899 (2001)
13. H. Everett III, The theory of the universal wavefunction, Ph.D. Thesis, Princeton University, 1955 (B. DeWitt, R. Neill Graham (eds.) *The Many-Worlds Interpretation of Quantum Mechanics* (Princeton University Press, Princeton, 1973), pp. 3–140
14. G. Lindblad, Entropy, information and quantum measurements. Commun. Math. Phys. **33**, 305 (1973)
15. G. Lindblad, Quantum entropy and quantum measurements, in *Quantum Aspects of Optical Communications*, ed. by C. Bendjaballah, et al. (Springer, Berlin, 1991), pp. 71–80
16. J. Oppenheim, M. Horodecki, P. Horodecki, R. Horodecki, Thermodynamical approach to quantifying quantum correlations. Phys. Rev. Lett. **89**, 180402 (2002)
17. M. Horodecki, P. Horodecki, R. Horodecki, J. Oppenheim, A. Sen(De), U. Sen, B. Synak-Radtke, Local versus nonlocal information in quantum-information theory: formalism and phenomena. Phys. Rev. A **71**, 062307 (2005)
18. A. Datta, A. Shaji, C.M. Caves, Quantum discord and the power of one qubit. Phys. Rev. Lett. **100**, 050502 (2008)
19. S. Luo, Quantum discord for two-qubit systems. Phys. Rev. A **77**, 042303 (2008)
20. A.R. Usha Devi, A.K. Rajagopa, Generalized information theoretic measure to discern the quantumness of correlations. Phys. Rev. Lett. **100**, 140502 (2008)
21. S. Luo, Q. Zhang, Observable correlations in two-qubit states. J. Stat. Phys. **136**, 165 (2009)
22. S. Wu, U.V. Poulsen, K. Mölmer, Correlations in local measurements on a quantum state, and complementarity as an explanation of nonclassicality. Phys. Rev. A **80**, 032319 (2009)
23. K. Modi, T. Paterek, W. Son, V. Vedral, M. Williamson, Unified view of quantum and classical correlations. Phys. Rev. Lett. **104**, 080501 (2010)
24. B. Dakic, V. Vedral, C. Brukner, Necessary and sufficient condition for nonzero quantum discord. Phys. Rev. Lett. **105**, 190502 (2010)
25. S. Luo, S. Fu, Geometric measure of quantum discord. Phys. Rev. A **82**, 034302 (2010)
26. J.-S. Xu, X.-Y. Xu, C.-F. Li, C.-J. Zhang, X.-B. Zou, G.-C. Guo, Experimental investigation of classical and quantum correlations under decoherence. Nature Commun. **1**, 7 (2010)
27. A. Brodutch, D.R. Terno, Quantum discord, local operations, and Maxwell's demons. Phys. Rev. A **81**, 062103 (2010)

28. P. Giorda, M.G.A. Paris, Gaussian quantum discord. Phys. Rev. Lett. **105**, 020503 (2010)
29. Q. Chen, C. Zhang, S. Yu, X.X. Yi, C.H. Oh, Quantum discord of two-qubit X states. Phys. Rev. A **84**, 042313 (2011)
30. D. Girolami, G. Adesso, Quantum discord for general two-qubit states: analytical progress. Phys. Rev. A **83**, 052108 (2011)
31. V. Madhok, A. Datta, Interpreting quantum discord through quantum state merging. Phys. Rev. A **83**, 032323 (2011)
32. F.F. Fanchini, M.F. Cornelio, M.C. de Oliveira, A.O. Caldeira, Conservation law for distributed entanglement of formation and quantum discord. Phys. Rev. A **84**, 012313 (2011)
33. D. Cavalcanti, L. Aolita, S. Boixo, K. Modi, M. Piani, A. Winter, Operational interpretations of quantum discord. Phys. Rev. A **83**, 032324 (2011)
34. A. Streltsov, H. Kampermann, D. Bruß, Linking quantum discord to entanglement in a measurement. Phys. Rev. Lett. **106**, 160401 (2011)
35. M. Piani, S. Gharibian, G. Adesso, J. Calsamiglia, P. Horodecki, A. Winter, All nonclassical correlations can be activated into distillable entanglement. Phys. Rev. Lett. **106**, 220403 (2011)
36. A. Streltsov, H. Kampermann, D. Bruß, Quantum cost for sending entanglement. Phys. Rev. Lett. **108**, 250501 (2012)
37. S. Luo, S. Maniscalco, K. Modi, G.M. Palma, M.G.A. Paris (eds.), Quantum correlations: entanglement and beyond, Int. J. Quantum Inf. **9**(7–8), (2011) (Special issues)
38. M. Piani, G. Adesso, Quantumness of correlations revealed in local measurements exceeds entanglement. Phys. Rev. A **85**, 040301(R) (2012)
39. T.K. Chuan, J. Maillard, K. Modi, T. Paterek, M. Paternostro, M. Piani, Quantum discord bounds the amount of distributed entanglement. Phys. Rev. Lett. **109**, 070501 (2012)
40. M. Gu, H.M. Chrzanowski, S.M. Assad, T. Symul, K. Modi, T.C. Ralph, V. Vedral, P.K. Lam, Operational significance of discord consumption: theory and experiment. Nature Phys. **8**, 671 (2012)
41. B. Dakic, Y.O. Lipp, X. Ma, M. Ringbauer, S. Kropatschek, S. Barz, T. Paterek, V. Vedral, A. Zeilinger, Č. Brukner, P. Walther, Quantum discord as resource for remote state preparation. Nature Phys. **8**, 666 (2012)
42. K. Modi, A. Brodutch, H. Cable, T. Paterek, V. Vedral, The classical-quantum boundary for correlations: discord and related measures. Rev. Mod. Phys. **84**, 1655 (2012)
43. M. Piani, Problem with geometric discord. Phys. Rev. A **86**, 034101 (2012)
44. B. Bellomo, G.L. Giorgi, F. Galve, R. Lo Franco, G. Compagno, R. Zambrini, Unified view of correlations using the square-norm distance. Phys. Rev. A **85**, 032104 (2012)
45. L. Chang, S. Luo, Remedying the local ancilla problem with geometric discord. Phys. Rev. A **87**, 062303 (2013)
46. D. Girolami, A.M. Souza, V. Giovannetti, T. Tufarelli, J.G. Filgueiras, R.S. Sarthour, D.O. Soares-Pinto, I.S. Oliveira, G. Adesso, Quantum discord determines the interferometric power of quantum states. Phys. Rev. Lett. **112**, 210401 (2014)
47. D. Spehner, M. Orszag, Geometric quantum discord with Bures distance. New J. Phys. **15**, 103001 (2013)
48. F. Ciccarello, T. Tufarelli, V. Giovannetti, Toward computability of trace distance discord. New J. Phys. **16**, 013038 (2014)
49. D. Spehner, Quantum correlations and distinguishability of quantum states. J. Math. Phys. **55**, 075211 (2014)
50. M.A. Nielsen, I.L. Chuang, *Quantum Computation and Quantum Information* (Cambridge University Press, Cambridge, 2000)
51. G. Alber, T. Beth, M. Horodecki, P. Horodecki, R. Horodecki, M. Rötteler, H. Weinfurter, R. Werner, A. Zeillinger, *Quantum Information: An Introduction to Basic Theoretical Concepts and Experiments* (Springer, Berlin, 2001)
52. R.F. Werner, Quantum states with Einstein-Podolsky-Rosen correlations admitting a hidden-variable model. Phys. Rev. A **40**, 4277 (1989)
53. C.H. Bennett, D.P. DiVincenzo, J.A. Smolin, W.K. Wootters, Mixed-state entanglement and quantum error correction. Phys. Rev. A **54**, 3824 (1996)

54. V. Vedral, M.B. Plenio, M.A. Rippin, P.L. Knight, Quantifying entanglement. Phys. Rev. Lett. **78**, 2275 (1997)
55. W.K. Wootters, Entanglement of formation of an arbitrary state of two qubits. Phys. Rev. Lett. **80**, 2245 (1998)
56. G. Vidal, Entanglement monotones. J. Mod. Opt. **47**, 355 (2000)
57. M. Christandl, A. Winter, Squashed entanglement-an additive entanglement measure. J. Math. Phys. **45**, 829 (2004)
58. R. Horodecki, P. Horodecki, M. Horodecki, K. Horodecki, Quantum entanglement. Rev. Mod. Phys. **81**, 865 (2009)
59. O. Gühne, G. Tóth, Entanglement detection. Phys. Rep. **474**, 1 (2009)
60. J.S. Bell, *Speakable and Unspeakable in Quantum Mechanics* (Cambridge University Press, Cambridge, 1987)
61. N. Brunner, D. Cavalcanti, S. Pironio, V. Scarani, S. Wehner, Bell nonlocality. Rev. Mod. Phys. **86**, 419 (2014)
62. C.E. Shannon, W. Weaver, *The Mathematical Theory of Communication* (University of Illinois Press, Urbana, 1949)
63. T.M. Cover, J.A. Thomas, *Elements of Information Theory* (Wiley, New York, 1991)
64. R.L. Stratonovich, The transmission rate for certain quantum communications channels. Probl. Inf. Transm. **2**, 35 (1966)
65. A. Wehrl, General properties of entropy. Rev. Mod. Phys. **50**, 221 (1978)
66. V. Vedral, The role of relative entropy in quantum information theory. Rev. Mod. Phys. **74**, 197 (2002)
67. B. Groisman, S. Popescu, A. Winter, Quantum, classical, and total amount of correlations in a quantum state. Phys. Rev. A **72**, 032317 (2005)
68. B. Schumacher, M.D. Westmoreland, Quantum mutual information and the one-time pad. Phys. Rev. A **74**, 042305 (2006)
69. S. Luo, N. Li, Total versus quantum correlations in quantum states. Phys. Rev. A **76**, 032327 (2007)
70. S. Luo, Using measurement-induced disturbance to characterize correlations as classical or quantum. Phys. Rev. A **77**, 022301 (2008)
71. M. Piani, P. Horodecki, R. Horodecki, No-local-broadcasting theorem for multipartite quantum correlations. Phys. Rev. Lett. **100**, 090502 (2008)
72. S. Luo, N. Li, Quantum correlations reduce classical correlations with ancillary systems. Chin. Phys. Lett. **27**, 120304 (2010)
73. S. Luo, W. Sun, Decomposition of bipartite states with applications to quantum no-broadcasting theorems. Phys. Rev. A **82**, 012338 (2010)
74. H. Barnum, C.M. Caves, C.A. Fuchs, R. Jozsa, B. Schumacher, Noncommuting mixed states cannot be broadcast. Phys. Rev. Lett. **76**, 2818 (1996)
75. S. Luo, N. Li, Classical and quantum correlative capacities of quantum systems. Phys. Rev. A **84**, 042124 (2011)
76. T.S. Cubitt, F. Verstraete, W. Dür, J.I. Cirac, Separable states can be used to distribute entanglement. Phys. Rev. Lett. **91**, 037902 (2003)
77. A. Kay, Using separable Bell-diagonal states to distribute entanglement. Phys. Rev. Lett. **109**, 080503 (2012)
78. A. Fedrizzi, M. Zuppardo, G. Gillett, M. Broome, M. Almeida, M. Paternostro, A. White, T. Paterek, Experimental distribution of entanglement with separable carriers. Phys. Rev. Lett. **111**, 230504 (2013)
79. S. Pirandola, Entanglement reactivation in separable environments. New J. Phys. **15**, 113046 (2013)
80. M. Koashi, A. Winter, Monogamy of quantum entanglement and other correlations. Phys. Rev. A **69**, 022309 (2004)
81. L.-X. Cen, X.-Q. Li, J. Shao, Y. Yan, Quantifying quantum discord and entanglement of formation via unified purifications. Phys. Rev. A **83**, 054101 (2011)
82. N. Li, S. Luo, Classical states versus separable states. Phys. Rev. A **78**, 024303 (2008)

83. G. Bellomo, A. Plastino, A.R. Plastino, Classical extension of quantum-correlated separable states. Int. J. Quantum Inf. **13**, 1550015 (2015)
84. A. Plastino, G. Bellomo, A.R. Plastino, Quantum state space dimension as a quantum resource. Int. J. Quantum Inform. **13**, 1550039 (2015)
85. G. Bellomo, A. Plastino, A.R. Plastino, Quantumness and the role of locality on quantum correlations. Phys. Rev. A **93**, 062322 (2016)

Monogamy of Quantum Correlations - A Review

Himadri Shekhar Dhar, Amit Kumar Pal, Debraj Rakshit, Aditi Sen(De) and Ujjwal Sen

Abstract Monogamy is an intrinsic feature of quantum correlations that gives rise to several interesting quantum characteristics which are not amenable to classical explanations. The monogamy property imposes physical restrictions on unconditional sharability of quantum correlations between different parts of a multipartite quantum system, and thus has a direct bearing on the cooperative properties of multipartite states, including those of large many-body systems. On the contrary, a certain party can be maximally classical correlated with an arbitrary number of parties. In recent years, the monogamy property of quantum correlations has been applied to understand several key aspects of quantum physics, including distribution of quantum resources, security in quantum communication, critical phenomena, and quantum biology. In this chapter, we look at some of the salient developments and applications in quantum physics that have been closely associated with the monogamy of quantum discord, and "discord-like" quantum correlation measures.

1 Introduction

Quantum correlations, shared between two or more parties [1, 2], boast of novel features, which are exclusive to the quantum world and are central to quantum information science. On one hand, they play a significant role in efficient quantum communication [3–5] and computational [6, 7] tasks, while on the other hand, they help us to understand cooperative phenomena in quantum many-body systems [8–10]. A key difference between classical and quantum correlations is the way they can be shared

H.S. Dhar · A.K. Pal · D. Rakshit · A. Sen(De) · U. Sen (✉)
Harish-Chandra Research Institute, Chhatnag Road, Jhunsi 211 019, Allahabad, India
e-mail: ujjwal@hri.res.in

H.S. Dhar
e-mail: dhar.himadri@gmail.com

H.S. Dhar · A.K. Pal · D. Rakshit · A. Sen(De) · U. Sen
Homi Bhabha National Institute, Training School Complex,
Anushaktinagar 400 085, Mumbai, India

F.F. Fanchini et al. (eds.), *Lectures on General Quantum Correlations and their Applications*, Quantum Science and Technology,
DOI 10.1007/978-3-319-53412-1_3

among various parts of a multiparty quantum system. Unlike classical correlations, which can be freely shared, the distribution of quantum correlations is restricted by the non-classical properties of the quantum system. For example, in a tripartite quantum state, ρ_{ABC}, if two parties A and B are maximally quantum correlated, then none of A and B can share any quantum correlation with the third party, C [4, 11–14] (for a social representation, see Fig. 1). However, there exist no such constraints for classical correlations, and the pairs of parties, say AB and AC, can simultaneously share maximum classical correlations. Similar situation arises in multiparty quantum systems of more than three parties. This exclusive trade-off between quantum correlations of different combinations of parties in a multiparty quantum system is known as *monogamy* of quantum correlations [4, 11–14]. The no-go theorems, like the no-cloning theorem [15–18], put restrictions on the available options in quantum cryptography [3]. Similarly, the monogamy of quantum correlations is a restriction on the sharability of quantum correlations, and yet can potentially help in obtaining advantages in a quantum system over their classical counterparts.

In the seminal work by Coffman, Kundu, and Wootters (CKW) [13], a monogamy relation for three-qubit pure states was established using an entanglement measure, namely, the squared concurrence [19, 20]. This scenario was later generalized by Osborne and Verstraete [21] for pure as well as mixed states of an arbitrary number of qubits. Since quantum correlations are not uniquely defined, even in the bipartite domain, the immediate question that follows is whether the monogamy inequality proposed by CKW is necessarily obeyed by all kinds of quantum correlation measures. In general, quantum correlations can be broadly classified into two categories – entanglement measures [1], and the information-theoretic measures of quantum correlations [2]. While entanglement of formation [12, 22, 23], concurrence [19, 20], distillable entanglement [24, 25], negativity [26–31],

Fig. 1 Monogamy: Two persons sharing an umbrella are unmindful of the presence of the third person

logarithmic negativity [28–31], relative entropy of entanglement [32–34], etc. belong to the first category, quantum discord [35, 36], and quantum work deficit [37–40] are examples of the second kind. Although quantum correlations are qualitatively monogamous, not all of them are limited to the form of monogamy constraint proposed by CKW. In particular, while the squared concurrence and negativity satisfy the CKW monogamy inequality for all three-qubit pure states [13, 21, 41], many others, such as logarithmic negativity and the information-theoretic measures, do not satisfy the same [41–45]. Deliberations on the monogamy of quantum correlations have led to important insights including a "conservation law" between entanglement and information-theoretic quantum correlations in multiparty quantum states [42]. In Ref. [46], the authors formulate the requirements for a bipartite entanglement measure to be monogamous for all quantum states, and show that additive and suitably normalized entanglement measures, which can faithfully describe the geometric structure of the fully antisymmetric state, are non-monogamous. However, it is also understood that all kinds of quantum correlations obey the CKW monogamy constraint for a given state when raised to a suitable power, provided they follow certain conditions [47].

Interestingly, the limitation imposed by the quantum mechanical principles, in the form of monogamy constraints, is not exclusive to quantum correlations. There exist no-go theorems which place parallel restrictions such as monogamy of Bell inequality violation [48, 49] and exclusion principle of classical information transmission over quantum channels [50] (cf. [51–54]). More precisely, within the consideration of a multiparty set-up, for example, of an editor with several reporters, if the shared quantum state between the editor and a single reporter violates a Bell inequality [55, 56] or is quantum dense codeable [57], then the rest of the channels shared between the editor and the other reporters are prohibited from possessing the same quantum advantage. Analogous monogamy constraints have also been addressed in the context of quantum steering [58, 59], quantum teleportation fidelity [60], and contextual inequalities [61, 62].

The monogamy properties of quantum correlations find potential applications in quantum information based protocols like quantum cryptography [3], entanglement distillation [22], quantum state and channel discrimination [44, 45, 63], and in characterizing quantum many-body systems [64–72] as well as in biological processes [73, 74]. The key concept of entanglement-based quantum cryptography essentially exploits the trade off in monogamy of quantum correlations, which limits the amount of information that an eavesdropper can extract about the secret key, shared between a sender and a receiver, obtained via measurement on both sides of an entangled state between the sender and the receiver [4, 11, 75]. The constraints on shareability of entanglement find further application in enhancing quantum privacy via entanglement purification [76]. Another importance of monogamy relations, arising due to the constraints that they put on the distribution of quantum correlation among many parties, is their ability to capture multipartite quantum correlations present in the system, the latter being, in general, a challenging task [13, 77, 78]. Moreover, they play a decisive role in designing the structure of eigenstates of quantum spin models, which are expected to obey the no-go principles arising from the monogamy constraints [67].

In this chapter, our main aim is to review the results on monogamy of quantum discord, and other "discord-like" measures of quantum correlations. We survey the properties of a proposed multiparty quantum correlation measure, called the "monogamy score", and its relations with other measures of quantum correlations. Finally, we also take a look at the usefulness of the monogamy score for quantum discord in quantum information science, and in quantum many-body physics. The structure of this chapter is as follows. In Sect. 2, we present a short review on the studies that have been carried out on the monogamy properties of entanglement. Section 3 consists of the definitions of several quantum correlation measures, such as quantum discord, quantum work deficit, and geometric quantum discord, that have been defined independent of the entanglement-separability paradigm. The monogamy properties of these measures are discussed in the following sections. In Sect. 4, we focus on the monogamy of quantum discord. The monogamy property of other information-theoretic quantum correlation measures are discussed in Sect. 5. Section 6 provides a report on the relation between the monogamy of quantum correlations with other multiparty measures. In Sect. 7, we review some noteworthy applications of the monogamy property. Section 8 contains some concluding remarks.

2 Monogamy Relations of Entanglement

In this section, we provide a brief discussion on the monogamy properties of different entanglement measures, such as concurrence, negativity, etc. Starting from tripartite quantum systems, we expand the discussion to the more complex cases in multiparty systems. For a review, see [14].

We also formally define the "monogamy score" corresponding to an arbitrary quantum correlation measure.

2.1 Tripartite System: CKW Inequality and Beyond

In this subsection, we review monogamy properties of entanglement in the tripartite scenario, using concurrence as the entanglement measure. For an arbitrary two-qubit quantum state ρ_{AB}, concurrence is defined [19, 20] as $\mathcal{C}_{AB} \equiv \mathcal{C}(\rho_{AB}) = \max\{0, \lambda_1 - \lambda_2 - \lambda_3 - \lambda_4\}$, where $\{\lambda_i\}, i = 1, \ldots, 4$, are square roots of the eigenvalues of the positive operator $\rho\tilde{\rho}$ in descending order, and the spin-flipped density matrix $\tilde{\rho}$ is given by $\tilde{\rho} = (\sigma_y \otimes \sigma_y)\rho^*(\sigma_y \otimes \sigma_y)$. Here and henceforth, σ_x, σ_y and σ_z denote the standard Pauli spin matrices. Note that $\mathcal{C}$ vanishes for the separable states and attains the value 1 for the maximally entangled states in $\mathbb{C}^2 \otimes \mathbb{C}^2$ systems. The physical significance of concurrence stems from the fact that the entanglement of formation is a monotonic function of concurrence, and vice versa, in $\mathbb{C}^2 \otimes \mathbb{C}^2$ [19, 20]. A formal definition of entanglement of formation is given later in this section.

Consider a three-qubit pure state, $\rho_{ABC} = (|\phi\rangle\langle\phi|)_{ABC}$. The concurrences corresponding to the reduced density matrices $\rho_{AB} = \mathrm{Tr}_C(\rho_{ABC})$ and $\rho_{AC} = \mathrm{Tr}_B(\rho_{ABC})$ satisfy the inequalities, $\mathcal{C}^2_{AB} \leq \mathrm{Tr}(\rho_{AB}\tilde{\rho}_{AB})$ and $\mathcal{C}^2_{AC} \leq \mathrm{Tr}(\rho_{AC}\tilde{\rho}_{AC})$, respectively. It can further be shown that $\mathrm{Tr}(\rho_{AB}\tilde{\rho}_{AB}) + \mathrm{Tr}(\rho_{AC}\tilde{\rho}_{AC}) = 4\ \mathrm{det}\rho_A$. This leads to following inequality [13]:

$$\mathcal{C}^2_{AB} + \mathcal{C}^2_{AC} \leq 4 \det \rho_A, \tag{1}$$

where $\rho_A = \mathrm{Tr}_{BC}(\rho_{ABC})$. Even though BC is a four-dimensional system, the support of $\rho_{BC} = \mathrm{Tr}_A(\rho_{ABC})$ is spanned by the eigenstates corresponding to at most two non-zero eigenvalues of the reduced density matrix ρ_{BC}, and hence is effectively defined on a two-dimensional space. This allows to treat the bipartite split of A and BC as an effective two-qubit system whose concurrence, $\mathcal{C}_{A:BC}$, is simply $2\sqrt{\det \rho_A}$. As a result, the inequality in (1) becomes

$$\mathcal{C}^2_{AB} + \mathcal{C}^2_{AC} \leq \mathcal{C}^2_{A:BC}, \tag{2}$$

and is referred to as the monogamy inequality for squared concurrence in the case of three-qubit pure states. For a three-qubit mixed state, the state ρ_{BC} may, in principle, have all four non-zero eigenvalues. However, a generalization of the above inequality for the mixed state is prescribed by replacing the right hand side of (2) by the minimum average concurrence squared over all possible pure state decompositions $\{p_i, |\psi_i\rangle\}$ of ρ_{BC}, and hence (2) also holds for arbitrary mixed three-qubit states.

In this context, let us define the tangle for a three-qubit pure state, ρ_{ABC}, expressed in terms of squared concurrences, as [13]

$$\tau_{ABC} = \mathcal{C}^2_{A:BC} - \mathcal{C}^2_{AB} - \mathcal{C}^2_{AC}. \tag{3}$$

It turns out that τ_{ABC}, which is also known as the *three-tangle*, or the *residual entanglement*, is independent of the choice of the "node", which is the site A here. The tangle has been argued to characterize three-qubit entanglement. The generalization of the tangle to mixed states can be obtained by the convex roof extension, the computation of which is difficult for arbitrary states.

Next, let us present the definition of the entanglement of formation (EoF) [20], a measure of bipartite entanglement, and its relation with concurrence for two-qubit states. Consider a bipartite quantum state ρ_{AB}, and the ensemble $\{p_i, |\psi_i\rangle\}$ denoting a possible pure state decomposition of ρ_{AB}, satisfying $\rho_{AB} = \sum_i p_i|\psi_i\rangle\langle\psi_i|$. The EoF is defined as

$$E_f(\rho_{AB}) = \min_{\{p_i, |\psi_i\rangle\}} \sum_i p_i S(\mathrm{Tr}_B[|\psi_i\rangle\langle\psi_i|]), \tag{4}$$

where $S(\text{Tr}_B[|\psi_i\rangle\langle\psi_i|])$ is the von Neumann entropy of the reduced density matrix corresponding to the A part of ρ_{AB}. A compact formula for the EoF is known for two-qubit systems in terms of the concurrence, $\mathcal{C}_{AB}$. For a two-qubit mixed state ρ_{AB}, $E_f(\rho_{AB}) = h\left((1+\sqrt{1-\mathcal{C}_{AB}^2})/2\right)$ [20], where $h(x) = -x\log_2 x - (1-x)\log_2(1-x)$. The EoF, being a concave function of squared concurrence, does not obey the CKW inequality. However, the EoF can also not be freely shared amongst the constituents of a multiparty system. In fact, the square of the EoF does obey the same relation as the squared concurrence for tripartite systems [79].

Several studies have addressed the monogamy property of concurrence within the tripartite scenario. We briefly mention some important findings in this direction. It is natural to ask whether the monogamy inequality is satisfied in tripartite systems consisting of higher-dimensional parties. The analysis understandably becomes complex with increasing dimension of the Hilbert space, as much less is known about the quantification of entanglement in higher dimensional cases. For example, the exact formula for EoF and concurrence are missing in higher dimensions. However, there have been definitive efforts to understand the monogamy constraints in higher dimensional systems. The results reported in Ref. [80] indicate that the CKW inequality cannot be directly extended to higher-dimensional states. However, it has been demonstrated that the squared concurrence satisfies the monogamy relation for arbitrary pure states in $\mathbb{C}^2 \otimes \mathbb{C}^2 \otimes \mathbb{C}^4$ [81] (cf. [82, 83]). The monogamy property of squared concurrence in higher-dimensional systems of more than three parties has also been addressed [84]. Other approaches to construct more monogamy inequalities for entanglement in tripartite states have made use of the generalized concurrence [85], which is a multipartite measure of entanglement.

2.2 *Monogamy Score*

Just like for squared concurrence, one can formulate the problem of monogamy for arbitrary quantum correlation measures. For a given bipartite quantum correlation measure, $\mathcal{Q}$, and a three-party quantum state, ρ_{ABC}, we call the state to be monogamous for the measure $\mathcal{Q}$ if

$$\mathcal{Q}(\rho_{A:BC}) \geq \mathcal{Q}(\rho_{AB}) + \mathcal{Q}(\rho_{AC}). \tag{5}$$

Otherwise, the state is non-monogamous for that measure. The measure is called monogamous for a given tripartite quantum system if it is monogamous for all tripartite states. Similar to the spirit of tangle defined in Eq. (3), one can define a quantity called *monogamy score* [77], of a bipartite quantum correlation measure, $\mathcal{Q}$, based on the monogamy relation in (5). It is given by

$$\delta_{\mathcal{Q}} = \mathcal{Q}(\rho_{ABC}) - \mathcal{Q}(\rho_{AB}) - \mathcal{Q}(\rho_{AC}), \tag{6}$$

where we call the party "A" as the nodal observer. With this notion, the tangle [13] can be called monogamy score for squared concurrence, and can also alternatively be denoted by $\delta_{\mathcal{C}^2}$. This definition can be extended to an N-party quantum state, $\rho_{12\ldots N}$. An arbitrary N-party state is said to be monogamous with respect to $\mathcal{Q}$, if

$$\mathcal{Q}(\rho_{j:rest}) \geq \sum_{k\neq j} \mathcal{Q}(\rho_{jk}), \tag{7}$$

and the corresponding monogamy score, for any quantum correlation measure, $\mathcal{Q}$, is defined as

$$\delta_{\mathcal{Q}}^{j}(\rho_{1:23\ldots N}) = \mathcal{Q}(\rho_{j:rest}) - \sum_{k\neq j} \mathcal{Q}(\rho_{jk}), \tag{8}$$

with j as the node. Here, ρ_{jk} represents the two-party density matrix, which can be obtained from $\rho_{12\ldots N}$ by tracing out all the other parties except j and k ($j, k = 1, 2, \ldots, N$). It has been argued [13, 77] that the monogamy score quantifies a multiparty quantum correlation measure, and hence constitutes a method of conceptualizing a multiparty measure using bipartite ones. A monogamous quantum correlation measure, for a given node j, have $\delta_{\mathcal{Q}}^{j} \geq 0$ for all multipartite states. Unless otherwise stated, we always use the first party as the nodal observer and in that case, we denote monogamy score as $\delta_{\mathcal{Q}}$, discarding the superscript. Henceforth, we shall always describe an N-party quantum state, pure or mixed, by $\rho_{12\ldots N}$, while for ease of notations, we denote a three-party quantum state by ρ_{ABC}.

In an N-party scenario, Coffman, Kundu, and Wootters conjectured a generalization of the CKW inequality to N-qubit pure states, $\rho_{12\ldots N}$, i.e., the inequality in (7), by replacing $\mathcal{Q}$ with $\mathcal{C}^2$. Some time later, the conjecture was proven for arbitrary N-qubit states by Osborne and Verstraete [21], by using an inductive strategy. Also, there have been several attempts to construct generalized monogamy inequalities for entanglement in qubit systems [85–93]. It has also been shown that in all multiqubit systems, there exist monogamy *equalities* for certain quantum correlations [94].

2.3 Monogamy of Negativity and Other Entanglement Measures

The negativity [26–31], $\mathcal{N}(\rho_{AB})$, corresponding to the quantum state ρ_{AB} defined on the Hilbert space $\mathbb{C}_A \otimes \mathbb{C}_B$ for two parties A and B, is defined by $\mathcal{N}_{AB} \equiv \mathcal{N}(\rho_{AB}) = (\|\rho_{AB}^{T_A}\|_1 - 1)/2$, where $\rho_{AB}^{T_A}$ is obtained by performing the partial transposition on the state ρ_{AB} with respect to the subsystem A [26, 27], i.e., $(\rho^{T_A})_{ij,kl} = (\rho)_{kj,il}$, and where $\|\rho\|_1 = \mathrm{Tr}\sqrt{\rho\rho^\dagger}$ denotes the trace norm of the matrix ρ. For systems in $\mathbb{C}^2 \otimes \mathbb{C}^2$ and $\mathbb{C}^2 \otimes \mathbb{C}^3$, $\mathcal{N} > 0$ is the necessary and sufficient for non-separability. In order to achieve a maximum value of unity in $\mathbb{C}^2 \otimes \mathbb{C}^2$, the negativity can be redefined as $\mathcal{N}_{AB} = \|\rho_{AB}^{T_A}\|_1 - 1$.

For any pure three-qubit state ρ_{ABC}, the monogamy inequality for squared negativity [41],

$$\mathcal{N}_{AB}^2 + \mathcal{N}_{AC}^2 \leq \mathcal{N}_{A:BC}^2, \tag{9}$$

holds, where A has been chosen as the nodal party, and in (5), Q is replaced by $\mathcal{N}^2$. For any three-qubit pure state, it turns out that $\mathcal{N}_{A:BC} = \mathcal{C}_{A:BC}$. Moreover, as shown in Ref. [95] for arbitrary two-qubit mixed states, we have

$$\mathcal{N}_{AB} = \|\rho_{AB}^{T_A}\|_1 - 1 \leq \mathcal{C}_{AB}. \tag{10}$$

Thus for any three-qubit pure state, we obtain $\mathcal{N}_{AB} \leq \mathcal{C}_{AB}$ and $\mathcal{N}_{AC} \leq \mathcal{C}_{AC}$, and consequently, the proof of (9) automatically follows from the corresponding one for squared concurrence. For an N-qubit pure state, the generalized monogamy inequality for squared negativity, given in Eq. (7) can be proven by using $\mathcal{N}_{1:23\ldots N} = \mathcal{C}_{1:23\ldots N}$ for pure states, and the relation in Eq. (10).

As mentioned in the previous subsection, there have been efforts to propose stronger versions of the monogamy relation for entanglement. Recently, a stronger monogamy inequality for negativity has been proposed [96], and a detailed study has been performed for four-qubit states. Another interesting proposal pitches the square of convex-roof extended negativity as an alternative candidate to characterize strong monogamy of multiparty quantum entanglement [97].

Monogamy has also been studied for other entanglement measures, including entanglement of assistance (EoA) [98], and squashed entanglement [99, 100]. The definition of EoA, which was originally introduced in terms of entropy of entanglement, can, in principle, be generalized for other measures of entanglement. For example, concurrence of assistance (CoA) is an entanglement monotone for pure tripartite states [101], and similar to the squared concurrence, monogamy properties of the squared CoA have been studied extensively [79, 101–103]. Ref. [104] found lower and upper bounds of EoA, among which the upper bound was shown to obey monogamy constraints for arbitrary N-qubit states. In Ref. [100], Koashi and Winter showed that for arbitrary tripartite states, the one-way distillable entanglement, the one-way distillable secret key [105], and the squashed entanglement [99] satisfy the monogamy relation, given in (5). For tripartite pure states, the entanglement of purification [106] was shown to be non-monogamous in general [107]. There have also been attempts to consider monogamy relations for entanglement in terms of Tsallis entropy [108–110].

The monogamy inequality of entanglement sharing has been investigated also for continuous variable systems. In Refs. [111, 112], monogamy inequality for the "continuous variable tangle", or the "contangle", has been provided for arbitrary three-mode Gaussian states and for symmetric arbitrary-mode Gaussian states, where contangle is defined as the convex roof of the square of the logarithmic negativity. Further generalization of the results has been achieved in [113–115].

3 Information-Theoretic Measures of Quantum Correlation

In this section, we define a few information-theoretic measures of quantum correlations, whose monogamy properties are discussed in the subsequent sections.

3.1 Quantum Discord

Let us consider a bipartite quantum state ρ_{AB}, for which the *uninterrogated*, or *unmeasured* quantum conditional entropy is defined as

$$\tilde{S}(\rho_{A|B}) = S(\rho_{AB}) - S(\rho_B), \tag{11}$$

where $\rho_B = \text{Tr}_A(\rho_{AB})$ is the reduced density matrix of the subsystem B, obtained by tracing over the subsystem A. One can also define an *interrogated* conditional entropy, given by

$$S(\rho_{A|B}) = \min_{\{\Pi_i^B\}} \sum_i p_i S(\rho_{A|i}), \tag{12}$$

where the minimization is performed over all complete sets of projective measurements, $\{\Pi_i^B\}$, performed on subsystem B. The corresponding post-measurement state for subsystem A is given by $\rho_{A|i} = \text{Tr}_B[(\mathbb{I}_A \otimes \Pi_i^B)\rho_{AB}(\mathbb{I}_A \otimes \Pi_i^B)]/p_i$, where $\mathbb{I}_A$ is the identity operator on the Hilbert space of the subsystem A, and $p_i = \text{Tr}_{AB}[(\mathbb{I}_A \otimes \Pi_i^B)\rho_{AB}(\mathbb{I}_A \otimes \Pi_i^B)]$ is the probability of obtaining the outcome i. These two definitions of quantum conditional entropy lead to two different but equivalent expressions of the classical mutual information. The former is used to define the uninterrogated quantum mutual information, given by

$$\tilde{I}(\rho_{AB}) = S(\rho_A) - \tilde{S}(\rho_{A|B}), \tag{13}$$

which is interpreted as the "total correlation" of ρ_{AB} [35, 36, 116–120]. On the other hand, the latter provides the definition of the interrogated quantum mutual information,

$$I^{\leftarrow}(\rho_{AB}) = S(\rho_A) - S(\rho_{A|B}), \tag{14}$$

also interpreted as the "classical correlation" present in the quantum state ρ_{AB} [35, 36]. The arrow in the superscript begins from the subsystem on which the measurement is performed. The quantum discord of the state ρ_{AB} is the difference between the uninterrogated and interrogated quantum mutual informations [35, 36], and is given by

$$D^{\leftarrow}(\rho_{AB}) = \tilde{I}(\rho_{AB}) - I^{\leftarrow}(\rho_{AB})$$
$$= S(\rho_{A|B}) - \tilde{S}(\rho_{A|B}). \quad (15)$$

Note here that one can also define quantum discord, $D^{\rightarrow}(\rho_{AB})$, by performing local measurement over the subsystem A instead of the subsystem B. In general, $D^{\leftarrow}(\rho_{AB}) \neq D^{\rightarrow}(\rho_{AB})$. Unless otherwise stated, here and throughout in this chapter, we consider $D^{\leftarrow}(\rho_{AB})$ as the measure of quantum discord. Note also that the definition of quantum discord is provided by using local projective measurement. However, quantum discord can also be defined in terms of positive operator valued measurements (POVMs). It has been shown that computation of quantum discord is NP-complete [121]. However, there has been efforts to determine quantum discord analytically as well as numerically for certain quantum states [122–128].

3.2 Quantum Work Deficit

The amount of extractable pure states from a bipartite state ρ_{AB}, under a set of global operations, called the "closed operations" (CO), is given by [37–40]

$$I_{\text{CO}} = \log_2 \dim(\mathcal{H}) - S(\rho_{AB}), \quad (16)$$

where the set of closed operations consists of *(i)* unitary operations, and *(ii)* dephasing the bipartite state by a set of projectors, $\{\Pi_k\}$, defined on the Hilbert space $\mathcal{H}$ of ρ_{AB}. On the other hand, considering the set of "closed local operations and classical communication" (CLOCC), the amount of extractable pure states from ρ_{AB} is given by [37–40]

$$I_{\text{CLOCC}} = \log_2 \dim(\mathcal{H}) - \min S\left(\rho'_{AB}\right). \quad (17)$$

Here, CLOCC consists of *(i)* local unitary operations, *(ii)* dephasing by local measurement on the subsystem, say, B, and *(iii)* communicating the dephased subsystem to the other party, A, via a noiseless quantum channel. The average quantum state, after the local projective measurement $\{\Pi_k^B\}$ is performed on B, can be written as $\rho'_{AB} = \sum_k p_k \rho^k_{AB}$ with ρ^k_{AB} and p_k being defined in a similar fashion as in the case of quantum discord. The minimization in I_{CLOCC} is achieved over all complete sets $\{\Pi_k^B\}$. The "one-way" quantum work deficit is then defined as [37–40]

$$W^{\leftarrow}(\rho_{AB}) = I_{\text{CO}} - I_{\text{CLOCC}}$$
$$= \min_{\{\Pi_k^B\}} \left[S\left(\rho'_{AB}\right) - S(\rho_{AB})\right], \quad (18)$$

where similar to quantum discord, the arrow in the superscript starts from the subsystem over which the measurement is performed.

3.3 Geometric Measure of Quantum Discord

Geometric quantum discord [129, 130] for a bipartite quantum state ρ_{AB} can be defined as the minimum squared Hilbert–Schmidt distance of ρ_{AB} from the set, $\mathcal{S}_{QC}$, of all "quantum-classical" states, given by $\sigma_{AB} = \sum_i p_i \sigma_A^i \otimes |i\rangle\langle i|$, where $\{|i\rangle\}$ forms a mutually orthonormal set of the Hilbert space of the subsystem B. Mathematically, geometric quantum discord is given by

$$D_G(\rho_{AB}) = \min_{\sigma_{AB} \in \mathcal{S}_{QC}} ||\rho_{AB} - \sigma_{AB}||_2^2, \tag{19}$$

where $||\rho - \sigma||_2 = \text{Tr}(\rho - \sigma)^2$, for two arbitrary density matrices ρ and σ (however, see [131]). Although geometric measures can be defined by using general Schatten p-norms [132] (see also [133, 134]), it has been shown in Ref. [135] that the geometric quantum discord can be consistently defined by using the one-norm, i.e., the trace-distance only. Note here that the asymmetry in the definition of quantum discord due to a local measurement over one of the parties also remains here in the choice of σ_{AB}. One can also consider "classical-quantum" states, $\tilde{\sigma}_{AB} = \sum_j p_j |j\rangle\langle j| \otimes \sigma_B^j$, to define the geometric quantum discord.

4 Monogamy of Quantum Discord

The monogamy score for quantum discord, defined after (6), is denoted by $\delta_D^{\leftarrow}$ or $\delta_D^{\rightarrow}$, depending on the subsystem over which the measurement is performed while computing the quantum discord. In Refs. [44, 45], it was found that quantum discord violates the monogamy relation already for certain three-qubit pure states.

Before discussing the monogamy properties of quantum discord in detail, let us first ask the question as to whether a measure of quantum correlation, chosen from the information-theoretic domain, can be monogamous. Although this is a difficult question to answer in its full generality, some insight can be obtained by noting that there is a marked difference between such a measure and the ones belonging to the entanglement-separability category. The former may have a non-zero value in the case of a separable state, while by definition, entanglement measures vanish for all unentangled states. Let us consider a general bipartite quantum correlation measure, $\mathcal{Q}$, which, for an arbitrary bipartite quantum state ρ_{AB}, obeys a set of basic properties, as enumerated below [43].

P1. Positivity: $\mathcal{Q}(\rho_{AB}) \geq 0$.

P2. Invariance under local unitary transformation: $\mathcal{Q}(\rho'_{AB}) = \mathcal{Q}(\rho_{AB})$, with $\rho'_{AB} = (U_A \otimes U_B)\rho_{AB}(U_A^\dagger \otimes U_B^\dagger)$. Here, U_A and U_B are unitary operators defined on the Hilbert spaces of the subsystems A and B.

P3. Non-increasing upon the introduction of a local pure ancilla: $\mathcal{Q}(\rho_{AB}) \geq \mathcal{Q}(\tilde{\rho}_{A:BC})$, with $\tilde{\rho}_{A:BC} = \rho_{AB} \otimes (|0\rangle\langle 0|)_C$.

Note here that the first two properties are standard requirements for any measure of quantum correlations, i.e., both entanglement and information-theoretic measures. The third property is satisfied, for example, by quantum discord, irrespective of whether the ancilla is attached to the measured, or the unmeasured side. The direction of the inequality in **P3** is interesting. It may seem that we should have $Q(\rho_{AB}) \leq Q_{A:BC}(\rho_{AB} \otimes (|0\rangle\langle 0|)_C)$, as throwing out the C-part, of a state in $A : BC$, may only "harm" (i.e., reduce Q), if at all. However, if we look at the definition of quantum discord, we find that *having* the extra C-part may do harm, as it is only the $I^{\leftarrow}$ term in $D^{\leftarrow}$ that can get affected due to the extra C-part. The extra C-part leads to a larger class of possible measurements that can be performed for the maximization in $I^{\leftarrow}_{A:BC}$, than in $I^{\leftarrow}_{AB}$.

A generic separable state of the AC system is given by

$$\rho_{AC} = \sum_i p_i P_A(|\psi_i\rangle) \otimes P_C(|\phi_i\rangle), \tag{20}$$

where $P(|\alpha\rangle) = |\alpha\rangle\langle\alpha|$. Let us now consider a special form of the tripartite separable state $\rho_{ABC} = \sum_i p_i P_A(|\psi_i\rangle) \otimes P_B(|i\rangle) \otimes P_C(|\phi_i\rangle)$ with $\{|i\rangle\}$ being a set of mutually orthonormal states. The quantum correlation, Q, in the $A : BC$ bipartition, has the same value as that in the unitarily connected state $\sigma_{ABC} = \sum_i p_i P_A(|\psi_i\rangle) \otimes P_B(|i\rangle) \otimes P_C(|0\rangle)$. Also, the amount of quantum correlation present in the state σ_{ABC} in the $A : BC$ bipartition can not be higher than that present in the $A : B$ bipartition in the state $\sigma_{AB} = \text{Tr}_C(\sigma_{ABC})$, i.e., $Q(\sigma_{AB}) \geq Q(\rho_{A:BC})$. If we now assume that Q satisfy the monogamy relation, then $Q(\sigma_{AB}) \geq Q(\rho_{AB}) + Q(\rho_{AC})$. Since $\rho_{AB} \equiv \sigma_{AB}$, we obtain $Q(\rho_{AC}) \leq 0$, which, due to the positivity of Q, implies $Q(\rho_{AC}) = 0$. Hence, a quantum correlation measure, Q, which is monogamous, and which obeys the properties **P1–P3**, must be zero for a generic separable state, ρ_{AC}. Contrapositively, a general bipartite quantum correlation measure, Q, which has a non-zero value for at least one separable state, and which obeys a set of basic properties must be non-monogamous [43].

The above discussion indicates that in the three-qubit scenario, monogamy for a general measure of quantum correlation, which belongs to the information-theoretic domain and which can be non-zero for a two-qubit separable state of rank 2, can be violated for pure three-qubit states. This includes quantum discord, and other "discord-like" measures [2]. As an example, let us consider the three-qubit generalized W states [136, 137], parametrized using two real parameters and given by

$$|\text{gW}\rangle = \sin\theta\cos\phi|100\rangle + \sin\theta\sin\phi|010\rangle + \cos\theta|001\rangle, \tag{21}$$

with $\theta, \phi\,(0 \leq \theta \leq \pi, 0 \leq \phi < 2\pi)$ being the real parameters. The plot of the negative of the quantity $\delta_D^{\leftarrow} = D^{\leftarrow}(\rho_{A:BC}) - D^{\leftarrow}(\rho_{AB}) - D^{\leftarrow}(\rho_{AC})$ as a function of θ and ϕ is presented in Fig. 2. Note that over the entire plane of (θ, ϕ), $-\delta_D^{\leftarrow}$ has positive values, indicating that generalized W states always violate monogamy of quantum discord [44, 45].

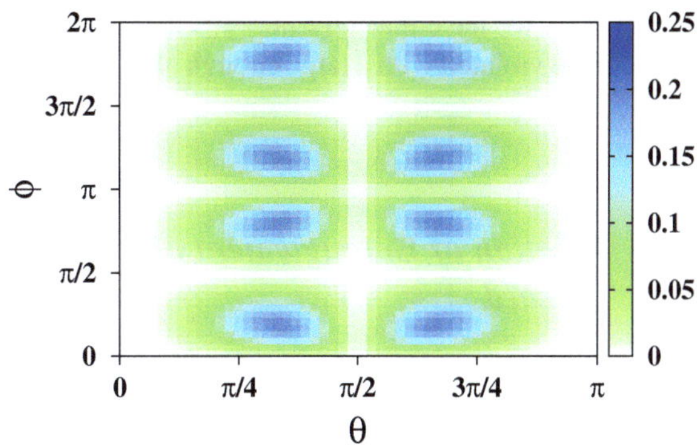

Fig. 2 Plot of $-\delta_D^{\leftarrow}$ as a function of θ and ϕ for three-qubit generalized W states given in Eq. (21). In agreement with the results reported in [44, 45], $-\delta_D^{\leftarrow}$ is positive over the entire plane of (θ, ϕ). Reproduced figure with permission from the Authors and the Publisher of Ref. [44]. Copyright (2012) of the American Physical Society

In this context, one must note that the monogamy inequality of quantum discord, as given by $D^{\rightarrow}(\rho_{A:BC}) \geq D^{\rightarrow}(\rho_{AB}) + D^{\rightarrow}(\rho_{AC})$, when the measurement is always performed over the nodal observer, can be shown to be equivalent to an inequality between the uninterrogated mutual information of the subsystem BC, and the EoF of the same, as given by [138]

$$E_f(\rho_{BC}) \leq \frac{\tilde{I}(\rho_{BC})}{2}, \tag{22}$$

for three-qubit pure states. Li and Luo [139] have shown that the inequality given in (22) does not hold for all tripartite pure states, thereby indicating that quantum discord can be both monogamous as well as non-monogamous, complementing the results obtained in [44, 45], where the measurements involved in the calculation of quantum discord are performed over the non-nodal observers.

The question that naturally arises next is whether a quantum correlation measure, Q, which violates one or more of the properties **P1–P3**, can be monogamous. In [43], it has been shown that a quantum correlation measure, Q, which is monogamous, and remains finite when the dimension of one of the subsystems, say, A, is fixed, i.e.,

$$Q(\rho_{AB}) \leq f(d_A) < \infty, \tag{23}$$

must be zero for all separable states. Here, d_A represents the dimension of A, and f is some function. To prove this, for a generic separable state ρ_{AB}, let us consider a symmetric extension of the form $\rho_{AB_1B_2\ldots B_N}$, where $\rho_{AB} = \rho_{AB_i} \forall i = 1, 2, \ldots, N$, N being an arbitrary positive integer, such that $Q(\rho_{AB}) = Q(\rho_{AB_i})$, $1 \leq i \leq N$ [140–143]. This implies $\sum_{i=1}^{N} Q(\rho_{AB_i}) = NQ(\rho_{AB})$. Now, the monogamy of Q implies

$$Q(\rho_{A:B_1B_2\ldots B_N}) \geq NQ(\rho_{AB}). \tag{24}$$

Using (23), we have that $Q(\rho_{A|B_1...B_N})$ is finite $\forall N$, including $N \to \infty$. Therefore, (24) can be violated with a large enough choice of N if $Q(\rho_{AB}) \neq 0$ for the separable state ρ_{AB}.

4.1 Relation of Entanglement of Formation to Quantum Discord

Let us consider a tripartite pure state ρ_{ABC}, which is a purification of the bipartite density matrices ρ_{AB} and ρ_{AC}, i.e., $\mathrm{Tr}_{B(C)}[\rho_{ABC}] = \rho_{AC(AB)}$. Let us now consider $I^{\leftarrow}(\rho_{AC})$, where the only difference with the quantity defined in Eq. (14) is that here we assume an optimization over POVMs [35, 144]. Let $\{p_i, |\psi_i\rangle\}$ be a pure state decomposition of ρ_{AB} that achieves the minimum in the definition of the EoF of ρ_{AB}. Now, there must exist a particular measurement setting $\{\tilde{M}_i\}$ corresponding to the subsystem C of the state ρ_{ABC}, for which the joint state of the rest of the system, AB, turns out to be $|\psi_i\rangle$ with probability p_i corresponding to the ith outcome [145]. This leaves the subsystem A in the state $\mathrm{Tr}_B(|\psi_i\rangle\langle\psi_i|)$. Following the definition of $I^{\leftarrow}(\rho_{AB})$, this implies [100]

$$I^{\leftarrow}(\rho_{AC}) \geq S(\rho_A) - \sum_i p_i S\left[\mathrm{Tr}_B(|\psi_i\rangle\langle\psi_i|)\right] \tag{25}$$

$$= S(\rho_A) - E_f(\rho_{AB}). \tag{26}$$

One may also consider an alternative approach, where a particular measurement $\{M_i\}$, when performed over the subsystem C, attains the maximum in $I^{\leftarrow}(\rho_{AC}) = S(\rho_A) - \sum_i p_i S(\rho_i)$. In general, $\{M_i\}$ may have a rank more than 1. It is now suggestive to decompose $\{M_i\}$ into rank-1 non-negative operators, M_{ij}, satisfying $M_i = \sum_j M_{ij}$. Let us assume that the action of M_{ij} on the subsystem C leaves the subsystem A in ρ_{ij} with probability p_{ij}, where the following relations hold: $p_i = \sum_j p_{ij}$ and $\rho_i = \sum_j \rho_{ij}$. Concavity of the von Neumann entropy implies that $S(\rho_A) - \sum_{ij} p_{ij} S(\rho_{ij}) \geq S(\rho_A) - \sum_i p_i S(\rho_i) = I^{\leftarrow}(\rho_{AC})$. However, this conflicts with the definition of $I^{\leftarrow}(\rho_{AC})$, unless $S(\rho_A) - \sum_{ij} p_{ij} S(\rho_{ij}) = I^{\leftarrow}(\rho_{AC})$. Now consider the action of $\{M_{ij}\}$ on the subsystem C. The state, $|\phi_{ij}\rangle$, of the subsystem AB, corresponding to the outcome ij, is a pure state, and the measurement, $\{M_{ij}\}$, therefore, leads to an ensemble $\{p_{ij}, |\phi_{ij}\rangle\}$, satisfying $\sum_{ij} p_{ij}|\phi_{ij}\rangle\langle\phi_{ij}| = \rho_{AB}$. This also signifies that $\rho_{ij} = \mathrm{Tr}_B\left[|\phi_{ij}\rangle\langle\phi_{ij}|\right]$. Consequently [100],

$$E_f(\rho_{AB}) \leq \sum_{ij} p_{ij} S(\rho_{ij}) \tag{27}$$

$$= S(\rho_A) - I^{\leftarrow}(\rho_{AC}). \tag{28}$$

Combining (26) and (28), we have [100]

$$E_f(\rho_{AB}) + I^{\leftarrow}(\rho_{AC}) = S(\rho_A). \tag{29}$$

This directly leads to a relation between EoF and quantum discord. The above relation turns out to be extremely important to prove several monogamy relations for different quantum correlation measures as can be seen in subsequent sections. However, note that the use of the relation implies that quantum discord is computed by performing POVMs.

Note here that $\rho_{ABC}^{\otimes n}$ is a purification of the states, $\rho_{AB}^{\otimes n}$ and $\rho_{AC}^{\otimes n}$. Moreover, the additivity of von Neumann entropy implies that $S(\rho_A^{\otimes n}) = nS(\rho_A)$. As a result, one obtains $E_f(\rho_{AB}^{\otimes n}) + I^{\leftarrow}(\rho_{AC}^{\otimes n}) = nS(\rho_A)$. Dividing both sides by n and taking the limit $n \to \infty$, Eq. (29) reduces to

$$E_C(\rho_{AB}) + C_D^{\leftarrow}(\rho_{AC}) = S(\rho_A), \tag{30}$$

where $E_C(\rho_{AB}) = \lim_{n\to\infty} \frac{1}{n} E_f(\rho_{AB}^{\otimes n})$ is the entanglement cost for creating ρ_{AB} by local operations and classical communication (LOCC) from a resource of shared singlets [146], and $C_D^{\leftarrow}(\rho_{AC}) = \lim_{n\to\infty} \frac{1}{n} I^{\leftarrow}(\rho_{AC})$ is the one-way distillable common randomness of ρ_{AC} [144].

4.2 *Conservation Law: Entanglement Versus Quantum Discord*

The above discussion directly leads to a conservation relation between EoF and quantum discord of an arbitrary three-qubit pure state, ρ_{ABC}. Note here that since ρ_{ABC} is pure, $S(\rho_A)$ is a good measure of quantum correlation of ρ_{ABC} in the $A : BC$ partition. Equation (29) implies that the amount of quantum correlation between the subsystem A and the rest of the system is the sum of the amount of quantum correlation present between A and B, and the amount of classical correlation present between A and C, thereby imposing a constraint over the distribution of the correlations between A and the rest of the system. Adding the uninterrogated mutual information between A and C, given by $\tilde{I}(\rho_{AC}) = S(\rho_A) + S(\rho_C) - S(\rho_{AC})$, to both sides of Eq. (29), one obtains

$$E_f(\rho_{AB}) = D^{\leftarrow}(\rho_{AC}) + \tilde{S}(\rho_{A|C}). \tag{31}$$

Proceeding in a similar fashion, one can write $E_f(\rho_{AB})$ and $E_f(\rho_{AC})$ as

$$E_f(\rho_{AB}) = D^{\leftarrow}(\rho_{BC}) + \tilde{S}(\rho_{B|C}),$$
$$E_f(\rho_{AC}) = D^{\leftarrow}(\rho_{AB}) + \tilde{S}(\rho_{A|B}). \tag{32}$$

Since the tripartite state is pure, $E_f(\rho_{C:AB}) = S(\rho_C)$ and $E_f(\rho_{B:AC}) = S(\rho_B)$, which, from Eq. (32), implies [42]

$$D^{\leftarrow}(\rho_{AB}) = E_f(\rho_{AC}) - E_f(\rho_{C:AB}) + E_f(\rho_{B:AC}). \tag{33}$$

Also, noticing that $\tilde{S}(\rho_{A|B}) = -\tilde{S}(\rho_{A|C})$ since the state ρ_{ABC} is pure, and using Eqs. (31) and (32), one obtains [42]

$$E_f(\rho_{AB}) + E_f(\rho_{AC}) = D^{\leftarrow}(\rho_{AB}) + D^{\leftarrow}(\rho_{AC}). \tag{34}$$

Note here that in the above discussion, we have considered the party A to be the nodal observer, and while computing quantum discord, the measurement is performed over the non-nodal observer. The above equation suggests that for an arbitrary tripartite pure state ρ_{ABC}, the sum of all possible bipartite entanglements shared by the nodal observer with the rest of the individual subsystems, as measured by the EoFs, can not be increased without increasing the sum of corresponding quantum discords by the same amount. This seems to indicate a "conservation relation" between EoF and quantum discord with respect to a fixed nodal observer in the case of a given tripartite pure state.

In this context, we point out that multipartite measures of quantum correlations have been defined as the sum of quantum correlations for all possible bipartitions in a multiparty quantum state, by using EoF, quantum discord, and geometric quantum discord as measures of bipartite quantum correlations. For tripartite pure states, the conservation law implies that certain such multipartite measures corresponding to EoF and quantum discord are equivalent [147, 148]. In the same vein, Ref. [149] investigates the above multipartite quantum correlations, in terms of EoFs and quantum discords for even and odd spin coherent states.

Since $D^{\leftarrow}(\rho_{A:BC}) = E_f(\rho_{A:BC}) = S(\rho_A)$ for an arbitrary tripartite pure state ρ_{ABC}, Eq. (34) implies an equivalence between the monogamy relation of EoF and quantum discord [45]. In the case of mixed states, the conservation law changes into [42]

$$D^{\leftarrow}(\rho_{AB}) + D^{\leftarrow}(\rho_{AC}) \geq E_f(\rho_{AC}) + E_f(\rho_{AB}) + \Delta, \tag{35}$$

where $\Delta = S(\rho_B) - S(\rho_{AB}) + S(\rho_C) - S(\rho_{AC})$.

4.3 Relation with Interrogated Information

We now derive a relation which gives a physical insight into the monogamy property of quantum correlation measures. For an arbitrary tripartite state ρ_{ABC}, an *uninterrogated* conditional mutual information is defined as

$$\tilde{I}(\rho_{A:B|C}) = \tilde{S}(\rho_{A|C}) - \tilde{S}(\rho_{A|BC}), \tag{36}$$

while the corresponding *interrogated* version can be expressed as

$$I(\rho_{A:B|C}) = S(\rho_{A|C}) - S(\rho_{A|BC}), \tag{37}$$

involving measurement over one or more of the subsystems. Here, $\tilde{I}(\rho_{A:B|C})$ and $I(\rho_{A:B|C})$ are non-negative, which is a direct consequence of the non-increasing nature of conditional entropy with an increase in the number of parties over which it is conditioned. The definitions of S and $\tilde{S}$ are as given in Sect. 3.1. Given a tripartite quantum state ρ_{ABC}, the *interaction information* [150], $I(\rho_{ABC})$, is defined as the difference between the information shared by the subsystem AB when C is present, and when C is traced out. Since $\tilde{S}(\rho_{A|C}) = S(\rho_{AC}) - S(\rho_C)$ and $\tilde{S}(\rho_{A|BC}) = S(\rho_{ABC}) - S(\rho_{BC})$, one can write an *uninterrogated* interaction information as [44]

$$\begin{aligned}\tilde{I}(\rho_{ABC}) &= \tilde{I}(\rho_{A:B|C}) - \tilde{I}(\rho_{AB}) \\ &= S(\rho_{AB}) + S(\rho_{BC}) + S(\rho_{AC}) - S(\rho_{ABC}) \\ &\quad -(S(\rho_A) + S(\rho_B) + S(\rho_C)).\end{aligned} \tag{38}$$

One can also define an *interrogated* interaction information, where the conditional entropies are defined so that a complete measurement has to be performed on one of the subsystems. In the case of the tripartite state ρ_{ABC}, an interrogated interaction information is given by

$$I(\rho_{ABC})_{\{\Pi_k^B,\Pi_i^C,\Pi_j^{BC}\}} = I(\rho_{A:B|C})_{\{\Pi_i^C,\Pi_j^{BC}\}} - I(\rho_{AB})_{\{\Pi_k^B\}}, \tag{39}$$

where the suffix on $I(\rho_{ABC})_{\{\Pi_k^B,\Pi_i^C,\Pi_j^{BC}\}}$ indicates that the measurements are performed over B, C, and BC. A similar notation is used to define $I(\rho_{A:B|C})_{\{\Pi_i^C,\Pi_j^{BC}\}} = S(\rho_{A|C})_{\{\Pi_i^C\}} - S(\rho_{A|BC})_{\{\Pi_j^{BC}\}}$ and $I(\rho_{AB})_{\{\Pi_k^B\}} = S(\rho_A) - S(\rho_{A|B})_{\{\Pi_k^B\}} \equiv S(\rho_A) - \sum_k p_k S(\rho_{A|k})$. For an arbitrary tripartite state ρ_{ABC}, the value of the interrogated interaction information is obtained by performing an optimization over the measurements. One can show that the quantum interaction information *(i)* can be either positive or negative, *(ii)* is invariant under local unitaries, and *(iii)* obeys the inequality $I(\rho_{ABC}) \geq \tilde{I}(\rho_{ABC})$ under unilocal measurements, which can be seen directly from the fact that quantum discord is non-negative [44].

If the optimization over the complete set of measurements is performed, the monogamy relation of quantum discord, i.e.,

$$D^{\leftarrow}(\rho_{AB}) + D^{\leftarrow}(\rho_{AC}) \leq D^{\leftarrow}(\rho_{A:BC}) \tag{40}$$

directly leads to

$$I(\rho_{A:B|C}) - I(\rho_{AB}) \leq \tilde{I}(\rho_{A:B|C}) - \tilde{I}(\rho_{AB}). \tag{41}$$

On the other hand, assuming the relation (41) implies that $\min_{\Pi_i^{BC}} S(\rho_{A|BC})_{\{\Pi_i^{BC}\}} - \tilde{S}(\rho_{A|BC}) \geq [\min_{\Pi_i^B} S(\rho_{A|B})_{\{\Pi_i^B\}} - \tilde{S}(\rho_{A|B})] + [\min_{\Pi_i^C} S(\rho_{A|C})_{\{\Pi_i^C\}} - \tilde{S}(\rho_{A|C})]$, which, in turn, implies the monogamy of quantum discord. Therefore, an arbitrary tripartite quantum state ρ_{ABC} is monogamous with respect to quantum discord if and only if $I(\rho_{ABC})_{\{\Pi_k^B,\Pi_i^C,\Pi_j^{BC}\}} \leq \tilde{I}(\rho_{ABC})$ [44]. For a tripartite pure state, since $\tilde{I}(\rho_{ABC}) = 0$, the interrogated interaction information is non-positive.

Consider the space of arbitrary three-qubit pure states, formed by the union of the GHZ and W classes, which are inequivalent under stochastic local operations and classical communication (SLOCC) [151]. The monogamy of quantum discord has been tested numerically for the GHZ and the W classes [44]. Evidence for both satisfaction and violation of monogamy relation of quantum discord in the former class has been found, while for the latter, monogamy of quantum discord is always found to be violated [44], if the measurement is performed over non-nodal observers. The violation of monogamy of quantum discord in the case of three-qubit pure states belonging to the W class can be proved analytically using the equivalence of monogamy relations of EoF and quantum discord in the case of tripartite pure states (see Sect. 4.2) [45]. Up to local operations, an arbitrary three-qubit pure state can be parametrized as [152, 153]

$$\begin{aligned}|\psi_{ABC}\rangle = {} & \lambda_0|000\rangle + \lambda_1 e^{i\gamma}|100\rangle + \lambda_2|101\rangle + \lambda_3|110\rangle \\ & +\lambda_4|111\rangle,\end{aligned} \tag{42}$$

where $\{\lambda_i : i = 1, \ldots, 4\}$ and γ are real parameters. For $\lambda_4 = 0$, Eq. (42) represents an arbitrary state from the W class, for which the tangle vanishes [13], i.e.,

$$\mathcal{C}_{AB}^2 + \mathcal{C}_{AC}^2 = \mathcal{C}_{A:BC}^2, \tag{43}$$

where $\mathcal{C}$ represents the concurrence [13, 20]. Since E_f $(0 \leq E_f \leq 1)$ is a concave function of $\mathcal{C}^2 (0 \leq \mathcal{C}^2 \leq 1)$, for two-qubit states, $E_f(\rho_{AB}) + E_f(\rho_{AC}) \geq E_f(\rho_{A:BC})$. Hence by using Eq. (34), we obtain the proof of violation of monogamy for quantum discord for the states from the W class.

4.4 Monogamy of Quantum Discord Raised to an Integer Power

Let us consider a bipartite quantum correlation measure, $\mathcal{Q}$, which is monotonically decreasing under discarding systems, and remains unchanged under discarding systems only for quantum states satisfying monogamy. Suppose that ρ_{ABC} is a state that violates the monogamy relation for $\mathcal{Q}$. This implies

$$\begin{aligned} Q(\rho_{A:BC}) &< Q(\rho_{AB}) + Q(\rho_{AC}), \\ Q(\rho_{A:BC}) &> Q(\rho_{AB}) > 0, \\ Q(\rho_{A:BC}) &> Q(\rho_{AC}) > 0. \end{aligned} \tag{44}$$

We have additionally assumed that $Q(\rho_{AB}) > 0$ and $Q(\rho_{AC}) > 0$. However, the vanishing Q cases can be handled separately. These directly lead to

$$\begin{aligned} \lim_{n\to\infty}\left[\frac{Q(\rho_{AB})}{Q(\rho_{A:BC})}\right]^n &= 0, \\ \lim_{n\to\infty}\left[\frac{Q(\rho_{AC})}{Q(\rho_{A:BC})}\right]^n &= 0. \end{aligned} \tag{45}$$

Therefore, for all values of $\epsilon > 0$, there exists two positive integers, $n_1(\epsilon)$ and $n_2(\epsilon)$, such that $[Q(\rho_{AB})/Q(\rho_{A:BC})]^m < \epsilon$ for all positive integers $m \geq n_1(\epsilon)$, and similarly for $[Q(\rho_{AB})/Q(\rho_{A:BC})]^m$ with respect to $n_2(\epsilon)$. With a choice of $\epsilon < 1/2$, one obtains $[Q(\rho_{AB})/Q(\rho_{A:BC})]^m, [Q(\rho_{AC})/Q(\rho_{A:BC})]^m < \epsilon$ for all positive integers $m \geq n(\epsilon)$, where $n(\epsilon) = \max[n_1(\epsilon), n_2(\epsilon)]$, leading to

$$\left[\frac{Q(\rho_{AB})}{Q(\rho_{A:BC})}\right]^m + \left[\frac{Q(\rho_{AC})}{Q(\rho_{A:BC})}\right]^m < 2\epsilon < 1, \tag{46}$$

for all positive integers $m \geq n(\epsilon)$. Therefore, considering the bipartite quantum correlation measure Q^m, we find that monogamy is obeyed for the state ρ_{ABC} [47].

In the case of quantum discord, monogamy property is obeyed for $m \geq 2$ [47, 154], while in the subsequent sections, we shall be discussing similar results regarding other measures of quantum correlations. Note also that if a quantum correlation measure Q is monogamous in the case of a three-party quantum state, any positive integer power of the measure is also monogamous for the same state, which can be shown directly by expanding $(Q(\rho_{AB}) + Q(\rho_{AC}))^m$, and considering the non-negativity of Q [47].

We illustrate this by considering the monogamy of squared quantum discord for two specific cases: **(1)** The three-qubit generalized W state, discussed in Sect. 4, and **(2)** a two-parameter three-qubit pure state, given by $|\psi(p,\epsilon)\rangle = \sqrt{p\epsilon}|000\rangle + \sqrt{p(1-\epsilon)}|111\rangle + \sqrt{(1-p)/2}(|101\rangle + |110\rangle)$, where p and ϵ are real. The variations of the monogamy score of squared quantum discord and quantum discord, as functions of the state parameters, are plotted in Fig. 3 [154], which clearly indicates that over the entire range of the state parameters, the squared quantum discord is monogamous, but the quantum discord itself is not.

The case of non-integer powers was considered in [155]. It was shown that a monogamous measure remains monogamous on raising its power, i.e., if $Q(\rho_{A:BC}) \geq Q(\rho_{A:B}) + Q(\rho_{AC})$, then $Q(\rho_{A:BC})^m \geq Q(\rho_{A:B})^m + Q(\rho_{AC})^m$, where $m \geq 1$. Similarly, one can also prove that a non-monogamous measure of quantum correlation remains non-monogamous with a lowering of the power [155]. Moreover, it has been pointed out in Ref. [155] that if a convex bipartite quantum correlation measure Q,

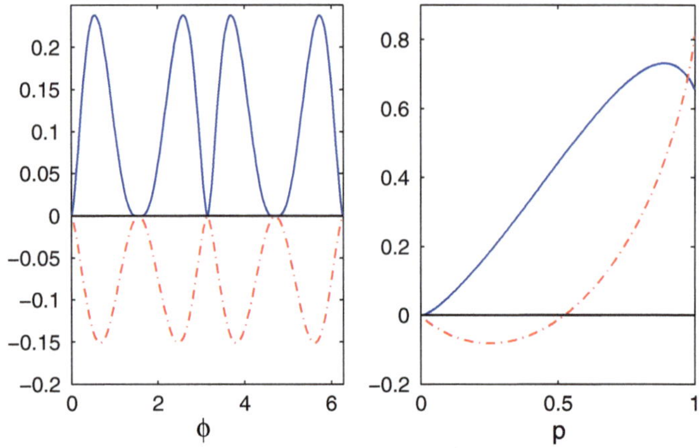

Fig. 3 Variation of the monogamy score for squared quantum discord (*blue solid line*) in comparison to that of quantum discord (*red dash-dotted line*) [154]. *Left* Variations of the monogamy score for squared quantum discord and quantum discord for the generalized W state, $|gW\rangle$, as functions of the parameter ϕ, where the parameter θ is set to $\pi/4$. *Right* Variations of the monogamy score for squared quantum discord and quantum discord in the case of the two-parameter state $|\psi(p, \epsilon)\rangle$, as a function of the state parameter p, where the other parameter is chosen to be $\epsilon = 0.5$. Reprinted figure with permission from the Authors and the Publisher of Ref. [154]. Copyright (2012) of the American Physical Society

when raised to a power $r = 1, 2$, is monogamous for pure tripartite states, then Q^r is also monogamous for the mixed states on the given Hilbert space. Note also that all the results mentioned in this subsection, except the monogamy of squared quantum discord, have been generalized to the N-partite case [47, 155].

4.5 N-Partite Quantum States

Research on the monogamy properties of quantum correlations belonging to either entanglement-separability or the information-theoretic paradigm predominantly deals with tripartite quantum states [13, 21, 41, 44, 45]. It is observed that "good" entanglement measures [156], which are known to be non-monogamous, in general, for tripartite quantum states, tend to obey monogamy, when considered for quantum states of a moderately large number of parties [157].

Irrespective of the genre of the measure used, it is possible to determine certain other independent sufficient conditions for an arbitrary bipartite quantum correlation measure to satisfy monogamy for arbitrary multiparty states. Let us consider an N-partite pure state $\rho_{12\ldots N}$, where each of the parties has a dimension d. As the first condition, we assume our chosen quantum correlation measure Q to be monogamous for all tripartite quantum states in dimension $d \otimes d \otimes d^m$ with $m \leq N - 2$. The monogamy of such an N-partite state can be expressed as

$$\mathcal{Q}(\rho_{1:23\ldots N}) \geq \mathcal{Q}(\rho_{12}) + \mathcal{Q}(\rho_{1:34\ldots N}), \tag{47}$$

where a partitioning of $\rho_{12\ldots N}$, given by $1:2:34\ldots N$, is assumed, and qubit 1 is chosen to be the nodal observer. One may, in turn, partition the state $\rho_{134\ldots N} = \text{Tr}_2[\rho_{12\ldots N}]$ as $1:3:4\ldots N$, and can continue to do so till the last couple of parties, labeled by $N-1$, and N. Recursively applying the tripartite monogamy relation, (47) can be reduced to

$$\begin{aligned}\mathcal{Q}(\rho_{1:23\ldots N}) &\geq \mathcal{Q}(\rho_{12}) + \mathcal{Q}(\rho_{1:34\ldots N}),\\ &\geq \mathcal{Q}(\rho_{12}) + \mathcal{Q}(\rho_{13}) + \mathcal{Q}(\rho_{1:45\ldots N}),\\ &\ldots\\ &\geq \sum_{k=1}^{N} \mathcal{Q}(\rho_{1k}),\end{aligned} \tag{48}$$

the intended monogamy of $\mathcal{Q}$ for the state $\rho_{12\ldots N}$ [21, 157].

As the second condition, let us assume the convexity of $\mathcal{Q}$, and a convex roof definition of $\mathcal{Q}$ in the case of a mixed state $\rho_{12\ldots N}$. For a tripartition $1:2:34\ldots N$ of an N-party mixed state $\rho_{12\ldots N}$, one obtains

$$\begin{aligned}\mathcal{Q}(\rho_{12\ldots N}) &= \mathcal{Q}\left(\sum_k p_k(|\psi\rangle\langle\psi|)^k_{12\ldots N}\right),\\ &= \sum_k p_k \mathcal{Q}\left(|\psi\rangle^k_{12\ldots N}\right),\end{aligned} \tag{49}$$

where the optimal convex roof decomposition providing $\mathcal{Q}(\rho_{12\ldots N})$ is $\{p_k, |\psi^k_{12\ldots N}\rangle\}$. If one additionally assumes that $\mathcal{Q}$ is monogamous for all tripartite pure states in dimension $d \otimes d \otimes d^m$, with $m \leq N-2$, then by using the convexity of $\mathcal{Q}$, from Eq. (49), one can write

$$\mathcal{Q}(\rho_{12\ldots N}) \geq \mathcal{Q}(\rho_{12}) + \mathcal{Q}(\rho_{1:34\ldots N}). \tag{50}$$

Continuing as before, the monogamy of the mixed state $\rho_{12\ldots N}$ with respect to $\mathcal{Q}$, i.e., the relation (48), can be proven.

The above result has been numerically tested in Ref. [157], by Haar-uniformly generating three-, four-, and five-qubit pure states. For several quantum correlation measures, it is found that the percentages of multiqubit pure states increase with increasing the number of parties. These measures include quantum discord and quantum work deficit. For example, the percentages of three-, four-, and five-qubit states, which are monogamous for quantum discord with measurement performed on the nodal observer, are respectively 90.5, 99.997, and 100%. This result indicates that in the case of a moderately large number of parties in the system, quantum correlation measures, which are known to be non-monogamous for tripartite quantum states, tend to obey monogamy for almost all states [157]. The adjective "almost" is necessary, since for a fixed number of parties, the set of Haar-uniformly generated states may exclude the sets of measure zero in the state space. Therefore, there may exist measure-zero non-monogamous multipartite states which can not be made monogamous for a specified quantum correlation measure by increasing the number of parties. Indeed, it is found that the family of N-qubit Dicke states [158] can not be made monogamous with respect to quantum discord by increasing the number of parties. More specifically, it has been shown that an N-partite pure state with vanishing tangle, i.e., $\mathcal{C}^2(\rho_{12\ldots N}) = \sum_{j=2}^{N} \mathcal{C}^2(\rho_{1j})$, violates the monogamy relation for quantum discord if the sum of the uninterrogated conditional entropy conditioned on all the non-nodal observers is a negative quantity [157]. In other words, $\mathcal{C}^2(\rho_{12\ldots N}) = \sum_{j=2}^{N} \mathcal{C}^2(\rho_{1j})$ implies $\delta_D^{\leftarrow} \leq \sum_{j=2}^{N} S(\rho_{1|j})$, and hence, for pure states having vanishing tangle and $\sum_{j=2}^{N} S(\rho_{1|j}) < 0$, the monogamy score for quantum discord is negative, and this is the case for the N-qubit Dicke states.

4.6 Monogamy of Quantum Discord in Open Quantum Systems

Although being of extreme importance, studies on the monogamy property of quantum correlations under noisy environments are limited, possibly due to the inherent mathematical difficulties. Recently, there has been experimental evidence, in photonic systems, of a flow of quantum correlations which occurs between a two-qubit system, AB, and its environment, E [159]. In this scenario, the initially present bipartite entanglement in the system AB decays with time, while multipartite entanglement and multipartite quantum discord emerge in the multipartite system consisting of the bipartite system and its environment. In another work, the dynamics of the monogamy property of quantum discord, in the case of global noise,[1] and dissipative and non-dissipative single-qubit quantum channels is discussed in [63], by using generalized GHZ [160] and generalized W [136, 137] states as input states to the noise. As a representative of the dissipative noise, amplitude-damping (AD) channel is used, while phase-damping (PD) and depolarizing (DP) channels are chosen as examples of non-dissipative channels.[2] In case of the three-qubit generalized GHZ

[1]The global noisy channel considered corresponds to the completely positive trace preserving (CPTP) map, $\rho \to \rho' = \zeta(\rho)$, given by $\rho' = \gamma\frac{\mathbb{I}}{d} + (1-\gamma)\rho$, where $\mathbb{I}$ is the identity matrix, d is the dimension of the Hilbert space on which ρ is defined, and $\gamma \in [0, 1]$ is the mixing factor.

[2]The CPTP maps, $\rho \to \rho' = \zeta(\rho)$, corresponding to these local noisy channels, can be given in the form of their respective Kraus operators, $\{E_k\}$, such that $\rho' = \sum_k E_k \rho E_k^\dagger$, where $\sum_k E_k^\dagger E_k = \mathbb{I}$. For the single-qubit AD, PD, and DP channels, the Kraus operators are given by $\{E_k^{ad}\}$, $\{E_k^{pd}\}$, and $\{E_k^{dp}\}$, respectively. The Kraus operators for the AD channel are given by

$$E_0^{ad} = \begin{pmatrix} 1 & 0 \\ 0 & \sqrt{1-\gamma} \end{pmatrix}, \quad E_1^{ad} = \begin{pmatrix} 0 & \sqrt{\gamma} \\ 0 & 0 \end{pmatrix},$$

state, given by $|\text{gGHZ}\rangle = a_0|000\rangle + a_1|111\rangle$, the monogamy score of quantum discord has been shown to be decaying monotonically with increasing strength of the noise parameter. On the other hand, in the case of three-qubit generalized W states given by $|\text{gW}\rangle = a_0|001\rangle + a_1|010\rangle + a_2|100\rangle$,[3] monogamy score for quantum discord exhibits non-monotonic behaviour when the value of the noise parameter is increased. Here, each qubit of the three-qubit states is used as input to the quantum channel under study. For example, when the DP channel is being studied, each qubit of the three-qubit state is fed into three independent DP channels. A characteristic value of the noise parameter, called the *dynamics terminal* [63], is introduced to quantify the robustness of the monogamy score against a particular type of noise applied to the input state, and depolarizing channel is identified as the one that destroys monogamy score faster than the other channels considered. A related statistics of three-qubit states belonging to the sets of generalized GHZ states and generalized W states is obtained numerically, which leads to a conclusive two-step distinguishing protocol to identify the type of noise applied to the three-qubit system. We discuss the details of the two-step channel discrimination protocol in Sect. 7.2. In [161], dynamics of quantum dissension [162], and the monogamy score of quantum discord in the case of amplitude-damping, dephasing, and depolarizing channels are discussed, when the input is from a set of three-qubit states including mixed GHZ states, mixed W states, and a certain mixture of separable and biseparable states. It was also found there that certain non-monogamous states become monogamous with the increase of noise.

5 Monogamy of Other Quantum Correlations

In this section, we discuss the monogamy properties of information-theoretic quantum correlation measures other than quantum discord. We will present results that are specific to the measures, while the results for generic measures of quantum

---

(Footnote 2 continued)
while for the PD and the DP channels, they are

$$E_0^{pd} = \sqrt{1-\gamma}\mathbb{I},\ E_1^{pd} = \frac{\sqrt{\gamma}}{2}(\mathbb{I}+\sigma_3),\ E_2^{pd} = \frac{\sqrt{\gamma}}{2},(\mathbb{I}-\sigma_3),$$

and

$$E_0^{dp} = \sqrt{1-\gamma}\mathbb{I},\ E_i^{dp} = \sqrt{\frac{\gamma}{3}}\sigma_i;\ i = 1, 2, 3,$$

respectively. Here, γ is the local noise parameter, with $\gamma \in [0, 1]$.

[3] Note that $|\text{gW}\rangle$, given in Eq. (21), has been parametrized in the spherical polar coordinates, and during numerical simulation, θ and ϕ are generated continuously, while in this case, a_0 and a_1 are chosen Haar-uniformly to simulate the $|\text{gW}\rangle$ state.

correlations, as discussed in the previous sections, remain valid. We start the discussion with quantum work deficit, which, apart from obeying the properties related to monogamy for general quantum correlation measures, has a direct relation with the monogamy of quantum discord in the case of tripartite pure states [47]. Assuming that the optimizations for both quantum discord and quantum work deficit of the bipartite state ρ_{AB} take place for the same ensemble $\{p_k, \rho^k_{AB}\}$, from the definition of quantum discord and quantum work deficit, one can show that

$$W^{\leftarrow}(\rho_{AB}) = D^{\leftarrow}(\rho_{AB}) - S(\rho_B) + H(\{p_k\}), \tag{51}$$

where $H(\{p_k\})$ is the Shannon entropy originating from the local measurement on the party B. Since $H(\{p_k\}) \geq S(\rho_B)$, $W^{\leftarrow}(\rho_{AB}) \geq D^{\leftarrow}(\rho_{AB})$. This implies that if quantum work deficit is monogamous, i.e., $W^{\leftarrow}(\rho_{A:BC}) \geq W^{\leftarrow}(\rho_{AB}) + W^{\leftarrow}(\rho_{AC})$, and since $W^{\leftarrow}(\rho_{A:BC}) = D^{\leftarrow}(\rho_{A:BC}) = S(\rho_A)$ [35–39] for pure states, we have

$$\begin{aligned} D^{\leftarrow}(\rho_{A:BC}) = W^{\leftarrow}(\rho_{A:BC}) &\geq W^{\leftarrow}(\rho_{AB}) + W^{\leftarrow}(\rho_{AC}) \\ &\geq D^{\leftarrow}(\rho_{AB}) + D^{\leftarrow}(\rho_{AC}), \end{aligned} \tag{52}$$

which implies monogamy of quantum discord. Note that the reverse is not true. Note also that although one can show that $W^{\leftarrow}(\rho_{AB}) + W^{\leftarrow}(\rho_{AC}) \geq D^{\leftarrow}(\rho_{AB}) + D^{\leftarrow}(\rho_{AC})$ for a three party mixed state under the same assumption of optimization.

In the case of arbitrary three-qubit pure states, quantum work deficit can be both monogamous and non-monogamous. As discussed in Sect. 4.4, similar to quantum discord, a state that is non-monogamous with respect to quantum work deficit can be made monogamous by raising quantum work deficit to an appropriate integer power, m. For quantum discord, one requires $m \geq 2$ to obtain monogamy, while for quantum work deficit, the percentage of non-monogamous three-qubit pure states with $m \geq 4$ is approximately 0.22, implying that a higher integer power of quantum work deficit is necessary to achieve monogamy for almost all three-qubit pure states.

Similar to the findings for quantum discord and quantum work deficit, and from the discussion in Sect. 4, geometric quantum discord is not monogamous in general, since it is non-zero in the case of some separable states. However, in contrast to quantum discord and quantum work deficit, geometric quantum discord is always monogamous for an arbitrary three-qubit pure state ρ_{ABC} [43]. Note that the geometric quantum discord, in the present case, is computed by performing the minimization over the set of all "classical-quantum states" instead of "quantum-classical states" as defined in Sect. 3.3. This is proved by showing the existence of a classical-quantum state, σ_{ABC}, for which $D_G(\rho_{A:BC}) \geq ||\rho_{AB} - \sigma_{AB}||_2^2 + ||\rho_{AC} - \sigma_{AC}||_2^2$, where $\sigma_{AB(AC)} = \text{Tr}_{C(B)}(\sigma_{ABC})$, whenever ρ_{ABC} is pure. Since the right hand side of the inequality is always bigger than $D_G(\rho_{AB}) + D_G(\rho_{AC})$, due to the minimization involved in geometric quantum discord, we obtain the claimed monogamy of D_G. Monogamy of geometric quantum discord is also considered in Ref. [163], while in the multiqubit scenario, the monogamy of D_G is addressed in Refs. [164, 165].

Ref. [166] investigates the monogamy of geometric quantum discord in photon added coherent states.

The monogamy of quantum correlations in the case of three-qubit pure symmetric states, in the Majorana representation [167], was addressed in [168], where the Rajagopal-Rendell quantum deficit (RRQD) [169, 170] is used as the quantum correlation measure. For a bipartite quantum state ρ_{AB}, RRQD is defined as the relative entropy distance [34, 171] of the state ρ_{AB} from $\rho^d_{AB} = \sum_{ij} p_{ij}|i\rangle\langle i| \otimes |j\rangle\langle j|$, which is diagonal in the eigenbasis of the marginals, ρ_A, and ρ_B, given by $\{|i\rangle\}$ and $\{|j\rangle\}$, respectively. Here, $p_{ij} = \langle j|\langle i|\rho_{AB}|i\rangle|j\rangle$, with $\sum_{i,j} p_{ij} = 1$. Mathematically,

$$D_{RR} = S(\rho_{AB}||\rho^d_{AB}), \tag{53}$$

where for two arbitrary density matrices ρ and σ, $S(\rho||\sigma) = \text{Tr}(\rho \log_2 \rho - \rho \log_2 \sigma)$. In the case of three-qubit pure symmetric states, RRQD was shown to be non-monogamous in general [168]. In particular, it was shown that although generalized W states can satisfy as well as violate the monogamy inequality for RRQD, the generalized GHZ states always satisfy the relation.

We conclude this section by mentioning the monogamy property of measurement-induced nonlocality, introduced by Luo and Fu [172], and defined as

$$N(\rho_{AB}) = \max_{\{\Pi_A\}} ||\rho_{AB} - \rho'_{AB}||_2^2, \tag{54}$$

where $\rho'_{AB} = \sum_i \Pi^i_A \otimes \mathbb{I}_B \rho_{AB} \Pi^i_A \otimes \mathbb{I}_B$, $\mathbb{I}_B$ is the identity matrix in the Hilbert space of B, and $\{\Pi^i_A\}$ is the set of elements of a projective measurement for which $\sum_i \Pi^i_A \rho_A \Pi^i_A = \rho_A$. For this measure, three-qubit pure states belonging to the GHZ and W classes can be non-monogamous in general [173], although unlike quantum discord, both the generalized GHZ and generalized W states satisfy the monogamy relation. The measurement-induced nonlocality has been also quantified from the perspective of relative entropy by Xi et al. [174]. Subsequently, Ref. [175] provides necessary and sufficient conditions for monogamy inequalities of this measure.

6 Relation with Other Multiparty Measures

An important perspective of the monogamy inequality of quantum correlations can be obtained by harvesting its inherent multipartite nature, and establish relations between the monogamy scores and the other quantum correlation measures including bipartite and multipartite entanglement. In several works, monogamy score of a given quantum correlation measure has been used as important markers of multipartite quantum correlations. In the seminal paper by Coffman, Kundu, and Wootters [13] introducing the monogamy inequality for tripartite states, tangle has been described as "essential three-qubit entanglement". Over the years, quantities such as monogamy scores for different quantum correlation measures, including quantum discord and quantum work deficit, have been used as bona-fide measures of multipartite quantum correlations [44, 77].

Recently, the relations of monogamy score for quantum discord with different multipartite quantum correlation measures, such as tangle [13], genuine multiparty entanglement measures quantified by generalized geometric measure (GGM) [176, 177], and global quantum discord [178], have been established [44, 179, 180]. Significant approaches to quantify multipartite entanglement and quantum correlations using the monogamy principle have also been undertaken [79, 154, 181, 182]. Monogamy scores of quantum discord and violation of Bell inequalities have also been connected [183, 184]. Tripartite dense coding capacities are also shown to have relations with the monogamy score of quantum discord [185, 186].

6.1 Monogamy Score Versus Other Multi-site Quantumness Measures

In Ref. [187], the monogamy score of a quantum correlation measure for an N-qubit pure multipartite state, $\rho_{12\ldots N} = |\Psi\rangle\langle\Psi|$, is shown to be intrinsically related to the genuine multipartite entanglement, as quantified by the GGM [176, 177]. An N-party pure quantum state $|\Psi\rangle$ is genuinely multipartite entangled if there exists no bipartition across which the state is product. The GGM, $\mathcal{G}$, of $|\Psi\rangle$ is defined as

$$\mathcal{G}(|\Psi\rangle) = 1 - \max_{\{|\Phi\rangle\}} |\langle\Phi|\Psi\rangle|^2, \tag{55}$$

where the maximization is over the set of states $\{|\Phi\rangle\}$, which are not genuinely multiparty entangled. The GGM is known to be an entanglement monotone [176]. The above expression for GGM can be simplified to the form, $\mathcal{G} = 1 - \max_{\mathrm{k}\in[1,\mathrm{N}/2]} \left[\{\xi_m(\rho^{(k)})\}\right]$, where $\{\xi_m(\rho^{(k)})\}$ is the set of highest eigenvalues of all possible reduced k-qubit states, where k ranges from 1 to $N/2$.

For three-qubit pure states, the maximum eigenvalue has to arise from single-qubit density matrices. Hence, an immediate relation between the monogamy score δ_Q and the GGM can be established in this case [44, 78]. Let $a = \max\{\xi_m(\rho^{(1)})\}$ be the maximum eigenvalue corresponding to the single-qubit reduced state, $\rho^{(1)}$ of $|\Psi\rangle$, so that $\mathcal{G} = 1 - a$. From Eq. (8), $\delta_Q^j \leq Q(\rho_{j:\mathrm{rest}})$, where j is the nodal qubit. Now, the quantity $Q(\rho_{j:\mathrm{rest}})$ is a function of a, say $F^Q(a)$. Moreover, we have $\mathcal{G} = 1 - a$, which gives us $F^Q(a) = F^Q(1 - \mathcal{G})$, and thus the bound, $\delta_Q^j \leq F^Q(a) = F^Q(1 - \mathcal{G})$. Now the monogamy score, δ_Q, is defined as the minimum score over all possible nodes, implying $\delta_Q \leq \delta_Q^j$. Hence, we obtain an upper-bound on the monogamy score in terms of a function of GGM, given by

$$\delta_Q(|\Psi\rangle) \leq F^Q(1 - \mathcal{G}(|\Psi\rangle)). \tag{56}$$

For quantum correlation measures that reduce to the von Neumann entropy for pure states, such as distillable entanglement [12], entanglement cost [188], entanglement of formation [22], squashed entanglement [99, 100], relative entropy of entanglement [32, 34], quantum discord [35, 36], and quantum work-deficit [37–40], the function $F^{\mathcal{Q}}(1-\mathcal{G}) = h(\mathcal{G})$, where $h(x)$ is the Shannon entropy. For the entanglement monotones, squared concurrence [13] and squared negativity [30, 41], the quantity $F^{\mathcal{Q}}(1-\mathcal{G})$ is equal to $z\mathcal{G}(1-\mathcal{G})$, where $z = 4$ and 1 for $\mathcal{C}^2$ and $\mathcal{N}^2$, respectively. We present a plot in Fig. 4 of the upper bound for the case of three-qubit pure states, with the quantum correlation being chosen as quantum discord. The above relation can be generalized to N-qubit pure states, where the upper bound on $\delta_{\mathcal{Q}}$ in terms of the entropic or quadratic functions of GGM, can be shown to exist for all states that satisfy a set of necessary conditions. For instance, for all N-qubit states $|\Psi\rangle$ with $\mathcal{G} = 1 - a$, the upper bound is universally valid. In other words, for N-qubit state $|\Psi\rangle$, if the maximum eigenvalue, among eigenvalues of all local density matrices, is obtained from a single-qubit density matrix, it has been shown that the upper-bound remains valid [187].

An important implication of the above bound is that it is even for those quantum correlation measures for which the corresponding $\delta_{\mathcal{Q}}$ can not be explicitly computed for arbitrary states. Examples of such measures include distillable entanglement, entanglement cost, and relative entropy of entanglement. The theorem implies that any possible value for these measures will always result in a $\delta_{\mathcal{Q}}$ that lies on or above the boundary.

Apart from entanglement measures, violations of Bell inequalities are important indicators of quantumness present in compound systems [56]. The two-point correlation function Bell inequality violations, and their monogamy properties [48, 49] have been connected with the monogamy scores of entanglement and quantum discord in three-qubit quantum systems [183, 184]. It has been shown that for three-qubit pure states, the monogamy scores for quantum correlations including quantum discord can be upper-bounded by a function of the monogamy score for Bell inequality violation [183]. Moreover, it was shown in Ref. [184] that if the monogamy score for quantum discord in the case of an arbitrary three-qubit pure state, $|\Psi\rangle$, is the same as that of the three-qubit generalized GHZ state, then the monogamy score corresponding to the Bell inequality violation of $|\Psi\rangle$ is bounded below by the same as that of the generalized GHZ state, when the measurements in quantum discord are performed on the non-nodal observer. In case the measurements are performed on the nodal observer, the role of the generalized GHZ state is replaced by the "special" GHZ state of N qubits [184], given by $|\text{sGHZ}\rangle = |00\ldots0\rangle_N + |11\rangle \otimes (\beta|00\ldots0\rangle + \sqrt{1-\beta^2}e^{-i\theta}|11\ldots1\rangle)_{N-2}$, where $\beta \in [0, 1]$, and θ is a phase.

6.2 Information Complementarity: Lower Bound on Monogamy Violation

For a multipartite system in a pure quantum state, $\rho_{12\ldots N}$, let us first divide the whole system into two parts, x and y, such that $x \cup y = \{1, 2, \ldots, N\}$. It is possible to derive an information-theoretic complementarity relation between the purity of the subsystem ρ_x, where $\rho_x = \text{Tr}_y \rho_{12\ldots N}$, and the bipartite quantum correlation shared between the subsystem x with the rest of the system, i.e., with y, and is given by [75]

$$P(\rho_x) + \mathcal{Q}(\rho_{xy}) \leq b \begin{cases} = 1, & \text{if } d_x \leq d_y, \\ = 2 - \frac{\log_2 d_y}{\log_2 d_x}, & \text{if } d_x > d_y, \end{cases}. \tag{57}$$

Here, $d_{x(y)}$ is the Hilbert-space dimension of $x(y)$, $P = \frac{\log_2 d_x - S(\rho_x)}{\log_2 d_x}$ is the normalized purity of the subsystem x, $S(\rho_x)$ is the von Neumann entropy of ρ_x, and $\mathcal{Q}(\rho_{xy}) = \frac{Q(\rho_{xy})}{\min\{\log_2 d_x, \log_2 d_y\}}$ is the normalized quantum correlation (with $Q(\rho_{xy})$ being the corresponding quantum correlation) shared between the subsystems x and y. The proof of the relation (57) requires that $Q(\rho_{xy})$ satisfies the conditions $Q(\rho_{xy}) \leq S(\rho_x)$. However, it is independently satisfied by several important quantum correlation measures [75, 189]. The above complementarity relation has useful application in quantum key distribution [75].

If we now consider a non-monogamous normalized bipartite quantum correlation measure, $\mathcal{Q}$, which could, for example, be normalized quantum discord or normalized quantum work deficit, we can obtain a useful lower bound on the monogamy score in terms of purity. For an N-qudit state, using (8) and (57), one obtains the relation [189],

$$\delta_{\mathcal{Q}} \geq -(N-2)\left(1 - P(\rho_{n_0}) + \frac{1 - x_0}{N-2}\right), \tag{58}$$

where n_0 is the nodal qubit and $x_0 = P(\rho_{n_0}) + \mathcal{Q}(\rho_{n_0:\text{rest}})$. For $x_0 \geq 1$ and large N, we obtain $\delta_{\mathcal{Q}} \geq -(N-2)(1 - P(\rho_{n_0}))$, which provides a nontrivial lower bound for the monogamy score of Q. For three-qubit states, the lower bound of $\delta_{\mathcal{Q}}$ reduces to $\delta_{\mathcal{Q}} \geq -S(\rho_{n_0})$. See Fig. 4.

6.3 Relation with Multiport Dense Coding Capacity

Another application of the monogamy inequality and the monogamy score is their role in estimating optimal classical information transfer in multiport dense coding protocols [57, 190–194]. A complementarity relation of the monogamy score for quantum correlation measures, such as squared concurrence and quantum discord, with the maximal dense coding capacity for pure tripartite quantum states was established in Ref. [185]. The dense coding capacity of a bipartite quantum state ρ_{12} is given by [190–194]

$$\mathbb{C}(\rho_{12}) = \max\left[\log_2 d_1, \log_2 d_1 + S(\rho_2) - S(\rho_{12})\right]. \tag{59}$$

Without a shared entangled resource, the capacity would be $\log_2 d_1$, and hence the quantum advantage is $\mathbb{C}_{\text{adv}} = \max\{0, S(\rho_2) - S(\rho_{12})\}$. One can consider a multiport communication with $\rho_{12\ldots N}$ as the resource state, where 1 is the sender, and the rests are the receivers. The quantum advantage in multiport dense coding for the transfer of classical information from 1 to $N-1$ individuals can be defined as $\mathbb{C}_{\text{adv}} = \max\left[\{S(\rho_i) - S(\rho_{1i})|\forall i = 2,\ldots,N\}, 0\right]$. Now, if one considers the set of pure tripartite quantum states, the monogamy score for squared concurrence and quantum discord are intrinsically related to the quantum advantage in dense coding. Specifically, it was shown that for any fixed monogamy score, the maximum quantum advantage is obtained from a single parameter family of three-qubit pure states, given by $|\psi_\alpha\rangle = |111\rangle + |000\rangle + \alpha(|101\rangle + |010\rangle)$ [185]. It is possible to derive a complementarity relation between the monogamy score of quantum discord and the quantum advantage. A similar complementarity relation exists between the monogamy score of squared concurrence and $\mathbb{C}_{\text{adv}}$.

Monogamy scores of entanglement and quantum discord have also been related to the multiparty dense coding capacity between several senders and a single receiver [186]. In particular, it was shown that in the noiseless scenario, among all multiqubit pure states with an arbitrary but fixed multiparty dense coding capacity, the generalized GHZ state has the maximum monogamy score for quantum discord, i.e., if $\tilde{C}(|\psi\rangle) = \tilde{C}(|gGHZ\rangle)$, it implies $\delta_D(|\psi\rangle) \le \delta_D(|gGHZ\rangle)$, where $\tilde{C}(\rho_{12\ldots N}) = \log_2 d_{1\ldots N-1} + S(\rho_N) - S(\rho_{1\ldots N})$, with $\rho_{1\ldots N}$ being a state shared between $N-1$ senders, $1, 2, \ldots, N-1$, and the receiver, N. Here, $d_{1\ldots N-1} = d_1 d_2 \ldots d_{N-1}$. We have suppressed the arrow in the superscript of δ_D here, as the result is true independent of the direction of the arrow. The above result is also true if δ_D is replaced by the tangle. Note also that Ref. [186] also considers the noisy channel case.

7 Physical Applications

In recent years, several works have been undertaken to elucidate the role of monogamy of quantum correlations in studying quantum systems and their dynamics. In particular, the concept of monogamy has been used to characterize quantum states [44, 45, 168, 173] and channels [63], and also to provide deeper understanding of many physical properties such as critical phenomena in many-body systems [68–72] involving complex quantum models such as frustrated spin lattices [72], and biological compounds [73, 74]. Moreover, monogamy also provides an important conceptual basis to quantify quantum correlations in multiparty mixed states, by using the concept of the monogamy score in situations where the usual measures of quantum correlations are neither easily accessible nor computable. More precisely, given a bipartite quantum correlation measure, the monogamy score for that measure defined for a given multiparty system, leads us to a measure of multipartite quantum correlation, without an increase in the complexity on both experimental and theoretical fronts as compared to those at the level of the bipartite measure.

7.1 State Discrimination

An important aspect in the study of the monogamy properties of quantum correlation measures, such as quantum discord and quantum work-deficit, is that these measures are not universally monogamous. In other words, for these quantum correlation measures, the monogamy inequality is not universally satisfied for all quantum states. This dichotomy allows the monogamy of quantum correlations to be an important figure of merit in state discrimination. In particular, for three-qubit pure states, it was shown that while the generalized GHZ states are always monogamous with respect to quantum discord, all generalized W states violate the monogamy inequality [44, 45]. The above results were extended to the more general sets, viz. the GHZ and the W class states, where it was shown that more than 80% of the Haar-uniformly generated GHZ class states satisfy monogamy, in contrast to *W* class states which are always non-monogamous (see Fig. 4). The monogamy score, therefore, plays

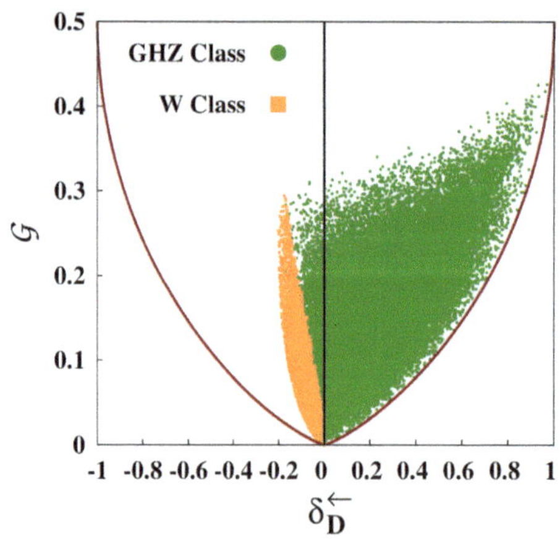

Fig. 4 *Upper* and *lower* bounds on the monogamy score of quantum discord for pure three-qubit GHZ- and W-class states. For three-qubit pure states, the *upper* bound using the GGM, $\mathcal{G}$, and the *lower* bound from the information complementarity, are given by the quantity $h(\mathcal{G})$ and $-h(\mathcal{G})$, respectively, where h is the binary entropy function. The scatter points in the figure correspond to 10^6 three-qubit pure states generated Haar uniformly. An equal number of W-class states are also Haar uniformly generated. The figure shows that all W-class states have a negative monogamy score that is weakly bounded below by $-S(\mathcal{G})$, whereas GHZ-class states can have both positive and negative monogamy scores. Reprinted figure with permission from the Authors and the Publisher of Ref. [78]. Copyright (2012) of the American Physical Society

a role that is akin to entanglement witnesses [27, 195–199]. Indeed a given linear entanglement witness allots values (real numbers) with a certain sign (say, negative), to all separable states while for entangled states, the same witness can have values of both signs. So, a positive value of the witness for a certain state immediately implies that the state is entangled. Similarly, a positive value of the monogamy score for quantum discord for a three-qubit pure state implies that the state is from the GHZ class. Subsequently, the comparative studies of the SLOCC inequivalent classes was discussed using the monogamy score of another measure of quantum correlation [168], namely, the quantum deficit [169, 170]. It was shown that while generalized W states may violate monogamy, generalized GHZ states always satisfy the monogamy inequality of quantum deficit. We therefore find that the state discrimination protocol using monogamy inequalities is dependent on the choice of the quantum correlation. For instance, using the monogamy properties of measurement induced non-locality [172], it was shown that both tripartite generalized GHZ and generalized W states are monogamous [173].

The monogamy inequality and the related monogamy scores have also been used to characterize pure tripartite quantum states [77], by finding the relation of the monogamy scores for those states with the corresponding values for measures of genuine multipartite entanglement, viz. the GGM [176, 177] and the multipartite Mermin-Klyshko Bell inequalities [55, 200–202]. In particular, tripartite states that have a vanishing monogamy score for quantum discord have been explored in this way [77]. Some of these aspects have been discussed in Sect. 6.

7.2 Channel Discrimination

Another application of the monogamy considerations of quantum correlations comes from the study of their behavior under the action of global and local noisy channels. It has been observed that an analysis of the dynamics of the monogamy scores of quantum discord and entanglement, quantified by negativity, for initial tripartite generalized GHZ and generalized W states can conclusively identify the noisy channel acting on the system [63]. By analyzing the monogamy scores for quantum discord (δ_D) and negativity ($\delta_{\mathcal{N}}$), with generalized GHZ and generalized W states as inputs, a two-step discrimination protocol to identify the above channels has been developed. To describe the protocol, let us consider an apriori unknown noisy channel, chosen from a set containing a global noise channel, and the AD, PD, and DP channels. See Sect. 4.6 for descriptions of these channels. In step 1 of the channel-discrimination protocol, one feeds generalized W state to the unknown channel with moderate noise, i.e., $\gamma \in [0.4, 0.6]$. The monogamy score of discord is the primary indicator in this step. For the global noise and DP channel, $\delta_D \geq 0$, while it is strictly negative for the AD and PD channels. In step 2 of the protocol, the same unknown channel, with moderate noise, is applied to an generalized GHZ state, and in this instance, both $\delta_{\mathcal{N}}$ and δ_D of the output state are estimated. It is observed that $\delta_{\mathcal{N}} > 0$ for the global noisy channel, but vanishes for the DP channel. Note that step 2 distinguishes between the

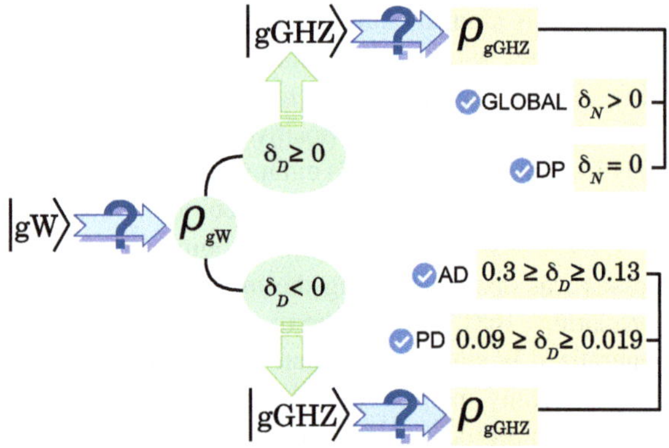

Fig. 5 Schematic representation of the two-step channel discrimination protocol proposed in [63]. Reproduced figure with permission from the Authors and the Publisher of Ref. [63]. Copyright (2016) of Physics Letter A (Elsevier)

instances which exhibit $\delta_D \geq 0$ in step 1. On the other hand, in step 2, $\delta_D \geq 0.13$ for the AD channel and $\delta_D \leq 0.09$ for the PD channel. Therefore, the values of δ_D and $\delta_{\mathcal{N}}$ together can discriminate between the global noise and the three local noisy channels [63]. A schematic representation of the two-step channel discrimination protocol can be found in Fig. 5.

Identification of quantum channels using the dynamics of monogamy scores of quantum correlations is potentially an important addition to the literature on channel identification [203] and estimation [204], which is a significant yet less discussed part of the vast literature on quantum state estimation [205].

7.3 *Characterization of Quantum Many-Body Systems*

Although several studies have attempted to characterize many-body quantum systems using quantum correlations beyond entanglement (for a review, see Ref. [2]), those engaging monogamy of quantum correlations to understand cooperative properties in strongly-correlated many-body systems are relatively scarce. In Ref. [68], monogamy property of quantum discord is used to characterize the ground state of the one dimensional bond-charge Hubbard model. A paradigmatic quantum spin model in one dimension is the XYZ model, represented by the Hamiltonian

$$H = \frac{J}{4}\sum_{i=1}^{N}\left\{(1+g)\sigma_i^x\sigma_{i+1}^x + (1-g)\sigma_i^y\sigma_{i+1}^y + \Delta\sigma_i^z\sigma_{i+1}^z\right\} + \frac{h_f}{2}\sum_{i=1}^{N}\sigma_i^z, \tag{60}$$

where J is the strength of the nearest-neighbour exchange interaction, g is the $x - y$ anisotropy parameter, Δ and h_f are the anisotropy, and the strength of the external

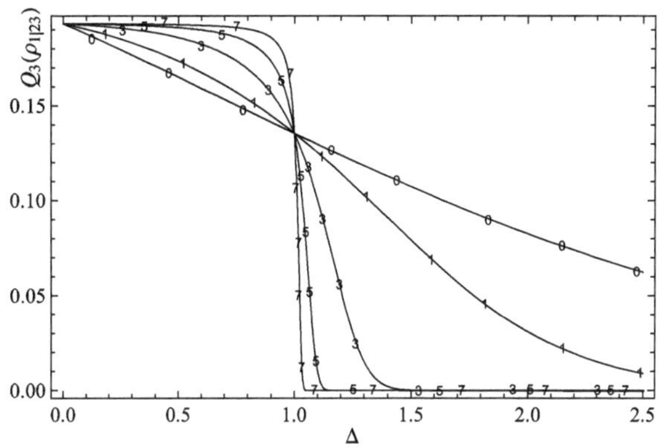

Fig. 6 Variation of the monogamy score of squared quantum discord as a function of Δ with increasing system-size in the case of the XXZ model. Reprinted figure with permission from the Authors and the Publisher of Ref. [70]. Copyright (2014) of the Europhysics Letters (Institute of Physics)

magnetic field, respectively, in the z direction. Several important one-dimensional quantum spin Hamiltonians emerge from Eq. (60). For example, the Hamiltonian in Eq. (60), with $g = 0$ and $h_f = 0$, represents the one-dimensional XXZ model, while for $\Delta = 0$ and $g = 1$, the model corresponds to the one-dimensional quantum Ising model in an external transverse magnetic field. In Ref. [70], the monogamy score of squared quantum discord is used to investigate the critical points of the one-dimensional XXZ model. See Fig. 6 for the variation of $\delta_{D^2}^{\leftarrow}$ (given by $\mathcal{Q}_3(\rho_{1|23})$ in Ref. [70]) against the anisotropy parameter, Δ. Monogamy properties of other quantum correlation measures such as geometric discord [69] and measurement induced disturbance [71] have also been investigated in the XXZ model. In an experimental study investigating the ground state of the one-dimensional quantum Ising model in a transverse magnetic field, by using a nuclear magnetic resonance (NMR) setup, monogamy scores of negativity and quantum discord are shown to distinguish between the cases of positive (frustrated phase) and negative (non-frustrated phase) values of J in the ground state of the system [72]. It is interesting to note that monogamy of entanglement has been used to constrain the bipartite entanglement of resonating valence bond states [64–66].

7.4 *Quantum Biological Processes*

An interesting and recent development in physics has been the investigation of quantum effects in certain complex biological processes. In particular, light-harvesting protein complexes have been modeled to investigate photosynthetic processes in certain bacteria, with specific interest in the role of quantum coherence and quantum correlations [206–209]. Several studies have investigated the well-known Fenna-Mathews-Olson (FMO) complex, which mediates energy transfer from the receiving chromophores to the central reaction center, and attempted to characterize the efficiency of the energy transfer in terms of quantum correlations [207, 208]. The role of monogamy of quantum correlations in the dynamics was recently investigated in Refs. [73, 74]. In Ref. [74], it is shown how the monogamy of quantum correlations, as quantified by negativity and quantum discord, is able to detect the arrangement of the different chromophore nodes in the FMO complex and support the predicted pathways for the transfer of excitation energy. The results also reiterate the predominance of multiparty quantum correlation measures over bipartite correlations between the nodes of the FMO complex.

8 Conclusions

Like other no-go theorems [15–18, 210–213] in quantum information science, in a multipartite domain, restrictions on sharability of quantum correlations, named as monogamy of quantum correlations, play a crucial role in achieving successes in, and in understanding of several quantum information processing tasks. In this chapter, we have discussed the monogamy properties of information-theoretic quantum correlations, specifically quantum discord, and highlighted their significant features. Computable multipartite quantum correlation measures are rare, although there are a handful of bipartite quantum correlation measures, including quantum discord and several "discord-like" measures, which are possible to calculate, at least numerically. The concept of monogamy opens up a new avenue where multipartite properties of a system can be studied via bipartite quantum correlations and becomes extremely useful to study different physical systems. This leads to another interesting aspect of monogamy, namely, its application in several key phenomena in quantum physics, ranging from quantum communication to the emerging research on quantum spin models, quantum biology, and open quantum systems, which is also reviewed.

References

1. R. Horodecki, P. Horodecki, M. Horodecki, K. Horodecki, Quantum entanglement. Rev. Mod. Phys. **81**, 865 (2009)
2. K. Modi, A. Brodutch, H. Cable, T. Patrek, V. Vedral, The classical-quantum boundary for correlations: discord and related measures. Rev. Mod. Phys. **84**, 1655 (2012)
3. N. Gisin, G. Ribordy, W. Tittel, H. Zbinden, Quantum cryptography. Rev. Mod. Phys. **74**, 145 (2002)
4. B.M. Terhal, Is entanglement monogamous? IBM J. Res. Dev. **48**, 71 (2004)
5. A. Sen(De) and U. Sen, Quantum advantage in communication networks. Phys. News **40**, 17 (2010)
6. R. Raussendorf, H.J. Briegel, A one-way quantum computer. Phys. Rev. Lett. **86**, 5188 (2001)
7. H.J. Briegel, D.E. Browne, W. Dür, R. Raussendorf, M. Van den Nest, Measurement-based quantum computation. Nat. Phys. **5**, 19 (2009)
8. M. Lewenstein, A. Sanpera, V. Ahufinger, B. Damski, A. Sen(De), U. Sen, Ultracold atomic gases in optical lattices: mimicking condensed matter physics and beyond. Adv. Phys. **56**, 243 (2007)

9. L. Amico, R. Fazio, A. Osterloh, V. Vedral, Entanglement in many-body systems. Rev. Mod. Phys. **80**, 517 (2008)
10. R. Augusiak, F.M. Cucchietti, M. Lewenstein, Many body physics from a quantum information perspective, in *Modern Theories of Many-Particle Systems in Condensed Matter Physics*, vol. 843, Lecture notes in physics, ed. by D.C. Cabra, A. Honecker, P. Pujol (Springer, Heidelberg, 2012), pp. 245–294
11. A.K. Ekert, Quantum cryptography based on Bell's theorem. Phys. Rev. Lett. **67**, 661 (1991)
12. C.H. Bennett, H.J. Bernstein, S. Popescu, B. Schumacher, Concentrating partial entanglement by local operations. Phys. Rev. A **53**, 2046 (1996)
13. V. Coffman, J. Kundu, W.K. Wootters, Distributed entanglement. Phys. Rev. A **61**, 052306 (2000)
14. J.S. Kim, G. Gour, B.C. Sanders, Limitations to sharing entanglement. Contemp. Phys. **53**, 417 (2012)
15. W.K. Wootters, W.H. Zurek, A single quantum cannot be cloned. Nature **299**, 802 (1982)
16. D. Dieks, Communication by EPR devices. Phys. Lett. A **92**, 271 (1982)
17. H.P. Yuen, Amplification of quantum states and noiseless photon amplifiers. Phys. Lett. A **113**, 405 (1986)
18. H. Barnum, C.M. Caves, C.A. Fuchs, R. Jozsa, B. Schumacher, Noncommuting mixed states cannot be broadcast. Phys. Rev. Lett. **76**, 2818 (1996)
19. S. Hill, W.K. Wootters, Entanglement of a pair of quantum bits. Phys. Rev. Lett. **78**, 5022 (1997)
20. W.K. Wootters, Entanglement of formation of an arbitrary state of two qubits. Phys. Rev. Lett. **80**, 2245 (1998)
21. T.J. Osborne, F. Verstraete, General monogamy inequality for bipartite qubit entanglement. Phys. Rev. Lett. **96**, 220503 (2006)
22. C.H. Bennett, D.P. DiVincenzo, J.A. Smolin, W.K. Wootters, Mixed-state entanglement and quantum error correction. Phys. Rev. A **54**, 3824 (1996)
23. P.M. Hayden, M. Horodecki, B.M. Terhal, The asymptotic entanglement cost of preparing a quantum state. J. Phys. A Math. Gen. **34**, 6891 (2001)
24. E.M. Rains, Rigorous treatment of distillable entanglement. Phys. Rev. A **60**, 173 (1999)
25. E.M. Rains, Bound on distillable entanglement. Phys. Rev. A **60**, 179 (1999)
26. A. Peres, Separability criterion for density matrices. Phys. Rev. Lett. **8**, 77 (1996)
27. M. Horodecki, P. Horodecki, R. Horodecki, Separability of mixed states: necessary and sufficient conditions. Phys. Lett. A **223**, 1 (1996)
28. K. Życzkowski, P. Horodecki, A. Sanpera, M. Lewenstein, Volume of the set of separable states. Phys. Rev. A **58**, 883 (1998)
29. J. Lee, M.S. Kim, Y.J. Park, S. Lee, Partial teleportation of entanglement in a noisy environment. J. Mod. Opt. **47**, 2151 (2000)
30. G. Vidal, R.F. Werner, Computable measure of entanglement. Phys. Rev. A **65**, 032314 (2002)
31. M.B. Plenio, Logarithmic negativity: a full entanglement monotone that is not convex. Phys. Rev. Lett. **95**, 090503 (2005)
32. V. Vedral, M.B. Plenio, M.A. Rippin, P.L. Knight, Quantifying entanglement. Phys. Rev. Lett. **78**, 2275 (1997)
33. V. Vedral, M.B. Plenio, Entanglement measures and purification procedures. Phys. Rev. A **57**, 1619 (1998)
34. V. Vedral, The role of relative entropy in quantum information theory. Rev. Mod. Phys. **74**, 197 (2002)
35. L. Henderson, V. Vedral, Classical, quantum and total correlations. J. Phys. A **34**, 6899 (2001)
36. H. Ollivier, W.H. Zurek, Quantum discord: a measure of the quantumness of correlations. Phys. Rev. Lett. **88**, 017901 (2002)
37. J. Oppenheim, M. Horodecki, P. Horodecki, R. Horodecki, Phys. Rev. Lett. **89**, 180402 (2002)
38. M. Horodecki, K. Horodecki, P. Horodecki, R. Horodecki, J. Oppenheim, A. Sen(De), U. Sen, Local information as a resource in distributed quantum systems. Phys. Rev. Lett. **90**, 100402 (2003)

39. M. Horodecki, P. Horodecki, R. Horodecki, J. Oppenheim, A. Sen(De), U. Sen, B. Synak-Radtke, Local versus nonlocal information in quantum-information theory: formalism and phenomena. Phys. Rev. A **71**, 062307 (2005)
40. I. Devetak, Distillation of local purity from quantum states. Phys. Rev. A **71**, 062303 (2005)
41. Y.-C. Ou, H. Fan, Monogamy inequality in terms of negativity for three-qubit states. Phys. Rev. A **75**, 062308 (2007)
42. F.F. Fanchini, M.F. Cornelia, M.C. de Oliveira, A.O. Caldeira, Conservation law for distributed entanglement of formation and quantum discord. Phys. Rev. A **84**, 012313 (2011)
43. A. Streltsov, G. Adesso, M. Piani, D. Bruß, Are general quantum correlations monogamous? Phys. Rev. Lett. **109**, 050503 (2012)
44. R. Prabhu, A.K. Pati, A. Sen(De), U. Sen, Conditions for monogamy of quantum correlations: Greenberger–Horne–Zeilinger versus W states. Phys. Rev. A **85**, 040102(R) (2012)
45. G.L. Giorgi, Monogamy properties of quantum and classical correlations. Phys. Rev. A **84**, 054301 (2011)
46. C. Lancien, S. Di Martino, M. Huber, M. Piani, G. Adesso, A. Winter, Should entanglement measures be monogamous or faithful? Phys. Rev. Lett. **117**, 060501 (2016)
47. K. Salini, R. Prabhu, A. Sen(De), U. Sen, Monotonically increasing functions of any quantum correlation can make all multiparty states monogamous. Ann. Phys. **348**, 297 (2014)
48. B. Toner, Monogamy of non-local quantum correlations. Proc. R. Soc. A **465**, 59 (2009)
49. P. Kurzyński, T. Paterek, R. Ramanathan, W. Laskowski, D. Kaszlikowski, Correlation complementarity yields bell monogamy relations. Phys. Rev. Lett. **106**, 180402 (2011)
50. R. Prabhu, A.K. Pati, A. Sen(De), U. Sen, Exclusion principle for quantum dense coding. Phys. Rev. A **87**, 052319 (2013)
51. T.R. de Oliveira, A. Saguia, M.S. Sarandy, Nonviolation of Bell's inequality in translation invariant systems. Europhys. Lett. **104**, 29901 (2013)
52. Z.-Y. Sun, Y.-Y. Wu, J. Xu, H.-L. Huang, B.-J. Chen, B. Wang, Violation of Bell inequality in perfect translation-invariant systems. Phys. Rev. A **88**, 054101 (2013)
53. R. Prabhu, A. Sen(De), U. Sen, Genuine multiparty quantum entanglement suppresses multiport classical information transmission. Phys. Rev. A **88**, 042329 (2013)
54. D. Sadhukhan, S. Singha Roy, D. Rakshit, A. Sen(De), U. Sen, Beating no-go theorems by engineering defects in quantum spin models. New J. Phys. **17**, 043013 (2015)
55. J.S. Bell, On the Einstein Podolsky Rosen paradox. Physics **1**, 195 (1964)
56. J.F. Clauser, M.A. Horne, A. Shimony, R.A. Holt, Proposed experiment to test local hidden-variable theories. Phys. Rev. Lett. **23**, 880 (1969)
57. C.H. Bennett, S.J. Wiesner, Communication via one- and two-particle operators on Einstein–Podolsky–Rosen states. Phys. Rev. Lett. **69**, 2881 (1992)
58. M.D. Reid, Monogamy inequalities for the Einstein–Podolsky–Rosen paradox and quantum steering. Phys. Rev. A **88**, 062108 (2013)
59. A. Milne, S. Jevtic, D. Jennings, H. Wiseman, T. Rudolph, Quantum steering ellipsoids, extremal physical states and monogamy. New J. Phys. **16**, 083017 (2014)
60. S. Lee, J. Park, Monogamy of entanglement and teleportation capability. Phys. Rev. A **79**, 054309 (2009)
61. R. Ramanathan, A. Soeda, P. Kurzynski, D. Kaszlikowski, Generalized monogamy of contextual inequalities from the no-disturbance principle. Phys. Rev. Lett. **109**, 050404 (2012)
62. P. Kurzyński, A. Cabello, D. Kaszlikowski, Fundamental monogamy relation between contextuality and nonlocality. Phys. Rev. Lett. **112**, 100401 (2014)
63. A. Kumar, S. Singha Roy, A.K. Pal, R. Prabhu, A. Sen(De), U. Sen, Conclusive identification of quantum channels via monogamy of quantum correlations. Phys. Lett. A **380**, 3588 (2016)
64. A. Chandran, D. Kaszlikowski, A. Sen(De), U. Sen, V. Vedral, Regional versus global entanglement in resonating-valence-bond states. Phys. Rev. Lett. **99**, 170502 (2007)
65. H.S. Dhar, A. Sen(De), Entanglement in resonating valence bond states: ladder versus isotropic lattices. J. Phys. A Math. Theor. **44**, 465302 (2011)
66. S. Singha Roy, H.S. Dhar, D. Rakshit, A. Sen(De), U. Sen, Response to defects in multi- and bipartite entanglement of isotropic quantum spin networks, arXiv:1607.05195 (2016)

67. D. Sadhukhan, S. Singha Roy, D. Rakshit, R. Prabhu, A. Sen(De), U. Sen, Quantum discord length is enhanced while entanglement length is not by introducing disorder in a spin chain. Phys. Rev. E **93**, 012131 (2016)
68. M. Allegra, P. Giorda, A. Montorsi, Quantum discord and classical correlations in the bond-charge Hubbard model: quantum phase transitions, off-diagonal long-range order, and violation of the monogamy property for discord. Phys. Rev. B **84**, 245133 (2011)
69. X.-K. Song, T. Wu, L. Ye, The monogamy relation and quantum phase transition in one-dimensional anisotropic XXZ model. Quantum Inf. Process. **12**, 3305 (2013)
70. L. Qiu, G. Tang, X.-Q. Yang, A.-M. Wang, Relating tripartite quantum discord with multisite entanglement and their performance in the one-dimensional anisotropic XXZ model. Europhys. Lett. **105**, 30005 (2014)
71. M. Qin, Z.-Z. Ren, X. Zhang, Renormalization of the global quantum correlation and monogamy relation in the anisotropic Heisenberg XXZ model. Quantum Inf. Process. **15**, 255 (2016)
72. K.R.K. Rao, H. Katiyar, T.S. Mahesh, A. Sen(De), U. Sen, A. Kumar, Multipartite quantum correlations reveal frustration in a quantum Ising spin system. Phys. Rev. A **88**, 022312 (2013)
73. J. Zhu, S. Kais, A. Aspuru-Guzik, S. Rodriques, B. Brock, P.J. Love, Multipartite quantum entanglement evolution in photosynthetic complexes. J. Chem. Phys. **137**, 074112 (2012)
74. T. Chanda, U. Mishra, A. Sen(De), U. Sen, Time dynamics of multiparty quantum correlations indicate energy transfer route in light-harvesting complexes, arXiv:1412.6519 (2014)
75. A. Bera, A. Kumar, D. Rakshit, R. Prabhu, A. Sen(De), U. Sen, Information complementarity in multipartite quantum states and security in cryptography. Phys. Rev. A **93**, 032338 (2016)
76. D. Deutsch, A. Ekert, R. Jozsa, C. Macchiavello, S. Popescu, A. Sanpera, Quantum privacy amplification and the security of quantum cryptography over noisy channels. Phys. Rev. Lett. **77**, 2818 (1996)
77. M.N. Bera, R. Prabhu, A. Sen(De), U. Sen, Characterization of tripartite quantum states with vanishing monogamy score. Phys. Rev. A **86**, 012319 (2012)
78. R. Prabhu, A.K. Pati, A. Sen(De), U. Sen, Relating monogamy of quantum correlations and multisite entanglement. Phys. Rev. A **86**, 052337 (2012)
79. Y.-K. Bai, Y.-F. Xu, Z.D. Wang, General monogamy relation for the entanglement of formation in multiqubit systems. Phys. Rev. Lett. **113**, 100503 (2014)
80. Y.-C. Ou, Violation of monogamy inequality for higher-dimensional objects. Phys. Rev. A **75**, 034305 (2007)
81. X.-J. Ren, W. Jiang, Entanglement monogamy inequality in a $2 \otimes 2 \otimes 4$ system. Phys. Rev. A **81**, 024305 (2010)
82. C.-S. Yu, H.-S. Song, Entanglement monogamy of tripartite quantum states. Phys. Rev. A **77**, 032329 (2008)
83. J.-H. Huang, S.-Y. Zhu, Entanglement monogamy in a three-qubit state. Phys. Rev. A **78**, 012325 (2008)
84. Y.-C. Ou, H. Fan, S.-M. Fei, Proper monogamy inequality for arbitrary pure quantum states. Phys. Rev. A **78**, 012311 (2008)
85. M.F. Cornelio, Multipartite monogamy of the concurrence. Phys. Rev. A **87**, 032330 (2013)
86. B. Regula, S.D. Martino, S. Lee, G. Adesso, Strong monogamy conjecture for multiqubit entanglement: the four-qubit case. Phys. Rev. Lett. **113**, 110501 (2014)
87. C. Eltschka, A. Osterloh, J. Siewert, Possibility of generalized monogamy relations for multipartite entanglement beyond three qubits. Phys. Rev. A **80**, 032313 (2009)
88. C. Eltschka, J. Siewert, Quantifying entanglement resources. J. Phys. A Math. Theor. **47**, 424005 (2014)
89. G. Gour, N.R. Wallach, All maximally entangled four qubits states. J. Math. Phys. **51**, 112201 (2010)
90. X.-N. Zhu, S.-M. Fei, Entanglement monogamy relations of qubit systems. Phys. Rev. A **90**, 024304 (2014)
91. Y. Luo, Y. Li, Monogamy of αth power entanglement measurement in qubit systems. Ann. Phys. **362**, 511 (2015)

92. B. Regula, A. Osterloh, G. Adesso, Strong monogamy inequalities for four qubits. Phys. Rev. A **93**, 052338 (2016)
93. Y.-K. Bai, D. Yang, Z.D. Wang, Multipartite quantum correlation and entanglement in four-qubit pure states. Phys. Rev. A **76**, 022336 (2007)
94. C. Eltschka, J. Siewert, Monogamy equalities for qubit entanglement from Lorentz invariance. Phys. Rev. Lett. **114**, 140402 (2015)
95. K. Chen, S. Albeverio, S.M. Fei, Concurrence of arbitrary dimensional bipartite quantum states. Phys. Rev. Lett. **95**, 040504 (2005)
96. S. Karmakar, A. Sen, A. Bhar, D. Sarkar, Strong monogamy conjecture in a four-qubit system. Phys. Rev. A **93**, 012327 (2016)
97. J.H. Choi, J.S. Kim, Negativity and strong monogamy of multiparty quantum entanglement beyond qubits. Phys. Rev. A **92**, 042307 (2015)
98. D.P. DiVincenzo, C.A. Fuchs, H. Mabuchi, J.A. Smolin, A. Thapliyal, A. Uhlmann, Entanglement of assistance, in *Quantum Computing and Quantum Communications*, vol. 1509, Lecture notes in computer science, ed. by C.P. Williams (Springer, Heidelberg, 1999), pp. 247–257
99. M. Christandl, A. Winter, "Squashed entanglement" - an additive entanglement measure. J. Math. Phys. (N.Y.) **45**, 829 (2004)
100. M. Koashi, A. Winter, Monogamy of quantum entanglement and other correlations. Phys. Rev. A **69**, 022309 (2004)
101. G. Gour, D.A. Meyer, B.C. Sanders, Deterministic entanglement of assistance and monogamy constraints. Phys. Rev. A **72**, 042329 (2005)
102. X.-N. Zhu, S.-M. Fei, Generalized monogamy relations of concurrence for N-qubit systems. Phys. Rev. A **92**, 062345 (2015)
103. T.R. de Oliveira, M.F. Cornelio, F.F. Fanchini, Monogamy of entanglement of formation. Phys. Rev. A **89**, 034303 (2014)
104. Z.-G. Li, S.-M. Fei, S. Albeverio, W.M. Liu, Bound of entanglement of assistance and monogamy constraints. Phys. Rev. A **80**, 034301 (2009)
105. I. Devetak, A. Winter, Distillation of secret key and entanglement from quantum states. Proc. R. Soc. Lond. A **461**, 207 (2005)
106. B.M. Terhal, M. Horodecki, D.W. Leung, D.P. DiVincenzo, The entanglement of purification. J. Math. Phys. **43**, 4286 (2002)
107. S. Bagchi, A.K. Pati, Monogamy, polygamy, and other properties of entanglement of purification. Phys. Rev. A **91**, 042323 (2015)
108. J.S. Kim, B.C. Sanders, Monogamy of multi-qubit entanglement in terms of Rnyi and Tsallis entropies, in *TQC10 Proceedings of the 5th Conference on Theory of Quantum Computation, Communication, and Cryptography* (Springer, Heidelberg, 2011), pp. 159–167
109. Y. Luo, T. Tian, L.-H. Shao, Y. Li, General monogamy of Tsallis q-entropy entanglement in multiqubit systems. Phys. Rev. A. **93**, 062340 (2016)
110. G.-M. Yuan, W. Song, M. Yang, D.-C. Li, J.-L. Zhao, Z.-L. Cao, Monogamy relation of multi-qubit systems for squared Tsallis-q entanglement. Sci. Rep. **6**, 28719 (2016)
111. G. Adesso, F. Illuminati, Continuous variable tangle, monogamy inequality, and entanglement sharing in Gaussian states of continuous variable systems. New J. Phys. **8**, 15 (2006)
112. G. Adesso, A. Serafini, F. Illuminati, Multipartite entanglement in three-mode Gaussian states of continuous-variable systems: quantification, sharing structure, and decoherence. Phys. Rev. A **73**, 032345 (2006)
113. G. Adesso, F. Illuminati, Strong monogamy of bipartite and genuine multipartite entanglement: the Gaussian case. Phys. Rev. Lett. **99**, 150501 (2007)
114. T. Hiroshima, G. Adesso, F. Illuminati, Monogamy Inequality for distributed Gaussian entanglement. Phys. Rev. Lett. **98**, 050503 (2007)
115. G. Adesso, F. Illuminati, Genuine multipartite entanglement of symmetric Gaussian states: strong monogamy, unitary localization, scaling behavior, and molecular sharing structure. Phys. Rev. A **78**, 042310 (2008)

116. W.H. Zurek, in *Quantum Optics, Experimental Gravitation and Measurement Theory*, ed. by P. Meystre, M.O. Scully (Plenum, New York, 1983)
117. S.M. Barnett, S.J.D. Phoenix, Entropy as a measure of quantum optical correlation. Phys. Rev. A **40**, 2404 (1989)
118. B. Schumacher, M.A. Nielsen, Quantum data processing and error correction. Phys. Rev. A **54**, 2629 (1996)
119. N.J. Cerf, C. Adami, Negative entropy and information in quantum mechanics. Phys. Rev. Lett. **79**, 5194 (1997)
120. B. Groisman, S. Popescu, A. Winter, Quantum, classical, and total amount of correlations in a quantum state. Phys. Rev. A **72**, 032317 (2005)
121. Y. Huang, Computing quantum discord in NP-complete. New J. Phys. **16**, 033027 (2014)
122. M. Ali, A.R.P. Rau, G. Alber, Quantum discord for two-qubit X states. Phys. Rev. A **81**, 042105 (2010); *ibid.* **82**, 069902(E) (2010)
123. M. Ali, Quantum discord for a two-parameter class of states in $2 \otimes d$ quantum systems. J. Phys. A Math. Theor. **43**, 495303 (2010)
124. X.-M. Lu, J. Ma, Z. Xi, X. Wang, Optimal measurements to access classical correlations of two-qubit states. Phys. Rev. A **83**, 012327 (2011)
125. D. Girolami, G. Adesso, Quantum discord for general two-qubit states: analytical progress. Phys. Rev. A **83**, 052108 (2011)
126. Q. Chen, C. Zhang, S. Yu, X.X. Yi, C.H. Oh, Quantum discord of two-qubit X states. Phys. Rev. A **84**, 042313 (2011)
127. S. Vinjanampathy, A.R.P. Rau, Calculation of quantum discord for qubit-qudit or N qubits. J. Phys. A Math. Theor. **45**, 095303 (2012)
128. T. Chanda, T. Das, D. Sadhukhan, A.K. Pal, A. Sen De, U. Sen, Reducing computational complexity of quantum correlations. Phys. Rev. A **92**, 062301 (2015)
129. S. Luo, S. Fu, Geometric measure of quantum discord. Phys. Rev. A **82**, 034302 (2010)
130. B. Dakić, V. Vedral, Č. Brukner, Necessary and sufficient condition for nonzero quantum discord. Phys. Rev. Lett. **105**, 190502 (2010)
131. M. Piani, Problem with geometric discord. Phys. Rev. A **86**, 034101 (2012)
132. T. Debarba, T.O. Maciel, R.O. Vianna, Witnessed entanglement and the geometric measure of quantum discord. Phys. Rev. A **86**, 024302 (2012)
133. S. Rana, P. Parashar, Comment on "witnessed entanglement and the geometric measure of quantum discord". Phys. Rev. A **87**, 016301 (2013)
134. T. Debarba, T.O. Maciel, R.O. Vianna, Reply to "comment on 'witnessed entanglement and the geometric measure of quantum discord' ". Phys. Rev. A **87**, 046301 (2013)
135. F.M. Paula, T.R. de Oliveira, M.S. Sarandy, Geometric quantum discord through the Schatten 1-norm. Phys. Rev. A **87**, 064101 (2013)
136. A. Zeilinger, M.A. Horne, D.M. Greenberger, in *Proceedings of Squeezed States and Quantum Uncertainty*, ed. by D. Han, Y.S. Kim, W.W. Zachary, vol. 3135 (NASA Conference Publication, 1992), p. 73
137. A. Sen(De), U. Sen, M. Wiesniak, D. Kaszlikowski, M. Zukowski, Multiqubit W states lead to stronger nonclassicality than Greenberger–Horne–Zeilinger states. Phys. Rev. A **68**, 062306 (2003)
138. X.-J. Ren, H. Fan, Non-monogamy of quantum discord and upper bounds for quantum correlation. Quant. Inf. Comp. **13**, 469 (2013)
139. N. Li, S. Luo, Total versus quantum correlations in quantum states. Phys. Rev. A **76**, 032327 (2007)
140. R.F. Werner, Quantum states with Einstein–Podolsky–Rosen correlations admitting a hidden-variable model. Phys. Rev. A **40**, 4277 (1989)
141. R.F. Werner, An application of Bell's inequalities to a quantum state extension problem. Lett. Math. Phys. **17**, 359 (1989)
142. A.C. Doherty, P.A. Parrilo, F.M. Spedalieri, Complete family of separability criteria. Phys. Rev. A **69**, 022308 (2004)
143. D. Yang, A simple proof of monogamy of entanglement. Phys. Lett. A **360**, 249 (2006)

144. I. Devetak, A. Winter, Distilling common randomness from bipartite quantum states. IEEE Trans. Inf. Theory **50**, 3183 (2004)
145. L.P. Hughston, R. Jozsa, W.K. Wootters, A complete classification of quantum ensembles having a given density matrix. Phys. Lett. A **183**, 14 (1993)
146. P.M. Hayden, M. Horodecki, B.M. Terhal, The asymptotic entanglement cost of preparing a quantum state. J. Phys. A **34**, 6891 (2001)
147. Z.H. Ma, Z.H. Chen, F.F. Fanchini, Multipartite quantum correlations in open quantum systems. New J. Phys. **15**, 043023 (2013)
148. F.F. Fanchini, L.K. Castelano, M.F. Cornelio, M.C. de Oliveira, Locally inaccessible information as a fundamental ingredient to quantum information. New J. Phys. **14**, 013027 (2012)
149. M. Daoud, R.A. Laamara, W. Kaydi, Multipartite quantum correlations in even and odd spin coherent states. J. Phys. A Math. Theor. **46**, 395302 (2013)
150. T.M. Cover, J.A. Thomas, *Elements of Information Theory* (Wiley, New Jersey, 2006)
151. W. Dür, G. Vidal, J.I. Cirac, Three qubits can be entangled in two inequivalent ways. Phys. Rev. A **62**, 062314 (2000)
152. A. Acín, A. Andrianov, L. Costa, E. Jané, J.I. Latorre, R. Tarrach, Generalized Schmidt decomposition and classification of three-quantum-bit states. Phys. Rev. Lett. **85**, 1560 (2000)
153. A. Acín, D. Bruss, M. Lewenstein, A. Sanpera, Classification of mixed three-qubit states. Phys. Rev. Lett. **87**, 040401 (2001)
154. Y.-K. Bai, N. Zhang, M.-Y. Ye, Z.D. Wang, Exploring multipartite quantum correlations with the square of quantum discord. Phys. Rev. A **88**, 012123 (2013)
155. A. Kumar, Conditions for monogamy of quantum correlations in multipartite systems. Phys. Lett. A **380**, 3044 (2016)
156. M. Horodecki, P. Horodecki, R. Horodecki, Limits for entanglement measures. Phys. Rev. Lett. **84**, 2014 (2000)
157. A. Kumar, R. Prabhu, A. Sen(de), U. Sen, Effect of a large number of parties on the monogamy of quantum correlations. Phys. Rev. A **91**, 012341 (2015)
158. R. Dicke, Coherence in spontaneous radiation processes. Phys. Rev. **93**, 99 (1954)
159. G.H. Aguilar, J. Farías, A. Valdés-Hernández, P.H. Souto Ribeiro, L. Davidovich, S.P. Walborn, Flow of quantum correlations from a two-qubit system to its environment. Phys. Rev. A **89**, 022339 (2014)
160. D.M. Greenberger, M.A. Horne, A. Zeilinger, in *Bells Theorem, Quantum Theory, and Conceptions of the Universe*, ed. by M. Kafatos (Kluwer Academic, Dordrecht, 1989)
161. P.K. Sarangi, I. Chakrabarty, A study of quantum correlation for three qubit states under the effect of quantum noisy channels, arXiv:1411.7579 (2014)
162. I. Chakrabarty, P. Agrawal, A.K. Pati, Quantum dissension: generalizing quantum discord for three-qubit states. Eur. Phys. J. D **65**, 605 (2011)
163. X.-K. Song, T. Wu, L. Ye, Monogamy properties of quantum discord for a three-qubit entangled state. Mod. Phys. Lett. B **27**, 1350049 (2013)
164. S. Rana, P. Parashar, Tight lower bound on geometric discord of bipartite states. Phys. Rev. A **85**, 024102 (2012)
165. Y.-K. Bai, T.-T. Zhang, L.-T. Wang, Z.D. Wang, Correlation evolution and monogamy of two geometric quantum discords in multipartite systems. Eur. Phys. J. D **68**, 274 (2014)
166. M. Daoud, W. Kaydi, H. El Hadfi, Distribution of geometric quantum discord in photon-added coherent states. Mod. Phys. Lett. B **29**, 1550239 (2015)
167. E. Majorana, Atomi orientati in campo magnetico variabile. Nuovo Cimento **9**, 43 (1932)
168. Sudha, A.R. Usha Devi, A.K. Rajagopal, Monogamy of quantum correlations in three-qubit pure states. Phys. Rev. A **85**, 012103 (2012)
169. A.K. Rajagopal, R.W. Rendell, Separability and correlations in composite states based on entropy methods. Phys. Rev. A **66**, 022104 (2002)
170. A.R. Usha Devi, A.K. Rajagopal, Generalized information theoretic measure to discern the quantumness of correlations. Phys. Rev. Lett. **100**, 140502 (2008)
171. A. Wehrl, General properties of entropy. Rev. Mod. Phys. **50**, 221 (1978)
172. S. Luo, S. Fu, Measurement-induced nonlocality. Phys. Rev. Lett. **106**, 120401 (2011)

173. A. Sen, D. Sarkar, A. Bhar, Monogamy of measurement induced non-locality. J. Phys. A: Math. Theor. **45**, 405306 (2012)
174. Z. Xi, X. Wang, Y. Li, Measurement-induced nonlocality based on the relative entropy. Phys. Rev. A **85**, 042325 (2012)
175. Y. Luo, Z.-J. Xi, Y.-M. Li, Monogamy of measurement-induced nonlocality based on relative entropy. Commun. Theor. Phys. **62**, 677 (2014)
176. A. Sen (De), U. Sen, Channel capacities versus entanglement measures in multiparty quantum states. Phys. Rev. A **81**, 012308 (2010)
177. A. Biswas, R. Prabhu, A. Sen (De), U. Sen, Genuine-multipartite-entanglement trends in gapless-to-gapped transitions of quantum spin systems. Phys. Rev. A **90**, 032301 (2014)
178. C.C. Rulli, M.S. Sarandy, Global quantum discord in multipartite systems. Phys. Rev. A **84**, 042109 (2011)
179. H.C. Braga, C.C. Rulli, T.R. de Oliveira, M.S. Sarandy, Monogamy of quantum discord by multipartite correlations. Phys. Rev. A **86**, 062106 (2012)
180. S.-Y. Liu, Y.-R. Zhang, L.-M. Zhao, W.-L. Yang, H. Fan, General monogamy property of global quantum discord and the application. Ann. Phys. **348**, 256 (2014)
181. Y.-K. Bai, Y.-F. Xu, Z.D. Wang, Hierarchical monogamy relations for the squared entanglement of formation in multipartite systems. Phys. Rev. A **90**, 062343 (2014)
182. F. Liu, F. Gao, S.-J. Qin, S.-C. Xie, Q.-Y. Wen, Multipartite entanglement indicators based on monogamy relations of n-qubit symmetric states. Sci. Rep. **6**, 20302 (2016)
183. P. Pandya, A. Misra, I. Chakrabarty, Complementarity between tripartite quantum correlation and bipartite bell inequality violation in three qubit pure states, arXiv:1512.01770 (2015)
184. K. Sharma, T. Das, A. Sen(De), Distribution of Bell-inequality violation versus multiparty-quantum-correlation measures. Phys. Rev. A **93**, 062344 (2016)
185. R. Nepal, R. Prabhu, A. Sen(De), U. Sen, Maximally-dense-coding-capable quantum states. Phys. Rev. A **87**, 032336 (2013)
186. T. Das, R. Prabhu, A. Sen(De), U. Sen, Multipartite dense coding versus quantum correlation: noise inverts relative capability of information transfer. Phys. Rev. A **90**, 022319 (2014)
187. A. Kumar, H.S. Dhar, R. Prabhu, A. Sen(De), U. Sen, Forbidden regimes in distribution of bipartite quantum correlations due to multiparty entanglement, arXiv:1505.01748 (2015)
188. C.H. Bennett, G. Brassard, S. Popescu, B. Schumacher, J.A. Smolin, W.K. Wootters, Purification of noisy entanglement and faithful teleportation via noisy channels. Phys. Rev. Lett. **76**, 722 (1996)
189. A. Kumar, H.S. Dhar, Lower bounds on violation of monogamy inequality for quantum correlation measures. Phys. Rev. A **93**, 062337 (2016)
190. S. Bose, M.B. Plenio, V. Vedral, Mixed state dense coding and its relation to entanglement measures. J. Mod. Opt. **47**, 291 (2000)
191. T. Hiroshima, Optimal dense coding with mixed state entanglement. J. Phys. A Math. Gen. **34**, 6907 (2001)
192. M. Ziman, V. Bužek, Correlation-assisted quantum communication. Phys. Rev. A **67**, 042321 (2003)
193. D. Bruß, G.M. DAriano, M. Lewenstein, C. Macchiavello, A. Sen(De), U. Sen, Distributed quantum dense coding. Phys. Rev. Lett. **93**, 210501 (2004)
194. D. Bruß, M. Lewenstein, A. Sen(De), U. Sen, G.M. DAriano, C. Macchiavello, Dense coding with multipartite quantum states. Int. J. Quant. Inf. **4**, 415 (2006)
195. B.M. Terhal, A family of indecomposable positive linear maps based on entangled quantum states. Lin. Alg. Appl. **323**, 61 (2001)
196. O. Gühne, P. Hyllus, D. Bruß, A. Ekert, M. Lewenstein, C. Macchiavello, A. Sanpera, Detection of entanglement with few local measurements. Phys. Rev. A **66**, 062305 (2002)
197. M. Lewenstein, B. Kraus, J.I. Cirac, P. Horodecki, Optimization of entanglement witnesses. Phys. Rev. A **62**, 052310 (2000)
198. D. Bruß, J.I. Cirac, P. Horodecki, F. Hulpke, B. Kraus, M. Lewenstein, A. Sanpera, Reflections upon separability and distillability. J. Mod. Opt. **49**, 1399 (2002)

199. O. Gühne, P. Hyllus, D. Bruß, A. Ekert, M. Lewenstein, C. Macchiavello, A. Sanpera, Experimental detection of entanglement via witness operators and local measurements. J. Mod. Opt. **50**, 1079 (2003)
200. N.D. Mermin, Extreme quantum entanglement in a superposition of macroscopically distinct states. Phys. Rev. Lett. **65**, 1838 (1990)
201. M. Ardehali, Bell inequalities with a magnitude of violation that grows exponentially with the number of particles. Phys. Rev. A **46**, 5375 (1992)
202. A.V. Belinskii, D.N. Klyshko, Interference of light and Bell's theorem. Phys. Usp. **36**, 653 (1993)
203. A. Fujiwara, Quantum channel identification problem. Phys. Rev. A **63**, 042304 (2001)
204. M. Sarovar, G.J. Milburn, Optimal estimation of one-parameter quantum channels. J. Phys. A Math. Gen. **39**, 8487 (2006)
205. M. Paris, J. Rehacek (eds.), *Quantum State Estimation*, vol. 649, Lecture notes in physics (Springer, Berlin, 2004)
206. G.S. Engel, T.R. Calhoun, E.L. Read, T.-K. Ahn, T.C.M. Caronal, Y.-C. Cheng, R.E. Blankenship, G.R. Fleming, Evidence for wavelike energy transfer through quantum coherence in photosynthetic systems. Nature **446**, 782 (2007)
207. F. Caruso, A.W. Chin, A. Datta, S.F. Huelga, M.B. Plenio, Highly efficient energy excitation transfer in light-harvesting complexes: the fundamental role of noise-assisted transport. J. Chem. Phys. **131**, 105106 (2009)
208. M. Sarovar, A. Ishizaki, G.R. Fleming, K.B. Whaley, Quantum entanglement in photosynthetic light-harvesting complexes. Nat. Phys. **6**, 462 (2010)
209. N. Lambert, Y.-N. Chen, Y.-C. Cheng, C.-M. Li, G.-Y. Chen, F. Nori, Quantum biology. Nat. Phys. **9**, 10 (2013)
210. J.S. Bell, On the problem of hidden variables in quantum mechanics. Rev. Mod. Phys. **38**, 447 (1966)
211. S. Kochen, E.P. Specker, The problem of hidden variables in quantum mechanics. J. Math. Mech. **17**, 59 (1967)
212. A.K. Pati, S.L. Braunstein, Impossibility of deleting an unknown quantum state. Nature **404**, 164 (2000)
213. A. Kalev, I. Hen, No-broadcasting theorem and its classical counterpart. Phys. Rev. Lett. **100**, 210502 (2008)

Measurement-Induced Nonlocality and Quantum Correlations Under Local Operations

Xueyuan Hu, Ming-Liang Hu and Heng Fan

Abstract One significant feature of quantum theory is the existence of non-local quantum correlations which have no classical counterpart. There are various measures quantifying quantum correlations from different view points. Here, we present some recent developments about the quantum correlation measures known as measurement-induced nonlocality, in the sense that quantum measurement may destroy the quantum correlations for quantum states resulting in measures of non-locality. Quantum correlations remain invariant under local unitary operations, they may decrease under general local operations, however, sometimes they can also show increasing for some local operations. We will review the properties of quantum correlations under local operations.

1 Introduction

Quantum mechanics is non-local. There exist non-local quantum correlations which have no classical counterpart. The study of quantum correlations can trace back to the well-known debate about whether quantum mechanics is complete, known as the

Xueyuan Hu and Ming-Liang Hu—those two authors contributed equally

X. Hu
School of Information Science and Engineering, Shandong University,
Jinan 250100, China
e-mail: xyhu@sdu.edu.cn

M.-L. Hu
School of Science, Xi'an University of Posts and Telecommunications,
Xi'an 710121, China
e-mail: mingliang0301@163.com

H. Fan (✉)
Institute of Physics, Chinese Academy of Sciences, Beijing 100190, China
e-mail: hfan@iphy.ac.cn

H. Fan
Collaborative Innovation Center of Quantum Matter, Beijing 100190, China

F.F. Fanchini et al. (eds.), *Lectures on General Quantum Correlations and their Applications*, Quantum Science and Technology,
DOI 10.1007/978-3-319-53412-1_4

Einstein-Podolsky-Rosen paradox [1]. It was proposed that there may exist hidden variables for quantum theory being complete. Later, various Bell-type inequalities were proposed which are derived based on the local hidden variable theory, see, e.g., Ref. [2] for an overview. It was found that the violation of Bell inequalities implies quantum entanglement in a system, while the opposite case is not always true [3, 4]. Entanglement also plays a critical role in many protocols of quantum information processing. Great progress has been made in studying quantum entanglement, which is one kind of quantum correlations showing non-locality of quantum mechanics. Entanglement is also believed to be the key resource for the advantages of quantum computation and protocols of quantum information processing. Very recently, it is realized that entanglement is not the only quantum correlation which has no classical counterpart. Other type of quantum correlations, such as the quantum discord and measurement-induced nonlocality, may also be responsible for the speedups in some quantum algorithms while entanglement may be vanishing or negligible, see review [5].

In general, most of the quantum correlations for pure quantum states may coincide, and sometimes may demonstrate similar behaviors for mixed states. However, there are also subtle differences for those quantum correlations and their physical interpretations are also different. All these indicate that the properties of quantum correlations, or nonlocality, of a system are intricate, and the characterizations of them from different aspects are in demanding. Here we will review some recent results of measurement-induced nonlocality and quantum correlations under local operations.

We would like to point out that measurement-induced nonlocality is one type of quantum correlations. However, we remark that nonlocality, for example in form of non-local correlation, which is non-classical from one side, may not always be possessed by quantum states from other side, such as the PR box [6, 7]. Here the measurement-induced nonlocality means the quantum correlation possessed by quantum states which is naturally quantum mechanical. For quantum correlations, we mean that some non-classical correlations possessed by quantum states. One may realize that there also exist classical correlation for quantum states. We remark that quantum correlation, as valuable resource, cannot be cloned (broadcasted) because of no-cloning theorem, in contrast with the classical correlation [8]. We know that entanglement cannot be created by local quantum operations even assisted by classical communications. However, some other quantum correlations may increase by local operations. Here, we will review results of quantum correlations under local operations.

Before we proceed, let us first introduce some notations. Quantum states are presented as the density operators ρ in the Hilbert space $\mathcal{H}_d$, where d denotes the dimension. A qubit is a two-dimension quantum system. Let $\vec{\sigma} = \{\sigma_1, \sigma_2, \sigma_3\}$ denote the Pauli basis and $\sigma_0 = I$ be the single-qubit identity operator. Any single-qubit state ρ^A can be written as $\rho^A = \frac{1}{2}\sum_{\mu=0}^{3} a_\mu \sigma_\mu^A$, where $a_\mu = \mathrm{tr}(\rho^A \sigma_\mu^A)$ and $\boldsymbol{a} \equiv (a_1, a_2, a_3)^{\mathrm{T}}$ is called the Bloch vector of the state ρ^A. We label $\tilde{\boldsymbol{a}} \equiv (a_0, a_1, a_2, a_3)^{\mathrm{T}}$ for later convenience. Similarly, a two-qubit state ρ^{AB} can be expanded in the Pauli basis as

$\rho^{AB} = \frac{1}{4}\sum_{\mu\nu}\Theta_{\mu\nu}\sigma_\mu^A \otimes \sigma_\nu^B$, where the coefficient matrix $\Theta_{\mu\nu} = \mathrm{tr}(\rho^{AB}\sigma_\mu^A \otimes \sigma_\nu^B)$ can be written in the block form $\Theta = \begin{pmatrix} 1 & \boldsymbol{b}^\mathrm{T} \\ \boldsymbol{a} & T \end{pmatrix}$.

Here, $\boldsymbol{a}$ and $\boldsymbol{b}$ are the Bloch vectors of the reduced density matrices ρ^A and ρ^B respectively, and the 3×3 matrix T represents the correlations.

The von Neumann entropy of a quantum state is denoted as $S(\rho) := -\mathrm{tr}(\rho \log_2 \rho)$. The relative entropy of two quantum states ρ and σ is $S(\rho||\sigma) := -\mathrm{tr}(\rho \log_2 \sigma) - S(\rho)$. When we consider the bipartite quantum state ρ^{AB}, the conditional entropy is $S_{A|B}(\rho^{AB}) := S(\rho^{AB}) - S(\rho^B)$.

A quantum channel is a trace-preserving completely positive (TPCP) linear map $\Lambda : \mathcal{D}(\mathcal{H}_d) \to \mathcal{D}(\mathcal{H}_{d'})$. Here $\mathcal{D}(\mathcal{H}_d)$ denotes the operator space defined on the Hilbert space $\mathcal{H}_d$. In the following context, we take $d = d'$. Any quantum channel can be presented as the Kraus decomposition $\Lambda(\cdot) = \sum_j E_j(\cdot)E_j^\dagger$, where E_j are called Kraus operators.

2 What are Quantum Correlations

A bipartite state ρ^{AB} is called quantum-classical (QC) if there exist a positive operator-valued measure (POVM) on B which does not disturb the whole state. The term "classical" is used to stress the nondisturbing property of classical measurements. Mathematically, a QC state can be written as

$$\rho^{\mathrm{QC}} = \sum_i p_i \rho_i^A \otimes |\phi_i\rangle_B \langle\phi_i|, \tag{1}$$

where $\{|\phi_i\rangle_B\}$ consist of an orthogonal basis for the Hilbert space of subsystem B, and ρ_i^A are density operators of A. The set of quantum-classical states are denoted as $\mathcal{QC}$. An equivalent expression for the QC states is

$$\rho^{\mathrm{QC}} = \sum_i p_i \zeta_i^A \otimes \xi_i^B, \tag{2}$$

where ζ_i^A are linearly independent, and ξ_i^B commute with each other.

A state is said to have nonzero quantum correlation on B if and only if it does not belong to $\mathcal{QC}$. Like entanglement, the amount of quantum correlation can be measured in various ways [5]. The measures of quantum correlation Q we discuss here satisfy the following three conditions:

(C1) $Q(\rho) = 0$ if and only if $\rho \in \mathcal{QC}$;
(C2) $Q(U_A \otimes U_B \rho U_A^\dagger \otimes U_B^\dagger) = Q(\rho)$, where U_A and U_B are arbitrary unitary operators on A and B;
(C3) $Q(\Lambda_A \otimes \mathbb{1}_B(\rho)) \leq Q(\rho)$.

Notice that the measures of quantum correlation are asymmetric for A and B, here and hereafter, we discuss only the quantum correlation defined on B. In the following, we list some quantum correlation measures which satisfy (C1–C3).

Quantum discord is defined as the minimum part of the mutual information shared between A and B that cannot be obtained by the measurement on B [9]:

$$\delta_{A|B}(\rho) := \min_{\{M_i^B\}} S_{A|B}\left(\sum_i I_A \otimes M_i^B \rho I_A \otimes M_i^{B\dagger}\right) - S_{A|B}(\rho), \tag{3}$$

where $\{M_i^B\}$ is a POVM on B. We point out here that the calculation of quantum discord is a hard task in general. Even for the two-qubit states, the analytical solutions of it exist only for certain special states [10].

Distance-based measure of quantum correlation is the minimum distance between the state ρ and the set of QC states [11]

$$Q_D(\rho) := \min_{\sigma \in QC} D(\rho, \sigma), \tag{4}$$

where the distance D does not increase under any quantum operation, such that Q_D satisfies (C3). When the relative entropy $S(\rho||\sigma) := \mathrm{tr}[\rho(\log_2 \rho - \log_2 \sigma)]$ is employed as the distance measure, we obtain the **one-way quantum deficit**

$$\Delta_{A|B} := \min_{\{\Pi_i^B\}} S\left(\sum_i I_A \otimes \Pi_i^B \rho I_A \otimes \Pi_i^{B\dagger}\right) - S(\rho), \tag{5}$$

where $\{\Pi_i^B\}$ is a projective measurement on B. $\Delta_{A|B}$ equals to the minimal distance between ρ and ρ^{QC}, and its operational connection with quantum entanglement has also been established [12, 13].

Measurement-induced disturbance is defined as the minimum disturbance caused by local projective measurements that do not change the reduced state $\rho^B \equiv \mathrm{tr}_A \rho$ [14]

$$Q_M(\rho) := \min_{\{\mathrm{E}_i^B\}} D\left(\rho, \sum_i I_A \otimes \mathrm{E}_i^B \rho I_A \otimes \mathrm{E}_i^{B\dagger}\right), \tag{6}$$

where $\{\mathrm{E}_i^B\}$ is a projective measurement on B which satisfies $\sum_i \mathrm{E}_i^B \rho^B \mathrm{E}_i^{B\dagger} = \rho^B$. This measure of quantum correlation is favored for its easy of calculation, and its generalization to continuous-variable systems has also been established [15].

3 Measurement-Induced Nonlocality

In the following, we recall the recently proposed measure of nonlocality which was termed as measurement-induced nonlocality (MIN) [16], as well as various forms of its extension [17–21]. They were all defined from the measurement perspective, and were motivated by those of the discord-like correlation measures [5]. We shall focus mainly on the bipartite systems described by the density operator ρ in the Hilbert space $\mathcal{H}_A \otimes \mathcal{H}_B$. But the related concepts and ideas can in fact be generalized to multipartite systems straightforwardly.

Motivated by the idea that the distance from a given state ρ to the set of states without the desired property is a measure of that property (e.g., the distance to the closest separable state is a measure of entanglement, and to the closest classical state is a measure of discord) [5], the MIN can be defined as the maximal distance that the considered state ρ to the set $\mathcal{L}$ of local quantum states, namely

$$N(\rho) = \max_{\delta \in \mathcal{L}} D(\rho, \delta), \tag{7}$$

where D denotes an arbitrary distance measure that does not increase under the action of TPCP map, while the maximum is taken over the full set of $\mathcal{L}$, which contains those of the quantum states δ obtained by the locally invariant measurements Π^A, that is, $\delta = \sum_k \Pi_k^A \rho \Pi_k^A$ for all $\Pi^A = \{\Pi_k^A\}$ satisfying $\sum_k \Pi_k^A \rho^A \Pi_k^A = \rho^A$, with $p_k = \mathrm{tr}(\Pi_k^A \rho \Pi_k^A)$, and $\rho^A = \mathrm{tr}_B \rho$ being the reduced state of ρ.

This definition of nonlocality measure was motivated by the consideration that a local state should not be disturbed by arbitrary locally invariant measurement Π^A on party A (or Π^B on B), while a nonlocal state may be disturbed by Π^A, and the maximal disturbance can be used to quantify the nonlocal property of it.

By adopting different distance measures D, one can define different MIN measures which possess distinct novel characteristics, and have been shown to play crucial role in many fields of quantum technology.

3.1 *MIN Quantified by the Hilbert–Schmidt Norm*

The notion of MIN was first introduced by Luo and Fu by using the Hilbert–Schmidt norm [16]. For a bipartite state ρ shared by two parties A and B, it was defined as

$$N_2(\rho) = \max_{\Pi^A} ||\rho - \Pi^A(\rho)||_2^2, \tag{8}$$

where Π^A denotes the locally invariant von Neumann measurements, and $||X||_2 = [\mathrm{tr}(X^\dagger X)]^{1/2}$ is the Hilbert–Schmidt norm.

Physically, $N_2(\rho)$ can be considered as the maximal global disturbance induced by the locally invariant measurements Π^A, or the maximal square Hilbert–Schmidt distance that the postmeasurement state $\Pi^A(\rho)$ departs from the premeasurement state ρ. From an applicative point of view, it is also hoped to be useful in the related field of quantum state steering, remote state control, superdense coding, and cryptography [16].

The MIN measure $N_2(\rho)$ has the following basic properties: (i) $N_2(\rho)$ is nonnegative, and equals zero for any product state $\rho = \rho^A \otimes \rho^B$. (ii) $N_2(\rho)$ is locally unitary invariant, namely, $N_2((U^A \otimes U^B)\rho(U^A \otimes U^B)^\dagger) = N_2(\rho)$ for any unitary operators U^A and U^B. (iii) If the reduced state ρ^A is nondegenerate with the spectral decomposition $\rho^A = \sum_k \lambda_k |k\rangle\langle k|$, then the optimal $\tilde{\Pi}^A$ for obtaining $N_2(\rho)$ is given by $\tilde{\Pi}^A(\rho) = \sum_k |k\rangle\langle k|\rho|k\rangle\langle k|$.

For the $(m \times n)$-dimensional bipartite states represented as

$$\rho = \sum_{ij} r_{ij} X_i \otimes Y_j, \tag{9}$$

with $\{X_i : i = 0, 1, \ldots, m^2 - 1\}$ ($X_0 = I_m/\sqrt{m}$) is the orthonormal operator base for subsystem A that satisfy $\mathrm{tr}(X_i^\dagger X_{i'}) = \delta_{ii'}$ (and likewise for Y_j), the MIN measure $N_2(\rho)$ has been shown to be upper bounded by

$$N_2(\rho) \le \sum_{i=1}^{m^2-m} \lambda_i, \tag{10}$$

where λ_i ($i = 1, 2, \ldots, m^2 - 1$) denote the eigenvalues of RR^T in nonincreasing order, $R = (r_{ij})$ with $i, j \ge 1$ is a real matrix, and the superscript T denotes transpose of vectors or matrices.

The MIN measure $N_2(\rho)$ is favored for its ease of calculation for a wide range of quantum states. First, for any bipartite pure state $|\psi\rangle$ with the Schmidt decomposition $|\psi\rangle = \sum_k \lambda_k |\phi_k^A\rangle \otimes |\phi_k^B\rangle$, one has

$$N_2(|\psi\rangle\langle\psi|) = 1 - \sum_k \lambda_k^2, \tag{11}$$

and for the $(2 \times n)$-dimensional states represented as Eq. (9), one has

$$N_2(\rho) = \begin{cases} ||R||_2^2 - \dfrac{1}{||\vec{x}||_2^2} \vec{x}^T RR^T \vec{x} & \text{if } \vec{x} \neq 0, \\ ||R||_2^2 - \lambda_{\min}(RR^T) & \text{if } \vec{x} = 0. \end{cases} \tag{12}$$

where $\lambda_{\min}(RR^T)$ is the smallest eigenvalue of RR^T, and $\vec{x} = (r_{10}, r_{20}, r_{30})^T$.

Moreover, for certain higher dimensional states with symmetry, $N_2(\rho)$ can also be calculated analytically [22], e.g., for the $(d \times d)$-dimensional Werner state ρ_W and isotropic state ρ_I of the following form

$$\begin{aligned}\rho_W &= \frac{d-x}{d^3-d}I_{d^2} + \frac{dx-1}{d^3-d}\sum_{ij}|ij\rangle\langle ji|, \quad x\in[-1,1],\\ \rho_I &= \frac{1-x}{d^2-1}I_{d^2} + \frac{d^2x-1}{d^3-d}\sum_{ij}|ii\rangle\langle jj|, \quad x\in[0,1],\end{aligned} \tag{13}$$

one has

$$N_2(\rho_W) = \frac{(dx-1)^2}{d(d+1)(d^2-1)}, \quad N_2(\rho_I) = \frac{(d^2x-1)^2}{d(d+1)(d^2-1)}. \tag{14}$$

The analytical solutions of $N_2(\rho)$ or its bound for certain bound entangled states [23] and other special states with degenerate ρ^A [24] have also been reported in the literature.

3.2 MIN Quantified by the Trace Norm

Although the MIN measure $N_2(\rho)$ is favored for its convenience of calculation, it is problematic as it can increase or decrease under trivial local reversible operations on the unmeasured subsystem B of ρ. For example, consider a map $\mathcal{E}_B$ which gives rise to $\mathcal{E}_B(\rho) = \rho \otimes \rho^C$, (i.e., it introduces a local ancilla to B), then by making use of the multiplicativity of the Hilbert–Schmidt norm under tensor products, we obtain $N_2(\rho^{A:BC}) = N_2(\rho)\mathrm{tr}(\rho^C)^2$. As the purity of a state is no larger than one, this equality means that the MIN is decreased by simply introducing an uncorrelated local ancillary sytem.

To avoid the aforementioned problem, another geometric measure of MIN based on the trace norm was introduced. It is given by [17]

$$N_1(\rho) = \max_{\Pi^A} ||\rho - \Pi^A(\rho)||_1, \tag{15}$$

where $||X||_1 = \mathrm{tr}\sqrt{X^\dagger X}$ is the trace norm, and Π^A denotes still the locally non-disturbing von Neumann measurements.

The new MIN measure can be interpreted as the maximal trace distance that the premeasurement state ρ departs from those of the postmeasurement states $\Pi^A(\rho)$ caused by the locally invariant measurements. In particular, it is nonincreasing under the action of any TPCP map $\mathcal{E}_B$ on the unmeasured party B [17], namely,

$$N_1(\rho) \geq N_1(\mathcal{E}_B(\rho)). \tag{16}$$

The proof is as follows. Let $\bar{\Pi}^A$ the optimal measurement for obtaining $N_1(\rho)$, and $\tilde{\Pi}^A$ be the optimal measurement for obtaining $N_1(\mathcal{E}_B(\rho))$, then as $\mathcal{E}_B$ and $\tilde{\Pi}^A$ commute, we obtain $\tilde{\Pi}^A(\mathcal{E}_B(\rho)) = \mathcal{E}_B(\tilde{\Pi}^A(\rho))$, and therefore

$$
\begin{aligned}
N_1(\rho) &= ||\rho - \bar{\Pi}^A(\rho)||_1 \\
&\geq ||\rho - \tilde{\Pi}^A(\rho)||_1 \\
&\geq ||\mathcal{E}_B(\rho) - \mathcal{E}_B[\tilde{\Pi}^A(\rho)]||_1 \\
&= N_1[\mathcal{E}_B(\rho)],
\end{aligned} \tag{17}
$$

where the first inequality comes from the fact that $\tilde{\Pi}^A \neq \bar{\Pi}^A$ in general, and the second inequality is due to the contractivity of the trace norm under TPCP map. Therefore, $N_1(\rho)$ circumvents successfully the problem incurred for $N_2(\rho)$.

For certain quantum states, analytical solutions of $N_1(\rho)$ can be obtained, e.g., for the $(2 \times n)$-dimensional pure state $|\psi\rangle$ with the Schmidt decomposition $|\psi\rangle = \sum_{k=1}^{2} \lambda_k |\phi_k^A\rangle \otimes |\phi_k^B\rangle$, the trace norm MIN is given by

$$
N_1(|\psi\rangle\langle\psi|) = 2\sqrt{\lambda_1 \lambda_2}, \tag{18}
$$

while for two-qubit state ρ of the form of Eq. (9) (i.e., $m = 2$) with the addition $r_{ij} = 0$ for $i \neq j$, we have

$$
N_1(\rho) = \begin{cases} \dfrac{\sqrt{\chi_+} + \sqrt{\chi_-}}{||\vec{x}||_1} & \text{if } \vec{x} \neq 0, \\ 2\max\{|r_{11}|, |r_{22}|, |r_{33}|\} & \text{if } \vec{x} = 0, \end{cases} \tag{19}
$$

where $\chi_\pm = \alpha \pm 4\sqrt{\beta}|\vec{x}|$, with $\alpha = |\vec{r}|^2|\vec{x}|^2 - |\vec{r} \cdot \vec{x}|^2$, $\vec{r} = (r_{11}, r_{22}, r_{33})$, $\beta = \sum_{\langle ijk\rangle} x_i^2 r_{jj}^2 r_{kk}^2$, and the summation runs over all the cyclic permutations of $\{1, 2, 3\}$.

Moreover, for the Werner state ρ_W and isotropic state ρ_I of Eq. (13), solutions of the the trace norm MIN are given, respectively, by

$$
N_1(\rho_W) = \frac{|dx - 1|}{d + 1}, \quad N_1(\rho_I) = \frac{2|d^2 x - 1|}{d(d + 1)}, \tag{20}
$$

which show qualitatively the same x-dependence as those of the MIN measure $N_2(\rho_W)$ and $N_2(\rho_I)$ with finite d. That is to say, for the symmetric states ρ_W and ρ_I, both the MIN measures N_1 and N_2 give the same descriptions of nonlocality.

3.3 *MIN Quantified by the Bures Distance*

The Bures distance $d_B(\rho, \chi) = [2(1 - F^{1/2}(\rho, \chi)]^{1/2}$ between two states ρ and χ, which is joint convexity, and is monotonous under the action of any TPCP map [25], can also be used to give a well-defined measure of MIN [17]. Without loss of generality, we define it as

$$
N_B(\rho) = \max_{\Pi^A}\{1 - \sqrt{F(\rho, \Pi^A(\rho))}\}, \tag{21}
$$

where Π^A is the locally invariant measurements on party A, and $F(\rho,\sigma)$ is the Uhlmann fidelity that is defined as

$$F(\rho,\sigma) = [\mathrm{tr}(\sqrt{\rho}\sigma\sqrt{\rho})^{1/2}]^2. \tag{22}$$

For states ρ with nondegenerate ρ^A, $N_B(\rho)$ can be obtained directly, as the optimal $\Pi^A = \{\Pi_i^A\}$ are induced by the spectral resolutions of $\rho^A = \sum_i p_i^A \Pi_i^A$. If ρ^A is degenerate, the calculation of $N_B(\rho)$ is difficult. But for the $(2\times n)$-dimensional states ρ, the minimum Uhlmann fidelity $F_{\min}(\rho, \Pi^A(\rho)) = \min_{\Pi^A} F(\rho, \Pi^A(\rho))$ can be calculated via

$$F_{\min}(\rho, \Pi^A(\rho)) = \frac{1}{2}\min_{\|\vec{u}\|=1}\left(1 - \mathrm{tr}\Lambda + 2\sum_{k=1}^{n}\lambda_k(\Lambda)\right), \tag{23}$$

with $\vec{u} = (\sin\theta\cos\phi, \sin\theta\sin\phi, \cos\theta)$ being a unit vector in $\mathbb{R}^3$, n the dimension of subsystem B, and $\lambda_k(\Lambda)$ eigenvalues of Λ arranged in non-increasing order. By denoting $\vec{\sigma} = (\sigma_1, \sigma_2, \sigma_3)$ the vector of the usual Pauli operators, and I_n the $n\times n$ identity matrix, we have

$$\Lambda = \sqrt{\rho}(\vec{u}\cdot\vec{\sigma}\otimes I_n)\sqrt{\rho}. \tag{24}$$

For the special case of the two-qubit Bell-diagonal states

$$\rho_{\mathrm{Bell}} = \frac{1}{4}\left(I_4 + \sum_{i=1}^{3} c_i\sigma_i\otimes\sigma_i\right), \tag{25}$$

as $\sqrt{\rho_{\mathrm{Bell}}}$ can be derived explicitly, $F_{\min}(\rho, \Pi^A(\rho))$ takes the form

$$F_{\min}(\rho, \Pi^A(\rho)) = \frac{1}{2}\left(1 + \min_{\{\theta,\phi\}}\sqrt{b_3^2 + [b_1^2 - b_3^2 + (b_2^2 - b_1^2)\sin^2\phi]\sin^2\theta}\right), \tag{26}$$

where $b_i = 8(t_0^2 + t_i^2) - 1$ $(i = 1, 2, 3)$, and by writing $c_{\mathrm{sum}} = c_1 + c_2 + c_3$, we have

$$\begin{aligned} t_0 &= \frac{1}{8}\sqrt{1 - c_{\mathrm{sum}}} + \frac{1}{8}\sum_{k=1}^{3}\sqrt{1 + c_{\mathrm{sum}} - 2c_k}, \\ t_i &= -\frac{1}{8}\sqrt{1 - c_{\mathrm{sum}}} + \frac{1}{8}\sum_{k=1}^{3}\sqrt{1 + c_{\mathrm{sum}} - 2c_k} - \frac{1}{4}\sqrt{1 + c_{\mathrm{sum}} - 2c_i}. \end{aligned} \tag{27}$$

From Eq. (26) one can see that $F_{\min}(\rho, \Pi^A(\rho)) = (1 + |b_1|)/2$ if $|b_1| \le \min\{|b_2|, |b_3|\}$, $F_{\min}(\rho, \Pi^A(\rho)) = (1 + |b_2|)/2$ if $|b_2| \leqslant \min\{|b_1|, |b_3|\}$, and $F_{\min}(\rho, \Pi^A(\rho)) = (1 + |b_3|)/2$ otherwise.

3.4 MIN Quantified by the von Neumann Entropy

Apart from the geometric measures, the MIN can also be quantified from the entropic perspective. In this respect, if we accept that the quantum mutual information (QMI) is a good measure of total correlations in ρ, and the entropic measure of MIN, in the spirit of its original definition [16], can be defined as the maximal discrepancy between the QMI of the pre- and post-measurement states as [18]

$$N_E(\rho) = I(\rho) - \min_{\Pi^A} I[\Pi^A(\rho)], \tag{28}$$

where $I(\rho) = S(\rho^A) + S(\rho^B) - S(\rho)$ is the QMI, and Π^A denotes still the locally measurements which do not disturb the reduced state ρ^A.

This measure of MIN quantifies in fact, the maximal loss of total correlations under locally non-disturbing measurements on party A. Moreover, as both ρ and $\Pi^A(\rho)$ have the same reduced states ρ^A and ρ^B, $N_E(\rho)$ defined above is equivalent to

$$N_E(\rho) = \max_{\Pi^A} S[\Pi^A(\rho)] - S(\rho), \tag{29}$$

which indicates that $N_E(\rho)$ quantifies in fact the maximal increment of von Neumann entropy induced by the locally invariant measurements. Moreover, as the entropy of a state measures how much uncertainty there is in it, $N_E(\rho)$ can also be interpreted as the maximal increment of our uncertainty about that system induced by the locally invariant measurements.

The entropic measure of MIN possesses the same basic properties (i), (ii), and (iii) as that of the Hilbert–Schmidt norm MIN. Furthermore, it is monotonous under the action of any TPCP map $\mathcal{E}_B$ on the unmeasured party B, i.e., $N_E(\mathcal{E}_B(\rho)) \leq N_E(\rho)$ [18], which shows that it is a well-defined measure of MIN. Moreover, $N_E(\rho)$ vanishes for the classical-quantum state $\rho_{CQ} = \sum_i p_i |i\rangle\langle i| \otimes \rho_i^B$ with nondegenerate reduced ρ_{CQ}^A, or ρ_{CQ} with degenerate ρ_{CQ}^A and $\rho_i^B = \rho_j^B$ for all i and j.

We point out here that $N_E(\rho)$ is also equivalent to the MIN measure defined based on the relative entropy, namely, $N_E(\rho) = N_{RE}(\rho)$, with

$$N_{RE}(\rho) = \max_{\Pi^A} S(\rho||\Pi^A(\rho)), \tag{30}$$

where $\Pi^A(\rho) = \sum_i \Pi_i^A \rho \Pi_i^A$, and $\{\Pi_i^A\}$ is the set of von Neumann measurements which do not disturb ρ^A locally. In fact, the relative entropy between two states can also be considered as a measure of their distance, although technically it does has a geometric interpretation.

For the state ρ with nondegenerate ρ^A, the optimal measurement operators $\tilde{\Pi}_k^A = |k\rangle\langle k|$ for obtaining $N_E(\rho)$ are induced by the spectral resolutions of $\rho^A = \sum_k \lambda_k |k\rangle\langle k|$. For general cases, $N_E(\rho)$ can be obtained numerically. It is lower bounded by $-S(A|B)$ and upper bounded by $\min\{I(\rho), S(\rho^A)\}$, with $S(A|B) = S(\rho) - S(\rho^B)$. As an example, we list here the analytical solution of $N_E(\rho)$ for the two-qubit Bell-diagonal state ρ_{Bell} of Eq. (25), which is given by

$$N_E(\rho_{\text{Bell}}) = H\left(\frac{1+c_{\min}}{2}\right) + \frac{1-c_{\text{sum}}}{4}\log_2\frac{1-c_{\text{sum}}}{4} + \sum_{k=1}^{3}\frac{1+c_{\text{sum}}-2c_k}{4}\log_2\frac{1+c_{\text{sum}}-2c_k}{4} + 1, \tag{31}$$

with $c_{\min} = \min\{|c_1|, |c_2|, |c_3|\}$, and $H(x) = -(1+x)\log_2(1+x) - (1-x)\log_2(1-x)$ is the binary Shannon entropy.

The quantitative relation between $N_E(\rho)$ and $N_2(\rho)$ has also been established, which is given by [19]

$$N_E(\rho) \geq \frac{1}{2\ln 2}N_2^2(\rho), \tag{32}$$

that is to say, the entropic MIN $N_E(\rho)$ is always greater than or equal to the square of the geometric MIN $N_2(\rho)$ divided by $2\ln 2$ for any state ρ. As the calculation of $N_2(\rho)$ is somewhat easy, the above inequality can serve as a lower bound of $N_E(\rho)$.

3.5 *MIN Quantified by the Wigner–Yanase Skew Information*

The Wigner–Yanase skew information is given by $\mathcal{I}(\rho, K) = -\text{tr}\{[\rho^{1/2}, K]^2\}/2$, with K being an observable [26]. $\mathcal{I}(\rho, K)$ is upper bounded by the variance of K, i.e., $\mathcal{I}(\rho, K) \leq \langle K^2\rangle_\rho - \langle K\rangle_\rho^2$, and vanishes iff the state and the observable commute. It has also been employed to quantify local quantum uncertainty and coherence.

The MIN based on Wigner–Yanase skew information is defined as [20]

$$N_{SI}(\rho) = \max_{\{\tilde{K}_i^A\}}\sum_{i=1}^{m}\mathcal{I}(\rho, \tilde{K}_i^A \otimes I_{d_B}), \tag{33}$$

where the local observables $\tilde{K}^A = \{\tilde{K}_i^A\}$ are restricted to rank-one projectors (i.e., $\tilde{K}_i^A = |i^A\rangle\langle i^A|$) which do not disturb the local state ρ^A, and I_{d_B} is the identity operator for subsystem B, with $d_B = \dim\mathcal{H}_B$. This MIN measure has been shown to be invariant under locally unitary operations, to be contractive under any TPCP map $\mathcal{E}_B$ on party B, and vanishes for the product states $\rho = \rho^A \otimes \rho^B$ and the classical-quantum states ρ_{CQ} with nondegenerate ρ^A.

If we decompose the bipartite state ρ as follows

$$\sqrt{\rho} = \sum_{ij}\gamma_{ij}X_i \otimes Y_j, \tag{34}$$

then it can be shown that $N_{SI}(\rho)$ is upper bounded by [20]

$$N_{SI}(\rho) \le 1 - \sum_{i=1}^{m-1} \mu_i, \tag{35}$$

with μ_i ($i = 1, 2, \ldots, m^2$) being the eigenvalues of $\Gamma\Gamma^T$ listed in decreasing order (counting multiplicity), and $\Gamma = (\gamma_{ij})$ is the $(m^2 \times n^2)$-dimensional correlation matrix.

For the pure states $|\psi\rangle$, $N_{SI}(|\psi\rangle\langle\psi|) = N_2(|\psi\rangle\langle\psi|)$, while for the $(2 \times n)$-dimensional states ρ, one has

$$N_{SI}(\rho) = \begin{cases} 1 - \mu_1 & \text{if } \vec{u} = 0, \\ 1 - \frac{1}{2}\text{tr}\left(\begin{pmatrix} 1 & \vec{u}_0 \\ 1 & -\vec{u}_0 \end{pmatrix} \Gamma\Gamma^T \begin{pmatrix} 1 & \vec{u}_0 \\ 1 & -\vec{u}_0 \end{pmatrix}^T\right) & \text{if } \vec{u} \neq 0. \end{cases} \tag{36}$$

where $\vec{u} = (u_1, u_2, u_3)$ with $u_i = \text{tr}(\rho^A \sigma_i)/\sqrt{2}$, and $\vec{u}_0 = \vec{u}/|\vec{u}|$.

Similarly, for the Werner state ρ_W and the isotropic state ρ_I, the skew information MIN are given, respectively, by

$$\begin{aligned} N_{SI}(\rho_W) &= \frac{1}{2}\left(\frac{d-x}{d+1} - \sqrt{\frac{d-1}{d+1}(1-x^2)}\right), \\ N_{SI}(\rho_I) &= \frac{1}{d}\left(\sqrt{(d-1)x} - \sqrt{\frac{1-x}{d+1}}\right)^2. \end{aligned} \tag{37}$$

The above measure of MIN is somewhat different from that of the MIN-like nonlocality measure defined as [27]

$$\mathcal{U}_{SI}(\rho) = \max_{K^A} \mathcal{I}(\rho, K^A \otimes I_{d_B}), \tag{38}$$

which was motivated by the notion of local quantum uncertainty [28], and was termed as uncertainty-induced nonlocality (UIN), with K^A being the Hermitian observable on A with non-degenerate spectrum and commuting with ρ^A.

The UIN $\mathcal{U}_{SI}(\rho)$ is invariant under locally unitary operation $U_A \otimes U_B$, and is contractive under TPCP map $\mathcal{E}_B$ on subsystem B. It also equals to the maximal squared Hellinger distance between ρ and $K^A \rho K^A$, namely, $\mathcal{U}_{SI}(\rho) = \max_{K^A} D_H^2(\rho, K^A \rho K^A)$, with $D_H^2(\rho, \chi) = \text{tr}\{(\rho^{1/2} - \chi^{1/2})^2\}/2$.

For the $(2 \times n)$-dimensional state of Eq. (9), the UIN is obtained explicitly as

$$\mathcal{U}_{SI}(\rho) = \begin{cases} 1 - \lambda_{\min}(W) & \text{if } \vec{x} = 0, \\ 1 - \frac{1}{|\vec{x}|^2}\vec{x}^T W \vec{x} & \text{if } \vec{x} \neq 0, \end{cases} \tag{39}$$

where $\vec{x} = (r_{10}, r_{20}, r_{30})^T$, and $\lambda_{\min}(W)$ is the smallest eigenvalue of the 3×3 matrix W, the elements of which is given by

$$W_{ij} = \mathrm{tr}\{\rho^{1/2}(\sigma_i \otimes I_{d_B})\rho^{1/2}(\sigma_j \otimes I_{d_B})\}, \tag{40}$$

and from Eq. (39) one can also obtain that for the pure $(2 \times n)$-dimensional state $|\psi\rangle$, $\mathcal{U}_{SI}(|\psi\rangle\langle\psi|)$ reduces to the linear entropy of entanglement $2[1 - \mathrm{tr}(\rho^A)^2]$.

3.6 Generalization of the MIN Measure to Multipartite States

The MIN measures presented above are all defined based on the one-sided locally invariant measurements Π^A on party A. They characterize in fact only partial information about the nonlocal properties of a state ρ. This is because a local state with respect to the subsystem A may be nonlocal with respect to the subsystem B.

The MIN measures can be extended to the cases with two-sided locally invariant measurements. Without loss of generality, we define it as

$$\tilde{N}(\rho) = \max_{\tilde{\delta} \in \mathcal{L}} D(\rho, \tilde{\delta}), \tag{41}$$

with $\tilde{\delta}$ being states that are obtained by the full set of locally invariant measurements $\Pi^A \otimes \Pi^B$, that is to say, $(\Pi^A \otimes \Pi^B)\tilde{\delta}(\Pi^A \otimes \Pi^B) = \tilde{\delta}$, and the measurement operators satisfy the equality $\sum_k \Pi_k^A \rho^A \Pi_k^A = \rho^A$ and $\sum_k \Pi_k^B \rho^B \Pi_k^B = \rho^B$ for arbitrary ρ. This definition of MIN reveals the genuine nonlocal characteristic of a bipartite state, and $\tilde{N}(\rho) = 0$ implies locality with respect to both the subsystems of A and B.

An example of the MIN over two-sided measurements is as follows [21]

$$\tilde{N}_2(\rho) = \max_{\Pi^A \otimes \Pi^B} ||\rho - \Pi^A \otimes \Pi^B(\rho)||_2^2, \tag{42}$$

which is locally unitary invariant, and vanishes for the product states $\rho = \rho^A \otimes \rho^B$.

For pure state $|\psi\rangle$, solution of $\tilde{N}_2(|\psi\rangle\langle\psi|)$ is completely the same as $N_2(|\psi\rangle\langle\psi|)$. For the special case that both ρ^A and ρ^B are nondegenerate, the optimal measurement operators are uniquely determined by the eigenvectors of ρ^A and ρ^B, while for more general case, it can be calculated using the numerical method.

Other measures of MIN presented in the above sections can be redefined in a similar way, namely, by replacing the original one-sided locally invariant measurements Π^A with the two-sided locally invariant measurements $\Pi^A \otimes \Pi^B$.

In fact, the MIN measure $\tilde{N}_2(\rho)$ can also be extended to the more general case of N-partite state ρ. The definition can be written in the same form of Eq. (41), with however the set $\mathcal{L}$ of local quantum states being obtained by performing all possible locally invariant measurements $\Pi^{A_1} \otimes \Pi^{A_2} \otimes \cdots \otimes \Pi^{A_N}$, with $\sum_k \Pi_k^{A_i} \rho^{A_i} \Pi_k^{A_i} =$

ρ^{A_i} for $i = \{1, 2, \ldots, N\}$, and ρ^{A_i} the reduced state of the subsystem A_i. But now the evaluation of their analytical expression becomes a hard work.

Here we summarize briefly the MIN. The notion of MIN is a recently introduced measure of nonlocality which is defined from a measurement perspective, and provides a better division between the local and nonlocal features of a system. In general, the MIN can be defined as distance between the quantum states before and after the measurement is performed.

Next, we consider the behaviors of quantum correlations under local operations.

4 Quantum Correlations Increased by Local Operations

The states that can be prepared by local operations and classical communications (LOCC) are called the separable states. The set of bipartite separable states can be written as $\mathcal{S} := \left\{\rho|\rho = \sum_i p_i \rho_i^A \otimes \rho_i^B\right\}$. By definition, $\mathcal{QC}$ is a strict subset of the separable states $\mathcal{S}$. There exist quantum correlated states which are separable and hence can be prepared via LOCC. Actually, the local operations (LO) alone can turn a QC state to a quantum correlated one. For example, consider a channel Λ with Kraus decompositions $K_1 = |0\rangle\langle 0|$ and $K_2 = |+\rangle\langle 1|$. When Λ is applied to B of the two-qubit QC state $\rho = \frac{1}{2}|00\rangle_{AB}\langle 00| + \frac{1}{2}|11\rangle_{AB}\langle 11|$, the output state $\rho_{out} = \frac{1}{2}|00\rangle_{AB}\langle 00| + \frac{1}{2}|1+\rangle_{AB}\langle 1+| \notin \mathcal{QC}$ has nonvanishing quantum correlations.

Then questions naturally arises:

(a) What kind of local operations have the ability to create quantum correlations?
(b) What is the power of a given local operation to create quantum correlations?
(c) What kind of states whose quantum correlations are more likely to be increased

In this section, we will give answers to these three questions.

4.1 Condition for Local Creation of Quantum Correlations

The main purpose of this subsection is to characterize the whole set of quantum operations satisfying

$$\mathbb{1}_A \otimes \Lambda_B(\rho) \in \mathcal{QC}, \ \forall \rho \in \mathcal{QC}. \tag{43}$$

Before solving the problem, let us first introduce a class of quantum channels, which we call the commutativity-preserving channels.

Definition 1 (*Commutativity-preserving channel*) A commutativity-preserving channel Λ^{CP} is the channel that can preserve the commutativity of any input density operators; i.e.

$$[\Lambda^{\text{CP}}(\xi), \Lambda^{\text{CP}}(\xi')] = 0 \tag{44}$$

holds for any density operators ξ and ξ' satisfying $[\xi, \xi'] = 0$.

When a commutativity-preserving channel acts on B of a QC state, the output state $\rho_{out} = \sum_i p_i \rho_i^A \otimes \Lambda^{\mathrm{CP}}(\phi_i^B)$ is still a QC state. This is because Bob's states $\Lambda^{\mathrm{CP}}(\phi_i^B)$ commute with each other and a projective measurement on their common eigenbasis does not change the state ρ_{out}.

Conversely, if Bob's channel Λ satisfies Eq. (43), it must be a commutativity-preserving channel. To see this, let us write the input state in the form of Eq. (2), and the output state $\rho_{out} = \sum_i p_i \zeta_i^A \otimes \Lambda(\xi_i^B)$ is still quantum-classical only when $[\Lambda(\xi_i^B), \Lambda(\xi_j^B)] = 0, \ \forall\, i, j$. As in Eq. (43), we have considered the whole set of QC states as input state, the channel Λ must preserves the commutativity of any two commutable states. Hence we arrive at the following theorem.

Theorem 1 *A local quantum channel Λ acting on a subsystem of a multipartite system can create quantum correlation if and only if it is not a commutativity-preserving channel.*

This theorem characterize the set of quantum channels which does not create quantum correlation in any QC states. The rest of this subsection will be devoted to provide the explicit form of the commutativity-preserving channels. We will see that when B is a qubit, the quantum correlations can never be created by a unital channel, but when B is a higher-dimension system, even unital channels can create quantum correlation.

Let us first consider the qubit case. In the Bloch presentation, any qubit state $\rho = \frac{I + \boldsymbol{r}\cdot\boldsymbol{\sigma}}{2}$ corresponds to a three-dimension real vector $\boldsymbol{r}$, where $\boldsymbol{\sigma} = \{\sigma_1, \sigma_2, \sigma_3\}$ are Pauli matrices and the Bloch vector $\boldsymbol{r}$ lives inside or on the surface of a unit ball, which is called the Bloch ball. The Bloch vectors of two commutative states are of the same or opposite orientation, so the necessary and sufficient condition for the commutativity-preserving channels is that they map radial segments onto radial segments in the Bloch ball. The unital channels are defined as those preserve the identity $\Lambda^u(I) = I$, i.e., the origin of the Bloch ball, and thus satisfies the above condition. Another set of channels which are apparently commutativity-preserving are the semiclassical channels, which map all input states onto states diagonal on the same basis. It can be strictly proved that a commutativity-preserving qubit channel is either a unital channel or a semiclassical channel. This leads to the following theorem.

Theorem 2 *A local quantum channel acting on a single qubit can create quantum correlations in a multiqubit system if and only if it is neither semiclassical nor unital.*

Now we turn to multipartite systems of higher-dimension. Apparently, the semiclassical channels do not have the ability to create quantum correlations. Here we propose another set of quantum channels, which we call the isotropic channels, that never create quantum correlations.

Definition 2 (*isotropic channel*) An isotropic channel is a channel $\Lambda : \mathcal{D}(\mathcal{H}_d) \to \mathcal{D}(\mathcal{H}_d)$ of the form

$$\Lambda^{\mathrm{iso}}(\rho) = p\Gamma(\rho) + (1 - p)\frac{I}{d}, \tag{45}$$

where Γ is any linear channel that preserves the eigenvalues of ρ, and the parameter p is chosen to make sure that Λ is a completely positive channel.

According to Ref. [29], Γ is either a unitary operation or unitarily equivalent to transpose. Direct calculations lead to $-1/(d-1) \leq p \leq 1$ when Γ is a unitary operation, and $-1/(d-1) \leq p \leq 1/(d+1)$ when Γ is unitarily equivalent to transpose.

Because the unitary operations and the transpose preserve the commutativity and the identity commutes with any state, the isotropic channels are commutativity preserving for arbitrary d. For $d = 3$, we have strictly proved that a commutativity channel is either semiclassical or isotropic.

Theorem 3 *A local quantum channel acting on a single qutrit of a multipartite system can create quantum correlations if and only if it is neither semiclassical nor isotropic.*

Since isotropic channels are a strict subset of unital channels, there exist unital channels that are able to locally create quantum correlations. Here we give an example to look more closely at why a unital channel can create quantum correlation in multipartite states of higher dimensions. Let us consider the unital channel $\Lambda(\cdot) = \sum_i E^{(i)}(\cdot)E^{(i)\dagger}$ with

$$E^{(0)} = |2\rangle\langle 2|, E^{(1)} = |0\rangle\langle 0| + |1\rangle\langle 1|. \tag{46}$$

It is *not* a commutativity-preserving channel, because when we choose the orthogonal pure state $|\psi\rangle = \frac{1}{\sqrt{3}}(|0\rangle + |1\rangle + |2\rangle)$ and $|\phi\rangle = \frac{1}{\sqrt{2}}(|0\rangle - |2\rangle)$ as input states, the output states do not commute. Two higher-dimension orthogonal states may become nonorthogonal when projected to subspaces. This is just the reason for creating quantum correlation using a local unital channel. Isotropic channels act on all of the states equivalently, so they are likely the only subset of unital channels which belongs to the set of commutativity-preserving channels. This observation leads to the following conjecture.

Conjecture *A local quantum channel acting on a single qubit with $d > 3$ can create quantum correlations in a multipartite system if and only if it is neither semiclassical nor isotropic.*

4.2 Quantum Correlating Power of Local Operations

So far, we have discussed the problem of *whether* a quantum channel can create quantum correlations. The problem of *how much* quantum correlations can be created by a quantum channel is the theme of this subsection. The quantum-correlating power of quantum channel is defined as the maximum amount of quantum correlations that can be created when the channel is applied locally to a subsystem of a multipartite system [31]. The formal definition is given as follows.

Definition 3 (*Quantum correlating power, QCP*) The quantum correlating power of a quantum channel Λ is defined as

$$\mathcal{Q}(\Lambda) := \max_{\rho \in \mathcal{QC}} Q(\mathbb{1} \otimes \Lambda(\rho)). \tag{47}$$

The QCP is an intrinsic attribute of a quantum channel, which quantifies the channel's ability to create quantum correlations. In the definition of QCP, the maximization is taken over the set of all quantum-classical states. The input states that correspond to the maximization are called the optimal input states, which are proved to be in the set of classical-classical (CC) states

$$\mathcal{CC} = \left\{ \rho | \rho = \sum_i p_i |\psi_i\rangle_A \langle \psi_i| \otimes |\phi_i\rangle_B \langle \phi_i| \right\}. \tag{48}$$

where $\{|\psi_i\rangle_A\}$ and $\{|\phi_i\rangle_B\}$ are orthogonal basis of $\mathcal{H}_A$ and $\mathcal{H}_B$ respectively. The proof can be easily sketched. For any output state ρ' that corresponds to a general QC input state, we can find a CC state, whose corresponding output state ρ can be transformed to ρ' by a local channel on A, i.e., $\rho' = \Lambda_A \otimes \mathbb{1}\rho$. From the condition (C3), we have $Q(\rho) \geq Q(\rho')$. Hence the definition of QCP can be optimized to

$$\mathcal{Q}(\Lambda) := \max_{\rho \in \mathcal{CC}} Q(\mathbb{1} \otimes \Lambda(\rho)). \tag{49}$$

A channel with larger amount of QCP is more quantum, in the sense of the ability to create quantum correlation. Hence it is of interest to find out the quantum channels with the most QCP. It can be proved that, the local single-qubit channel which maximum QCP can be found in the set of measuring-preparing channels

$$\mathcal{MP} = \left\{ \Lambda | \Lambda(\cdot) = \sum_{i=1}^{1} |\alpha_i\rangle\langle \phi_i| \cdot |\phi_i\rangle\langle \alpha_i| \right\}, \tag{50}$$

where $|\alpha_0\rangle$ and $|\alpha_1\rangle$ are two nonorthogonal pure states.

When two channels used paralleled, the QCP of the composed channel is no less than the sum of the QCPs of the two channels. We call this property the superadditivity of QCP [32]. We here give an example of phase-damping (PD) channel to show exactly how this property works. The Kraus operators of PD channel are $E_0^{\text{PD}} = |0\rangle\langle 0| + \sqrt{1-p}|1\rangle\langle 1|$ and $E_1^{\text{PD}} = \sqrt{p}|1\rangle\langle 1|$. Here we consider the nontrivial case where $0 < p < 1$. Clearly, PD channel is a unital channel and thus has vanishing QCP.

Now consider a four-qubit initial state shared between Alice and Bob

$$\rho_{AA'BB'} = \frac{1}{4} \sum_{i,j} |ij\rangle_{AA'} \langle ij| \otimes |\psi_{ij}\rangle_{BB'} \langle \psi_{ij}|, \tag{51}$$

where $|\psi_{00}\rangle = \frac{1}{\sqrt{2}}(|00\rangle + |11\rangle)$, $|\psi_{11}\rangle = \frac{1}{\sqrt{2}}(|0+\rangle + |1-\rangle)$, $|\psi_{01}\rangle = \frac{1}{\sqrt{2}}(|01\rangle - |10\rangle)$, and $|\psi_{10}\rangle = \frac{1}{\sqrt{2}}(|0-\rangle - |1+\rangle)$. Here qubits AA' belong to Alice and BB' belong to Bob. Since $|\psi_{ij}\rangle$ are orthogonal to each other, the quantum correlation on Bob is zero. Then qubits B and B' each transmits through a PD channel, and the output state becomes $\rho'_{AA'BB'} = \mathbb{1}_{AA'} \otimes \Lambda_B^{\text{PD}} \otimes \Lambda_{B'}^{\text{PD}}(\rho_{AA'BB'})$. Because $[\Lambda^{\text{PD}} \otimes \Lambda^{\text{PD}}(\psi_{00}), \Lambda^{\text{PD}} \otimes \Lambda^{\text{PD}}(\psi_{11})] = \frac{1}{8}\tilde{i}p\sqrt{1-p}(\sigma^y \otimes \sigma^z + \sigma_z \otimes \sigma^y) \neq 0$, the output state $\rho'_{AA'BB'}$ is not a QC state. Therefore, the quantum correlation on Bob's qubits BB' is created by the channel $\Lambda_B^{\text{PD}} \otimes \Lambda_{B'}^{\text{PD}}$.

The super-activation of QCP is a collective effect. For both the input state $\rho_{AA'BB'}$ and the output state $\rho'_{AA'BB'}$, any two-qubit marginal is a completely mixed state. In other words, no correlation exists between any two qubits of the four-qubit state $\rho'_{AA'BB'}$. Therefore, we suppose that the effect of super-additivity of QCP is due to the genuine quantum correlation.

4.3 States Whose Quantum Correlations Can Be Increased

Our aim is now to characterize the quantum states whose quantum correlations can be increased locally. This is a less studied subject than the condition on quantum channels to locally create quantum correlations. Obviously, the quantum correlation of all the QC state can be increased locally, by the quantum channels which are not commutativity preserving. However, it is not obvious whether the quantum correlation of a discordant state can be increased locally.

Before study the problem, we first introduce the quantum steering ellipsoids (QSE), which provides a natural geometric presentation of two-qubit states. The quantum steering ellipsoid of a two-qubit state ρ_{AB} is the whole set of Bloch vectors that the qubit A can be collapsed to by a positive-operator valued measurement (POVM) on qubit B. When the Bloch vector $\boldsymbol{b}$ of ρ_B satisfies $b = 1$, ρ_B is a pure state which is not correlated to A; hence the QSE at A reduces to a single point $\boldsymbol{a}$.

Now we consider the case with $b \in [0, 1)$. Suppose the qubit B is projected to a pure state ρ_x with Bloch vector $\boldsymbol{x}$. The state of A is steered to $\rho_A^S = \text{tr}_B[\rho_{AB}(I \otimes \rho_x)]/\text{tr}[\rho_{AB}(I \otimes \rho_x)]$, whose Bloch vector is $\boldsymbol{a}^S = \frac{\boldsymbol{a}+T\boldsymbol{x}}{1+\boldsymbol{b}\cdot\boldsymbol{x}}$. Let $\boldsymbol{x}$ varies through the Bloch ball, the set of corresponding $\boldsymbol{a}^S$ forms an ellipsoid

$$\mathcal{E}_A = \left\{ \frac{\boldsymbol{a} + T\boldsymbol{x}}{1 + \boldsymbol{b} \cdot \boldsymbol{x}} | x \leq 1 \right\}. \tag{52}$$

To obtain $\mathcal{E}_B$, one only need to make the substitution $\boldsymbol{a} \to \boldsymbol{b}, \boldsymbol{b} \to \boldsymbol{a}, T \to T^{\text{T}}$. It is worth mentioning that the QSE $\mathcal{E}_A$ and $\mathcal{E}_B$ of state ρ_{AB} have the same dimension, which equals to $\text{rank}(\Theta) - 1$.

The state ρ_{AB} is a QC state if and only if $\mathcal{E}_B$ is a radial line segment. Local channels on qubit B can create B-side quantum discord from the above quantum-classical state. The output discordant state can be written as

$$\rho \equiv \mathbb{I}_A \otimes \Lambda_B(\rho_{AB}) = p_0\rho_0^A \otimes \rho_0^B + p_1\rho_1^A \otimes \rho_1^B. \tag{53}$$

Here $\rho_i^B \equiv \Lambda_B(|\phi_i\rangle_B\langle\phi_i|)$ $(i = 0, 1)$ do not commute with each other [30] and thus are linearly independent. The following statement builds the connection between locally created discordant states and the states with needle-shape QSE. *A B-side discordant two-qubit state can be created from a classical state by a trace-preserving local channel on B if and only if its QSE at qubit B $\mathcal{E}_B$ is a non-radial line segment* [33]. It means that all of the quantum states with pancake-shape or obese-shape QSE, even though not entangled, can not be prepared by local operations.

Next we focus on the Bell diagonal states and study the relation between the effect of locally increased quantum discord and the shape and position of QSE. For a Bell diagonal two-qubit state, the density matrix can be written as

$$\tilde{\rho} = \frac{1}{4}\left(\sigma_0 \otimes \sigma_0 + \sum_{i=1}^{3} c_i\sigma_i \otimes \sigma_i\right). \tag{54}$$

For such a state, both $\mathcal{E}_A$ and $\mathcal{E}_B$ are unit spheres shrunk by c_1, c_2 and c_3 in the x, y and z direction, respectively. After the action of an amplitude damping channel on B, the QSE at B becomes

$$\mathcal{E}_B^{\mathrm{AD}} = \left\{ \begin{pmatrix} 0 \\ 0 \\ p \end{pmatrix} + \begin{pmatrix} \sqrt{1-p}c_1x_1 \\ \sqrt{1-p}c_2x_2 \\ (1-p)c_3x_3 \end{pmatrix} \middle| x \le 1 \right\}. \tag{55}$$

The effect of Λ_B^{AD} on $\mathcal{E}_B$ is to translate it by p in the z direction and meanwhile shrink the ellipsoid on three directions. Notice that when $p > \frac{c_3}{1+c_3}$, the ellipsoid does not contain the origin point any more.

Increase of quantum discord by local amplitude damping channel on B can occur for all of the three cases when the initial QSE is a needle, a pancake and an obese. For the last two cases, the local increase of discord occurs when $|c_1| \gg |c_2|, |c_3|$, which means that the shape of the plate or the ball is like a baguette perpendicular to the z axis. It is worth noticing that, the local quantum operation can increase the quantum discord of an entangled state.

5 Summary

A local quantum channel acting on a subsystem of a multipartite system can create quantum correlation if and only if it is not a commutativity-preserving channel. A qubit channel is commutativity-preserving if and only if it is unital or semiclassical. For the high-dimension case, some unital channels have the ability to create quantum correlations. In order to characterise the maximum quantum correlation that a quantum channel can create, the quantum-correlating power is defined. It is an intrinsic property of quantum channels. The superactivity of QCP is proved. Concerning the two-qubit states whose quantum correlation can be increased locally, it is observed that the quantum correlation of those states possesses baguette-like quantum steering ellipsoids.

We also presented that nonlocality of a quantum state can be described from different aspects. We provided a short review about the recently introduced quantification of MIN by considering the distances for states before and after the measurement is taken. The MIN in terms of distance can be defined based on the Hilbert–Schmidt norm, the trace norm, the Bures distance, the von Neumann entropy, and the Wigner–Yanase skew information. The basic formulae for their respective definitions, the analytical solutions of them for certain special states, and a comparison of their similarities and differences, are given in detail. We have also provided a outlook for its further development such as its generalization to multipartite systems.

Acknowledgements This work is supported by MOST (2016YFA0302104, 2016YFA0300600), NSFC (91536108, 11504205, 11205121), CAS (XDB01010000, XDB21030300), New Star Project of Science and Technology of Shaanxi Province (2016KJXX-27), the Fundamental Research Funds of Shandong University under Grant No. 2014TB018.

References

1. A. Einstein, B. Podolsky, N. Rosen, Can quantum-mechanical description of physical reality be considered complete? Phys. Rev. **47**, 777–780 (1935)
2. M. Genovese, Research on hidden variable theories: a review of recent progresses. Phys. Rep. **413**, 319–396 (2005)
3. R.F. Werner, Quantum states with Einstein-Podolsky-Rosen correlations admitting a hidden-variable model. Phys. Rev. A **40**, 4277–4281 (1989)
4. L. Masanes, Y.C. Liang, A.C. Doherty, All bipartite entangled states display some hidden nonlocality. Phys. Rev. Lett. **100**, 090403 (2008)
5. K. Modi, A. Brodutch, H. Cable, Z. Paterek, V. Vedral, The classical-quantum boundary for correlations: discord and related measures. Rev. Mod. Phys. **84**, 1655–1707 (2012)
6. S. Popescu, R. Rohrlich, Nonlocality as an axiom. Found. Phys. **24**, 379 (1994)
7. S. Popescu, Nonlocality beyond quantum mechanics. Nat. Phys. **10**, 264 (2014)
8. H. Fan, Y.N. Wang, L. Jing, J.D. Yue, H.D. Shi, Y.L. Zhang, L.Z. Mu, Quantum cloning machines and the applications. Phys. Rep. **544**, 241–322 (2014)
9. H. Ollivier, W.H. Zurek, Quantum discord: a measure of the quantumness of correlations. Phys. Rev. Lett. **88**, 017901 (2002)
10. D. Girolami, G. Adesso, Quantum discord for general two-qubit states: analytical progress. Phys. Rev. A **83**, 052108 (2011)
11. K. Modi, T. Paterek, W. Son, V. Vedral, M. Williamson, Unified view of quantum and classical correlations. Phys. Rev. Lett. **104**, 080501 (2010)
12. M. Piani, S. Gharibian, G. Adesso, J. Calsamiglia, P. Horodecki, A. Winter, All nonclassical correlations can be activated into distillable entanglement. Phys. Rev. Lett. **106**, 220403 (2011)
13. M.F. Cornelio, M.C. de Oliveira, F.F. Fanchini, Entanglement irreversibility from quantum discord and quantum deficit. Phys. Rev. Lett. **107**, 020502 (2011)
14. S. Luo, Using measurement-induced disturbance to characterize correlations as classical or quantum. Phys. Rev. A **77**, 022301 (2008)
15. L. Mista Jr., R. Tatham, D. Girolami, N. Korolkova, G. Adesso, Measurement-induced disturbances and nonclassical correlations of Gaussian states. Phys. Rev. A **83**, 042325 (2011)

16. S. Luo, S. Fu, Measurement-induced nonlocality. Phys. Rev. Lett. **106**, 120401 (2011)
17. M.L. Hu, H. Fan, Measurement-induced nonlocality based on the trace norm. New J. Phys. **17**, 033004 (2015)
18. M.L. Hu, H. Fan, Dynamics of entropic measurement-induced nonlocality in structured reservoirs. Ann. Phys. **327**, 2343–2353 (2012)
19. Z. Xi, X. Wang, Y. Li, Measurement-induced nonlocality based on the relative entropy. Phys. Rev. A **85**, 042325 (2012)
20. L. Li, Q.W. Wang, S.Q. Shen, M. Li, Measurement-induced nonlocality based on Wigner-Yanase skew information. Europhys. Lett. **114**, 10007 (2016)
21. Y. Guo, Measurement-induced nonlocality over two-sided projective measurements. Int. J. Mod. Phys. B **27**, 1350067 (2013)
22. S. Luo, S. Fu, Global effects of quantum states induced by locally invariant measurements. Europhys. Lett. **92**, 20004 (2010)
23. S. Rana, P. Parashar, Geometric discord and measurement-induced nonlocality for well known bound entangled states. Quantum Inf. Process. **12**, 2523–2534 (2013)
24. S.Y. Mirafzali, I. Sargolzahi, A. Ahanj, K. Javidan, M. Sarbishaei, Measurement-induced nonlocality for an arbitrary bipartite state. Quantum Inf. Comput. **13**, 479–489 (2013)
25. R. Jozsa, Fidelity for mixed quantum states. J. Modern Opt. **41**, 2315–2323 (1994)
26. E.P. Wigner, M.M. Yanase, Information contents of distributions. Proc. Natl. Acad. Sci. U. S. A. **49**, 910–918 (1963)
27. S. Wu, J. Zhang, C. Yu, H. Song, Uncertainty-induced quantum nonlocality. Phys. Lett. A **378**, 344–347 (2014)
28. D. Girolami, T. Tufarelli, G. Adesso, Characterizing nonclassical correlations via local quantum uncertainty. Phys. Rev. Lett. **110**, 240402 (2013)
29. M. Marcus, B.N. Moyls, Linear transformations on algebras of matrices. Can. J. Math. **11**, 61 (1959)
30. X.Y. Hu, H. Fan, D.L. Zhou, W.M. Liu, Necessary and sufficient conditions for local creation of quantum correlation. Phys. Rev. A **85**, 032102 (2012)
31. X. Hu, H. Fan, D.L. Zhou, W.M. Liu, Quantum correlating power of local quantum channels. Phys. Rev. A **87**, 032340 (2013)
32. X. Hu, H. Fan, D.L. Zhou, W.M. Liu, Superadditivity of quantum-correlating power. Phys. Rev. A **88**, 012315 (2013)
33. X. Hu, H. Fan, Effect of local channels on quantum steering ellipsoids. Phys. Rev. A **91**, 022301 (2015)

Quantum Correlations in Multipartite Quantum Systems

Thiago R. de Oliveira

Abstract We review some concepts and properties of quantum correlations, in particular multipartite measures, geometric measures and monogamy relations. We also discuss the relation between classical and total correlations.

1 Introduction

Entanglement is usually said to be the characteristic trait of quantum mechanics. All started with the recognition by Einstein, Podolsky and Rosen [1] that two-qubit states such as the superposition

$$|\psi\rangle = |00\rangle + |11\rangle, \tag{1}$$

where $|0\rangle$ and $|1\rangle$ are the eigenstates of σ^z, have some kind of non-local "action at a distance" since a measure of the first qubit somehow "changes" the state of the second qubit, no matter how far away it is: if I measure one qubit and obtain $|0\rangle$ ($|1\rangle$), I know immediately that a measurement on the other, in the same basis, will also return the state $|0\rangle$ ($|1\rangle$). However it was later realized that such states alone do not allow communication at a distance and therefore do not violate the principle of special relativity. But such states do allow for stronger correlations than allowed by a classical theory, as seen in the violation of Bell inequalities. One way to see such stronger correlations is to note that the perfect correlations between measurements of the spin, are not true only for measurements along the z direction, but actually in any direction. As far as one deals with pure states, the situation is clear. However, the generalization of the concept of entanglement to mixed states is more complicated. Werner in 1989 [2] proposed non-entangled, or separable mixed states, to be the ones which can be written as

T.R. de Oliveira (✉)
Instituto de Física, Universidade Federal Fluminense, Niterói, RJ 24210-346, Brazil
e-mail: tro@if.uff.br

F.F. Fanchini et al. (eds.), *Lectures on General Quantum Correlations and their Applications*, Quantum Science and Technology,
DOI 10.1007/978-3-319-53412-1_5

$$\rho = \sum_i p_i \, \rho_A^i \otimes \rho_B^i. \tag{2}$$

This definition is motivated by the fact that these are the states which can be created by two separated labs using local quantum operations and classical communication: they contain only classical correlations due to the classical communication. Entanglement was then rigorously defined as a property of quantum states which can not be created by local operation and classical communication (LOCC). This framework of LOCC created the basis for entanglement theory which defines good entanglement measures as the ones which do not increase under LOCC. But already Werner [2] noted that such definition allowed for entangled states which do not violated any Bell-type inequality, opening a gap between the concept of entanglement and non-locality.[1]

And in 2002, studying the correlation between apparatus and system in a measurement, Ollivier and Zurek realized [4] that separable states, as defined by Werner, may still have some quantumness in the sense that they can be perturbed by local measurements. Let's focus on "perfect" von Neumman measurements, defined by a set of one-dimensional orthogonal projectors $\{\Pi_j^B\}$ on system B, the apparatus. The state of A after the outcome corresponding to Π_j^B has been detected is

$$\rho_{A|\Pi_j^B} = \frac{\Pi_j^B \rho_{AB} \Pi_j^B}{\mathrm{Tr}[\Pi_j^B \rho_{AB}]}, \tag{3}$$

and this outcome happens with probability $p_j = \mathrm{Tr}[\Pi_j^B \rho_{AB}]$. It can be shown that the only way that the state of A is not perturbed by this measurement is if it can be written as

$$\chi_{AB} = \sum_i \rho_A^i \otimes \Pi_i^B. \tag{4}$$

This is a separable state with only fully-distinguishable states (orthogonal ones) for B and some indistinguishable states for A. Such states are called quantum-classical since there are measurements on B which do not perturb the state; but measurements on A may perturb it. States which can not be written in such form are perturbed by all local measurements on B. This perturbation of the state of A by local measurements on B is a quantum aspect of the correlations between A and B that goes beyond entanglement and that can be quantified in many ways. The perturbation will also in general decrease the correlation between the parts.

There are many possibilities to quantify this quantum aspect of the correlation. Lets consider the conditional entropy of A given B:

$$S(\rho_A|\rho_B) = S(\rho_{AB}) - S(\rho_B). \tag{5}$$

---

[1]There is a vast literature studying this gap and looking for more general types of Bell inequality which may close the gap; see [3].

Considering that entropy measures the uncertainty about the system, the conditional entropy is the remaining uncertainty about A after we learn the state of B and it is associated with our uncertainty, on average, about A, given that we know the state of B; we measured it. But as we mentioned, the acting of measuring the system can perturb it and thus change the conditional entropy. Therefore for quantum states we can define the conditional entropy in a alternative way as

$$S(\rho_{AB}|\Pi^B) = \sum_j p_j \; S(\rho_{A|\Pi_j^B}). \tag{6}$$

It still has the interpretation of the average uncertainty about A given that we measured B. Classically these two definitions are equivalent, but for some quantum states they can differ.[2] This difference in the definition also propagates for other entropy measures of correlation. The mutual information, for example, can be written as

$$I(\rho_{AB}) = S(\rho_A) + S(\rho_B) - S(\rho_{AB}) \tag{7}$$

or in terms of the conditional entropy

$$J_{\Pi^B}(\rho_{AB}) = S(\rho_A) - S(\rho_{AB}|\Pi^B), \tag{8}$$

where we used the alternative definition for the conditional entropy and another symbol, J, since the two definitions of the mutual information may not be equivalent. The first expression suggests the interpretation of the mutual information as the common information between A and B and therefore as the measure of its total correlation. The second expression suggests that the mutual information is a measure of the decrease on uncertainty, or gain in information, about A as a result of a measurement on B. The second definition was introduced in [6] as a measure of the classical correlation.

Based on these considerations the discord was defined by Ollivier and Zurek [4] as

$$D_{\Pi^B}(\rho_{AB}) = I(\rho_{AB}) - J_{\Pi^B}(\rho_{AB}). \tag{9}$$

It measures how much common information, or correlation, was lost in the measurement. In other words, it measures the information about A that exists in the correlation but can not be extracted locally by reading the state of B. It can also be written as the difference between the two definitions of conditional entropy: $S(\rho_A|\rho_B) - S(\rho_{AB}|\Pi^B)$. One should minimize over all possible measurements on B to find the one which disturbs the least A and allows us to extract the most information about A by measuring B. Thus the measurement independent discord was defined as

$$D_B(\rho_{AB}) = \min_{\Pi^B} D_{\Pi^B}(\rho_{AB}). \tag{10}$$

---

[2] Actually $S(\rho_A|\rho_B)$ is always positive in classical setting, but can be negative for entangled states and took it a long time to understand this negativity; see [5].

The discord has the following properties: (i) it is not symmetric under the change of A for B; (ii) it is non-negative; (iii) it vanishes if and only if the state is quantum-classical; (iv) it is invariant under local unitary transformations. Unfortunately it does not have an important property for correlations measures: to not increase under local operations.[3] It may increase by simple local operations and therefore is not a bona fine measure of correlations. In sum, discord does indicate that the correlation in the state has a quantum aspect, but it is not a bona fide quantifier of the amount of such correlation.

There are also many other possibilities to quantify this quantumness in separable states. And in fact many discord-like measures were proposed, and still are being proposed (see [7] for a review). We will just mention another one since it will be of our interest later, and actually is closely related to the original definition of discord. The idea is to consider local von Neumann measurements on both parties. The state after the non-selective measurement is

$$\Pi^{AB}(\rho_{AB}) = \sum_{i,j}(\Pi_i^A \otimes \Pi_j^B)\rho_{AB}(\Pi_i^A \otimes \Pi_j^B) \tag{11}$$

and has the general form

$$\chi_{AB} = \sum_{i,j} p_{ij}\; \Pi_i^A \otimes \Pi_j^B. \tag{12}$$

Such states are called classical-classical, since they are the ones which are not perturbed by the local measurement on A or B. Thus the probability p_{ij} can be regarded as a classical joint probability of the random variables i and j. One then defines the symmetric discord as

$$D_S(A:B) = \min_{\Pi^{AB}} [I(\rho_{AB}) - I(\Pi^{AB}(\rho_{AB}))]. \tag{13}$$

This idea was originally proposed as a measurement-induced disturbance and without the optimization over measurements. It was then redefined with the optimization and studied by several authors (see Sect. II.E of [7]). Note also that it may be argued that the asymmetric discord is not a good measure of the quantum correlation since it could be null from one side and not the other. In other words, a quantum-classical state still has some quantumness in its correlations when measurements are made on part A. It was also realized that the original asymmetric discord is equivalent to the difference between the mutual information before and after a local measurement is made on one of the parts.

For the sake of completeness we should mention that a natural question is what would happen if one considers positive operator valued measurement (POVM), which

---

[3]We should stress that it is a natural requirement that correlation measurements should not increase under local operations: one should no be able to increase their correlation with someone far away only acting on their own system.

are more general than von Neumann projective measures. And the question is if the minimum discord is attained with von Neumann measurements, in which case considering POVM would not be necessary. It can be shown that von Neumann are not always optimum but extremal rank-one POVM are sufficient (see Sect. II.I of [7]).

2 Multipartite Quantum Discord

One important and difficult question when dealing with correlations is how to extend them beyond the two-part scenario. In this multipartite scenario there is not even a single conceptual framework, not to mention measures. One can for example consider many different bipartitions of the multipartite system. For three qubits we could consider the correlation between one of the particles and the rest. We could then average over all possible bipartitions of one with the other two. Or we could take the minimum, or the maximum. Actually we could consider any function of the possible combinations. And for more than three particles there are even more options, since besides the bipartition of one with the rest, we could still have two with the rest, three with the rest and so on. Thus we have many possible bipartitions and can still combine the correlations in many different ways. Another possibility would be not to use a single number but build a correlation vector (or matrix) to characterize the multipartite correlation. It is clear then that the problem is very complex, something already realized in the quantification of multipartite entanglement where a zoo of measures exist, but still very little is well understood.

For discord we also have many possibilities. We could consider measures only on single particles or in groups of particles. This is equivalent to the many possible ways to write the mutual information in terms of conditional entropies. For three particles the mutual information can be written as

$$I(\rho_{ABC}) = S(\rho_A) + S(\rho_B) + S(\rho_C). - S(\rho_{ABC}) \tag{14}$$

But in terms of the conditional entropy there are many possible combinations. These would be the classical correlation and two possibilities are

$$S(\rho_{AB}) - S(\rho_B|\rho_A) - S(\rho_A|\rho_B) - S(\rho_A|\rho_C) - S(\rho_B|\rho_C) + S(\rho_{AB}|\rho_C) \tag{15}$$

and

$$S(\rho_A) + S(\rho_B) + S(\rho_C) - S(\rho_{AB}) - S(\rho_{AC}) + S(\rho_A|\rho_{BC}) \tag{16}$$

Note that the first case involves only single-particle measures, while the second one involves only two-particle measures. One of the first works on multipartite discord proposed to use the two expressions above to define multipartite discord measures, which they called quantum dissension for one- and two-particle measures [8]. And of course one could also combine all these quantities in many ways to define another multipartite measure or construct a correlation vector as proposed in [9].

One proposal, which gained some attention [10], started by defining a symmetric version of discord as the minimum between the asymmetric discord in relation to A and in relation to B: $D_{AB} = \min\{D_A, D_B\}$. In the same way one can define the symmetric classical correlation as the maximum between the asymmetric ones: $J_{AB} = \max\{J_A, J_B\}$. It then considers that the correlation in a tripartite system can be decomposed in a bipartite part and a genuine tripartite part. This division should be true for the total, classical and quantum correlation. The next step is to use conditional entropies involving both single and two particles measures as

$$J_{BC,B}(\rho_{ABC}) = S(\rho_A) + S(\rho_B) - S(\rho_A|\rho_{BC}) - S(\rho_C|\rho_B) \tag{17}$$

to define the total classical correlation (we are assuming a maximization over all possible measures on BC and B). Actually, there are six possible definitions similar to the above with difference only on the single or two parties being measured. And the total classical correlation is defined as the maximum among them: $J(\rho_{ABC}) = \max_{i,j,k}\{J_{ij,k}(\rho_{ABC})\}$. The bipartite part of this classical correlation is defined as $J^{(2)} = \max\{J_{AB}, J_{AC}, J_{BC}\}$. The genuine tripartite classical correlation is then the difference between the total and the bipartite classical correlation: $J^{(3)}(\rho_{ABC}) = J(\rho_{ABC}) - J^{(2)}(\rho_{ABC})$. In the same way we can define the total, bipartite and genuine tripartite correlations. While the definitions may seem arbitrary they are interesting as they have nice properties and are related to the relative entropy (see [10] for more details).

There is also the option to consider sequential single-particle measures. In the bipartite case, one first makes the optimal measurement $\tilde{\Pi}^B$ on B to get the discord $D_{\tilde{\Pi}^B}(\rho_{AB})$. One then makes the optimal measurement on part A of $\tilde{\Pi}^B(\rho_{AB})$. We thus have a symmetric discord as the sum:

$$D_{\tilde{\Pi}^B}(\rho_{AB}) + D_{\tilde{\Pi}^A}(\tilde{\Pi}^B(\rho_{AB})). \tag{18}$$

The generalization to the N-partite system is straightforward, one realizes the sequential optimal measurements on each particle adding the corresponding discords (see [11] for more details).

A natural generalization of discord for multipartite systems is to extend the symmetric discord as defined by the mutual information before and after a local measurement on both parts $\Pi^A \otimes \Pi^B$. This was named global quantum discord and defined as [12]

$$D(\rho_{AB}) = \min_{\Pi^A \otimes \Pi^B} [I(\rho_{AB}) - I(\Pi^A \otimes \Pi^B(\rho_{AB}))]. \tag{19}$$

Note that here measures can be done all together or sequentially since they are local and commute. But one does not add the partial discords after each measurement. For an arbitrary multipartite state of N parts the natural generalization is

$$D(\rho_{A_1...A_N}) = \min_{\Pi^{A_1}\otimes...\otimes\Pi^{A_N}} [I(\rho_{A_1...A_N}) - I(\Pi^{A_1}\otimes...\otimes\Pi^{A_N}(\rho_{A_1...A_N}))] \tag{20}$$

It can be shown that this measure is non-negative. It takes value one for the tripartite GHZ state. And when one considers a mixture of the tripartite GHZ state with a fully mixed state (the identity), one can show that the global discord decreases with the decrease in the weight of the GHZ in the mixture, becoming null only for a zero contribution of the GHZ.

It is also possible to define multipartite geometric measures of discord. One just generalizes the definition of the set of product and classical states to states of the form $\pi_1 \otimes ... \otimes \pi_N$ and $\sum p_{i_1...i_N}\pi_1 \otimes ... \otimes \pi_N$. Then we just chooses a distance measure to define the discord, and even other correlations. These measures may be viewed as true multipartite measures, since one does not appeal to the use of many bipartitions.

3 Geometric Correlations

As we mentioned before, one can use different figures of merit to quantify Discord. Most of these measures are related to entropy measures. A different[4] approach is the geometric one: to use the distance of a given state to the set of classical states (see Fig. 1). Mathematically the geometric Discord of a given state ρ is given by

$$D_G(\rho) = \min_{\chi\in\mathcal{C}} ||\rho - \chi||^2 \tag{21}$$

where $||X||$ is a operator distance in the Hilbert space and $\mathcal{C}$ is the set of classical states (the ones with zero discord), which are mixtures of locally distinguishables states

$$\chi = \sum_{i,j} p_{ij}\ \Pi_i^A \otimes \Pi_j^B, \tag{22}$$

with p_{ij} a joint probability distribution, $\Pi_i = |k_i\rangle\langle k_i|$ with local states $|k_i\rangle$ spanning a local orthonormal basis. Here one also has in principle many possible geometric measures using different distance measures. These measures have the appeal of a geometric interpretation and for some choices of distances can be interpreted as the distinguishability between states.

The first proposal of a geometric measure appeared in 2010 and used the Hilbert-Schmidt norm [13]. The Hilbert-Schmidt norm, also known as the 2-norm, of an operator X is given by

$$||X||_2^2 = \mathrm{Tr}(XX^\dagger). \tag{23}$$

[4]Note that the relative entropy can also be understood as a distance measure, even though technically it is not a genuine distance since it is not symmetric.

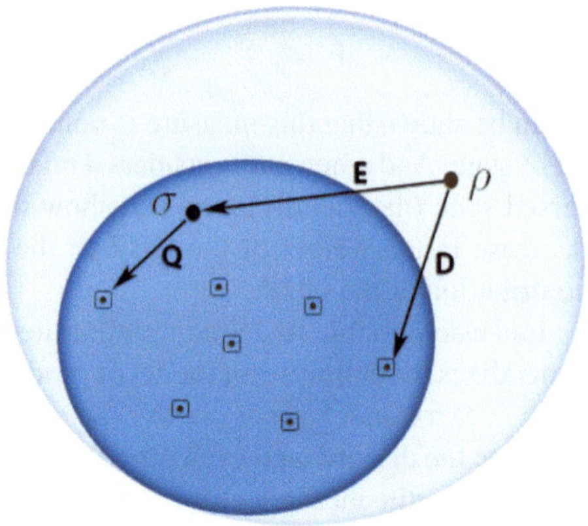

Fig. 1 From [32]. The *large ball* represents the set of all quantum states and the *inside ball* the set of separable states, which is convex. The *squares* represents the set of classical states and the point the set of product states, neither of them are convex. Note that our representation is just a sketch; actually the set of zero discord states, the classical ones, has null measure

This norm is the most used one since it is easy to evaluate. It thus allows for a closed expression of the geometrical discord of two qubits (see [13]), what is an important feature since there are no analytical expressions for most entropic discord measures. As it was easy to calculate the geometric discord using the Hilber-Schmidt norm was used in many works, including experimental papers, and it was even related to the performance of remote state preparation [14, 15]. Nonetheless, it was found that it was not a proper measure of quantumness of correlations, since it can increase under trivial local reversible operations on the unmeasured party [16]. To see this, consider a simple map (channel) $\Gamma^{\sigma} : X \to X \otimes \sigma$: it introduces an ancilla which can be noisy but is uncorrelated. Under such operation we have

$$||X||_2 \to ||X \otimes \sigma||_2 = ||X||_2||\sigma||_2 = ||X||_2\sqrt{\mathrm{Tr}(\sigma^2)}, \tag{24}$$

where in the second equality we used that the Hilbert-Schmidt norm is multiplicative on tensor products. With this we have that

$$D_G(\Gamma_B^{\sigma}(\rho_{AB})) = D_G(\rho_{AB})\mathrm{Tr}[\sigma^2]. \tag{25}$$

Therefore just adding, or removing, a local, uncorrelated and noisy ancilla ($\mathrm{Tr}[\sigma^2] < 1$) in the unmeasured part B, a reversible operation, can change the Discord. But, if one remembers that Discord measures are not monotonic under local operations anyway, this can be considered not a fundamental problem. But on the other side this is a very trivial local operation: we are just adding or removing an uncorrelated ancilla, which can actually always be there, and in the unmeasured part. We should also note that the map increases the dimension of part B, but there are also examples showing that the geometric discord can increase even for local maps which preserve the dimension [17].

The origin of the problem with the Hilbert-Schmidt geometric discord lies in the fact that the Hilbert-Schmidt norm can increase under completely positive trace-preserving (CPTP) maps; it can increase under quantum evolution. Actually this problem was already recognized after the proposals of geometric measures for entanglement. In the beginning of the development of entanglement theory, Vedral et al. proposed three necessary conditions that any entanglement measure should satisfy. [18]. They then showed that the distance between a state and the set of separable states is a good entanglement measure (satisfying their conditions) if the given distance has the property of not increasing under CPTP maps: $D(\Gamma(\rho), \Gamma(\sigma)) \leq D(\rho, \sigma)$. But later it was shown that it was not the case for the Hilbert-Schmidt norm [19]. Thus a possible solution is to use a contractive norm, as usually one calls norms which do not increase under CPTP maps.

We should also mention that there are other possibilities to fix the problem raised in [16]. One could redefine the measure taking the supremum over all maps on the unmeasured part [16]. Besides seeming a bit artificial, this measure is in principle much more difficult to calculate. Another possibility it to rescale the measure by the state purity [20]. However in both cases problems are still expected to appear from the non-contractive property of the Hilbert-Schmidt norm.

Soon after the problem was raised, [21] proposed to use the trace norm for the geometric discord and obtained an analytical expression for a class of states: the Bell-diagonal states. Actually they considered an general Schatten p-norm Discord. The Schatten norm[5] of an operator is given by $||X||_p = \mathrm{Tr}[(X^\dagger X)^{\frac{p}{2}}]^{\frac{1}{p}}$ and they are multiplicative under tensor products: $||\Gamma\sigma(X)||_p = ||X||_p||\sigma||_p$. We then define the p-Schatten geometric Discord as

$$D_p(\rho) = \min_{\chi \in C} ||\rho - \chi||_p^p. \tag{26}$$

It is trivial to note that

$$D_p(\Gamma_B^\sigma(\rho_{AB})) = D_p(\rho_{AB})||\sigma||_p^p. \tag{27}$$

As density operators are Hermitian we have that $||\sigma||_p = \mathrm{Tr}[\sigma^p]^{1/p}$. And as $\mathrm{Tr}[\sigma] = 1$ we have that $||\sigma||_p = 1$ if and only if $p = 1$. Therefore the only p-Schatten geometric Discord which does not increase under the removal or addition of local ancillas is the trace norm. Even more, as the trace norm is contractive under the CPTP maps, the trace distance geometric Discord can not increase under local operations in the

[5]There are many different ways to define a norm for a matrix (or operator). One should first consider the p-norms of a vector $\vec{v}$ given by $||\vec{v}|| = (\sum_i |v_i|^p)^{1/p}$ with v_i being the components of $\vec{v}$ in some basis and $p \geq 1$. For $p = 2$ we have the Euclidean norm. One can then define the induced norm for the matrix as the maximum norm the matrix can induce in a unit vector: $||X|| = \sup_{|\vec{v}|=1} ||X\vec{v}||$. Then given a vector p-norm vector we get a operator p-norm. Another possibility is to consider an m x n matrix as a mn vector and use an vector norm. These are usually called "entrywise" norms. A third possibility, the Schatten norms, is to apply the vector p-norm to the singular values of the matrix (the singular values are the square root of the eingenvalues of $X^\dagger X$). For $p = 2$ we have the Hilbert-Schmidt, also called Frobenius, norm which is equivalent to the $p = 2$ entrywise norm mentioned before. For $p = 1$ we have the trace norm and for $p = \infty$ we have the spectral norm which is equivalent to the induced $p = 2$ norm and also called operator norm and given by the largest singular value.

unmeasured system. This would also be the case for other contractive distance measures as the Bures and Hellinger distances.

The trace norm geometric Discord is then a bona fide measure of correlations. It is also related to the probability of distinguishing between two states via a single measurement[6] and to another measure of quantumness of correlation, the negativity of quantumness, when the measured part is a qubit [23]. In this measure the quantum correlation is defined as the minimum entanglement created between a system and a measurement apparatus by a local measurement. One drawback of the trace distance discord is that is not as simple to calculate as the Hilbert-Schmidt norm. In fact there is still no closed analytical expression for general two-qubit states, but only for some classes. An expression for the Bell diagonal states was presented first in [21], but assuming that the closest classical state also has the Bell diagonal form. Such assumption was confirmed numerically for random states. Later, using a different approach, the same formula was obtained without any assumption [23]; they also obtained a closed expression for Werner, isotropic states and for all two-qubit states for which the reduced state of the measured systems is maximally mixed. More recently, the optimization problem for general two-qubit states was shown to be equivalent to the minimization of a two-variable function (but which parametrically depends on the Bloch vectors of the reduced density matrix and the singular values of the correlation matrix) and a closed expression for X states was also obtained [24]. We should also mention that a general analytical expression exists for the geometric discord using the Hellinger distance [25].

It is also possible to define a measurement-induced geometric measure of discord as [26]

$$D_{MG}(\rho) = \min_{\Pi^B} D(\rho - \Pi^B(\rho)), \tag{28}$$

where the minimum is over all von Neumann measurements (rank one projectors) $\Pi^B = \{\Pi_i^B\}$ on the part B and $\Pi^B(\rho) = \sum_i (\mathbf{1^A} \otimes \Pi_\mathbf{i}^\mathbf{B})\rho(\mathbf{1^A} \otimes \Pi_\mathbf{i}^\mathbf{B})$ is the post measurement state in the absence of readout. Here again there are many possibilities of distance measure to use. When using the same distance it is clear that $D_{MG}(\rho) \geq D_G(\rho)$ since in the measurement-induced one needs to optimize only over classical states generated by von Neumann measurements. The two measurements are equivalent when using the Hilbert-Schmidt distance. For the trace distance the equivalence is true only if the system A is a qubit. They are generally different when using the Bures and Hellinger distances [25].

One last possibility we should mention is to use the relative entropy as a measure of distance. The quantum relative entropy, is defined as

$$S(\rho||\sigma) = \mathrm{Tr}(\rho \log \rho - \rho \log \sigma). \tag{29}$$

[6]This follows directly from the fact that the trace distance itself has a interpretation in terms of distinguishability: Suppose Alice prepares a quantum system in state ρ with probability 1/2 and in state σ with probability 1/2. She then gives the system to Bob, who performs a POVM measurement to distinguish the two states. It can be shown that Bob's probability of correctly identifying which state Alice prepared is $1/2 + 1/2||\rho - \sigma||_1$ (see sec. 9.2 of [22]).

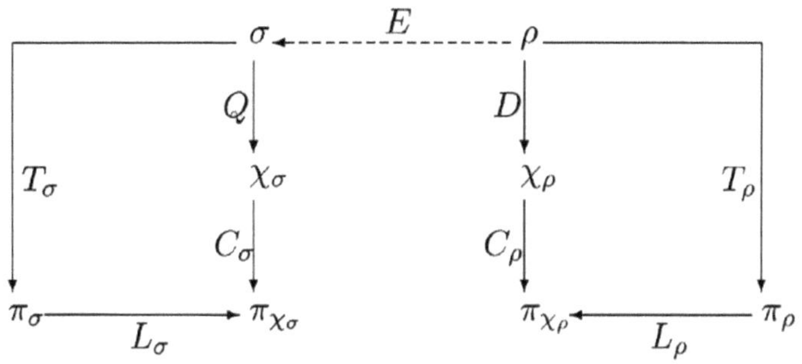

Fig. 2 From [7]. The *arrows* represent the closest state using the relative entropy. ρ is a entangled state, σ a separable state, χ a classical state and π a product state. E is the entanglement, D the Discord, Q the quantum dissonance, T_σ and T_ρ the total mutual information, C_σ and C_ρ the classical correlation. L_σ and L_ρ have no physical interpretation but allow for an additivity relation between the quantities: $T_\rho = D + C_\rho - L_\rho$ and analogously for σ

This entropy is also named the Kullback-Leibler divergence and is often used to distinguish two probability distributions or density operators. It resembles a distance measure but strictly it is not one since it is not symmetric. However, it was originally proposed as a possible unified view on the quantum and classical correlations. Besides the set of separable and classical states one also defines the set of product states,

$$\pi = \pi_1 \otimes \ldots \otimes \pi_N \tag{30}$$

as the states having no correlation at all. The separable states are mixtures of general product states and classical states are mixtures of product states which are orthogonal. Then discord is defined as

$$D(\rho) = \min_{\chi \in \mathcal{C}} S(\rho || \chi) \tag{31}$$

and entanglement as

$$E(\rho) = \min_{\sigma \in \mathcal{S}} S(\rho || \sigma). \tag{32}$$

For an entangled state ρ, the discord $D(\rho)$ can also contain some entanglement. So we can look at the closest classical state to the closest separable state σ, denoted by χ_σ. This distance contains the non-classical correlation excluding entanglement and was named quantum dissonance (see Figs. 1 and 2)

$$Q(\sigma) = \min_{\chi \in \mathcal{C}} S(\sigma || \chi) \tag{33}$$

Further, we can compute the classical correlation as the minimal distance between a classically correlated state and a product state

$$C(\chi) = \min_{\pi \in \mathcal{P}} S(\chi || \pi) \tag{34}$$

and the total correlations as the distance of the original state to the closest product state

$$T(\rho) = \min_{\pi \in \mathcal{P}} S(\rho || \pi). \tag{35}$$

These distances are illustrated in Fig. 2. Note that for an entangled state ρ one has both $T(\rho)$ and $T(\sigma)$, and that given a state ρ one can first find the closest classical state χ_ρ and then look for the closest product state to χ_ρ, which is π_{χ_ρ}. Or one can directly look at the closest product state to ρ, π_ρ. The two states are not equal and we then define

$$L(\rho) = S(\pi_\rho || \pi_{\chi_\rho}) \tag{36}$$

and similarly $L(\sigma)$. These two last quantities allow additivity conditions for correlations as illustrated in Fig. 2.

The advantage of using the relative entropy is that it is possible to find the closest product states and then a closed expression for some of the correlation measures. First, it is easy to show (as done in [27]) that "the closest product state to any state ρ, as measured by the relative entropy, is its reduced state in the product form, i.e., $\pi_\rho = \pi_1 \otimes ... \otimes \pi_N$". Thus, using the linearity of trace and additivity of log we easily obtain

$$T_\rho \equiv S(\rho || \pi_\rho) = -\mathrm{tr}[\rho \log(\pi_1 \otimes ... \otimes \pi_N) + \rho \log \rho] \tag{37}$$

$$= \sum_i -\mathrm{tr}(\pi_i \log \pi_i) + \mathrm{tr}(\rho \log \rho) \tag{38}$$

$$= S(\pi_\rho) - S(\rho). \tag{39}$$

This also allows us to write

$$C_\rho = S(\pi_{\chi_\rho}) - S(\chi_\rho) \tag{40}$$

and

$$C_\sigma = S(\pi_{\chi_\sigma}) - S(\chi_\sigma). \tag{41}$$

It is also possible to show that the closest classical state to a generic state ρ is $\chi_\rho = \sum_{\vec{k}} |\vec{k}\rangle\langle\vec{k}|\rho|\vec{k}\rangle\langle\vec{k}|$ with $\{|\vec{k}\rangle\}$ the eigeinbasis of χ_ρ. Then it is possible to show [27] that the quantum correlations are also a difference between entropies

$$D = S(\chi_\rho) - S(\rho) \tag{42}$$

$$Q = S(\chi_\sigma) - S(\sigma) \tag{43}$$

with $S(\chi_\rho) = \min_{\vec{k}} S(\sum_{\vec{k}} |\vec{k}\rangle\langle\vec{k}|\rho|\vec{k}\rangle\langle\vec{k}|)$ and analogously to $S(\chi_\sigma)$. The minimization involved in D and Q is only over the possible local basis $\vec{k}$. However, this is equivalent to minimizing over all rank-one POVM measurements and is still a very difficult problem. Note that finding the closest separable state to obtain the relative entropy of entanglement is also a hard task. Finally it is possible to show that

$$L_\rho = S(\pi_{\chi_\rho}) - S(\pi_\rho) \tag{44}$$

and

$$L_\sigma = S(\pi_{\chi_\sigma}) - S(\pi_\sigma). \tag{45}$$

So all the quantities can be written as the difference of entropies, the entropic cost, of performing operations that bring the states to the closest one without the desired property; it is the entropic cost of destroying the given correlation. This simple relation also allows us to obtain the following additivity inequalities:

$$T_\rho = D + C_\rho - L_\rho \tag{46}$$

$$T_\sigma = Q + C_\sigma - L_\sigma \tag{47}$$

which justify the diagram of Fig. 2. We should also mention a relation with the original entropic discord. For this we should consider the set of classical-quantum states instead of the classical-classical states. In this case it is possible to show that the original entropic discord is equal to $D - L_\rho$ or equivalently to $T_\rho - C_\rho$, but only when considering a minimization over only projective measurements for the original entropic discord.

It is also possible to define the geometric total and classical correlation using other distance measures. This was studied for the trace distance [28, 29], for the Hilbert-Schmidt norm in [30] and for the Bures distance in [31], but the separation of the total correlation in a classical and quantum part does not obey the additivity relation anymore. Even the first property that the closest product state is the product of the marginals is not true in general, which makes the computation of the total correlation non-trivial.

4 Monogamy of Quantum Correlations

One of the most important properties of entanglement is the fact that it can not be freely shared between many parties. If two systems, A and B, for example, are maximally entangled, neither of them can be entangled with a third system C. This comes from the fact that for A and B two be maximally entangled they should be in a pure state. On the other hand, if any of them is entangled it C, it should have non-zero entropy and therefore be in a mixed state. But how about when A and B

are not maximally entangled. In this situation A and B can be entangled with system C, but there are limits in the amount of entanglement they can share. These relations are named monogamy inequalities. The first and most well-know one was given by Cofmman, Kundu and Wootters in 2000 [33], the CKW inequality,

$$C^2_{A|BC} \geq C^2_{A|B} + C^2_{A|C}, \tag{48}$$

with $C^2_{A|BC}$ representing the squared concurrence between A and BC. So we can see that given the amount of entanglement between A and BC, the amount of entanglement that A and B can share with C is restricted. And if C increases its entanglement with A or B, it has to decrease the entanglement with the other. This relation can also be used to define a measure of genuine tripartite entanglement as the difference: $C^2_{A|BC} - C^2_{A|B} + C^2_{A|C}$. This measure is usually named tangle or three tangle. It gives the intuition that the entanglement between A and BC is composed of the sum of bipartite entanglement of A with B and of A with C, plus a genuine tripartite entanglement. The relation is also valid for N qubits [34]:

$$C^2_{A_1|A_2 \ldots A_N} \geq C^2_{A_1|A_2} + C^2_{A_1|A_3} + \ldots + C^2_{A_1|A_N} \tag{49}$$

Despite its appeal, the monogamy inequality above is not universal: not true for all measures of entanglement. In particular it is not true for the entanglement of formation but only for its square. And when it is valid for qubits it usually breaks down in higher dimensions, with the only known exception being the squashed entanglement. And in fact, recently it has been shown that an important class of entanglement measures may not obey a general monogamy relation for arbitrary dimension [35].

Given the importance of monogamy relations, a natural question is if such a property is also true for other measures of quantum correlation. It was first noted that discord itself did not obey the CKW inequality for three qubits; it was violated even for the W state. This could lead one to say that discord is not monogamous. But we should be careful, as not obeying the specific CKW inequality does not mean it can be freely shared. It may obey other inequalities. And actually not even the concurrence or the entanglement of formation obey the CKW inequality, but only their squares. And it was later realized that the square of discord does obey the CKW inequality [36], being in this sense as monogamous as entanglement. In fact there is a stronger relation between entanglement of formation and discord for three qubits. This comes from an important monogamy relation between entanglement and other correlations, obtained by Koashi and Winter in 2004 [37]:

$$E_F(AB) + J_A(AC) \leq S(A), \tag{50}$$

with E_F being the entanglement of formation, $J_A(AC)$ the classical correlation between A and C with measure in A and $S(A)$ the usual von Neumann entropy of A. The equality holds if ρ_{ABC} is a pure state. By adding the mutual information between A and C on both sides we have that for pure states

$$E_F(AB) = D_A(AC) + S(A|C), \tag{51}$$

with $S(A|C) = S(AC) - S(C)$ the unmeasured conditional entropy; a formula which can be used to obtain the discord. And as for pure states $S(AC) = S(B)$ and $S(A|C) = S(B) - S(C)$ we have

$$D_A(AC) = S(C) - S(B) + E_F(AB) \tag{52}$$

And with some further manipulation it is possible to show that [38]

$$D_A(AB) + D_A(AC) = E_F(AB) + E_F(AC) \tag{53}$$

which is a monogamy relation between the sum of bipartite Discord and bipartite entanglement of formation in a three qubit system. It can also be seen as a conservation law between bipartite discord and entanglement. This also shows a relationship between the CKW inequality for discord and entanglement of formation. Actually, since for pure states discord and entanglement are equivalent, so are the two inequalities [39]. But this is true only for pure states, with the distributed discord exceeding the distributed entanglement for mixed states. This same relationship is used to show that the squared discord does obey the CKW inequality.

The general question of which measure of quantum correlation obeys the CKW inequality for general three-qudit states was addressed in [40]. It was shown that all measures of quantum correlation beyond entanglement, which are nonzero on at least one separable state, and obey some basic properties of a bona fide measure are not monogamous in general. So there is no good measure of quantum correlation which obey the CKW inequality for all states in any dimension. This can be true only for some class of states or for some specific dimension, as the squared Discord. We should also mention that the Hilbert-Schmidt geometric discord also obeys the CKW inequality for three-qubit pure states. In [40] is is also shown that any bona fide measure of quantum correlation which obeys the CKW inequality can not increase under local operations. And it is known that in general quantum correlation can increase under local operations, which can thus be connected with the lack of monogamy for such measures. Recently it has also been shown that an important class of entanglement measures can not satisfy an CKW type of inequality independent of the dimension [35].

Another type of monogamy inequality was proposed in [41], by replacing the bipartite quantum correlation between A and BC by a multiparite measure between ABC. More specifically, the following inequality was proven

$$D(A:B:C) \geq D(A:B) + D(A:C) \tag{54}$$

with $D(A:B:C)$ being the global quantum discord. However the inequality can only be proven for quantum states whose conditional mutual information does not increase under measurement.

5 Conclusions

In sum, it is now clear that entanglement is just one of many interesting and intriguing characteristics of quantum mechanics. This opens the possibility of new phenomena and applications based not on entanglement but in these other forms of correlations. And even tough these measures are not strictly correlation measures, they may have some operational meaning, as already showed in some particular situations, but still a question being explored. And the characterization of discord-like measures in condensed matter systems and dissipative system has also been a very active field; see [7] for more details of such works.

But as we mentioned before there are many possibilities to quantify the quantumness of correlations. And many of them give qualitative different behaviors when characterizing physical systems, or just ordering the quantum states by degree of quantumness. Even the choice of distance in geometric measures can give such distinct behavior. This is actually not very surprising, given that it is a known result that different entanglement measures may also present different behaviors and induce different ordering in the state space. But, while entanglement theory is well developed, although with still important open questions, the discord and related measures have just started to be explored.

Acknowledgements I would like to thanks Marcelo Sarandy for many discussion about discord and quantum correlation and Ernesto Galvão for carefully reading the first draft. I also acknowledge financial support from the Brazilian agencies CNPq, CAPES, FAPERJ, and the Brazilian National Institute of Science and Technology for Quantum Information (INCT-IQ).

References

1. A. Einstein, B. Podolsky, N. Rosen, Phys. Rev. **47**, 777 (1935)
2. R.F. Werner, Phys. Rev. A **40**, 4277 (1989)
3. N. Brunner, D. Cavalcanti, S. Pironio, V. Scarani, S. Wehner, Rev. Mod. Phys. **86**, 419 (2014)
4. H. Ollivier, W.H. Zurek, Phys. Rev. Lett. **88**, 017901 (2001)
5. M. Horodecki, J. Oppenheim, A. Winter, Nature **436**, 673 (2005)
6. L. Henderson, V. Vedral, J. Phys. A **34**, 6899 (2001)
7. K. Modi, A. Brodutch, H. Cable, T. Paterek, V. Vedral, Rev. Mod. Phys. **84**, 1655 (2012)
8. I. Chakrabarty, P. Agrawal, A.K. Pati, Eur. Phys. J. D **65**, 605 (2011)
9. S. Sazim, P. Agrawal. arXiv:1607.05155 [quant-ph]
10. G.L. Giorgi, B. Bellomo, F. Galve, R. Zambrini, Phys. Rev. Lett. **107**, 190501 (2011)
11. M. Okrasa, Z. Walczak, EPL **96**, 60003 (2011)
12. C.C. Rulli, M.S. Sarandy, Phys. Rev. A **84**, 042109 (2011)
13. B. Dakić, V. Vedral, Č. Brukner, Phys. Rev. Lett. **105**, 190502 (2010)
14. B. Dakić, Y.O. Lipp, X. Ma, M. Ringbauer, S. Kropatschek, S. Barz, T. Paterek, V. Vedral, A. Zeilinger, Č. Brukner, P. Walther, Nat. Phys. **8**, 666 (2012)
15. T. Tufarelli, D. Girolami, R. Vasile, S. Bose, G. Adesso, Phys. Rev. A **86**, 052326 (2012)
16. M. Piani, Phys. Rev. A **86**, 034101 (2012)
17. X. Hu, H. Fan, D.L. Zhou, W.-M. Liu, Phys. Rev. A **87**, 032340 (2013)
18. V. Vedral, M.B. Plenio, M.A. Rippin, P.L. Knight, Phys. Rev. Lett. **78** (1997)
19. M. Ozawa, Phys. Lett. A **268**, 158 (2000)

20. T. Tufarelli, T. MacLean, D. Girolami, R. Vasile, G. Adesso, J. Phys. A Math. Theor. **46**, 275308 (2013)
21. F.M. Paula, T.R. de Oliveira, M.S. Sarandy, Phys. Rev. A **87**, 064101 (2013)
22. M.A. Nielsen, I.L. Chuang, *Quantum Computation and Quantum Information* (Cambridge University Press, Cambridge, 2000)
23. T. Nakano, M. Piani, G. Adesso, Phys. Rev. A **88**, 012117 (2013)
24. F. Ciccarello, T. Tufarelli, V. Giovannetti, New J. Phys. **16**, 013038 (2014)
25. W. Roga, D. Spehner, F. Illuminati, J. Phys. A Math. Theor. **49** (2016)
26. S. Luo, S. Fu, Phys. Rev. A **82**, 034302 (2010)
27. K. Modi, T. Paterek, W. Son, V. Vedral, M. Williamson, Phys. Rev. Lett. **104**, 080501 (2010)
28. F.M. Paula, A. Saguia, T.R. de Oliveira, M.S. Sarandy, EPL **108**, 10003 (2014)
29. F.M. Paula, J.D. Montealegre, A. Saguia, T.R. de Oliveira, M.S. Sarandy, EPL **103**, 50008 (2013)
30. B. Bellomo, G.L. Giorgi, F. Galve, R.L. Franco, G. Compagno, R. Zambrini, Phys. Rev. A **85**, 032104 (2012)
31. T.R. Bromley, M. Cianciaruso, R.L. Franco, G. Adesso, J. Phys. A Math. Theor. **47**, 405302 (2012)
32. K. Modi, V. Vedral, AIP Conf. Proc. **1384**, 69 (2011)
33. V. Coffman, J. Kundu, W.K. Wootters, Phys. Rev. A **61**, 052306 (2000)
34. T.J. Osborne, F. Verstraete, Phys. Rev. Lett. **96**, 220503 (2006)
35. C. Lancien, S. Di Martino, M. Huber, M. Piani, Gerardo Adesso, Andreas Winter, Phys. Rev. Lett. **117**, 060501 (2016)
36. Y.-K. Bai, N. Zhang, M.-Y. Ye, Z.D. Wang, Phys. Rev. A **88**, 012123 (2013)
37. M. Koashi, A. Winter, Phys. Rev. A **69**, 022309 (2004)
38. F.F. Fanchini, M.F. Cornelio, M.C. de Oliveira, A.O. Caldeira, Phys. Rev. A **84**, 012313 (2011)
39. G.L. Giorgi, Phys. Rev. A **84**, 054301 (2011)
40. A. Streltsov, G. Adesso, M. Piani, D. Bruß, Phys. Rev. Lett. **109**, 050503 (2012)
41. H.C. Braga, C.C. Rulli, T.R. de Oliveira, M.S. Sarandy, Phys. Rev. A **86**, 062106 (2012)

Geometric Measures of Quantum Correlations with Bures and Hellinger Distances

D. Spehner, F. Illuminati, M. Orszag and W. Roga

1 Introduction

Quantum correlations in composite quantum systems are at the origin of the most peculiar features of quantum mechanics such as the violation of Bell's inequalities and non-locality. In quantum information theory, they are viewed as quantum resources used by quantum algorithms and communication protocols to outperform their classical analogs. If the composite system is in a mixed state, classical correlations between the parties – arising e.g. from a random state preparation – may be present at the same time as quantum correlations. In two seminal papers, Ollivier and Zurek [66] and Henderson and Vedral [41] proposed a way to separate in bipartite systems classical from quantum correlations and introduced the quantum discord as a quantifier of the latter. For pure states, this quantifier coincides with the entanglement of formation, in agreement with the fact that quantum correlations in pure

D. Spehner (✉)
Université Grenoble Alpes, Institut Fourier, 38000 Grenoble, France
e-mail: dominique.spehner@univ-grenoble-alpes.fr

D. Spehner
CNRS, Laboratoire de Physique et Modélisation des Milieux Condensés,
38000 Grenoble, France

F. Illuminati
Dipartimento di Ingegneria Industriale, Università degli Studi di Salerno,
Via Giovanni Paolo II 132, 84084 Fisciano (SA), Italy

F. Illuminati
INFN, Sezione di Napoli, Gruppo collegato di Salerno, 84084 Fisciano (SA), Italy

M. Orszag
Pontificia Universidad Católica, Instituto de Física, Casilla 306, Santiago 22, Chile

W. Roga
Department of Physics, University of Strathclydeb,
John Anderson Building, 107 Rottenrow, Glasgow G4 0NG, UK

F.F. Fanchini et al. (eds.), *Lectures on General Quantum Correlations and their Applications*, Quantum Science and Technology,
DOI 10.1007/978-3-319-53412-1_6

states are synonymous to entanglement. For mixed states, however, the states with a vanishing discord, i.e. those states which possess only classical correlations, form a small (zero-measure) subset of the set of separable states. It has been argued that a non-zero discord could be responsible for the quantum speed-up of the DQC1 algorithm [26, 27]. Furthermore, the discord can be interpreted as the cost of quantum communication in certain protocols such as quantum state merging [20, 56, 60] and can be related to the distillable entanglement between one subsystem and a measurement apparatus [74, 86]. On the other hand, the evaluation of the quantum discord remains a difficult challenge, even in the simplest case of two qubits (see [35, 60] and references therein).

In this chapter, we study alternative measures of quantum correlations which share many of the properties of the quantum discord while being easier to compute and enabling for operational interpretations in terms of state distinguishability. Such measures are related to the geometry of the set of quantum states $\mathcal{E}(\mathcal{H}_{AB})$ of the bipartite system AB. Actually, they are defined in terms of a distance on $\mathcal{E}(\mathcal{H}_{AB})$. Apart from easier computability and operational interpretations, a notable advantage of the geometric approach is that it provides additional tools going beyond the quantification of correlations. In particular, one can determine the closest separable and closest classically-correlated states to a given state ρ, as well as the geodesics linking ρ to those states. These tools may be useful when studying dissipative dynamical evolutions. For instance, one can gain some insight on the efficiency of a dynamical process in changing the amount of entanglement or quantum correlations by comparing the physical trajectory $t \mapsto \rho_t$ in $\mathcal{E}(\mathcal{H}_{AB})$ with the geodesics connecting ρ_t to its closest separable or classically-correlated state(s).

The aim of what follows is to introduce and review the main properties of a few geometric measures of quantum correlations depending on the choice of a distance on $\mathcal{E}(\mathcal{H}_{AB})$. Instead of discussing the (huge amount of) different measures present in the literature, we shall restrict our attention to three quantities. We will mainly focus on (1) the *geometric discord* [25], defined as the minimal distance between the bipartite state ρ and a classically-correlated state. We compare this discord with two other measures characterizing the sensitivity of the state to local measurements and unitary perturbations on one subsystem, namely (2) the *measurement-induced geometric discord* [55], defined as the minimal distance between ρ and the corresponding post-measurement state after an arbitrary local measurement on subsystem A, and (3) the *discord of response* [32, 34], defined as the minimal distance between ρ and its time-evolved version after an arbitrary local unitary evolution on A implemented by a unitary operator with a fixed non-degenerate spectrum. As indicated in the title of the chapter, we will only consider two distinguished distances on the set of quantum states, namely the Bures and Hellinger distances. The discord of response for these two distances corresponds (in a sense that will become clear below) to well known measures of quantum correlations having clear operational interpretations, called the *interferometric power* [37] and *Local Quantum Uncertainty (LQU)* [36]. We will show that the geometric discord with Bures and Hellinger distances are related to a quantum state discrimination task, thereby establishing an explicit link between quantum correlations and state distinguishability. We will also demonstrate that the geometric discord and discord of response with the Hellinger distance are almost as easy to compute as their analogs for the Hilbert–Schmidt distance (for instance, an explicit formula valid for arbitrary qubit-qudit states, which involves the coefficients of the expansion of the square root of the state in terms of generalized Pauli matrices, will be derived in Sect. 6.4). We point out that for the Bures and Hellinger distances, the measures (1)–(3) obey all the basic axioms of *bona fide* measures of quantum correlations, in contrast to what happens for the Hilbert–Schmidt distance [75]. Hence, the geometric discord and discord of response with the Hellinger distance offer the advantage of easy computability while being physically reliable.

The material of this chapter is to a large extend self-contained. The proofs of most results save for basic theorems related in textbooks (e.g. in Ref. [64]) are included. A few technical proofs are, however, omitted. We apologize to the authors of many papers related to geometric measures of quantum correlations for not citing their works, either because they are not directly related to the results presented here or because we are not aware of them.

The remaining of the chapter is organized as follows. We recall in Sect. 2 the definitions of the entropic quantum discord and classically correlated states and formulate the basic postulates on measures of quantum correlations. The three measures outlined above are defined properly in Sect. 3. Sufficient conditions on the distance insuring that they obey the basic postulates are given in this section. A detailed review on the Bures and Hellinger distances and their metrics is provided in Sect. 4. Sections 5 and 6 are devoted to the geometric discord with the Bures and Hellinger distances, respectively. We present without proofs in Sect. 7 some results on the other two measures (2)–(3), in particular some bounds involving these measures and the geometric discord. The last Sect. 8 contains a few concluding remarks.

2 Quantum Versus Classical Correlations

2.1 Entropic Quantum Discord

In all what follows, we consider a bipartite quantum system *AB*, formed by putting together two systems *A* and *B*, with Hilbert space $\mathcal{H}_{AB} = \mathcal{H}_A \otimes \mathcal{H}_B$, $\mathcal{H}_A$ and $\mathcal{H}_B$ being the Hilbert spaces of the two subsystems. In the whole chapter, we only consider systems with finite dimensional Hilbert spaces, $n_A = \dim \mathcal{H}_A < \infty$ and $n_B = \dim \mathcal{H}_B < \infty$. Let us recall that a state of *AB* is given by a density matrix ρ, that is, a non-negative operator on $\mathcal{H}_{AB}$ with unit trace $\operatorname{tr} \rho = 1$. We write $\mathcal{E}(\mathcal{H})$ the convex set formed by all density matrices on the Hilbert space $\mathcal{H}$. The extreme points of this convex set are the pure states $\rho_\psi = |\psi\rangle\langle\psi|$, with $|\psi\rangle \in \mathcal{H}$, $\|\psi\| = 1$. We often abusively write $|\psi\rangle$ instead of ρ_ψ. Given a state $\rho \in \mathcal{E}(\mathcal{H}_{AB})$ of the bipartite system *AB*, the reduced states of *A* and *B* are defined by partial tracing over the other subsystem, that is, $\rho_A = \operatorname{tr}_B(\rho) \in \mathcal{E}(\mathcal{H}_A)$ and $\rho_B = \operatorname{tr}_A(\rho) \in \mathcal{E}(\mathcal{H}_B)$. They correspond to the marginals of a joint probability in classical probability theory.

[45, 60]. For a large number of qubits, the computation of the discord is an NP-complete problem [46].

2.2 Classical-Quantum and Classical States

States of a bipartite system AB with a vanishing quantum discord with respect to A possess only classical correlations. They are usually called classical-quantum states, but we shall prefer here the terminology "A-classical states". One can show that a state $\sigma_{A-\mathrm{cl}}$ is A-classical if and only if it is left unchanged by a local von Neumann measurement on A with rank-one projectors Π_i^A, i.e., $\sigma_{A-\mathrm{cl}} = \mathcal{M}_A^\Pi \otimes 1(\sigma_{A-\mathrm{cl}})$, where $\mathcal{M}_A^\Pi$ is defined by (6).[4] Therefore, all A-classical states are of the form

$$\sigma_{A-\mathrm{cl}} = \sum_{i=1}^{n_A} q_i |\alpha_i\rangle\langle\alpha_i| \otimes \rho_{B|i} \,, \tag{9}$$

where $\{|\alpha_i\rangle\}_{i=1}^{n_A}$ is an orthonormal basis of $\mathcal{H}_A$, $\{q_i\}$ is a probability distribution (i.e., $q_i \geq 0$ and $\sum_i q_i = 1$), and $\rho_{B|i}$ are arbitrary states of B. Equation (9) means that the zero-discord states are mixtures of locally discernable states, that is, of states which can be perfectly discriminated by local measurements on A.

The A-classical states form a non-convex set $\mathcal{C}_A$, the convex hull of which is the set of all separable states $\mathcal{S}_{AB}$ of the bipartite system. It is clear from (9) that a pure state $|\Phi\rangle$ is A-classical if and only if it is a product state $|\Phi\rangle = |\alpha\rangle|\beta\rangle$. Thus, for pure states classicality is equivalent to separability, as already evidenced by the relation (7). In contrast, most separable mixed states are not A-classical.

The B-classical states are defined analogously as the states with a vanishing discord with respect to subsystem B. They are of the form (9) with $\{|\alpha_i\rangle\}$ replaced by an orthonormal basis $\{|\beta_i\rangle\}$ of $\mathcal{H}_B$ and $\rho_{B|i}$ by arbitrary states $\rho_{A|i}$ of A. The states which are both A- and B-classical are called *classical states*. They are of the form

$$\sigma_{\mathrm{clas}} = \sum_{i,j=1}^{n_A,n_B} q_{ij} |\alpha_i\rangle\langle\alpha_i| \otimes |\beta_j\rangle\langle\beta_j| \,. \tag{10}$$

We denote by $\mathcal{C}_B$ and $\mathcal{C}_{AB} = \mathcal{C}_A \cap \mathcal{C}_B$ the sets of B-classical states and of classical states, respectively. An illustrative picture of these subsets of the set of quantum states is displayed in Fig. 1. Note that this picture does not reflect all geometrical aspects (in particular, $\mathcal{C}_A$, $\mathcal{C}_B$, and $\mathcal{C}_{AB}$ typically have a lower dimensionality than $\mathcal{E}(\mathcal{H}_{AB})$ and $\mathcal{S}_{AB}$).

---

[4] This can be justified by using the identity (5) and a theorem due to Petz, which gives a necessary and sufficient condition on ρ such that $I_{A:B}(\rho) = I_{A:B}(\mathcal{M}_A \otimes 1(\rho))$ for a fixed quantum operation $\mathcal{M}_A$ on A (saturation of the data processing inequality) [39, 71]. We refer the reader to [81] for more detail. Note that the proof originally given in Ref. [66] is not correct.

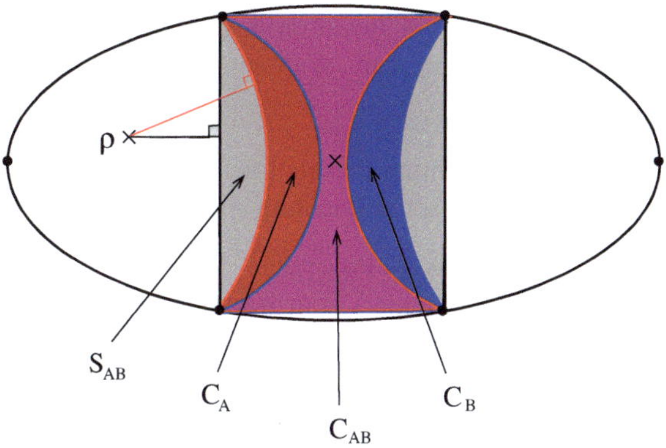

Fig. 1 Schematic view of the set of quantum states $\mathcal{E}(\mathcal{H}_{AB})$ of a bipartite system AB. The subset $\mathcal{C}_{AB}$ of classical states (in *magenta*) is the intersection of the subsets $\mathcal{C}_A$ and $\mathcal{C}_B$ of A- and B-classical states (in *red* and *blue*). The convex hull of $\mathcal{C}_A$ (or $\mathcal{C}_B$) is the subset $\mathcal{S}_{AB}$ of separable states (*gray rectangle*). All these subsets intersect the border of $\mathcal{E}(\mathcal{H}_{AB})$ (which contains all pure states of AB) at the pure product states, represented by the four vertices of the *rectangle*. The maximally mixed state $\rho = 1/(n_A n_B)$ lies at the center (*cross*). The two points at the *left* and *right* extremities of the *ellipse* represent the maximally entangled pure states, which are the most distant states from $\mathcal{S}_{AB}$ (and also from $\mathcal{C}_A$, $\mathcal{C}_B$, and $\mathcal{C}_{AB}$). The square distances between a given state ρ and $\mathcal{S}_{AB}$ (*black line*) and between ρ and $\mathcal{C}_A$ (*red line*) define the geometric entanglement $E^{\mathrm{G}}_{AB}(\rho)$ and the geometric discord $D^{\mathrm{G}}_A(\rho)$, respectively

2.3 *Axioms on Measures of Quantum Correlations*

Before proceeding further, let us briefly recall the definition of a quantum operation (or quantum channel). We denote by $\mathcal{B}(\mathcal{H})$ the C^*-algebra of bounded linear operators from $\mathcal{H}$ into itself, that is, $n \times n$ complex matrices with $\dim \mathcal{H} = n$ in our finite dimensional setting. Mathematically, a quantum operation is a completely positive (CP)[5] trace-preserving linear map $\mathcal{M} : \mathcal{B}(\mathcal{H}) \to \mathcal{B}(\mathcal{H}')$. Physically, quantum operations represent either quantum evolutions or changes in the system state due to measurements without readout like in (6). More precisely, let a quantum system S initially in state ρ_S be coupled at time $t = 0$ to its environment E, with which it has never interacted at prior times. If the joint state $\rho_{SE}(t)$ of SE either evolves unitarily according to the Schrödinger equation or is modified by a measurement process, then the reduced state of S at time $t > 0$ is given by $\rho_S(t) = \mathcal{M}_t(\rho_S)$ where $\mathcal{M}_t$ is a quantum operation.

[5]A linear map $\mathcal{M} : \mathcal{B}(\mathcal{H}) \to \mathcal{B}(\mathcal{H}')$ is positive if it transforms a non-negative operator $\rho \geq 0$ into a non-negative operator $\mathcal{M}(\rho) \geq 0$. It is CP if the map

$$\mathcal{M} \otimes 1 : X \in \mathcal{B}(\mathcal{H} \otimes \mathbb{C}^m) \mapsto \sum_{k,l=1}^{m} \mathcal{M}(X_{kl}) \otimes |k\rangle\langle l| \in \mathcal{B}(\mathcal{H}' \otimes \mathbb{C}^m) \tag{11}$$

is positive for any integer $m \geq 1$.

By studying the properties of the quantum discord, one is led to define the following axioms for a *bona fide* measure of quantum correlations [24, 36, 37, 76, 81].

Definition 1 A measure of quantum correlations on the bipartite system AB is a non-negative function D_A on the set of quantum states $\mathcal{E}(\mathcal{H}_{AB})$ such that:

(i) $D_A(\rho) = 0$ if and only if ρ is A-classical;
(ii) D_A is invariant under local unitary transformations, i.e., $D_A(U_A \otimes U_B \rho U_A^\dagger \otimes U_B^\dagger) = D_A(\rho)$ for any unitaries U_A and U_B acting on $\mathcal{H}_A$ and $\mathcal{H}_B$;
(iii) D_A is monotonically non-increasing under quantum operations on B, i.e., $D_A(1 \otimes \mathcal{M}_B(\rho)) \leq D_A(\rho)$ for any quantum operation $\mathcal{M}_B : \mathcal{B}(\mathcal{H}_B) \to \mathcal{B}(\mathcal{H}'_B)$;
(iv) D_A reduces to an entanglement monotone on pure states.[6]

These axioms are satisfied in particular by the entropic discord.[7]

It can be shown[8] that axioms (iii) and (iv) imply that, if the space dimensions of $\mathcal{H}_A$ and $\mathcal{H}_B$ are such that $n_A \leq n_B$, D_A is maximum on maximally entangled pure states $|\Psi_{\mathrm{me}}\rangle$, i.e., if $\rho = \rho_{\Psi_{\mathrm{me}}}$ then $D_A(\rho) = D_{\max}$ [75]. It is argued in Ref. [78] that a proper measure of quantum correlations D_A must actually be such that the maximally entangled states are the *only* states satisfying $D_A(\rho) = D_{\max}$. We will thus consider the following additional axiom, fulfilled in particular by the entropic discord D_A^{ent} [81]:

(v) if $n_A \leq n_B$ then $D_A(\rho)$ is maximum if and only if ρ is maximally entangled, that is, ρ has maximal entanglement of formation $E_{\mathrm{EoF}}(\rho) = \ln n_A$.

Many authors have looked for functions $D_A \neq D_A^{\mathrm{ent}}$ on $\mathcal{E}(\mathcal{H}_{AB})$ fulfilling axioms (i–iv), which can be used as D_A^{ent} to quantify quantum correlations in bipartite systems while being easier to compute and having operational interpretations. Among such measures, the distance-based measures studied in this chapter are especially appealing since they provide a geometric understanding of quantum correlations not limited to their quantification, as stressed in the Introduction.

---

[6] Recall that an entanglement monotone E on pure states is a function which does not increase under Local Operations and Classical Communication (LOCC), i.e., $E(|\Phi\rangle) \leq E(|\Psi\rangle)$ whenever $|\Psi\rangle$ can be transformed into $|\Phi\rangle$ by a LOCC operation [44, 64].

[7] Actually, D_A^{ent} obeys axiom (i) by definition. Axiom (ii) follows from the unitary invariance of the entropy S. Axiom (iv) is a consequence of (7) and the entanglement monotonicity of the entanglement of formation. The proof of axiom (iii) is given e.g. in [81].

[8] This follows from the facts that a function D_A on $\mathcal{E}(\mathcal{H}_{AB})$ satisfying (iii) is maximal on pure states if $n_A \leq n_B$ [88] and that any pure state can be obtained from a maximally entangled pure state via a LOCC.

3 Geometric Measures of Quantum Correlations

3.1 Contractive Distances on the Set of Quantum States

A fundamental issue in quantum information theory is the problem of distinguishing quantum states, that is, quantifying how "different" or how "far from each other" are two given states ρ and σ. A natural way to deal with this problem is to endow the set of quantum states $\mathcal{E}(\mathcal{H})$ with a distance d. One has a priori the choice between many distances. The most common ones are the L^p-distances

$$d_p(\rho, \sigma) = \|\rho - \sigma\|_p = \left[\mathrm{tr}(|\rho - \sigma|^p)\right]^{\frac{1}{p}} \tag{12}$$

with $p \geq 1$ (here $|X|$ denotes the non-negative operator $|X| = \sqrt{X^\dagger X}$).

In quantum information theory, it is important to consider distances d having the following property: if two identical systems in states ρ and σ undergo the same quantum evolution or are subject to the same measurement described by a quantum operation $\mathcal{M}$, then the time-evolved or post-measurement states $\mathcal{M}(\rho)$ and $\mathcal{M}(\sigma)$ cannot be farther from each other than the initial states ρ and σ. In other words, the two states are less distinguishable after the evolution or the measurement, because some information has been lost in the environment or in the measurement apparatus. Distances d on the sets of quantum states satisfying this property are said to be *contractive under quantum operations* (or "contractive" for short). More precisely, d is contractive if for any finite-dimensional Hilbert spaces $\mathcal{H}$ and $\mathcal{H}'$, any quantum operation $\mathcal{M} : \mathcal{B}(\mathcal{H}) \to \mathcal{B}(\mathcal{H}')$, and any $\rho, \sigma \in \mathcal{E}(\mathcal{H})$, it holds

$$d(\mathcal{M}(\rho), \mathcal{M}(\sigma)) \leq d(\rho, \sigma) . \tag{13}$$

Note that a contractive distance is in particular unitary invariant, i.e.,

$$d\big(U\rho U^\dagger, U\sigma U^\dagger\big) = d(\rho, \sigma) \quad \text{if } U \text{ is unitary on } \mathcal{H} \tag{14}$$

(in fact, $\rho \mapsto U\rho U^\dagger$ is an invertible quantum operation on $\mathcal{B}(\mathcal{H})$).

The relative von Neumann entropy $S(\rho||\sigma) = \mathrm{tr}[\rho(\ln\rho - \ln\sigma)]$ is a prominent example of contractive function on $\mathcal{E}(\mathcal{H}) \times \mathcal{E}(\mathcal{H})$ and has a fundamental interpretation in terms of information. However, $S(\rho||\sigma)$ is not a distance (it is not symmetric under the exchange of ρ and σ). It is desirable to work with contractive functions d on $\mathcal{E}(\mathcal{H}) \times \mathcal{E}(\mathcal{H})$ which can be interpreted like S in terms of information while being true distances. It turns out that the L^p-distances d_p are *not* contractive, with the notable exception of the trace distance d_1 [67, 69, 79]. Hence d_p, $p > 1$ (and in particular the Hilbert–Schmidt distance d_2) cannot be reliably used to distinguish quantum states. We will focus in what follows on two particular distances, called the Bures and Hellinger distances, defined by

$$d_{\mathrm{Bu}}(\rho,\sigma) = \left(2 - 2\sqrt{F(\rho,\sigma)}\right)^{\frac{1}{2}} \tag{15}$$

$$d_{\mathrm{He}}(\rho,\sigma) = \left(2 - 2\,\mathrm{tr}\,\sqrt{\rho}\sqrt{\sigma}\right)^{\frac{1}{2}} , \tag{16}$$

where the Uhlmann fidelity is given by

$$F(\rho,\sigma) = \|\sqrt{\rho}\sqrt{\sigma}\|_1^2 = \left(\mathrm{tr}\big[(\sqrt{\sigma}\rho\sqrt{\sigma})^{\frac{1}{2}}\big]\right)^2 . \tag{17}$$

These distances will be studied in Sect. 4. We will show that they are contractive, enjoy a number of other nice properties, and are related to the Rényi relative entropies.

3.2 Distances to Separable, Classical-Quantum, and Product States

From a geometrical viewpoint, it is quite natural to quantify the amount of quantum correlations in a state ρ of a bipartite system AB by the distance $d(\rho, \mathcal{C}_A)$ of ρ to the subset $\mathcal{C}_A$ of A-classical states, i.e., by the minimal distance between ρ and an arbitrary A-classical state (see Fig. 1). This idea goes back to Vedral and Plenio [94, 95], who proposed to define the entanglement in AB by the (square) distance from ρ to the set of separable states $\mathcal{S}_{AB}$,

$$E_{AB}^{\mathrm{G}}(\rho) = d(\rho, \mathcal{S}_{AB})^2 = \min_{\sigma_{\mathrm{sep}} \in \mathcal{S}_{AB}} d(\rho, \sigma_{\mathrm{sep}})^2 . \tag{18}$$

These authors have shown that E_{AB}^{G} is an entanglement monotone if the distance d is contractive. By analogy, Dakić, Vedral, and Brukner [25] introduced the geometric discord

$$D_A^{\mathrm{G}}(\rho) = d(\rho, \mathcal{C}_A)^2 = \min_{\sigma_{A-\mathrm{cl}} \in \mathcal{C}_A} d(\rho, \sigma_{A-\mathrm{cl}})^2 . \tag{19}$$

Unfortunately, the distance d was chosen in Ref. [25] to be the Hilbert–Schmidt distance d_2, which is not a good choice because d_2 is not contractive (Sect. 3.1). Further works have studied the geometric discords based on the more physically reliable Bures distance (see [1, 2, 82, 83, 87]), Hellinger distance (see [57, 78]), and trace distance (see [24, 63, 68] and references therein).

The discord D_B^{G} relative to subsystem B is defined by replacing $\mathcal{C}_A$ by $\mathcal{C}_B$ in (19). Like for the entropic discord, one has in general $D_A^{\mathrm{G}} \neq D_B^{\mathrm{G}}$. Symmetric measures of quantum correlations are obtained by considering the square distance $D_{AB}^{\mathrm{G}}(\rho) = d(\rho, \mathcal{C}_{AB})^2$ to the set of classical states $\mathcal{C}_{AB} = \mathcal{C}_A \cap \mathcal{C}_B$.

We emphasize that since $\mathcal{C}_{AB} \subset \mathcal{C}_A \subset \mathcal{S}_{AB}$ (see Fig. 1), the geometric measures are ordered as

$$E_{AB}^{\mathrm{G}}(\rho) \leq D_A^{\mathrm{G}}(\rho) \leq D_{AB}^{\mathrm{G}}(\rho) . \tag{20}$$

This ordering is a nice feature of the geometrical approach. In contrast, depending on ρ, the entanglement of formation $E_{\rm EoF}(\rho)$ can be larger or smaller than the entropic discord $D_A^{\rm ent}(\rho)$.

It is easy to show that $E_{AB}^{\rm G}$ is an entanglement monotone if the distance d is contractive (this follows from the invariance of $\mathcal{S}_{AB}$ under LOCC operations, see [81, 94]) and that $E_{AB}^{\rm G}(\rho) = 0$ if and only if ρ is separable (since a distance d separates points). Hence $E_{AB}^{\rm G}$ qualifies as a reliable measure of entanglement.[9] Similarly, one may ask whether the geometric discord $D_A^{\rm G}$ satisfies axioms (i–iv) of Definition 1. If d is contractive, one easily shows[10] that $D_A^{\rm G}$ obeys the first three axioms (i–iii). Finding general conditions on d insuring the validity of the last axiom (iv) is still an open question. We will show below that (see Sects. 5.1, 5.2, and 6.1)

Proposition 2 *$D_A^{\rm G}$ is a* bona fide *measure of quantum correlations when d is the Bures or the Hellinger distance. Furthermore, if $d = d_{\rm Bu}$ then $D_A^{\rm G}$ satisfies the additional axiom (v).*

It can be proven that $D_A^{\rm G}$ obeys axiom (v) also for the Hellinger distance $d_{\rm He}$ when A is a qubit or a qutrit ($n_A = 2$ or $n_A = 3$) [78], and we believe that this is still true for higher dimensional spaces $\mathcal{H}_A$. It is conjectured by several authors that $D_A^{\rm G}$ is a *bona fide* measure of quantum correlations for the trace distance $d = d_1$, but as far as we are aware the justification of axiom (iv) is still open (however, this axiom holds for $n_A = 2$, see e.g. [78]). In contrast, $D_A^{\rm G}$ is *not* a measure of quantum correlations for the Hilbert–Schmidt distance $d = d_2$. Indeed, as one might expect from the fact that d_2 is not contractive, $D_A^{\rm G}$ does not fulfill axiom (iii) (an explicit counter-example is given in Ref. [73]).

One can replace the square distance by the relative entropy S in formulas (18)–(19). Since S is contractive under quantum operations and satisfies $S(\rho||\sigma) = 0$ if and only if $\rho = \sigma$, one shows in the same way as for contractive distances that the corresponding entanglement measure $E_{AB}^{\rm S}$ is entanglement monotone [95] and that the discord $D_A^{\rm S}$ obeys axioms (i–iii). Furthermore, one finds that the closest separable state to a pure state $|\Psi\rangle$ for the relative entropy is a classical state and that $E_{AB}^{\rm S}(|\Psi\rangle)$ is equal to the entanglement of formation $E_{\rm EoF}(|\Psi\rangle)$ [94]. Hence $D_A^{\rm S}(|\Psi\rangle) = E_{AB}^{\rm S}(|\Psi\rangle) - E_{\rm EoF}(|\Psi\rangle)$ for any pure state $|\Psi\rangle \in \mathcal{H}_{AB}$. As a result, $D_A^{\rm S}$ is a *bona fide* measure of quantum correlations.

The mutual information (1) quantifying the total amount of correlations between A and B is equal to[11]

---

[9] Furthermore, $E_{AB}^{\rm G}$ is convex if d is the Bures or the Hellinger distance since then d^2 is jointly convex, see Sect. 4.3. Convexity is sometimes considered as another axiom for entanglement measures, apart from entanglement monotonicity and vanishing for separable states and only for those states.

[10] Actually, $D_A^{\rm G}$ clearly obeys axiom (i), irrespective of the choice of the distance. It satisfies axiom (ii) for any unitary-invariant distance, thus in particular for contractive distances. One shows that it fulfills axiom (iii) by using the contractivity of d and the fact that the set of A-classical states $\mathcal{C}_A$ is invariant under quantum operations acting on B, as is evident from (9).

[11] This identity follows from the relations $I_{A:B}(\rho) = S(\rho||\rho_A \otimes \rho_B)$ and $S(\rho||\sigma_A \otimes \sigma_B) - S(\rho||\rho_A \otimes \rho_B) = S(\rho_A||\sigma_A) + S(\rho_B||\sigma_B) \geq 0$. It means in particular that the "closest" product state to ρ for the relative entropy is the product $\rho_A \otimes \rho_B$ of the marginals of ρ [59].

$$I_{A:B}(\rho) = \min_{\sigma_{\rm prod} \in \mathcal{P}_{AB}} S(\rho || \sigma_{\rm prod}) , \tag{21}$$

where $\mathcal{P}_{AB} = \{\sigma_A \otimes \sigma_B; \sigma_S \in \mathcal{E}(\mathcal{H}_S), S = A, B\}$ is the set of product (i.e., uncorrelated) states. In analogy with (21), one can define a geometrical mutual information $I_{AB}^{\rm G}$ and a measure $C_A^{\rm G}$ of classical correlations by [18, 59]

$$I_{AB}^{\rm G}(\rho) = d(\rho, \mathcal{P}_{AB})^2 \quad , \quad C_A^{\rm G}(\rho) = \min_{\sigma_\rho \in \mathcal{C}_A} d(\sigma_\rho, \mathcal{P}_{AB})^2 , \tag{22}$$

where the minimum is over all[12] closest A-classical states σ_ρ to ρ. Unlike in the case of the entropic discord, the total correlations $I_{AB}^{\rm G}(\rho)$ is not the sum of the quantum and classical correlations $D_A^{\rm G}(\rho)$ and $C_A^{\rm G}(\rho)$ [59]. However, the triangle inequality yields $I_{AB}^{\rm G}(\rho) \leq (\sqrt{D_A^{\rm G}(\rho)} + \sqrt{C_A^{\rm G}(\rho)})^2$.

3.3 *Response to Local Measurements and Unitary Perturbations*

An alternative way to quantify quantum correlations with the help of a distance d is to consider the sensitivity of the state $\rho \in \mathcal{E}(\mathcal{H}_{AB})$ to local measurements or local unitary perturbations.

(1) The distinguishability of ρ with the corresponding post-measurement state after a local projective measurement on subsystem A is characterized by the *measurement-induced geometric discord*, defined by [55]

$$D_A^{\rm M}(\rho) = \min_{\{\Pi_i^A\}} d\big(\rho, \mathcal{M}_A^{\Pi} \otimes 1(\rho)\big)^2 , \tag{23}$$

where the minimum is over all measurements on A with rank-one projectors Π_i^A and $\mathcal{M}_A^{\Pi}$ is the associated quantum operation (6). Since the outputs of such measurements are always A-classical, one has $D_A^{\rm G}(\rho) \leq D_A^{\rm M}(\rho)$ for any state ρ. This inequality is an equality if $d = d_2$ is the Hilbert–Schmidt distance [55]. For the Bures and Hellinger distances, $D_A^{\rm G}$ and $D_A^{\rm M}$ are in general different, even if A is a qubit [78]. For the trace distance, $D_A^{\rm G} = D_A^{\rm M}$ when A is a qubit but this is no longer true for $n_A > 2$ [63].

(2) The distinguishability of ρ with the corresponding state after a local unitary evolution on subsystem A is characterized by the *discord of response* [32, 34, 76]

$$D_A^{\rm R}(\rho) = \frac{1}{\mathcal{N}} \min_{U_A, {\rm sp}(U_A) = e^{{\rm i}\Lambda}} d\big(\rho, U_A \otimes 1 \, \rho \, U_A^\dagger \otimes 1\big)^2 , \tag{24}$$

[12] As we shall see below, ρ may have an infinite family of closest A-classical states.

where the minimum is over all unitary operators U_A on $\mathcal{H}_A$ separated from the identity by the condition of having a fixed spectrum $\mathrm{sp}(U_A) = e^{\mathrm{i}\Lambda} = \{e^{2\mathrm{i}\pi j/n_A}; j = 0, \dots, n_A - 1\}$ given by the roots of unity.[13] The normalization factor $\mathcal{N}$ in (24) depends on the distance and is chosen such that $D_A^{\mathrm{R}}(\rho)$ has maximal value equal to unity. For instance, $\mathcal{N} = 2$ for the Bures, Hellinger, and Hilbert–Schmidt distances and $\mathcal{N} = 4$ for the trace distance [78].

The measurement-induced geometric discord and discord of response are special instances of measures of quantumness given by

$$Q_{\delta,\mathcal{F}_A}(\rho) = \inf_{\mathcal{M}_A \in \mathcal{F}_A} \delta(\rho, \mathcal{M}_A \otimes 1(\rho)) , \tag{25}$$

where $\mathcal{F}_A$ is a family of quantum operations on A and δ is a (square) distance or a relative entropy. The following result of Piani, Narasimhachar, and Calsamiglia [75] is useful to check that such measures are *bona fide* measures of quantum correlations.

Theorem 1 ([75]) *For all spaces $\mathcal{H}$ with* $\dim \mathcal{H} < \infty$, *let $\delta(\rho, \sigma)$ be non-negative functions on $\mathcal{E}(\mathcal{H}) \times \mathcal{E}(\mathcal{H})$ which are contractive under quantum operations and satisfy the 'flags' condition*

$$\delta\Big(\sum_i \eta_i \rho_i \otimes |i\rangle\langle i|, \sum_i \eta_i \sigma_i \otimes |i\rangle\langle i|\Big) = \sum_i \eta_i \delta(\rho_i, \sigma_i) , \tag{26}$$

where $\{|i\rangle\}$ is an orthonormal basis of an ancilla Hilbert space $\mathcal{H}_E$. Assume that $n_A \leq n_B$. Let the family $\mathcal{F}_A$ of quantum operations on $\mathcal{B}(\mathcal{H}_A)$ be closed under unitary conjugations. Then the measure of quantumness (25) satisfies axioms (ii–iv) of Definition 1.

Proof We first show that $Q_{\delta,\mathcal{F}_A}$ is an entanglement monotone when restricted to pure states. It is known (see e.g. [64]) that when $n_A \leq n_B$, a LOCC acting on a pure state $|\Psi\rangle$ may always be simulated by a one-way communication protocol involving only three steps: (1) Bob first performs a measurement with Kraus operators $\{B_i\}$ on subsystem B; (2) he sends his measurement result to Alice; (3) Alice performs a unitary evolution U_i on subsystem A conditional to Bob's result. Therefore, it is enough to show that for any pure state $|\Psi\rangle \in \mathcal{H}_{AB}$, any family $\{B_i\}$ of Kraus operators on $\mathcal{H}_B$ (satisfying $\sum_i B_i^\dagger B_i = 1$), and any family $\{U_i\}$ of unitaries on $\mathcal{H}_A$, it holds

$$\sum_i \eta_i Q_{\delta,\mathcal{F}_A}(|\Phi_i\rangle) \leq Q_{\delta,\mathcal{F}_A}(|\Psi\rangle) , \tag{27}$$

where $\eta_i = \|1 \otimes B_i |\Psi\rangle\|^2$ is the probability that Bob's outcome is i and $|\Phi_i\rangle = \eta_i^{-\frac{1}{2}} U_i \otimes B_i |\Psi\rangle$ is the corresponding conditional post-measurement state after Alice's unitary evolution. The inequality (27) is proven by considering the following quantum operation $\mathcal{M} : \mathcal{B}(\mathcal{H}_{AB}) \to \mathcal{B}(\mathcal{H}_{ABE})$

[13] See [76] for a discussion on the choice of the non-degenerate spectrum $e^{\mathrm{i}\Lambda}$.

$$\mathcal{M}(\rho) = \sum_i U_i \otimes B_i \, \rho \, U_i^\dagger \otimes B_i^\dagger \otimes |i\rangle\langle i| \,. \tag{28}$$

From the contractivity of δ and the flags condition, one gets

$$\begin{aligned} Q_{\delta,\mathcal{F}_A}(|\Psi\rangle) &\geq \inf_{\mathcal{M}_A\in\mathcal{F}_A} \delta\big(\mathcal{M}(|\Psi\rangle\langle\Psi|)\,,\, \mathcal{M}\circ\mathcal{M}_A\otimes 1(|\Psi\rangle\langle\Psi|)\big) \\ &= \inf_{\mathcal{M}_A\in\mathcal{F}_A} \delta\Big(\sum_i \eta_i|\Phi_i\rangle\langle\Phi_i|\otimes|i\rangle\langle i|\,,\, \sum_i \eta_i \mathcal{M}_A^{(i)}\otimes 1(|\Phi_i\rangle\langle\Phi_i|)\otimes|i\rangle\langle i|\Big) \\ &= \inf_{\mathcal{M}_A\in\mathcal{F}_A} \sum_i \eta_i\, \delta\Big(|\Phi_i\rangle\langle\Phi_i|\,,\, \mathcal{M}_A^{(i)}\otimes 1(|\Phi_i\rangle\langle\Phi_i|)\Big) \end{aligned} \tag{29}$$

with $\mathcal{M}_A^{(i)}(\cdot) = U_i \mathcal{M}_A(U_i^\dagger \cdot U_i)U_i^\dagger$. Bounding the infimum of the sum by the sum of the infima and using the assumption $U_i\mathcal{F}_A U_i^\dagger = \mathcal{F}_A$, one is led to the desired result

$$Q_{\delta,\mathcal{F}_A}(|\Psi\rangle) \geq \sum_i \eta_i \inf_{\mathcal{M}_A^{(i)}\in\mathcal{F}_A} \delta\Big(|\Phi_i\rangle\langle\Phi_i|, \mathcal{M}_A^{(i)}\otimes 1(|\Phi_i\rangle\langle\Phi_i|)\Big) = \sum_i \eta_i Q_{\delta,\mathcal{F}_A}(|\Phi_i\rangle) \,. \tag{30}$$

In particular, if the pure state $|\Psi\rangle$ can be transformed by a LOCC operation into the pure state $|\Phi\rangle$, which means that $|\Phi_i\rangle = |\Phi\rangle$ for all outcomes i, then $Q_{\delta,\mathcal{F}_A}(|\Psi\rangle) \geq Q_{\delta,\mathcal{F}_A}(|\Phi\rangle)$. Axiom (ii) follows from a similar argument and the unitary invariance of δ (which is a consequence of the contractivity assumption, see Sect. 3.1). Finally, one easily verifies that $Q_{\delta,\mathcal{F}_A}$ fulfills axiom (iii) by exploiting the contractivity of δ. □

Proposition 3 *D_A^{M} and D_A^{R} are* bona fide *measures of quantum correlations if the distance d is contractive and d^2 satisfies the flag condition (24).*

It is easy to show that the square Bures and Hellinger distances d_{Bu}^2 and d_{He}^2 satisfy the flags condition, so that Proposition 3 applies in particular to these distances. The result applies to the trace distance d_1 as well, see [75].

Proof Let us first discuss axiom (i). For D_A^{M}, its validity comes from the fact that a state is A-classical if and only if it is invariant under a von Neumann measurement on A with rank-one projectors (Sect. 2.2). Note that this axiom would not hold if the minimization in (23) was performed over projectors Π_i^A with ranks larger than one. For D_A^{R}, one uses an equivalent characterization of A-classical states as the states ρ which are left invariant by a local unitary transformation on A for some unitary U_A on $\mathcal{H}_A$ having a non-degenerate spectrum [76]. Actually, $U_A\otimes 1\rho U_A^\dagger\otimes 1 = \rho$ if and only if ρ commutes with $U_A\otimes 1$, or, equivalently, with all its spectral projectors Π_i^A. This means that $\mathcal{M}_A^\Pi\otimes 1(\rho) = \rho$ with $\mathcal{M}_A^\Pi$ defined by (6). Since $\mathrm{sp}(U_A)$ is not degenerate, the spectral projectors Π_i^A have rank one. Consequently, the above condition on unitary transformations is equivalent to the invariance of ρ under a measurement on A with rank-one projectors and thus to ρ being A-classical. This proves that D_A^{R} satisfies axiom (i). The fact that D_A^{M} and D_A^{R} obey the other axioms (ii–iv) is a consequence of Theorem 1. □

As for the geometric discord, we do not have a general argument implying that D_A^{M} and D_A^{R} fulfill the additional axiom (v) under appropriate assumptions on the distance. However, one can show that D_A^{R} obeys axiom (v) for the Bures, Hellinger, and trace distances, and this is also true for D_A^{M} for the Bures distance [78].

3.4 Speed of Response to Local Unitary Perturbations

All distance-based measures of quantum correlations defined above are global geometric quantities, in the sense that they depend on the distance between ρ and states that are not in the neighborhood of ρ (excepted of course when the measure vanishes). It is natural to look for quantifiers of quantum correlations involving local geometric quantities[14] depending only on the properties of $\mathcal{E}(\mathcal{H}_{AB})$ in the vicinity of ρ. The idea of the sensitivity to local unitary perturbations sustaining the definition of the discord of response is well suited for this purpose. Indeed, instead of considering the minimal distance between ρ and its perturbation by a local unitary with a fixed spectrum, one may consider the minimal speed at time $t = 0$ of the time-evolved states

$$\rho_{\mathrm{out}}(t) = e^{-\mathrm{i}tH_A \otimes 1} \rho\, e^{\mathrm{i}tH_A \otimes 1} . \tag{31}$$

This leads to the definition of the *discord of speed of response*

$$D_A^{\mathrm{SR}}(\rho) = \min_{H_A, \mathrm{sp}(H_A) = \Lambda} \lim_{t \to 0} t^{-2} d\big(\rho, \rho_{\mathrm{out}}(t)\big)^2 , \tag{32}$$

where the minimum is over all self-adjoint operators H_A on $\mathcal{H}_A$ with a fixed non-degenerate spectrum $\Lambda = \{2\pi j/n_A; j = 0, \ldots, n_A - 1\}$. This local geometric version of the discord of response has apparently not been defined before in the literature. The results of Propositions 4 and 5 below have up to our knowledge not been published elsewhere.

Proposition 4 *If the distance d is contractive and Riemannian, and if d^2 satisfies the flags condition (26), then D_A^{SR} is a* bona fide *measure of quantum correlations. Furthermore, one has*

$$\mathcal{N} D_A^{\mathrm{R}}(\rho) \leq D_A^{\mathrm{SR}}(\rho) . \tag{33}$$

Proof If d is a Riemannian distance then the limit in (32) exists and is equal to $g_\rho(-i[H_A \otimes 1, \rho], -i[H_A \otimes 1, \rho])$ where g is the metric associated to d (see Sect. 4.6). Since g_ρ is a scalar product, $D_A^{\mathrm{SR}}(\rho) = 0$ if and only if $[H_A \otimes 1, \rho] = 0$ for some observable H_A of A with a non-degenerate spectrum $\mathrm{sp}(H_A) = \Lambda$. As in the proof of Proposition 3, this is equivalent to ρ being A-classical. Hence axiom (i) holds true. One easily convinces oneself that D_A^{SR} fulfills axioms (ii) and (iii) by

[14]The word "local" refers here to the geometry on $\mathcal{E}(\mathcal{H}_{AB})$ and should not be confused with the usual notion of locality in quantum mechanics.

using the unitary invariance and the contractivity of d, respectively. One deduces from the flags condition (26) for $\delta = d^2$ that the metric g satisfies

$$g_{\rho_{ABE}}\Big(\sum_i \eta_i \dot{\rho}_i \otimes |i\rangle\langle i|\,,\,\sum_i \eta_i \dot{\rho}_i \otimes |i\rangle\langle i|\Big) = \sum_i \eta_i g_{\rho_i}(\dot{\rho}_i, \dot{\rho}_i) \quad \text{for } \rho_{ABE} = \sum_i \eta_i \rho_i \otimes |i\rangle\langle i|\,. \quad (34)$$

Similarly, one infers from the contractivity of d that the metric g satisfies the inequality (66) below. By repeating the arguments in the proof of Theorem 1, one finds that $D_A^{\rm SR}$ is an entanglement monotone for pure states, i.e., it obeys axiom (iv).

The bound (33) is a consequence of the triangle inequality and the unitary invariance of d. Actually, one has

$$d(\rho, U_A \otimes 1 \rho U_A^\dagger \otimes 1)^2 \leq \lim_{N\to\infty} \Big\{ \sum_{n=1}^{N} d\Big(U_A^{\frac{n-1}{N}} \otimes 1 \rho (U_A^{\frac{n-1}{N}})^\dagger \otimes 1, U_A^{\frac{n}{N}} \otimes 1 \rho (U_A^{\frac{n}{N}})^\dagger \otimes 1\Big)\Big\}^2$$
$$= \lim_{N\to\infty} N^2 d\Big(\rho, U_A^{\frac{1}{N}} \otimes 1 \rho (U_A^{\frac{1}{N}})^\dagger \otimes 1\Big)^2 = \lim_{t\to 0} t^{-2} d(\rho, \rho_{\rm out}(t))^2\,, \quad (35)$$

where $\rho_{\rm out}(t)$ is given by (31) and $U_A = e^{{\rm i}H_A}$. □

Note that when subsystem A is a qubit ($n_A = 2$), the dependence of $D_A^{\rm SR}$ on the choice of the spectrum Λ reduces to a multiplication factor.[15] It follows from the physical interpretations of the Bures and Hellinger metrics (see Sect. 4.5 below) that

Proposition 5 *The discord of speed of response (32) coincides when ρ is invertible with*

- *the* interferometric power *[37] if $d = d_{\rm Bu}$ is the Bures distance:*

$$D_A^{\rm SR}(\rho) = \mathcal{P}_\Lambda(\rho) = \frac{1}{4} \min_{H_A, {\rm sp}(H_A)=\Lambda} \mathcal{F}_Q(\rho, H_A \otimes 1)\,, \quad (36)$$

  where $\mathcal{F}_Q(\rho, H_A \otimes 1)$ is the quantum Fisher information giving the best precision in the estimation of the unknown parameter t from arbitrary measurements on the output states (31).
- *(twice the)* Local Quantum Uncertainty (LQU) *[36] if $d = d_{\rm He}$ is the Hellinger distance:*

$$D_A^{\rm SR}(\rho) = 2LQU_\Lambda(\rho) = 2 \min_{H_A, {\rm sp}(H_A)=\Lambda} \mathcal{I}_{\rm skew}(\rho, H_A \otimes 1)\,, \quad (37)$$

[15] This is a consequence of the following observations [36]: (a) any $H \in \mathcal{B}(\mathbb{C}^2)_{\rm sa}$ with spectrum $\{\lambda_-, \lambda_+\}$ has the form $(\lambda_+ - \lambda_-)\sigma_{\vec{u}}/2 + (\lambda_+ + \lambda-)/2$, where $\sigma_{\vec{u}} = \sum_{m=1}^{3} u_m \sigma_m$, $\vec{u}$ is a unit vector in $\mathbb{R}^3$, and σ_1, σ_2, and σ_3 are the three Pauli matrices; (b) as noted in the proof of Proposition 4, the limit in the r.h.s. of (32) is equal to $g_\rho(-{\rm i}[H_A \otimes 1, \rho], -{\rm i}[H_A \otimes 1, \rho])$ where g_ρ is a scalar product. Hence changing the spectrum Λ from $\{0, \pi\}$ to $\{\lambda_-, \lambda_+\}$ amounts to multiply $D_A^{\rm SR}$ by the constant factor $[(\lambda_+ - \lambda_-)/\pi]^2$.

where $\mathcal{I}_{\rm skew}(\rho, H) = -\frac{1}{2}\,{\rm tr}([\sqrt{\rho}, H]^2)$ *is the skew information [97] describing the amount of information on the values of observables not commuting with H inferred from measurements on a system in state* ρ*.*

This proposition shows that $D_A^{\rm SR}$ has operational interpretations related to parameter estimation and to quantum uncertainty in local measurements for the Bures and Hellinger distances, respectively.

If $n_A = 2$ and $d = d_{\rm He}$ is the Hellinger distance, one finds that $D_A^{\rm R}$ and $D_A^{\rm SR}$ (i.e., the LQU) are proportional to each other,

$$LQU_{\{0,\pi\}}(\rho) = \frac{\pi^2}{4} D_A^{\rm R}(\rho) = \frac{\pi^2}{4} \inf_{\|\vec{u}\|=1} \mathcal{I}_{\rm skew}(\rho, \sigma_{\vec{u}} \otimes 1) \qquad (38)$$

(this follows from the identity $2\mathcal{I}_{\rm skew}(\rho, U) = d_{\rm He}(\rho, U\rho U^\dagger)^2$ for $U = U^\dagger = U^{-1}$ and from the aforementioned property of $D_A^{\rm SR}$ with respect to changes in the spectrum Λ).

3.5 Different Quantum Correlation Measures Lead to Different Orderings on $\mathcal{E}(\mathcal{H})$

It may appear as an unpleasant fact that the orderings on the states of $\mathcal{E}(\mathcal{H}_{AB})$ defined by the different measures of quantum correlations are in general different, in particular they depend on the choice of the distance d. This means that, for instance, it is possible to find two states ρ and σ which satisfy $D_A^{\rm R}(\rho) < D_A^{\rm R}(\sigma)$ for the Bures distance and the reverse inequality $D_A^{\rm R}(\rho) > D_A^{\rm R}(\sigma)$ for the Hellinger distance. This is illustrated in Fig. 2. This figure displays the distributions in the planes defined by pairs of discords of response based on different distances for randomly generated two-qubit states ρ with a fixed purity $P = {\rm tr}(\rho^2)$ (a similar figure would be obtained if the purity was not fixed). The different orderings translate into the non-vanishing area of the plane covered by the distribution, which in turn reflects the absence of a functional relation between the two discords. It is quite analogous to the different orderings on quantum states established by different entanglement measures. More strikingly, the states with a fixed purity $P < 1$ which are maximally quantum correlated for one discord are not maximally quantum correlated for another discord, as is also illustrated in Fig. 2. The distance-dependent families of such maximally quantum correlated two-qubit states have been determined in Refs. [76, 78] as a function of the purity P for the Bures, Hellinger, and trace distances. Note that if the purity is not fixed, axiom (v) precisely makes sure that the family of maximally quantum correlated states is universal and is composed of the maximally entangled states.

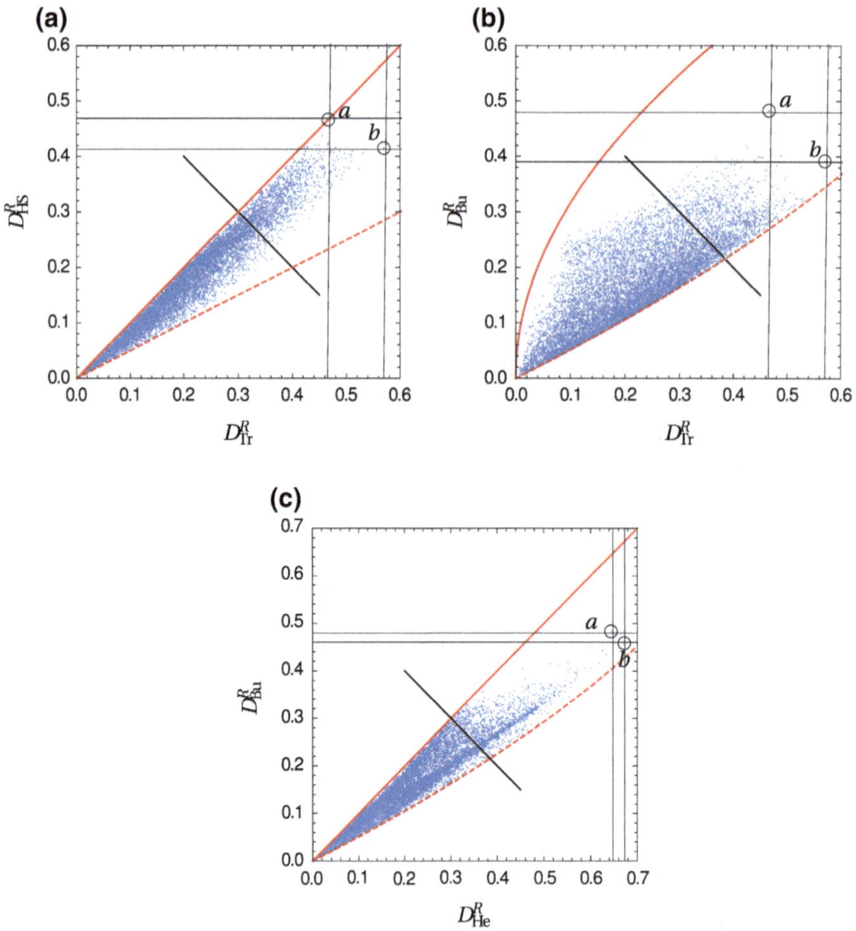

Fig. 2 **a** Values of the Hilbert–Schmidt and trace discords of response $D^{\rm R}_{\rm HS}$ and $D^{\rm R}_{\rm tr}$ for 10^4 random two-qubit states with the same fixed purity $P = 0.6$. These states are generated from random spectra and eigenvectors given by the column vectors of a unitary matrix distributed according to the Haar measure. States on the *thick black line* have a hierarchy with respect to $D^{\rm R}_{\rm HS}$ that is reversed compared to the hierarchy with respect to $D^{\rm R}_{\rm tr}$. The points a and b represent, respectively, some states with purity P maximizing $D^{\rm R}_{\rm HS}$ and $D^{\rm R}_{\rm tr}$. Note that a has not maximal trace discord of response $D^{\rm R}_{\rm tr}$, and similarly for b and $D^{\rm R}_{\rm HS}$. **b** Same for the Bures and trace discords of response $D^{\rm R}_{\rm Bu}$ and $D^{\rm R}_{\rm tr}$. **c** Same for the Bures and Hellinger discords of response $D^{\rm R}_{\rm Bu}$ and $D^{\rm R}_{\rm He}$. The *solid* and *dashed lines* are the borders of the regions delimited by bounds on $D^{\rm R}$ derived from Table 3. This figure is taken from Ref. [78]

4 Bures and Hellinger Distances

In this section we review the properties of the Bures and Hellinger distances between quantum states.

4.1 Bures Distance

The Bures distance was first introduced by Bures in the context of infinite products of von Neumann algebras [19] (see also [4]) and was later studied in a series of papers by Uhlmann [92, 93]. Uhlmann used it to define parallel transport and related it to the fidelity generalizing the usual fidelity $|\langle\psi|\phi\rangle|^2$ between pure states. Indeed, the Bures distance is an extension to mixed states of the Fubini–Study distance for pure states. Recall that the pure states $\rho_\psi = |\psi\rangle\langle\psi|$ of a quantum system with Hilbert space $\mathcal{H}$ can be identified with elements of the projective space $P\mathcal{H}$, that is, the set of equivalence classes of normalized vectors in $\mathcal{H}$ modulo a phase factor. The vectors $|\psi_\theta\rangle = e^{i\theta}|\psi\rangle \in \mathcal{H}$ with $0 \leq \theta < 2\pi$ are called the representatives of ρ_ψ. The Fubini–Study distance on $P\mathcal{H}$ is defined by

$$d_{\rm FS}\big(\rho_\psi, \sigma_\phi\big) = \inf_{\|\psi_\theta\|=\|\phi_\delta\|=1} \big\| |\psi_\theta\rangle - |\phi_\delta\rangle \big\| = \big(2 - 2|\langle\psi|\phi\rangle|\big)^{\frac{1}{2}} , \qquad (39)$$

where the infimum in the second member is over all representatives $|\psi_\theta\rangle$ of ρ_ψ and $|\phi_\delta\rangle$ of σ_ϕ. Observe that the third member depends on the equivalent classes ρ_ψ and σ_ϕ only.

For two mixed states ρ and $\sigma \in \mathcal{E}(\mathcal{H})$, one can define analogously [47, 93]

$$d_{\rm Bu}(\rho, \sigma) = \inf_{R,S} d_2(R, S) , \qquad (40)$$

where d_2 is the Hilbert–Schmidt distance and the infimum is over all Hilbert–Schmidt matrices R and $S \in \mathcal{B}(\mathcal{H})$ satisfying $RR^\dagger = \rho$ and $SS^\dagger = \sigma$. Such matrices are given by $R = \sqrt{\rho}V$ and $S = \sqrt{\sigma}W$ for some unitaries V and W on $\mathcal{H}$ (polar decompositions).

Proposition 6 $d_{\rm Bu}$ *defines a distance on the set of quantum states* $\mathcal{E}(\mathcal{H})$, *which coincides with the Fubini–Study distance for pure states.*

Proof It is clear on (40) that $d_{\rm Bu}(\rho, \sigma)$ is symmetric, non-negative, and vanishes if and only if $\rho = \sigma$. To prove the triangle inequality, let us first observe that by the polar decomposition and the invariance property $d_2(RV, SV) = d_2(R, S)$ of the Hilbert–Schmidt distance for any unitary V, one has $d_{\rm Bu}(\rho, \sigma) = \inf_U d_2(\sqrt{\rho}, \sqrt{\sigma}U)$ with an infimum over all unitaries U. Let ρ, σ, and τ be three states in $\mathcal{E}(\mathcal{H})$. The triangle inequality for d_2 and the aforementioned invariance property yield

$$\begin{aligned} d_{\rm Bu}(\rho,\tau) &\leq \inf_{U,V}\big\{d_2(\sqrt{\rho},\sqrt{\sigma}V)+d_2(\sqrt{\sigma}V,\sqrt{\tau}U)\big\} \\ &= \inf_V d_2(\sqrt{\rho},\sqrt{\sigma}V)+\inf_W d_2(\sqrt{\sigma},\sqrt{\tau}W) \\ &= d_{\rm Bu}(\rho,\sigma)+d_{\rm Bu}(\sigma,\tau)\ . \end{aligned} \tag{41}$$

Hence $d_{\rm Bu}$ defines a distance on $\mathcal{E}(\mathcal{H})$. For pure states $\rho_\psi = |\psi\rangle\langle\psi|$ and $\sigma_\phi = |\phi\rangle\langle\phi|$, the Hilbert–Schmidt operators are of the form $R = |\psi\rangle\langle\mu|$ and $S = |\phi\rangle\langle\nu|$ with $\|\mu\| = \|\nu\| = 1$. A simple calculation then shows that the r.h.s. of (39) and (40) coincide. □

By using the polar decompositions and the formula $\|O\|_1 = \sup_U \mathrm{Re}\ \mathrm{tr}(UO)$ for the trace norm $\|\cdot\|_1$ (the supremum is over all unitaries U), one finds

$$d_{\rm Bu}(\rho,\sigma) = \big(2-2\sup_U \mathrm{Re}\ \mathrm{tr}(U\sqrt{\rho}\sqrt{\sigma})\big)^{\frac{1}{2}} = \big(2-2\sqrt{F(\rho,\sigma)}\big)^{\frac{1}{2}}\ , \tag{42}$$

where $F(\rho,\sigma) = \|\sqrt{\rho}\sqrt{\sigma}\|_1^2$ is the Uhlmann fidelity. Furthermore, the infimum in (40) is attained if and only if the parallel transport condition $R^\dagger S \geq 0$ holds.

Since the fidelity $F(\rho,\sigma)$ belongs to $[0,1]$, $d_{\rm Bu}(\rho,\sigma)$ takes values in the interval $[0,\sqrt{2}]$. Two states ρ and σ have a maximal distance $d_{\rm Bu}(\rho,\sigma) = \sqrt{2}$ (i.e., a vanishing fidelity $F(\rho,\sigma)$) if and only if they have orthogonal supports, $\mathrm{ran}\,\rho \perp \mathrm{ran}\,\sigma$. Such orthogonal states are thus perfectly distinguishable.

Comparing (39) and (42), one sees that the Uhlmann fidelity F is a generalization of the usual pure state fidelity $F(|\psi\rangle,|\phi\rangle) = |\langle\psi|\phi\rangle|^2$. More generally, if σ_ϕ is pure, then it follows from (17) that

$$F(\rho,\sigma_\phi) = \langle\phi|\rho|\phi\rangle \tag{43}$$

for any $\rho \in \mathcal{E}(\mathcal{H})$. A very useful result due to Uhlmann shows that for any states ρ and σ, $F(\rho,\sigma)$ is equal to the fidelity between two pure states $|\Psi\rangle$ and $|\Phi\rangle$ belonging to an enlarged space $\mathcal{H}\otimes\mathcal{K}$ and having marginals $\rho = \mathrm{tr}_{\mathcal{K}}(|\Psi\rangle\langle\Psi|)$ and $\sigma = \mathrm{tr}_{\mathcal{K}}(|\Phi\rangle\langle\Phi|)$. Such states $|\Psi\rangle$ and $|\Phi\rangle$ are called *purifications* of ρ and σ on $\mathcal{H}\otimes\mathcal{K}$. More precisely, one has

Theorem 2 ([92]) *Let ρ, $\sigma \in \mathcal{E}(\mathcal{H})$ and $|\Psi\rangle$ be a purification of ρ on the Hilbert space $\mathcal{H}\otimes\mathcal{K}$, with* $\dim\mathcal{K} \geq \dim\mathcal{H}$. *Then*

$$F(\rho,\sigma) = \max_{|\Phi\rangle} |\langle\Psi|\Phi\rangle|^2\ , \tag{44}$$

where the maximum is over all purifications $|\Phi\rangle$ of σ on $\mathcal{H}\otimes\mathcal{K}$.

A simple proof of this theorem for finite-dimensional Hilbert spaces $\mathcal{H}$ has been given in Ref. [50] (see also [64]).

As the fidelity satisfies $F(\rho_A\otimes\rho_B,\sigma_A\otimes\sigma_B) = F(\rho_A,\sigma_A)F(\rho_B,\sigma_B)$, the Bures distance increases by taking tensor products, i.e.,

$$d_{\mathrm{Bu}}(\rho_A \otimes \rho_B, \sigma_A \otimes \sigma_B) \geq d_{\mathrm{Bu}}(\rho_A, \sigma_A) \tag{45}$$

with equality if and only if $\rho_B = \sigma_B$. Note that the trace distance does not enjoy this property.

4.2 *Classical and Quantum Hellinger Distances*

Let $\mathcal{E}_{\mathrm{clas}} = \{\mathbf{p} \in \mathbb{R}_+^n ; \sum_k p_k = 1\}$ be the simplex of classical probability distributions on the finite sample space $\{1, 2, \ldots, n\}$. The restriction of a distance d on $\mathcal{E}(\mathcal{H})$ to all density matrices commuting with a given state ρ_0 defines a distance on $\mathcal{E}_{\mathrm{clas}}$. In particular, if ρ and σ are two commuting states with spectral decompositions $\rho = \sum_k p_k |k\rangle\langle k|$ and $\sigma = \sum_k q_k |k\rangle\langle k|$, then

$$d_{\mathrm{Bu}}(\rho, \sigma) = d_{\mathrm{clas}}(\mathbf{p}, \mathbf{q}) \equiv \Big(2 - 2\sum_{k=1}^{n} \sqrt{p_k q_k}\Big)^{\frac{1}{2}} = \Big(\sum_{k=1}^{n} (\sqrt{p_k} - \sqrt{q_k})^2\Big)^{\frac{1}{2}} \tag{46}$$

reduces to the classical Hellinger distance d_{clas} on $\mathcal{E}_{\mathrm{clas}}$. One can of course define other distances on $\mathcal{E}(\mathcal{H})$ which coincide with d_{clas} for commuting density matrices, by choosing a different ordering of the operators inside the trace in the definition (17) of the fidelity. For the "normal ordering", one obtains the quantum Hellinger distance

$$d_{\mathrm{He}}(\rho, \sigma) = \big(2 - 2\,\mathrm{tr}\,\sqrt{\rho}\sqrt{\sigma}\big)^{\frac{1}{2}} = d_2(\sqrt{\rho}, \sqrt{\sigma})\ . \tag{47}$$

Since d_2 is a distance on $\mathcal{E}(\mathcal{H})$, this is also the case for d_{He}. In the sequel, d_{He} will be referred to as the Hellinger distance when it is clear from the context that one works with quantum states and not probability distributions.

Comparing (40) and (47), one immediately sees that $d_{\mathrm{He}}(\rho, \sigma) \geq d_{\mathrm{Bu}}(\rho, \sigma)$ for any states $\rho, \sigma \in \mathcal{E}(\mathcal{H})$. Like d_{Bu}, the Hellinger distance satisfies the monotonicity (45) under tensor products. A notable difference between d_{Bu} and d_{He} is that the latter does not coincide with the Fubini–Study distance for pure states (in fact, $d_{\mathrm{He}}(\rho_\psi, \sigma_\phi) = (2 - 2|\langle\psi|\phi\rangle|^2)^{\frac{1}{2}} > d_{\mathrm{FS}}(\rho_\psi, \sigma_\phi)$ if ρ_ψ and σ_ϕ are distinct and non-orthogonal).

One can associate to two non-commuting states ρ and σ the outcome probabilities $\mathbf{p} = (p_1, \ldots, p_m)$ and $\mathbf{q} = (q_1, \ldots, q_m)$ of a measurement performed on the system respectively in states ρ and σ. A natural question is whether $d_{\mathrm{Bu}}(\rho, \sigma)$ or $d_{\mathrm{He}}(\rho, \sigma)$ coincide with the supremum of the classical distance $d_{\mathrm{clas}}(\mathbf{p}, \mathbf{q})$ over all such measurements.

Proposition 7 *For any $\rho, \sigma \in \mathcal{E}(\mathcal{H})$, one has*

$$d_{\mathrm{Bu}}(\rho, \sigma) = \sup_{\{M_i\}} d_{\mathrm{clas}}(\mathbf{p}, \mathbf{q})\ , \tag{48}$$

where the supremum is over all POVMs $\{M_i\}_{i=1}^m$ *and* $p_i = \operatorname{tr} M_i \rho$ *(respectively* $q_i = \operatorname{tr} M_i \sigma$*) is the probability of the measurement outcome i in the state* ρ *(respectively* σ*). The supremum is achieved for von Neumann measurements with rank-one projectors* $M_i = |i\rangle\langle i|$.

A proof of this result and references to the original works can be found in Nielsen and Chuang's book [64]. Note that a similar statement also holds for the trace distance (with $d_{\rm clas}$ replaced by the ℓ^1-distance). In contrast, while $d_{\rm clas}(\mathbf{p}, \mathbf{q}) \leq d_{\rm Hel}(\rho, \sigma)$ for any POVM, the maximum over all POVMs is strictly smaller than $d_{\rm Hel}(\rho, \sigma)$, except when $d_{\rm He}(\rho, \sigma) = d_{\rm Bu}(\rho, \sigma)$.

4.3 Contractivity and Joint Convexity

Proposition 8 *The Bures and Hellinger distances* $d_{\rm Bu}$ *and* $d_{\rm He}$ *are contractive under quantum operations. Moreover,* $d_{\rm Bu}^2$ *and* $d_{\rm He}^2$ *are jointly convex, that is,*

$$d_{\rm Bu}^2\Big(\sum_i p_i \rho_i, \sum_i p_i \sigma_i\Big) \leq \sum_i p_i d_{\rm Bu}^2(\rho_i, \sigma_i)\ , \tag{49}$$

with a similar inequality for $d_{\rm He}$.

The relative entropy $S(\rho||\sigma)$ is also jointly convex. This mathematical property is interpreted as follows. Given two ensembles $\{\rho_i, p_i\}$ and $\{\sigma_i, p_i\}$ of states in $\mathcal{E}(\mathcal{H})$ with the same probabilities p_i, by erasing the information about which state of the ensemble is chosen, the state of the system becomes $\rho = \sum_i p_i \rho_i$ or $\sigma = \sum_i p_i \sigma_i$. The joint convexity means that the entropy between the two ensembles after the loss of information provoked by the state mixing is smaller or equal to the average of the entropies $S(\rho_i||\sigma_i)$. Note that the L^p-distances d_p also fulfill this requirement. According to Proposition 8, the same is true for the squares of the Bures and Hellinger distances, but not for the distances themselves.

The contractivity of $d_{\rm He}$ will be deduced from the following more general result, known as Lieb's concavity theorem [51] (see e.g. [64] for a proof).[16] We denote by $\mathcal{B}(\mathcal{H})_+$ the set of all non-negative operators on $\mathcal{H}$.

Theorem 3 ([51]) *For any fixed operator* $K \in \mathcal{B}(\mathcal{H})$, $\beta \in [-1, 0]$, *and* $q \in [0, 1+\beta]$, *the function* $(\rho, \sigma) \mapsto \operatorname{tr}(K^\dagger \rho^q K \sigma^{-\beta})$ *on* $\mathcal{B}(\mathcal{H})_+ \times \mathcal{B}(\mathcal{H})_+$ *is jointly concave in* (ρ, σ).

Proof of Proposition 8. Let us first show that $d_{\rm Bu}^2$ is jointly convex. This is a consequence of the bound

[16] The justification by Lieb and Ruskai [52] of the strong subadditivity of the von Neumann entropy is based on this important theorem.

$$\sqrt{F\Big(\sum_i p_i\rho_i, \sum_i q_i\sigma_i\Big)} \geq \sum_i \sqrt{p_i q_i}\sqrt{F(\rho_i, \sigma_i)} \,. \tag{50}$$

To establish (50), we use Theorem 2 and introduce some purifications $|\Psi_i\rangle$ of ρ_i and $|\Phi_i\rangle$ of σ_i on $\mathcal{H}\otimes\mathcal{H}$ such that $\sqrt{F(\rho_i, \sigma_i)} = |\langle\Psi_i|\Phi_i\rangle| = \langle\Psi_i|\Phi_i\rangle$. Let us define the vectors

$$|\Psi\rangle = \sum_i \sqrt{p_i}|\Psi_i\rangle|i\rangle \quad , \quad |\Phi\rangle = \sum_i \sqrt{p_i}|\Phi_i\rangle|i\rangle \tag{51}$$

in $\mathcal{H}\otimes\mathcal{H}\otimes\mathcal{H}_E$, where $\mathcal{H}_E$ is an auxiliary Hilbert space with orthonormal basis $\{|i\rangle\}$. Then $|\Psi\rangle$ and $|\Phi\rangle$ are purifications of $\rho = \sum_i p_i\rho_i$ and $\sigma = \sum_i q_i\sigma_i$, respectively. Using Theorem 2 again, one finds

$$\sqrt{F(\rho, \sigma)} \geq |\langle\Psi|\Phi\rangle| = \sum_i \sqrt{p_i q_i}\langle\Psi_i|\Phi_i\rangle = \sum_i \sqrt{p_i q_i}\sqrt{F(\rho_i, \sigma_i)} \,. \tag{52}$$

We have thus proven that $d_{\rm Bu}^2$ is jointly convex. The joint convexity of $d_{\rm He}^2$ is a corollary of Theorem 3, which insures that $(\rho, \sigma) \mapsto \mathrm{tr}(\sqrt{\rho}\sqrt{\sigma})$ is jointly concave.

The following general argument shows that the contractivity of $d_{\rm Bu}$ and $d_{\rm He}$ is a consequence of the joint convexity proven above and of Stinespring's theorem [84] on CP maps [31, 91, 99]. Recall that if $\mu_{\rm H}$ is the normalized Haar measure on the group $U(n)$ of $n \times n$ unitary matrices, then $\int \mathrm{d}\mu_{\rm H}(U)\, UBU^\dagger = n^{-1}\,\mathrm{tr}\, B$ for any $B \in \mathcal{B}(\mathcal{H})$ (in fact, all diagonal matrix elements of $O = \int \mathrm{d}\mu_{\rm H}(U)\, UBU^\dagger$ in an arbitrary basis are equal, as a result of the left-invariance of the Haar measure, $\mathrm{d}\mu_{\rm H}(VU) = \mathrm{d}\mu_{\rm H}(U)$ for any $V \in U(n)$; thus O is proportional to the identity matrix). Let $\mathcal{M}$ be a quantum operation on $\mathcal{B}(\mathcal{H})$. One infers from the Stinespring theorem that there exists a pure state $|\epsilon_0\rangle$ of an ancilla system E and a unitary U on $\mathcal{H}\otimes\mathcal{H}_E$ such that

$$\begin{aligned}\mathcal{M}(\rho)\otimes(1/n_E) &= \mathrm{tr}_E(U\rho\otimes|\epsilon_0\rangle\langle\epsilon_0|U^\dagger)\otimes(1/n_E)\\ &= \int \mathrm{d}\mu_{\rm H}(U_E)\,(1\otimes U_E)U\rho\otimes|\epsilon_0\rangle\langle\epsilon_0|U^\dagger(1\otimes U_E^\dagger)\,.\end{aligned} \tag{53}$$

By using the property $d_{\rm Bu}(\rho\otimes\tau, \sigma\otimes\tau) = d_{\rm Bu}(\rho, \sigma)$, see (45), and the joint convexity and unitary invariance of $d_{\rm Bu}^2$, one gets

$$\begin{aligned}d_{\rm Bu}^2(\mathcal{M}(\rho), \mathcal{M}(\sigma)) &= d_{\rm Bu}^2\big(\mathcal{M}(\rho)\otimes(1/n_E)\,,\ \mathcal{M}(\sigma)\otimes(1/n_E)\big)\\ &\leq \int \mathrm{d}\mu_{\rm H}(U_E) d_{\rm Bu}^2\big((1\otimes U_E)U\rho\otimes|\epsilon_0\rangle\langle\epsilon_0|U^\dagger(1\otimes U_E^\dagger)\,,\\ &\quad (1\otimes U_E)U\sigma\otimes|\epsilon_0\rangle\langle\epsilon_0|U^\dagger(1\otimes U_E^\dagger)\big)\\ &= \int \mathrm{d}\mu_{\rm H}(U_E) d_{\rm Bu}^2(\rho, \sigma) = d_{\rm Bu}^2(\rho, \sigma)\,.\end{aligned} \tag{54}$$

A similar reasoning applies to $d_{\rm He}$. □

4.4 Riemannian Metrics

In Riemannian geometry, a metric on a smooth manifold X is a (smooth) map g associating to each point x in X a scalar product g_x on the tangent space T_xX at x. A metric g induces a Riemannian distance d, which is such that the square distance $\mathrm{d}s^2 = d(x, x + \mathrm{d}x)^2$ between two infinitesimally close points x and $x + \mathrm{d}x$ is equal to $g_x(\mathrm{d}x, \mathrm{d}x)$. For the manifold $X = \mathcal{E}(\mathcal{H})$ of quantum states, the tangent spaces $T_\rho\mathcal{E}(\mathcal{H})$ can be identified with the (real) vector space $\mathcal{B}(\mathcal{H})^0_{\text{s.a.}}$ of self-adjoint operators on $\mathcal{H}$ with zero trace. A curve Γ in $\mathcal{E}(\mathcal{H})$ joining two states ρ_0 and ρ_1 is a (continuously differentiable) map $\Gamma : t \in [0, 1] \mapsto \rho(t) \in \mathcal{E}(\mathcal{H})$ with $\Gamma(0) = \rho_0$ and $\Gamma(1) = \rho_1$ (see Fig. 3). Its length $\ell(\Gamma)$ is

$$\ell(\Gamma) = \int_\Gamma \mathrm{d}s = \int_0^1 \mathrm{d}t \sqrt{g_{\rho(t)}(\dot{\rho}(t), \dot{\rho}(t))} \,, \tag{55}$$

where $\dot{\rho}(t)$ stands for the time derivative $\mathrm{d}\rho/\mathrm{d}t$. A curve $\Gamma_{\mathrm{g}}(\rho, \sigma)$ with the shortest length joining ρ and σ, or more generally, a curve $\Gamma_{\mathrm{g}}(\rho, \sigma) \in C(\rho, \sigma) = \{\Gamma \in C^1([0, 1], \mathcal{E}(\mathcal{H}); \Gamma(0) = \rho, \Gamma(1) = \sigma\}$ at which the map $\Gamma \in C(\rho, \sigma) \to \ell(\Gamma)$ has a stationary point. The distance between two states ρ and σ is the length of the shortest geodesic joining these two states, $d(\rho, \sigma) = \min\{\ell(\Gamma_{\mathrm{g}}(\rho, \sigma))\} = \min_{\Gamma \in C(\rho,\sigma)} \ell(\Gamma)$. Thanks to this formula, a distance d on $\mathcal{E}(\mathcal{H})$ can be associated to any metric g. Conversely, one can associate a metric g to a distance d if the following condition is satisfied (we ignore here the regularity assumptions): for any $\rho \in \mathcal{E}(\mathcal{H})$ and $\dot{\rho} \in \mathcal{B}(\mathcal{H})^0_{\text{s.a.}}$, the square distance between ρ and $\rho + t\dot{\rho}$ has a small time Taylor expansion of the form

$$\mathrm{d}s^2 = d(\rho, \rho + t\dot{\rho})^2 = g_\rho(\dot{\rho}, \dot{\rho})t^2 + \mathcal{O}(t^3) \,. \tag{56}$$

Needless to say, determining the metric induced by a given distance d is much simpler than finding an explicit formula for $d(\rho, \sigma)$ for arbitrary states $\rho, \sigma \in \mathcal{E}(\mathcal{H})$ from the expression of the metric g.

A trivial example of metric on $\mathcal{E}(\mathcal{H})$ is

$$g_\rho(O, O') = \langle O, O' \rangle = \mathrm{tr}(OO') \quad , \quad O, O' \in \mathcal{B}(\mathcal{H})^0_{\text{s.a.}} \,, \tag{57}$$

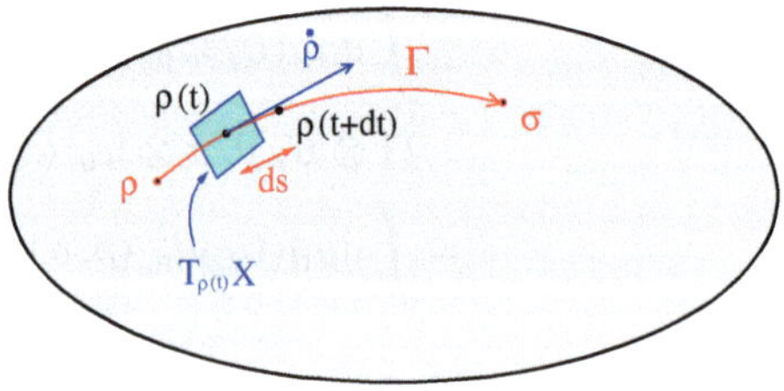

Fig. 3 Curve Γ joining two states ρ and σ in the set of quantum states $X = \mathcal{E}(\mathcal{H})$

i.e., g_ρ is independent of ρ and given by the Hilbert–Schmidt scalar product for matrices. Introducing an orthonormal basis $\{|i\rangle\}_{i=1}^n$ of $\mathcal{H}$, one finds that $g_\rho(O, O') = \sum_{i,j=1}^n \overline{O_{ij}} O'_{ij}$ is nothing but the Euclidean scalar product. Thus the geodesics are straight lines, $\Gamma_g(\rho, \sigma) : t \in [0, 1] \mapsto (1 - t)\rho + t\sigma$, and the distance between two arbitrary states ρ and σ is the Hilbert–Schmidt distance $d_2(\rho, \sigma) = \langle -\rho + \sigma, -\rho + \sigma\rangle^{\frac{1}{2}} = (\mathrm{tr}[(\rho - \sigma)^2])^{\frac{1}{2}}$.

It is not difficult to show (see [81]) that the Bures and Hellinger distances are Riemannian and have metrics given by

$$\left.\begin{aligned} (g_{\mathrm{Bu}})_\rho(O, O) &= \frac{1}{2}\sum_{k,l=1}^n \frac{|\langle k|O|l\rangle|^2}{p_k + p_l} \\ (g_{\mathrm{He}})_\rho(O, O) &= \sum_{k,l=1}^n \frac{|\langle k|O|l\rangle|^2}{(\sqrt{p_k} + \sqrt{p_l})^2} \end{aligned}\right\} , \quad O \in \mathcal{B}(\mathcal{H})_{\mathrm{s.a.}}^0 \, , \quad \rho > 0 \, , \tag{58}$$

where $\{|k\rangle\}$ is an orthonormal basis of eigenvectors of ρ with eigenvalues p_k. In contrast, the trace distance d_1 is not Riemannian. One deduces from (58) that

$$(g_{\mathrm{Bu}})_\rho(O, O) \leq (g_{\mathrm{He}})_\rho(O, O) \leq 2(g_{\mathrm{Bu}})_\rho(O, O) \, . \tag{59}$$

The volume of $\mathcal{E}(\mathcal{H})$ and the area of its boundary for the Bures metric have been determined in Ref. [80].

4.5 *Physical Interpretations of the Bures and Hellinger Metrics*

The metrics g_{Bu} and g_{He} have interpretations in quantum metrology and quantum hypothesis testing. Let us first discuss the link with quantum metrology. Consider the curve in $\mathcal{E}(\mathcal{H})$ given by the unitary evolution of the state $\rho(0) = \rho$ under the Hamiltonian $H \in \mathcal{B}(\mathcal{H})_{\mathrm{s.a.}}$,

$$\rho(t) = e^{-\mathrm{i}tH} \rho \, e^{\mathrm{i}tH} \, . \tag{60}$$

Then $\dot{\rho}(t) = -\mathrm{i}[H, \rho(t)]$. Assuming that ρ is invertible, the speed of the state evolution, $v(t_0) = \lim_{t\to 0} t^{-1} d_{\mathrm{Bu}}(\rho(t_0), \rho(t_0 + t))$, is given by $\sqrt{\mathcal{F}_Q(\rho(t_0), H)}/2 = \sqrt{\mathcal{F}_Q(\rho, H)}/2$, where

$$\mathcal{F}_Q(\rho, H) = 4(g_{\mathrm{Bu}})_\rho\big(-\mathrm{i}[H, \rho], -\mathrm{i}[H, \rho]\big) = 2 \sum_{k,l,p_k+p_l>0} \frac{(p_k - p_l)^2}{p_k + p_l} |\langle k|H|l\rangle|^2 \tag{61}$$

is the *quantum Fisher information*. This quantity is related to the smallest error Δt that can be achieved when estimating the unknown parameter t by performing measurements on the output states $\rho(t)$. Indeed, optimizing over all measurements and all unbiased statistical estimators (that is, all functions $t_{\rm est}(i_1, \ldots, i_N)$ depending on the measurement outcomes $i_1, \ldots, i_N$ and such that $\langle t_{\rm est}\rangle = t$), the best precision is given by [17]

$$(\Delta t)_{\rm best} = \frac{1}{\sqrt{N}\sqrt{\mathcal{F}_Q(\rho, H)}}\,, \tag{62}$$

where N is the number of measurements.[17] Note that for pure states $\mathcal{F}_Q(|\psi\rangle, H) = 4\langle(\Delta H)^2\rangle_\psi$ reduces to the square quantum fluctuation $\langle(\Delta H)^2\rangle_\psi = \langle\psi|H^2|\psi\rangle - \langle\psi|H|\psi\rangle^2$ up to a factor of four. Hence (62) takes the form of a generalized uncertainty relation $(\Delta t)^2\langle(\Delta H)^2\rangle_\psi \geq 1/4$ (here we take $N = 1$), in which H plays the role of the variable conjugated to the parameter t. We remark that the second equality in (61) is only valid when $\rho > 0$. The quantum Fisher information is, however, given by the last expressionin (61) for any state ρ.

The analog of (61) for the Hellinger metric is the *skew information* [97]

$$\mathcal{I}_{\rm skew}(\rho, H) = \frac{1}{2}(g_{\rm He})_\rho\big(-{\rm i}[H,\rho], -{\rm i}[H,\rho]\big) = -\frac{1}{2}\,{\rm tr}\left([\sqrt{\rho}\,,\,H]^2\right). \tag{63}$$

It describes the amount of information on the values of observables not commuting with H in a system in state ρ. The Fisher and skew informations have the following properties [53, 97]:

(a) they are non-negative and vanish if and only if $[\rho, H] = 0$ (this follows from the fact that $(g_{\rm Bu})_\rho$ and $(g_{\rm He})_\rho$ are scalar products);
(b) they are convex in ρ (this follows from the joint convexity of $d_{\rm Bu}^2$ and $d_{\rm He}^2$).[18]
(c) they are additive, i.e., $\mathcal{F}_Q(\rho_A \otimes \rho_B, H_A \otimes 1 + 1 \otimes H_B) = \mathcal{F}_Q(\rho_A, H_A) + \mathcal{F}_Q(\rho_B, H_B)$, with a similar identity for $\mathcal{I}_{\rm skew}$;
(d) the Fisher information is given by [30, 90]

$$\frac{1}{4}\mathcal{F}_Q(\rho, H) = \inf_{\{|\psi_i\rangle,\eta_i\}}\left\{\sum_i \eta_i\langle(\Delta H)^2\rangle_{\psi_i}\right\}, \tag{64}$$

where the infimum is over all pure state decompositions $\rho = \sum_i \eta_i|\psi_i\rangle\langle\psi_i|$ of ρ;
(e) they obey the bounds[19]

$$\frac{1}{8}\mathcal{F}_Q(\rho, H) \leq \mathcal{I}_{\rm skew}(\rho, H) \leq \frac{1}{4}\mathcal{F}_Q(\rho, H) \leq \langle(\Delta H)^2\rangle_\rho\,, \tag{65}$$

---

[17] More precisely, the error $\Delta t = \langle(t_{\rm est} - t)^2\rangle^{1/2}$ in the parameter estimation is always larger or equal to $(\Delta t)_{\rm best}$ and equality is reached asymptotically as $N \to \infty$ by using the maximum-likelihood estimator and an optimal measurement.

[18] Actually, if a Riemannian distance d with metric g is such that $d^2(\rho, \sigma)$ is jointly convex, then $g_\rho(\sum_i p_i O_i, \sum_i p_i O_i) \leq \sum_i p_i g_{\rho_i}(O_i, O_i)$ for any $O_i \in \mathcal{B}(\mathcal{H})^0_{\rm s.a.}$ and any $\rho = \sum p_i \rho_i$. In view of their expressions (61) and (63) in terms of $g_{\rm Bu}$ and $g_{\rm He}$, this implies that the Fisher and skew informations are convex in ρ.

[19] This follows from (59) and, for the last bound, from (64) and the concavity of $\rho \mapsto \langle(\Delta H)^2\rangle_\rho$.

where $\langle(\Delta H)^2\rangle_\rho = \mathrm{tr}(\rho H^2) - (\mathrm{tr}\,\rho H)^2$ is the variance of H. The second and third inequalities are equalities for pure states.

It can be shown that if the system is composed of N_p particles, H is the sum of the same single particle Hamiltonian $H_{1\mathrm{p}}$ acting on each particle, and Δh is the half difference between the maximal and minimal eigenvalues of $H_{1\mathrm{p}}$, then $\mathcal{F}_Q(\rho, H) > 4(\Delta h)^2 N_\mathrm{p}$ is a sufficient (but not necessary) condition for particle entanglement [72, 81]. Furthermore, high values of $\mathcal{F}_Q(\rho, H)$ imply multipartite entanglement between a large number of particles [48, 89].

Let us now discuss the link with the hypothesis testing problem. This problem consists in discriminating two probability measures μ_1 and μ_2 given the outcomes of N independent identically distributed random variables with laws given by either μ_1 or μ_2. In the quantum setting, this is rephrased as a discrimination of two states ρ and σ given N independent copies of ρ and σ, by means of measurements on the N copies either in state $\rho^{\otimes N}$ or $\sigma^{\otimes N}$. One decides among the two alternatives according to the two possible measurement outcomes. According to the quantum Chernoff bound [6, 65], the probability of error decays exponentially in the limit $N \to \infty$, with a rate given by a contractive function $\xi(\rho, \sigma)$, which is equal to $g_{\mathrm{He}}(\mathrm{d}\rho, \mathrm{d}\rho)/2$ for two infinitesimally close states ρ and $\sigma = \rho + \mathrm{d}\rho$.

4.6 Characterization of All Riemannian Contractive Distances

In Ref. [70], Petz has determined the general form of all Riemannian contractive distances on $\mathcal{E}(\mathcal{H})$ for finite-dimensional Hilbert spaces $\mathcal{H}$. Such distances are induced by metrics g satisfying

$$g_{\mathcal{M}(\rho)}\big(\mathcal{M}(O), \mathcal{M}(O)\big) \leq g_\rho(O, O) \quad , \quad O \in \mathcal{B}(\mathcal{H})^0_{\mathrm{s.a.}} \, , \tag{66}$$

for any $\rho \in \mathcal{E}(\mathcal{H})$ and any quantum operation $\mathcal{M} : \mathcal{B}(\mathcal{H}) \to \mathcal{B}(\mathcal{H}')$. We recall that a real function $f : \mathbb{R}_+ \to \mathbb{R}$ is *operator monotone-increasing* if for any space dimension $n = \dim \mathcal{H} < \infty$ and any $A, B \in \mathcal{B}(\mathcal{H})_+$, one has $A \leq B \Rightarrow f(A) \leq f(B)$ (see e.g. [16]).

Theorem 4 ([70]) *Any continuous contractive metric g on $\mathcal{E}(\mathcal{H})$ has the form*

$$g_\rho(O, O) = \sum_{k,l=1}^{n} c(p_k, p_l) |\langle k|O|l\rangle|^2 \quad , \quad O \in \mathcal{B}(\mathcal{H})^0_{\mathrm{s.a.}} \, , \tag{67}$$

where $\rho = \sum_k p_k |k\rangle\langle k|$ is a spectral decomposition of ρ,

$$c(p, q) = \frac{pf(q/p) + qf(p/q)}{2pqf(p/q)f(q/p)} \, , \tag{68}$$

and $f : \mathbb{R}_+ \to \mathbb{R}_+$ is an operator monotone-increasing function satisfying $f(x) = xf(x^{-1})$. Conversely, the metric defined by (67) are contractive for any function f with these properties. The Bures distance is the smallest of all contractive Riemannian distances with metrics satisfying the normalization condition $g_\rho(1, 1) = \mathrm{tr}(\rho^{-1})/4$.

This theorem is of fundamental importance in geometrical approaches to quantum information. It relies on the fact that from the classical side, there exists only one (up to a normalization factor) contractive metric[20] on the probability simplex $\mathcal{E}_{\rm clas}$, namely the Fisher metric $\mathrm{d}s^2_{\rm Fisher} = \sum_{k=1}^n \mathrm{d}p_k^2/p_k$ [21]. The metric $\mathrm{d}s^2_{\rm Fisher}$ plays a crucial role in statistics. It induces the Hellinger distance (46) up to a factor of one fourth. Therefore, all contractive Riemannian distances on $\mathcal{E}(\mathcal{H})$ satisfying the normalization condition $g_\rho(1, 1) = \mathrm{tr}(\rho^{-1})/4$ coincide with the classical Hellinger distance for commuting density matrices.

It can be shown that the following functions are operator monotone-increasing:

$$f_{\rm KM}(x) = 4\frac{x-1}{\ln x} \;\leq\; f_{\rm He}(x) = (1+\sqrt{x})^2 \;\leq\; f_{\rm Bu}(x) = 2(x+1)\,. \tag{69}$$

Substituting them into the formula (68), we get

$$c_{\rm KM}(p,q) = \frac{\ln p - \ln q}{4(p-q)} \;\geq\; c_{\rm He}(p,q) = \frac{1}{(\sqrt{p}+\sqrt{q})^2} \;\geq\; c_{\rm Bu}(p,q) = \frac{1}{2(p+q)}\,. \tag{70}$$

In view of (58), the last choice $f_{\rm Bu}$ gives the Bures metrics and $f_{\rm He}$ gives the Hellinger metric.

The first choice in (69) corresponds to the so-called Kubo-Mori (or Bogoliubov) metric, which is associated to the relative entropy. In fact, an explicit calculation gives [81]

$$S(\rho + t\dot\rho||\rho) = \frac{t^2}{2}(\widetilde{g}_{\rm KM})_\rho(\dot\rho,\dot\rho) + \mathcal{O}(t^3) = S(\rho||\rho+t\dot\rho) + \mathcal{O}(t^3)\,, \tag{71}$$

where we defined for convenience $\widetilde{g}_{\rm KM} = 4g_{\rm KM}$, which satisfies the normalization condition $(\widetilde{g}_{\rm KM})_\rho(1,1) = \mathrm{tr}(\rho^{-1})$. As noted in [7, 8], the Kubo-Mori metric is quite natural from a physical viewpoint because $d\widetilde{s}^2_{\rm KM} = -\mathrm{d}^2 S$, where S is the von Neumann entropy (since S is concave, its second derivative is non-positive and defines a scalar product on $\mathcal{B}(\mathcal{H})$). Actually, one easily deduces from (71) that[21]

---

[20] Here, the contractivity of the classical metrics refers to Markov mappings $\mathbf{p} \mapsto \mathcal{M}^{\rm clas}\mathbf{p}$ on $\mathcal{E}_{\rm clas}$, with stochastic matrices $\mathcal{M}^{\rm clas}$ having non-negative elements $\mathcal{M}^{\rm clas}_{ij}$ such that $\sum_i \mathcal{M}^{\rm clas}_{ij} = 1$ for any $j = 1, \dots, n$.

[21] The first equality is a consequence of (71) and the identity $S(\rho + t\dot\rho) = S(\rho) - S(\rho + t\dot\rho||\rho) - t\,\mathrm{tr}(\dot\rho\ln\rho)$, and the second expression follows from $\ln(\rho + t\dot\rho) = \ln\rho + t\int_0^\infty \mathrm{d}u\,(\rho+u)^{-1}\dot\rho(\rho+u)^{-1} + \mathcal{O}(t^2)$.

$$(\widetilde{g}_{\rm KM})_\rho(\dot{\rho},\dot{\rho}) = -\frac{{\rm d}^2 S(\rho+t\dot{\rho})}{{\rm d}t^2}\bigg|_{t=0} = {\rm tr}\,\dot{\rho}\,\frac{{\rm d}\ln(\rho+t\dot{\rho})}{{\rm d}t}\bigg|_{t=0} . \tag{72}$$

Let us consider the exponential mapping $\rho \in \mathcal{E}(\mathcal{H}) \mapsto O \in \mathcal{B}(\mathcal{H})_{\rm s.a.}$ defined by

$$\rho = \frac{e^O}{{\rm tr}(e^O)} \quad \Leftrightarrow \quad O - F(O) = \ln\rho \quad \text{with} \quad F(O) = \ln({\rm tr}\, e^O) . \tag{73}$$

Note that $F(O) - {\rm tr}\,\rho O = {\rm tr}\,\rho(F(O) - O) = S(\rho)$, hence F is the Legendre transform of the von Neumann entropy. As a result, ${\rm d}^2 F = {\rm d}^2 S + 2\,{\rm tr}\,{\rm d}\rho\,{\rm d}O = {\rm d}\widetilde{s}^2_{\rm KM}$ (the last equality follows from ${\rm d}\widetilde{s}^2_{\rm KM} = -{\rm d}^2 S$, the last expression of $(\widetilde{g}_{\rm KM})_\rho(\dot{\rho},\dot{\rho})$ in (72), and ${\rm tr}({\rm d}\rho) = 0$). Hence the metric $\widetilde{g}_{\rm KM}$ can also be viewed as the Hessian of the free energy F [7]. A physical interpretation of the Kubo-Mori metric in terms of information losses in state mixing is as follows: the loss of information when mixing the two states $\rho_t = \rho_0 + t\dot{\rho}$ and $\rho_{-t} = \rho_0 - t\dot{\rho}$ with the same weight $p = 1/2$, $\Delta S = S(\rho_0) - S(\rho_t)/2 + S(\rho_{-t})/2$, equals $(t^2/2)(\widetilde{g}_{\rm KM})_{\rho_0}(\dot{\rho},\dot{\rho})$ in the small t limit. We point out that the explicit expression of the Kubo-Mori distance between two arbitrary states ρ and σ is unknown, except in the case of a single qubit [7].

4.7 *Comparison of the Bures, Hellinger, and Trace Distances*

One can find explicit bounds between the Bures, trace, and Hellinger distances showing that these distances define equivalent topologies.

Proposition 9 *For any $\rho,\sigma \in \mathcal{E}(\mathcal{H})$, one has*

$$d_{\rm Bu}(\rho,\sigma) \leq d_{\rm He}(\rho,\sigma) \;\leq\; \sqrt{2}\, d_{\rm Bu}(\rho,\sigma) \tag{74}$$

$$d_{\rm He}(\rho,\sigma)^2 \leq d_1(\rho,\sigma) \;\leq\; 2\Big\{1 - \Big(1 - \frac{1}{2} d_{\rm Bu}(\rho,\sigma)^2\Big)^2\Big\}^{\frac{1}{2}} . \tag{75}$$

The last inequality in (75) is saturated for pure states.

The bounds $d_{\rm Bu}(\rho,\sigma)^2 \leq d_1(\rho,\sigma)$ and $d_{\rm He}(\rho,\sigma)^2 \leq d_1(\rho,\sigma)$, which are consequences of (74) and (75), have been first proven in the C^*-algebra setting by Araki [4] and Holevo [42], respectively. An upper bound on $d_1(\rho,\sigma)$ similar to the one in (75) but with $d_{\rm Bu}$ replaced by $d_{\rm He}$ (which is weaker than the bound in (75) because of (74)) has been also derived by Holevo. Lower and upper bounds on the fidelity $F(\rho,\sigma)$ in terms of traces of polynomials in ρ and σ, which are easier to compute than the trace distance and the fidelity itself, have been derived in [58].

Proof The inequalities in (74) are consequences of the bounds (59) on the Bures and Hellinger metrics. The first bound in (75) can be obtained as follows [42]. We set $A = \sqrt{\rho} - \sqrt{\sigma}$ and $B = \sqrt{\rho} + \sqrt{\sigma}$ and consider the polar decomposition $A = U|A|$ with the unitary $U = P_+ - P_-$, where P_+ and $P_- = 1 - P_+$ are the spectral projectors of

A on $[0,\infty)$ and $(-\infty,0)$, respectively. Noting that $\rho-\sigma=(AB+BA)/2$, $UA=AU=|A|$, and $|A|P_\pm=P_\pm|A|$, we obtain by using $|\operatorname{tr} UO|\leq\|O\|_1$ that

$$\|\rho-\sigma\|_1\geq\operatorname{tr} U(\rho-\sigma)=\operatorname{tr}|A|B=\operatorname{tr}|A|^{\frac{1}{2}}(P_+BP_++P_-BP_-)|A|^{\frac{1}{2}}\ . \tag{76}$$

Now $-B\leq A\leq B$, so that

$$-AP_-=-P_-AP_-\leq P_-BP_-\quad,\quad AP_+=P_+AP_+\leq P_+BP_+\ . \tag{77}$$

Hence the r.h.s. of (76) is bounded from below by $\operatorname{tr}|A|^{\frac{1}{2}}A(P_+-P_-)|A|^{\frac{1}{2}}=\operatorname{tr}A^2$. This yields $\|\rho-\sigma\|_1\geq\|\sqrt{\rho}-\sqrt{\sigma}\|_2^2$, that is, $d_1(\rho,\sigma)\geq d_{\rm He}(\rho,\sigma)^2$.

To prove the last bound in (75), we first argue that if $\rho_\psi=|\psi\rangle\langle\psi|$ and $\sigma_\phi=|\phi\rangle\langle\phi|$ are pure states, then $d_1(\rho_\psi,\sigma_\phi)=2\sqrt{1-F(\rho_\psi,\sigma_\phi)}$, showing that this bound holds with equality. Actually, let $|\phi\rangle=\cos\theta|\psi\rangle+e^{{\rm i}\delta}\sin\theta|\psi^\perp\rangle$, where $\theta,\delta\in[0,2\pi)$ and $|\psi^\perp\rangle$ is a unit vector orthogonal to $|\psi\rangle$. Since $\rho_\psi-\sigma_\phi$ has non-vanishing eigenvalues $\pm\sin\theta$, one has $d_1(\rho_\psi,\sigma_\phi)=2|\sin\theta|$. But $F(\rho_\psi,\sigma_\phi)=\cos^2\theta$, hence the aforementioned statement is true. It then follows from Theorem 2 and from the contractivity of the trace distance under partial traces that for arbitrary ρ and $\sigma\in\mathcal{E}(\mathcal{H})$,

$$d_1(\rho,\sigma)\leq 2\sqrt{1-F(\rho,\sigma)}\ . \tag{78}$$

This concludes the proof. □

4.8 Relations with the Quantum Relative Rényi Entropies

The Rényi entropies $S_\alpha(\rho)=(1-\alpha)^{-1}\ln\operatorname{tr}(\rho^\alpha)$ depending on a parameter $\alpha>0$ are generalizations of the von Neumann entropy $S(\rho)$. For indeed, $S_\alpha(\rho)$ converges to $S(\rho)$ when $\alpha\to1$. Moreover, $S_\alpha(\rho)$ is a non-increasing function of α. Similarly, the relative Rényi entropies generalize the relative entropy $S(\rho||\sigma)=\operatorname{tr}[\rho(\ln\rho-\ln\sigma)]$. Different definitions have been proposed in the literature. The "sandwiched" relative entropies studied in [62, 98] seem to have the nicer properties. A family of relative Rényi entropies depending on two parameters (α,z), which includes the sandwiched entropies (obtained for $z=\alpha$) as special cases, has been introduced in the context of fluctuation relations in quantum statistical physics [14, 49] and was later on studied from a quantum information perspective [5]. These entropies are defined when $\ker\sigma\subset\ker\rho$ by

$$S_{\alpha,z}(\rho||\sigma)=-\frac{1}{2(1-\alpha)}\ln F_{\alpha,z}(\rho||\sigma)\quad,\quad F_{\alpha,z}(\rho||\sigma)=\Big(\operatorname{tr}\big[(\sigma^{\frac{1-\alpha}{2z}}\rho^{\frac{\alpha}{z}}\sigma^{\frac{1-\alpha}{2z}})^z\big]\Big)^2\ . \tag{79}$$

Taking $\alpha = z \to 1$, one recovers the von Neumann relative entropy $S(\rho||\sigma)$ [62]. The max-entropy is obtained in the limit $\alpha = z \to \infty$ [62]. For commuting matrices ρ and σ with eigenvalues $\mathbf{p}$ and $\mathbf{q}$, $S_{\alpha,z}(\rho||\sigma)$ reduces to the classical Rényi divergence $S_\alpha^{\rm clas}(\mathbf{p}||\mathbf{q}) = (\alpha-1)^{-1}\ln(\sum_k p_k^\alpha q_k^{1-\alpha})$.

It is known that $S_{\alpha,z}(\rho||\sigma)$ is contractive and jointly convex when $\alpha \in (0,1]$ and $z \geq \max\{\alpha, 1-\alpha\}$ (see [5] and references therein) and is contractive when $\alpha = z \geq 1/2$ (see [81] and references therein). For those values of (α, z), it is easy to show[22] that $S_{\alpha,z}(\rho||\sigma) \geq 0$ with equality if and only if $\rho = \sigma$. Furthermore, the following monotonicity properties hold: for any $\rho, \sigma \in \mathcal{E}(\mathcal{H})$, $S_{\alpha,\alpha}(\rho||\sigma)$ is non-decreasing in α on $(0, \infty)$ [62] and for any fixed $\alpha \in (0,1)$, $S_{\alpha,z}(\rho||\sigma)$ is non-decreasing in z on $(0,\infty)$ (this follows from the Lieb–Thirring–Araki trace inequality).

We observe that the Bures and Hellinger distances are functions of the generalized Rényi relative entropies $S_{\alpha,z}$ for $(\alpha, z) = (1/2, 1/2)$ and $(1/2, 1)$, respectively. In fact,

$$
\begin{aligned}
d_{\rm Bu}(\rho,\sigma)^2 &= 2 - 2\exp\Big\{-\frac{1}{2}S_{1/2,1/2}(\rho||\sigma)\Big\} \ , \\
d_{\rm He}(\rho,\sigma)^2 &= 2 - 2\exp\Big\{-\frac{1}{2}S_{1/2,1}(\rho||\sigma)\Big\} \ .
\end{aligned}
\tag{80}
$$

Thus, $S_{\alpha,z}$ connects monotonously and continuously to each other the von Neumann relative entropy S, the Bures distance $d_{\rm Bu}$, and the Hellinger distance $d_{\rm He}$.

5 Bures Geometric Discord

In this section we study the Bures geometric discord, obtained by choosing the Bures distance $d = d_{\rm Bu}$ in (19),

$$
D_{\rm Bu}^{\rm G}(\rho) = d_{\rm Bu}(\rho, \mathcal{C}_A)^2 = 2(1-\sqrt{F(\rho,\mathcal{C}_A)}) \quad , \quad F(\rho,\mathcal{C}_A) = \max_{\sigma_{A-{\rm cl}}\in\mathcal{C}_A} F(\rho, \sigma_{A-{\rm cl}}) \ , \tag{81}
$$

where F is the fidelity (17). Hereafter, we omit the lower subscript A on all discords, as we will always take A as the reference subsystem. Instead, the chosen distance is indicated as a lower subscript. The main result of this section is Theorem 5 below, which shows that the determination of $D_{\rm Bu}^{\rm G}(\rho)$ and of the closest A-classical state(s) to ρ are related to a minimal-error quantum state discrimination problem.

---

[22] This follows from the contractivity of $S_{\alpha,z}(\rho||\sigma)$ applied to a measurement with rank-one projectors $\{|k\rangle\langle k|\}$ and the fact that $S_\alpha^{\rm clas}(\mathbf{p}||\mathbf{q}) \geq 0$ with equality if and only if $\mathbf{p} = \mathbf{q}$. The property is actually true for any $\alpha = z > 0$ (see e.g. [81]) and, probably, for other values of (α, z).

5.1 The Case of Pure States

Let us first restrict our attention to pure states $\rho_\Psi = |\Psi\rangle\langle\Psi|$, for which a simple formula for the geometric discord in terms of the Schmidt coefficients μ_i of $|\Psi\rangle$ can be obtained. We recall that any pure state $|\Psi\rangle \in \mathcal{H}_A \otimes \mathcal{H}_B$ admits a Schmidt decomposition

$$|\Psi\rangle = \sum_{i=1}^{n} \sqrt{\mu_i} |\varphi_i\rangle \otimes |\chi_i\rangle \;, \tag{82}$$

where $\{|\varphi_i\rangle\}_{i=1}^{n_A}$ (respectively $\{|\chi_j\rangle\}_{j=1}^{n_B}$) is an orthonormal basis of $\mathcal{H}_A$ ($\mathcal{H}_B$) and $n = \min\{n_A, n_B\}$. The basis $\{|\varphi_i\rangle\}$ (respectively $\{|\chi_j\rangle\}$) and Schmidt coefficients $\mu_i \geq 0$ are the eigenbasis and eigenvalues of the reduced state $[\rho_\Psi]_A$ (respectively $[\rho_\Psi]_B$).

Let us show that $D^{\mathrm{G}}_{\mathrm{Bu}}(|\Psi\rangle)$ is equal to the geometric entanglement $E^{\mathrm{G}}_{\mathrm{Bu}}(|\Psi\rangle)$. In order to calculate the latter, we write the decomposition of separable states into pure product states, $\sigma_{\mathrm{sep}} = \sum_m q_m |\phi_A^m \otimes \phi_B^m\rangle\langle\phi_A^m \otimes \phi_B^m|$ and use the expression (43) of the fidelity and $\sum_m q_m = 1$ to get

$$\begin{aligned} F(\rho_\Psi, \mathcal{S}_{AB}) \equiv \max_{\sigma_{\mathrm{sep}} \in \mathcal{S}_{AB}} F(\rho_\Psi, \sigma_{\mathrm{sep}}) &= \max_{\{|\phi_A^m\rangle, |\phi_B^m\rangle, q_m\}} \Big\{ \sum_m q_m |\langle \phi_A^m \otimes \phi_B^m |\Psi\rangle|^2 \Big\} \\ &= \max_{\|\phi_A\| = \|\phi_B\| = 1} \big\{ |\langle \phi_A \otimes \phi_B |\Psi\rangle|^2 \big\} \;. \end{aligned} \tag{83}$$

For any fixed normalized vectors $|\phi_A\rangle \in \mathcal{H}_A$ and $|\phi_B\rangle \in \mathcal{H}_B$, one deduces from (82) and the Cauchy–Schwarz inequality that

$$\begin{aligned} |\langle \phi_A \otimes \phi_B |\Psi\rangle| &\leq \sqrt{\mu_{\max}} \sum_{i=1}^{n} \big| \langle \phi_A|\varphi_i\rangle\langle\phi_B|\chi_i\rangle \big| \\ &\leq \sqrt{\mu_{\max}} \Big(\sum_{i=1}^{n} |\langle\phi_A|\varphi_i\rangle|^2 \Big)^{1/2} \Big(\sum_{j=1}^{n} |\langle\phi_B|\chi_j\rangle|^2 \Big)^{1/2} \\ &\leq \sqrt{\mu_{\max}} \;, \end{aligned} \tag{84}$$

where $\mu_{\max} = \max_i \mu_i$ is the largest Schmidt eigenvalue. All bounds are saturated by taking $|\phi_A\rangle$ and $|\phi_B\rangle$ equal respectively to the eigenvectors $|\varphi_{\max}\rangle$ and $|\chi_{\max}\rangle$ of $[\rho_\Psi]_A$ and $[\rho_\Psi]_B$ with maximal eigenvalue $\mu_{\max}$. Thus $F(\rho_\Psi, \mathcal{S}_{AB}) = \mu_{\max}$. Furthermore, the pure product state $|\varphi_{\max}\rangle|\chi_{\max}\rangle$ is a closest separable state to $|\Psi\rangle$. Now, a product state is also an A-classical state. Since $d_{\mathrm{Bu}}(|\Psi\rangle, \mathcal{C}_A) \geq d_{\mathrm{Bu}}(|\Psi\rangle, \mathcal{S}_{AB})$ (because $\mathcal{C}_A \subset \mathcal{S}_{AB}$, see Fig. 1), $|\varphi_{\max}\rangle|\chi_{\max}\rangle$ is also a closest A-classical state to $|\Psi\rangle$ and $D^{\mathrm{G}}_{\mathrm{Bu}}(\rho_\Psi) = E^{\mathrm{G}}_{\mathrm{Bu}}(\rho_\Psi)$, as claimed above.

Proposition 10 ([82]) *The Bures geometric discord is given for pure states* $|\Psi\rangle \in \mathcal{H}_{AB}$ *by*

$$D^{\mathrm{G}}_{\mathrm{Bu}}(|\Psi\rangle) = E^{\mathrm{G}}_{\mathrm{Bu}}(|\Psi\rangle) = 2(1 - \sqrt{\mu_{\max}}) \;. \tag{85}$$

(1) *If the maximal Schmidt eigenvalue $\mu_{\max}$ is non-degenerate, then the closest A-classical (respectively classical, separable) state to ρ_Ψ for the Bures distance is unique and given by the pure product state $|\varphi_{\max}\rangle|\chi_{\max}\rangle$.*

(2) *If $\mu_{\max}$ is r-fold degenerate, say $\mu_{\max} = \mu_1 = \cdots = \mu_r > \mu_{r+1}, \ldots, \mu_n$, then ρ has infinitely many closest A-classical (respectively classical, separable) states. These closest states are convex combinations of the pure product states $|\widehat{\varphi}_l\rangle|\widehat{\chi}_l\rangle$, with*

$$|\widehat{\varphi}_l\rangle = \sum_{i=1}^{r} u_{il}|\varphi_i\rangle \quad , \quad |\widehat{\chi}_l\rangle = \sum_{i=1}^{r} \overline{u_{il}}|\chi_i\rangle \quad , \quad l = 1, \ldots, r \; , \tag{86}$$

where $\{|\varphi_i\rangle\}_{i=1}^r$ and $\{|\chi_i\rangle\}_{i=1}^r$ are some fixed orthonormal families of eigenvectors of $[\rho_\Psi]_A$ and $[\rho_\Psi]_B$ with eigenvalue $\mu_{\max}$ and $(u_{il})_{i,l=1}^r$ is an arbitrary $r \times r$ unitary matrix.

The relation (85) is analogous to the equality between the entropic discord and the entanglement of formation for pure states (Sect. 2.1). It comes here from the existence of a pure product state which is closer or at the same distance from the pure state $|\Psi\rangle$ than any other separable state. This property is a special feature of the Bures distance.

We refer the reader to Refs. [81, 82] for a proof of statements (1) and (2). It should be noticed that when $\mu_{\max}$ is degenerate, the vectors (86) provide together with $|\varphi_i\rangle$, $|\chi_i\rangle$, $i = r+1, \ldots, n$, a Schmidt decomposition of $|\Psi\rangle$ (in that case this decomposition is not unique). Conversely, disregarding degeneracies among the other eigenvalues $\mu_i < \mu_{\max}$, all Schmidt decompositions of $|\Psi\rangle$ are of this form for some unitary matrix $(u_{il})_{i,l=1}^r$. Thus, the existence of an infinite family of closest A-classical states to $|\Psi\rangle$ is related to the non-uniqueness of the Schmidt vectors associated to $\mu_{\max}$. This shows in particular that the maximally entangled pure states (for which $\mu_{\max}$ is n-fold degenerate) are the pure states with the largest family of closest states.[23]

The properties of the Bures geometric entanglement E_{Bu}^{G} have been investigated in [85, 94, 96]. We have already argued above that E_{Bu}^{G} is an entanglement monotone (Sect. 3.2). Hence, in view of (85), the geometric discord D_{Bu}^{G} fulfills axiom (iv) of Definition 1 and is thus a *bona fide* measure of quantum correlations (recall that axioms (i–iii) hold for any contractive distance). One can deduce from the Uhlmann theorem (Theorem 2) and the one-to-one correspondence between purifications and pure state decompositions of a state ρ that $F(\rho, \mathcal{S}_{AB})$ is equal to $\max \sum_i \eta_i F(|\Psi_i\rangle, \mathcal{S}_{AB})$, the maximum being over all pure state decompositions $\rho = \sum_i \eta_i |\Psi_i\rangle\langle\Psi_i|$ of ρ (convex roof) [85].

[23] This family forms a $(n^2 + n - 2)$ real-parameter submanifold of $\mathcal{E}(\mathcal{H}_{AB})$.

5.2 Link with Quantum State Discrimination

As for all other measures of quantum correlations, determining $D_{\rm Bu}^{\rm G}(\rho)$ is harder for mixed states ρ than for pure states. Interestingly, this problem is related to an ambiguous quantum state discrimination task.

The objective of quantum state discrimination is to distinguish states taken randomly from a known ensemble of states [15, 40, 81]. If these states are not orthogonal, any measurement devised to distinguish them cannot succeed to identify exactly which state from the ensemble has been chosen. The quantum state discrimination problem is to find the optimal measurement leading to the smallest probability of equivocation. More precisely, a receiver is given a state $\rho_i \in \mathcal{E}(\mathcal{H})$ drawn from a known ensemble $\{\rho_i, \eta_i\}_{i=1}^{n_A}$ with a prior probability η_i. In order to determine which state he has received, he performs a measurement given by a POVM $\{M_i\}$ and concludes that the state is ρ_j when he gets the measurement outcome j. The probability of this outcome given that the state is ρ_i is $p_{j|i} = \operatorname{tr} M_j \rho_i$. In the ambiguous (or minimal-error) strategy, the number of measurement outcomes is chosen to be equal to the number of states in the ensemble $\{\rho_i, \eta_i\}$. The maximal success probability of the receiver reads

$$P_{\rm S}^{\rm opt}(\{\rho_i, \eta_i\}) = \max_{\text{POVM } \{M_i\}} \sum_{i=1}^{n_A} \eta_i \operatorname{tr} M_i \rho_i \ . \tag{87}$$

If the ρ_i span $\mathcal{H}$ and are linearly independent, in the sense that their eigenvectors $|\xi_{ij}\rangle$ with nonzero eigenvalues form a linearly independent family $\{|\xi_{ij}\rangle\}_{i=1,\dots,n_A}^{j=1,\dots,n_B}$ of vectors in $\mathcal{H}$, it is known that the optimal POVM is a von Neumann measurement with projectors of rank $r_i = \operatorname{rank}(\rho_i)$ [28]. In that case, the maximal success probability $P_{\rm S}^{\rm opt}(\{\rho_i, \eta_i\})$ is equal to

$$P_{\rm S}^{\text{opt v.N.}}(\{\rho_i, \eta_i\}) = \max_{\{\Pi_i\}} \sum_{i=1}^{n_A} \eta_i \operatorname{tr} \Pi_i \rho_i \ , \tag{88}$$

the maximum being over all projective measurements with projectors Π_i of rank r_i.

Theorem 5 ([82]) *For any state ρ of the bipartite system AB, the largest fidelity between ρ and an A-classical state reads*

$$F(\rho, \mathcal{C}_A) = \max_{\{|\alpha_i\rangle\}} P_{\rm S}^{\text{opt v.N.}}(\{\rho_i, \eta_i\}) \ , \tag{89}$$

where the maximum is over all orthonormal bases $\{|\alpha_i\rangle\}_{i=1}^{n_A}$ of $\mathcal{H}_A$ and $P_{\rm S}^{\text{opt v.N.}}(\{\rho_i, \eta_i\})$ is the maximal success probability in discriminating the states ρ_i by von Neumann measurements on AB with projectors of rank n_B. Here, the states ρ_i and probabilities η_i depend on $\{|\alpha_i\rangle\}_{i=1}^{n_A}$ and are given by

$$\eta_i = \langle \alpha_i | \rho_A | \alpha_i \rangle \quad , \quad \rho_i = \eta_i^{-1} \sqrt{\rho} |\alpha_i\rangle\langle\alpha_i| \otimes 1 \sqrt{\rho} \quad , \quad i = 1, \dots, n_A \ . \tag{90}$$

Furthermore, the closest A-classical states to ρ are given by

$$\sigma_{\mathrm{Bu},\rho} = \frac{1}{F(\rho, \mathcal{C}_A)} \sum_{i=1}^{n_A} |\alpha_i^{\mathrm{opt}}\rangle\langle\alpha_i^{\mathrm{opt}}| \otimes \langle\alpha_i^{\mathrm{opt}}|\sqrt{\rho}\,\Pi_i^{\mathrm{opt}}\sqrt{\rho}|\alpha_i^{\mathrm{opt}}\rangle \,, \tag{91}$$

where $\{|\alpha_i^{\mathrm{opt}}\rangle\}$ is an orthonormal basis of $\mathcal{H}_A$ maximizing the r.h.s. of (89) and $\{\Pi_i^{\mathrm{opt}}\}$ is an optimal measurement with projectors of rank n_B maximizing the success probability in (88).

We postpone the proof of this theorem to Sect. 5.5 and proceed with a few comments and consequences of the theorem. Firstly, the ρ_i are quantum states because $\rho_i \geq 0$ and η_i is chosen such that $\operatorname{tr}\rho_i = 1$ (if $\eta_i = 0$ then ρ_i is not defined but does not contribute to the sum in (88)). Secondly, the η_i are the outcome probabilities of a measurement on A with rank-one projectors $\Pi_i^A = |\alpha_i\rangle\langle\alpha_i|$, see (3). Denoting by $\rho_{AB|i} = \eta_i^{-1}\Pi_i^A \otimes 1\rho\,\Pi_i^A \otimes 1$ the corresponding conditional states of AB and by $\mathcal{M}_A^{\Pi}$ the associated quantum operation on A, see (6), we remark that $\rho_i = \mathcal{R}_{\mathcal{M}_A^{\Pi},\rho}(\rho_{AB|i})$ is the image of $\rho_{AB|i}$ under the Petz transpose operation $\mathcal{R}_{\mathcal{M}_A^{\Pi},\rho}$, that is, the approximate reversal operation of $\mathcal{M}_A^{\Pi} \otimes 1$ with respect to ρ (see [81] for more detail). Now, $\mathcal{M}_A^{\Pi} \otimes 1(\rho) = \sum_i \eta_i\rho_{AB|i}$ and, by definition of the transpose operation, $\mathcal{R}_{\mathcal{M}_A^{\Pi},\rho} \circ \mathcal{M}_A^{\Pi} \otimes 1(\rho) = \rho$. Thus $\rho = \sum_i \eta_i\rho_i$, so that the ensemble $\{\rho_i, \eta_i\}_{i=1}^{n_A}$ gives a convex decomposition of ρ (this can also be checked directly on (90)). Another notable property of this ensemble is that the least square measurement[24] associated to it, defined by the POVM $\{M_i^{\mathrm{lsm}}\}$ with

$$M_i^{\mathrm{lsm}} = \eta_i\rho^{-1/2}\rho_i\rho^{-1/2} \quad , \quad i = 1, \ldots, n_A \,, \tag{92}$$

coincides with $\{|\alpha_i\rangle\langle\alpha_i| \otimes 1\}$.

Corollary 1 *If ρ is invertible then one can substitute $P_S^{\mathrm{opt\,v.N.}}(\{\rho_i, \eta_i\})$ in (89) by the maximal success probability $P_S^{\mathrm{opt}}(\{\rho_i, \eta_i\})$ over all POVMs, given by (87).*

Proof If $\rho > 0$ then the states ρ_i defined in (90) are linearly independent, thus the optimal measurement to discriminate them is a von Neumann measurement with projectors of rank $r_i = \operatorname{rank}(\rho_i)$ (see above). The linear independence can be justified as follows. Let us first notice that ρ_i has rank $r_i = n_B$ (for indeed, it has the same rank as $\eta_i\rho^{-1/2}\rho_i = |\alpha_i\rangle\langle\alpha_i| \otimes 1\sqrt{\rho}$). A necessary and sufficient condition for $|\xi_{ij}\rangle$ to be an eigenvector of ρ_i with eigenvalue $\lambda_{ij} > 0$ is $|\xi_{ij}\rangle = (\lambda_{ij}\eta_i)^{-1}\sqrt{\rho}|\alpha_i\rangle \otimes |\zeta_{ij}\rangle$, where $|\zeta_{ij}\rangle \in \mathcal{H}_B$ is an eigenvector of $R_i = \langle\alpha_i|\rho|\alpha_i\rangle$ with eigenvalue $\lambda_{ij}\eta_i > 0$. For any i, the Hermitian invertible matrix R_i admits an orthonormal eigenbasis $\{|\zeta_{ij}\rangle\}_{j=1}^{n_B}$.

---

[24]This measurement bears several other names: it is referred to as the "pretty good measurement" in [38] and is sometimes also called "square-root measurement" [29]. For a pure state ensemble $\{|\psi_i\rangle, \eta_i\}$, it is given by $\{M_i^{\mathrm{lsm}} = |\widetilde{\mu}_i\rangle\langle\widetilde{\mu}_i|\}$ and the vectors $|\widetilde{\mu}_i\rangle = \sqrt{\eta_i}(\sum_j \eta_j|\psi_j\rangle\langle\psi_j|)^{-\frac{1}{2}}|\psi_i\rangle$ are such that they minimize the sum of the square norms $\||\widetilde{\mu}_i\rangle - \sqrt{\eta_i}|\psi_i\rangle\|^2$ under the constraint that $\{M_i^{\mathrm{lsm}}\}$ is a POVM, i.e., $\sum_i |\widetilde{\mu}_i\rangle\langle\widetilde{\mu}_i| = 1$ [43].

Thanks to the invertibility of $\sqrt{\rho}$, $\{|\xi_{ij}\rangle\}_{i=1,\dots,n_A}^{j=1,\dots,n_B}$ is a basis of $\mathcal{H}_{AB}$ and thus the states ρ_i are linearly independent and span $\mathcal{H}_{AB}$. □

5.3 *Quantum Correlations and Distinguishability of Quantum States*

We give in this subsection a physical interpretation of Theorem 5. We start by discussing the state discrimination problem in the special cases where ρ is either pure or A-classical. Of course, the values of $D^{\rm G}_{\rm Bu}(\rho)$ are already known in these cases (they are given by (85) and by $D^{\rm G}_{\rm Bu}(\rho) = 0$, respectively), but it is instructive to recover that from Theorem 5.

(a) If $\rho = \rho_\Psi$ is pure then all states ρ_i with $\eta_i > 0$ are identical and equal to ρ_Ψ, so that $P_{\rm S}^{\rm opt\,v.N.} = \max_{\{\Pi_i\}}\{\sum_i \eta_i\langle\Psi|\Pi_i|\Psi\rangle\} = \eta_{\max}$. One gets $F(\rho_\Psi, \mathcal{C}_A) = \mu_{\max}$ by optimization over the basis $\{|\alpha_i\rangle\}$.

(b) If ρ is an A-classical state, i.e., if it can be decomposed as in (9), then the optimal basis $\{|\alpha_i^{\rm opt}\rangle\}$ coincides with the basis appearing in this decomposition. With this choice one obtains $\eta_i = q_i$ and $\rho_i = |\alpha_i\rangle\langle\alpha_i| \otimes \rho_{B|i}$ for all i such that $q_i > 0$. The states ρ_i are orthogonal and can thus be perfectly discriminated by von Neumann measurements. This yields $F(\rho, \mathcal{C}_A) = 1$ and $D^{\rm G}_{\rm Bu}(\rho) = 0$ as it should be. Reciprocally, if $F(\rho, \mathcal{C}_A) = 1$ then $P_{\rm S}^{\rm opt\,v.N.}(\{\rho_i, \eta_i\}) = 1$ for some basis $\{|\alpha_i\rangle\}$ and the corresponding ρ_i must be orthogonal. Hence one can find an orthonormal family $\{\Pi_i\}$ of projectors with rank n_B such that $\rho_i = \Pi_i\rho_i\Pi_i$ for any i with $\eta_i > 0$. It is an easy exercise to show that this implies that $\Pi_i = |\alpha_i\rangle\langle\alpha_i| \otimes 1$ if $\rho|_{\Pi_i\mathcal{H}_{AB}}$ is invertible. Thus $\rho = \sum_i \eta_i\rho_i$ is A-classical, in agreement with axiom (i).

These special cases help us to interpret Theorem 5 in the following way. The discordant states ρ are characterized by ensembles $\{\rho_i, \eta_i\}$ of non-orthogonal states, which are thereby not perfectly distinguishable for any orthonormal basis $\{|\alpha_i\rangle\}$ of the reference system.[25] This means that the transpose operation $\mathcal{R}_{\mathcal{M}_A^\Pi,\rho}$ transforms the ensemble of orthogonal states $\{\rho_{AB|i}, \eta_i\}$ into a non-orthogonal ensemble $\{\rho_i, \eta_i\}$. Furthermore, *the less distinguishable are the ρ_i for the optimal basis $\{|\alpha_i^{\rm opt}\rangle\}$, the most distant is ρ from the set of A-classical states, i.e., the most quantum-correlated is the state ρ.*

The states ρ for which the discrimination of the ensemble $\{\rho_i^{\rm opt}, \eta_i^{\rm opt}\}$ is the most difficult are the maximally entangled states. Actually, with the help of Theorem 5 one can show (see [81, 82]) that $D^{\rm G}_{\rm Bu}$ satisfies axiom (v) of Sect. 2.3, as already anticipated in Proposition 2. More precisely, one has

---

[25]Note that the entropic discord can also be interpreted in terms of state distinguishability, but for states of subsystem B. Actually, the measure of classical correlations $J_{B|A}(\rho)$ is the maximum over all orthonormal bases $\{|\alpha_i\rangle\}$ of the Holevo quantity $\chi(\{\rho_{B|i}, \eta_i\})$ (see (4) and the footnote after this equation). The latter is related to the problem of decoding a message encoded in the post-measurement states $\rho_{AB|i}$ when one has access to subsystem B only.

Corollary 2 *If $n_A \leq n_B$ then the maximal value of $D^{\rm G}_{\rm Bu}(\rho)$ is equal to $D^{\rm G}_{\rm max} = 2(1 - 1/\sqrt{n_A})$ and $D^{\rm G}_{\rm Bu}(\rho) = D^{\rm G}_{\rm max}$ if and only if ρ is a maximally entangled state.*

Proof of the value of $D^{\rm G}_{\rm max}$ One deduces from (85) and the bound $\mu_{\rm max} \geq 1/n$ (which follows from $\sum_{i=1}^{n} \mu_i = 1$) that for any pure state $|\Psi\rangle \in \mathcal{H}_{AB}$,

$$D^{\rm G}_{\rm Bu}(|\Psi\rangle) \leq 2\Big(1 - \frac{1}{\sqrt{n}}\Big) \quad , \quad n = \min\{n_A, n_B\} \,. \tag{93}$$

The inequality is saturated when $\mu_i = 1/n$ for any i, i.e., for the maximally entangled states. Assuming that $n_A \leq n_B$, since a measure of quantum correlations is maximal for pure maximally entangled states (Sect. 2.3), one has $D^{\rm G}_{\rm Bu}(\rho) \leq D^{\rm G}_{\rm max}$ for any state $\rho \in \mathcal{E}(\mathcal{H}_{AB})$. □

It is worth mentioning that finding the optimal measurement and success probability for discriminating an ensemble of $n_A > 2$ states is highly non-trivial and is still an open problem, even though it has been solved for particular ensembles.[26] However, the Helstrom formula [40] provides a celebrated solution for any ensemble with $n_A = 2$ states. Thus, as we shall see in the next subsection, Theorem 5 can be used to compute $D^{\rm G}_{\rm Bu}(\rho)$ when the reference subsystem A is a qubit. Despite our belief that this should not be hopeless, we have not succeeded so far to solve the discrimination problem for the ensemble given in (90) when $n_A > 2$.

5.4 Computability for Qubit-Qudit Systems

If subsystem A is a qubit then the ensemble $\{\rho_i, \eta_i\}$ in Theorem 5 contains only $n_A = 2$ states and the optimal probability and measurement to discriminate the ρ_i are easy to determine. One starts by writing the projector Π_1 as $1 - \Pi_0$ in the expression of the success probability,

$$P_{\rm S}^{\{\Pi_i\}}(\{\rho_i, \eta_i\}) = \eta_0 \,{\rm tr}\, \Pi_0 \rho_0 + \eta_1 \,{\rm tr}\, \Pi_1 \rho_1 = \frac{1}{2}\big(1 - {\rm tr}\, \Lambda\big) + {\rm tr}\, \Pi_0 \Lambda \tag{94}$$

with $\Lambda = \eta_0 \rho_0 - \eta_1 \rho_1$. The maximum of ${\rm tr}\, \Pi_0 \Lambda$ over all projectors Π_0 of rank n_B is achieved when Π_0 projects onto (the direct sum of) the eigenspaces associated to the n_B highest eigenvalues $\lambda_1 \geq \cdots \geq \lambda_{n_B}$ of the Hermitian matrix Λ. The maximal success probability is thus given by a variant of Helstrom's formula [40],

$$P_{\rm S}^{\rm opt\, v.N.}(\{\rho_i, \eta_i\}) = \frac{1}{2}\big(1 - {\rm tr}\, \Lambda\big) + \sum_{l=1}^{n_B} \lambda_l \,. \tag{95}$$

[26] In particular, if the states $\rho_i = U^{i-1} \rho_1 (U^{i-1})^\dagger$ are related between themselves through conjugations by powers of a single unitary operator U satisfying $U^m = +1$, one can show that the least square measurement is optimal [9, 10, 23, 29].

For the states ρ_i associated to the orthonormal basis $\{|\alpha_i\rangle\}_{i=0}^1$ of $\mathbb{C}^2$ via formula (90), one has $\Lambda = \sqrt{\rho}\,(|\alpha_0\rangle\langle\alpha_0| - |\alpha_1\rangle\langle\alpha_1|) \otimes 1\,\sqrt{\rho}$. The operator inside the parenthesis is equal to $\sigma_{\vec{u}} = \vec{u}\cdot\vec{\sigma}$ for some unit vector $\vec{u} \in \mathbb{R}^3$ depending on $\{|\alpha_i\rangle\}$ (here $\vec{\sigma}$ is the vector formed by the three Pauli matrices). Conversely, one can associate to any unit vector $\vec{u} \in \mathbb{R}^3$ the eigenbasis $\{|\alpha_i\rangle\}_{i=0}^1$ of $\sigma_{\vec{u}}$. Thus, according to Theorem 5, $F(\rho, \mathcal{C}_A)$ is obtained by maximizing the r.h.s. of (95) over all Hermitian matrices

$$\Lambda(\vec{u}) = \sqrt{\rho}\,\sigma_{\vec{u}} \otimes 1\,\sqrt{\rho} \tag{96}$$

with $\vec{u} \in \mathbb{R}^3$, $\|\vec{u}\| = 1$. One can show [81] that $\Lambda(\vec{u})$ has at most n_B positive eigenvalues $\lambda_l(\vec{u}) > 0$ and at most n_B negative eigenvalues $\lambda_l(\vec{u}) < 0$, counting multiplicities. This yields to the following formula, which shows that the computation of $D_{\rm Bu}^{\rm G}(\rho)$ for qubit-qudit states reduces to an optimization problem of a trace norm.

Corollary 3 ([81]) *If A is a qubit ($n_A = 2$) and B is an arbitrary system with a n_B-dimensional Hilbert space (qudit), the fidelity between ρ and the set of A-classical states is given by*

$$F(\rho, \mathcal{C}_A) = \frac{1}{2} \max_{\|\vec{u}\|=1} \left\{1 + \|\Lambda(\vec{u})\|_1\right\}, \tag{97}$$

where $\Lambda(\vec{u})$ is the $2n_B \times 2n_B$ matrix defined in (96).

One can also conclude from the arguments above that the closest A-classical state(s) to ρ is (are) given by (91) where $\Pi_0^{\rm opt}$ is a spectral projector associated to the n_B largest eigenvalues of $\Lambda(\vec{u}^{\,\rm opt})$ and $\vec{u}^{\,\rm opt} \in \mathbb{R}^3$ is a unit vector achieving the maximum in (97). Using Corollary 3, an analytical expression for $D_{\rm Bu}^{\rm G}(\rho)$ can be derived for Bell-diagonal two-qubit states ρ, and the closest A-classical states to such Bell-diagonal states can be determined explicitly [83]. The same result for $D_{\rm Bu}^{\rm G}(\rho)$ has been found independently in Ref. [1] by another method. Analytical expressions for the geometric total and classical correlations $I_{AB}^{\rm G}(\rho)$ and $C_A^{\rm G}(\rho)$ for Bell-diagonal two-qubit states ρ have been obtained in Ref. [18].

The properties of the Bures geometric discord established in this section are summarized in the second column of Table 1.

5.5 *Proof of Theorem 5*

To establish Theorem 5, we rely on a slightly more general statement summarized in the following lemma.

Lemma 1 *For a fixed family $\{\sigma_{A|i}\}_{i=1}^n$ of states $\sigma_{A|i} \in \mathcal{E}(\mathcal{H}_A)$ having orthogonal supports and spanning $\mathcal{H}_A$, with $1 \leq n \leq n_A$, let us define*

Table 1 Properties of the geometric discords with the Bures, Hellinger, trace, and Hilbert–Schmidt distances. Here n_A is the Hilbert space dimension of the reference subsystem A, $\mu_{\max} = \max\{\mu_i\}$ is the maximal Schmidt coefficient, and $K = (\sum_i \mu_i^2)^{-1}$ is the Schmidt number of a pure state. The question marks "?" indicate unsolved problems. The results quoted in this table have been obtained in Refs. [1, 24, 25, 78, 82, 83]. The table is taken from [78]

Distance	Geometric discord D^{G}			
	Bures	Hellinger	Trace	Hilbert Schmidt
Bona fide measure of quantum correlations	✓	✓	Proved for $n_A = 2$	No
Satisfies axiom (v)	✓	Proved for $n_A = 2, 3$	Proved for $n_A = 2$	
Maximal value if $n_A \le n_B$	$2 - 2/\sqrt{n_A}$	$2 - 2/\sqrt{n_A}$	1 for $n_A = 2$	
Value for pure states	$2 - 2\sqrt{\mu_{\max}}$	$2 - 2K^{-\frac{1}{2}}$	?	$1 - K^{-1}$
Relations and cross inequalities	$2 - 2\sqrt{1 - D^{\mathrm{G}}_{\mathrm{He}}(\rho)/2} \le D^{\mathrm{G}}_{\mathrm{Bu}}(\rho) \le D^{\mathrm{G}}_{\mathrm{He}}(\rho) = 2 - 2\sqrt{1 - D^{\mathrm{G}}_{\mathrm{HS}}(\sqrt{\rho})}$			
Computability for two qubits	Bell-diagonal states	All states	{ X-states; *B*-classical states	All states

$$\mathcal{C}_A(\{\sigma_{A|i}\}) = \left\{\sigma = \sum_{i=1}^{n} q_i \sigma_{A|i} \otimes \sigma_{B|i}\ ;\ \{q_i, \sigma_{B|i}\}_{i=1}^{n} \text{ is a state ensemble on } \mathcal{H}_B \right\}. \tag{98}$$

Then

$$F\big(\rho, \mathcal{C}_A(\{\sigma_{A|i}\})\big) \equiv \max_{\sigma \in \mathcal{C}_A(\{\sigma_{A|i}\})} \big\{F(\rho,\sigma)\big\} = \max_U \left\{ \sum_{i=1}^{n} \|W_i(U)\|_2^2 \right\}, \tag{99}$$

where the last maximum is over all unitaries U on $\mathcal{H}_{AB}$, $\|\cdot\|_2$ is the Hilbert–Schmidt norm, and

$$W_i(U) = \operatorname{tr}_A\big(\sqrt{\sigma_{A|i}} \otimes 1\, \sqrt{\rho}\, U\big)\,. \tag{100}$$

Moreover, there exists a unitary U_{opt} achieving the maximum in (99) which satisfies $W_i(U_{\mathrm{opt}}) \ge 0$. The states σ_{opt} satisfying $F(\rho, \sigma_{\mathrm{opt}}) = F(\rho, \mathcal{C}_A(\{\sigma_{A|i}\}))$ are given in terms of this unitary by

$$\sigma_{\mathrm{opt}} = \frac{1}{F(\rho, \mathcal{C}_A(\{\sigma_{A|i}\}))} \sum_{i=1}^{n} \sigma_{A|i} \otimes W_i(U_{\mathrm{opt}})^2\,. \tag{101}$$

Proof Using the spectral decompositions of the states $\sigma_{B|i}$, any $\sigma \in \mathcal{C}_A(\{\sigma_{A|i}\})$ can be written as

$$\sigma = \sum_{i=1}^{n} \sum_{j=1}^{n_B} q_{ij} \sigma_{A|i} \otimes |\beta_{j|i}\rangle\langle\beta_{j|i}| \quad \text{with} \quad q_{ij} \geq 0\,, \; \sum_{ij} q_{ij} = 1\,, \tag{102}$$

where $\{|\beta_{j|i}\rangle\}_{j=1}^{n_B}$ is an orthonormal basis of $\mathcal{H}_B$ for any i. By assumption, if $i \neq i'$ then $\operatorname{ran} \sigma_{A|i} \perp \operatorname{ran} \sigma_{A|i'}$, so that $\sqrt{\sigma} = \sum_{i,j} \sqrt{q_{ij}} \sqrt{\sigma_{A|i}} \otimes |\beta_{j|i}\rangle\langle\beta_{j|i}|$. We start by evaluating the trace norm in the definition (17) of the fidelity by means of the formula $\|O\|_1 = \max_U |\operatorname{tr} UO|$ to obtain

$$\begin{aligned} F\big(\rho, \mathcal{C}_A(\{\sigma_{A|i}\})\big) &= \max_{\sigma \in \mathcal{C}_A(\{\sigma_{A|i}\})} \max_U \Big\{ \big|\operatorname{tr} U^\dagger \sqrt{\rho}\sqrt{\sigma}\big|^2 \Big\} \\ &= \max_U \Bigg\{ \max_{\{q_{ij}\},\{|\beta_{j|i}\rangle\}} \Bigg| \sum_{i,j} \sqrt{q_{ij}} \langle\beta_{j|i}| W_i(U)^\dagger |\beta_{j|i}\rangle \Bigg|^2 \Bigg\} \,. \end{aligned} \tag{103}$$

The square modulus can be bounded by invoking twice the Cauchy–Schwarz inequality and $\sum_{ij} q_{ij} = 1$,

$$\begin{aligned} \Bigg| \sum_{i,j} \sqrt{q_{ij}} \langle\beta_{j|i}| W_i(U)^\dagger |\beta_{j|i}\rangle \Bigg|^2 &\leq \sum_{i,j} \big| \langle\beta_{j|i}| W_i(U)^\dagger |\beta_{j|i}\rangle \big|^2 \\ &\leq \sum_{i,j} \big\| W_i(U) |\beta_{j|i}\rangle \big\|^2 = \sum_i \| W_i(U) \|_2^2 \,. \end{aligned} \tag{104}$$

The foregoing inequalities are equalities if the following conditions are satisfied:

(1) $W_i(U) = W_i(U)^\dagger \geq 0$;
(2) $q_{ij} = \langle\beta_{j|i}| W_i(U) |\beta_{j|i}\rangle^2 / (\sum_{i,j} \langle\beta_{j|i}| W_i(U) |\beta_{j|i}\rangle^2)$;
(3) $\{|\beta_{j|i}\rangle\}_{j=1}^{n_B}$ is an eigenbasis of $W_i(U)$ for any i.

Therefore, (99) holds true provided that there is a unitary U on $\mathcal{H}_{AB}$ satisfying (1). For a given U, let us define $U_{\text{opt}} = U \sum_i \Pi_i^A \otimes V_i^\dagger$, where Π_i^A is the projector onto $\operatorname{ran} \sigma_{A|i}$ and V_i is a unitary on $\mathcal{H}_B$ such that $W_i(U) = |W_i(U)^\dagger| V_i$ (polar decomposition). Then U_{opt} is unitary since by hypothesis $\Pi_i^A \Pi_{i'}^A = \delta_{ii'} \Pi_i^A$ and $\sum_i \Pi_i^A = 1$. Furthermore, one readily shows that $W_i(U_{\text{opt}}) = W_i(U_{\text{opt}})^\dagger = |W_i(U)^\dagger| \geq 0$. As $\sum_i \|W_i(U)\|_2^2 = \sum_i \|W_i(U_{\text{opt}})\|_2^2$, the identity (99) follows from (103) and (104). From condition (3) one has $W_i(U_{\text{opt}}) |\beta_{j|i}^{\text{opt}}\rangle = w_{ji} |\beta_{j|i}^{\text{opt}}\rangle$ with $\sum_{i,j} w_{ji}^2 = F(\rho, \mathcal{C}_A(\{\sigma_{A|i}\}))$, see (104). Condition (2) entails

$$\sum_j q_{ij}^{\text{opt}} |\beta_{j|i}^{\text{opt}}\rangle\langle\beta_{j|i}^{\text{opt}}| = \frac{W_i(U_{\text{opt}})^2}{F(\rho, \mathcal{C}_A(\{\sigma_{A|i}\}))} \,, \tag{105}$$

which together with (102) leads to (101). □

Proof of Theorem 5. Let $\{|\alpha_i\rangle\}_{i=1}^{n_A}$ be an orthonormal basis of $\mathcal{H}_A$. Applying Lemma 1 with $\sigma_{A|i} = |\alpha_i\rangle\langle\alpha_i|$ one gets

$$F\big(\rho, \mathcal{C}_A(\{|\alpha_i\rangle\})\big) = \max_U \left\{ \sum_{i=1}^{n_A} \operatorname{tr} U|\alpha_i\rangle\langle\alpha_i| \otimes 1\, U^\dagger \sqrt{\rho}\, |\alpha_i\rangle\langle\alpha_i| \otimes 1 \sqrt{\rho} \right\} ,$$
$$= \max_{\{\Pi_i\}} \left\{ \sum_{i=1}^{n_A} \operatorname{tr} \Pi_i \sqrt{\rho} |\alpha_i\rangle\langle\alpha_i| \otimes 1 \sqrt{\rho} \right\} = P_{\mathrm{S}}^{\mathrm{opt\,v.N.}}(\{\rho_i, \eta_i\}) . \tag{106}$$

The last maximum is over all orthonormal families $\{\Pi_i\}_{i=1}^{n_A}$ of projectors of rank n_B and the success probability $P_{\mathrm{S}}^{\mathrm{opt\,v.N.}}(\{\rho_i, \eta_i\})$ is given by (88). Since the fidelity $F(\rho, \mathcal{C}_A)$ is the maximum of $F(\rho, \mathcal{C}_A(\{|\alpha_i\rangle\}))$ over all bases $\{|\alpha_i\rangle\}$, this leads to (89) and (91). □

6 Hellinger Geometric Discord

In this section we study the geometric discord for the Hellinger distance, given by (see (16) and (19))

$$D_{\mathrm{He}}^{\mathrm{G}}(\rho) = 2 - 2 \max_{\sigma_{A-\mathrm{cl}} \in \mathcal{C}_A} \operatorname{tr} \sqrt{\rho}\sqrt{\sigma_{A-\mathrm{cl}}} . \tag{107}$$

6.1 *Values for Pure States, General Expression, and Closest A-Classical States*

Theorem 6 ([78])

(a) *If $|\Psi\rangle \in \mathcal{H}_{AB}$ is a pure state, then*

$$D_{\mathrm{He}}^{\mathrm{G}}(|\Psi\rangle) = 2 - 2K(|\Psi\rangle)^{-\frac{1}{2}} , \tag{108}$$

where $K(|\Psi\rangle) = (\sum_i \mu_i^2)^{-1}$ is the Schmidt number of $|\Psi\rangle$. Furthermore, the closest A-classical state to $|\Psi\rangle$ for the Hellinger distance is the classical state

$$\sigma_{\mathrm{He},\Psi} = K(|\Psi\rangle) \sum_{i=1}^{n} \mu_i^2 |\varphi_i\rangle\langle\varphi_i| \otimes |\chi_i\rangle\langle\chi_i| , \tag{109}$$

where $|\varphi_i\rangle$ and $|\chi_i\rangle$ are the eigenvectors of $[\rho_\Psi]_A$ and $[\rho_\Psi]_B$ in the Schmidt decomposition (82).

(b) *If ρ is a mixed state, then*

$$D_{\rm He}^{\rm G}(\rho) = 2 - 2 \max_{\{|\alpha_i\rangle\}} \left\{ \sum_{i=1}^{n_A} \operatorname{tr}_B[\langle \alpha_i | \sqrt{\rho} | \alpha_i \rangle^2] \right\}^{\frac{1}{2}} = 2 - 2 \max_{\{|\alpha_i\rangle\}} \sqrt{P_{\rm S}^{\rm lsm}(\{\rho_i, \eta_i\})} \,, \tag{110}$$

where the maximum is over all orthonormal bases $\{|\alpha_i\rangle\}$ *for A and* $P_{\rm S}^{\rm lsm}(\{\rho_i, \eta_i\})$ *is the success probability in discriminating the ensemble* $\{\rho_i, \eta_i\}$ *defined in (90) by the least–square measurement. Let the maxima in (110) be reached for the basis* $\{|\alpha_i^{\rm opt}\rangle\}$. *Then the closest A-classical state(s) to* ρ *for the Hellinger distance is (are)*

$$\sigma_{{\rm He},\rho} = \left(1 - \frac{D_{\rm He}^{\rm G}(\rho)}{2}\right)^{-2} \sum_{i=1}^{n_A} |\alpha_i^{\rm opt}\rangle\langle\alpha_i^{\rm opt}| \otimes \langle\alpha_i^{\rm opt}|\sqrt{\rho}|\alpha_i^{\rm opt}\rangle^2 \,. \tag{111}$$

As the Schmidt number $K(|\Psi\rangle)$ is an entanglement monotone, one infers from (a) that $D_{\rm He}^{\rm G}$ satisfies Axiom (iv) of Definition 1 and is thus a *bona fide* measure of quantum correlations, as claimed in Proposition 2. Moreover, if $n_A \leq n_B$ then $D_{\rm He}^{\rm G}$ has the same maximal value $D_{\rm max}^{\rm G} = 2 - 2/\sqrt{n_A}$ as the Bures geometric discord (in fact, $D_{\rm He}^{\rm G}(\rho)$ is maximum for maximally entangled pure states which have Schmidt numbers equal to n_A).

Proof Let us first prove part (b) of the theorem. By using the spectral decompositions of the states $\rho_{B|i}$ in (9), any A-classical state can be written as

$$\sigma_{A-{\rm cl}} = \sum_{i=1}^{n_A} \sum_{j=1}^{n_B} q_{ij} |\alpha_i\rangle\langle\alpha_i| \otimes |\beta_{j|i}\rangle\langle\beta_{j|i}| \,, \tag{112}$$

where $\{q_{ij}\}$ is a probability distribution, $\{|\alpha_i\rangle\}_{i=1}^{n_A}$ is an orthonormal basis for A and, for any i, $\{|\beta_{j|i}\rangle\}_{j=1}^{n_B}$ is an orthonormal basis for B (note that the $|\beta_{j|i}\rangle$ need not be orthogonal for distinct i's). The square root of $\sigma_{A-{\rm cl}}$ is obtained by replacing q_{ij} by $\sqrt{q_{ij}}$ in the r.h.s. of (112). Hence, in the same way as in the proof of Sect. 5.5,

$$\operatorname{tr} \sqrt{\rho}\sqrt{\sigma_{A-{\rm cl}}} = \sum_{i,j} \sqrt{q_{ij}} \langle \alpha_i \otimes \beta_{j|i} | \sqrt{\rho} | \alpha_i \otimes \beta_{j|i} \rangle \leq \Big(\sum_{i,j} \langle \alpha_i \otimes \beta_{j|i} | \sqrt{\rho} | \alpha_i \otimes \beta_{j|i} \rangle^2 \Big)^{\frac{1}{2}} . \tag{113}$$

The last bound follows from the Cauchy–Schwarz inequality and the identity $\sum_{i,j} q_{ij} = 1$. It is saturated when

$$q_{ij} = \frac{\langle \alpha_i \otimes \beta_{j|i} | \sqrt{\rho} | \alpha_i \otimes \beta_{j|i} \rangle^2}{\sum_{i,j} \langle \alpha_i \otimes \beta_{j|i} | \sqrt{\rho} | \alpha_i \otimes \beta_{j|i} \rangle^2} \,. \tag{114}$$

Therefore,

$$\max_{\{q_{ij}\}} \operatorname{tr} \sqrt{\rho}\sqrt{\sigma_{A-{\rm cl}}} = \Big(\sum_{i,j} \langle \beta_{j|i} | B_i | \beta_{j|i} \rangle^2 \Big)^{\frac{1}{2}} \tag{115}$$

with $B_i = \langle \alpha_i | \sqrt{\rho} | \alpha_i \rangle \in \mathcal{B}(\mathcal{H}_B)_{\rm s.a.}$. Now, for any fixed i, one has

$$\sum_j \langle \beta_{j|i} | B_i | \beta_{j|i} \rangle^2 \leq \operatorname{tr}[B_i^2] \,. \tag{116}$$

This inequality is saturated when $\{|\beta_{j|i}\rangle\}$ is an eigenbasis of B_i. Since maximizing over all A-classical states in (107) amounts to maximize over all $\{q_{ij}\}$, $\{|\alpha_i\rangle\}$, and $\{|\beta_{j|i}\rangle\}$, this gives

$$\left(1 - \frac{D_{\rm He}^{\rm G}(\rho)}{2}\right)^2 = \max_{\{|\alpha_i\rangle\}} \sum_{i=1}^{n_A} \operatorname{tr}_B \left[\langle \alpha_i | \sqrt{\rho} | \alpha_i \rangle^2 \right] . \tag{117}$$

It has been observed in Sect. 5.2 that the least square measurement for the ensemble $\{\rho_i, \eta_i\}$ defined in Theorem 5 is the projective measurement $\{|\alpha_i\rangle\langle\alpha_i| \otimes 1\}_{i=1}^{n_A}$. Thus

$$P_{\rm S}^{\rm lsm}(\{\rho_i, \eta_i\}) = \sum_{i=1}^{n_A} \eta_i \operatorname{tr} \rho_i |\alpha_i\rangle\langle\alpha_i| \otimes 1 = \sum_{i=1}^{n_A} \operatorname{tr}_B \langle \alpha_i | \sqrt{\rho} | \alpha_i \rangle^2 . \tag{118}$$

Equation (110) follows from (117) and (118). The closest A-classical state is given by (112) in which $|\alpha_i\rangle = |\alpha_i^{\rm opt}\rangle$ are the vectors realizing the maximum in (117), $|\beta_{j|i}\rangle = |\beta_{j|i}^{\rm opt}\rangle$ are the eigenvectors of $B_i^{\rm opt} = \langle \alpha_i^{\rm opt} | \sqrt{\rho} | \alpha_i^{\rm opt} \rangle$, and (see (114)):

$$q_{ij} = \frac{\langle \beta_{j|i}^{\rm opt} | (B_i^{\rm opt})^2 | \beta_{j|i}^{\rm opt} \rangle}{\sum_i \operatorname{tr}(B_i^{\rm opt})^2} . \tag{119}$$

The expression (111) readily follows.

We now establish part (a) of the theorem. Let $\rho = |\Psi\rangle\langle\Psi|$ be a pure state with reduced state $\rho_A = \operatorname{tr}_B |\Psi\rangle\langle\Psi|$. Then $B_i = |\beta_i\rangle\langle\beta_i|$, where $|\beta_i\rangle = \langle\alpha_i|\Psi\rangle$ has square norm $\|\beta_i\|^2 = \langle\alpha_i|\rho_A|\alpha_i\rangle$. Thus (117) yields

$$\left(1 - \frac{D_{\rm He}^{\rm G}(|\Psi\rangle)}{2}\right)^2 = \max_{\{|\alpha_i\rangle\}} \sum_{i=1}^{n_A} \langle \alpha_i | \rho_A | \alpha_i \rangle^2 . \tag{120}$$

In analogy with (116), the sum in the r.h.s. is bounded from above by $\operatorname{tr} \rho_A^2 = K(|\Psi\rangle)^{-1}$, the bound being saturated when $\{|\alpha_i\rangle\}$ is an eigenbasis of ρ_A. This leads to (108). The closest A-classical state to $|\Psi\rangle$ is given by (111) with $|\alpha_i^{\rm opt}\rangle = |\varphi_i\rangle$, which gives (109). □

6.2 Link with the Hilbert–Schmidt Geometric Discord

In view of the definition $d_{\rm He}(\rho,\sigma) = d_2(\sqrt{\rho},\sqrt{\sigma})$ of the Hellinger distance, it should not come as a surprise that $D^{\rm G}_{\rm He}(\rho)$ is related to the Hilbert–Schmidt geometric discord $D^{\rm G}_{\rm HS}(\sqrt{\rho})$ of the square root of ρ.

Proposition 11 ([78]) *For any $\rho \in \mathcal{E}(\mathcal{H}_{AB})$, one has*

$$D^{\rm G}_{\rm He}(\rho) = 2 - 2\big(1 - D^{\rm G}_{\rm HS}(\sqrt{\rho})\big)^{\frac{1}{2}} . \tag{121}$$

Note that the Hilbert–Schmidt geometric discord is evaluated for the square root of ρ, which is not a state but is nevertheless a non-negative operator. Thus $\sigma = \sqrt{\rho}\,/\,{\rm tr}\,\sqrt{\rho}$ is a density operator and $D^{\rm G}_{\rm HS}(\sqrt{\rho})$ is defined as $D^{\rm G}_{\rm HS}(\sqrt{\rho}) \equiv ({\rm tr}\,\sqrt{\rho})^2 D^{\rm G}_{\rm HS}(\sigma)$.

Proof The following expression of $D^{\rm G}_{\rm HS}(\rho)$ has been found by Luo and Fu [55]:

$$D^{\rm G}_{\rm HS}(\rho) = {\rm tr}\,\rho^2 - \max_{\{|\alpha_i\rangle\}} \sum_{i=1}^{n_A} {\rm tr}_B \langle\alpha_i|\rho|\alpha_i\rangle^2 = \min_{\{|\alpha_i\rangle\}} \sum_{i\neq j}^{n_A} {\rm tr}_B\, |\langle\alpha_i|\rho|\alpha_j\rangle|^2 . \tag{122}$$

For completeness, let us give a simple derivation of (122). By definition,

$$D^{\rm G}_{\rm HS}(\rho) = \min_{\sigma_{A-{\rm cl}}\in\mathcal{C}_A} \|\rho - \sigma_{A-{\rm cl}}\|_2^2 = {\rm tr}\,\rho^2 + \min_{\sigma_{A-{\rm cl}}\in\mathcal{C}_A} {\rm tr}(\sigma_{A-{\rm cl}}^2 - 2\rho\sigma_{A-{\rm cl}}) . \tag{123}$$

Thanks to (112), the last trace is equal to

$$\sum_{i,j} \Big\{ \big(q_{ij} - \langle\alpha_i\otimes\beta_{j|i}|\rho|\alpha_i\otimes\beta_{j|i}\rangle\big)^2 - \langle\alpha_i\otimes\beta_{j|i}|\rho|\alpha_i\otimes\beta_{j|i}\rangle^2 \Big\} . \tag{124}$$

The minimum over the probability distribution $\{q_{ij}\}$ is obviously achieved for $q_{ij} = \langle\alpha_i\otimes\beta_{j|i}|\rho|\alpha_i\otimes\beta_{j|i}\rangle$. Minimizing also over the orthonormal bases $\{|\alpha_i\rangle\}$ and $\{|\beta_{j|i}\rangle\}$ and using (116) again, one finds the first equality in (122). The second equality follows from the relation ${\rm tr}\,\rho^2 = \sum_{i,j} {\rm tr}_B\,|\langle\alpha_i|\rho|\alpha_j\rangle|^2$. The result of Proposition 11 is now obtained by comparing (110) and (122). □

Remark 1 By using similar arguments as in the proof of Theorem 6, one finds that the closest A-classical state to ρ for the Hilbert–Schmidt distance coincides with the post-measurement state $\mathcal{M}_A^\Pi \otimes 1(\rho)$, where $\mathcal{M}_A^\Pi$ is the quantum operation (6) associated to a measurement on A with projectors $\Pi_i^A = |\alpha_i^{\rm opt}\rangle\langle\alpha_i^{\rm opt}|$, $\{|\alpha_i^{\rm opt}\rangle\}$ being the orthonormal basis maximizing the first sum in (122). Therefore, as already observed in Ref. [55], for the Hilbert–Schmidt distance the geometric and measurement-induced geometric discords are equal, $D^{\rm G}_{\rm HS} = D^{\rm M}_{\rm HS}$. Furthermore, the known value $D^{\rm G}_{\rm HS}(|\Psi\rangle) = 1 - K(|\Psi\rangle)^{-1}$ for pure states [22] is recovered by noting that (121) implies $D^{\rm G}_{\rm He}(|\Psi\rangle) = 2 - 2(1 - D^{\rm G}_{\rm HS}(|\Psi\rangle))^{\frac{1}{2}}$ and by comparing with (108).

Table 2 Properties of the measurement-induced geometric discords with the Bures, Hellinger, trace, and Hilbert–Schmidt distances. The function g is given by (126). The remaining notations are the same as in the caption of Table 1. The results quoted in this table have been obtained in Refs. [24, 25, 55, 63, 75] and [78]. This table is taken from [78]

Distance	Measurement-induced geometric discord D^{M}			
	Bures	Hellinger	Trace	Hilbert-Schmidt
Bona fide measure of quantum correlations	✓	✓	✓	No
Satisfies axiom (v)	✓	For $n_A = 2$ (conjecture)	Proved for $n_A = 2$	
Maximal value if $n_A \leq n_B$	$2 - 2/\sqrt{n_A}$	$2 - 2/\sqrt{n_A}$	$(2 - 2/n_A)^2$	
Value for pure states	$2 - 2K^{-\frac{1}{2}}$	$2 - 2\sum_i \mu_i^{\frac{3}{2}}$	See Theorem 3.3 in [75]	$1 - K^{-1}$
Comparison with the geometric discord	$D_{\mathrm{Bu}}^{\mathrm{G}} \leq D_{\mathrm{Bu}}^{\mathrm{M}} \leq g(D_{\mathrm{Bu}}^{\mathrm{G}})$	$D_{\mathrm{He}}^{\mathrm{G}} \leq D_{\mathrm{He}}^{\mathrm{M}} \leq g(D_{\mathrm{He}}^{\mathrm{G}})$	$\begin{cases} D_{\mathrm{tr}}^{\mathrm{M}} = D_{\mathrm{tr}}^{\mathrm{G}} & \text{for } n_A = 2 \\ D_{\mathrm{tr}}^{\mathrm{M}} \geq D_{\mathrm{tr}}^{\mathrm{G}} & \text{for } n_A > 2 \end{cases}$	$D_{\mathrm{HS}}^{\mathrm{M}} = D_{\mathrm{HS}}^{\mathrm{G}}$
Computability for two qubits	?	?	{ X-states *B*-classical states	All states

6.3 *Comparison Between the Bures and Hellinger Geometric Discords*

As pointed out in Sect. 3.5, the Bures and Hellinger geometric discords are not functions of each other and thereby define different orderings on $\mathcal{E}(\mathcal{H}_{AB})$. A large number of inequalities enabling to compare D^{G}, D^{M}, and D^{R} for the Bures, Hellinger, trace, and Hilbert Schmidt distances have been established in Ref. [78] (some of these inequalities are given in Tables 1, 2 and 3). A particular bound is as follows.

Proposition 12 ([78]) *The Bures and Hellinger geometric discords satisfy*

$$g^{-1}(D_{\mathrm{He}}^{\mathrm{G}}(\rho)) \leq D_{\mathrm{Bu}}^{\mathrm{G}}(\rho) \leq D_{\mathrm{He}}^{\mathrm{G}}(\rho) \, , \tag{125}$$

where the increasing function $g(d)$ and its inverse are defined by

$$g(d) = 2d - \frac{1}{2}d^2 \quad , \quad g^{-1}(d) = 2 - 2\sqrt{1 - d/2} \, . \tag{126}$$

If A is a qubit, the stronger bound $D_{\mathrm{He}}^{\mathrm{G}}(\rho) \leq g^{-1} \circ h(D_{\mathrm{Bu}}^{\mathrm{G}}(\rho))$ holds and is saturated for pure states, with $h(d) = 2g(d) - g(d)^2$.

Table 3 Properties of the discords of response with the Bures, Hellinger, trace, and Hilbert-Schmidt distances. Here E^{R} is the entanglement of response [33, 61] and the function g is given by (126). Inequalities denoted by the symbol $\precsim$ instead of $\leq$ are saturated for pure states. The remaining notations are the same as in the caption of Table 1. The results quoted in this table have been obtained in Refs. [76–78]. This table is taken from [78]

Distance		Discord of response D^{R}			
		Bures	Hellinger	Trace	Hilbert Schmidt
Bona fide measure of quantum correlations		✓	✓	✓	No
Satisfies axiom (v)		✓	✓	✓	No if $n_B \geq 2n_A$
Maximal value if $n_A \leq n_B$		1	1	1	1
Value for pure states		$1-\sqrt{1-E^{\mathrm{R}}}$	E^{R}	E^{R}	E^{R}
Functional relation with D^{G}	$n_A = 2$	$D^{\mathrm{R}}_{\mathrm{Bu}} = g(D^{\mathrm{G}}_{\mathrm{Bu}})$	$D^{\mathrm{R}}_{\mathrm{He}} = g(D^{\mathrm{G}}_{\mathrm{He}})$	$D^{\mathrm{R}}_{\mathrm{tr}} = D^{\mathrm{G}}_{\mathrm{tr}}$	$D^{\mathrm{R}}_{\mathrm{HS}} = 2D^{\mathrm{G}}_{\mathrm{HS}}$
	$n_A = 3$	No	$D^{\mathrm{R}}_{\mathrm{He}} = \frac{3}{4}g(D^{\mathrm{G}}_{\mathrm{He}})$	No	$D^{\mathrm{R}}_{\mathrm{HS}} = \frac{3}{2}D^{\mathrm{G}}_{\mathrm{HS}}$
	$n_A > 3$		No		No
Comparison with D^{G} and D^{M}	$n_A = 2$	$D^{\mathrm{M}}_{\mathrm{Bu}} \precsim 2-\sqrt{2}\sqrt{1+(1-D^{\mathrm{R}}_{\mathrm{Bu}})^2}$	$\sin^2\left(\frac{\pi}{n_A}\right)g(D^{\mathrm{G}}_{\mathrm{He}}) \leq D^{\mathrm{R}}_{\mathrm{He}} \leq g(D^{\mathrm{G}}_{\mathrm{He}})$	$D^{\mathrm{R}}_{\mathrm{tr}} = D^{\mathrm{M}}_{\mathrm{tr}} = D^{\mathrm{G}}_{\mathrm{tr}}$	$D^{\mathrm{R}}_{\mathrm{HS}} = 2D^{\mathrm{M}}_{\mathrm{HS}}$
	$n_A = 3$	$D^{\mathrm{M}}_{\mathrm{Bu}} \precsim 2-\frac{2}{\sqrt{3}}\sqrt{1+2(1-D^{\mathrm{R}}_{\mathrm{Bu}})^2}$		$\frac{1}{n_A n_B}\sin^2\left(\frac{\pi}{n_A}\right)D^{\mathrm{G}}_{\mathrm{tr}} \leq D^{\mathrm{R}}_{\mathrm{tr}} \leq n_A n_B D^{\mathrm{G}}_{\mathrm{tr}}$	$D^{\mathrm{R}}_{\mathrm{HS}} = \frac{3}{2}D^{\mathrm{M}}_{\mathrm{HS}}$
	$n_A > 3$	$D^{\mathrm{M}}_{\mathrm{Bu}} \leq 2-\frac{2}{\sqrt{n_A}}(1-D^{\mathrm{R}}_{\mathrm{Bu}})$			$2\sin^2\left(\frac{\pi}{n_A}\right)D^{\mathrm{G}}_{\mathrm{HS}} \leq D^{\mathrm{R}}_{\mathrm{HS}} \leq 2D^{\mathrm{G}}_{\mathrm{HS}}$
Cross inequalities and relations		$D^{\mathrm{R}}_{\mathrm{Bu}} \leq D^{\mathrm{R}}_{\mathrm{He}} \precsim 1-(1-D^{\mathrm{R}}_{\mathrm{Bu}})^2$, $(D^{\mathrm{R}}_{\mathrm{He}})^2 \leq D^{\mathrm{R}}_{\mathrm{tr}} \leq 1-(1-D^{\mathrm{R}}_{\mathrm{Bu}})^2$, $D^{\mathrm{R}}_{\mathrm{He}}(\rho) = D^{\mathrm{R}}_{\mathrm{HS}}(\sqrt{\rho})$			
Computability for two qubits		Bell-diagonal states	All states	X-states *B*-classical states	All states

Proof The first statement is a consequence of Theorem 5 and of an upper bound on the probability of success in quantum state discrimination due to Barnum and Knill [11]. According to such bound, the maximum probability of success $P_{\rm S}^{\rm opt\,v.N.}(\{\rho_i,\eta_i\})$ is at most equal to the square root of the probability of success $P_{\rm S}^{\rm lsm}(\{\rho_i,\eta_i\})$ obtained by discriminating the states ρ_i with the least–square measurement. Hence

$$\max_{\{|\alpha_i\rangle\}} P_{\rm S}^{\rm lsm}(\{\rho_i,\eta_i\}) \leq F(\rho,\mathcal{C}_A) \leq \max_{\{|\alpha_i\rangle\}} P_{\rm S}^{\rm lsm}(\{\rho_i,\eta_i\})^{\frac{1}{2}} \,. \tag{127}$$

The second inequality together with (81) and (110) yields to the first bound in (125). The second bound in (125) is an immediate consequence of the fact that the Bures distance is always smaller or equal to the Hellinger distance[27] (Proposition 9). The stronger bound when A is a qubit follows from the inequality $D_{\rm He}^{\rm R}(\rho) \leq 1-(1-D_{\rm Bu}^{\rm R}(\rho))^2$ on the discords of response[28] and from the identities $D_{\rm Bu}^{\rm R}(\rho) = g(D_{\rm Bu}^{\rm G}(\rho))$ and $D_{\rm He}^{\rm R}(\rho) = g(D_{\rm He}^{\rm G}(\rho))$, see Table 3 and Ref. [78]. □

Proposition 9 also yields bounds on $D_{\rm He}^{\rm G}$ and $D_{\rm Bu}^{\rm G}$ in terms of the trace geometric discord $D_{\rm tr}^{\rm G}$:

$$[D_{\rm He}^{\rm G}(\rho)]^2 \leq D_{\rm tr}^{\rm G}(\rho) \leq 2g(D_{\rm Bu}^{\rm G}(\rho)) \,. \tag{128}$$

Similar bounds hold for the measurement-induced geometric discord $D^{\rm M}$ and discord of response $D^{\rm R}$ (but one has to take care of the different normalization factors in the definition of $D^{\rm R}$, see Sect. 3.3).

6.4 *Computability for Qubit-Qudit Systems*

We show in this subsection that the Hellinger geometric discord is an easily computable quantity, at least when A is a qubit. For indeed, we will determine with the help of (110) an explicit expression for $D_{\rm He}^{\rm G}(\rho)$ for arbitrary qubit-qudit states ρ.

Let us introduce the vector $\vec{\gamma}$ formed by the (n_B^2-1) self-adjoint operators γ_p on $\mathcal{H}_B$ satisfying $\operatorname{tr}\gamma_p = 0$ and $\operatorname{tr}\gamma_p\gamma_q = n_B\delta_{pq}$ for any $p,q = 1,\dots,n_B^2-1$ (this means that $\{1/\sqrt{n_B},\gamma_p/\sqrt{n_B}\}$ is an orthonormal basis of the Hilbert space of all $n_B\times n_B$ matrices). This vector is the analog for B of the vector $\vec{\sigma}$ formed by the three Pauli matrices for A. The square root of ρ can be decomposed as

$$\sqrt{\rho} = \frac{1}{\sqrt{2n_B}}\Big(t_0 1\otimes 1 + \vec{x}\cdot\vec{\sigma}\otimes 1 + 1\otimes\vec{y}\cdot\vec{\gamma} + \sum_{m=1}^{3}\sum_{p=1}^{n_B^2-1} t_{mp}\,\sigma_m\otimes\gamma_p\Big) \tag{129}$$

[27] We remark that by exploiting (81) and (110), this second bound is equivalent precisely to the lower bound in (127).

[28] This inequality follows from the definitions of $D_{\rm He}^{\rm R}$ and $D_{\rm Bu}^{\rm R}$ and from the trace inequality $F(\rho, U_A\otimes 1\,\rho\,U_A^\dagger\otimes 1) = \|\sqrt{\rho}\,U_A\otimes 1\,\sqrt{\rho}\|_1^2 \leq \operatorname{tr}(\sqrt{\rho}\,U_A\otimes 1\,\sqrt{\rho}\,U_A^\dagger\otimes 1)$. It is saturated for pure states (see [78] for more detail).

with $t_0 \in [-1, 1]$, $\vec{x} \in \mathbb{R}^3$, and $\vec{y} \in \mathbb{R}^{n_B^2-1}$. We denote by T the $3 \times (n_B^2 - 1)$ complex matrix with coefficients t_{mp}. The condition $\operatorname{tr}(\sqrt{\rho})^2 = 1$ entails $t_0^2 + \|\vec{x}\|^2 + \|\vec{y}\|^2 + \operatorname{tr}(TT^{\mathrm{T}}) = 1$ (here T^{T} stands for the transpose of T). For any orthonormal basis $\{|\alpha_i\rangle\}_{i=0,1}$ for qubit A, one finds

$$\sum_{i=0,1} \operatorname{tr}\langle\alpha_i|\sqrt{\rho}|\alpha_i\rangle^2 = t_0^2 + \|\vec{y}\|^2 + \vec{u}^{\mathrm{T}}(\vec{x}\vec{x}^{\mathrm{T}} + TT^{\mathrm{T}})\vec{u}\;, \tag{130}$$

where we have introduced the unit vector $\vec{u} = \langle\alpha_0|\vec{\sigma}|\alpha_0\rangle = -\langle\alpha_1|\vec{\sigma}|\alpha_1\rangle$. Maximizing over all such vectors and using (110), we have [78]

$$D_{\mathrm{He}}^{\mathrm{G}}(\rho) = 2 - 2\sqrt{t_0^2 + \|\vec{y}\|^2 + k_{\max}}\;, \tag{131}$$

where $k_{\max}$ is the largest eigenvalue of the 3×3 matrix $K = \vec{x}\vec{x}^{\mathrm{T}} + TT^{\mathrm{T}}$. Therefore, the calculation of $D_{\mathrm{He}}^{\mathrm{G}}(\rho)$ is straightforward once one has determined the decomposition (129) of the square root of ρ.

A formula for the Hilbert–Schmidt geometric discord for two-qubit states has been given in Ref. [25]. An alternative derivation of (131) consists in using this formula and Proposition 11. The trace geometric discord $D_{\mathrm{tr}}^{\mathrm{G}}$ seems harder to compute than $D_{\mathrm{He}}^{\mathrm{G}}$ and $D_{\mathrm{Bu}}^{\mathrm{G}}$, but analytical expressions have been found in Ref. [24] for two-qubit X-states and two-qubit B-classical states.

The results of this section are summarized in the third column of Table 1.

7 Measurement-Induced Geometric Discord and Discord of Response

The properties of the measurement-induced geometric discord D^{M} and discord of response D^{R} for the Bures, Hellinger, trace, and Hilbert–Schmidt distances are summarized in Tables 2 and 3. We refer the reader to Ref. [78] for the proofs and references to the original works. For any $\rho \in \mathcal{E}(\mathcal{H}_{AB})$, the following general expressions and bounds on D^{M} and D^{R} can be derived. For the Bures distance, one has (compare with (89)) [78]

$$\begin{aligned} D_{\mathrm{Bu}}^{\mathrm{G}}(\rho) \quad &\leq D_{\mathrm{Bu}}^{\mathrm{M}}(\rho) = 2 - 2\max_{\{|\alpha_i\rangle\}} \operatorname{tr}\sqrt{\sum_{i=1}^{n_A} \eta_i^2 \rho_i^2} \leq g(D_{\mathrm{Bu}}^{\mathrm{G}}(\rho)) \\ 1 - \sqrt{1 - D_{\mathrm{He}}^{\mathrm{R}}(\rho)} &\leq D_{\mathrm{Bu}}^{\mathrm{R}}(\rho) = 1 - \max_{\{|\alpha_i\rangle\}} \operatorname{tr}\left|\sum_{i=1}^{n_A} \eta_i e^{-\mathrm{i}\frac{2\pi i}{n_A}} \rho_i\right| \leq \; D_{\mathrm{He}}^{\mathrm{R}}(\rho) \end{aligned}\;, \tag{132}$$

where $\{\rho_i, \eta_i\}$ is the state ensemble defined in (90) and g is the function (126). Similarly, one finds for the Hellinger distance (compare with (110)) [78]

$$\begin{array}{llr} D_{\rm He}^{\rm G}(\rho) & \leq D_{\rm He}^{\rm M}(\rho) = 2 - 2 \max_{\{|\alpha_i\rangle\}} \sum_{i=1}^{n_A} {\rm tr}_B \langle \alpha_i | \sqrt{\rho} | \alpha_i \rangle \sqrt{\langle \alpha_i | \rho | \alpha_i \rangle} & \leq g(D_{\rm He}^{\rm G}(\rho)) \\ \sin^2\left(\frac{\pi}{n_A}\right) g(D_{\rm He}^{\rm G}(\rho)) & \leq D_{\rm He}^{\rm R}(\rho) = 2 \min_{\{|\alpha_i\rangle\}} \sum_{i,j=1}^{n_A} \sin^2\left(\frac{\pi(i-j)}{n_A}\right) {\rm tr}_B \big| \langle \alpha_i | \sqrt{\rho} | \alpha_j \rangle \big|^2 & \leq g(D_{\rm He}^{\rm G}(\rho)) \end{array} \quad (133)$$

The first inequality in the last line is an equality when $n_A = 2$ or 3. Thus, for the Hellinger distance the discord of response is a function of the geometric discord when A is a qubit or a qutrit. This is also true for the Bures and trace distances when A is a qubit (see Table 3). In that case, $D_{\rm He}^{\rm R}(\rho) = g(D_{\rm He}^{\rm G}(\rho))$ can be evaluated analytically by relying on the formula (131), showing that $D_{\rm He}^{\rm R}$ is an easily computable measure of quantum correlations. In fact, when $n_A = 2$ then $D_{\rm He}^{\rm R}(\rho)$ is related to the LQU (see (38)), which has been determined for arbitrary qubit-qudit states in Ref. [36].

8 Conclusion

We have presented the properties of three classes of geometric measures of quantum correlations, namely the geometric discord $D^{\rm G}$, the measurement-induced geometric discord $D^{\rm M}$, and the discord of response $D^{\rm R}$, for two distinguished distances on the set of quantum states, the Bures and Hellinger distances. These measures satisfy all the axiomatic criteria for *bona fide* measures of quantum correlations while being easier to compute than the entropic quantum discord and having operational interpretations. Indeed, we have found that the geometric discord may be interpreted in terms of a probability of success in a quantum state discrimination task. The discords of response for the Hellinger and Bures distances are related respectively to the Local Quantum Uncertainty (LQU) [36] and the interferometric power [37]. The latter are in fact local geometrical versions of $D^{\rm R}$ (called here the discords of speed of response) and enjoy clear interpretations in local measurements and quantum metrology. The geometric measures $D^{\rm G}$, $D^{\rm M}$, and $D^{\rm R}$ are likely to appear as figures of merit in other protocols of quantum information and quantum technologies (for instance, $D^{\rm R}$ provides upper and lower bounds on the probability of error in quantum reading [77]). We have addressed the issue of the explicit evaluation of the geometric measures when the reference subsystem A is a qubit. We have found in particular that the Hellinger geometric discord and Hellinger discord of response are easily computable for any qubit-qudit states. When A is a qubit or a qutrit, different measures may be linked by algebraic relations. This is what happens for instance for the Hellinger geometric discord, Hellinger discord of response, and LQU. When A has a higher dimensional Hilbert space, however, each geometric measure defines its own ordering on the set of quantum states. In this sense, the different measures are not equivalent. Some bounds enabling to compare them have been given.

From a broader perspective, we have tried in this chapter to show that the study of the geometry on the set of quantum states defined by contractive Riemannian distances sheds new light on quantum correlations in bipartite systems and, more generally, on the whole field of quantum information theory.

Acknowledgements We acknowledge support from the French ANR project No. ANR-13-JS01-0005-01, the EU FP7 Cooperation STREP Projects iQIT No. 270843 and EQuaM No. 323714, the Italian Minister of Scientific Research (MIUR) national PRIN programme, and the Chilean Fondecyt project No. 1140994.

References

1. B. Aaronson, R.L. Franco, G. Adesso, Comparative investigation of the freezing phenomena for quantum correlations under nondissipative decoherence. Phys. Rev. A **88**, 012120 (2013)
2. T. Abad, V. Karimipour, L. Memarzadeh, Power of quantum channels for creating quantum correlations. Phys. Rev. A **86**, 062316 (2012)
3. M. Ali, A.R.P. Rau, G. Alber, Quantum discord for two-qubit X states. Phys. Rev. A **81**, 042105 (2010)
4. H. Araki, A remark on Bures distance function for normal states. Publ. RIMS Kyoto Univ. **6**, 477–482 (1970)
5. K.M.R. Audenaert, N. Datta, α-z-relative Rényi entropies. J. Math. Phys. **56**, 022202 (2015)
6. K.M.R. Audenaert, J. Calsamiglia, R. Muñoz-Tapia, E. Bagan, L.I. Masanes, A. Acin, F. Verstraete, Discriminating States: The Quantum Chernoff Bound. Phys. Rev. Lett. **98**, 160501 (2007)
7. R. Balian, The entropy-based quantum metric. Entropy 2014 **16**(7), 3878–3888 (2014)
8. R. Balian, Y. Alhassid, H. Reinhardt, Dissipation in many-body systems: a geometric approach based on information theory. Phys. Rep. **131**, 1 (1986)
9. M. Ban, K. Kurokawa, R. Momose, O. Hirota, Optimum measurements for discrimination among symmetric quantum states and parameter estimation. Int. J. Theor. Phys. **36**, 1269–1288 (1997)
10. S.M. Barnett, Minimum error discrimination between multiply symmetric states. Phys. Rev. A **64**, 030303 (2001)
11. H. Barnum, E. Knill, Reversing quantum dynamics with near-optimal quantum and classical fidelity. J. Math. Phys. **43**, 2097–2106 (2002)
12. C.H. Bennett, H.J. Bernstein, S. Popescu, B. Schumacher, Concentrating partial entanglement by local operations. Phys. Rev. A **53**, 2046 (1996)
13. C.H. Bennett, D.P. DiVincenzo, J.A. Smolin, W.K. Wootters, Mixed-state entanglement and quantum error correction. Phys. Rev. A **54**, 3824 (1996)
14. T. Benoist, V. Jakšić, Y. Pautrat, C.-A. Pillet, On entropy production of repeated quantum measurements I. General theory, arXiv:1607.00162 [math-ph]
15. J.A. Bergou, U. Herzog, M. Hillery, Discrimination of quantum states, in *Quantum State Estimation*, vol. 649, Lecture Notes in Physics, ed. by M. Paris, J. Rehacek (Springer, Berlin, 2004), pp. 417–465
16. R. Bhatia, *Matrix Analysis* (Springer, Berlin, 1991)
17. S.L. Braunstein, C.M. Caves, Statistical distance and the geometry of quantum states. Phys. Rev. Lett **72**, 3439–3443 (1994)
18. T.R. Bromley, M. Cianciaruso, R. Lo Franco, G. Adesso, Unifying approach to the quantification of bipartite correlations by Bures distance. J. Phys. A: Math. Theor. **47**, 405302 (2014)
19. D. Bures, An extension of Kakutani's theorem on infinite product measures to the tensor product of semifinite w^*-algebras. Trans. Am. Math. Soc. **135**, 199–212 (1969)
20. D. Cavalcanti, L. Aolita, S. Boixo, K. Modi, M. Piani, A. Winter, Operational interpretations of quantum discord. Phys. Rev. A **83**, 032324 (2011)

21. N.N. Cencov, *Statistical Decision Rules and Optimal Interferences*, vol. 53, Translations of Mathematical Monographs (American Mathematical Society, Providence, 1982)
22. L. Chang, S. Luo, Remedying the local ancilla problem with geometric discord. Phys. Rev. A **87**, 062303 (2013)
23. C.-L. Chou, L.Y. Hsu, Minimal-error discrimination between symmetric mixed quantum states. Phys. Rev. A **68**, 042305 (2003)
24. F. Ciccarello, T. Tufarelli, V. Giovannetti, Towards computability of trace distance discord. New J. Phys. **16**, 013038 (2014)
25. B. Dakić, V. Vedral, C. Brukner, Necessary and sufficient condition for nonzero quantum discord. Phys. Rev. Lett. **105**, 190502 (2010)
26. A. Datta, S.T. Flammia, C.M. Caves, Entanglement and the power of one qubit. Phys. Rev. A **72**, 042316 (2005)
27. A. Datta, A. Shaji, C.M. Caves, Quantum discord and the power of one qubit. Phys. Rev. Lett. **100**, 050502 (2008)
28. Y.C. Eldar, von Neumann measurement is optimal for detecting linearly independent mixed quantum states. Phys. Rev. A **68**, 052303 (2003)
29. Y.C. Eldar, G.D. Forney Jr., On quantum detection and the square-root measurement. IEEE Trans. Inf. Theory **47**, 858–872 (2001)
30. B.M. Escher, R.L. de Matos Filho, L. Davidovish, General framework for estimating the ultimate precision limit in noisy quantum-enhanced metrology. Nat. Phys. **7**, 406–411 (2011)
31. R.L. Frank, E.H. Lieb, Monotonicity of a relative Rényi entropy. J. Math. Phys. **54**, 122201 (2013)
32. S. Gharibian, Quantifying nonclassicality with local unitary operations. Phys. Rev. A **86**, 042106 (2012)
33. S.M. Giampaolo, F. Illuminati, Characterization of separability and entanglement in $(2 \times D)$- and $(3 \times D)$-dimensional systems by single-qubit and single-qutrit unitary transformations. Phys. Rev. A **76**, 042301 (2007)
34. S.M. Giampaolo, A. Streltsov, W. Roga, D. Bruß, F. Illuminati, Quantifying nonclassicality: global impact of local unitary evolutions. Phys. Rev. A **87**, 012313 (2013)
35. D. Girolami, G. Adesso, Quantum discord for general two-qubit states: analytical progress. Phys. Rev. A **83**, 052108 (2011)
36. D. Girolami, T. Tufarelli, G. Adesso, Characterizing nonclassical correlations via local quantum uncertainty. Phys. Rev. Lett. **110**, 240402 (2013)
37. D. Girolami, A.M. Souza, V. Giovannetti, T. Tufarelli, J.G. Filgueiras, R.S. Sarthour, D.O. Soares-Pinto, I.S. Oliveira, G. Adesso, Quantum discord determines the interferometric power of quantum states. Phys. Rev. Lett. **112**, 210401 (2014)
38. P. Hausladen, W.K. Wootters, A "pretty good" measurement for distinguishing quantum states. J. Mod. Opt. **41**, 2385–2390 (1994)
39. P. Hayden, R. Jozsa, D. Petz, A. Winter, Structure of states which satisfy strong subadditivity of quantum entropy with equality. Commun. Math. Phys. **246**, 359–374 (2004)
40. C.W. Helstrom, *Quantum Detection and Estimation Theory* (Academic Press, New York, 1976)
41. L. Henderson, V. Vedral, Classical, quantum and total correlations. J. Phys. A: Math. Gen. **34**, 6899–6905 (2001)
42. A.S. Holevo, On quasiequivalence of locally normal states. Theor. Math. Phys. **13**(2), 1071–1082 (1972)
43. A.S. Holevo, On asymptotically optimal hypothesis testing in quantum statistics. Theory Probab. Appl. **23**, 411–415 (1979)
44. R. Horodecki, P. Horodecki, M. Horodecki, K. Horodecki, Quantum entanglement. Rev. Mod. Phys. **81**, 865–942 (2009)
45. Y. Huang, Quantum discord for two-qubit X states: analytical formula with very small worst-case error. Phys. Rev. A **88**, 014302 (2013)
46. Y. Huang, Computing quantum discord is NP-complete. New J. Phys. **16**, 033027 (2014)
47. M. Hübner, Explicit computation of the Bures distance for density matrices. Phys. Lett. A **163**, 239–242 (1992)

48. P. Hyllus, W. Laskowski, R. Krischek, C. Schwemmer, W. Wieczorek, H. Weinfurter, L. Pezzé, A. Smerzi, Fisher information and multiparticle entanglement. Phys. Rev. A **85**, 022321 (2012)
49. V. Jakšić, C.-A. Pillet, Entropic Functionals in Quantum Statistical Mechanics, in *Proceedings of XVIIth International Congress of Mathematical Physics (Aalborg 2012)* (World Scientific, Singapore, 2013), 336–343
50. R. Jozsa, Fidelity for mixed quantum states. J. Mod. Opt. **41**, 2315–2323 (1994)
51. E.H. Lieb, Convex trace functions and the Wigner-Yanase-Dyson conjecture. Adv. Math. **11**, 267–288 (1973)
52. E.H. Lieb, M.B. Ruskai, Proof of the strong subadditivity of quantum-mechanical entropy. J. Math. Phys. **14**, 1938–1941 (1973)
53. S. Luo, Wigner-Yanase skew information and uncertainty relations. Phys. Rev. Lett. **91**, 180403 (2003)
54. S. Luo, Quantum discord for two-qubit systems. Phys. Rev. A **77**, 042303 (2008)
55. S. Luo, S. Fu, Geometric measure of quantum discord. Phys. Rev. A **82**, 034302 (2010)
56. V. Madhok, A. Datta, Interpreting quantum discord through quantum state merging. Phys. Rev. A **83**, 032323 (2011)
57. P. Marian, T.A. Marian, Hellinger distance as a measure of gaussian discord. J. Phys. A: Math. Theor. **48**, 115301 (2015)
58. J.A. Miszczak, Z. Puchala, P. Horodecki, A. Uhlmann, K. Życzkowski, Sub- and super-fidelity as bounds for quantum fidelity. Quantum Inf. Comput. **9**(1–2), 0103–0130 (2009)
59. K. Modi, T. Parerek, W. Son, V. Vedral, M. Williamson, Unified view of quantum and classical correlations. Phys. Rev. Lett. **104**, 080501 (2010)
60. K. Modi, A. Brodutch, H. Cable, T. Paterek, V. Vedral, The classical-quantum boundary for correlations: discord and related measures. Rev. Mod. Phys. **84**, 1655–1707 (2012)
61. A. Monras, G. Adesso, S.M. Giampaolo, G. Gualdi, G.B. Davies, F. Illuminati, Entanglement quantification by local unitary operations. Phys. Rev. A **84**, 012301 (2011)
62. M. Müller-Lennert, F. Dupuis, O. Szehr, S. Fehr, M. Tomamichel, On quantum Rényi entropies: a new generalization and some properties. J. Math. Phys. **54**, 122203 (2013)
63. T. Nakano, M. Piani, G. Adesso, Negativity of quantumness and its interpretations. Phys. Rev. A **88**, 012117 (2013)
64. N.A. Nielsen, I.L. Chuang, *Quantum Computation and Information* (Cambridge University Press, Cambridge, 2000)
65. M. Nussbaum, A. Szkola, The Chernoff lower bound for symmetric quantum hypothesis testing, vol. 37, The Annals of Statistics (Institute of Mathematical Statistics, 2009), pp. 1040–1057
66. H. Ollivier, W.H. Zurek, Quantum discord: a measure of the quantumness of correlations. Phys. Rev. Lett. **88**, 017901 (2001)
67. M. Ozawa, Entanglement measures and the Hilbert-Schmidt distance. Phys. Lett. A **268**, 158–160 (2000)
68. F.M. Paula, T.R. de Oliveira, M.S. Sarandy, Geometric quantum discord through the Schatten 1-norm. Phys. Rev. A **87**, 064101 (2013)
69. D. Pérez-Garcia, M.M. Wolf, D. Petz, M.B. Ruskai, Contractivity of positive and trace-preserving maps under L^p-norms. J. Math. Phys. **47**, 083506 (2006)
70. D. Petz, Monotone metrics on matrix spaces. Lin. Alg. Appl. **244**, 81–96 (1996)
71. D. Petz, Monotonicity of quantum relative entropy revisited. Rev. Math. Phys. **15**, 79–91 (2003)
72. L. Pezzé, A. Smerzi, Entanglement, nonlinear dynamics, and the Heisenberg limit. Phys. Rev. Lett. **102**, 100401 (2009)
73. M. Piani, Problem with geometric discord. Phys. Rev. A **86**, 034101 (2012)
74. M. Piani, S. Gharibian, G. Adesso, J. Calsamiglia, P. Horodecki, A. Winter, All nonclassical correlations can be activated into distillable entanglement. Phys. Rev. Lett. **106**, 220403 (2011)
75. M. Piani, V. Narasimhachar, J. Calsamiglia, Quantumness of correlations, quantumness of ensembles and quantum data hiding. New J. Phys. **16**, 113001 (2014)
76. W. Roga, S.M. Giampaolo, F. Illuminati, Discord of response. J. Phys A: Math. Theor. **47**, 365301 (2014)

77. W. Roga, D. Buono, F. Illuminati, Device-independent quantum reading and noise-assisted quantum transmitters. New J. Phys. **17**, 013031 (2015)
78. W. Roga, D. Spehner, F. Illuminati, Geometric measures of quantum correlations: characterization, quantification, and comparison by distances and operations. J. Phys. A: Math. Theor. **49**, 235301 (2016)
79. M.B. Ruskai, Beyond strong subadditivity: improved bounds on the contraction of the generalized relative entropy. Rev. Math. Phys. **6**(5a), 1147–1161 (1994)
80. H.-J. Sommers, K. Życzkowski, Bures volume of the set of mixed quantum states. J. Phys. A: Math. Gen. **36**, 10083–10100 (2003)
81. D. Spehner, Quantum correlations and distinguishability of quantum states. J. Math. Phys. **55**, 075211 (2014)
82. D. Spehner, M. Orszag, Geometric quantum discord with Bures distance. New J. Phys. **15**, 103001 (2013)
83. D. Spehner, M. Orszag, Geometric quantum discord with Bures distance: the qubit case. J. Phys. A: Math. Theor. **47**, 035302 (2014)
84. W.F. Stinespring, Positive functions on C^*-algebras. Proc. Am. Soc. **6**, 211–216 (1955)
85. A. Streltsov, H. Kampermann, D. Bruß, Linking a distance measure of entanglement to its convex roof. New J. Phys. **12**, 123004 (2010)
86. A. Streltsov, H. Kampermann, D. Bruß, Linking quantum discord to entanglement in a measurement. Phys. Rev. Lett. **106**, 160401 (2011)
87. A. Streltsov, H. Kampermann, D. Bruß, Behavior of quantum correlations under local noise. Phys. Rev. Lett. **107**, 170502 (2011)
88. A. Streltsov, G. Adesso, M. Piani, D. Bruß, Are general quantum correlations monogamous? Phys. Rev. Lett. **109**, 050503 (2012)
89. G. Toth, Multipartite entanglement and high-precision metrology. Phys. Rev. A **85**, 022322 (2012)
90. G. Tóth, D. Petz, Extremal properties of the variance and the quantum Fisher information. Phys. Rev. A **87**, 032324 (2013)
91. A. Uhlmann, Endlich-dimensionale Dichtematrizen II. Wiss. Z. Karl-Marx-Univ. Leipzig, Math.-Nat R. **22**, 139–177 (1973)
92. A. Uhlmann, The "transition probability" in the state space of a $*$-algebra. Rep. Math. Phys. **9**, 273–279 (1976)
93. A. Uhlmann, Parallel transport and "quantum holonomy" along density operators. Rep. Math. Phys. **24**, 229–240 (1986)
94. V. Vedral, M.B. Plenio, Entanglement measures and purifications procedures. Phys. Rev. A **57**, 1619–1633 (1998)
95. V. Vedral, M.B. Plenio, M.A. Rippin, P.L. Knight, Quantifying entanglement. Phys. Rev. Lett. **78**, 2275–2279 (1997)
96. T.C. Wei, P.M. Goldbart, Geometric measure of entanglement and applications to bipartite and multipartite quantum states. Phys. Rev. A **68**, 042307 (2003)
97. E.P. Wigner, M.M. Yanase, Information contents of distributions. Proc. Natl. Acad. Sci. U.S.A. **49**, 910–918 (1963)
98. M.M. Wilde, A. Winter, D. Yang, Strong converse for the classical capacity of entanglement-breaking and Hadamard channels via a sandwiched Renyi relative entropy. Commun. Math. Phys. **331**, 593–622 (2014)
99. M.M. Wolf, *Quantum Channels and Operations Guided Tour* (2002). http://www-m5.ma.tum.de/foswiki/pub/M5/Allgemeines/MichaelWolf/QChannelLecture.pdf

Metrological Measures of Non-classical Correlations

Pieter Bogaert and Davide Girolami

1 Introduction

In this work, we will review studies showing that non-classical, discord-like correlations do not necessarily describe a statistical dependence between measurements performed by non-communicating parties. We will explain how they yield the impossibility of global observers to obtain full knowledge of local properties of quantum systems. This apparently detrimental feature translates, on the other hand, in an increased capability of an observer to acquire information about a quantum perturbation by establishing correlations between its probe and an unchanged ancillary system. The phenomenon is undoubtedly not explicable by classical physics, being a direct consequence of quantum complementarity. We will present our arguments by following a two-step line of thinking.

First, we will point out that quantum coherence manifests in the intrinsic quantum randomness of measurement outcomes (Sect. 2). Genuinely quantum uncertainty differs from classical randomness. We will explain how to discriminate between them and quantify the quantum uncertainty from experimental data. Non-classical correlations in a bipartite system will be defined as the degree of irreducible coherence, i.e. quantum uncertainty or randomness, experienced when measuring local observables. The result links a local property as quantum uncertainty to a global feature as non-classical correlations. The proof is given by showing that a quantity called Local Quantum Uncertainty, which quantifies the minimum local quantum randomness in a bipartite state, satisfies the very same properties enjoyed by entropic measures of discord-like correlations (Sect. 2.2).

Then, we will show the link between quantum-induced uncertainty and supraclassical measurement precision (Sect. 3). A measurement can be thought as an informa-

P. Bogaert · D. Girolami (✉)
Department of Atomic and Laser Physics, University of Oxford,
Parks Road, Oxford OX1 3PU, UK
e-mail: davegirolami@gmail.com

F.F. Fanchini et al. (eds.), *Lectures on General Quantum Correlations and their Applications*, Quantum Science and Technology,
DOI 10.1007/978-3-319-53412-1_7

tion processing task where knowledge encoded in a physical systems is transmitted to an apparatus. Specifically, a measurement requires a preliminary step in which the probe is prepared in an input configuration. In a second stage, the information we want to access is imprinted in the probe state through a quantum dynamics. The final part is the information decoding by collection and statistical analysis of the data. We will focus here on the first step, i.e. input state preparation. Arguably, the probe state has to be sensitive to the perturbation. We will explain why quantum systems displaying non-classical correlations are intrinsically more sensitive probes. The key observation is that quantum uncertainty entails sensitivity to quantum dynamics. Consequently, non-classical correlations guarantee non-vanishing sensitivity to local quantum perturbations. We will explain how the concept of Interferometric Power captures genuinely quantum sensitivity in a standard measurement setting, and how this leads to non-classical performances for phase estimation (Sect. 3.3). Remarkably, the minimum precision for local measurements will be shown to be a measure of non-classical correlations. A third interesting correlation quantifier, the Discriminating Strength (Sect. 3.4), will be shown to evaluate the worst case precision in another important metrological task, state discrimination.

It is our hope to highlight the main merit, in our opinion, of the metrological approach to characterizing non-classical correlations. That is, giving a physical meaning to an information-theoretic construction, providing an operational interpretation which goes truly beyond the original one [1]. Quantum discord is a concept developed to study environmentally-induced decoherence [2], and the limit to information transmission established by classical correlations [3]. Other concurrent studies characterized non-classical correlations in the context of quantum Shannon information theory [4]. While Quantum Mechanics is somehow a theory of information itself, we owe its postulate and structure to key experimental observations of low energy light and atomic structure in the beginning of 20th century [5]. For example, the somehow elusive concept of Entanglement was originally discussed by means of carefully designed thought experiments. It is therefore reassuring to make real the concept of non-classical correlations by linking it with observable experimental effects.

2 Local Quantum Uncertainty

2.1 Quantum Uncertainty

Quantum Mechanics predicts the existence of coherent superpositions of quantum states [5]. The first experimental evidence which suggested such possibility was the wave-like probability distribution of measurement outcomes observed in low energy optical experiments [6]. The intuition linking coherence and non-classical outcome statistics can be formalized. By focusing on finite dimensional quantum systems, let us suppose to measure the observable being represented by a non-degenerate Hermitian operator with spectral decomposition $O = \sum_i o_i |i_o\rangle\langle i_o|$. The information about O in a state represented by a density matrix ρ, $\mathrm{tr}[\rho] = 1$, $\rho = \rho^\dagger$, $\rho \geq 0$, can be quantified by the state change due to the measurement (without postselection) of O. For our purposes, we focus on the von Neumann measurement model $\rho \rightarrow \rho' = \sum_i |i_o\rangle\langle i_o|\rho|i_o\rangle\langle i_o|$ [7]. If and only if state and observable commute, there is no change in the state, $\rho = \rho'$. This is easily proven to happen if and only if the state is an eigenstate or a mixture of eigenstates of the observable, taking the form $\rho_O = \sum_i p_i |i_o\rangle\langle i_o|$. That is, if and only if the state is incoherent in the observable eigenbasis, the measurement output statistics will be classical. Without coherence, the measurement uncertainty is only due to incomplete knowledge of the system state, which is a classical error source. In the more general case of states displaying coherence, the contribution to the measurement uncertainty is then twofold. Apart from the classical randomness, there is an additional quantum component, which manifests in the interference pattern of the outcome statistics.

Let us now quantify quantum uncertainty. The first quantity that has in many ways become almost synonymous with uncertainty, at least in undergraduate Physics textbooks, is the variance $V(\rho, O) = \mathrm{tr}[\rho O^2] - (\mathrm{tr}[\rho O])^2$. The variance enjoys both a simple expression and a close tie to experimental practice. However, for mixed states the variance includes a contribution of classical uncertainty due to the mixedness of the state. It is easy to see that the variance does not vanish even if ρ and O commute, but the case in which the state is an observable eigenstate. The variance is therefore not suitable to quantify quantum uncertainty. A way to solve the issue is to formally split the variance, which captures the total measurement uncertainty, into quantum and classical contributions: $V = V_q + V_c$ [8]. Note that the idea can be extended to entropic uncertainty quantifiers [9]. A good measure of quantum uncertainty V_q should be zero if and only if ρ and O commute. Yet, an arbitrary norm of their commutator is not the finest choice. Additionally, a measure of quantum uncertainty should be convex, i.e. non-increasing under classical mixing, as this only generates classical uncertainty, $V_q\left(\sum_i p_i\rho_i, O\right) \leq \sum_i p_i V_q(\rho_i, O)$. A suitable candidate is the (Wigner–Yanase) skew information [10], given by

$$\mathcal{I}(\rho, O) := -\frac{1}{2}\mathrm{tr}[[\rho^{1/2}, O]^2]. \tag{1}$$

The skew information is upper bounded by the variance, being equal to it for pure states: $\mathcal{I}(\rho, O) \leq V(\rho, O)$. This can be shown as follows [8]. By defining $O_0 = O - \mathrm{tr}[\rho O]$, one has $V(\rho, O) = \mathrm{tr}[\rho O_0^2]$ and $\mathcal{I}(\rho, O) = \mathrm{tr}[\rho O_0^2] - \mathrm{tr}[\rho^{1/2} O_0 \rho^{1/2} O_0] - V(\rho, O) - \mathrm{tr}[\rho^{1/2} O_0 \rho^{1/2} O_0]$. It is easy to see that the second term is non-negative as it equals $\mathrm{tr}[(\rho^{1/4} O_0 \rho^{1/4})(\rho^{1/4} O_0 \rho^{1/4})]$ and noting that $\rho^{1/4} O_0 \rho^{1/4}$ is self-adjoint. Whilst being just one of the potential choices, the skew information is a consistent yet sufficiently manageable measure of quantum uncertainty. We illustrate the interplay between classical and quantum uncertainty by a simple example presented in Fig. 1.

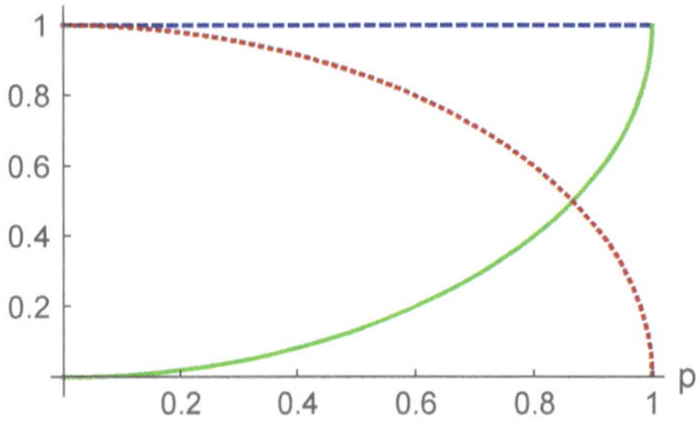

Fig. 1 Quantum uncertainty disclosed. We calculate the uncertainty on the measurement outcome of the observable $\sigma_z = |0\rangle\langle 0| - |1\rangle\langle 1|$ in the state $\rho = (1-p)\mathbb{I}_2/2 + p|\phi\rangle\langle\phi|$, $|\phi\rangle = 1/\sqrt{2}(|0\rangle + |1\rangle)$, $p \in [0, 1]$. The *blue dashed line* is the variance, the *green blue continuous curve* is the skew information. The *red dotted curve* depicts the difference between the two quantities, being an heuristic mixedness quantifier. As expected by a measure of quantum uncertainty, the skew information monotonically increases with the purity parameter p

2.2 *Discord Triggers Local Quantum Uncertainty*

The Heisenberg uncertainty principle states that complementary properties of quantum systems cannot be measured with arbitrary precision, in the sense that, regardless our experimental ability, the product of the experimental uncertainties about their values in a given state is non-negative [11]. In the original form of the uncertainty relations, the non-commutativity between observables captures such ineludible quantum randomness. However, it may seem that any single physical quantity, such as one spin or position component, could be measured with arbitrary precision. We are going to show that this is not true in general. We identified the truly quantum uncertainty of the measurement, and, not surprisingly, quantified it by a measure of state-observable non-commutativity. Zero quantum uncertainty implies that the measurement performed by a flawless experimental implementation, i.e. whenever there is no even classical uncertainty, has a deterministic outcome. Yet, a non-negotiable intrinsic quantum uncertainty on single observable measurements appears whenever the system of interest shares non-classical correlations.

Let us examine the quantum uncertainty in local quantum measurements on a bipartite system. For example, it is given a two-qubit system prepared in a maximally entangled state $|\varphi\rangle_{AB} = (|00\rangle + |11\rangle)/\sqrt{2}$. It is immediate to observe that this is an eigenstate of the global observable $\sigma_z \otimes \sigma_z$, which means that there is no quantum uncertainty when measuring that observable. Any *local* spin measurement, however, will have intrinsic uncertainty. The only vector $\boldsymbol{n}$ for which $\boldsymbol{n} \cdot \boldsymbol{\sigma}_A \otimes \mathbb{I}_B|\varphi\rangle_{AB} = k|\varphi\rangle_{AB}$, $k \in \mathbb{R}$, where $\boldsymbol{\sigma}$ are the Pauli matrices, is indeed $\boldsymbol{n} = \boldsymbol{0}$. More generally, only product states (e.g. $|11\rangle$) can be eigenstates of local observables.

By extending the argument to mixed states, it is clear that one does not want to associate quantum uncertainty to state mixedness (which quantifies the incomplete knowledge about the state). Given a local complete measurement, we still require that performing the measurement leaves the mixed state ρ_{AB} invariant if and only if it commutes with the observable. Supposing without loss of generality that the measurement is performed on A, this means that it must be possible to express the state in the following form:

$$\rho_{AB} = \sum_i p_i |i\rangle\langle i|_A \otimes \sigma_B^i, \tag{2}$$

where the elements $\{|i\rangle\}$ form an orthonormal basis. Such density matrices are called classical-quantum (CQ) states, and they are precisely the states with zero quantum discord [4]. Therefore, non-classical correlations imply local quantum uncertainty. In other words, for any CQ state there is *at least* one local measurement which does not alter it, while for other states quantum uncertainty always appears. However, the interplay between local randomness and non-local quantum effects turns out to be deeper. The minimum quantum uncertainty on local measurements is a quantifier of non-classical correlations. To prove that, let us quantify the quantum uncertainty of an observable O_A in a state ρ_{AB} by the skew information $\mathcal{I}(\rho_{AB}, O_A \otimes \mathbb{I}_B)$. By reminding the definition in Eq. (1), we note that the quantity depends on the state and the observable, while non-classical correlations are a property of the state only. It is sensible to introduce the Local Quantum Uncertainty (LQU) [12], defined as the minimum skew information between the state and a local observable. To be more precise, let us define the set of local observables $\{K_A^\Lambda := K_A^\Lambda \otimes \mathbb{I}_B\}$, where the K_A^Λ are Hermitian operators with spectrum Λ, which we demand to be non-degenerate, as this would represent an additional classical uncertainty source. Thus, the LQU with respect to the subsystem A is given by

$$\mathcal{U}_A^\Lambda(\rho_{AB}) := \min_{K^\Lambda} \mathcal{I}(\rho_{AB}, K_A^\Lambda), \tag{3}$$

with an optimisation over the previously defined set of local observables with non-degenerate spectrum Λ. We rewrite them as $K_A^\Lambda = U_A \mathrm{diag}(\Lambda) U_A^\dagger$, $U_A \in SU(d)$, where d is the dimension of subsystem A and $\mathrm{diag}(\Lambda)$ is a diagonal matrix with the observable eigenvalues being the diagonal entries. The minimisation then runs over all the possible unitary transformations U_A. The LQU is still dependent on the spectrum Λ, and this can be interpreted as fixing a "ruler" for the measurement. The non-degeneracy condition ensures the quality of the ruler, namely that there exist states for which a measurement will be maximally informative (i.e. states which commute with the observable and hence do not exhibit quantum uncertainty for it). Any spectrum choice identifies a different measure of non-classical correlations. On the other hand, the LQU is by no means dependent on the measurement basis, as U_A is varied over SU(d).

2.2.1 Local Quantum Uncertainty as a Measure of Non-classical Correlations

We here review the proof that the LQU is a measure for non-classical correlations, i.e. it meets the criteria identifying discord-like quantifiers [4]. We shall always work with the LQU defined by measurements on A.

1. The LQU is zero if and only if the state is CQ. If ρ_{AB} is CQ, then one can pick a K_A^Λ which is diagonal in the local basis of A, which means that the LQU vanishes. Conversely, if the LQU is zero, then there exists a local observable K_A^Λ which is simultaneously diagonalisable with ρ_{AB}. Since Λ is non-degenerate, this defines a basis on A which is unique up to phases (let us call it $\{|k_i\rangle\}$). An eigenvector basis for K_A^Λ must then be of the form $\{|k_i\rangle_A \otimes |\varphi_{ij}\rangle_B\}$, and the state must therefore be of the form $\rho_{AB} = \sum_{ij} p_{ij} |k_i\rangle\langle k_i|_A \otimes |\varphi_{ij}\rangle\langle\varphi_{ij}|_B$, i.e. it must be CQ.
2. The LQU is invariant under local unitary transformations. Few algebra steps give

$$\begin{aligned}\mathcal{U}_A^\Lambda((U_A \otimes U_B)\rho_{AB}(U_A \otimes U_B)^\dagger) &= \min_{K^\Lambda} \mathcal{I}((U_A \otimes U_B)\rho_{AB}(U_A \otimes U_B)^\dagger, K_A^\Lambda \otimes \mathbb{I}_B) \\ &= \min_{K^\Lambda} \mathcal{I}(\rho_{AB}, (U_A \otimes U_B)^\dagger K_A^\Lambda \otimes \mathbb{I}_B (U_A \otimes U_B)) \\ &= \min_{K^\Lambda} \mathcal{I}(\rho_{AB}, (U_A^\dagger K_A^\Lambda U_A) \otimes \mathbb{I}_B) = \mathcal{U}_A^\Lambda(\rho_{AB}), \qquad (4)\end{aligned}$$

   where the second and third lines follow from the definition of the skew information. The last equality holds because minimising over K_A^Λ is equivalent to minimising over the observable $U_A^\dagger K_A^\Lambda U_A$.
3. The LQU is contractive under completely positive trace-preserving (CPTP) maps on the non-measured subsystem B. The skew information is contractive under CPTP maps Φ_B: $\mathcal{I}(\rho_{AB}, K_A \otimes \mathbb{I}_B) \geq \mathcal{I}((\mathbb{I}_A \otimes \Phi_B)\rho_{AB}, K_A \otimes \mathbb{I}_B)$. This can be easily proved by writing the CPTP map Φ_B in a Stinespring representation and noting that the skew information is contractive under partial trace: $\mathcal{I}(\sigma_{AB}, X_A \otimes \mathbb{I}_B) \geq \mathcal{I}(\sigma_A, X_A)$. Let us suppose now that $\tilde{K}_A$ is the local observable minimising the skew information. The LQU takes the form

$$\mathcal{U}_A^\Lambda(\rho_{AB}) = \mathcal{I}(\rho_{AB}, \tilde{K}_A \otimes \mathbb{I}_B) \geq \mathcal{I}((\mathbb{I}_A \otimes \Phi_B)\rho_{AB}, \tilde{K}_A \otimes \mathbb{I}_B) \geq \mathcal{U}_A^\Lambda((\mathbb{I}_A \otimes \Phi_B)\rho_{AB}). \quad (5)$$

4. The LQU reduces to an entanglement monotone for pure states. For the full proof of this property, we refer to [12], presenting here just a sketch of it. Given the contractivity and invariance under CPTP and unitary maps respectively, we only need to prove that the LQU cannot increase on average under local operations on A:

$$\sum_i p_i \mathcal{U}_A^\Lambda(|\phi_i\rangle\langle\phi_i|_{AB}) \leq \mathcal{U}_A^\Lambda(|\psi\rangle\langle\psi|_{AB}), \qquad (6)$$

   where $\{p_i, |\phi_i\rangle$ is the output ensemble after a channel with Kraus operators $\{M_i\}$ is applied on A: $M_{i,A}|\psi\rangle_{AB} = \sqrt{p_i}|\phi_i\rangle_{AB}$. It is possible to prove two auxiliary lemmas. First, one can always assume $d_A \geq d_B$. Then, one shows that the LQU is not affected when measuring B instead of A, where $\Lambda(K_B)$ is a subset of $\Lambda(K_A)$. Suppose that the minimum is achieved for $\tilde{K}_B^\Lambda$. Since the skew information is equal to the variance for pure states, and the latter is concave, one finally has

$$
\begin{aligned}
\sum_i p_i \mathcal{U}_A^\Lambda(|\phi_i\rangle\langle\phi_i|_{AB}) &\leq \sum_i p_i \min_{K_B^\Lambda} \mathcal{I}(|\phi_i\rangle\langle\phi_i|_{AB}, K_B^\Lambda) \leq \sum_i p_i \mathcal{I}(|\phi_i\rangle\langle\phi_i|_{AB}, \tilde{K}_B^\Lambda) \\
&= \sum_i p_i V(|\phi_i\rangle\langle\phi_i|_{AB}, \tilde{K}_B^\Lambda) \leq V\left(\sum_i p_i |\phi_i\rangle\langle\phi_i|_{AB}, \tilde{K}_B^\Lambda\right) \\
&= \sum_i p_i \langle\phi_i|(\tilde{K}_B^\Lambda)^2|\phi_i\rangle_{AB} - \left(\sum_i p_i \langle\phi_i|\tilde{K}_B^\Lambda|\phi_i\rangle_{AB}\right)^2 \\
&= \sum_i \langle\psi|M_i(\tilde{K}_B^\Lambda)^2 M_i^\dagger|\psi\rangle_{AB} - \left(\sum_i \langle\psi|M_i \tilde{K}_B^\Lambda M_i^\dagger|\psi\rangle_{AB}\right)^2 \\
&= \langle\psi|(\tilde{K}_B^\Lambda)^2|\psi\rangle_{AB} - \left(\langle\psi|\tilde{K}_B^\Lambda|\psi\rangle_{AB}\right)^2 \\
&= \mathcal{I}(|\psi\rangle\langle\psi|_{AB}, \tilde{K}_B^\Lambda) = \min_{K_A^\Lambda} \mathcal{I}(|\psi\rangle\langle\psi|_{AB}, K_A^\Lambda) \\
&= \mathcal{U}_A^\Lambda(|\psi\rangle\langle\psi|_{AB}).
\end{aligned} \tag{7}
$$

2.2.2 Restriction to $\mathbb{C}^2 \otimes \mathbb{C}^d$

We now consider the case where system A is a qubit and B a qudit, i.e. with states defined on an Hilbert space $\mathbb{C}^2 \otimes \mathbb{C}^d$. A question that remains to be answered is in which way the LQU depends on the choice of non-degenerate spectrum Λ. It is straightforward to show that, since A is a qubit, all Λ-dependent $\mathcal{U}^\Lambda(\rho_{AB})$ are equivalent up to a multiplicative factor. This is because a general local observable K_A^Λ with non-degenerate spectrum $\Lambda = \{\lambda_1, \lambda_2\}$ can be parametrised as

$$
K_A^\Lambda = U_A\left(\frac{\lambda_1 - \lambda_2}{2}\sigma_{zA} + \frac{\lambda_1 + \lambda_2}{2}\mathbb{I}_A\right)U_A^\dagger - \frac{\lambda_1 - \lambda_2}{2}\boldsymbol{n}\cdot\boldsymbol{\sigma}_A + \frac{\lambda_1 + \lambda_2}{2}\mathbb{I}_A, \tag{8}
$$

where $\boldsymbol{n}$ is a unit vector. From the definition of the skew information, it follows that $\mathcal{I}(\rho_{AB}, K_A^\Lambda) = \frac{(\lambda_1-\lambda_2)^2}{4}\mathcal{I}(\rho_{AB}, \boldsymbol{n}\cdot\boldsymbol{\sigma}_A)$. Therefore, for qubit-qudit systems the choice of the spectrum Λ does not affect the quantification of non-classical correlations (we shall therefore drop the Λ superscript from here onwards), and without loss of generality, we assume the local observables to be of the form $K_A = \boldsymbol{n}\cdot\boldsymbol{\sigma}_A$.

Having simplified the form of the observables over which we need to optimise (the minimisation runs over $\boldsymbol{n}$ now), we can write the LQU in the following fashion:

$$
\mathcal{U}_A(\rho_{AB}) = 1 - \lambda_{\max}(W_{AB}), \tag{9}
$$

being $\lambda_{\max}(W_{AB})$ the maximum eigenvalue of the 3×3 symmetric matrix W with entries

$$(W_{AB})_{ij} = \mathrm{tr}[\rho_{AB}^{1/2}(\sigma_{iA} \otimes \mathbb{I}_B)\rho_{AB}^{1/2}(\sigma_{jA} \otimes \mathbb{I}_B)], \quad (10)$$

where i, j label the Pauli matrices. Finally, for pure states $|\psi\rangle\langle\psi|_{AB}$, this further reduces to (two times) the linear entropy of entanglement

$$\mathcal{U}_A(|\psi\rangle\langle\psi|_{AB}) = 2(1 - \mathrm{tr}[\rho_A^2]) = 1 - (\sigma_0 - \sigma_1)^2, \quad (11)$$

where we used the Schmidt coefficients $\rho_A = \sigma_1|\psi_1\rangle\langle\psi_1|_A + \sigma_2|\psi_2\rangle\langle\psi_2|_A$. We observe that with our choice of observables K_A the LQU equals one for pure, maximally entangled states.

2.2.3 Geometric Insight

Finally, we provide a geometric interpretation of the LQU in qubit-qudit states. The (squared) Hellinger distance between two states ρ and σ is defined as $D_H^2(\rho, \sigma) = (1/2)\,\mathrm{tr}[\rho^{1/2} - \sigma^{1/2}]^2 = 1 - \mathrm{tr}[\rho^{1/2}\sigma^{1/2}]$. Since $K_A = \boldsymbol{n} \cdot \boldsymbol{\sigma}$ is a root-of-unity unitary, for every function f and any bipartite state one has $K_A f(\rho_{AB}) K_A = f(K_A \rho_{AB} K_A)$. Hence, the skew information takes the form

$$\begin{aligned}\mathcal{I}(\rho_{AB}, K_A) &= 1 - \mathrm{tr}[\rho_{AB}^{1/2} K_A \rho_{AB}^{1/2} K_A] = 1 - \mathrm{tr}[\rho_{AB}^{1/2}(K_A \rho_{AB} K_A)^{1/2}] \\ &= D_H^2(\rho_{AB}, K_A \rho_{AB} K_A).\end{aligned} \quad (12)$$

The LQU then represents the minimum distance between the state before and after a local root-of-unity unitary operations is applied.

3 Interferometric Power and Discriminating Strength

3.1 Quantum Metrology

We discussed a measure of discord-like correlations, the LQU, linked to the uncertainty in a given measurement. Perhaps surprisingly, in this section we will show that non-classical correlations yield measurement precision! We will explain how the two apparently contradictory viewpoints are consistently related to each other in the context of quantum metrology, which we briefly introduce here.

Metrology is the study of measurement strategies and tools. The term can be used in a variety of contexts related to measurements, for example to denote the establishment of units of measurement, or the technological application of measurement instruments and related issues such as calibration. For our purposes, however, metrology denotes the study of parameter estimation schemes and the strategies to reach the highest possible precision in them. Many of the concepts in metrology were first defined for classical systems, but we shall only discuss the ones which are useful for the extension to the quantum realm. For enjoyable reviews on quantum metrology, we refer the Reader to Refs. [13, 14]. It is indeed possible to take advantage of quantumness to increase the precision of measurement schemes. The reason is that quantum systems are more sensitive probes in a number of situations. Quantum metrology is the research line that studies what properties of quantum systems are responsible for this. Results in quantum metrology have a wide applicability in optical interferometry, atomic spectroscopy, and even gravitometry. A metrology task usually consists of three steps. First, the preparation of a probe in an input state. Second, an interaction or perturbation of the probe, which encodes information in it. Third, a measurement on the probe followed by data analysis. We here focus on the first step, and we investigate how non-classical correlations in the input help in two important metrology protocols: interferometric phase estimation and state discrimination.

3.2 Quantum Phase Estimation

We here focus on the important metrology primitive of parameter estimation [15]. The goal is to assign a probability function $p_\theta(x)$ to the independent measurement outcomes x of a random variable X. The parameter θ, which is unknown and unmeasurable, acts as a coordinate in the probability function space. The task is then to extract an observable estimator $\hat{\theta}(x)$ from the measurement outcomes, such that $p_{\hat{\theta}}(x)$ characterizes well the observed data. We require the estimator to be unbiased, i.e. its average value does equal the real value of the parameter, $\int(\theta - \hat{\theta}(x))p_\theta(x)dx = 0$. The quality of the estimation can be then quantified by the variance of the estimator $\hat{\theta}$.

It is possible to establish a fundamental limit to parameter estimation. By employing the maximum likelihood method, the best estimator $\hat{\theta}_{\text{best}}$ is defined as the one maximising the log-likelihood function $\max_{\hat{\theta}} \ln l(\hat{\theta}|x) = \ln l(\hat{\theta}_{\text{best}}|x)$, $l(\hat{\theta}|x) \equiv p_{\hat{\theta}}(x)$, where the logarithm is just a convention. This means that $p_{\hat{\theta}_{\text{best}}}(x)$ is the best function to describe the measurement outcomes. The information about θ which can be obtained by the data x is quantified by the rate of change of the likelihood function with the parameter value. A measure of such information is the zero mean value score function $\frac{\partial \ln l(\theta|x)}{\partial\theta}$. The second moment of the score is called the Fisher Information:

$$F(\theta) = \int \left(\frac{\partial}{\partial\theta} \log p(x,\theta)\right)^2 p(x,\theta)dx. \tag{13}$$

An important result in classical statistics is the Cramér–Rao bound, which gives a lower bound on the variance of $\hat{\theta}$:

$$V(p_\theta, \hat{\theta}) \geq \frac{1}{nF(\theta)}, \tag{14}$$

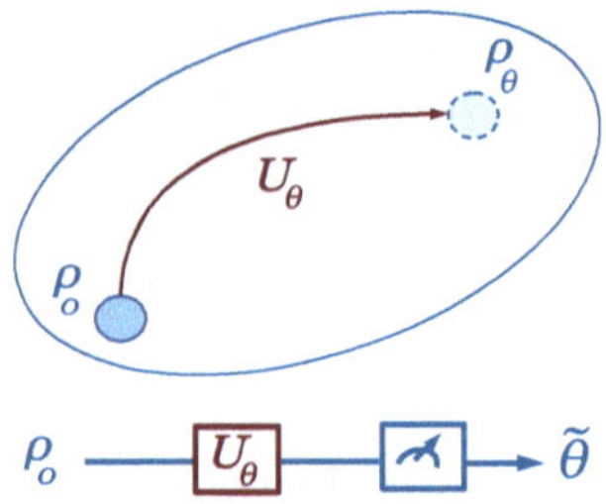

Fig. 2 Quantum phase estimation. A system initialized in the state ρ_0 is perturbed through a unitary transformation U_θ. A measurement and statistical processing of outcomes give an estimated value $\tilde{\theta}$ of the phase shift. The perturbation can be represented both as a geometric path in the parametrized space of quantum states, where θ is a coordinate (*top*), or as a logic transformation by applying a unitary gate (*bottom*). The resource is found to be the speed of evolution of the state during the phase shift, as quantified by the quantum Fisher information

for n repetitions of the measurement. Hence, the Fisher information is a key figure of merit of a parameter estimation protocol. We observe that, under the assumptions of single parameter, unbiased estimation, the best estimator $\hat{\theta}_{\text{best}}$ saturates the bound.

Let us now discuss the quantum case. The state of the system under study is represented by a parametrized density matrix ρ_θ. Let us assume that the parameter represents the information about a unitary perturbation $\rho_\theta = U_\theta \rho_0 U_\theta^\dagger$, $U_\theta = e^{-iH\theta}$ (Fig. 2). An estimator is built up by a generalized positive operator value measurement (POVM) $\{\Pi_x\}$ on the output state ρ_θ, where the Π_x denote the operators corresponding to the measurement outcomes x, thus obtaining $p_\theta(x) = \text{tr}[\rho_\theta \Pi_x]$. The expression of the Fisher information for an arbitrary POVM is

$$F(\rho_\theta) := \int dx \frac{1}{\text{tr}[\rho_\theta \Pi_x]} \left(\text{tr}[\partial_\theta \rho_\theta \Pi_x]\right)^2 . \tag{15}$$

However, the quantum scenario implies a further optimization of the measurement [15, 16]. One can prove that the optimal estimator is given by a projective measurement into the eigenbasis of the symmetric logarithmic derivative (SLD) L, defined implicitly as $\frac{\partial}{\partial\theta}\rho_\theta = \frac{1}{2}(\rho_\theta L + L\rho_\theta)$. In particular, an upper bound is obtained: $F(\rho_\theta) \leq \text{tr}[\rho_\theta L^2]$. The quantum Fisher information (QFI, from now on) is then given by the optimal measurement strategy:

$$\mathcal{F}(\rho, H) := \text{tr}[\rho L^2], \tag{16}$$

where we dropped the parameter label as the QFI is independent of its value. The quantum extension of the Cramér–Rao bound reads:

$$V(\rho, \hat{\theta}) \geq 1/[n\mathcal{F}(\rho, H)], \tag{17}$$

which alike the classical case is saturated asymptotically by the best estimator. The QFI enjoys a peculiar compact expression:

$$\mathcal{F}(\rho, H) = 4\sum_{k<l} \frac{(\lambda_k - \lambda_l)^2}{\lambda_k + \lambda_l} |\langle k|H|l\rangle|^2. \tag{18}$$

where we have used the eigendecomposition of the state, $\rho = \sum_k \lambda_k |k\rangle\langle k|$. The formula highlights that the sensitivity of a probe, and therefore its usefulness for phase estimation, is quantified by the non-commutativity of its state with the Hamiltonian. In fact, the QFI measures the sensitivity of the state ρ to the unitary evolution $e^{-iH\theta}$, or, in other words, the speed of evolution of the probe under such dynamics. If and only if H is diagonal in the eigenbasis of ρ, the transformation leaves ρ invariant. It is easy to see that in that case $\mathcal{F}(\rho, H) = 0$.

3.2.1 Properties of the Quantum Fisher Information

Finally, we mention a non-exhaustive list of properties of the QFI, which will be useful in proofs later in this Section.

1. Up to a constant factor, the QFI is upper bounded by the variance, $\mathcal{F}(\rho, H) \leq 4V(\rho, H)$, where the equality is reached for pure states. More precisely, the QFI is the variance convex roof, $\mathcal{F}(\sum_i p_i|\psi_i\rangle, H) = 4 \inf_{\{p_i, |\psi_i\rangle\}} \sum_i p_i V(|\psi_i\rangle, H)$ [14].
2. The QFI is convex: $\mathcal{F}(p\rho_1 + (1-p)\rho_2, H) \leq p\mathcal{F}(\rho_1, H) + (1-p)\mathcal{F}(\rho_2, H)$.
3. For unitaries U, $\mathcal{F}(U\rho U^\dagger, H) = \mathcal{F}(\rho, U^\dagger H U)$.
4. The QFI is non-increasing under CPTP maps Φ which do not depend on the parameter: $\mathcal{F}(\Phi(\rho), H) \leq \mathcal{F}(\rho, H)$.

3.3 Interferometry and Non-classical Correlations

An important phase estimation scenario is represented by estimation through interferometric measurements (Fig. 3). Such template has been the testbed of the first observations of quantum phenomena, and it is still the standard textbook example to introduce students to quantum laws. Apart from the historical and pedagogical value, interferometry plays a premier role in modern quantum sensing schemes [13]. The architecture of an interferometric measurement is extremely simple. A bipartite system AB in the input state $\rho_{AB,0}$ is injected into a two-arm channel. Subsystem A undergoes a phase shift $U_A = e^{-iH_A^\Lambda \theta}$, generated by an Hamiltonian with non-degenerate spectrum Λ. This restriction is useful for understanding the role of non-classical correlations in this scenario. We remind that the phase θ represents the unknown perturbation we want to estimate, being not directly measurable. Its value is a function of the output visibility, i.e. the outcome statistics of a polarization

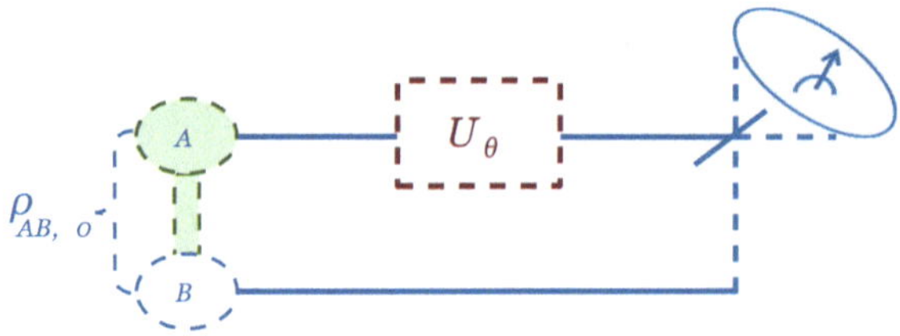

Fig. 3 Non-classical correlations guarantee non-vanishing precision in interferometric phase estimation. A bipartite system is prepared in an input state $\rho_{AB,0}$, and it is injected into a two-arm interferometer. A unitary transformation U_θ is applied to subsystem A. The Hamiltonian eigenbasis, i.e. the phase direction, is just revealed after the interaction. The value of the imprinted phase is estimated by a measurement at the output. The minimum precision of the estimation, as quantified by the Interferometric Power (IP), is a measure of non-classical correlations in the input. That is, non-classical correlations ensure non-vanishing precision for any Hamiltonian

measurement into the output $\rho_{AB,\theta} = (U_A^\Lambda \otimes \mathbb{I}_B)\rho_{AB,0}(U_A^\Lambda \otimes \mathbb{I}_B)^\dagger$. We here focus on the optimization of the input. If the Hamiltonian H_A^Λ is fully known, then coherence of the reduced state ρ_A in its eigenbasis, also called asymmetry in literature [17], is the necessary and sufficient resource of the phase estimation. In fact, the QFI $\mathcal{F}(\rho, H)$ is a measure of asymmetry of the state with respect to a unitary transformation generated by H. Here correlations seem to not play any role, a single party estimation is sufficient and the interferometric configuration appears redundant. Let us now introduce a further difficulty. We suppose that the estimation is blind, in the sense that only the spectrum of the Hamiltonian generating the phase imprinting is known during the input preparation. There is no prior information about the Hamiltonian eigenbasis. We allow to disclose the phase direction at the output, so that the measurement step can be still optimized, and the best estimator is reached. It is easy to see that there is no possible single system input ρ_A guaranteeing an arbitrary degree of precision for every possible Hamiltonian. In other words, the estimation by a single party relies on pure luck as the key information about the phase direction is missing. Let us consider what happens if instead we implement the interferometer to perform the estimation. One can prove that a classically correlated probe AB, or even a CQ state, are still insufficient to ensure precision for any Hamiltonian. On the other hand, by employing non-classically correlated states one can overcome the lack of knowledge about the phase direction [18]. Similarly to what happens for the LQU, it is possible to show that a quantifier for the worst-case precision is a bona fide measure of non-classical correlations. The optimal estimator is the one that saturates the Cramér–Rao bound in the limit of very large n, and in that case the quality of the input is determined by the QFI. The worst-case QFI for a given state reads

$$\mathcal{P}_A^\Lambda(\rho_{AB}) := \frac{1}{4}\min_{H_A}\mathcal{F}(\rho_{AB}, H_A^\Lambda), \tag{19}$$

where the minimisation is over all Hamiltonians with the given non-degenerate spectrum Λ (and where the factor 1/4 is chosen such that it cancels out the one in Eq. (18)

for the QFI under unitary dynamics). This quantity is called Interferometric Power (IP) of the state ρ_{AB} [18]. It quantifies the minimum sensitivity in interferometric phase estimation.

3.3.1 Interferometric Power as a Discord-Like Quantity

One can prove that the IP enjoys the same properties of the measures of non-classical correlations, as discussed in Sect. 2.2.1 for the LQU.

1. The IP is zero if and only if ρ_{AB} is CQ. If ρ_{AB} is CQ, then one can choose a Hamiltonian H_A^Λ which is diagonal in the local basis of A so that the QFI and hence the IP vanish. If, on the other hand, the IP is zero, the LQU has to be zero as well. Then we use the fact that the LQU vanishes if and only if ρ_{AB} is CQ.
2. The IP is invariant under local unitary transformations. It is clear from the expression in Eq. (18) that the QFI for Hamiltonians on A is invariant under local unitaries on B. On the same hand, local unitaries on A are absorbed in the definition of H_A^Λ, thus they do not affect the minimisation.
3. The IP is contractive under CPTP maps on the non-affected party B. This is easy to prove from the properties of the QFI itself, see Sect. 3.2.1, or alternatively by the following, more intuitive proof. Since any map Φ_B acting on B commutes with H_A^Λ, it can be included in the measurement process. Next, we note that the QFI quantifies the maximum precision that is achievable by picking the optimal estimation strategy. Since this maximum precision can only decrease when applying an extra map on B, we have that $\mathcal{F}(\rho_{AB}, H_A^\Lambda) \geq \mathcal{F}((\mathbb{1}_A \otimes \Phi_B)\rho_{AB}, H_A^\Lambda)$.
4. The IP reduces to an entanglement monotone for pure states. For pure states, the QFI is proportional to the variance of H_A^Λ, and the IP becomes equal to the LQU. The latter is known to be an entanglement for pure states.

3.3.2 Restriction to $\mathbb{C}^2 \otimes \mathbb{C}^d$

We report a simplified formula for the IP in the case A is a qubit, making it a computable measure of non-classical correlations for qubit-qudit systems. From the definition of QFI, one has $\mathcal{F}(\rho_{AB}, aH_A^\Lambda + b\mathbb{I}_A) = a^2\mathcal{F}(\rho_{AB}, H_A^\Lambda)$. By setting the spectrum Λ to be $\{1, -1\}$ one has $H_A = \boldsymbol{n} \cdot \boldsymbol{\sigma}$. The IP then becomes the minimisation of a quadratic form over the unit sphere, which leads to the following expression (like the one for the LQU of a qubit-qudit system):

$$\mathcal{P}_A^\Lambda(\rho_{AB}) = \lambda_{\min}(M_{AB}). \tag{20}$$

So the IP is the minimal eigenvalue of the 3×3-matrix M_{AB} with the following elements:

$$(M_{AB})_{mn} = \frac{1}{2} \sum_{i,j:p_i+p_j \neq 0} \frac{(p_i - p_j)^2}{p_i + p_j} \langle \psi_i | \sigma_{mA} \otimes \mathbb{1}_B | \psi_j \rangle_{AB} \langle \psi_j | \sigma_{nA} \otimes \mathbb{1}_B | \psi_i \rangle_{AB}, \tag{21}$$

where again we have used the eigendecomposition of the state $\rho_{AB} = \sum_i p_i |\psi_i\rangle \langle \psi_i|_{AB}$.

3.3.3 Interplay Between LQU and IP

We have shown how the LQU characterizes the minimum quantum uncertainty obtained upon measuring local observables. We here point out that the skew information and the QFI, and therefore the LQU and the IP, are closely related quantities. Both the skew information $\mathcal{I}(\rho, H)$ and the QFI given by $\mathcal{F}(\rho, H)$ measure the speed of evolution of a quantum state undergoing a unitary dynamics $e^{-iH\theta}$. In particular, they are associated with two metrics included in the Fisher metrics family, which is proven to be the only class of Riemannian metrics in the space of quantum states which is contractive under noisy maps [19]. Any speed measure obtained from such metrics is a measure of asymmetry. For classical probability distributions and stochastic processes, they all reduce to the classical Fisher information given in Eq. (15).

We observe that the following chain of inequalities holds:

$$\mathcal{I}(\rho, H) \leq \frac{1}{4}\mathcal{F}(\rho, H) \leq 2\mathcal{I}(\rho, H), \forall \rho, H. \tag{22}$$

This implies $\mathcal{U}_A^\Lambda(\rho_{AB}) \leq \mathcal{P}_A^\Lambda(\rho_{AB})$, and it makes possible to give a metrological interpretation to the LQU as well, by deriving an upper bound for the minimum variance in the interferometric scheme presented in Fig. 3. In order to estimate the parameter θ, we can optimise the input state ρ_{AB}, the Hamiltonian H_A, and the final measurement. As mentioned before, the quantum Cramér–Rao bound is saturated asymptotically by employing the most informative measurement, $V(\rho, \hat{\theta}_{\text{best}}) = 1/(n\mathcal{F}(\rho, H))$ [13, 14]. Therefore, few algebra steps show that for interferometric phase estimations one has

$$V(\rho_{AB}, \hat{\theta}_{\text{best}}) = \frac{1}{n\mathcal{F}(\rho_{AB}, H_A^\Lambda)} \leq \frac{1}{n\mathcal{P}_A^\Lambda(\rho_{AB})} \leq \frac{1}{4n\mathcal{U}_A^\Lambda(\rho_{AB})}. \tag{23}$$

Hence, (the inverse of) non-classical correlations upper bound the smallest possible variance of the estimator. In other words, it is guaranteed the existence of an Hamiltonian and a measurement such that the parameter θ can be estimated with a variance lower than a value determined by the amount of discord-like correlations and the number of the experiment repetitions. Note that in this set-up we assume perfect unitary evolution and ideal measurements, but that we allow for noise in the prepared input state ρ_{AB}.

3.4 Discriminating Strength

We here discuss a third measure of non-classical correlations which represents the worst-case precision in another metrology task, state discrimination [20]. We also show how it relates to the LQU and therefore the IP.

Suppose that we want to establish if n copies of a quantum system are prepared in a state ρ_1 or ρ_2, where each occurs with equal probability. It is allowed to obtain information by measuring the system. According to the Holevo–Helstrom theorem, the minimum error probability after optimising over all possible POVMs is given by

$$P^{(n)}_{\text{err,min}} := \frac{1}{2}\left(1 - \frac{1}{2}||\rho_1^{\otimes n} - \rho_2^{\otimes n}||_1\right), \tag{24}$$

where the optimal POVM discriminates the positive and negative eigenspaces of $\rho_1^{\otimes n} - \rho_2^{\otimes n}$. In the asymptotic limit of large n, the minimum error probability follows an exponential decay law

$$P^{(n)}_{\text{err,min}} \approx e^{-n\xi(\rho_1,\rho_2)}, \tag{25}$$

where the decay constant is given by

$$\xi(\rho_1,\rho_2) := -\lim_{n\to\infty}\frac{\ln P^{(n)}_{\text{err,min}}}{n} = -\ln\left(\min_{0\leq s\leq 1}\text{tr}[\rho_1^s\rho_2^{1-s})\right]. \tag{26}$$

Such limit is called quantum Chernoff bound [21]. Finally, we define the quantity

$$Q(\rho_1,\rho_2) := e^{-\xi(\rho_1,\rho_2)} = \min_{0\leq s\leq 1}\text{tr}[\rho_1^s\rho_2^{1-s}]. \tag{27}$$

It is immediately clear that $0 \leq Q(\rho_1,\rho_2) \leq \text{tr}[\rho_1^{1/2}\rho_2^{1/2}] \leq 1$, and, if at least one of the two states is pure, $Q(\rho_1,\rho_2)$ reduces to Uhlmann's fidelity $F(\rho_1,\rho_2) := \left(\text{tr}[\sqrt{\sqrt{\rho_1}\rho_2\sqrt{\rho_1}}]\right)^2$.

A state discrimination problem represents the discretized version of a phase estimation scenario, where instead of a continuous parameter θ one wishes to know the value of a two-value label identifying one of the two options $\rho_{1,2}$. It is then not surprising that non-classical correlations play a role in an interferometric state discrimination scheme called quantum illumination [22, 23]. The protocol runs as follows. An experimentalist Alice prepares n copies of a bipartite state ρ_{AB}, where A is the probe part and B is a reference system. A second player Charlie chooses an undisclosed unitary C_A from a given set of allowed transformations $\mathcal{S}$. Then, Alice sends her n copies to Charlie who is free to either leave the n copies unaltered, or rotate all of them by implementing C_A. Finally, Alice has to decide which of the two actions Charlie has chosen, being allowed to perform any POVM on the n copies. That means that she has to discriminate between $\rho_1^{\otimes n} = \rho_{AB}^{\otimes n}$ and $\rho_2^{\otimes n} = (C_A\rho_{AB}C_A^\dagger)^{\otimes n}$. The Discriminating Strength (DS) of the probe state ρ_{AB} is defined as thc Alice discriminating ability in the worst possible case:

$$\mathcal{D}_A^{\mathcal{S}}(\rho_{AB}) := 1 - \max_{C_A \in \mathcal{S}} Q\left(\rho_{AB}, C_A \rho_{AB} C_A^\dagger\right). \tag{28}$$

From the definition of the quantum Chernoff bound, it is clear that A is able to perform better if the DS is higher. Note that we meet again a context in which there is a clear asymmetry between the role played by the parts of a bipartite system.

So far, we have not specified what the set of allowed transformations $\mathcal{S}$ is, and the DS of course depends heavily on the choice of this set. A first observation is that if $\mathcal{S}$ were chosen to be the whole group of unitaries on A, the DS would always be zero as this group includes the identity. Clearly, we need to avoid such pathological case. We restrict the Charlie's choice within the set of unitaries $C_A = \exp(i H_A^\Lambda)$. In this parametrisation, H_A^Λ is an Hamiltonian acting on A, with non-degenerate spectrum Λ (notice the similarity with the LQU case): $H_A^\Lambda = U_A \text{diag}(\Lambda) U_A^\dagger$, where $U_A \in \text{U}(d_A)$. The DS is then define as

$$\mathcal{D}_A^\Lambda(\rho_{AB}) := 1 - \max_{H_A^\Lambda} Q\left(\rho_{AB}, e^{i H_A^\Lambda} \rho_{AB} e^{-i H_A^\Lambda}\right). \tag{29}$$

A crucial point to discuss is to what extent the DS depends on the choice of the spectrum Λ. Although in Ref. [20] the authors mention that it is tempting to conjecture that the harmonic spectrum (i.e. $\lambda_i - \lambda_{i+1}$ is constant, $\forall i$) should be optimal, no clear answer to this question is given. One obvious property of the DS is the invariance under constant shifts, $\mathcal{D}_A^\Lambda(\rho) = \mathcal{D}_A^{\Lambda+b}(\rho)$, $\forall \rho, b \in \mathbb{R}$.

3.4.1 Discriminating Strength as a Measure of Non-classical Correlations

The intuition behind the DS is that establishing whether the state has undergone a local rotation should be easier the more the part A potentially affected by the rotation is non-classically correlated with an unaffected part B. We here report the proof that the DS is a bona fide measure for non-classical correlations, as it has the same properties of the LQU and the IP discussed in Sects. 2.2.1 and 3.3.1 respectively.

1. The DS is zero if and only if ρ is CQ. The DS is zero if and only if there is a C_A such that $Q(\rho, C_A \rho C_A^\dagger) = 1$, which is the case if and only if $\rho = C_A \rho C_A^\dagger$. Since C_A has a non-degenerate spectrum, this is equivalent to require that ρ and H_A^Λ are diagonal in the same basis, i.e. ρ is CQ.
2. The DS is invariant under local unitary transformations. First we note that $(U \rho U^\dagger)^s = U \rho^s U^\dagger$ for any unitary U. Using this property and the cyclicity of the trace, it follows that Q is invariant under local unitaries on B. For local unitaries on A, we can use the same property and absorb the transformation in the Hamiltonian, since maximising over $U_A^\dagger H_A^\Lambda U_A$ is equivalent to maximising over H_A^Λ.

3. The DS is contractive under CPTP maps on the unchanged party B. Since any local map Φ_B commutes with the transformation on A induced by H_A^Λ, Φ_B can be absorbed in the POVM. The minimum error probability is obtained by minimising the error probability over all POVMs on $\rho_{AB}^{\otimes n}$. Absorbing the extra local map Φ_B can only increase the error probability, and hence Q is monotonically increasing: $\mathcal{D}_A^\Lambda(\Phi_B(\rho_{AB})) \leq \mathcal{D}_A^\Lambda(\rho_{AB})$.
4. The DS reduces to an entanglement monotone for pure states. If $|\psi\rangle_{AB}$ is transformed to $|\phi\rangle_{AB}$ under LOCC operations, we can write

$$|\phi\rangle\langle\phi|_{AB} = \sum_i M_{i,A} V_{i,B} |\psi\rangle\langle\psi|_{AB} M_{i,A}^\dagger V_{i,B}^\dagger, \tag{30}$$

where $\{M_{i,A}\}$ are Kraus operators on A and $\{V_{i,B}\}$ are unitaries on B. One has $M_{i,A} V_{i,B} |\psi\rangle_{AB} = \sqrt{p_i}|\phi\rangle_{AB}$. Similarly to the case of the LQU, one can prove that maximising over Hamiltonians on A is equivalent to maximising over Hamiltonians on B. Assume that $\tilde{H}_B^\Lambda$ achieves that maximum. Then one obtains

$$\begin{aligned} D_A^\Lambda(|\phi\rangle\langle\phi|_{AB}) &= 1 - \max_{H_B^\Lambda} \left| \langle\phi| e^{iH_B^\Lambda} |\phi\rangle_{AB} \right|^2 = 1 - \sum_i \frac{1}{p_i} \max_{H_B^\Lambda} \left| \langle\psi| M_{i,A}^\dagger e^{iH_B^\Lambda} M_{i,A} |\psi\rangle_{AB} \right|^2 \\ &\leq 1 - \sum_i \frac{1}{p_i} \left| \langle\psi| M_{i,A}^\dagger e^{i\tilde{H}_B^\Lambda} M_{i,A} |\psi\rangle_{AB} \right|^2 \\ &\leq 1 - \left| \langle\psi| \sum_i M_{i,A}^\dagger M_{i,a} e^{iH_B^\Lambda} |\psi\rangle_{AB} \right|^2 \\ &= 1 - \left| \langle\psi| e^{i\tilde{H}_B^\Lambda} |\psi\rangle_{AB} \right|^2 = D_A^\Lambda(|\psi\rangle\langle\psi|_{AB}). \end{aligned} \tag{31}$$

Note that in the first line $V_{i,B}$ and $V_{i,B}^\dagger$ are included into the maximisation over H_B^Λ; in the second line, we rely on the fact that the maximum of a function is lower bounded by the function evaluated at any given point, and the Cauchy–Schwarz inequality.

3.4.2 Interplay with the Local Quantum Uncertainty

The DS is related to the LQU. To show it, we remind that, for any given density matrix ρ and Hermitian operator O, the following result holds:

$$\min_{0\leq s\leq 1} \mathrm{tr}[\rho^s O \rho^{1-s} O] = \mathrm{tr}[\rho^{1/2} O \rho^{1/2} O]. \tag{32}$$

This is clear by writing ρ in terms of its eigenvectors $\{|\psi_i\rangle\}$ and by employing a non-increasing order for the eigenvalues λ_i:

$$\min_{0\le s\le 1}\text{tr}[\rho^s O\rho^{1-s}O] = \sum_i \lambda_i|\langle\psi_i|O|\psi_i\rangle|^2 + \min_{0\le s\le 1}\sum_{i<i'}(\lambda_i^s\lambda_{i'}^{1-s} + \lambda_{i'}^s\lambda_i^{1-s})|\langle\psi_i|O|\psi_i'\rangle|^2. \tag{33}$$

It is then easy to see that for each term in the second sum the minimum is achieved for $s = 1/2$, which proves the result. The link between LQU and DS is manifest by Taylor expanding $e^{iH_A^\Lambda}$ with respect to Λ:

$$\begin{aligned}\mathcal{D}_A^\Lambda(\rho_{AB}) &= 1 - \max_{\{H_A^\Lambda\}}\min_{0\le s\le 1}\text{tr}[\rho_{AB}^s e^{iH_A^\Lambda}\rho_{AB}^{1-s}e^{-iH_A^\Lambda}] \\ &= -\max_{\{H_A^\Lambda\}}\min_{0\le s\le 1}\text{tr}[\rho_{AB}^s H_A^\Lambda\rho_{AB}^{1-s}H_A^\Lambda - H_A^\Lambda\rho_{AB}H_A^\Lambda] + O(\Lambda^3) \\ &= -\max_{\{H_A^\Lambda\}}\text{tr}[\rho_{AB}^{1/2}H_A^\Lambda\rho_{AB}^{1/2}H_A^\Lambda - H_A^\Lambda\rho_{AB}H_A^\Lambda] + O(\Lambda^3) \\ &= \min_{\{H_A^\Lambda\}}\text{tr}[H_A^\Lambda\rho_{AB}H_A^\Lambda - \rho_{AB}^{1/2}H_A^\Lambda\rho_{AB}^{1/2}H_A^\Lambda] + O(\Lambda^3) \\ &= \mathcal{U}_A^\Lambda(\rho_{AB}) + O(\Lambda^3).\end{aligned} \tag{34}$$

where we have used Eq. (32) in the third line.

We observe that for small Λ the LQU can be interpreted as the DS in a discrimination task. In this statement, small Λ means that the local transformations should be close to the identity, i.e. only small perturbations are allowed.

3.4.3 Computable Expressions of the DS

In Ref. [20], the authors present expressions for the DS in a few special cases. We only give details about the derivation of the formula for qubit-qudit states, in analogy with the LQU, but other cases are mentioned for the sake of completeness.

First, let us consider pure bipartite states $|\psi\rangle_{AB}$. The Schmidt decomposition is given by $|\psi\rangle_{AB} = \sum_{i=1}^{\min\{d_A,d_B\}}\sqrt{\sigma_i}|i\rangle_A|i\rangle_B$ with Schmidt coefficients $\{\sigma_i\}$. Then, the DS is given by the following expression:

$$\mathcal{D}_A^\Lambda(|\psi\rangle\langle\psi|_{AB}) = 1 - \max_{\pi_\alpha}\left|\sum_k \sigma_{\pi_\alpha[k]}e^{i\lambda_k}\right|^2. \tag{35}$$

where we now have a maximisation over the group of permutations π_α on the Schmidt coefficients $\{\sigma_i\}$, instead of the maximisation over all Hamiltonians H_A^Λ with spectrum Λ (which is an infinite set). If $d_A > d_B$, the set of Schmidt coefficients should be extended with zeros to obtain a set of size d_A.

We mentioned before that it is tempting to hypothesise that the DS obtained by fixing an harmonic spectrum would yield the most accurate measure for non-classical correlations. Even though it is not clear if this is true, it explains why it is interesting to calculate the expression of the DS in such a case. By defining the fundamental frequency $\omega := |\lambda_i - \lambda_{i+1}| \leq 2\pi/d_A$, we can further simplify the previous formula. The permutation which maximises the second term gives the following values: σ_1=0, $\sigma_2 = \omega$, $\sigma_3 = -\omega$, $\sigma_4 = 2\omega$, $\sigma_5 = -2\omega$, etcetera. The resulting expression for the DS is then

$$\mathcal{D}_A^{\Lambda}(|\psi\rangle\langle\psi|_{AB}) = 1 - \left| \sum_{n=0}^{[(d_A+1)/2]-1} \sigma_{2n+1} e^{in\omega} + \sum_{n=1}^{d_A-[(d_A+1)/2]} \sigma_{2n} e^{in\omega} \right|^2 . \quad (36)$$

The precise details of this expression are not very relevant to our discussion, but it is noteworthy that we have managed to get rid of the optimisation over the unitaries.

We now analyse the qubit-qudit case (where the subsystem A is the qubit). The DS is invariant under constant shifts, thus we can parametrise the spectrum as $\{-\lambda, \lambda\}$. The Hamiltonian takes the form $H_A^{\Lambda} = \lambda \boldsymbol{n} \cdot \boldsymbol{\sigma}_A$. For conciseness of notation, we introduce $\boldsymbol{\sigma}_{A,n} := \boldsymbol{n} \cdot \boldsymbol{\sigma}_A$. The quantum Chernoff bound then reads

$$\begin{aligned} Q(\rho_1, \rho_2) &= \min_{0\leq s\leq 1} \text{tr}\left[\rho_{AB}^s e^{i\lambda\boldsymbol{\sigma}_{A,n}} \rho_{AB}^{1-s} e^{-i\lambda\boldsymbol{\sigma}_{A,n}}\right] \\ &= \cos^2\lambda + \min_{0\leq s\leq 1} \text{tr}[\rho_{AB}^s \boldsymbol{\sigma}_{A,n} \rho_{AB}^{1-s} \boldsymbol{\sigma}_{A,n}] \sin^2\lambda \\ &= \cos^2\lambda + \text{tr}[\rho_{AB}^{1/2} \boldsymbol{\sigma}_{A,n} \rho_{AB}^{1/2} \boldsymbol{\sigma}_{A,n}] \sin^2\lambda. \end{aligned} \quad (37)$$

Using this expression, we finally get the formula

$$\begin{aligned} \mathcal{D}_A^{\Lambda}(\rho_{AB}) &= \min_{\boldsymbol{n}} \left(1 - \text{tr}[\rho_{AB}^{1/2} \boldsymbol{\sigma}_{A,n} \rho_{AB}^{1/2} \boldsymbol{\sigma}_{A,n}]\right) \sin^2\lambda \\ &= \mathcal{U}_A^{\Lambda}(\rho_{AB}) \frac{\sin^2\lambda}{\lambda^2}. \end{aligned} \quad (38)$$

To summarise, there is a proportionality relation between the DS and the LQU for qubit-qudit systems, which turns out to be an equality when λ approaches zero, as $\frac{\sin^2\lambda}{\lambda^2} \to 1$.

4 Conclusion

We here reviewed recent works providing a metrological interpretation to non-classical correlations. Our understanding of an elusive, information-theoretic concept has been shaped by linking it to experimentally testable effects. State-observable complementarity implies genuine quantum uncertainty. Such uncertainty corresponds to sensitivity to a quantum evolution. The state rate of change triggers measurement precision of a complementary property. The peculiar asymmetry of non-classical correlations finds an operational interpretation in metrology, when such an argument is extended to compound systems. If and only if the state of a bipartite system shows non-classical correlations, Quantum Mechanics dictates sensitivity to local perturbations, which translates into a guaranteed minimum performance in paradigmatic scenarios as parameter estimation and state discrimination. The LQU, the IP and the DS are parent discord-like measures which capture this distinctive feature of quantum states. An interesting question is to establish if the metrologic measures of discord, which have been introduced to catch bipartite statistical dependence, can be extended to quantify multipartite correlations. We are actively working on the problem and we are able to anticipate that the answer is positive, while a complete study on the topic will be published in the near future. Such extension relies on employing non-unitary evolutions, where the information is imprinted by noisy channels. The scenario will provide an operational interpreation of multipartite non-classical correlations in more realistic scenarios, taking in account non-negligible errors in both state and gate preparations, and the presence of an environment.

Finally, we would like to point the Reader to further results on metrological measures of non-classical correlations. An experimental comparison of classical and quantum resources in interferometric phase estimation has been implemented in a room temperature NMR (Nuclear Magnetic Resonance) system [18]. Extensions of the reported results to continuous variable systems have been obtained [24–26]. Other geometric measures of non-classical correlations inspired by metrological tasks have also been proposed [27, 28].

References

1. D. Girolami, J. Phys. Conf. Ser. **626**, 012042 (2015)
2. H. Ollivier, W.H. Zurek, Phys. Rev. Lett. **88**, 017901 (2002)
3. L. Henderson, V. Vedral, J. Phys. A Math. Gen. **34**, 6899 (2001)
4. K. Modi, A. Brodutch, H. Cable, T. Paterek, V. Vedral, Rev. Mod. Phys. **84**, 1655 (2012)
5. P.A.M. Dirac, *The Principles of Quantum Mechanics* (Oxford University Press, Oxford, 1930)
6. G.I. Taylor, Proc. Cam. Phil. Soc. **15**, 114 (1909)
7. J. von Neumann, *Mathematical Foundations of Quantum Mechanics* (Princeton University Press, New Jersey, 1955)
8. S. Luo, Phys. Rev. Lett. **91**, 180403 (2003)
9. F. Herbut, J. Phys. A **38**, 2959 (2005)
10. E.P. Wigner, M.M. Yanase, Proc. Natl. Acad. Sci. USA **49**, 910 (1963)
11. W. Heisenberg, Z. Phys. **43**, 172 (1927)
12. D. Girolami, T. Tufarelli, G. Adesso, Phys. Rev. Lett. **110**, 240402 (2013)
13. V. Giovannetti, S. Lloyd, L. Maccone, Nature Photon. **5**, 222 (2011)
14. G. Tóth, I. Apellaniz, J. Phys. A Math. Theor. **47**, 424006 (2014)
15. C.W. Helstrom, *Quantum Detection and Estimation Theory* (Academic Press, Cambridge, 1976)
16. S.L. Braunstein, C.M. Caves, Phys. Rev. Lett. **72**, 3439 (1994)
17. S.D. Bartlett, T. Rudolph, R.W. Spekkens, Rev. Mod. Phys. **79**, 555 (2007)
18. D. Girolami, A.M. Souza, V. Giovannetti, T. Tufarelli, J.G. Filgueiras, R.S. Sarthour, D.O. Soares-Pinto, I.S. Oliveira, G. Adesso, Phys. Rev. Lett. **112**, 210401 (2014)
19. I. Bengtsson, K. Zyczkowski, *Geometry of Quantum States* (Cambridge University Press, Cambridge, 2007)
20. A. Farace, A. De Pasquale, L. Rigovacca, V. Giovannetti, New J. Phys. **16**, 073010 (2014)
21. K.M.R. Audenaert, J. Calsamiglia, R. Munoz-Tapia, E. Bagan, Ll Masanes, A. Acin, F. Verstraete, Phys. Rev. Lett. **98**, 160501 (2007)

22. S. Lloyd, Science **321**, 1463 (2008)
23. C. Weedbrook, S. Pirandola, J. Thompson, V. Vedral, M. Gu, New J. Phys. **18**, 043027 (2016)
24. G. Adesso, Phys. Rev. A **90**, 022321 (2014)
25. L.A.M. Souza, H.S. Dhar, M.N. Bera, P. Liuzzo-Scorpo, G. Adesso, Phys. Rev. A **92**, 052122 (2015)
26. L. Rigovacca, A. Farace, A. De Pasquale, V. Giovannetti, Phys. Rev. A **92**, 042331 (2015)
27. D. Spehner, M. Orszag, New J. Phys. **15**, 103001 (2013)
28. W. Roga, S.M. Giampaolo, F. Illuminati, J. Phys. A Math. Theor. **47**, 365301 (2014)

Part II
Operational Interpretations and Applications

Why Should We Care About Quantum Discord?

Aharon Brodutch and Daniel R. Terno

Abstract Entanglement is a central feature of quantum theory. Mathematical properties and physical applications of pure state entanglement make it a template to study quantum correlations. However, an extension of entanglement measures to mixed states in terms of separability does not always correspond to all the operational aspects. Quantum discord measures allow an alternative way to extend the idea of quantum correlations to mixed states. In many cases these extensions are motivated by physical scenarios and quantum information protocols. In this chapter we discuss several settings involving correlated quantum systems, ranging from distributed gates to detectors testing quantum fields. In each setting we show how entanglement fails to capture the relevant features of the correlated system, and discuss the role of discord as a possible alternative.

1 Introduction

Entanglement has been hailed as the quintessential feature of quantum mechanics. In Schrödinger's words it is not "*one* but rather *the* characteristic trait of quantum mechanics, the one that enforces its entire departure from classical lines of thought"

A. Brodutch (✉)
Center for Quantum Information and Quantum Control, University of Toronto,
Toronto, ON, Canada
e-mail: brodutch@physics.utoronto.ca

A. Brodutch
Department of Physics, Institute for Optical Sciences, University of Toronto,
Toronto, ON, Canada

A. Brodutch
The Edward S. Rogers Department of Electrical and Computer Engineering,
University of Toronto, Toronto, ON, Canada

D.R. Terno
Department of Physics and Astronomy, Macquarie University,
Sydney, NSW 2109, Australia
e-mail: daniel.terno@mq.edu.au

F.F. Fanchini et al. (eds.), *Lectures on General Quantum Correlations and their Applications*, Quantum Science and Technology,
DOI 10.1007/978-3-319-53412-1_8

[1]. While its role as the only characteristic trait of quantum mechanics has been challenged, it is clear that pure bipartite entangled states play an essential role in uniquely quantum phenomena such as Bell non-locality, steering and teleportation [2, 3]. These phenomena are not restricted to pure bipartite states, and their relation to entanglement becomes less trivial as we move to mixed states or ensembles of pure states. For example, it is known that not all entangled states are Bell nonlocal, steerable or useful for teleportation. Moreover, the quantification of entanglement becomes more complicated as we step away from the pure bipartite scenario where all measures of entanglement are functions of the spectrum of the reduced states [2]. For mixed states there is a multitude of entanglement measures, matching different information processing tasks and, while all vanish on separable states, some vanish for specific entangled states. A well known example is the distillable entanglement which vanishes for bound entangled states [2, 4].

The fact that some mixed entangled states do not always exhibit properties that are directly related to the entanglement in pure states is the first hint that it does not fully capture the departure from classicality. The second hint in this direction is that some separable mixed states exhibit properties that are associated with entanglement for pure states. In pure states, entanglement and classical correlation are synonymous and some properties may be mistakenly identified with the former instead of the latter. However, a number of phenomena are related to correlations on the one hand, but seem to be outside the scope of classical correlations on the other, can be observed with separable mixed states. It has been argued that entanglement is only a special case of more general types of quantum correlations. These ideas have led to a great amount of work in trying to quantify these quantum correlations using various measures that have become known under the collective name of quantum discord (see, e.g., [5] and references therein).[1]

In this chapter we present an overview of some scenarios where quantum correlations in bipartite systems are not synonymous with entanglement. We begin with a brief discussion of entanglement in pure and mixed states, pointing out some examples where entangled mixed states do not have all the properties associated with entangled pure states. We continue with a brief introduction to discord, focusing on one particular discord measure. We then move on to three examples of phenomena that involve quantum correlations and are in some sense related to measurement disturbance. In the first example, we examine the ability to distinguish between orthogonal pure bipartite product states. In the second example, we discuss more general scenarios where (the lack of) entanglement in the input and output states fails to indicate the non-local nature of a quantum protocol. In the final example we consider a scenario where discord is a better figure of merit than entanglement for capturing a non-classical nature of the physical system.

---

[1] In some cases quantum correlations and quantum discord have been used interchangeably, in other cases quantum correlations have been used as a synonym for entanglement. Here we use the term *discord* when referring to discord-like quantities and *quantum correlations* when referring to a more operational aspect which may or may not relate to either discord and/or entanglement.

2 Pure State Entanglement, Mixed State Entanglement and Discord

2.1 Mathematical Preliminaries and Notation

We consider quantum states that are shared between two distant parties Alice and Bob.[2] Subscripts (e.g., A, B, AB) denote subsytems: for example we will consider a bipartite state ρ_{AB} whose local reduced states $\rho_A = \mathrm{tr}_B \rho_{AB}$ and $\rho_B = \mathrm{tr}_B \rho_{AB}$ are controlled by Alice and Bob, respectively. Here tr_K means a partial trace over the subsystem K.

A classical probability distribution represented by the set of probabilities $\{p_k\}$ can be encoded in the quantum state $\sum_k p_k |k\rangle\langle k|$, where $|k\rangle$ are the normalized orthogonal *computational basis* states. If the probability distribution is bipartite it can be encoded in the state $\sum_{k,l} p_{kl} |k\rangle\langle k|_A \otimes |k\rangle\langle k|_B$.

We use entropic measures to quantify most of our information theoretic quantities [4, 6, 7]. These will be based on the von Neumann entropy, that reduces to the (classical) Shannon entropy when the states represent classical distributions. The von Neumann entropy of a state ρ is defined as $S(\rho) = -\mathrm{tr}\rho \log \rho$ (all logarithms used here are base 2). It is non-negative and vanishes only for a pure state $S(|\psi\rangle\langle\psi|) = 0$. The state of maximal entropy on a d dimensional system is the d dimensional maximally mixed state I_d, $S(I_d) = \log d$.

An entropic measure of correlations in a quantum state ρ is given by the quantum mutual information

$$I(\rho_{AB}) = S(\rho_A) + S(\rho_B) - S(\rho_{AB}), \tag{1}$$

which is one particular way to extend the corresponding classical quantity [4, 5]. The original motivation for discord was based on the difference between various ways of extending the classical mutual information [7] to quantum states. A reader who is unfamiliar with the original motivation for discord is encouraged to read the original papers [8, 9] or one of the reviews on the subject [5, 10].

Fidelity [4] is a measure of closeness between quantum states. It is defined as $F(\rho, \sigma) = \mathrm{tr}\sqrt{\sqrt{\rho}\sigma\sqrt{\rho}}$ and has the following properties:

$$F(|\psi\rangle\langle\psi|, |\phi\rangle\langle\phi|) = |\langle\psi|\phi\rangle|, \tag{2a}$$

$$F(\rho_1 \otimes \sigma_1, \rho_2 \otimes \sigma_2) = F(\rho_1, \rho_2)F(\sigma_1\sigma_2), \tag{2b}$$

$$F(\rho, \sigma) = F(U\rho U^\dagger, U\sigma U^\dagger), \qquad \text{For all unitaries } U \tag{2c}$$

$$F(\rho, \sigma) \leq F(\Phi(\rho), \Phi(\sigma)), \qquad \text{For all quantum channels } \Phi. \tag{2d}$$

[2] We assume that the identity of the subsystems is unambiguous. The extension to systems of identical particles (where the position wave-function has to be accounted for explicitly) is mentioned in Sect. 6.

In the last equation a quantum channel Φ is represented by a completely positive trace preserving map [4].

2.2 Pure State Entanglement and LOCC

A pure bipartite state $|\psi\rangle_{AB} \in \mathcal{H}_A \otimes \mathcal{H}_B$ can always be brought into the Schmidt form $|\psi\rangle_{AB} = \sum_k \lambda_k |\alpha_k\rangle_A |\beta_k\rangle_B$ where λ_k are unique positive numbers (the Schmidt coefficients), and $\{|\alpha_k\rangle_A\}$, $\{|\beta_k\rangle_B\}$ are complete orthogonal bases for $\mathcal{H}_A$ and $\mathcal{H}_B$ respectively. The state $|\psi\rangle_{AB}$ is separable (and also a product state) if and only if it can be decomposed as $|\psi\rangle_{AB} = |\alpha_1\rangle_A |\beta_1\rangle_B$, i.e. it only has one non-zero Schmidt coefficient. Pure states that are not separable are called entangled.

The amount of entanglement in a pure state can be quantified in various ways that depend only on the Schmidt coefficients [2, 4]. Noting that $\rho_A = \sum_k \lambda_k^2 |\alpha_k\rangle\langle\alpha_k|$ and $\rho_B = \sum_k \lambda_k^2 |\beta_k\rangle\langle\beta_k|$ we see that the local states contain all the relevant information about entanglement. Direct product states are parameterized by strictly fewer parameters than arbitrary pure states in the same bipartite Hilbert space. Consequently, the direct product states are of measure zero in the set of all pure states.

A pure state can be described as a state of the maximal knowledge, i.e. zero entropy. If a pure state is entangled, its reduced states are no longer in such a state of maximal knowledge, i.e. the local entropies are non-zero. A pure state is maximally entangled when the knowledge about the local states is minimal, i.e. these states are completely mixed and thus have maximal entropy. In general the entanglement entropy

$$E(|\psi\rangle\langle\psi|_{AB}) = S(\rho_A) = S(\rho_B), \tag{3}$$

is a preferred measure of a pure bipartite entanglement [2]. It equals to the Shannon entropy of the Schmidt coefficients. Since the entropy of a pure state is zero, the mutual information is $I\left(|\psi\rangle\langle\psi|_{AB}\right) = S(\rho_A) + S(\rho_B) = 2S(\rho_A) = 2E(|\psi\rangle\langle\psi|_{AB})$.

To further study the properties of pure entangled states we will describe their role in two operational tasks: Bell inequality violations and distillation. The Bell-type experiment can be used to verify that a given state shared by Alice and Bob does not have a local realistic description in terms of hidden variables [3, 11]. A state that violates a Bell inequality is known as Bell non-local. A pure state is Bell non-local if and only if it is entangled [3].

In the Bell-type experiments Alice and Bob cannot communicate. Scenarios where Alice and Bob can perform arbitrary local quantum operations on their subsystems and communicate classically, but cannot send quantum information to each other belong to the paradigm of *local operations and classical communications* (LOCC). If Alice and Bob share some maximally entangled pairs they can use LOCC to perform tasks that cannot be performed locally, e.g. by using teleportation to send quantum information to each other. Consequently, if they share an unlimited supply of maximally entangled pairs they can perform any quantum operation in a finite amount of time. If, on the other hand, they have a finite amount of partially entangled pairs, they can use LOCC to distill them into maximally entangled pairs and use them for teleportation or other tasks. A supply of entangled (but not maximally entangled) pure states can always be distilled into a smaller supply of states that are more entangled [4].

Before moving on to mixed states, we recap a few properties of pure state bipartite entanglement that (as shown below) do not carry over to mixed states:

- All separable pure states are product states (correlations $\Leftrightarrow$ entanglement).
- Local mixed states imply a global entangled pure state (and the local states have the same spectrum).
- Pure product states are zero measure in the set of all pure states.
- All pure entangled states are distillable and can be used to violate a Bell inequality.

2.3 Mixed State Entanglement

A generic state ρ_{AB} (i.e. a trace 1 positive-semidefinite operator on $\mathcal{H}_A \otimes \mathcal{H}_B$) is a product state if it can be represented as $\rho_A \otimes \rho_B$. It is a separable state if it can be decomposed as

$$\rho_{AB} = \sum_k \alpha_k \tau_A^k \otimes \omega_B^k \tag{4}$$

(here $\{\tau_A^k\}$ and $\{\omega_B^k\}$ are sets of local states and $\{\alpha^k\}$ is a set of probabilities). If a state is not separable it is entangled.

Unlike pure states, not all separable states are product states. If a state is not a product state, it is correlated as can be verified using mutual information and the fact that $S(\rho_{AB}) = S(\rho_A) + S(\rho_B) \Leftrightarrow \rho_{AB} = \rho_A \otimes \rho_B$. If a mixed state is correlated it is not necessarily entangled, but if it is entangled it must be correlated. It is easy to verify whether a state is correlated or not, but it is usually difficult to verify whether it is separable or entangled.

The set of separable mixed states is dense. The simplest way to see this is by showing that for small enough $p > 0$ the states of the form

$$\rho^{p,\psi} = p|\psi\rangle\langle\psi|_{AB} + \frac{(1-p)}{4}\mathbb{1}_n \tag{5}$$

(where $\mathbb{1}_n$ is the n qubit identity) are separable for any normalized $|\psi\rangle_{AB}$. A state of this type is called pseudo pure and is a natural state in various implementations of quantum computing.

One interesting family of bipartite pseudo pure states is the family of two qubit Werner states [2, 12]. Denote the maximally entangled singlet state $|\Psi^-\rangle = \frac{1}{\sqrt{2}}[|01\rangle - |10\rangle]$. The two-qubit Werner state is

$$\rho^{W,p} = p|\Psi^-\rangle\langle\Psi^-| + \frac{(1-p)}{4}\mathbb{1}_2 \tag{6}$$

We can think of this $\rho^{W,p}$ as a depolarized singlet state. This state is entangled for $p > 1/3$ [2], but not Bell non-local for $p < 0.66$ [13].

Another difference from pure states is that not all entangled mixed states can be distilled. States that are entangled but cannot be distilled are called *bound entangled.* There are no bound entangled states for a pair of qubits or a qubit and a qutrit. A pair of qutrits provides a simple example using the so-called tile basis and a stopper tile [14]. The tile basis is formed by the orthogonal basis states

$$|\psi_{\frac{1}{2}}\rangle = \tfrac{1}{\sqrt{2}}|0\rangle \otimes \big(|0\rangle \pm |1\rangle\big), \tag{7a}$$

$$|\psi_{\frac{3}{4}}\rangle = \tfrac{1}{\sqrt{2}}|2\rangle \otimes \big(|1\rangle \pm |2\rangle\big), \tag{7b}$$

$$|\psi_5\rangle = |1\rangle \otimes |1\rangle, \tag{7c}$$

$$|\psi_{\frac{6}{7}}\rangle = \tfrac{1}{\sqrt{2}}\big(|0\rangle \pm |1\rangle\big) \otimes |2\rangle, \tag{7d}$$

$$|\psi_{\frac{8}{9}}\rangle = \tfrac{1}{\sqrt{2}}\big(|1\rangle \pm |2\rangle\big) \otimes |0\rangle, \tag{7e}$$

and the stopper tile

$$|\psi_S\rangle = \tfrac{1}{3}\big(|0\rangle + |1\rangle + |2\rangle\big] \otimes \big[|0\rangle + |1\rangle + |2\rangle\big). \tag{8}$$

It is possible to show that the state

$$\rho_{AB} = \tfrac{1}{4}\Big(\mathbb{1}_9 - \sum_{i\in\{2,4,7,9,S\}} |\psi_i\rangle\langle\psi_i|\Big), \tag{9}$$

is bound entangled [14].

The fact that some mixed entangled states are not distillable and some cannot be used to violate a Bell inequality suggests that at least some of the properties associated with pure state entanglement are not shared by all (mixed) entangled states. In the following we discuss the opposite scenario, i.e. situations where a property that we would intuitively associate with entangled states carries over to correlated separable states.

2.4 Discord

The idea of quantifying quantum correlation beyond entanglement originally appeared in the studies of decoherence [15]. Within this framework it was noted that entanglement is not sufficient for capturing all quantum correlations and that some separable states retain some quantum properties. At around the same time, a number of different versions of quantum discord and a similar idea called the information deficit were used to quantify non-classicality in various scenarios (for a review see [5]). These quantities usually vanish for one of three families of classical states, often called Quantum-Classical, Classical-Quantum and Classical-Classical (although a few vanish for more general families such as product-basis states). A state ρ_{AB} is called Classical-Quantum if there is a basis on $\{|a\rangle\}$ for $\mathcal{H}_A$ and a set of states $\{\tau^a\}$ on $\mathcal{H}_B$ such that

$$\rho_{AB} = \sum_a \alpha_a |a\rangle\langle a| \otimes \tau^a. \tag{10}$$

where α_a are probabilities that sum up to 1. The state is Quantum-Classical if it has the same structure with A and B swapped and it is Classical-Classical if it is both Classical-Quantum and Quantum-Classical. These families are all measure zero in the set of all states. Various versions of discord can be described as different ways of quantifying the 'distance' from the desired family of classical states. One way to introduce them is by calculating the difference between the quantum mutual information $I(\rho_{AB})$ that is given by Eq. (1) and different measurement-dependent quantum generalizations of the classically equivalent expression [7–9],

$$J^{\Pi^A} := S(\rho_B) - S(\rho_B|\Pi^A), \tag{11}$$

where the conditional entropy depends on the measurement on A that is described by a positive operator-valued measure [4] Π^A via

$$S(\rho_B|\Pi^A) = \sum_a p_a S(\rho_{B|a}), \tag{12}$$

where the probability p_a of the outcome a is $p_a = \mathrm{tr}\rho_A \Pi_a$, and $\rho_{B|a}$ is the state of B conditioned on obtaining the outcome a. Different choices of optimization condition that determines the measurement selection lead to different versions of discord [5],

$$D(\rho_{AB}) = I(\rho_{AB}) - J(\rho_{AB}). \tag{13}$$

In this work we focus on a specific version of discord which we call D_3 [16]. It has the advantage of being easy to calculate and providing an upper bound on some other discord measures. Most importantly it vanishes if and only if the states are the Classical-Quantum states of Eq. (10).

Given a state ρ_{AB} with marginals ρ_A and ρ_B we define the local basis $\{|l_a\rangle\}$ to be the basis where ρ_A is diagonal (note that this is not well defined when ρ_A has a degenerate spectrum). The dephasing channel Φ_l is defined as[3]

$$\Phi_l(\rho_{AB}) = \sum_a |l_a\rangle\langle l_a|\rho_{AB}|l_a\rangle\langle l_a| \tag{14}$$

The quantity D_3 is the loss of correlations under this channel

[3]It should be noted that this channel depends on the state, and is therefore not linear [17].

$$D_3(\rho_{AB}) = I(\rho_{AB}) - I\left[\Phi_l(\rho_{AB})\right] \tag{15}$$

As a simple example we can consider the Werner state (6). If it was classically correlated, the correlations would, in principle, be immune to decoherence, however it can easily be verified that the mutual information for a Werner state gets degraded when one of the qubits is decohered. In this sense the Werner state is always (for $p > 0$) non-classically correlated. This is in-fact true for any pseudo pure state with $|\psi\rangle$ entangled.

3 Local Distinguishability and the Failure of Discord

One of the first hints that separability does not imply classicality in the context of correlations was the discovery of *non-locality without entanglement* [14]. Consider a bipartite system of two qutrits and the set of nine orthonormal basis states of Eq. (7). Imagine the following task: Alice and Bob are given one of these orthogonal states and are asked to identify which one it is, they can communicate but cannot use any shared entanglement. Despite the fact that these are orthogonal product states the task cannot be completed deterministically. Any LOCC protocol used to identify out one of these nine states will misidentify some states with some probability. In other words, any protocol that can perfectly identify all the nine states must include quantum communication and is in that sense non-local. We can also say that these states are non-classically correlated although they are separable.

Now, let us assume that the a priori probability for each of the nine states is $1/9$, in such a case we can construct a density matrix ρ_{AB} that represents Alice and Bob's knowledge about the unknown state. Since these states are an orthonormal basis, their equal mixture is the maximally mixed state, $\rho_{AB} = \mathbb{1}_9$. In that sense, we can see that the non-classical correlations in this scenario cannot be captured by discord in the average state since the maximally mixed state is not correlated [16]. A natural approach for correcting this problem is to quantify quantum correlations in a different way for ensembles. Here we will consider a simple definition of classical ensembles which is motivated by other approaches [18, 19], but requires fewer formalities.

An ensemble $\{\rho^i_{AB}\}$ is classical if and only if for any choice of non-negative coefficients $\{\alpha_i\}$, such that $\sum_i \alpha_i = 1$ the state $\rho^{\{\alpha_i\}}_{AB} = \sum_i \alpha_i \rho^i_{AB}$ is classical. It is clear that the ensemble $\{|\psi_i\rangle\}$ is not classical in this sense. However, neither is an ensemble that consists of two orthogonal maximally entangled pure states [16]. Now on the one hand an ensemble of two orthogonal maximally entangled states is not a classical ensemble (by the above definition), on the other hand, it is well known that any two orthogonal states can be distinguished using LOCC. Consequently, the notion of non-classically correlated ensembles which we described above does not seem to play a role in locally distinguishing between orthogonal states.

4 Restricted Distributed Gates

The process of identifying an unknown state $|\phi_k\rangle_{AB}$ from the set of orthogonal states $\{|\phi_i\rangle\}$ can be described as an isometry that takes the state $|\phi_i\rangle$ from the space $\mathcal{H}_A \otimes \mathcal{H}_B$ to the state $|i\rangle \otimes |i\rangle$ on a different space $\mathcal{H}_{A'} \otimes \mathcal{H}_{B'}$, where the orthogonal states $|i\rangle$ are quantum pointers to the 'classical' labels. The *restricted, distributed gates* paradigm [20, 21] is set up along the same lines but with different restrictions.

Consider a unitary operation $\mathcal{G}(\rho) = U\rho U^\dagger$ (a quantum gate) and a subset of states $\mathbb{S} = \{\rho^i_{AB}\}$. Now consider the family of channels $\mathcal{G}_\mathbb{S}$ defined through

$$\mathcal{G}_\mathbb{S}(\rho^i_{AB}) = \mathcal{G}(\rho^i_{AB}), \qquad \forall \rho^i_{AB} \in \mathbb{S} \tag{16}$$

We call such a channel $\mathcal{G}_\mathbb{S}$ a distributed gate if it can be implemented using LOCC. There are situations where $\mathcal{G}_\mathbb{S}$ cannot be distributed without shared entanglement resources, even when both $\mathbb{S}$ and $\mathbb{S}' = \{\mathcal{G}(\rho^i_{AB})|\rho^i_{AB} \in \mathbb{S}\}$ contain only separable states. This restriction holds even when the set $\mathbb{S}$ is very small —- in fact it can contain only two states [20, 21].[4]

We begin with the simplest case [20] where $\mathbb{S} = \{|\psi_1\rangle, |\psi_2\rangle\}$ contains two non-orthogonal pure product states $|\psi_i\rangle = |a_i\rangle|b_i\rangle$, $\langle\psi_1|\psi_2\rangle \neq 0$. In such a case $\mathcal{G}_\mathbb{S}$ can be implemented using LOCC if and only if $\mathcal{G}_\mathbb{S} = \mathcal{G}'_A \otimes \mathcal{G}'_B$, where $\mathcal{G}'_A$ and $\mathcal{G}'_B$ are unitary gates. In other words if $\mathcal{G}_\mathbb{S}$ changes correlations (classical or quantum) for any convex combination of the two states, $\rho_{AB} = \alpha|\psi_1\rangle\langle\psi_1| + (1-\alpha)|\psi_2\rangle\langle\psi_2|$, then it cannot be implemented using LOCC.

The proof of this statement is as follows. Denote $|\psi_i^f\rangle\langle\psi_i^f| = \mathcal{G}(|\psi_i\rangle\langle\psi_i|)$. The execution by LOCC of a unitary gate implies that the output states are pure and separable,

$$|\psi_i^f\rangle = |a_i^f\rangle|b_i^f\rangle, \tag{17}$$

as well as the states at all intermediate steps [20]. Now, consider the protocol Alice and Bob need to use to implement the gate. Without loss of generality, we can assume that the protocol is broken into rounds where one party performs an operation and sends the classical outcomes of their measurement to the other party. We can also assume that the classical measurement results are recorded as quantum states. Since fidelity is non-decreasing under quantum channels, Eq. (2d), it also cannot increase at any point due to unitarity of $\mathcal{G}$ which implies it is unchanged at the end of the process.

Assume that Alice acts first by performing some operation, possibly including a measurement on her input state that corresponds to a classical message k that she sends to Bob. When averaged over many implementations of the protocol, it results in a channel Φ_A. As a result of Eq. (2b) the fidelity

$$F\big(|a_1\rangle\langle a_1|, |a_2\rangle\langle a_2|\big) = |\langle a_1|a_2\rangle| = F\big(\Phi_A(|a_1\rangle\langle a_1|), \Phi_A(|a_2\rangle\langle a_2|)\big), \tag{18}$$

[4]Note that if $\mathbb{S}'$ contains only one state and this state is separable then the transformation $\mathcal{G}_\mathbb{S}$ is trivial in LOCC.

is preserved. Alice's state is now is either of

$$\Phi_A(|a_1\rangle\langle a_1|) = \sum_k p_1^k |a_1^k\rangle\langle a_1^k| \otimes |k\rangle\langle k|^{A'}, \tag{19a}$$

$$\Phi_A(|a_2\rangle\langle a_2|) = \sum_k p_2^k |a_2^k\rangle\langle a_2^k| \otimes |k\rangle\langle k|^{A'}, \tag{19b}$$

where p_i^k are the probabilities of obtaining the outcome k given the state $i = 1, 2$, and the subsystem A' holds the classical information to be sent to Bob. Since the pointer states on A' are orthogonal, the fidelity satisfies

$$|\langle a_1|a_2\rangle| = F\big(\Phi_A(|a_1\rangle\langle a_1|), \Phi_A(|a_2\rangle\langle a_2|)\big) = \sum_k \sqrt{p_1^k p_2^k}\, \big|\langle a_1^k|a_2^k\rangle\big|, \tag{20}$$

However, since

$$\sum_k \sqrt{p_1^k p_2^k} \leq 1, \tag{21}$$

Equation (20) cannot be satisfied unless either the probability distributions coincide (which implies that Alice has no relevant information to send Bob[5]) and

$$\big|\langle a_1^k|a_2^k\rangle\big| = |\langle a_1|a_2\rangle|, \qquad \forall k, \tag{22}$$

or there must be some l with

$$\big|\langle a_1^l|a_2^l\rangle\big| > |\langle a_1|a_2\rangle|. \tag{23}$$

However, if Alice gets the result l (that she will send to Bob) and then the two parties proceed with the successful implementation of the protocol, they must deterministically decrease the fidelity in at least one stage on the way to the final state. This contradicts the non-decreasing of fidelity in quantum channels, Eq. (2d). The conclusion is that Alice gets no useful information during the measurement and has nothing to send Bob. Consequently the overall transformation must be implemented by local unitary operations.

In the general case, it can be shown [21] that if $\mathbb{S}$ contains only two states: ρ and the maximally mixed state, then an LOCC $\mathcal{G}_\mathbb{S}$ cannot change the correlations in ρ unless there is some measurement that leaves ρ invariant. This suggests that the maximally mixed state may play an important role in increasing the quantum resources required by a quantum protocol.

[5]If Alice has no relevant information to send Bob, then by symmetry Bob cannot have any relevant information to send Alice, and the protocol should not involve any communication.

5 Discord and Unruh-DeWitt Detectors

Various discord-like quantities were calculated in a number of problems of relativistic quantum information [5, 22]. The scenario we consider below is interesting from several points of view. The state ρ_{AB} of the two detectors that are used to characterize the vacuum entanglement belongs to the family of the X-states at all orders of the perturbation theory; the discord D_3 is a natural quantity to characterize quantumness of correlations; correlations and discord persist in the region of strictly zero entanglement.

5.1 The Model

From the point of view of local observers the vacuum state of any quantum field is entangled, and thus localized vacuum fluctuations are correlated [22]. It was demonstrated that vacuum correlations measured by local inertial observers can, in principle, violate Bell-type inequalities [23]. Further, it is known that localized particle detectors can extract entanglement form the vacuum state of a quantum field, even while remaining spacelike separated [24–26].

An Unruh-DeWitt detector is a two-level quantum systems that interacts with (a real massless) scalar field ϕ via a monopole coupling [27]. It is a popular tool in analysis of entanglement in quantum fields. The time-dependent interaction Hamiltonian in the interaction picture is given by

$$H_I(\tau) = \lambda(\tau)\left(e^{i\Omega\tau}\sigma^+ + e^{-i\Omega\tau}\sigma^-\right)\phi\left[x(\tau)\right], \tag{24}$$

where τ is the proper time of the detector, $\lambda(\tau)$ is a weak time-dependent coupling parameter that controls the strength and length of the interaction, Ω is the energy gap between the ground state $|0\rangle_d$ of the detector and its excited state $|1\rangle_d$, $\sigma^\pm$ are SU(2) ladder operators that act on the state of the detector according to $\sigma^+|0\rangle_d = |1\rangle_d$, $\sigma^-|1\rangle_d = |0\rangle_d$, $(\sigma^\pm)^2 = 0$, and $\phi(x(\tau))$ is the field evaluated along the trajectory of the detector.

It is convenient to parameterize the time evolution by the common coordinate time t [28]. We express the coupling parameter as $\lambda(t) = \epsilon_0\epsilon(t)$, where $\epsilon_0 \ll 1$ is the coupling strength and $\epsilon(t) = e^{-t^2/2\sigma^2}$ is a Gaussian switching function.

Prior to the interaction the detectors had been in their ground states $|0\rangle_A$ and $|0\rangle_B$, and the field in the vacuum state $|0\rangle$, hence the initial joint state of the two detectors and field was given by $|\Psi\rangle = |0\rangle_A|0\rangle_B|0\rangle$. The unitary evolution of the detectors-field system is given by

$$U = \hat{T}e^{-i\int dt[H_A(t)+H_B(t)]}, \tag{25}$$

where $\hat{T}$ denotes time ordering and the Hamiltonians H_A and H_B (that are given by Eq. (24) each) describe the field interaction with detectors A and B, respectively.

The joint state of the two detectors is $\rho_{AB} = \mathrm{tr}_\phi[U|\Psi\rangle\langle\Psi|U^\dagger]$, where the trace is over the field degrees of freedom. It is possible to show that [28] at all orders of perturbation theory the density matrix has the form of an X-state [29]

$$\rho_{AB} = \begin{pmatrix} r_{11} & 0 & 0 & r_{14}e^{-i\xi} \\ 0 & r_{22} & r_{23}e^{-i\zeta} & 0 \\ 0 & r_{23}e^{i\zeta} & r_{22} & 0 \\ r_{14}e^{i\xi} & 0 & 0 & r_{44}, \end{pmatrix}, \tag{26}$$

in the basis $\{|00\rangle, |01\rangle, |10\rangle, |11\rangle\}$ where $|ij\rangle = |i\rangle_A|j\rangle_B$, and all the coefficients r_{ij} are positive. Since ρ_{AB} is a valid density matrix, the normalization condition $\sum_i r_{ii} = 1$, and the following two positivity conditions must be satisfied:

$$r_{11}r_{44} \geq r_{14}^2, \qquad r_{22}r_{33} \geq r_{23}^2. \tag{27}$$

A useful parametrization of this matrix that explicitly separates the local and nonlocal quantities is

$$\rho_{AB} = \begin{pmatrix} 1-A-B+E & 0 & 0 & X \\ 0 & B-E & C & 0 \\ 0 & C^* & A-E & 0 \\ X^* & 0 & 0 & E \end{pmatrix}, \tag{28}$$

where A and B are the probabilities that either detector A or B are excited after the interaction with the field, and the other parameters are functions of the properties of both detectors. Indeed, tracing out either of the detectors in the state ρ_{AB}, say detector B, results in the state ρ_A of detector A

$$\rho_A = \begin{pmatrix} 1-A & 0 \\ 0 & A \end{pmatrix}, \tag{29}$$

in the basis $\{|0\rangle_A, |1\rangle_A\}$.

To simplify the exposition we consider the case of two identical detectors, i.e. $H_A = H_B$, at rest at the distance L from each other in the Minkowski spacetime. We find [28] the matrix elements of ρ_{AB} to be

$$A = \frac{\epsilon_0^2}{4\pi}\left[e^{-\sigma^2\Omega^2} - \sqrt{\pi}\sigma\Omega\,\mathrm{erfc}\,(\sigma\Omega)\right] + \mathcal{O}(\epsilon_0^4), \tag{30}$$

$$X = \frac{\epsilon_0^2}{4\sqrt{\pi}}\frac{\sigma}{L}ie^{-\sigma^2\Omega^2-\frac{L^2}{4\sigma^2}}\left[1+\mathrm{erf}\left(i\frac{L}{2\sigma}\right)\right] + \mathcal{O}(\epsilon_0^4), \tag{31}$$

$$C = \frac{\epsilon_0^2}{4\sqrt{\pi}}\frac{\sigma}{L}e^{-\frac{L^2}{4\sigma^2}}\left(\mathrm{Im}\left[e^{i\Omega L}\,\mathrm{erf}\left(i\frac{L}{2\sigma}+\sigma\Omega\right)\right] - \sin(\Omega L)\right) + \mathcal{O}(\epsilon_0^4), \tag{32}$$

$$E = |X|^2 + A^2 + 2C^2 + \mathcal{O}\left(\epsilon_0^6\right), \tag{33}$$

where $\mathrm{erf}(z)$ is the error function, $\mathrm{erfc}(z) = 1 - \mathrm{erf}(z)$. When the distance L between detectors increases the total state approaches the direct product of the density matrices of the individual detectors, that is

$$X \to 0, \qquad C \to 0, \qquad E \to A^2. \tag{34}$$

5.2 *Information-Theoretical Properties of the Joint State*

Application of the Peres–Horodecki criterion [2, 4] shows that the X-states are entangled if and only if either of the alternatives

$$r_{14}^2 > r_{22}r_{23}, \qquad r_{23}^2 > r_{11}r_{44}, \tag{35}$$

holds.

For two identical detectors in the state ρ_{AB} given in Eq. (28), these conditions are equivalent to

$$|X| - A + \mathcal{O}(\epsilon_0^4) > 0, \qquad |C| - \sqrt{E} + \mathcal{O}\left(\epsilon_0^4\right) > 0, \tag{36}$$

respectively. However, Eqs. (32) and (33) ensure that the second condition is never satisfied. The concurrence [2, 4]

$$\mathcal{C} = 2\max\left(0, |X| - A + \mathcal{O}\left(\epsilon_0^4\right)\right), \tag{37}$$

is non-zero if and only if $r_{14} > r_{22}$. This is the area below the red line on Fig. 1.

Concurrence and other entanglement measures are inaccessible by local measurements. Instead we focus on the correlation between the detectors A and B. We characterize the measurement results by random variables r_A and r_B, respectively, with $r_A,\ r_B \in \{0, 1\}$. The correlation between these variables is given by

$$\mathsf{corr}_{AB} = \frac{\mathsf{cov}_{AB}}{\sigma_A\sigma_B} = \frac{E - AB}{\sqrt{A(1-A)B(1-B)}} = \frac{|X|^2 + 2C^2}{A} + \mathcal{O}\left(\epsilon_0^4\right), \tag{38}$$

where $\mathsf{cov}_{AB} := \langle r_A r_B\rangle - \langle r_A\rangle\langle r_B\rangle$ is the covariance between r_A and r_B and $\sigma_A^2 = \mathsf{cov}_{AA}$ and $\sigma_B^2 = \mathsf{cov}_{BB}$ are the variances associated with r_A and r_B.

Equation (36) implies that the state ρ_{AB} is not entangled when the local terms dominate the bi-local effects, i.e. $A > |X|$. However, the state still contains non-classical correlations that are characterized by quantum discord.

The measurement on the Unruh-DeWitt detector precisely selects the eigenbasis of Eq. (29), making the discord D_3 the preferred measure. The measurement of A is

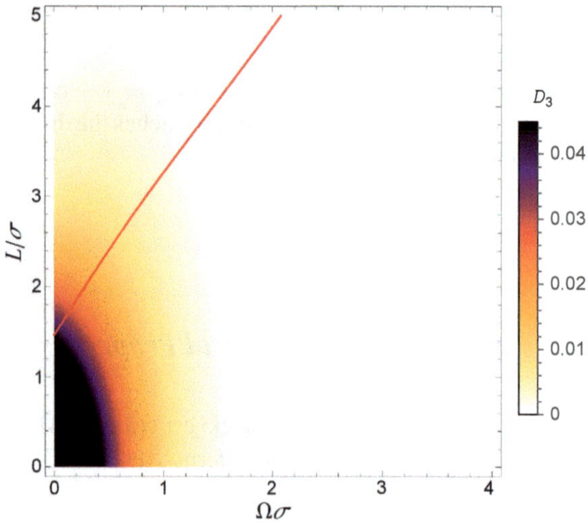

Fig. 1 The discord $D(\rho_{AB})/\epsilon_0^2$ as given by Eq. (41). The domain of zero entanglement lies above the *red line*, but there is no reason to suspect any qualitative change in the physics in the vicinity of this line

a standard projective measurement in the eigenbasis of the reduced state ρ_A, and

$$D_3(\rho_{AB}) = S(\rho^*_{AB}) - S(\rho_{AB}), \tag{39}$$

where $S(\rho)$ is the von Neumann entropy of the state ρ and

$$\rho^*_{AB} = \sum_a \rho_{B|a} \otimes |a\rangle\langle a|, \tag{40}$$

i.e. ρ^*_{AB} is the average state of the joint system after the eigenbasis projective measurement on A [16].

The discord D_3 stands out for the same reason that the entanglement measures are unobservable. The optimization procedures required by other measures are unavailable: no basis other than the standard $\{|0\rangle, |1\rangle\}$ is accessible. A straightforward calculation shows that

$$D_3(\rho_{AB}) = \big(A + C\big)\log\big(A + C\big) + \big(A - C\big)\log\big(A - C\big) - 2A\log A + \mathcal{O}(\epsilon_0^4), \tag{41}$$

hence quantum correlations persist for any finite separation of the detectors.

6 Discussion

While entanglement is a central feature of quantum mechanics and a fundamental resource in quantum information processing, it is not always directly related to interesting qualitative features of correlated quantum systems, be it open quantum systems correlated with the environment or quantum phase transitions in correlated many body systems. Extensions of the quantitative measures of pure state entanglement into mixed states, while very useful in many scenarios,[6] do not capture all the richness of non-classical correlations. These are the settings were the discord-like quantities find their use.

The three examples described are only a small sample of the vast work regarding the role of correlations (quantum or classical) in various non-classical scenarios (see [5] for a review). In many cases the role of correlations is still being explored, and in others the relation seems to have failed (see for example the work on correlations and complete positivity [30–32]).

Another area that generated a burst of interest with respect to quantum correlations is quantum computing and the difficulty of simulating large quantum systems. It is known that in general it is possible to efficiently simulate the dyanamics of a many-body system when the state is pure and the entanglement (as measured by the Schmidt rank over all bipartitions) scales at most logarithmically with the system size at all times [33]. Similarly it is known that a quantum circuit can be efficiently simulated (classically) when the correlations between the qubits are restricted to blocks of a constant size [34]. Both of these results imply that pure state dynamics should be easy to simulate when the systems are separable. It has, however been speculated [34] that this is not true for mixed states. In general there is growing evidence [35–37] that mixed state dynamics may be difficult to simulate even when the systems are separable over most bi-partitions.

One particularly unexplored, but potentially important area, is the study of quantum correlations in the system of identical particles. Entanglement of identical particles, particularly in many-body systems [38], has features distinct from that of the identical particles, and poses more open questions. Discord-like quantities are also much less understood beyond few-fermion systems [39]. On the other hand, quantum simulations of many-body system [40], may resolve the problem of the exponential scaling of resources needed to calculate, e.g., energies of atoms or molecules, with their size [41]. While correlations seem to be at the root of the requirement for exponential resources, it is unclear how to best quantify the correlations in order to provide the best figure of merit for the difficulty of simulating the many-body system. This still remains to be investigated, particularly in light of the classical cumulant-based methods of approximating many-electron wave functions that raise the possibility of breaking this exponential wall [41].

It is clear that ideas regarding pure state entanglement do not always carry forward to mixed state entanglement and that in many cases a more general (and sometimes more restrictive) class of states must be considered. The question we should be asking is therefore not *why discord?* but rather *when discord?*. We hope that this brief review further stimulates work in that direction.

---

[6]Mixed state entanglement monotones are a particularly good quantities in scenarios where entanglement is consumed as a resource.

Acknowledgements A part of this work was done when AB was at the Institute for Quantum Computing and the Department of Physics and Astronomy at the University of Waterloo. AB was supported by NSERC, Industry Canada, CIFAR and a fellowship from the Center for Quantum Information and Quantum Control.

References

1. E. Schrödinger, Naturwissenschaften, **23**, 844 (1935); English translation in J.A. Wheeler, W.H. Zurek (eds.), *Quantum Theory and Measurement* (Princeton University Press, Princeton, 1983), p. 152
2. R. Horodecki, P. Horodecki, M. Horodecki, K. Horodecki, Rev. Mod. Phys. **81**, 865 (2009)
3. N. Brunner, D. Cavalcanti, S. Pironio, V. Scarani, S. Wehner, Rev. Mod. Phys. **86**, 419 (2014)
4. D. Bruß, G. Leuchs, *Lectures on Quantum Information* (Wiley-VCH, Weinheim, 2007); M.W. Wilde, *Quantum Information Theory* (Cambridge University Press, Cambridge, 2013)
5. K. Modi, A. Brodutch, H. Cable, T. Paterek, V. Vedral, Rev. Mod. Phys. **84**, 1655 (2012)
6. A. Wehrl, Rev. Mod. Phys. **50**, 221 (1978)
7. T.M. Cover, J.A. Thomas, *Elements of Information Theory*, 2nd edn. (Wiley-Interscience, Hoboken, 2006)
8. H. Ollivier, W.H. Zurek, Phys. Rev. Lett. **88**, 017901 (2002)
9. L. Henderson, V. Vedral, J. Phys. A **34**, 6899 (2001)
10. L.C. Celeri, J. Maziero, R.M. Serra, Int. J. Quanum Inform. **9**, 1837 (2011)
11. A. Peres, *Quantum Theory: Concepts and Methods* (Kluwer, Dordrecht, 1993)
12. R.F. Werner, Phys. Rev. A **40**, 4277 (1989)
13. D. Cavalcanti, M.L. Almeida, V. Scarani, A. Acin, Nat. Commun. **2**, 184 (2011)
14. C.H. Bennett, D.P. DiVincenzo, T. Mor, P.W. Shor, J.A. Smolin, B.M. Terhal, Phys. Rev. Lett. **82**, 5385 (1999)
15. W.H. Zurek, Ann. Phys. **9**, 855 (2000)
16. A. Brodutch, D.R. Terno, Phys. Rev. A **81**, 062103 (2010)
17. Z.-W. Liu, X. Hu, S. Lloyd (2016), Phys. Rev. Lett. **118**, 060502 (2017)
18. S. Luo, N. Li, S. Fu, Theor. Math. Phys. **169**, 1724 (2011)
19. M. Piani, V. Narasimhachar, J. Calsamiglia, New J. Phys. **16**, 113001 (2014)
20. A. Brodutch, D.R. Terno, Phys. Rev. A **83**, 010301(R) (2011)
21. A. Brodutch, Phys. Rev. A **88**, 022307 (2013)
22. A. Peres, D.R. Terno, Rev. Mod. Phys. **76**, 93 (2004); R.B. Mann, T.C. Ralph (eds.), Class. Quantum Gravity **29**(22) (2012)
23. S.J. Summers, R. Werner, Phys. Lett. A **110**, 257 (1985); J. Math. Phys. **28**, 2440 (1987)
24. A. Valentini, Phys. Lett. A **153**, 321 (1991)
25. B. Reznik, Found. Phys. **33**, 167 (2003)
26. B.L. Hu, S.Y. Lin, J. Louko, Class. Quantum Gravity **29**, 224005 (2012)
27. L.C.B. Crispino, Rev. Mod. Phys. **80**, 787 (2008)
28. E. Martín-Martínez, A.R.H. Smith, D.R. Terno, Phys. Rev. D. **93**, 044001 (2016)
29. M. Ali, A.R.P. Rau, G. Alber, Phys. Rev. A **81**, 042105 (2010)
30. A. Brodutch, A. Datta, K. Modi, A. Rivas, C. Rodriguez-Rosario, Phys. Rev. A **87**, 010301 (2013)
31. A. Brodutch, A. Gilchrist, D.R. Terno, C.J. Wood, J. Phys. Conf. Ser. **306**, 012030 (2011)
32. F. Buscemi, Phys. Rev. Lett. **113**, 140502 (2014)
33. G. Vidal, Phys. Rev. Lett. **91**, 147902 (2003); G. Vidal, Phys. Rev. Lett. **93**, 040502 (2004); J.I. Latorre, E. Rico, G. Vidal, Quant. Inf. Comput. **4**, 48 (2004)
34. R. Jozsa, N. Linden, Proc. R. Soc. A **459**, 2011 (2003)
35. A. Datta, G. Vidal, Phys. Rev. A **75**, 042310 (2007)
36. A. Datta, A. Shaji, C. Caves, Phys. Rev. Lett. **100**, 050502 (2008)

37. M. Boyer, A. Brodutch, T. Mor, Phys. Rev. A **95**, 022330 (2017)
38. L. Amico, R. Fazio, A. Osterloh, V. Vedral, Rev. Mod. Phys. **80**, 517 (2008)
39. A.P. Majtey, C. Zander, A.R. Plastino, Eur. Phys. J. D **67**, 79 (2013)
40. A. Aspuru-Guzik, A.D. Dutoi, P.J. Love, Science **309**, 1704 (2005); R. Babbush, D.W. Berry, I.D. Kivlichan, A.Y. Wei, P.J. Love, A. Aspuru-Guzik, New J. Phys. **18**, 033032 (2016)
41. P. Fulde, Nat. Phys. **12**, 107 (2016)

Local Broadcasting of Quantum Correlations

Marco Piani

Operations that are trivial in the classical world, like accessing information without introducing any change or disturbance, or like copying information, become non-trivial in the quantum world. In this chapter we will discuss several limitations in the local redistributing correlations, when it comes to dealing with bipartite quantum states. In particular, we focus on the task of local broadcasting, by discussing relevant no-go theorems, and by quantifying the non-classicality of correlations in terms of the degree to which local broadcasting is possible in an approximate fashion.

1 Cloning, Broadcasting, and Local Broadcasting

1.1 Cloning

A key aspect of quantum information, which strongly differentiates it from classical information, is the inability to freely copy quantum states. Suppose we are given a quantum system S_1 in an unknown state $|\psi\rangle$, and that we want to put – equivalently, prepare – another system S_2 (let us say, of similar physical nature, and initially in some fiducial state $|0\rangle$) in the same state, without changing the state of the system that was given to us. Thinking in terms of a unitary evolution U, what we want to accomplish is the following:

$$U|\psi\rangle_{S_1}|0\rangle_{S_2}|0\rangle_E = |\psi\rangle_{S_1}|\psi\rangle_{S_2}|\xi_\psi\rangle_E, \tag{1}$$

M. Piani (✉)
SUPA and Department of Physics, University of Strathclyde, Glasgow G4 0NG, UK
e-mail: marco.piani@strath.ac.uk

F.F. Fanchini et al. (eds.), *Lectures on General Quantum Correlations and their Applications*, Quantum Science and Technology,
DOI 10.1007/978-3-319-53412-1_9

where E is an ancillary system that we may want to use in the process, initially prepared in some fiducial state $|0\rangle_E$ independent of $|\psi\rangle$ and ending up in a state $|\xi_\psi\rangle$, which potentially depends on $|\psi\rangle$. Consider now two known states $|\psi\rangle$ and $|\psi'\rangle$, for both of which we assume (1) to hold for the same unitary U. By taking the inner product of the left-hand and of the right-hand sides of the two occurrencies (one for ψ, and one for ψ') of (1), and using the fact that U preserves inner products, we arrive at the relation

$$\langle\psi'|\psi\rangle = \langle\psi'|\psi\rangle^2\langle\xi_{\psi'}|\xi_\psi\rangle. \tag{2}$$

Given that the modulus of the inner product of two normalized vector states is always less or equal to 1, the latter relation can be satisfied only if $|\langle\psi'|\psi\rangle|$ is either 0 (ψ and ψ' are orthogonal) or 1 (ψ and ψ' are the same). This is the content of the no-cloning theorem [1, 2], which says that, within the quantum formalism, there is no physical process able to clone pure quantum states that are not orthogonal.

In the general case where one adopts the formalism of density matrices and channels, the cloning of a state ρ of a system S by means of an $S \to S_1S_2$ channel Λ corresponds to the request

$$\Lambda[\rho_S] = \rho_{S_1} \otimes \rho_{S_2}, \tag{3}$$

where the introduction of a system onto which to copy the state and the possibility of using an ancillary system in the process are already taken into account by the quantum channel formalism. It is useful to recall that every quantum channel $\Lambda_{S\to S'}$ from a system S to a system S' (of potentially different dimensionality) can be seen as the result of an isometry $V_{S\to S'E}$ from S to a combined system $S'E$, followed by the tracing out of E:

$$\Lambda_{S\to S'}[\rho_S] = \mathrm{Tr}_E\left(V_{S\to S'E}\,\rho_S\,V^\dagger_{S\to S'E}\right) \tag{4}$$

This is known as Stinespring or isometric dilation of quantum channels [3]. In the case of cloning, the output system S' consists of two copies of the input system S, that is, $S' = S_1S_2$.

Consider the fidelity between two states ρ and σ [4],

$$F(\rho,\sigma) := \mathrm{Tr}\left(\sqrt{\sqrt{\rho}\,\sigma\sqrt{\rho}}\right) = \|\sqrt{\rho}\sqrt{\sigma}\|_1. \tag{5}$$

In the rightmost expression, $\|\cdot\|_1$ indicates the 1-norm (also called trace norm), $\|X\|_1 := \mathrm{Tr}\sqrt{XX^\dagger} = \mathrm{Tr}\sqrt{X^\dagger X}$. Such an expression shows explicitly that the fidelity is symmetric in ρ and σ. The fidelity satisfies $0 \leq F(\rho,\sigma) \leq 1$, with $F(\rho,\sigma) = 0$ if and only if ρ and σ are orthogonal, and $F(\rho,\sigma) = 1$ if and only if $\rho = \sigma$. Furthermore, it is multiplicative on tensor states,

$$F(\rho\otimes\rho',\sigma\otimes\sigma') = F(\rho,\sigma)F(\rho',\sigma'), \tag{6}$$

and it is monotone under quantum channels, i.e.,

$$F(\Gamma[\rho], \Gamma[\sigma]) \geq F(\rho, \sigma),$$

for any pair of states ρ, σ and any quantum channel Γ. Suppose now that ρ and ρ' can each be cloned by the action of the same Λ. Then we have

$$F(\rho, \rho') \leq F(\Lambda[\rho], \Lambda[\rho']) = F(\rho \otimes \rho, \rho' \otimes \rho') = F(\rho, \rho')F(\rho, \rho'). \tag{7}$$

The inequality is due to the monotonicity of the fidelity under quantum channels; the first equality is just the hypothesis that Λ is able to clone both ρ and ρ'; the second equality is due to the multiplicativity of fidelity on tensor states. Since $0 \leq F(\rho, \rho') \leq 1$, Eq. (7) can hold only in two situations: either $F(\rho, \rho') = 0$ (the states are orthogonal) or $F(\rho, \rho') = 1$ (the two states are actually the same state). Notice that in both latter circumstances the two states commute, $[\rho, \rho'] = 0$, and they do so in a trivial way. Thus, we find that, when it comes to cloning, the conditions for it to be possible do not vary when moving from pure states to the consideration of general mixed states.

1.2 Broadcasting

In the general framework of density operators, though, one can relax the condition of cloning to that of 'broadcasting', for which we only require that $\tilde{\rho}_{S_1S_2} = \Lambda_{S\to S_1S_2}[\rho_S]$ is such that its marginals $\tilde{\rho}_{S_1} = \mathrm{Tr}_{S_2}(\tilde{\rho}_{S_1S_2})$ and $\tilde{\rho}_{S_2} = \mathrm{Tr}_{S_1}(\tilde{\rho}_{S_1S_2})$ satisfy

$$\tilde{\rho}_{S_1} = \tilde{\rho}_{S_2} = \rho_S. \tag{8}$$

This means that we require that 'having two copies of ρ' is only achieved at the level of reduced states of the output systems, and the copies are not necessarily independent, that is, we allow $\tilde{\rho}_{S_1S_2} \neq \tilde{\rho}_{S_1} \otimes \tilde{\rho}_{S_1}$. It should be clear that mixed density matrices that are the convex combinations of a fixed set of orthonormal pure states, can be broadcast in the above sense. Indeed, for any fixed set $\{|\psi_i\rangle\}$ of states respecting $|\langle\psi_i|\psi_j\rangle| = \delta_{ij}$, there is a channel Λ such that

$$\Lambda\left[\sum_i p_i|\psi_i\rangle\langle\psi_i|_S\right] = \sum_i p_i|\psi_i\rangle\langle\psi_i|_{S_1} \otimes |\psi_i\rangle\langle\psi_i|_{S_2}, \tag{9}$$

for any *any* probability distribution $\{p_i\}$. Indeed, it is enough to consider a channel Λ that clones the orthogonal pure states $\{|\psi_i\rangle\}$—something we well know to be possible—and to exploit the linearity of the action of a channel. One immediately checks that the state on the right-hand side of (9) is such that the condition (8) is satisfied with respect to $\rho_S = \sum_i p_i|\psi_i\rangle\langle\psi_i|_S$. What is less trivial is that this is the only case where broadcasting of mixed quantum states is possible; this fact is captured by the no-broadcasting theorem [5].

Theorem 1 *Two mixed states ρ and ρ' can be broadcast simultaneously if and only if they admit a spectral decomposition with the same eigenvectors, that is, if and only if they commute,* $[\rho, \rho'] = 0$.

Notice that, this theorem is immediately extended to a collection of states, since pairwise commutation of Hermitian operators implies joint commutation.

1.3 Local Broadcasting

It is worth stressing that, for a fixed and known state ρ, there is no problem with 'cloning', as knowing the state allows one to create as many copies of it as one wants. Nonetheless, limitations kick in again even for a single state if we consider restrictions on the operations, like allowing only local operations in a distributed setting. Then, the broadcasting of a single state of a distributed system cannot be performed arbitrarily. For example, consider the case where Alice and Bob would like to clone the maximally entangled state $|\psi^+\rangle_{AB} = \frac{1}{\sqrt{2}}(|00\rangle_{AB} + |11\rangle_{AB})$. Notice that, since we are considering a pure state, we are dealing with actual cloning, not with broadcasting, that is, the target state is composed of independent copies. If Alice and Bob could use global operations, with no limit on what they can do across their laboratories, then Alice and Bob could certainly produce two copies of $|\psi^+\rangle_{AB}$—even from scratch, without using the fact that they shared one copy of the state to begin. If instead they can only implement fully local quantum channels of the form $\Lambda_{AB} = \Lambda_A \otimes \Lambda_B$, then they cannot transform one copy of $|\psi^+\rangle$ into two copies of it, because they cannot increase the entanglement they share. Actually, the task would be impossible even if they were allowed to communicate classically, that is, allowed to apply Local Operations aided by Classical Communication (LOCC) [6].

Is there a general no-go theorem for local broadcasting, that goes beyond considerations related to entanglement? Yes, there is! In Ref. [7] the following was proved.

Theorem 2 *Let ρ_{AB} be a bipartite state. There exist local maps $\Lambda_{A\to A_1A_2}$ and $\Gamma_{B\to B_1B_2}$ such that*

$$\tilde{\rho}_{A_1A_2B_2B_2} = (\Lambda_{A\to A_1A_2} \otimes \Gamma_{B\to B_1B_2})[\rho_{AB}] \tag{10}$$

satisfies

$$\tilde{\rho}_{A_1B_1} = \tilde{\rho}_{A_2B_2} = \rho_{AB}, \tag{11}$$

if and only if ρ_{AB} is classical-classical, that is,

$$\rho_{AB} = \sum_{ij} p_{ij} |a_i\rangle\langle a_i|_A \otimes |b_j\rangle\langle b_j|_B, \tag{12}$$

with $\{|a_i\rangle_A\}$ *(*$\{|b_j\rangle_B\}$*) some orthonormal basis for* A *(*B*).*

The previous result can be considered a no-local-broadcasting theorem, more precisely a two-sided no-local broadcasting theorem, that spells out a limitation about broadcasting in a distributed setting, when only local operations are allowed. In the next sections we will prove it, and even prove a one-sided version of it. Before we do that, we will recall some entropic quantifiers of distinguishability of quantum states and of correlations, and their relation with quantum recoverability.

2 Entropy, Relative Entropy, Mutual Information, and Conditional Mutual Information

2.1 Entropy

We begin by recalling the notion of von Neumann entropy [4].

Definition 1 The von Neumann Entropy a quantum state ρ is a quantifier of how mixed ρ is, and is defined as

$$S(\rho) := -\operatorname{Tr}(\rho \log_2 \rho). \tag{13}$$

In the following we will sometimes consider the entropy of subsystems. The entropy $S(X)_\rho$ of a subsystem X of a bipartite system XY in a global state $\rho = \rho_{XY}$ is defined as

$$S(X)_\rho := S(\rho_X), \tag{14}$$

where ρ_X is the reduced state of X. Notice that, when considering a multipartite system, we can always think of the bipartition into, on one side, the system of interest and, on the other side, all the other systems.

2.2 Relative Entropy

Relative entropy is a quantifier of the distinguishability of two quantum states [4].

Definition 2 The relative entropy $S(\rho\|\sigma)$ between a quantum state ρ and a quantum state σ is defined as

$$S(\rho\|\sigma) = \begin{cases} \operatorname{Tr}(\rho \log_2 \rho) - \operatorname{Tr}(\rho \log_2 \sigma) & \operatorname{supp}(\rho) \subseteq \operatorname{supp}(\sigma) \\ +\infty & \text{otherwise.} \end{cases} \tag{15}$$

Notice that the relative entropy is not symmetric in the two arguments ρ and σ. In the following we will always consider the first case, $\operatorname{supp}(\rho) \subseteq \operatorname{supp}(\sigma)$.

It holds that $S(\rho\|\sigma) \geq 0$, with equality, $S(\rho\|\sigma) = 0$, if and only if $\rho = \sigma$. Furthermore the relative entropy is monotone under channels, that is,

$$S(\rho\|\sigma) \geq S(\Gamma[\rho]\|\Gamma[\sigma]), \tag{16}$$

for any pair of states ρ, σ and any quantum channel Γ. The latter relation is often called the 'data-processing inequality' for relative entropy; operationally, it follows directly form the interpretation of relative entropy as measure of distinguishability between the two states [8].

2.3 *Mutual Information*

Mutual information is a quantifier of correlations encoded in a bipartite quantum state, that can be understood as the relative entropy between the state and the product of its marginals [4, 8].

Definition 3 The mutual information $I(A : B)_\rho$ between systems A and B in a state $\rho = \rho_{AB}$ can be defined as

$$I(A : B)_\rho := S(\rho_{AB}\|\rho_A \otimes \rho_B) = S(A)_\rho + S(B)_\rho - S(AB)_\rho. \tag{17}$$

The rightmost side of (17) comes from computing the relative entropy for the specific choice of states ρ and σ in (15), and proves that mutual information is symmetric under the exchange of A and B. A consequence of the data-processing inequality for relative entropy is that mutual information is monotone under fully local quantum channels of the form $\Lambda_{A\to A'} \otimes \Lambda_{B\to B'}$: if $\rho' = \rho'_{A'B'} = (\Lambda_{A\to A'} \otimes \Lambda_{B\to B'})[\rho_{AB}]$, then $I(A : B)_{\rho'} \leq I(A : B)_\rho$. Notice that, since $\Lambda_{A\to A'} \otimes \Lambda_{B\to B'} = (\Lambda_{A\to A'} \otimes \mathrm{id}_B) \circ (\mathrm{id}_A \otimes \Lambda_{B\to B'})$, monotonicity of mutual information under fully local quantum channel is equivalent to the monotonicity of mutual information under both channels that act non-trivially only on A, i.e., of the form $\Lambda_{A\to A'} \otimes \mathrm{id}_B$, and channels that act non-trivially only on B.

2.4 *Conditional Mutual Information*

Let us consider a tripartite system ABC in a state $\rho = \rho_{ABC}$. It is useful to introduce the notion of conditional mutual information between A and C, conditioned on B, defined as [3, 4]

$$\begin{aligned} I(A : C|B)_\rho &:= I(A : BC)_\rho - I(A : B)_\rho \\ &= S(AB)_\rho + S(BC)_\rho - S(ABC)_\rho - S(B)_\rho. \end{aligned} \tag{18}$$

The last expression for the conditional mutual information in (18) proves that $I(A : C|B)_\rho$ is symmetric between A and C; indeed, it holds

$$I(A : C|B)_\rho = I(A : BC)_\rho - I(A : B)_\rho = I(AB : C)_\rho - I(C : B)_\rho.$$

Conditional mutual information $I(A : C|B)_\rho$ is non-negative, a fact known also as the strong subadditivity of von Neumann entropy [3, 4], and of the foremost importance in quantum information theory. It has an interpretation as the amount of correlations, as measured by mutual information, lost between A and BC when C gets discarded.

Notice that, if we consider the tripartite state $\rho_{AB'E} = V_{B\to B'E}\rho_{AB}V^\dagger_{B\to B'E}$, that arises from the isometry step in the implementation of an arbitrary local channel $\Lambda_{B\to B'}$, the mutual information between A and $B'E$ is the same as the mutual information between A and B, before the action of the channel. On the other hand, the mutual information after the action of the channel is the mutual information between A and B', after having discarded E. Thus we see that the fact that conditional mutual information is non-negative is equivalent to the fact that mutual information is monotone under fully local operations.

3 Quantum Recoverability

The data processing inequality (16) can be refined, and linked to the issue of the recoverability of the action of the quantum channel Γ on ρ [9, 10].

Theorem 3 *Given two states ρ and σ, and a channel Γ, there is a recovery channel $R = R_{\sigma,\Gamma}$ that depends only on σ and Γ such that*

$$S(\rho||\sigma) - S(\Gamma[\rho]\|\Gamma[\sigma]) \geq -\log_2 F^2(\rho, (R \circ \Gamma)[\rho]) \tag{19}$$

$$(R \circ \Gamma)[\sigma] = \sigma. \tag{20}$$

Notice that the right-hand side of (19) is non-negative, so that indeed (19) constitutes a strengthening of (16). In Theorem 3, the recovery channel R always recovers σ from $\Gamma[\sigma]$ perfectly. On the other hand, how well R recovers ρ from $\Gamma[\rho]$—that is, how large the fidelity $F(\rho, (R \circ \Gamma)[\rho])$ can be—depends on the decrease of the relative entropy under the action of Γ: the fidelity is large—close to 1—if the decrease in the relative entropy is small, since (19) is equivalent to

$$F(\rho, (R \circ \Gamma)[\rho]) \geq 2^{-\frac{1}{2}(S(\rho||\sigma)-S(\Gamma[\rho]\|\Gamma[\sigma]))}.$$

Notice that the state ρ is perfectly recovered by R from $\Gamma[\rho]$—that is, the fidelity of $(R \circ \Gamma)[\rho]$ with ρ is equal to 1—if there is no decrease of the relative entropy. The case of equality in the left-hand side of (19) had already been considered by Petz, who was able to provide an explicit form for a perfect recovery map $R^P = R^P_{\sigma,\Gamma}$ [11–13] for such a case:

$$R^P_{\sigma,\Gamma}[\tau] = \sigma^{1/2}\Gamma^\dagger\left[(\Gamma[\sigma])^{-1/2}\tau(\Gamma[\sigma])^{-1/2}\right]\sigma^{1/2}. \tag{21}$$

Here $\Gamma^\dagger$ is the map dual to Γ, i.e., such that $\mathrm{Tr}(X^\dagger\Gamma[Y]) = \mathrm{Tr}((\Gamma^\dagger[X])^\dagger Y)$ for all X, Y. In particular, since Γ is a channel that admits a Kraus decomposition, $\Gamma[Y] = \sum_i K_i Y K_i^\dagger$, with Kraus operators K_i, then $\Gamma^\dagger$ acts as follows: $\Gamma^\dagger[X] = \sum_i K_i^\dagger X K_i$. One verifies that R^P is completely positive and trace-preserving, hence a channel. In the general case covered by Theorem 3, which considers a non-vanishing decrease in the relative entropy, it has been proven that the recovery channel R can be chosen to have some close connection with the structure of the Petz recovery channel [9, 10], but a discussion of the present status in the study of the expression of the best recovery map—in the general case of imperfect recovery—goes beyond the scope of these notes.

3.1 *Quantum Recoverability and Mutual Information*

Given that mutual information is as a special case of relative entropy, it should be no surprise that Theorem 3 can be specialized to the case of mutual information. Actually, Theorem 3 can be seen as a generalization of a theorem previously derived by Fawzi and Renner about mutual information [14].

Theorem 4 *For any state $\rho = \rho_{ABC}$ there is a recovery channel $R_{B\to BC}$ such that*

$$F(\rho_{ABC}, R_{B\to BC}[\rho_{AB}]) \geq 2^{-\frac{1}{2}I(A:C|B)_\rho}.$$

Theorem 4 says that, if correlations between A and BC do not decrease too much because of the loss (that is, the tracing out) of C, then the total state ρ_{ABC} can be recovered pretty well by means of a channel $R_{B\to BC}$ acting on ρ_{AB}; more in detail, it is possible to choose such a map so that it only depends on ρ_{BC}, and not on the full state ρ_{ABC} [15].

4 Proof of the No-Local-Broadcasting Theorem

We will use what we recalled about quantum recoverability in Sect. 3 to prove the no-local-broadcasting Theorem 2; we will do so by leveraging the no-broadcasting Theorem 1. We will use an intermediate step that one could call the no-unilocal-broadcasting theorem [16, 17].

4.1 No-Unilocal-Broadcasting

We say that a bipartite state ρ_{AB} can be locally broadcast on B if there exists a local map $\Gamma_{B\to B_1B_2}$ such that

$$\tilde{\rho}_{AB_1B_2} = (\mathrm{id}_A \otimes \Gamma_{B\to B_1B_2})[\rho_{AB}] \tag{22}$$

satisfies

$$\tilde{\rho}_{AB_1} = \tilde{\rho}_{AB_2} = \rho_{AB}. \tag{23}$$

The following holds [16, 17].

Theorem 5 *A bipartite state ρ_{AB} can be locally broadcast on B if and only if ρ_{AB} is classical on B, that is,*

$$\rho_{AB} = \sum_j p_j \rho_j^A \otimes |b_j\rangle\langle b_j|_B \tag{24}$$

with $\{|b_j\rangle_B\}$ some orthonormal basis for B.

In order to prove Theorem 5 we will need the following lemma, for which we provide a more direct proof than the one given in Ref. [17].

Lemma 1 *Any bipartite state ρ_{AB} admits a decomposition of the form*

$$\rho_{AB} = \sum_i p_i F_i^A \otimes \rho_i^B \tag{25}$$

with $\{p_i\}$ a probability distribution, $\{F_i^A\}$ a collection of linearly independent operators on A, and ρ_i^B normalized states on B.

Notice that the operators $\{F_i^A\}$ in (25) are in general non-positive; this must be the case, because otherwise the lemma would claim that every bipartite state can be written as the convex combination of tensor products of positive operators, while we know that the latter applies—by definition—only to unentangled states.

Proof (*of Lemma* 1) We will consider a minimal informationally complete POVM on A, which means a collection of linearly independent operators $\{E_i^A\}$ that form a valid POVM, i.e., $E_i^A \geq 0$, $\sum_i E_i^A = I_A$, and at the same time constitute a basis for the space of operators on A [18]. A minimal informationally complete POVM constitutes a quantum frame [19]. We can then consider a dual frame $\{F_i\}$ to it, that is, a collection of linearly independent operators such that

$$X = \sum_i \mathrm{Tr}(E_i X) F_i \quad \forall X.$$

Then, one has

$$\rho_{AB} = \sum_i p_i F_i^A \otimes \rho_i^B,$$

with

$$\rho_i^B = \frac{1}{p_i} \mathrm{Tr}_A(E_i^A \otimes I_B \rho_{AB}) \quad p_i = \mathrm{Tr}_{AB}(E_i^A \otimes I_B \rho_{AB}). \tag{26}$$

Notice that the ρ_i^B's are normalized states, thanks to the fact that $\{E_i\}$ is a POVM; for the same reason, $\{p_i\}$ is a valid probability distribution.

We are now in the position to prove Theorem 5.

Proof (*of Theorem* 5) That a state classical on B can be locally broadcast on B is trivial. To prove the other direction, let us consider a map $\Gamma_{B\to B_1B_2}$ that achieves the local broadcasting of ρ_{AB} on B. We are going to prove that $\Gamma_{B\to B_1B_2}$ broadcasts the individual states ρ_i^B of (26); then, the no-broadcasting Theorem 1 will allow us to conclude that all the ρ_i^B can be diagonalized in a same basis, which is equivalent to saying that ρ_{AB} is classical on B.

That $\Gamma_{B\to B_1B_2}$ broadcasts the individual states ρ_i^B is true, since we have (we focus on the B_1 output for concreteness, but the same goes for B_2):

$$\begin{aligned}
\mathrm{Tr}_{B_2}(\Gamma_{B\to B_1B_2}[\rho_i^B]) &= \frac{1}{p_i} \mathrm{Tr}_{AB_2}((E_i^A \otimes I_{B_1B_2})(\mathrm{id}_A \otimes \Gamma_{B\to B_1B_2})[\rho_{AB}]) \\
&= \frac{1}{p_i} \mathrm{Tr}_A((E_i^A \otimes I_{B_1})\tilde{\rho}_{AB_1}) \\
&= \frac{1}{p_i} \mathrm{Tr}_A((E_i^A \otimes I_B)\rho_{AB}) \\
&= \rho_i^B.
\end{aligned}$$

The first equality is due to the definition of conditional state (26), and the fact that the POVM $\{E_i^A\}$ and the map $\Gamma_{B\to B_1B_2}$ operate on different systems; the second and third equalities are due to the broadcasting conditions. □

4.2 No-Local-Broadcasting

We are now in the position to give a straightforward proof of Theorem 2.

Proof (*of Theorem* 2) That a classical-classical state can be locally broadcast is trivial. To prove the other direction, let us assume that ρ_{AB} can be locally broadcast, and let $\Lambda_{A\to A_1A_2}$ and $\Gamma_{B\to B_1B_2}$ be the locally broadcasting maps, so that $\tilde{\rho}_{A_1A_2B_2B_2} = (\Lambda_{A\to A_1A_2} \otimes \Gamma_{B\to B_1B_2})[\rho_{AB}]$ satisfies the broadcasting conditions (11). We will prove that ρ_{AB} can also be locally broadcast on both A and B, and hence it is classical on both A and B, which means it is classical-classical. For the sake of concreteness we will focus on proving that ρ_{AB} can be locally broadcast on B. A similar proof can be followed to prove classicality on A.

Besides ρ_{AB} and $\tilde{\rho}_{A_1A_2B_1B_2}$ defined above, it is convenient to consider also

$$\tilde{\rho}'_{AB_1B_2} := (\mathrm{id}_A \otimes \Gamma_{B\to B_1B_2})[\rho_{AB}].$$

Notice that it holds

$$\tilde{\rho}_{A_1A_2B_1B_2} = (\Lambda_{A\to A_1A_2} \otimes \mathrm{id}_B)[\tilde{\rho}'_{AB_1B_2}].$$

The key point is that one goes from ρ_{AB}, to $\tilde{\rho}'_{AB_1B_2}$, to $\tilde{\rho}_{A_1A_2B_1B_2}$ by a sequence of local operations. This, together to monotonicity of mutual information under local operations (including partial trace), implies

$$I(A_1 : B_1)_{\tilde{\rho}} \leq I(A_1A_2 : B_1)_{\tilde{\rho}} \leq I(A : B_1)_{\tilde{\rho}'} \leq I(A : B_1B_2)_{\tilde{\rho}'} \leq I(A : B)_{\rho}.$$

Notice that, because of the broadcasting conditions, the leftmost quantity is actually equal to the rightmost quantity, hence, all the mutual information quantities in the latter equation are equal. In particular, $I(A : B_1)_{\tilde{\rho}'} = I(A : B)_{\rho}$. Moreover,

$$\tilde{\rho}'_{AB_1} = \left(\mathrm{id}_A \otimes \Gamma'_{B\to B_1}\right)[\rho_{AB}],$$

with $\Gamma'_{B\to B_1} = (\mathrm{id}_{B_1} \otimes \mathrm{Tr}_{B_2}) \circ \Gamma_{B\to B_1B_2}$. It follows then from Theorem 4 that there is a recovery channels $R^{(1)}_{B_1\to B}$ such that

$$(\mathrm{id}_A \otimes (R^{(1)}_{B_1\to B} \circ \Gamma'_{B\to B_1}))[\rho_{AB}] = \rho_{AB}$$

One can argue similarly about B_2. We arrive at the conclusion that there are two channels $R^{(1)}_{B_1\to B_1}$ and $R^{(2)}_{B_2\to B_2}$ such that

$$(R^{(1)}_{B_1\to B_1} \otimes R^{(2)}_{B_2\to B_2}) \circ \Gamma_{B\to B_1B_2}$$

locally broadcasts ρ_{AB} on B. □

5 Broadcasting Mutual Information

The characterization of recoverability in terms of mutual information captured by Theorem 4 is such—one could say, 'strong enough'—that the local (or unilocal) broadcasting of quantum states, understood in a structural sense—that is, in terms the density matrices—is fully equivalent to the broadcasting of the correlations contained in the states, as quantified by mutual information. More precisely, one can state a no-go theorem for local broadcasting, which, at least at face value, is more general than Theorem 2.

Theorem 6 *Let ρ_{AB} be a bipartite state. There exist local maps $\Lambda_{A\to A_1A_2}$ and $\Gamma_{B\to B_1B_2}$ such that*

$$\tilde{\rho}_{A_1A_2B_2B_2} = (\Lambda_{A\to A_1A_2} \otimes \Gamma_{B\to B_1B_2})[\rho_{AB}] \tag{27}$$

satisfies

$$I(A_1 : B_1)_{\tilde{\rho}} = I(A_2 : B_2)_{\tilde{\rho}} = I(A : B)_{\rho_{AB}} \tag{28}$$

if and only if ρ_{AB} is classical-classical.

This version of the no-local-broadcasting theorem would indeed appears to be more general than Theorem 2 because if $\tilde{\rho}_{A_1A_2B_2B_2}$ satisfies the broadcasting conditions (11), then it satisfies also conditions (27), but the opposite is not immediately evident. Nonetheless, it is true—and we know it thanks to Theorem 4. Similarly, one can have an alternative version of the no-unilocal-broadcasting theorem.

Theorem 7 *Let ρ_{AB} be a bipartite state. There exists a local map $\Gamma_{B\to B_1B_2}$ such that*

$$\tilde{\rho}_{AB_1B_2} = (\mathrm{id}_A \otimes \Gamma_{B\to B_1B_2})[\rho_{AB}] \tag{29}$$

satisfies

$$I(A : B_1)_{\tilde{\rho}} = I(A : B_2)_{\tilde{\rho}} = I(A : B)_{\rho}. \tag{30}$$

if and only if ρ_{AB} is classical on B.

6 Quantifying Non-classical Correlations Through Broadcasting

So far we have characterized quantum correlations only qualitatively, via no-go theorems. Moreover, we have made use of Theorem 4 only for the case of exact recoverability. In this section, we delineate some ways in which local and unilocal broadcasting can be used to *quantify* the degree of non-classicality of correlations of the state ρ_{AB} under scrutiny. Interestingly, we will connect broadcasting to the well established quantifier of non-classical correlations known as quantum discord. The latter is an asymmetric quantity, based on the notion of minimal loss of correlations, as measured by mutual information, when one tries to 'extract' such correlations and map them into a classical register. More explicitly, consider quantum-to-classical channels $\mathcal{M}_{B\to B'}[\sigma_B] = \sum_i \mathrm{Tr}(M_i\sigma_B)|i\rangle\langle i|_{B'}$, where $\{M_i\}$ is a POVM on B, and $\{|i\rangle\}$ is an orthonormal basis for the 'classical register' B'. We define $\rho'_{AB'} = (\mathrm{id}_A \otimes \mathcal{M}_{B\to B'})[\rho_{AB}]$. Then, the discord of ρ_{AB} on B is equal to

$$D(A|B)_\rho := I(A:B)_\rho - \max_{\mathcal{M}_{B\to B'}} I(A:B')_{\rho'}, \tag{31}$$

Here the maximum is over all quantum-to-classical channels. Notice that discord can be rewritten as

$$D(A|B)_\rho := \min_{V_{B\to B'E}} I(A:E|B')_{V\rho V^\dagger}, \tag{32}$$

where $V = V_{B\to B'E}$ is any isometry that realizes a quantum-to-classical channel $\mathcal{M}_{B\to B'}$. Discord vanishes only for the quantum-classical states [20–22]; in all other cases, there is a loss of correlations—a measured by quantum mutual information—in the local quantum-to-classical mapping. Nonetheless, one can try to recover ρ_{AB} from the quantum-classical state $\rho_{AB'}$, via a recovery channel $R_{B'\to B}$. It is easy to check that, without loss of generality, such a recovery map consists of a preparation procedure, so that the composition of measurement $\mathcal{M}$ and preparation/recovery R gives rise to a so-called entanglement breaking map, $R_{B'\to B}\circ\mathcal{M}_{B\to B'}[\sigma_B] = \Lambda_B^{\mathrm{EB}}[\sigma_B] = \sum_i \mathrm{Tr}(M_i\sigma_B)\tau_B^i$, for all σ_B, with $\{\tau_B^i\}$ a collection of states. Let us define [23]

$$F_B^{\mathrm{EB}}(\rho_{AB}) := \max_{\Lambda_B^{\mathrm{EB}}} F\big(\rho_{AB}, (\mathrm{id}_A\otimes\Lambda_B^{\mathrm{EB}})[\rho_{AB}]\big). \tag{33}$$

Then, Theorem 4 implies [23]

$$D(A|B)_\rho \geq -2\log_2 F_B^{\mathrm{EB}}(\rho_{AB}). \tag{34}$$

6.1 *Imperfect Structural Local Broadcasting*

Although cloning and broadcasting of general unknown states is not possible, as formalized by the no-cloning and no-broadcasting theorems, one can consider how well the task can be achieved, at least in an approximate sense. This corresponds to the relative large topic of optimal (albeit not perfect) quantum cloners (see, e.g., [24, 25]). The same applies to local or unilocal broadcasting. For example, one can consider the state-dependent[1] maximally achievable fidelity

$$F_{B,1\to 2}^{\max}(\rho_{AB}) := \max_{\Lambda_{B\to B_1B_2}} F\big(\rho_{AB}, \mathrm{Tr}_{B_1}((\mathrm{id}_A\otimes\Lambda_{B\to B_1B_2})[\rho_{AB}])\big), \tag{35}$$

where the maximum is over maps $\Lambda_{B\to B_1B_2}$ whose output is invariant under swap of B_1B_2. Alternatively, one could consider the average of the fidelities, for a general map that does not have symmetric output, but one can argue that a symmetric output is always optimal, thanks to the (joint) concavity of the fidelity in each of its

[1] A large part of the study about optimal cloners deals with single system, where it is natural to discuss the cloning of a set of states, or universal cloners, where the figure of merit is either an average, or state-independent (for pure states).

arguments [4]. Notice that, exactly because of the symmetry of the output of $\Lambda_{B\to B_1B_2}$, on the right side of (35) we can indifferently consider the trace over B_1 or B_2. The no-unilocal-broadcasting Theorem 5 ensures that $F^{\max}_{B,1\to 2}(\rho_{AB}) < 1$ (strictly) as soon as ρ_{AB} is not classical on B.

We now observe that any entanglement breaking map can be seen as the composition of a map with symmetric output followed by a partial trace, because

$$\sum_i \mathrm{Tr}(M_i\sigma_B)\tau^i_B = \mathrm{Tr}_{B_1}\left(\sum_i \mathrm{Tr}(M_i\sigma_B)\tau^i_{B_1}\otimes\tau^i_{B_2}\right).$$

This implies that $F^{\max}_{B,1\to 2}(\rho_{AB}) \geq F^{\mathrm{EB}}_B(\rho_{AB})$, which combined with (34), gives

$$D(A|B)_\rho \geq -2\log_2 F^{\max}_{B,1\to 2}(\rho_{AB}).$$

Thus, we see that discord can be bounded in terms of the quality of approximate broadcasting. In Ref. [26] this relation is further explored, showing, on one hand, that considering a larger number of outputs leads to more stringent bound on the discord, and, on the other hand, that each quantities like $F^{\max}_{B,1\to 2}(\rho_{AB})$ can be computed numerically in an efficient and reliable way, since such a quantity can be calculated by semidefinite programming [27, 28].

6.2 Imperfect Local Broadcasting of Mutual Information

In the same way in which we have cast the no-local-broadcasting and no-unilocal-broadcasting theorems in terms of broadcasting mutual information, so we can approach the issue of approximate broadcasting. We will focus on unilocal broadcasting. Then, for any channel $\Lambda_{B\to B'^n}$, where $B'^n = B'_1B'_2\ldots B'_n$, and each B'_1, $B'_2,\ldots,B'_n$ could potentially have a dimensionality different from that of B, one can define the average loss of correlations in broadcasting as

$$I(A:B)_{\rho_{AB}} - \frac{1}{n}\sum_{i=1}^{n} I(A:B'_i)_{(\mathrm{id}_A\otimes\Lambda_{B\to B'^n})[\rho_{AB}]}. \tag{36}$$

The mutual information version of the no-unilocal-broadcasting theorem, Theorem 7, says that such an average loss is strictly positive for all $n\geq 2$, if ρ_{AB} is not quantum-classical. One proves that such an average loss actually converges to exactly the quantum discord (31), with n going to infinity [29].

7 Conclusions

It is hard to overestimate the importance of the no-cloning theorem in our understanding of quantum mechanics and quantum information, and the pivotal role it has played in the latter field. For example, born from the attempts to reconcile entanglement with a principle of no-faster-than-light signalling, it contributed to the development of quantum cryptography. We have seen that the counterpart of no-cloning in the scenario where one considers mixed states is no-broadcasting.

When it comes to distributed system, and to the study of the limitations in the local manipulation of correlations, other no-go theorems can be derived, like the no-local-broadcasting theorem or the no-unilocal-broadcasting theorem. It is worth emphasizing one last time that the latter no-go results apply to single distributed quantum states, contrary to the no-cloning and no-broadcasting theorems, which instead deal with multiple states.

The quantification of the limits in the local manipulation of correlations provide a sound and physically meaningful way to quantify the non-classicality of correlations. Interestingly, this field of research connects directly to recent and exciting advances in our understanding of quantum recoverability. The latter connection can also be exploited for an efficient numerical approach to the quantification of the quantumness of correlations, allowing, for example, the derivation of reliable numerical bounds to quantum discord.

Acknowledgements The productions of these notes was supported by the European Union's Horizon 2020 research and innovation programme under the Marie Sklodowska-Curie grant agreement No 661338. I would like to thank the Institute for Quantum Computing at the University of Waterloo for its hospitality during the completion of these notes.

References

1. W.K. Wootters, W.H. Zurek, Nature **299**(5886), 802 (1982)
2. D. Dieks, Phys. Lett. A **92**(6), 271 (1982). doi:10.1016/0375-9601(82)90084-6
3. M.M. Wilde, *Quantum Information Theory* (Cambridge University Press, Cambridge, 2013)
4. M.A. Nielsen, I.L. Chuang, *Quantum Computation and Quantum Information* (Cambridge University Press, Cambridge, 2010)
5. H. Barnum, C.M. Caves, C.A. Fuchs, R. Jozsa, B. Schumacher, Phys. Rev. Lett. **76**(15), 2818 (1996)
6. R. Horodecki, P. Horodecki, M. Horodecki, K. Horodecki, Rev. Mod. Phys. **81**, 865 (2009). doi:10.1103/RevModPhys.81.865
7. M. Piani, P. Horodecki, R. Horodecki, Phys. Rev. Lett. **100**(9), 090502 (2008)
8. V. Vedral, Rev. Mod. Phys. **74**, 197 (2002). doi:10.1103/RevModPhys.74.197
9. M.M. Wilde, Proc. R. Soc. Lond. A: Math. Phys. Eng. Sci. **471**(2182) (2015). doi:10.1098/rspa.2015.0338
10. M. Junge, R. Renner, D. Sutter, M.M. Wilde, A. Winter (2015). arXiv:1509.07127
11. D. Petz, Commun. Math. Phys. **105**(1), 123 (1986)
12. D. Petz, Q. J. Math. **39**(1), 97 (1988)
13. P. Hayden, R. Jozsa, D. Petz, A. Winter, Commun. Math. Phys. **246**(2), 359 (2004)

14. O. Fawzi, R. Renner, Commun. Math. Phys. **340**(2), 575 (2015)
15. D. Sutter, O. Fawzi, R. Renner, Proc. R. Soc. Lond. A: Math. Phys. Eng. Sci. **472**(2186) (2016). doi:10.1098/rspa.2015.0623
16. S. Luo, Lett. Math. Phys. **92**(2), 143 (2010). doi:10.1007/s11005-010-0389-1
17. S. Luo, W. Sun, Phys. Rev. A **82**(1), 012338 (2010)
18. C.M. Caves, C.A. Fuchs, R. Schack, J. Math. Phys. **43**(9), 4537 (2002). doi:10.1063/1.1494475
19. C. Ferrie, J. Emerson, New J. Phys. **11**(6), 063040 (2009). http://stacks.iop.org/1367-2630/11/i=6/a=063040
20. H. Ollivier, W.H. Zurek, Phys. Rev. Lett. **88**(1), 017901 (2001)
21. L. Henderson, V. Vedral, J. Phys. A: Math. Gen. **34**(35), 6899 (2001)
22. M. Hayashi, *Quantum Information* (Springer, Berlin, 2006)
23. K.P. Seshadreesan, M.M. Wilde, Phys. Rev. A **92**, 042321 (2015). doi:10.1103/PhysRevA.92.042321
24. N. Gisin, S. Massar, Phys. Rev. Lett. **79**(11), 2153 (1997)
25. R.F. Werner, Phys. Rev. A **58**, 1827 (1998). doi:10.1103/PhysRevA.58.1827
26. M. Piani, Phys. Rev. Lett. **117**, 080401 (2016). doi:10.1103/PhysRevLett.117.080401
27. S. Boyd, L. Vandenberghe, *Convex Optimization* (Cambridge University Press, Cambridge, 2009)
28. J. Watrous (2012). arXiv:1207.5726
29. F.G. Brandão, M. Piani, P. Horodecki, Nat. Commun. **6**, 7908 (2015)

Entanglement Distribution and Quantum Discord

Alexander Streltsov, Hermann Kampermann and Dagmar Bruß

Abstract Establishing entanglement between distant parties is one of the most important problems of quantum technology, since long-distance entanglement is an essential part of such fundamental tasks as quantum cryptography or quantum teleportation. In this lecture we review basic properties of entanglement and quantum discord, and discuss recent results on entanglement distribution and the role of quantum discord therein. We also review entanglement distribution with separable states, and discuss important problems which still remain open. One such open problem is a possible advantage of indirect entanglement distribution, when compared to direct distribution protocols.

1 Introduction

This lecture presents an overview of the task of establishing entanglement between two distant parties (Alice and Bob) and its connection to quantum discord [1–5]. Surprisingly, it is possible for the two parties to perform this task successfully by exchanging an ancilla which has never been entangled with Alice and Bob. This puzzling quantum protocol was already suggested in [6], but a thorough study [7–12] and experimental verification [13–15] (see also [16]) had to wait for almost ten years until recently, when interest in general quantum correlations arose and led to insights about their role in the entanglement distribution protocol.

A composite quantum system does not need to be in a product state for the subsystems, but it can also occur as a superposition of product states, or as a mixture of such superpositions. This feature does not exist in the classical world, and a state

A. Streltsov (✉)
Dahlem Center for Complex Quantum Systems, Freie Universität Berlin,
D-14195 Berlin, Germany
e-mail: streltsov.physics@gmail.com

H. Kampermann · D. Bruß
Institut für Theoretische Physik III, Heinrich-Heine-Universität Düsseldorf,
D-40225 Düsseldorf, Germany

F.F. Fanchini et al. (eds.), *Lectures on General Quantum Correlations and their Applications*, Quantum Science and Technology,
DOI 10.1007/978-3-319-53412-1_10

exhibiting it is called entangled. In general, a state is said to contain entanglement if it cannot be written as a mixture of projectors onto product states. It is said to contain quantum correlations, if it cannot be written as a mixture of projectors onto product states with local orthogonality properties. And it is said to contain correlations (classical or quantum), if it cannot be written as a product state.

Let us formalize these notions. In the following definitions we will consider for simplicity only bipartite quantum systems (with superscripts A and B for Alice and Bob, respectively); the generalization to composite quantum systems with more than two subsystems is straightforward. Let us denote by $\{|e_i\rangle\}$ a complete set of orthogonal basis states (which could also be interpreted as classical states), i.e. $\langle e_i|e_j\rangle = \delta_{ij}$, while Greek letters indicate quantum states which are not necessarily orthogonal, i.e., for the ensemble $\{|\psi_i\rangle\}$ in general $\langle\psi_i|\psi_j\rangle \neq \delta_{ij}$ holds.

A *separable* state ρ_{sep} can be written as [17]

$$\rho_{\text{sep}}^{AB} = \sum_{i,j} p_{ij} \, |\psi_i\rangle\langle\psi_i|^A \otimes |\phi_j\rangle\langle\phi_j|^B , \tag{1}$$

where p_{ij} are probabilities with $\sum_{i,j} p_{ij} = 1$. The set of all separable states will be denoted by $\mathcal{S}$. Any separable state can be produced with local operations and classical communication (LOCC). An *entangled* state cannot be written as in Eq. (1). In order to produce an entangled state, a non-local operation is needed. In Sect. 2 we will review different ways to quantify the amount of entanglement in a given state.

A state is called *classically correlated* (CC) if it can be written as [18]

$$\rho_{\text{cc}}^{AB} = \sum_{i,j} p_{ij} \, |e_i\rangle\langle e_i|^A \otimes |e_j\rangle\langle e_j|^B , \tag{2}$$

with $\langle e_i|e_j\rangle = \delta_{ij}$. Measuring ρ_{cc} in the local bases $\{|e_i\rangle^A\}$ and $\{|e_j\rangle^B\}$ does not change the state, i.e.,

$$\Pi^A \otimes \Pi^B[\rho_{\text{cc}}^{AB}] = \rho_{\text{cc}}^{AB} , \tag{3}$$

where the von Neumann measurement Π is defined as

$$\Pi[\sigma] = \sum_i |e_i\rangle\langle e_i| \, \sigma \, |e_i\rangle\langle e_i| . \tag{4}$$

The set of all classically correlated states will be denoted by $\mathcal{CC}$. A *quantum correlated* state cannot be written as in Eq. (2). The eigenbasis of a quantum correlated state is not a product basis with the property that the sets of local states are orthogonal ensembles.

It is also possible to combine the aforementioned frameworks of separability and classicality, thus arriving at *classical-quantum* (CQ) states [18]:

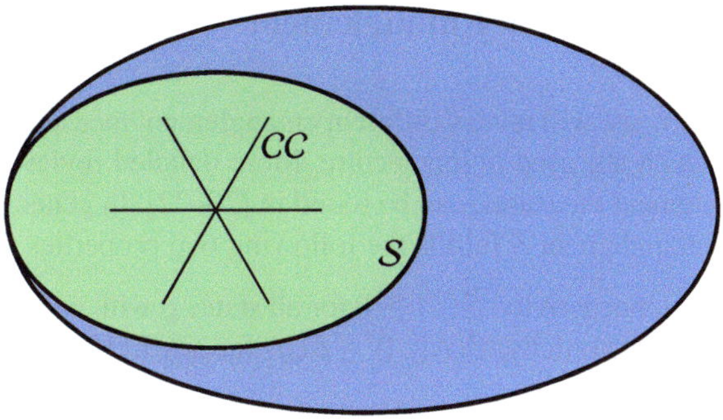

Fig. 1 State space for composite quantum systems [19]: classically correlated states form a connected set $\mathcal{CC}$ of measure zero; separable states form a convex set $\mathcal{S}$ (*green*, containing $\mathcal{CC}$). Quantum correlated states are all states outside of $\mathcal{CC}$, and entangled states (*blue*) are all states outside of $\mathcal{S}$

$$\rho_{\mathrm{cq}}^{AB} = \sum_{i,j} p_{ij} \left|e_i\right\rangle \left\langle e_i\right|^A \otimes \left|\psi_j\right\rangle\left\langle\psi_j\right|^B . \tag{5}$$

The set of all classical-quantum states will be denoted by $\mathcal{CQ}$. For any CQ state, there exists a local von Neumann measurement on the subsystem A which leaves the state unchanged, i.e.,

$$\Pi^A \otimes \mathbb{1}^B[\rho_{\mathrm{cq}}^{AB}] = \rho_{\mathrm{cq}}^{AB}, \tag{6}$$

where the von Neumann measurement Π is given in Eq. (4). If a state cannot be written as in Eq. (5), we say that the state has nonzero *quantum discord* with respect to the subsystem A. Measuring a state with nonzero discord in any orthogonal basis on the subsystem A necessarily changes the state. In Sect. 3 we will present different ways to quantify the amount of discord in a given state.

From the above definitions it is clear that classically correlated states are a subset of separable states, and that entangled states are a subset of quantum correlated states. These different types of states for composite quantum systems therefore form a nested structure [19] which is sketched in Fig. 1. Note that separable states form a convex set, due to their definition in Eq. (1). However, classically correlated states do *not* form a convex set: one can produce a quantum correlated state by mixing two classically correlated states.

Those states which are not entangled, but nevertheless possess quantum correlations a la discord, may exhibit puzzling features. They can be produced via LOCC, but nevertheless they carry quantum properties. Namely, in order to produce them one has to create quantumness in the form of non-orthogonality. This makes them a potential resource for quantum information processing protocols. Counterintuitively, even though they do not carry entanglement, they may be used for the distribution of entanglement, as we will see below.

The structure of this lecture is as follows: in Sect. 2 we review different measures of quantum entanglement, discord quantifiers are reviewed in Sect. 3. In Sect. 4 we review recent results on entanglement distribution and discuss the role of quantum discord therein. Conclusions in Sect. 5 complete our lecture.

2 Quantum Entanglement

Here, we will review different entanglement measures, mainly focusing on measures which are used in this lecture. More detailed reviews, also containing other entanglement measures, can be found in [20–22]. In general, we require that a measure of entanglement $\mathcal{E}$ fulfills the following two properties [23, 24]:

- Nonnegativity: $\mathcal{E}(\rho) \geq 0$ for all states ρ with equality for all separable states [25],
- Monotonicity: $\mathcal{E}(\Lambda[\rho]) \leq \mathcal{E}(\rho)$ for any LOCC operation Λ.

Many entanglement measures also have additional properties such as strong monotonicity in the sense that entanglement does not increase on average under selective LOCC operations [23, 24]: $\sum_i p_i \mathcal{E}(\sigma_i) \leq \mathcal{E}(\rho)$, where the states σ_i are obtained from the state ρ by the means of LOCC with the corresponding probabilities p_i. Moreover, many entanglement measures are also convex in the state, i.e., $\mathcal{E}(\sum_i p_i \rho_i) \leq \sum_i p_i \mathcal{E}(\rho_i)$ [23, 24].

From now on we will focus on the bipartite scenario with two parties A and B of the same dimension d. In this case, any entanglement measure is maximal on states of the form

$$\left|\phi_d^+\right\rangle = \frac{1}{\sqrt{d}} \sum_{i=0}^{d-1} |ii\rangle , \tag{7}$$

since from this state any quantum state can be created via LOCC operations [22]. Of particular importance is the two-qubit *singlet* state $(|01\rangle - |10\rangle)/\sqrt{2}$, which can be obtained from the state $\left|\phi_2^+\right\rangle$ via local unitaries. In entanglement theory local unitaries do not change the properties of a state, and thus we will refer to the state $\left|\phi_2^+\right\rangle$ as a singlet.

Operational measures of entanglement are *distillable entanglement* and *entanglement cost*. Distillable entanglement quantifies the maximal rate for extracting singlets from a state via LOCC operations [21, 22]:

$$\mathcal{E}_{\mathrm{d}}(\rho) = \sup\left\{R : \lim_{n\to\infty}\left(\inf_{\Lambda}\left\|\Lambda\left[\rho^{\otimes n}\right] - \left(\phi_2^+\right)^{\otimes nR}\right\|_1\right) = 0\right\}, \tag{8}$$

where $||M||_1 = \mathrm{Tr}\sqrt{M^\dagger M}$ is the trace norm, ϕ_2^+ is the projector onto the state $\left|\phi_2^+\right\rangle$ [26], and the infimum is performed over all LOCC operations Λ. Entanglement cost on the other hand quantifies the minimal singlet rate required for creating a state via LOCC operations [21, 22]:

$$\mathcal{E}_{\mathrm{c}}(\rho) = \inf\left\{R : \lim_{n\to\infty}\left(\inf_{\Lambda}\left\|\Lambda\left[\left(\phi_2^+\right)^{\otimes nR}\right] - \rho^{\otimes n}\right\|_1\right) = 0\right\}. \tag{9}$$

For pure states $|\psi\rangle = |\psi\rangle^{AB}$ these two quantities coincide and are equal to the von Neumann entropy of the reduced state [27]: $\mathcal{E}_{\mathrm{d}}(\psi) = \mathcal{E}_{\mathrm{c}}(\psi) = S(\rho^A) = -\mathrm{Tr}[\rho^A \log_2 \rho^A]$. This implies that the resource theory of entanglement is reversible for pure states [21, 22]. In general, it holds that $\mathcal{E}_{\mathrm{d}}(\rho) \leq \mathcal{E}_{\mathrm{c}}(\rho)$, and there exist states which have zero distillable entanglement but nonzero entanglement cost. This phenomenon is also known as *bound entanglement* [28].

An important family of entanglement measures is obtained by taking the minimal distance to the set of separable states $\mathcal{S}$ [23, 24]:

$$\mathcal{E}(\rho) = \inf_{\sigma \in \mathcal{S}} D(\rho, \sigma). \tag{10}$$

Here, $D(\rho, \sigma)$ can be an arbitrary functional which is nonnegative and monotonic under quantum operations, i.e., $D(\Lambda[\rho], \Lambda[\sigma]) \leq D(\rho, \sigma)$ for any quantum operation Λ [29]. Examples for such distances are the trace distance $||\rho - \sigma||_1/2$, the infidelity $1 - F(\rho, \sigma)$ with fidelity $F(\rho, \sigma) = ||\sqrt{\rho}, \sqrt{\sigma}||_1^2$, and the quantum relative entropy $S(\rho||\sigma) = \mathrm{Tr}[\rho \log_2 \rho] - \mathrm{Tr}[\rho \log_2 \sigma]$. In the latter case, the corresponding measure is known as the *relative entropy of entanglement* [23, 24]:

$$\mathcal{E}_{\mathrm{r}}(\rho) = \min_{\sigma \in \mathcal{S}} S(\rho||\sigma). \tag{11}$$

The second important family of measures are convex roof measures defined as [30]

$$\mathcal{E}(\rho) = \inf \sum_i p_i \mathcal{E}(\psi_i), \tag{12}$$

where the infimum is taken over all pure state decompositions of $\rho = \sum_i p_i \psi_i$. If for pure states entanglement is defined as the von Neumann entropy of the reduced state $\mathcal{E}(\psi) = S(\rho^A)$, the corresponding convex roof measure is known as the *entanglement of formation* [31]:

$$\mathcal{E}_{\mathrm{f}}(\rho) = \min \sum_i p_i S\left(\mathrm{Tr}_A\left[\psi_i\right]\right). \tag{13}$$

In general, the relative entropy of entanglement is between the distillable entanglement and the entanglement of formation [32]:

$$\mathcal{E}_{\mathrm{d}}(\rho) \leq \mathcal{E}_{\mathrm{r}}(\rho) \leq \mathcal{E}_{\mathrm{f}}(\rho). \tag{14}$$

Moreover, the regularized entanglement of formation is equal to the entanglement cost [33]: $\mathcal{E}_{\mathrm{c}}(\rho) = \lim_{n\to\infty} \mathcal{E}_{\mathrm{f}}(\rho^{\otimes n})/n$. We also mention that the geometric measure of entanglement defined as

$$\mathcal{E}_{\mathrm{g}}(\rho) = 1 - \max_{\sigma \in \mathcal{S}} F(\rho, \sigma) \tag{15}$$

is a distance-based and a convex roof measure simultaneously [34, 35].

Another important entanglement measure which will be used in this lecture is the *logarithmic negativity*. For a bipartite state $\rho = \rho^{AB}$ it is defined as [36, 37]

$$\mathcal{E}_{\mathrm{n}}(\rho) = \log_2 \left\| \rho^{T_A} \right\|_1 \tag{16}$$

with the partial transposition T_A. The logarithmic negativity is zero for states which have positive partial transpose, and thus there exist entangled states which have zero logarithmic negativity [38]. Nevertheless, these states cannot be distilled into singlets [28]. Interestingly, the logarithmic negativity is not convex [39], and is related to the entanglement cost under quantum operations preserving the positivity of the partial transpose [40].

Several entanglement measures discussed above are subadditive, i.e., they fulfill the inequality

$$\mathcal{E}\left(\rho \otimes \sigma\right) \leq \mathcal{E}\left(\rho\right) + \mathcal{E}\left(\sigma\right) \tag{17}$$

for any two states ρ and σ. Examples for subadditive measures are entanglement cost, entanglement of formation, and relative entropy of entanglement. The logarithmic negativity is additive, i.e., it fulfills Eq. (17) with equality. It is conjectured [41] that the distillable entanglement violates Eq. (17).

3 Quantum Discord

Quantum discord was introduced in [1, 2] as a quantifier for correlations different from entanglement. In the modern language of quantum information theory, quantum discord of a state $\rho = \rho^{AB}$ can be expressed in the following compact way [42, 43]:

$$\delta(\rho) = I(\rho) - \sup_{\Lambda_{\mathrm{eb}}} I\left(\Lambda_{\mathrm{eb}} \otimes \mathbb{1}\left[\rho\right]\right). \tag{18}$$

Here, $I(\rho^{AB}) = S(\rho^A) + S(\rho^B) - S(\rho^{AB})$ is the quantum mutual information and the supremum is performed over all entanglement breaking channels Λ_{eb} [44]. Quantum discord vanishes on CQ-states and is larger than zero otherwise [45]. The quantity $I(\rho) - \delta(\rho)$ was initially introduced in [2] as a measure of classical correlations. Interestingly, quantum discord is closely related to the entanglement of formation via the Koashi-Winter relation [46, 47]:

$$\delta(\rho^{AB}) = E_{\mathrm{f}}(\rho^{BC}) - S(\rho^{AB}) + S(\rho^A), \tag{19}$$

where the total state ρ^{ABC} is pure [48].

Similar as for entanglement, we can define distance-based measures of discord [49]:

$$\mathcal{D}(\rho) = \inf_{\Pi} D(\rho, \Pi \otimes \mathbb{1}[\rho]), \tag{20}$$

where the infimum is performed over all local von Neumann measurements Π and $D(\rho, \sigma)$ is a suitable distance between ρ and σ, such as the relative entropy. In the latter case, the corresponding quantity is called *relative entropy of discord* [50]:

$$\mathcal{D}_{\mathrm{r}}(\rho) = \min_{\Pi} S(\rho||\Pi \otimes \mathbb{1}[\rho]), \tag{21}$$

and has also been studied earlier in the context of thermodynamics [51, 52]. If the distance is chosen to be the squared Hilbert-Schmidt distance $\mathrm{Tr}(\rho - \sigma)^2$, the corresponding measure is known as the *geometric discord* [53, 54]. Interestingly, the geometric discord can increase under local operations on any of the subsystems [55]. It was also shown to play a role for remote state preparation [56].

The role of quantum discord in quantum information theory has been studied extensively in the last years [3, 4]. Several alternative quantifiers of discord have been presented [5], and criteria for good discord measures have also been discussed [57]. As an important example, we mention the *interferometric power* [58], which is a computable measure of discord and a figure of merit in the task of phase estimation with bipartite states. Further results on the role of quantum discord in quantum metrology have been presented in [59, 60]. The relation between quantum discord and entanglement creation in the quantum measurement process has also been investigated, both theoretically [61, 62] and experimentally [63]. Monogamy of quantum discord [64, 65] and its behavior under local noise [66] and non-Markovian dynamics [67] have also been studied. Experimentally friendly measures of discord were presented in [68, 69], and the possibility of local detection of discord has been reported in [70]. As we will see in the next section, quantum discord also plays an important role for entanglement distribution [7, 8].

4 Entanglement Distribution

In the following discussion we will distinguish between *direct* and *indirect* entanglement distribution between two parties (Alice and Bob) [11, 12]. Direct entanglement distribution is achieved if Alice prepares two particles in an entangled state ρ and sends one of them to Bob. The amount of distributed entanglement is then given by $\mathcal{E}(\mathbb{1} \otimes \Lambda[\rho])$, where Λ describes the corresponding quantum channel, and $\mathcal{E}$ is a suitable entanglement measure.

Indirect entanglement distribution is a more general scenario where Alice and Bob already share correlations initially. In this case the total initial state is a tripartite state $\rho = \rho^{ABC}$, where Alice is in possession of the particles A and C, and the particle B is in Bob's hands. Entanglement distribution is then achieved by sending the particle C from Alice to Bob, see Fig. 2. The amount of distributed entanglement is then given by $\mathcal{E}^{A|BC}(\mathbb{1}^{AB} \otimes \Lambda_C[\rho]) - \mathcal{E}^{AC|B}(\rho)$. In the following, we will discuss recent results on these two types of entanglement distribution [11, 12].

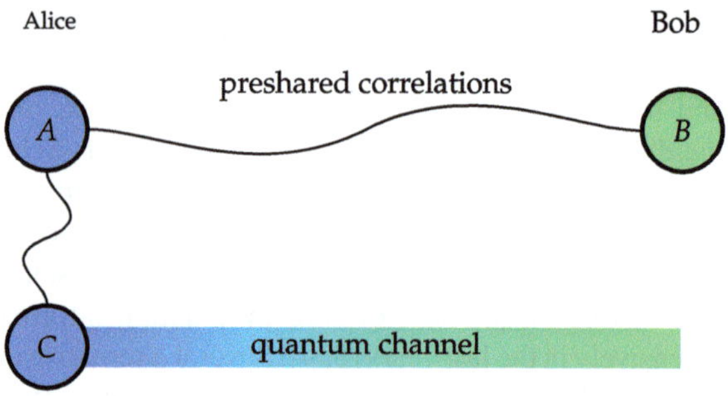

Fig. 2 Indirect entanglement distribution. Alice and Bob initially share the state $\rho = \rho^{ABC}$, where Alice holds the particles A and C, and Bob holds the particle B. Entanglement distribution is achieved by sending the particle C from Alice to Bob via a (possibly noisy) quantum channel. The figure is taken from [11]

4.1 Direct Entanglement Distribution

What is the maximal amount of entanglement that can be directly distributed via a given quantum channel Λ? For answering this question, we first introduce the corresponding figure of merit:

$$\mathcal{E}_{\rm direct}(\Lambda) = \sup_{\sigma} \mathcal{E}\left(\mathbb{1} \otimes \Lambda[\sigma]\right). \tag{22}$$

In general, the supremum is performed over all bipartite quantum states σ. However, if the entanglement quantifier $\mathcal{E}$ is convex, we can restrict ourselves to pure states.

If the distribution channel is noiseless, i.e., $\Lambda = \mathbb{1}$, then Eq. (22) reduces to

$$\mathcal{E}_{\rm direct}(\mathbb{1}) = \mathcal{E}(\phi_d^+), \tag{23}$$

where d is the dimension of the carrier particle. It is tempting to believe that this also extends to noisy channels, i.e., that for any noisy channel the optimal performance is achieved by sending one half of a maximally entangled state. Quite surprisingly, this procedure is not optimal in general [11, 71, 72]. In particular, for any convex entanglement measure $\mathcal{E}$ there exists a noisy channel Λ and a bipartite state ρ such that [71]

$$\mathcal{E}(\mathbb{1} \otimes \Lambda[\rho]) > \mathcal{E}(\mathbb{1} \otimes \Lambda[\phi_d^+]). \tag{24}$$

Even more, if entanglement is quantified via the logarithmic negativity, then states with arbitrary little entanglement can outperform maximally entangled states for some noisy channels [11]. Nevertheless, maximally entangled states are still optimal in various scenarios, e.g. if the carrier particle is a qubit and entanglement quantifier is the entanglement of formation or the geometric entanglement [11]. If the distribution channel is a single-qubit Pauli channel, i.e.,

$$\Lambda_{\rm p}[\rho] = \sum_{i=0}^{3} p_i \sigma_i \rho \sigma_i, \tag{25}$$

where σ_i are Pauli matrices with $\sigma_0 = \mathbb{1}$, then maximally entangled states are optimal for entanglement distribution, regardless of the particular entanglement measure [11]:

$$\mathcal{E}_{\mathrm{direct}}(\Lambda_{\mathrm{p}}) = \mathcal{E}(\mathbb{1} \otimes \Lambda_{\mathrm{p}}[\phi_2^+]). \tag{26}$$

This result also holds if entanglement distribution is performed via a combination of (possibly different) Pauli channels, also in this case sending one half of a maximally entangled state is the best strategy. Finally, if entanglement is quantified via the logarithmic negativity, maximally entangled states are optimal for all unital single-qubit channels [73].

This completes our discussion on direct entanglement distribution, and we will present the more general scenario in the following.

4.2 *Indirect Entanglement Distribution*

Can Alice and Bob gain an advantage if they share some correlations initially? To answer this question, we first introduce a figure of merit for indirect entanglement distribution:

$$\mathcal{E}_{\mathrm{indirect}}(\Lambda) = \sup_{\rho} \left\{ \mathcal{E}^{A|BC}(\mathbb{1}^{AB} \otimes \Lambda_C[\rho]) - \mathcal{E}^{AC|B}(\rho) \right\}, \tag{27}$$

where the supremum is taken over all tripartite states $\rho = \rho^{ABC}$. In particular, we are interested in the question if $\mathcal{E}_{\mathrm{indirect}}$ is larger than $\mathcal{E}_{\mathrm{direct}}$ for some noisy channel and some entanglement measure.

Note that so far no general answer to this question is known, and partial results have been presented in [11, 12]. In particular, if the channel used for entanglement distribution is a single-qubit Pauli channel given in Eq. (25) and entanglement is quantified via a subadditive measure $\mathcal{E}$, then indirect entanglement distribution does not provide any advantage [11]:

$$\mathcal{E}_{\mathrm{indirect}}(\Lambda_{\mathrm{p}}) = \mathcal{E}_{\mathrm{direct}}(\Lambda_{\mathrm{p}}) = \mathcal{E}(\mathbb{1} \otimes \Lambda_{\mathrm{p}}[\phi_2^+]). \tag{28}$$

This means that in this case sending one half of a singlet state is the optimal distribution strategy. This result can be generalized to the case where entanglement is distributed via a combination of (possibly different) Pauli channels [11].

However, not all entanglement measures are subadditive. An important example is the distillable entanglement $\mathcal{E}_{\mathrm{d}}$ which was defined in Eq. (8) and is conjectured [41] to violate subadditivity. Interestingly, if this conjecture is true, then indirect entanglement distribution provides an advantage for the distribution of distillable entanglement [11].

Finally, we note that entanglement breaking channels cannot be used for entanglement distribution for any entanglement measure $\mathcal{E}$ [12]:

$$\mathcal{E}_{\text{indirect}}(\Lambda_{\text{eb}}) = \mathcal{E}_{\text{direct}}(\Lambda_{\text{eb}}) = 0 \tag{29}$$

for any entanglement breaking channel Λ_{eb}. This can be seen by noting that any entanglement breaking channel is equivalent to an LOCC protocol [74].

4.3 Entanglement Distribution with Separable States

Entanglement can also be distributed by sending a carrier particle which is not entangled with the rest of the system. In particular, there exist tripartite states $\rho = \rho^{ABC}$ such that

$$\mathcal{E}^{AC|B}(\rho) = \mathcal{E}^{AB|C}(\rho) = 0, \quad \mathcal{E}^{A|BC}(\rho) > 0. \tag{30}$$

The first example for a state fulfilling Eq. (30) was presented in [6], and can be written as

$$\eta = \frac{1}{3}|\Psi_{\text{GHZ}}\rangle\langle\Psi_{\text{GHZ}}| + \sum_{i,j,k=0}^{1} \beta_{ijk}\Pi_{ijk}, \tag{31}$$

with $|\Psi_{\text{GHZ}}\rangle = (|000\rangle + |111\rangle)/\sqrt{2}$, $\Pi_{ijk} = |ijk\rangle\langle ijk|$, and all β_{ijk} are zero apart from $\beta_{001} = \beta_{010} = \beta_{101} = \beta_{110} = 1/6$. These results were extended to Gaussian states in [75], and experiments verifying this phenomenon have also been reported [13–15].

Motivated by this result, Zuppardo et al. [12] proposed a classification of entanglement distribution protocols. In particular, a noiseless distribution protocol is called *excessive* if the amount of distributed entanglement is larger than the amount of entanglement between the carrier and the rest of the system, i.e.,

$$\mathcal{E}^{A|BC}(\rho) - \mathcal{E}^{AC|B}(\rho) > \mathcal{E}^{AB|C}(\rho). \tag{32}$$

Otherwise, the protocol is called nonexcessive. As discussed above, the state η in Eq. (31) gives rise to an excessive distribution protocol.

It is natural to ask if such entanglement distribution with separable states can provide an advantage when compared to scenarios where the carrier particle is entangled with the rest of the system. In particular, one might ask if a separable state can show a better performance for entanglement distribution when compared to maximally entangled states. This question could be especially relevant if the distribution channel is noisy. Despite attempts by several authors [9, 13], the question has not yet been settled.

Finally, we mention that rank two separable states are not useful for entanglement distribution if entanglement is quantified via logarithmic negativity [10].

4.4 Role of Quantum Discord for Entanglement Distribution

As was shown in [7, 8], the amount of entanglement that can be distributed via a noiseless channel by using a tripartite quantum state $\rho = \rho^{ABC}$ is bounded above by the discord between the carrier particle C and the rest of the system:

$$\mathcal{E}^{A|BC}(\rho) - \mathcal{E}^{AC|B}(\rho) \leq \mathcal{D}^{C|AB}(\rho). \tag{33}$$

This inequality is true for any distance-based measure of entanglement and discord given in Eqs. (10) and (20) if the corresponding distance does not increase under quantum operations and fulfills the triangle inequality. Moreover, it is also true for the relative entropy of entanglement and discord [7, 8].

The inequality (33) immediately implies that zero-discord states cannot be used for entanglement distribution. Moreover, this result can also be used to bound the amount of entanglement in one cut of a tripartite state $\rho = \rho^{ABC}$ in terms of entanglement and discord in the other cuts [7, 8]:

$$\mathcal{E}^{AC|B}(\rho) + \mathcal{D}^{C|AB}(\rho) \geq \mathcal{E}^{A|BC}(\rho) \geq \mathcal{E}^{AC|B}(\rho) - \mathcal{D}^{C|AB}(\rho). \tag{34}$$

For the relative entropy of entanglement and discord, the inequality (33) is saturated for pure states of the form $|\psi\rangle^{AC} \otimes |\phi\rangle^{B}$ and also for the state η given in Eq. (31) [7].

If the channel used for entanglement distribution is noisy, we get the following generalized inequality [11]:

$$\mathcal{E}^{A|BC}(\rho') - \mathcal{E}^{AC|B}(\rho) \leq \min\left\{\mathcal{D}^{C|AB}(\rho), \mathcal{D}^{C|AB}(\rho')\right\}. \tag{35}$$

Here, we used the notation $\rho' = \mathbb{1}^{AB} \otimes \Lambda_C[\rho]$, and $\mathcal{E}$ and $\mathcal{D}$ are any measures of entanglement and discord which fulfill Eq. (33).

5 Conclusions

In this lecture we discussed recent results on entanglement distribution and the role of quantum discord in this task. Despite substantial progress in recent years, several important questions in this research field still remain open. In particular, it is still unclear if indirect entanglement distribution can provide an advantage in comparison to direct distribution protocols. The question also concerns entanglement distribution with separable states: also in this case it remains unclear if such scheme can be more useful than any direct distribution procedure.

We also mention that studying entanglement distribution in relation to the resource theory of coherence [76–78] and its extension to distributed scenarios [79–86] could potentially shed new light on these questions, and also lead to new independent results.

Acknowledgements We thank Remigiusz Augusiak, Maciej Demianowicz, Jens Eisert, and Maciej Lewenstein for discussion. This work was supported by the Alexander von Humboldt-Foundation, Bundesministerium für Bildung und Forschung, and Deutsche Forschungsgemeinschaft.

References

1. H. Ollivier, W.H. Zurek, Phys. Rev. Lett. **88**, 017901 (2001)
2. L. Henderson, V. Vedral, J. Phys. A **34**, 6899 (2001)
3. K. Modi, A. Brodutch, H. Cable, T. Paterek, V. Vedral, Rev. Mod. Phys. **84**, 1655 (2012)
4. A. Streltsov, *Quantum Correlations Beyond Entanglement and their Role in Quantum Information Theory* (SpringerBriefs in Physics, 2015). arXiv:1411.3208
5. G. Adesso, T.R. Bromley, M. Cianciaruso, J. Phys. A **49**, 473001 (2016). arXiv:1605.00806
6. T.S. Cubitt, F. Verstraete, W. Dür, J.I. Cirac, Phys. Rev. Lett. **91**, 037902 (2003)
7. A. Streltsov, H. Kampermann, D. Bruß, Phys. Rev. Lett. **108**, 250501 (2012)
8. T.K. Chuan, J. Maillard, K. Modi, T. Paterek, M. Paternostro, M. Piani, Phys. Rev. Lett. **109**, 070501 (2012)
9. A. Kay, Phys. Rev. Lett. **109**, 080503 (2012)
10. A. Streltsov, H. Kampermann, D. Bruß, Phys. Rev. A **90**, 032323 (2014)
11. A. Streltsov, R. Augusiak, M. Demianowicz, M. Lewenstein, Phys. Rev. A **92**, 012335 (2015)
12. M. Zuppardo, T. Krisnanda, T. Paterek, S. Bandyopadhyay, A. Banerjee, P. Deb, S. Halder, K. Modi, M. Paternostro, Phys. Rev. A **93**, 012305 (2016)
13. A. Fedrizzi, M. Zuppardo, G.G. Gillett, M.A. Broome, M.P. Almeida, M. Paternostro, A.G. White, T. Paterek, Phys. Rev. Lett. **111**, 230504 (2013)
14. C.E. Vollmer, D. Schulze, T. Eberle, V. Händchen, J. Fiurášek, R. Schnabel, Phys. Rev. Lett. **111**, 230505 (2013)
15. C. Peuntinger, V. Chille, L. Mišta, N. Korolkova, M. Förtsch, J. Korger, C. Marquardt, G. Leuchs, Phys. Rev. Lett. **111**, 230506 (2013)
16. C. Silberhorn, Physics **6**, 132 (2013)
17. R.F. Werner, Phys. Rev. A **40**, 4277 (1989)
18. M. Piani, P. Horodecki, R. Horodecki, Phys. Rev. Lett. **100**, 090502 (2008)
19. A. Ferraro, L. Aolita, D. Cavalcanti, F.M. Cucchietti, A. Acín, Phys. Rev. A **81**, 052318 (2010)
20. D. Bruß, J. Math. Phys. **43**, 4237 (2002)
21. M.B. Plenio, S. Virmani, Quantum Inf. Comput. **7**, 1 (2007). arXiv.org/abs/quant-ph/9701014
22. R. Horodecki, P. Horodecki, M. Horodecki, K. Horodecki, Rev. Mod. Phys. **81**, 865 (2009)
23. V. Vedral, M.B. Plenio, M.A. Rippin, P.L. Knight, Phys. Rev. Lett. **78**, 2275 (1997)
24. V. Vedral, M.B. Plenio, Phys. Rev. A **57**, 1619 (1998)
25. Note that some entanglement measures (like distillable entanglement and logarithmic negativity) also vanish for some entangled states
26. For a general pure state $|\psi\rangle$ we will denote the corresponding projector by ψ, i.e., $\psi = |\psi\rangle\langle\psi|$
27. C.H. Bennett, H.J. Bernstein, S. Popescu, B. Schumacher, Phys. Rev. A **53**, 2046 (1996)
28. M. Horodecki, P. Horodecki, R. Horodecki, Phys. Rev. Lett. **80**, 5239 (1998)
29. Note that $D(\rho, \sigma)$ does not have to be a distance in the mathematical sense, since it does not necessarily fulfill the triangle inequality
30. A. Uhlmann, Open Syst. Inf. Dyn. **5**, 209 (1998). arXiv.org/abs/quant-ph/9701014
31. C.H. Bennett, D.P. DiVincenzo, J.A. Smolin, W.K. Wootters, Phys. Rev. A **54**, 3824 (1996)

32. M. Horodecki, P. Horodecki, R. Horodecki, Phys. Rev. Lett. **84**, 2014 (2000)
33. P.M. Hayden, M. Horodecki, B.M. Terhal, J. Phys. A **34**, 6891 (2001)
34. T.-C. Wei, P.M. Goldbart, Phys. Rev. A **68**, 042307 (2003)
35. A. Streltsov, H. Kampermann, D. Bruß, New J. Phys. **12**, 123004 (2010)
36. K. Życzkowski, P. Horodecki, A. Sanpera, M. Lewenstein, Phys. Rev. A **58**, 883 (1998)
37. G. Vidal, R.F. Werner, Phys. Rev. A **65**, 032314 (2002)
38. P. Horodecki, Phys. Lett. A **232**, 333 (1997)
39. M.B. Plenio, Phys. Rev. Lett. **95**, 090503 (2005)
40. K. Audenaert, M.B. Plenio, J. Eisert, Phys. Rev. Lett. **90**, 027901 (2003)
41. P.W. Shor, J.A. Smolin, B.M. Terhal, Phys. Rev. Lett. **86**, 2681 (2001)
42. A. Streltsov, W.H. Zurek, Phys. Rev. Lett. **111**, 040401 (2013)
43. A. Streltsov, S. Lee, G. Adesso, Phys. Rev. Lett. **115**, 030505 (2015)
44. An entanglement breaking channel Λ_{eb} has the property that $\Lambda_{eb} \otimes \mathbb{1}[\rho]$ is not entangled for any bipartite input state ρ. We refer to [74] for more details
45. A. Datta (2010). arXiv:1003.5256
46. M. Koashi, A. Winter, Phys. Rev. A **69**, 022309 (2004)
47. F.F. Fanchini, M.F. Cornelio, M.C. de Oliveira, A.O. Caldeira, Phys. Rev. A **84**, 012313 (2011)
48. Interestingly, Eq. (19) implies that a simple formula for quantum discord for all quantum states is out of reach, since such an expression would also allow for an exact evaluation of entanglement of formation. Nevertheless, analytical progress on the evaluation of discord for particular families of states has been presented in [87–89]
49. Many authors also consider the minimal distance to the set of CQ states, i.e., $\Delta(\rho) = \inf_{\sigma \in \mathcal{CQ}} D(\rho, \sigma)$. We note that Δ and $\mathcal{D}$ coincide for the quantum relative entropy, and $\Delta(\rho) \leq \mathcal{D}(\rho)$ in general [3]
50. K. Modi, T. Paterek, W. Son, V. Vedral, M. Williamson, Phys. Rev. Lett. **104**, 080501 (2010)
51. J. Oppenheim, M. Horodecki, P. Horodecki, R. Horodecki, Phys. Rev. Lett. **89**, 180402 (2002)
52. M. Horodecki, P. Horodecki, R. Horodecki, J. Oppenheim, A. Sen(De), U. Sen, B. Synak-Radtke, Phys. Rev. A **71**, 062307 (2005)
53. B. Dakić, V. Vedral, Č. Brukner, Phys. Rev. Lett. **105**, 190502 (2010)
54. S. Luo, S. Fu, Phys. Rev. A **82**, 034302 (2010)
55. M. Piani, Phys. Rev. A **86**, 034101 (2012)
56. B. Dakić, Y.O. Lipp, X. Ma, M. Ringbauer, S. Kropatschek, S. Barz, T. Paterek, V. Vedral, A. Zeilinger, Č. Brukner, P. Walther, Nat. Phys. **8**, 666 (2012)
57. A. Brodutch, K. Modi, Quantum Inf. Comput. **12**, 0721 (2012). arXiv:1108.3649
58. D. Girolami, A.M. Souza, V. Giovannetti, T. Tufarelli, J.G. Filgueiras, R.S. Sarthour, D.O. Soares-Pinto, I.S. Oliveira, G. Adesso, Phys. Rev. Lett. **112**, 210401 (2014)
59. K. Modi, H. Cable, M. Williamson, V. Vedral, Phys. Rev. X **1**, 021022 (2011)
60. D. Girolami, T. Tufarelli, G. Adesso, Phys. Rev. Lett. **110**, 240402 (2013)
61. A. Streltsov, H. Kampermann, D. Bruß, Phys. Rev. Lett. **106**, 160401 (2011)
62. M. Piani, S. Gharibian, G. Adesso, J. Calsamiglia, P. Horodecki, A. Winter, Phys. Rev. Lett. **106**, 220403 (2011)
63. G. Adesso, V. D'Ambrosio, E. Nagali, M. Piani, F. Sciarrino, Phys. Rev. Lett. **112**, 140501 (2014)
64. A. Streltsov, G. Adesso, M. Piani, D. Bruß, Phys. Rev. Lett. **109**, 050503 (2012)
65. H.S. Dhar, A.K. Pal, D. Rakshit, A. Sen(De), U. Sen (2016). arXiv:1610.01069
66. A. Streltsov, H. Kampermann, D. Bruß, Phys. Rev. Lett. **107**, 170502 (2011)
67. F.F. Fanchini, T. Werlang, C.A. Brasil, L.G.E. Arruda, A.O. Caldeira, Phys. Rev. A **81**, 052107 (2010)
68. R. Auccaise, J. Maziero, L.C. Céleri, D.O. Soares-Pinto, E.R. deAzevedo, T.J. Bonagamba, R.S. Sarthour, I.S. Oliveira, R.M. Serra, Phys. Rev. Lett. **107**, 070501 (2011)
69. D. Girolami, G. Adesso, Phys. Rev. Lett. **108**, 150403 (2012)
70. M. Gessner, M. Ramm, T. Pruttivarasin, A. Buchleitner, H.-P. Breuer, H. Häffner, Nat. Phys. **10**, 105 (2014)
71. M. Ziman, V. Bužek (2007). arXiv:0707.4401

72. R. Pal, S. Bandyopadhyay, S. Ghosh, Phys. Rev. A **90**, 052304 (2014)
73. The authors of [72] proved this statement for negativity $\mathcal{N}$, which is related to the logarithmic negativity as $\mathcal{E}_n = \log_2(2\mathcal{N} + 1)$. Since $\mathcal{N}$ is a nondecreasing function of $\mathcal{E}_n$, it follows that the statement is also true for the logarithmic negativity
74. M. Horodecki, P.W. Shor, M.B. Ruskai, Rev. Math. Phys. **15**, 629 (2003)
75. L. Mišta Jr., N. Korolkova, Phys. Rev. A **77**, 050302 (2008)
76. T. Baumgratz, M. Cramer, M.B. Plenio, Phys. Rev. Lett. **113**, 140401 (2014)
77. A. Winter, D. Yang, Phys. Rev. Lett. **116**, 120404 (2016)
78. A. Streltsov, G. Adesso, M.B. Plenio (2016). arXiv:1609.02439
79. T.R. Bromley, M. Cianciaruso, G. Adesso, Phys. Rev. Lett. **114**, 210401 (2015)
80. A. Streltsov, U. Singh, H.S. Dhar, M.N. Bera, G. Adesso, Phys. Rev. Lett. **115**, 020403 (2015)
81. E. Chitambar, A. Streltsov, S. Rana, M.N. Bera, G. Adesso, M. Lewenstein, Phys. Rev. Lett. **116**, 070402 (2016)
82. J. Ma, B. Yadin, D. Girolami, V. Vedral, M. Gu, Phys. Rev. Lett. **116**, 160407 (2016)
83. E. Chitambar, M.-H. Hsieh, Phys. Rev. Lett. **117**, 020402 (2016)
84. J.M. Matera, D. Egloff, N. Killoran, M.B. Plenio, Quantum Sci. Technol. **1**, 01LT01 (2016)
85. A. Streltsov, S. Rana, M.N. Bera, M. Lewenstein, Phys. Rev. X **7**, 011024 (2017). arXiv:1509.07456
86. B. Yadin, J. Ma, D. Girolami, M. Gu, V. Vedral, Phys. Rev. X **6**, 041028 (2016). arXiv:1512.02085
87. M. Ali, A.R.P. Rau, G. Alber, Phys. Rev. A **81**, 042105 (2010)
88. M. Ali, A.R.P. Rau, G. Alber, Phys. Rev. A **82**, 069902 (2010)
89. D. Girolami, G. Adesso, Phys. Rev. A **83**, 052108 (2011)

Discord, Quantum Knowledge and Private Communications

Mile Gu and Stefano Pirandola

Abstract In this brief review, we discuss the role that quantum correlations, as quantified by quantum discord, play in two interesting settings. The first one is discerning which unitaries have been applied on a quantum system, by taking advantage of knowledge regarding its initial configuration. Here discord captures the 'quantum' component of this knowledge, useful only when we have access to a quantum memory. In particular, discord can be used to detect whether an untrusted party has certain quantum capabilities. The second setting is quantum cryptography. Here discord represents an important resource for trusted-noise quantum key distribution and also provides a general upper bound for the optimal secret key rates that are achievable by ideal protocols. In particular, the (two-way assisted) secret key capacity of a lossy bosonic channel exactly coincides with the maximum discord that can be distributed between the remote parties at the two ends of the channel.

1 Knowledge, Correlations, and Guessing Channels

The 1962 James Bond's movie 'Dr. No' taught children around the world a valuable lesson in how to detect whether nosy siblings are snooping into their rooms. You stick a small piece hair across the door and the doorframe. When the door is opens, the hair

M. Gu
School of Physical and Mathematical Sciences, Nanyang Technological University, Singapore 639673, Singapore

M. Gu
Complexity Institute, Nanyang Technological University, Singapore 637723, Singapore

M. Gu
Centre for Quantum Technologies, National University of Singapore, Singapore 117543, Singapore

S. Pirandola (✉)
Computer Science & York Centre for Quantum Technologies, University of York, York YO10 5GH, UK
e-mail: Stefano.pirandola@york.ac.uk

F.F. Fanchini et al. (eds.), *Lectures on General Quantum Correlations and their Applications*, Quantum Science and Technology,
DOI 10.1007/978-3-319-53412-1_11

falls to the floor. The unsuspecting perpetrator has unwittingly communicated to you their rather unscrupulous action. This trick demonstrates the power of knowledge; by knowing how a system is initially configured (the location of hair), one can gain information about actions that have affected the system (opening the door).

This phenomena can be described by information theory. We denote a system of interest to be A, and knowledge about the system to be encoded within some memory B - an approach previously adopted to understand uncertainty relations under quantum memory [1]. If B contains information about A, the two systems will be correlated, such that $I(A, B) > 0$.

The classical one time pad provides a simple example. Here Alice and Bob wish to communicate some secret message in the future. To do this, Alice and Bob gather in some secure location, where Alice generates a string of random bits that Bob commits them to memory. That is, they share many copies of the classically correlated state

$$\rho = |00\rangle\langle 00| + |11\rangle\langle 11|. \tag{1}$$

Should Alice choose to flip some of her bits and give the resulting string to Bob, Bob is able to discern exactly which bits have been flipped by comparing the resulting string with the one stored in his memory. In contrast, anyone without access to Bob's memory would gain no information about Alice's actions. The optimal such scheme would allow Alice to communicate $\delta_I = 1$ bit per copy of ρ_{AB} shared. Thus possession of B allows exclusive knowledge of how the system was manipulated. One notes that here, $I(A, B) = 1$, which is equal to δ_I. This is in fact, not a coincidence.

Consider the following general "channel guessing game".

1. Alice and Bob initially share a state ρ distributed over the system of interest A, and the memory B. This initial state is publicly known.
2. Alice applies some unitary operator U_k onto her subsystem A with probability p_k. She publicly announces her protocol (e.g. the unitaries U_k and their probability of application), but not the specific k she selects in each run.
3. Alice gives A to Bob, so that Bob is now in possession of $\rho_{AB}^{(k)} = U_k \rho_{AB} U_k^\dagger$. Without knowledge of k, Bob sees the ensemble state $\tilde{\rho}_{AB} = \sum_k p_k \rho_{AB}^{(k)}$.
4. Alice challenges Bob to guess which U_k she has applied, i.e., to estimate the value of k.

This game captures a communication channel between Alice and Bob, where Alice has encoded a random variable K that takes the value k with probability p_k, onto corresponding codewords $\rho_{AB}^{(k)}$. The maximum information rate of this channel is then bounded above by the Holevo quantity

$$I_q = \tilde{S}(A, B) - S(A, B), \tag{2}$$

where $S(A, B)$ and $\tilde{S}(A, B)$ represent the respective entropies of ρ_{AB} and $\tilde{\rho}_{AB}$. Here we consider the i.i.d. limit of many trials, where Alice repeats this game a large number of times; the performance of Bob, as quantified by the maximum information per trial, then saturates I_q.

This relation has a nice interpretation. In fact $\tilde{\rho}$ describes the state of the bipartite system after encoding, as viewed by an observer who is unaware of which k was encoded in each run. Therefore $\tilde{S}(A, B) - S(A, B)$ captures the gain in entropy (or alternatively, the cost in negentropy) of encoding K from their perspective. Thus Eq. (2) tells us that communication of k bits of data necessarily incurs a minimum entropic cost of k.

Suppose Bob cannot access his memory (e.g. it was lost), the effective codewords would now be be $\rho_A^{(k)}$, with associated Holevo quantity

$$I_0 = \tilde{S}(A) - S(A). \tag{3}$$

The impact of having memory on Bob's in performance at the i.i.d. limit is then

$$\Delta_q \equiv I_q - I_0 = I(A, B) - \tilde{I}(A, B), \tag{4}$$

where $I(A, B)$ and $\tilde{I}(A, B)$ are the respective mutual information of ρ_{AB} and $\tilde{\rho}_{AB}$. The quantity $I(A, B) - \tilde{I}(A, B)$ then represents the cost, in terms of total correlations between A and B of encoding K. Meanwhile Δ_q represents information about K that is exclusively available to Bob due to his possession of B.

If we consider $I(A, B)$ to capture the amount of knowledge Bob knows about A, we find an interesting resource based view of knowledge: *Bob can expend k bits of knowledge about a system A to learn at most k bits of information about actions on A; in the i.i.d limit, this bound can be saturated*. That is, knowing k bits about some system A, as captured by possessing a system B such that $I(A, B) = k$, implies that one can gain up to k extra bits about actions on A. Thus for the one-time pad, a shared mutual information of 1 allows Alice to securely communicate a single bit to Bob. Meanwhile in quantum dense coding, Alice and Bob initially share a Bell state - such that $I(A, B) = 2$. Thus, Bob can harness his memory to gain 2 exclusive bits about Alice's actions on A.

1.1 The Role of Discord

Recall that we can separate correlations into two components, i.e., we can write $I(A, B) = J(A|B) + \delta(A|B)$, where $J(A|B)$ and $\delta(A|B)$ respectively represent the purely-classical correlations and the quantum correlations (discord [2]). This fits well with the channel guessing game. The approach is to relate quantum and classical correlations with the quantum and classicality of Bob's memory. Specifically let us consider the three scenarios:

1. *Memoryless Bob*: Bob's memory is completely faulty. That is, Bob cannot access B at all. Bob's resulting performance is then given by I_0 (as defined above).

2. *Classical Bob*: Bob's memory is classical. That is, Bob is required to measure any ρ_b given to him with respect to some orthogonal basis, and stored the measurement results in place of ρ_b. Denote Bob's resulting performance by I_c.
3. *Quantum Bob*: Bob has unrestricted quantum information processing, and can (i) store ρ_b without error, and (ii) coherently interact his memory with the system of interest. Bob's resulting performance is given by I_q.

Cases 1 and 3 have been outlined above. Our focus here is thus case 2. The rationale is that a classical memory should be able to make use of purely classical correlations, but not quantum correlations. Therefore, we would expect discord to be related with the performance gap, $I_q - I_c$, between quantum and classical Bob. This problem was studied in Gu et al. [3], where they established that

$$J(A|B) - \tilde{J}(A|B) \leq I_c - I_0 \leq J(A|B). \tag{5}$$

Here $\tilde{J}(A|B)$ and $\tilde{J}(A|B)$ represent the classical correlations in ρ_{AB} and $\tilde{\rho}_{AB}$. The equation describes an interesting connection between discord and the performance advantage of having quantum-over-classical memory. That is, introducing $\delta(A|B) - \tilde{\delta}(A|B)$ as the discord difference before and after encoding, we get

$$\Delta\delta(A|B) - \tilde{I}(A, B) \leq I_q - I_c \leq \Delta\delta(A|B), \tag{6}$$

where $\tilde{I}(A|B)$ is the mutual information of $\tilde{\rho}$. Consider now any encoding that attempts to communicate the maximum amount of information (known as a maximal encoding). In this scenario, $\tilde{\rho}_A$ is maximally mixed, $\tilde{\delta} = \tilde{I} = 0$, and thus we have:

1. $I_0 = 1 - S(A)$: a memoryless Bob can only access the local memory available on A. That is, the maximum amount of information Bob can learn about what happens to A is exactly the negentropy of A.
2. $I_c = I_0 + J(A|B)$: a classical Bob can learn an additional $\Delta_c = J(A|B)$ bits of information about actions on A. That is, he can exactly take advantage of the classical correlations between A and B.
3. $I_q = I_0 + I(A, B)$: a quantum Bob can take advantage of the full correlations between A and B. As such, his performance advantage over the classical case is exactly $\delta(A|B)$, the discord between B and A.

These relations capture an operational interpretation of discord $\delta(A|B)$ as how much purely quantum mechanical knowledge B has about A. An example if given in Fig. 1.

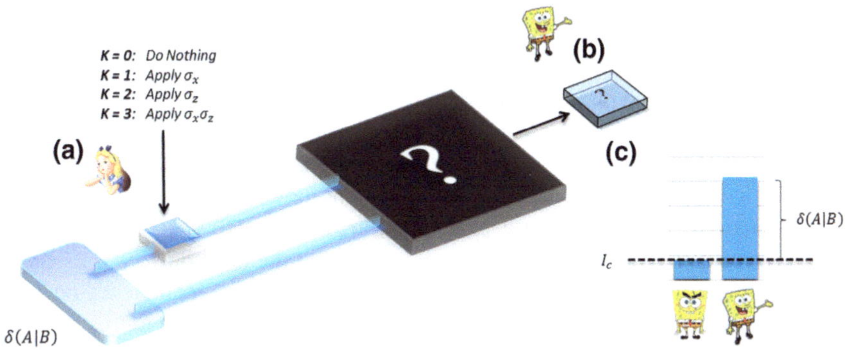

Fig. 1 **Example on Two Qubits**. Consider the special case where Alice and Bob share a correlated state on two qubits, A and B, with discord $\delta(A|B)$. Alice then encodes a random variable K governed by a uniform distribution over $\{0, 1, 2, 3\}$ by applying one of four possible unitaries, I, σ_x, σ_z or $\sigma_x\sigma_z$ and challenges Bob to estimate K. In this scenario, the encoding is maximal, and Bob's performance gain when using quantum in the place of classical memory is given exactly by $\delta(A|B)$. This protocol has been experimentally implemented by Almeida et.al. [4]

1.2 Example: Certifying Entangling Gates Without Entanglement

The interpretation of discord as quantum knowledge can be applied to verify whether someone is in possession of entangling gates, as also experimentally realized by using polarization photons [4]. Consider the case where Bob claims that he is capable of building entangling two-qubit gates. How can Alice verify that Bob is telling the truth - without being able to generate entanglement herself?

The inability for classical processors to harness quantum knowledge suggests an immediate solution. Suppose now Alice prepares some discord, two-qubit state, ρ_{AB}. She can then perform the protocol above, using a specific encoding scheme that encodes two bits, $a, b \in \{0, 1\}$, onto A, by applying the unitary $U = X^a Z^b$, where X and Z are standard Pauli operators. This corresponds to a scenario where $\tilde{\rho}_A = I/2$ is maximally mixed. Alice then challenges Bob guess a and b. Bob's performance is then characterized by the mutual information between the encoded bits, and that of Bob's guess.

In the 2 qubit case, it can be shown that if Bob is incapable of synthesizing entangling two-qubit gates, then he cannot exceed the performance level of I_c. As such any performance exceeding I_c implies that Bob is capable of some entangling operations. Thus, discord can be used as a way of certifying entanglement without entangling gates.

2 Discord in Quantum Key Distribution

Quantum discord also plays an important role in private communications and quantum key distribution (QKD) [5–7]. The fact that it must be non-zero is intuitive: Quantum discord and its geometric formulation are connected with the concept of non-orthogonality, which is the essential ingredient for QKD. A scenario where this is particularly evident is device-dependent (or trusted-device) QKD. This includes all those realistic situations where the noise affecting the local devices is assumed to be trusted. For instance this can be detection noise (genuine inefficiency or noise added by the parties [8, 9]) or preparation noise, as in the settings of untrusted-relay QKD [10, 11] and thermal-QKD [12–16]. Such trusted noise may be so high to prevent any entanglement distribution, but still a secure key can be extracted due to non-zero discord.

Any QKD protocol can be recast into a measurement-based scheme, where Alice sends Bob part of a bipartite state, then subject to local detections. Let us describe a device-dependent protocol in this representation. In her private space, Alice prepares two systems, A and a, in a generally mixed state ρ_{Aa}. This state is purified into a 3-partite state Φ_{PAa} with the ancillary system P being inaccessible to Alice, Bob or Eve. This system accounts for the trusted noise in Alice's side. Then, system b is sent to Bob, who gets the output B after the channel (eavesdropping). Bob's output B is assumed to be affected by other local trusted noise in Bob's private space (denoted as P as before). Finally, from the shared state ρ_{AB}, Alice and Bob extract two correlated variables by applying suitable measurements. On the output data, they perform error correction and privacy amplification with the help of one-way classical communication (CC), which can be either forward (direct reconciliation, $\blacktriangleright$), or backward (reverse reconciliation, $\blacktriangleleft$).

They finally extract a key at a rate $R = \max\{R_\blacktriangleright, R_\blacktriangleleft\}$, maximised between the reconciliations. Now we have [17]

$$E_D(A, B) \leq R \leq E_D(A, B) + I(AB, P), \tag{7}$$

where $E_D(A, B)$ is the one-way distillable entanglement for systems A and B, as quantified by the maximum between the coherent [18, 19] and reverse coherent information [20, 21], while $I(AB, P)$ is the quantum mutual information between AB and the trusted-noise system P. From Eq. (7), we see that the existence of P is necessary in order to have $R > 0$ in the absence entanglement (i.e., for $E_D = 0$). Indeed it is easy to find discord-based Gaussian QKD protocols for which this is possible [17]. According to Eq. (7), the absence of P implies $R = E_D$, so that secure key distribution becomes equivalent to entanglement distillation [22].

In the absence of trusted noise, we have ideal QKD protocols where all the noise in the global output state is partly controlled by the parties and partly by Eve. In this setting, quantum discord becomes a simple upper bound for the key rate. In fact, for any ideal QKD protocol in direct or reverse reconciliation, we may write [17]

$$R \leq \max\{\delta(A|B), \delta(B|A)\}, \tag{8}$$

where $\delta(A|B)$ and $\delta(B|A)$ are the two types of discord. Surprisingly, for the important practical case of a lossy channel [7] with transmissivity η, such as an optical fiber or a free-space link, the previous bound becomes tight. This is due to a combination of elements. First of all, we may always write [17]

$$R_{\blacktriangleleft} = \delta(B|A) - E_F(B, E), \tag{9}$$

where $E_F(B, E)$ is the entanglement of formation between Bob and Eve. Second, the Stinespring dilation of a lossy channel is a beam splitter with transmissivity η, mixing the Alice's input state with a vacuum environmental mode. For this reason, Bob and Eve's output state is not entangled, i.e., $E_F(B, E) = 0$. Therefore, in a lossy channel, we always have

$$R_{\blacktriangleleft} = \delta(B|A). \tag{10}$$

Most importantly, one can prove [23] that the maximum discord $\delta_{max}(B|A)$ that can be distributed to the parties through the lossy channel coincides with the secret-key capacity K of the lossy channel (where this capacity is generally defined assuming the most general feedback-assisted protocols for key generation, based on unlimited two-way CC and adaptive local operations). In fact, Ref. [23] showed that

$$K(\eta) = \delta_{max}(B|A) = -\log_2(1-\eta), \tag{11}$$

which provides the ultimate rate-loss scaling for bosonic secure communications, approximately 1.44η secret bits per channel use for high loss (i.e., at long distances)

The proof Eq. (11) is based on several ingredients. First of all, it exploits the technique of teleportation stretching, devised in Ref. [23] for point-to-point quantum/private communications, and then extended in Ref. [24] to quantum repeaters and communication networks. In this technique, an arbitrary adaptive protocol for quantum/private communication is simplified into a much simpler non-adaptive form, providing the same output state as the original one. The advantage is that such output state is now decomposed in the form $\bar{\Lambda}(\rho_{\mathcal{E}}^{\otimes n})$, where $\bar{\Lambda}$ is a trace-preserving LOCC, $\rho_{\mathcal{E}}$ is the Choi matrix [25] of the channel $\mathcal{E}$ (to be defined as suitable limit for a lossy channel), and n is the number of uses of the channel. This decomposition is possible because the lossy channel is covariant with respect to the displacement operators and therefore can be simulated by means of continuous variable quantum teleportation [26, 27]. In other words, the lossy channel is a specific example of teleportation-covariant channel [23].

The second ingredient is introduction of the channel's relative entropy of entanglement $E_R(\mathcal{E})$, which extends the original definition for quantum states [28–30] to quantum channels. Ref. [23] proved that, for any channel $\mathcal{E}$, the secret-key capacity satisfies the bound $K(\mathcal{E}) \leq E_R(\mathcal{E})$. For the specific case of the lossy channel, one may combine the Choi-decomposition of the output $\bar{\Lambda}(\rho_{\mathcal{E}}^{\otimes n})$ together with the properties of the relative entropy of entanglement to prove that $K(\eta) \leq E_R(\rho_{\mathcal{E}})$. The latter term is the relative entropy of entanglement of the asymptotic Choi matrix of the lossy channel and must be computed as a limit over a sequence of two-mode squeezed vacuum states [23]. This procedure leads to the upper bound

$$K(\eta) \leq -\log_2(1-\eta). \tag{12}$$

Since the upper bound is achievable by a suitable Gaussian protocol in reverse reconciliation [17, 20], we then achieve Eq. (11). The proof can be easily extended to include the two-way quantum capacity, so that we also have $K(\eta) = Q_2(\eta)$ [23].

3 Conclusions

In this brief review, we have discussed the role that quantum discord plays in two interesting settings. First of all, we considered the scenario of a bipartite system consisting of a system of interest, A, and a memory system B, such that their correlations, $I(A, B)$, represent knowledge B has about A. This knowledge can be harnessed by a person in possession of B to gain extra information about what performed on A. In this context, we outlined how discord is captured in the quantum component of such knowledge - measuring the component of $I(A, B)$ that is useful only when B can be stored in quantum memory.

We then reviewed how quantum discord can be seen as a primitive for quantum cryptography, where it plays a double role. It is the bipartite resource which is exploited in trusted-noise QKD, where the presence of such noise may prevent the exploitation of quantum entanglement but not the distribution of a secret key. Then, quantum discord provides a general upper bound to the key rate in the ideal case when trusted noise is absent. In particular, this bound is achievable in the important case of lossy bosonic communications. In this setting, the maximum discord that two remote parties can generate at the two ends of a lossy channel corresponds exactly to the maximum number of secret bits that they can generate through the channel by means of the most general adaptive protocols for QKD.

Acknowledgements Authors acknowledge financial support from the National Research Foundation of Singapore (NRF), NRF-Fellowship (Reference No: NRF-NRFF2016-02), the John Templeton Foundation (Grant No 54914) and the EPSRC via the 'UK Quantum Communications Hub' (EP/M013472/1).

References

1. M. Berta et al., Nat. Phys. **6**, 659–662 (2011)
2. K. Modi et al., Rev. Mod. Phys. **84**, 1655 (2012)
3. M. Gu et al., Nat. Phys. **8**, 671–675 (2012)
4. M. Almeida et al., Phys. Rev. A **89**, 042323 (2014)

5. N. Gisin, G. Ribordy, W. Tittel, H. Zbinden, Rev. Mod. Phys. **74**, 145 (2002)
6. V. Scarani et al., Rev. Mod. Phys. **81**, 1301 (2009)
7. C. Weedbrook, S. Pirandola, R. Garcia-Patron, N.J. Cerf, T.C. Ralph, J.H. Shapiro, S. Lloyd, Rev. Mod. Phys. **84**, 621 (2012)
8. R. Renner, N. Gisin, B. Kraus, Phys. Rev. A **72**, 012332 (2005)
9. S. Pirandola, R. García-Patrón, S.L. Braunstein, S. Lloyd, Phys. Rev. Lett. **102**, 050503 (2009)
10. S.L. Braunstein, S. Pirandola, Phys. Rev. Lett. **108**, 130502 (2012)
11. S. Pirandola, C. Ottaviani, G. Spedalieri, C. Weedbrook, S.L. Braunstein, S. Lloyd, T. Gehring, C.S. Jacobsen, U.L. Andersen, Nat. Photonics **9**, 397–402 (2015)
12. R. Filip, Phys. Rev. A **77**, 022310 (2008)
13. V.C. Usenko, R. Filip, Phys. Rev. A **81**, 022318 (2010)
14. C. Weedbrook, S. Pirandola, S. Lloyd, T.C. Ralph, Phys. Rev. Lett. **105**, 110501 (2010)
15. C. Weedbrook, S. Pirandola, T.C. Ralph, Phys. Rev. A **86**, 022318 (2012)
16. C. Weedbrook, C. Ottaviani, S. Pirandola, Phys. Rev. A **89**, 012309 (2014)
17. S. Pirandola, Quantum discord as a resource for quantum cryptography. Sci. Rep. **4**, 6956 (2014)
18. B. Schumacher, M.A. Nielsen, Phys. Rev. A **54**, 2629 (1996)
19. S. Lloyd, Phys. Rev. A **55**, 1613 (1997)
20. S. Pirandola, R. García-Patrón, S.L. Braunstein, S. Lloyd, Phys. Rev. Lett. **102**, 050503 (2009)
21. R. García-Patrón, S. Pirandola, S. Lloyd, J.H. Shapiro, Phys. Rev. Lett. **102**, 210501 (2009)
22. A.K. Ekert, Phys. Rev. Lett. **67**, 661 (1991)
23. S. Pirandola, R. Laurenza, C. Ottaviani, L. Banchi, *Fundamental Limits of Repeaterless Quantum Communications* (2015). Preprint arXiv:1510.08863
24. S. Pirandola, *Capacities of Repeater-Assisted Quantum Communications* (2016). Preprint arXiv:1601.00966
25. C. Choi, Linear Algebra Appl. **10**, 285–290 (1975)
26. S.L. Braunstein, H.J. Kimble, Phys. Rev. Lett. **80**, 869–872 (1998)
27. S. Pirandola et al., Nat. Photon **9**, 641–652 (2015)
28. V. Vedral, M.B. Plenio, M.A. Rippin, P.L. Knight, Phys. Rev. Lett. **78**, 2275–2279 (1997)
29. V. Vedral, M.B. Plenio, Phys. Rev. A **57**, 1619 (1998)
30. V. Vedral, Rev. Mod. Phys. **74**, 197 (2002)

Quantum Discord in Quantum Communication Protocols

Animesh Datta and Vaibhav Madhok

Abstract We review an operational interpretation of quantum discord by quantifying it as the difference in the yield of the noisy and noiseless fully quantum Slepian-Wolf (FQSW) protocol and the closely related Mother protocol. The fully quantum Slepian-Wolf protocol is the most general form of unidirectional and bipartite quantum communication protocols and hence we provided an operational interpretation of discord for the entire spectrum of such protocols. We discuss examples giving specific scenarios for the role of discord in quantum communication. We provide examples of how the properties of discord can be derived intuitively from our operational view point.

1 Introduction

The universe, as we know, is quantum mechanical. Yet, classical mechanics gives an excellent description of the macroscopic world. What aspects of quantum mechanics make the quantum world different than the classical world is still an open question. Historically, this question has been asked in several forms, quantum-to-classical transition and the Bohr correspondence [1], quantum signatures of chaos, semi-classical treatment of quantum mechanics [2] etc. Here the key goal is two fold-to explain how classical mechanics arises out of underlying quantum theory, and to explain quantum aspects by connecting it to some classical properties of dynamical systems. Emergence of classicality through weak continuous measurement and the issues of decoherence as a way into the classical world and the entire framework of open quantum systems are intimately related to the first goal. Gutzwiller's Trace

A. Datta (✉)
Department of Physics, University of Warwick, Coventry CV4 7AL, UK
e-mail: animesh.datta@warwick.ac.uk

V. Madhok (✉)
Department of Physics, Indian Institute of Technology Madras, Chennai 600036, India
e-mail: vmadhok@gmail.com

F.F. Fanchini et al. (eds.), *Lectures on General Quantum Correlations and their Applications*, Quantum Science and Technology,
DOI 10.1007/978 3 319 53412 1_12

Formula and Semiclassical description of quantum mechanics [2–4], coherent state representations and Wigner functions and WKB approximations are the crowning accomplishments of the latter.

Information theory provides another window in exploring this connection. As Landauer first pointed out [5], there is an intimate connection between information processing and the physical properties of a device. "Information is Physical". After all, information processing has to respect the laws of physics-classical and quantum. Does information processing governed by quantum mechanics fundamentally different than it's classical counterpart? The answer has been a resounding yes! Over the last three decades, there has been a revolution as a result of exploring this question. This has led to the advent of quantum information science.

Two overarching motivations drive the field of quantum information science. The first is the to uncover the fundamental limits to classical information processing. The second is to overcome these limits by using the laws of quantum mechanics, in the process uncovering the *more* fundamental limits, and attaining them in practice. Between these two motivations lies a very likely chance of delineating the quantum-classical boundary.

Quantum information science has added a radically new viewpoint to the investigation of quantum mechanics. This has brought about a superior comprehension of quantum aspects like entanglement and decoherence, and given us the devices to see certain quantum properties of physical systems as an asset. This has additionally empowered us to address the key inquiries in quantum computation from another viewpoint. From a more foundational point of view, quantum information speculation has a more significant message for us. How physical systems process and exchange information is crucial for better understanding of the workings of our universe. For example, the relationship between entropy, information and thermodynamics are fundamental to understanding of statistical physics.

Quantum information theory has a – perhaps more revolutionary – message for us: Devices employing the laws of quantum physics have superior information processing capabilities than their classical counterparts. What aspects of quantum mechanics makes this possible? This, in the context of communication, is the focus of our work.

Two aspects of quantum mechanics set it apart from its classical counterpart. The first is the existence of nonclassical states with correlations of a strength impossible classically [6–8]. The second is the ability of quantum measurements to disturb the states that are measured. Both of these are uniquely quantum, and their harnessing in numerous ways underlies quantum information science. For quantum teleportation, quantum computation [9], and some protocols of quantum cryptography using pure states, quantum entanglement [10, 11] has been shown to be the reason behind the quantum enhancement.

However, much less is known about the necessity of quantum entanglement in mixed-state quantum information processing. In 2008, quantum discord was proposed as a possible resource [12] behind mixed-state quantum computation [13]. In spite of several early investigations [14–17], and some recent partial successes [18], a complete proof on the necessity of quantum discord as a resource for mixed-state quantum computation remains elusive. Quantum discord has also been studied in numerous other contexts, as reviewed in Ref. [19].

In this work, we review the role of quantum discord in quantum communication. We show that quantum discord plays a vital role in *all* unidirectional, memoryless, bipartite quantum communication.

The remainder of the paper is organized as follows. In Sect. 2 we review the definition and a few properties of quantum discord. In Sect. 3 we provide a very brief background in quantum Shannon theory that leads to the fully quantum Slepian Wolf (FQSW) protocol. This protocol is the unification of *all* unidirectional, bipartite and memoryless quantum communication protocols. Next, in Sect. 4, we discuss the role of quantum discord in the efficiency of the FQSW protocol. Thus, a role of quantum discord in the FQSW protocol implies its role in all quantum information processing protocols and therefore establishing the role of quantum correlations in quantum information theory. We discuss this in Sect. 5, connecting quantum discord to the efficiency of quantum communication protocols. We give examples of specific bipartite communication protocols and demonstrate how discord makes them more efficient. The significance of discord in device dependent cryptography and entanglement transfer is discussed next in Sects. 6 and 7 respectively.

We conclude with a discussion on the significance of our results in the broader context of quantum information theory in Sect. 8.

2 The Ingredient: Quantum Discord

Quantum discord was first defined in 2002 [20, 21]. It remained a quantity of marginal interest until its proposed possibility as the resource of quantum enhancement in mixed-state quantum computation [12] in 2008. Originally defined to quantitatively separate bipartite quantum states from classical probability distributions of two random variables, quantum discord attempts to demarcate the border between classical and quantum for bipartite systems that can be extended to multi-party scenarios. It aims to capture essentially all the "quantumness" in an information theoretic way, including entanglement [20, 21].

In order to characterize "quantumness" or quantum correlations, we first look at two ways to define mutual information. For a bipartite systems XY, the mutual information is defined as $I(X:Y) = H(X) + H(Y) - H(X,Y)$, where $H(\cdot)$ stands for the Shannon entropy, as defined below.

Let X be a random variable that is distributed according to the probability distribution function (PDF) $P(x)$. Y is the output random variable distributed according to $P(y)$. Y carries information about X and the conditional PDF is given by $P(y|x)$.

$$P(y) = \sum_i P(y|x_i)P(x_i) \tag{1}$$

The MI is defined as

$$I(X:Y) = h(X) - h(X|Y) \\ = h(Y) - h(Y|X) \tag{2}$$

or it can be defined as,

$$I(X:Y) = h(X) + h(Y) - h(X,Y), \tag{3}$$

where h represents the Shannon entropy function for the random variable with a given probability distribution. Writing h explicitly in terms of the PDF, we get

$$h(X) = -\sum P_i(x) \ln P(x_i); \\ h(X|Y) = -\sum\sum P(x_i, y_j) \ln P(x_i I|y_j); \\ h(X,Y) = -\sum\sum P(x_i, y_j) \ln P(x_i, y_j) \tag{4}$$

For a classical probability distribution, Eqs. (2) and (3) are equivalent definitions of the mutual information as $I(X:Y) = h(X) - h(X|Y)$, where the conditional entropy $h(X|Y)$ is an average of the Shannon entropies of X, conditioned on the outcomes of Y. This is enabled by Bayes' rule, a vital cog in proving numerous results in classical information theory. As we will now show, the definition of conditional probabilities is nontrivial in quantum mechanics due to the role of quantum measurements. Quantum discord thus captures the nonclassical effects of quantum measurements in information processing.

In order to determine outcomes of Y to compute conditional probabilities in quantum mechanics, one needs to make measurements. In general, these measurements disturb the quantum systems and lead to different values of mutual information. For a quantum system, this depends on the measurements that are made on B. For a POVM given by the set $\{\Pi_i\}$, the state of A after the measurement corresponding to the outcome i is given by

$$\rho_{A|i} = \text{Tr}_B(\Pi_i \rho_{AB})/p_i, \quad p_i = \text{Tr}_{A,B}(\Pi_i \rho_{AB}). \tag{5}$$

Therefore, if we define the conditional quantum entropy as, $\tilde{H}_{\{\Pi_i\}}(A|B) \equiv \sum_i p_i H(\rho_{A|i})$, and an alternative version of the quantum mutual information can now be defined as $\mathcal{J}_{\{\Pi_i\}}(A:B) = H(A) - \tilde{H}_{\{\Pi_i\}}(A|B)$, which depends on the chosen set of measurements $\{\Pi_i\}$. To capture all the classical correlations present in ρ_{AB}, we maximize $\mathcal{J}_{\{\Pi_i\}}(A:B)$ over all $\{\Pi_i\}$, arriving at a measurement independent quantity $\mathcal{J}(A:B) = \max_{\{\Pi_i\}}(H(A) - \tilde{H}_{\{\Pi_i\}}(A|B)) \equiv H(A) - \tilde{H}(A|B)$, where $\tilde{H}(A|B) = \min_{\{\Pi_i\}} \tilde{H}_{\{\Pi_i\}}(A|B)$. Since the conditional entropy is concave over the set of POVMs, which is convex, the minimum is attained on the boundary points of the set of POVMs, which are rank 1 [22]. Then, quantum discord is finally defined as

$$\mathcal{D}(A:B) = I(A:B) - \mathcal{J}(A:B) \qquad (6)$$
$$= H(A) - H(A,B) + \min_{\{\Pi_i\}} \tilde{H}_{\{\Pi_i\}}(A|B),$$

where $\{\Pi_i\}$ are now, and henceforth in the paper, rank 1 POVMs. Quantum discord is non-negative for all quantum states [21, 22], and that quantum discord is sub-additive [23].

3 The Recipe: Quantum Shannon Theory

The study of inter-conversions between non-local information-processing resources is the mandate of quantum Shannon theory. Within the resource framework [24, 25], resources are classified as static (e.g. entanglement, shared randomness) or dynamic (e.g. communication channels), noisy or noiseless, finite or asymptotic. All protocols involving static resources emanate from a single parent protocol, a quantum communication-assisted entanglement distillation protocol, known as the 'mother' protocol. The 'father' on the other hand is an entanglement-assisted quantum communication protocol, the parent of protocols involving dynamic resources [24]. A protocol which unifies the family tree described above would be at the heart of quantum Shannon theory, and that is what the fully quantum Slepian-Wolf (FQSW) achieves [26]. It derives its name from its applicability to distributed compression, the classical case of which was solved by Slepian and Wolf [27, 28]. FQSW is a generalization of the 'mother' protocol since, in addition to quantum communication-assisted entanglement distillation, it accomplishes state transfer from the sender to the receiver. Moreover, the FQSW can also be transformed into the 'father' protocol by employing the Schmidt symmetry [26]. FQSW therefore lies at the heart of quantum information theory, and sometimes referred to as state transfer or the merging mother [29].

This review builds upon on the protocol of quantum state merging. Importance of the quantum state merging protocol to the field of quantum information theory is perhaps underappreciated. It provides an operational interpretation for conditional probabilities in quantum mechanics [30]. Indeed, conditional entropies in quantum mechanics can be negative, marking a fundamental point of departure of the statistics of quantum systems from that of classical systems. The state merging protocol is itself inspired by the classical Slepian-Wolf protocol, which provides an operational interpretation of classical conditional probabilities [27], via its analysis of distributed compression. In the classical version, the receiver, Bob, has access to some information Y, and the sender Alice possesses the missing information X. Here X and Y as random variables. If Bob wishes to learn X fully, how much information must Alice send to him? One way she can do this is by sending the entire $H(X)$ bits to Bob. However, Slepian and Wolf showed that she can this more efficiently, by just sending $H(X|Y) = H(X,Y) - H(Y)$, the conditional information [28]. Since

$H(X|Y) \leq H(X)$, Alice can take advantage of correlations between X and Y to reduce the communication cost needed to accomplish the given task.

An interpretation of quantum discord through quantum state merging therefore reflects this departure, and provides a distinctive measure of what it means and takes for a system to be quantum. Our result is a consequence of the strong subadditivity of the Von-Neumann entropy [31]. Strong subadditivity is one of the centerpieces of information theory and statistical mechanics. Our work considers a bipartite quantum system of A and B, with C being the ancilla (initially in a pure state) with B. The information that is lost in the process of measurement on B (which is equivalent to discarding C after a unitary interaction between B and C) results in making it more expensive for A to merge her state with B. This enhancement in the cost is quantum discord [23].

3.1 The "Mother" and the FQSW Protocol

The mother protocol is a transformation on asymptotically many copies of a purified quantum state $|\Psi^{ABR}\rangle^{\otimes n}$. The reference system R purifies ρ_{AB} and does not actively participate in the protocol. To start with, Alice and Bob share the state ρ_{AB}. The mother protocol can be viewed as an entanglement distillation protocol between A and B when the only type of communication permitted is the ability to send qubits from Alice to Bob. The transformation can be expressed concisely in the resource inequality formalism as [25]

$$\langle \Psi^{AB} \rangle + \frac{1}{2} I(A:R)[q \to q] \geq \frac{1}{2} I(A:B)[qq], \tag{7}$$

which means that n copies of the state $|\Psi^{ABR}\rangle$ can be converted to $\frac{1}{2}I(A:B)$ EPR pairs per copy, provided Alice is allowed to communicate with Bob by sending him qubits at the rate $\frac{1}{2}I(A:R)$ per copy.

The mother protocol generalizes to a more general FQSW protocol, which lies at the heart of quantum Shannon theory. Starting with the same $|\Psi^{ABR}\rangle^{\otimes n}$, and using $\frac{1}{2}I(A:R)$ bits of quantum communication from Alice to Bob, they can distill $\frac{1}{2}I(A:B)$ EPR pairs per copy, and in addition Alice can accomplish merging her state with Bob. In other words, they create a state $|\Phi^{R\hat{B}}\rangle^{\otimes n}$, where $\hat{B}$ is a register held with B and $|\Psi^{R}\rangle = |\Phi^{R\hat{B}}\rangle$. As purifications are equivalent upto local unitaries, Bob can convert $|\Phi^{\hat{B}}\rangle$ to $|\Psi^{AB}\rangle$ at his end and thus complete the state merging with Alice. In the state merging task, as described above, Alice is able to successfully transfer her entanglement with the reference system R to Bob. This is expressed as

$$\langle \Psi^{AB} \rangle + \frac{1}{2} I(A:R)[q \to q] \geq \frac{1}{2} I(A:B)[qq] + \text{State Merging}. \tag{8}$$

To start with, A and B share n copies (asymptotic limit) of a pure state $|\Psi^{ABR}\rangle^{\otimes n}$. Here R is simply the purification of the system and does not participate actively in the protocol. At the end of the protocol, the aim is for A to its entanglement with R to B and in addition A and B distill EPR pairs between them. The exact quantitative expression showing the resource transfer is given by

$$\langle \mathcal{U}^{S\to AB} : \Psi^S\rangle + \frac{1}{2} I(A:R)[q \to q] \geq \frac{1}{2} I(A:B)[qq] + \langle \mathbb{I}^{S\to B} : \Psi^S\rangle. \tag{9}$$

The above inequality is another way of expressing the FQSW protocol, where we accomplish state merging as well as entanglement distillation. The state S on the left-hand side of the inequality, is distributes to Alice and Bob, while on the right-hand side, that same state is given to Bob aone. $\mathcal{U}$ is the channel that accomplishes this "handing over" of the state to the concerned parties. The FQSW protocol is valid asymptotically in the limit of a large number of copies and we denote this by the symbol $\geq$. The final state is of the form $|\Phi^{A_1 B_1}\rangle \otimes |\Psi^{\tilde{B}BR}\rangle^{\otimes n}$, where A_1 is held by Alice, and B, $\tilde{B}$, B_1 are held by Bob, and $|\Phi^{A_1 B_1}\rangle$ is a maximally entangled state with $|A_1| = nI(A:B)_\Psi - o(n)$. The aim is to have $|\Psi^{\tilde{B}BR}\rangle$ identical to $|\Psi^{ABR}\rangle$ with system A now in Bob's possession. Therefore, the initial entanglement between A and R has been transferred to B.

In the FQSW protocol, Alice and Bob share the quantum state $\rho_{AB}^{\otimes n}$, with each party having the marginal density operators $\rho_A^{\otimes n}$ and $\rho_B^{\otimes n}$ respectively. It was shown that in the limit of $n \to \infty$, and asymptotically vanishing errors, the answer is given by the quantum conditional entropy: $S(A|B) = S(A, B) - S(B)$. When $S(A|B)$ is negative, Bob obtains the full state with just local operations and classical communication, and distill $-S(A|B)$ ebits with Alice, which can be used to transfer additional quantum information in the future. The accounting of resources can be seen from the resource inequality (9) in a straightforward way. If the bell pairs, quantified by $\frac{1}{2} I(A:B)[qq]$, recovered at the end of the FQSW protocol can be used for teleportation, the net cost of quantum communication for state merging is given by $(I(A:R) - I(A:B))/2$ is exactly equal to the quantum conditional entropy $S(A|B) = S(A, B) - S(B)$.

The central feature to note is that all bipartite, unidirectional, memoryless can be derived from the FQSW protocol. In the next section, we explore this further and discuss the role of discord in a few specific examples of such communication.

4 The Result: Quantum Discord in the FQSW Protocol

In this section we present the main result on the role of quantum discord in quantum communication protocols. Since any practical implementation of a quantum communication protocol will be affected by noise, we now show that quantum discord assumes its relevance.

We consider loss of information and coherence only at Bob's end. This can be effected by considering a global unitary between Bob's system B and an ancillary environment system, say C, and then tracing C out. Physically, such a quantum operation will emulate environmental decoherence.

We begin by expanding the size of the Hilbert space so that an arbitrary pre-measurement (or any other quantum operation) can be modeled by coupling to the auxiliary subsystem and then discarding it. We assume C to initially be in a pure state $|\mathbf{0}\rangle$, and a unitary interaction U between B and C. Letting primes denote the state of the system after U has acted we have $H(A, B) = H(A, BC)$ as C starts out in a product state with AB. We also have $I(A : BC) = I(A' : B'C')$. As discarding quantum systems cannot increase the mutual information, we get $I(A' : B') \leq I(A' : B'C')$. Now consider the FQSW protocol between A and B in the presence of C. We can always view the yield of the FQSW protocol on the system AB to be the same as that of performing the protocol between systems A and BC, where C is some ancilla (initially in a pure state) with which B interacts coherently through a unitary U. Such an operation does not change the cost or yield of the FQSW protocol, as shown, but helps us in counting resources. Discarding system C yields

$$I(A' : B') \leq I(A' : B'C') = I(A : BC) = I(A : B). \tag{10}$$

Now consider a protocol which we call as $FQSWD_B$ (fully quantum Slepian-Wolf after decoherence), where the subscript refers to the decoherence at B. The resource inequality for $FQSWD_B$ is

$$\langle \mathcal{U}^{S \to A'B'} : \Psi^S \rangle + \frac{1}{2} I(A' : R')[q \to q] \geq \frac{1}{2} I(A' : B')[qq] + \langle \mathbb{I}^{S \to \hat{B}} : \Psi^S \rangle. \tag{11}$$

As in the fully coherent version, the Alice is able to transfer her entanglement with the reference system R', and is able to distill $\frac{1}{2} I(A' : B')$ EPR pairs ($[qq]$) with Bob. Since $I(A' : R') = I(A : R)$, the FQSW protocol without decoherence has a net gain of $\frac{1}{2} I(A : B) - \frac{1}{2} I(A' : B') = \frac{1}{2} D$ EPR pairs. Since the cost of quantum communication is same for both (coherent and decoherent) versions, we regard D as the metric of how coherently the FQSW protocol operates.

The minimum of D over all possible measurements is the quantum discord. The state ρ_{AB}, under measurement of subsystem B, changes to $\rho'_{AB} = \sum_j p_j \rho_{A|j} \otimes \pi_j$, where $\{\pi_j\}$ are orthogonal projectors resulting from a Neumark extension of the POVM elements. The unconditioned post measurement states of A and B are

$$\rho'_A = \sum_j p_j \rho_{A|j} = \rho_A, \quad \rho'_B = \sum_j p_j \pi_j.$$

Invoking these relations, we get

$$
\begin{aligned}
I(A' : B') &= H(A') + H(B') - H(A', B'), \\
&= H(A') + S(p) - \big\{S(p) + \sum_j p_j H(\rho_{A|j})\big\}, \\
&= H(A) - \sum_j p_j H(\rho_{A|j}),
\end{aligned} \tag{12}
$$

where $S(\cdot)$ is the Shannon entropy. After maximization, it reduces to $\mathcal{J}(\rho_{AB})$, as defined earlier. The reduction to rank 1 POVMs follows as stated earlier. Quantum discord quantifies the loss in yield of the FQSW protocol due to decoherence, and therefore $\mathcal{D}$ serves as a valid metric for the net loss in the number of EPR pairs in the noisy version of the protocol. This result shows that quantum discord is a vital quantifier of the performance of noisy quantum communication protocols.

In the following section, we show that discord plays the role of quantifying the role of noise in essentially all the children protocols that are the special cases of the FQSW protocol [32, 33]. This is to aid in the appreciation of this result in the context of a few commonly encountered and studied quantum communication protocols. They are also chosen for their central role quantum computation and quantum error correction.

5 Quantum Discord in the Children Protocols

5.1 *Noisy Teleportation*

The resource inequality for teleportation in the presence of noise is given by

$$
\langle \Psi^{AB} \rangle + I(A : B)[c \to c] \geq I(A\rangle B)[q \to q]. \tag{13}
$$

This resource inequality is a combination of the mother protocol with teleportation [25], and $I(A\rangle B)$, the coherent information [34], is equal to $-S(A|B)$. The above resource inequality states that A can accomplish teleporting $I(A\rangle B)$ bits of quantum information using the state $\langle \Psi^{AB} \rangle$ and some classical resources.

Under environmental decoherence at Bob, B, the protocol is given by

$$
\langle \Psi^{A'B'} \rangle + I(A' : B')[c \to c] \geq I(A'\rangle B')[q \to q]. \tag{14}
$$

Therefore, we see that the amount of bits teleported gets reduced by $I(A\rangle B) - I(A'\rangle B')$ or $S(A'|B') - S(A|B)$.

5.2 Noisy Super-Dense Coding

Superdense coding is a process that is the reverse of teleportation. Here, the goal is to transmit classical information using entanglement. The resource inequality for superdense coding is given by:

$$[qq] + [q \to q] \succeq 2[c \to c], \tag{15}$$

showing that one can employ a shared entangled bit and a single bit of quantum communication to communicate 2 bits of classical information. The symbol $\succeq$ is used to denote exact attainability as compared to $\geq$ which is to denote asymptotic attainability.

Combining the FQSW protocol with superdense coding, we get,

$$\langle \Psi^{AB} \rangle + H(A)[q \to q] \geq I(A:B)[c \to c]. \tag{16}$$

When the party B is undergoing decoherence, the noisy superdense coding can be expressed as,

$$\langle \Psi^{A'B'} \rangle + H(A')[q \to q] \geq I(A':B')[c \to c]. \tag{17}$$

We note that $H(A) = H(A')$. Thus, due to decoherence, the number of classical bits communicated through this protocol gets reduced by the amount $I(A:B) - I(A':B')$. This difference shows up as discord of the state $\langle \Psi^{AB} \rangle$.

5.3 Entanglement Distillation

The one-way entanglement distillation can be expressed as

$$\langle \Psi^{AB} \rangle + I(A:R)[c \to c] \geq I(A\rangle B)[qq]. \tag{18}$$

This inequality can be derived by combining the FQSW protocol Eq. (9) and recycling the $\frac{1}{2}I(A:R)$ ebits out of the total $\frac{1}{2}I(A:B)$ produced for teleportation, as shown in [25]. Decoherence at Bob's end B provides

$$\langle \Psi^{A'B'} \rangle + I(A':R')[c \to c] \geq I(A'\rangle B')[qq]. \tag{19}$$

The net change in entanglement distillation is equal to $I(A'\rangle B') - I(A\rangle B) = H(A|B) - H(A'|B')$, which is the negative of the quantum discord of the original state. As is well known, classical communication between parties cannot enhance entanglement, and we can neglect the overhead of $I(A:R) - I(A':R')$ classical bits.

6 Quantum Discord in Device Dependent Cryptography

Recently, quantum discord has also been shown to play a vital role in quantum cryptography [35].

Consider the scenario where two parties, Alice and Bob, want to communicate secret messages between them using quantum states. Alice prepares a bipartite quantum state ρ_{AB}, whose purification is given by ρ_{ABR}. Then she sends one half of the state, ρ_B, to Bob. Alice and Bob then make measurements on their respective subsystems two extract correlated variables, X and Y. More specifically, Alice and Bob both make rank-1 POVM measurements on their respective subsystems. Alice extracts the random variable X with the distribution $\{x, p_x\}$. Similarly, Bob recovers Y with distribution $\{y, p_y\}$. After the procedure is repeated a number of times, Alice and Bob compare their data. After classical procedures like privacy amplification and error correction, they extract the secret key at a rate of $K(x, y) \leq I(X, Y)$.

The role of discord becomes evident when we consider the above protocol in the presence of an eavesdropper, Eve. It was shown in [35], that nonzero discord is a necessary condition for such a device dependent secure quantum cryptography. Let us begin with states having zero quantum discord. Suppose, Alice prepares a "quantum-classical" state, $\rho_{Aa} = \sum_i p_i \rho_A(k) \otimes |k\rangle\langle k|$, where $|k\rangle$ are all orthogonal. The classical part of this state is perfectly cloneable by Eve due to this orthogonality. The joint state of Alice, Bob and Eve can be represented as,

$$\rho^{ABE} = \sum_i p_i \rho_A(k) \otimes |k\rangle_B\langle k| \otimes |k\rangle_E\langle k|. \tag{20}$$

As the above state is symmetric to perturbations of B and E, Eve is able to decode the information in variable A of Alice with the same accuracy as Bob. We can also see that secure communication in the presence of eavesdropper also fails in the reverse direction. After measurement by Bob, the joint state of the system is $\rho_{AE|y} = \sum_i p_i|y\rho_A(k) \otimes |k\rangle_E\langle k|.$, where $p_{k|y} = \langle k|M_y|k\rangle$. The Eve makes measurement using POVM to get $K = \{k, p_{k|y}\}$ and Alice recovers variable X with the distribution Tr $[M_x \rho_A(k)]$. Since $Y \rightarrow K \rightarrow X$ form a Markov chain, the data processing inequality states that $I(Y, K) \geq I(Y, X)$. Therefore, Eve recovers more information than Alice.

Thus, non-zero discord, that is $\mathcal{D}(A : B) > 0$, is essential to device dependent cryptography in the above scenario. Essentially, the same features of traditional cryptography, the information encoded in non-orthogonal states, is manifested as non-zero discord in a bipartite setting and this enables secure communication. It is important to notice that entanglement can be completely absent and yet cryptography is successful due to quantum discord. It was shown in [35] that any prepare and measure protocol based on non-orthogonal quantum states can be recast into an entanglement free device dependent with non zero discord as a necessary condition for it to be successful.

7 Quantum Discord in Entanglement Transfer

It was shown in 2003 that no entanglement is necessary to distribute entanglement. In other words, two distant particles can be entangled by sending a third particle that is never entangled with the other two [36]. The absence of entanglement in transferring it was somewhat of a paradox. A primary analysis[1] was presented discussing the role of bound entanglement and quantum discord was first presented in 2008 [22].

In 2012, it was shown that the amount by which the entanglement can be increased between a sender and a receiver is, in fact, bounded by the amount of quantum discord between them and the intermediate carrier [37]. To get an overview of the scheme, consider Alice and Bob, with their respective quantum systems A and B. They aim to increase their entanglement via local operations and sending an auxiliary system C between them. The difference between entanglement between two partitions $A : BC$ and $AC : B$ can be bound as [37]

$$\mathcal{E}_{A:CB}(\rho) - \mathcal{E}_{AC:B}(\rho) \leq \mathcal{D}_{AB|C}(\rho) \tag{21}$$

Let α be the initial states of A, B and C. Let $\beta = \mathcal{M}_{AC}(\alpha)$ be the state obtained through local operations on the system AC. Now, C is send over to B. The above bound implies,

$$\mathcal{E}_{A:CB}(\rho(\beta)) \leq \mathcal{E}_{AC:B}(\rho(\alpha)) + \mathcal{D}_{AB|C}(\rho(\beta)). \tag{22}$$

The above shows that the entanglement between two remote subsystems cannot increase if the discord between the bipartite state cut $AB|C$ is zero. In particular, states of the form, $\rho(\beta) = \sum_i p_i \rho_A(k) \otimes |k\rangle\langle k|_C$, where $|k\rangle$ are orthogonal, known as the classical-quantum states, cannot be useful for entanglement gain, as they are equivalent to local operations and classical communication between A and B.

8 Discussions and Outlook

The study we presented in this paper quantifies quantum discord as the cost of quantum communication under environmental noise. We were mainly concerned with quantum communication protocols. We hope this will lead to further progress in understanding the quantum resources in quantum information processing. Alternatively, can this picture help us understand quantum resources in quantum computation and shed new light on such issues? Our work on quantum discord raises many interesting questions and opens new directions.

The first in the extension of the formalism of the fully-quantum Slepian-Wolf protocol to multipartite scenarios. This would enable us to show more rigorously the role of quantum discord in noisy implementations of quantum cryptography and entanglement. A multipartite generalization has existed since 2008 [38], but a

[1] It was first brought to the attention of one of the authors (AD) by W. Zurek during 2007-8.

complete unification into a 'mother of all protocols' seems lacking. This unification should be a worthwhile endeavor as quantum technologies make possible applications such as multi-party cryptography, anonymous communication, voting, and message passing.

The next extension would be extend the fully-quantum Slepian-Wolf beyond memoryless channels and bi- and then multidirectional scenarios. One of the challenge seems to that of deriving the rates of classical randomness distillation with 2-way communication [39], and its possible connection of multipartite quantum discord. All of these would be commendable advances in the area of quantum Shannon theory in their own right, and quantum information theory in general.

At a more foundational level, one may wonder whether a state with zero discord be treated as "classical"? Are states with vanishing quantum discord useful for certain tasks which are not possible classically? For example, the question whether *concordant* computations can be simulated classically was investigated by Eastin [17]. A concordant computation is one in which after each stage of computation, the resulting quantum state is diagonal in a product basis, and hence has zero discord. The entire simulation is finding the right product basis, yet Eastin's findings suggest that it might be difficult to simulate such computations efficiently. Cable and Browne recently showed that this is possible in a large number of cases [18]. This represents not only a big step towards a full classical simulation of concordant computations, but also in the understanding of the role of non-classical correlations in quantum computation [40]. With greater understanding that quantum discord has provided, it would appear that the boundary between the quantum and the classical world might not be that simple after all.

At another level, the role of quantum discord in quantum-to-classical transition could be vital. Quantum discord is also closely related to the measurement problem and, indeed, quantum discord arose in the context of pointer states and environment induced decoherence. Under what conditions does a quantum trajectory that tracks a measurement of a given observable follow the classical trajectory? A related question is the role of generic quantum correlations like quantum discord in the emergence of the classical world. What is the role of quantum discord in the interfering paths a many-body quantum system takes from one point in spacetime to another? Further, the role of discord in characterizing quantum chaotic dynamics [41] has come to the fore. In [41], a multiqubit system that can be viewed as a quantum kicked top was employed to study this connection. As the initial quantum state, a coherent wave packet on the entire multi-qubit Hilbert space, is varied from the regular to the chaotic regions of the corresponding classical phase space, a contour plot of the long time averaged discord remarkably reproduces the structures of the classical stroboscopic map. The key message here suggests as classical chaos produces classical information, captured by uncertainty and Kolmogorov-Sinai entropy, quantum chaos seems to produce quantum information that is manifest is superposition of quantum states of a multipartite system. Discord is one way to quantify this quantum information and superpositions and hence characterize chaotic dynamics. Such studies have put in light the role of discord in efficient simulation quantum chaotic dynamics.

Acknowledgements This work was supported, in part, by the UK EPSRC (EP/K04057X/2) and, in part, by the start up faculty grant to VM by IIT Madras.

References

1. N. Bohr, Z. Phys. **2**(5), 423–478 (1920)
2. F. Haake, *Quantum Signatures of Chaos* (Spring, Berlin, 1991)
3. M.C. Gutzwiller, *Chaos in Classical and Quantum Mechanics* (Springer, New York, 1990)
4. M.C. Gutzwiller, J. Math. Phys. **11**, 1791 (1970)
5. R. Landauer, Irreversibility and heat generation in the computing process. IBM J. Res. Dev. **5**, 183–191 (1961)
6. J.S. Bell, On the Einstein–Podolsky–Rosen paradox. Physics **1**, 195–200 (1964)
7. D.M. Greenberger, M.A. Horne, A. Zeilinger, in *Bell's Theorem: Quantum Theory, and Conceptions of the Universe*, Going beyond Bell's theorem (Kluwer, Dordrecht, 1989)
8. A. Aspect, P. Gragier, G. Roger, Experimental test of Bell's inequalities using time varying analyzers. Phys. Rev. Lett. **49**, 1804–1808 (1993)
9. R. Jozsa, N. Linden, Proc. R. Soc. A **459**, 2011 (2003)
10. R. Horodecki, P. Horodecki, M. Horodecki, K. Horodecki, Rev. Mod. Phys. **81**, 865–942 (2007)
11. M.B. Plenio, S. Virmani, Quantum Inf. Comput. **1**, 1–51 (2007)
12. A. Datta, A. Shaji, C.M. Caves, Phys. Rev. Lett **100**, 050502 (2008)
13. E. Knill, R. Laflamme, Phys. Rev. Lett. **81**, 5672 (1998)
14. B. Dakic, V. Vedral, C. Bruckner, Phys. Rev. Lett. **105**, 190502 (2010)
15. A. Datta, A. Shaji, Int. J. Quantum Inf. **9**, 1787–1805 (2011)
16. F.F. Fanchini, L.K. Castelano, M.F. Cornelio, M.C. de Oliveira. arXiv:1106.0289
17. B. Eastin (2010). arXiv:1006.4402
18. H. Cable, D.E. Browne, New J. Phys. **17**, 113049 (2015)
19. K. Modi, A. Brodutch, H. Cable, T. Paterek, V. Vedral, Rev. Mod. Phys. **84**, 1655 (2012)
20. L. Henderson, V. Vedral, J. Phys. A: Math. Gen. **34**, 6899 (2001)
21. W.H. Zurek, Ann. Phys. (Leipzig), **9**, 855 (2000); H. Ollivier, W.H. Zurek, Phys. Rev. Lett. **88**, 017901 (2002)
22. A. Datta, Studies on the role of entanglement in mixed-state quantum computation, Ph.D. thesis, University of New Mexico (2008). arXiv:0807.4490
23. V. Madhok, A. Datta, Phys. Rev. A **83**, 032323 (2011)
24. I. Devetak, A. Harrow, A. Winter, Phys. Rev. Lett. **88**, 017901 (2002)
25. I. Devetak, A. Harrow, A. Winter, IEEE Trans. Inf. Theory **54**, 4587 (2008)
26. A. Abeyesinghe, I. Devetak, P. Hayden, A. Winter, Proc. R. Soc. A **465**, 2537 (2009)
27. D. Slepian, J.K. Wolf, IEEE Trans. Inf. Theory **19**, 471 (1971)
28. T. Cover, J. Thomas, *Elements of Information Theory* (Wiley, New York, 2006)
29. J. Oppenheim (2008). arXiv:0805.1065
30. M. Horodecki, J. Oppenheim, A. Winter, Nature **436**, 673 (2005)
31. M. Horodecki, J. Oppenheim, A. Winter, Commun. Math. Phys. **268**, 107 (2007)
32. V. Madhok, A. Datta, Quantum discord in quantum information theory, from strong subadditivity to the Mother protocol, in *6th Conference on Theory of Quantum Computation, Communication and Cryptography*. Lecture Notes in Computer Science (Springer, 2011)
33. V. Madhok, A. Datta, Role of quantum discord in quantum information theory. Int. J. Mod. Phys. B (2012)
34. J. Yard, I. Devetak, P. Hayden, IEEE Trans. Inf. Theory **54**, 3091 (2008)
35. S. Pirandola, Sci. Rep. **4**, 6956 (2014)
36. T. Cubitt, F. Verstraete, W. Dur, J.I. Cirac, Phys. Rev. Lett. **91**, 37902 (2003)
37. T.K. Chuan, J. Maillard, K. Modi, T. Paterek, M. Paternostro, M. Piani, Phys. Rev. Lett **109**, 070501 (2012)

38. I. Devetak, J. Yard, Phys. Rev. Lett. **100**, 230501 (2008)
39. I. Devetak, A. Winter, IEEE Trans. Inf. Theory **50**, 3183–3196 (2004)
40. B. Eastin, New J. Phys. **18**, 021003 (2016)
41. V. Madhok, V. Gupta, D. Trottier, S. Ghose, Phys. Rev. E **91**, 032906 (2015)

Non-Classical Correlations in Information Processing

Anil Shaji

1 Introduction

In purely functional terms, a computation transforms a human readable bit string referred to as the input or question into another human readable bit string referred to as the output or answer. We are assuming here that irrespective of the nature of the input or output, information can always be expressed in the simplest possible language with only two elements in its alphabet; namely binary. In the process of getting from the input to the output, the sequence of steps - the algorithm - would employ additional resources like computational space (memory) computational time etc. Quantum information processing brings new resources into the mix; entanglement and quantum coherence being the most prominent among them. These additional resources are known to shorten the path from input to output, as measured in terms of the conventional resources like computational space and computational time, for several interesting and useful computational problems. However there is reason to believe that the full potential of the use of quantum resources in information processing is only beginning to be understood.

A substantial amount of literature exists exploring the role and relevance of quantum entanglement as a key resource in quantum information processing. These studies have provided valuable insights into the role of non-classicality as a useful resource [1, 2]. Josza and Linden [3] showed that the presence of multi-partite entanglement involving a number of partics increasing unboundedly with input size is necessary for a quantum algorithm *operating on pure quantum states* to offer an exponential speed-up over classical computation. If the entanglement does not increase unboundedly in a computational process then each step of the quantum algorithm can, to any desired accuracy, be simulated on a classical computer.

A. Shaji (✉)
School of Physics, Indian Institute of Science Education and Research Thiruvananthapuram, Maruthamala PO, Vithura, Thiruvananthapuram, India
e-mail: shaji@iisertvm.ac.in

F.F. Fanchini et al. (eds.), *Lectures on General Quantum Correlations and their Applications*, Quantum Science and Technology,
DOI 10.1007/978-3-319-53412-1_13

An explicit construction for such a simulation scheme is also available [4]. It is also known that just having large amounts of entanglement by itself need not lead to a computational advantage. For instance, quantum computations involving a restricted set of highly entangled states like the stabilizer states can be simulated efficiently classically, as shown by the Gottesman-Knill theorem [5].

While exploring the role of entanglement in quantum information processing with exponentially enhanced efficiencies relative to classical algorithms [3], Jozsa and Linden make it quite clear that the resource(s) responsible for the computational power of mixed states which may or may not be entangled remains an open question. The realisation that nonClassical correlations in quantum states - especially mixed ones - are not limited to entanglement has given some pointers towards unraveling the resources responsible for enhanced efficiencies in quantum information processing using mixed states. In this lecture I put together what is known along these lines and give some pointers on problems that are still open and directions that remain unexplored.

2 DQC1 and Quantum Discord

The DQC1 model of a quantum computation using mixed states [6] provided the first indication that nonClassical correlations other than entanglement may be a useful resource for quantum information processing. In the DQC1 model, the normalised trace of a random unitary matrix is evaluated to any desired accuracy in a fixed number of steps independent of the size of the unitary. The quantum circuit corresponding to the DQC1 computation is given below:

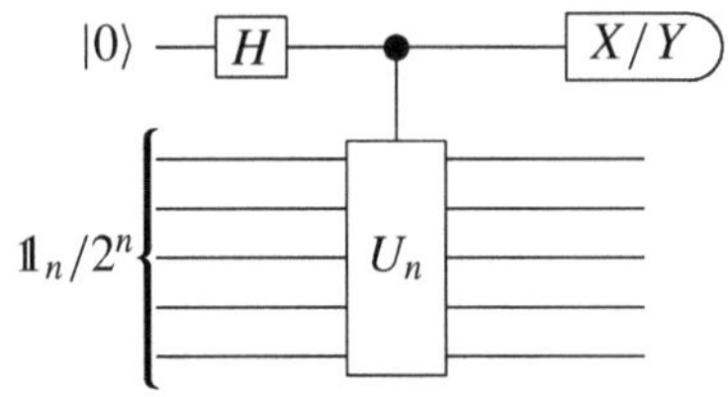

This circuit evaluates the normalized trace of U_n, $\tau = \mathrm{tr}(U_n)/2^n$ to any desired accuracy. The real and imaginary parts of the normalized trace is obtained from the measurement statistics of the sole pure qubit in the circuit. The measurement accuracy therefore depends only on the number of repetitions of the computation and is crucially independent of n. In contrast, the best known classical algorithm to compute the trace of U_n, which is a $2^n \times 2^n$ matrix, requires resources (computational steps) that grows as 2^n.

The DQC1 circuit transforms the highly-mixed initial state $\rho_0 \equiv |0\rangle\langle 0| \otimes \mathbb{1}_n/2^n$ into the final state ρ_{n+1},

$$\rho_{n+1} = \frac{1}{2^{n+1}}\Big(|0\rangle\langle 0| \otimes I_n + |1\rangle\langle 1| \otimes I_n + \ \alpha|0\rangle\langle 1| \otimes U_n^{\dagger} + \alpha|1\rangle\langle 0| \otimes U_n\Big). \quad (1)$$

The pure qubit at the top of the circuit is separable from the bottom mixed qubits at all times as can be seen easily by inserting the eigen-decomposition of the unitary $U_n = \sum_j e^{i\varphi_j}|\varphi_j\rangle\langle\varphi_j|$ into the state above leading to the explicitly separable form,

$$\rho_{n+1} = \frac{1}{2^n}\sum_j |\psi_j\rangle\langle\psi_j \otimes |\varphi_j\rangle\langle\varphi_j|, \qquad |\psi_j\rangle = \frac{1}{\sqrt{2}}(|0\rangle + e^{i\varphi_j}|1\rangle).$$

The final state also has vanishingly small entanglement, as measured by the negativity [7] across any split that groups the top qubit with some of the mixed qubits. Instead of looking for entanglement in the DQC1 state, in [8] nonClassical correlations as measured by the quantum discord across the natural bipartite split of the state between the top qubit and the rest was computed and found to be present.

To recap briefly, the notion of quantum discord was introduced by Ollivier and Zurek [9]. A closely related quantity was introduced independently by Henderson and Vedral [10] around the same time. Quantum discord captures all the quantum correlations including entanglement across a bipartite split of a quantum state. Quantum discord [9, 10] is defined in terms of the mutual information, which is an entropic measure of correlations between two systems A and B, in a joint state ρ_{AB}, defined as,

$$I(A:B) = S(\rho_A) + S(\rho_B) - S(\rho_{AB}), \quad (2)$$

where $\rho_{A,B} = \mathrm{tr}_{B,A}(\rho_{AB})$ are the reduced sub-system density matrices and $S(\rho) = -\mathrm{tr}(\rho\log\rho)$ is the vonNeumann entropy of the quantum state ρ. Based on a measurement of subsystem A we can quantify the correlations between A and B alternatively as

$$J(A:B) = S(\rho_B) - S(B|A), \quad (3)$$

where the conditional entropy $S(B|A)$ is defined with respect to the measurement $\mathcal{M}$ performed on system A with possible results labelled by a_k as

$$S(B|A) = \sum_k p_k S(\rho_B|a_k). \quad (4)$$

Here $p_k \equiv p(a_k)$ is the probability of obtaining the measurement result a_k. Note that if A and B are assumed to the classical random variables with associated probability distributions then the classical analogues of the mutual informations $I(A:B)$ and $J(A:B)$ defined by replacing the vonNeumann entropies with the corresponding Shannon entropies in Eqs. (2) and (3) are identical as a consequence of Bayes' theorem. Subtracting J from I quantifies the correlations in the AB system that are not revealed by the measurement $\mathcal{M}$ on A. Removing the ambiguity in $I - J$ stemming from the choice of measurement $\mathcal{M}$ by maximising over all possible local measurements, one defines discord as

$$\mathcal{D} = I(A:B) - \max_{\mathcal{M}} J(A:B). \tag{5}$$

In principle the maximisation over $\mathcal{M}$ includes all possible measurements including POVMs. Often discord is computed and/or operationally defined using a simpler maximisation over all possible projective measurements.

In [8] the DQC1 circuit above was generalised to allow the top qubit to have an initial state of arbitrary purity given by $(\mathbb{1} + \alpha Z)/2$, where $|\alpha| \leq 1$ and Z is the Pauli spin-z operator. A closed form expression for the quantum discord of the DQC1 state across the natural bipartite split between the top qubit and the rest assuming a uniform distribution for the eigen-phases of U_n, was obtained as

$$\mathcal{D}_{\text{DQC1}} = 2 - H_2\Big(\frac{1-\alpha}{2}\Big) - \log\Big(1 + \sqrt{1-\alpha^2}\Big) - \Big(1 - \sqrt{1-\alpha^2}\Big)\log e. \tag{6}$$

For $\alpha = 1$ which corresponds to the circuit above, we obtain $\mathcal{D}_{\text{DQC1}} = 0.5573$.

The assumption that the eigen-phases of U_n are distributed uniformly around the unit circle implies that the normalized trace is zero. So the analysis in [8] is open to the criticism that one is finding the trace using DQC1 for a case where the answer is already known to be zero and then claiming that the quantum advantage may possibly be attributed to the discord in the circuit. Numerical investigations however suggest that the expression in Eq. (6) continues to be a good approximation as well as an upper bound to the value of discord in DQC1 even if the eigen-phases are not uniformly distributed. For example the assumption of uniform distribution of phases breaks down as soon as we assume that the bottom register of qubits in the DQC1 circuit is made of a small number of qubits. With only a small number of qubits, unitaries U_n can be numerically generated and the corresponding discord in the DQC1 final state can be computed.

To generate random unitaries while sampling faithfully from the set of all unitaries, we use the method proposed in [11]. For a random unitary on n qubits, one starts with n independent 2×2 SU(2) matrices constructed as

$$U_2(\theta, \phi, \xi) = \begin{pmatrix} e^{i\xi}\cos\theta & e^{i\phi}\sin\theta \\ -e^{-i\phi}\sin\theta & e^{-i\xi}\cos\theta \end{pmatrix},$$

where ξ, θ and ϕ are three random numbers between 0 and 2π. Taking the tensor product of n such matrices, we get a $2^n \times 2^n$ unitary matrix,

$$W = U_2(\theta_1, \phi_1, \xi_1) \otimes U_2(\theta_2, \phi_2, \xi_2) \otimes \cdots \otimes U_2(\theta_n, \phi_n, \xi_n),$$

which is further "mixed up" by multiplying it by the unitary matrix

$$V = \exp\left[i\frac{\pi}{4}\sum_{j=1}^{n-1}\sigma_z^j \otimes \sigma_z^{j+1}\right],$$

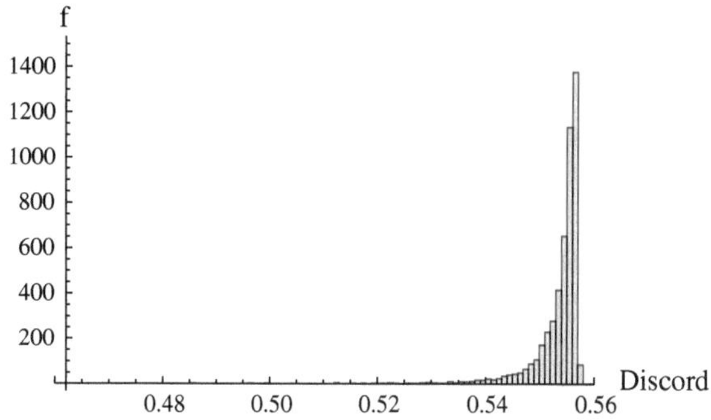

Fig. 1 Histogram showing the distribution of quantum discord when 5000 instances of random unitary matrices are used in a DQC1 circuit in which the bottom half of the circuit has nine qubits and the top (control) qubit is initially in a pure state

The resulting unitary,

$$U_{\text{PR}} = VW,$$

is a pseudo-random unitary "that can reproduce those statistical properties of random unitary operators that are most relevant for quantum information processing tasks" [11].

Numerically computing the discord from five thousand instances of randomly picked unitaries yields a distribution for the discord in the DQC1 state after the conditional application of the unitary shown in Fig. 1. The distribution has mean $\bar{\mathcal{D}} = 0.553261$ and standard deviation 0.006032. The minimum value for discord that was found for this trial was 0.462253 and the maximum value was 0.557196. Numerical computations of the same kind done after changing the number of qubits in the lower half of the circuit indicate a similar behavior. The discord values cluster together close to $\mathcal{D}_{\text{DQC1}}$. However the values seems to be always less than $\mathcal{D}_{\text{DQC1}}$.

The results in [8] suggested that nonClassical correlations other than entanglement may play a role in the exponentially enhanced ability of the DQC1 circuit in finding the normalised trace of a random unitary U_n. This suggestion triggered a veritable deluge of results and papers on quantum discord; too numerous to summarise here. An overview of this flurry of activity may be found in [12] and references therein. We will focus on a few pertinent lines of thought that came out of these investigations.

2.1 *Discord: Resource or Not?*

A state ρ has a zero discord if it can be written as $\rho = \sum_j p_j |\psi_i\rangle\langle\psi_j| \otimes \rho_j$ where $|\psi_j\rangle$ form an orthonormal basis. This state has a one way zero discord, since one can find a projective measurement on subsystem A for which $\rho = \sum_j (\Pi_j^A \otimes \mathbb{1}^B)\rho(\Pi_j^{A\dagger} \otimes \mathbb{1}^B)$. In [13] the condition for zero discord is shown to be equivalent to the case where the state has the form $\sum_j c_j S_j^A \otimes F_j^B$ with $[S_j^A, S_k^A] = 0$, where S_j^A form a suitable operator basis for the Hilbert space of subsystem A. The commutativity of all the S_j^A that appear in the expansion of ρ ensure set of common eigenvectors and the state can be re-written in the zero discord form explicitly in terms of the projectors on to these eigenvectors. Using this condition for zero discord states it can be shown that if the random unitary in the DQC1 circuit has the form $U_n = e^{i\phi} A_n$ where A_n is a binary observable ($A_n^2 = A_n^2$), then the final DQC1 state will have no discord across the division between the top qubit and the rest.

In [13] the authors point out that an efficient classical algorithm for simulating the DQC1 circuit may not exist even for such an unitary which does not produce any discord, since the eigenvectors of U_n can effectively be random in any fixed basis. In order to compute the phase picked up by the top qubit as a result of the action of U_n on the bottom qubits, one has to represent the fully mixed state of the bottom qubits in the eigenbasis of U_n. However since the eigenbasis of U are typically arbitrary and entangled the resources required to simulate these states classically can be exponential in the number of qubits.

The results in [13] does legitimately pose the question whether non-Classical correlations as quantified by discord is indeed a resource in mixed state quantum computation and in particular in the DQC1 model. It does not necessarily imply that discord is not a resource because it remains to be seen whether the particular family of unitaries that leave the final DQC1 state with no discord has further structure that allows for efficient computation of the trace of such unitaries on a classical computer. There are indications this may be case. For one, it is clear from the structure of these unitaries that their eigenvalues are restricted to being ± 1, or equivalently they are Hermitian unitaries. In this case finding the trace reduces to finding the number of positive (or negative) eigenvalues and there may exist an efficient classical algorithm for doing the same. Such an algorithm would by implication then lend further support to the view that the power of DQC1 model of quantum computation obtains at least partially because of the nonClassical correlations that are generated between the qubits during the course of the computation.

A further clue that nonClassical correlations may be the key resource in mixed state quantum computation comes from Eastin's result on concordant quantum computations [14]. Concordant states are those that have zero discord across any bipartition and in a concordant computation, the states remain concordant at every step. Eastin gives a classical algorithm for efficiently simulating a conventional concordant computation provided that the gate set acting on the qubits is restricted to one qubit or two qubit gates. The computation is conventional in the sense that the input state is diagonal in the computational basis and can therefore be read-off as a bit string. This result was further extended recently by Cable et al. [15]. The results by Eastin and Cable et al. indicate that there may be transition in terms of the amount and type of nonClassical correlations generated in the computational qubit registers during the course of a quantum information processing protocol that decides whether the quantum computation can be classically simulated or not. Role of discord in DQC1 will then be decided depending on which side of the transition the case of the family of unitaries in [13] lies.

3 Other Measures of NonClassical Correlations

Discord is but one way of quantifying the nonClassical correlations in a quantum state and it is not clear whether discord is the best measure of such correlations to use when trying to relate it to the enhanced efficiencies of quantum information processing protocols. Several alternate measures have been proposed as well [16–18]. In [19], several of these measures were be brought together under a unified picture following the general strategy of constructing measures of nonClassical correlations for bipartite states as the difference between a quantum entropic measure, $\mathcal{Q}(\rho_{AB})$, and its classical counterpart, $\mathcal{C}(\rho_{AB})$, which is derived from the probabilities for results of local measurements on one or both of the subsystems. Here $\mathcal{Q}$ quantifies appropriately all the correlations in the system, whereas the corresponding classical $\mathcal{C}$ captures only the corresponding classical correlations. The difference, $\mathcal{M} = \mathcal{Q} - \mathcal{C}$, is therefore a way of quantifying the nonclassical correlations in the quantum state.

In [19] three types of entropic measures of correlations are chosen, namely (1) mutual information, (2) conditional entropy and (3) Joint entropy. The joint-entropy based measures (3) are equivalently thought of measures based on the differences in abilities of 'quantum' and 'classical' Maxwell's demons to extract work from the same quantum state. Within broad type of measure of nonClassical correlations, finer sub-divisions can be thought of based on the type of measurements done on one or more of the subsystems, the statistics of which, in turn, is used to compute the classical quantity $\mathcal{C}$. We imagine that there are classical observers A and B—demons or otherwise—who have access to the two parts of the bipartite system in the state ρ_{AB}. We allow these observers to employ one of three measurement strategies: (a) Local, rank-one-projector measurements in the eigenbases of the marginal density operators, (b) Unconditioned local measurements and (c) Conditioned local measurements. Measurement strategy (a) allows for no freedom in the choice of measurements performed by the local observers. In the other two cases, $\mathcal{C}$ has to be optimised, as in the case of discord, over all allowed measurements. Further the first two measurement strategies are symmetric between the two observers while the third one is not.

Putting together the three types of correlation measures with the three possible measurement strategies of the local classical observers, we obtain nine possible measures of nonClassical correlations labeled as $\mathcal{M}_{j\alpha}$, where the index $j = 1, 2, 3$ labels the type of correlation measure used and the index $\alpha = a, b, c$ denotes the particular measurement strategy employed. The nine measures and the inter-relationships between them are summarised in the following array reproduced from [19]:

$$
\begin{array}{ccccccc}
 & & \mathcal{M}_{1a}^{(\mathrm{MID})} & \geq & \mathcal{M}_{1b}^{(\mathrm{WPM})} & & \\
 & & \| & & \wedge\!| & & \\
 & & \mathcal{M}_{2a}^{(\mathrm{MID})} & \geq & \mathcal{M}_{2b} & \geq & \mathcal{M}_{2c}^{(\mathrm{discord})} \\
 & & \| & & \wedge\!| & & \wedge\!| \\
S(A,B) = & \mathcal{M}_{3(\mathrm{i})} \geq & \mathcal{M}_{3a}^{(\mathrm{MID})} & \geq & \mathcal{M}_{3b} & \geq & \mathcal{M}_{3c}^{(\mathrm{dd})}
\end{array}
\tag{7}
$$

In the array (7) above, previously known measures belonging to the family of nine measures we consider are indicated as superscripts within brackets. In particular $\mathcal{M}_{1a}, \mathcal{M}_{2a}$ and $\mathcal{M}_{3a}$ are all equal to the *measurement induced disturbance* introduced by Luo [16], while $\mathcal{M}_{1b}$ is the symmetric measure of nonClassical correlations introduced by Wu, Poulsen and Molmer (WPM) [17]. $\mathcal{M}_{2c}$ is the discord that we have already considered while $\mathcal{M}_{3c}$ is the Maxwell demon based measure, (dd), explored by Zurek in [18]. One of the quantities $\mathcal{M}_{1c}$, it turns out, is ill defined and can be arbitrarily large and negative and hence is omitted from the list. The joint-entropy based measures (demon based) have one extra entry because one can further consider work extraction with or without subsequent erasure of the demons' memory. From the table above we also see that all the measures are upper bounded by the joint entropy of the bipartite system, $S(A, B)$.

From the hierarchy of measures we see that the quantum discord is typically the most parsimonious one when it comes to quantifying nonClassical correlations and in this regard it may be the best suited to quantify nonClassical correlations when discussing their role in quantum information processing.

4 Discord in Entanglement Distribution

Quantum information processing tasks other than computations per se, also leverages nonClassical correlations in quantum states to achieve objectives that may not be possible using classical means. In [20], Cubitt, Verstraete, Dür and Cirac showed that entanglement can be generated between two qubits A and B using a third qubit C as an intermediary. In the "Cubitt" protocol, joint operations between the A and C qubits and separate joint operations between the B and C qubits are allowed. However, there is no entanglement between A and C or B and C at any point in the protocol but still entanglement is generated between A and B even though they started off in a non-entangled state.

In a pair of closely related papers [21, 22], the idea that in order to distribute entanglement between two parties A and B using a third quantum system C as in the Cubitt protocol, there has to be at least some non-classical correlations between A and C has been elaborated upon. In [22] this point is expressed in inequality (6):

$$E_{\text{final}} - E_{\text{initial}} = \sum_{i=1}^{n} \Delta_i, \tag{8}$$

where E denotes entanglement quantified using any suitable measure and Δ_i denotes the amount of quantum correlations between the sent particle (C) and the remaining particle (A) in the i-th application of the quantum channel (that sends C across from A to B). A similar statement is made in [21] wherein the entanglement gain is found to be bounded by the amount of quantum discord between the two parties and the carrier. This idea has been previously illustrated using the example from [20] and using quantum discord as the measure of quantum correlations in [23]. The same is outlined in the following as an example of the role of nonClassical correlation in information processing tasks other than computing.

The starting point in the example given in [20] is a tripartite state of the three qubits where, as above, A and B are the qubits between which entanglement is distributed using the carrier/channel qubit C. The initial state of the three qubits given in an obviously separable form amongst the three qubits is

$$\rho^{(0)}_{ABC} = \frac{1}{6}\sum_{k=0}^{3} |\Psi_k, \Psi_{-k}, 0\rangle\langle\Psi_k, \Psi_{-k}, 0| + \frac{1}{6}\sum_{i=0}^{1} |i, i, 1\rangle\langle i, i, 1|, \quad |\Psi_k\rangle = \frac{|0\rangle + e^{ik\pi/2}|1\rangle}{\sqrt{2}}.$$

In the computational basis,

$$\rho^{(0)}_{ABC} = \frac{1}{6}\Big(|000\rangle\langle 000| + |010\rangle\langle 010| + |100\rangle\langle 100| + |110\rangle\langle 110| + |000\rangle\langle 110| + |110\rangle\langle 000| + |001\rangle\langle 001| + |111\rangle\langle 111|\Big).$$

By construction this is a separable state among all three qubits A, B and C. A CNOT operation between qubits A and C produces the state

$$\rho^{(1)}_{ABC} = \frac{1}{6}\Big[\big(|000\rangle + |111\rangle\big)\big(\langle 000| + \langle 111|\big) + |001\rangle\langle 001| + |010\rangle\langle 010| + |101\rangle\langle 101| + |110\rangle\langle 110|\Big]. \tag{9}$$

To complete the entanglement distribution protocol qubit C is brought together with B and another CNOT operation is performed between them yielding the state

$$\rho^{(2)}_{ABC} = \frac{1}{3}|\Phi^+\rangle_{AB}\langle\Phi^+| \otimes |0\rangle_C\langle 0| + \frac{2}{3}\mathbb{1}_{AB} \otimes |1\rangle_C\langle 1|, \qquad |\Phi^+\rangle = \frac{1}{\sqrt{2}}\big(|00\rangle + |11\rangle\big). \tag{10}$$

Qubit C is separable from AB and by measuring qubit C one can get with probability $1/3$ a maximally entangled state of AB. The entanglement distribution protocol produces a state from which $1/3$ ebits of entanglement can be extracted starting from a state with no entanglement between A and B. In this case

$$E_{\text{final}} - E_{\text{initial}} = \frac{1}{3},$$

when the entanglement is measured in ebits.

Let us now look at the intermediate state $\rho^{(1)}_{ABC}$ in some more detail so as to quantify the resources that are required for distributing the $1/3$ ebits of entanglement between qubits A and B using this two step protocol. Since the first CNOT gate acts on state vectors of the form,

$$|\Psi_k\rangle_A \otimes |0\rangle_C = \frac{1}{\sqrt{2}}(|0\rangle_A + e^{ik\pi/2}|1\rangle_A) \otimes |0\rangle_C,$$

in the first four terms of $\rho^{(0)}_{ABC}$, it is easy to see that $\rho^{(1)}_{ABC}$ can be written as

$$\rho^{(1)}_{ABC} = \frac{1}{6}\Bigg[\sum_{k=0}^{3} |\Phi^{AC}_k, \Psi^B_{-k}\rangle\langle\Phi^{AC}_k, \Psi^B_{-k}| + |001\rangle\langle 001| + |110\rangle\langle 110|\Bigg], \tag{11}$$

where

$$\Phi_k = \frac{|00\rangle + e^{ik\pi/2}|11\rangle}{\sqrt{2}}.$$

Clearly, $\rho^{(1)}_{ABC}$ is separable with respect to the $B - AC$ split. From Eq. 9, we see that the state $\rho^{(1)}_{ABC}$ is symmetric under the exchange of B and C and it follows that the $C - AB$ split has to be separable as well. In fact,

$$\rho^{(1)}_{ABC} = \frac{1}{6}\Bigg[\sum_{k=0}^{3} |\Phi^{AB}_k, \Psi^C_{-k}\rangle\langle\Phi^{AB}_k, \Psi^C_{-k}| + |010\rangle\langle 010| + |101\rangle\langle 101|\Bigg]. \tag{12}$$

However, the partial transpose of $\rho^{(1)}_{ABC}$ with respect to qubit A has eigenvalues

$$\left\{-\frac{1}{6}, \frac{1}{6}, \frac{1}{6}, \frac{1}{6}, \frac{1}{6}, \frac{1}{6}, \frac{1}{6}, \frac{1}{6}\right\}.$$

This shows that across the $A - BC$ split, $\rho^{(1)}_{ABC}$ is entangled. It us useful to note that even if the CNOT was done between A and C, the reduced state

$$\rho^{(1)}_{AC} = \mathrm{tr}_B\big[\rho^{(1)}_{ABC}\big] = \frac{1}{3}|00\rangle\langle 00| + \frac{1}{3}|11\rangle\langle 11| + \frac{1}{6}|01\rangle\langle 01| + \frac{1}{6}|10\rangle\langle 10|,$$

does not have any entanglement. The reduced states $\rho^{(1)}_{AB}$ and $\rho^{(1)}_{BC}$ are also separable.

In spite of the entanglement between A and BC one cannot extract an entangled state probabilistically from $\rho^{(1)}_{ABC}$ by measuring C unlike in the case of $\rho^{(2)}_{ABC}$. The post measurement state after projecting qubit C on to an arbitrary state,

$$|M_1\rangle = \cos\theta|0\rangle + e^{-i\phi}\sin\theta|1\rangle$$

is

$$\rho^{(M_1)}_{AB} = \frac{1}{3}\Bigg(|00\rangle\langle 00| + \cos^2\theta|01\rangle\langle 01| + \sin^2\theta|10\rangle\langle 10| + |11\rangle\langle 11| \\ + \frac{1}{2}e^{-i\phi}\sin 2\theta|00\rangle\langle 11| + \frac{1}{2}e^{i\phi}\sin 2\theta|11\rangle\langle 00|\Bigg). \tag{13}$$

The eigenvalues of the partial transpose of $\rho^{(M)}_{AB}$ with respect to either A or B are $(1/3, 1/3, 1/3, 0)$ independent of θ and ϕ. This means that one cannot extract an entangled state of qubits A and B out of $\rho^{(1)}_{ABC}$ by measuring C while after the CNOT operation between B and C, this is possible.

Two facts; namely that there is no entanglement between qubits A and C in the intermediate state $\rho^{(1)}_{ABC}$ and that there is no known way of quantifying the entanglement between A and BC in terms of ebits prompts one to look for alternate ways of identifying and quantifying the resources that are essential for this entanglement distribution protocol. The form of $\rho^{(1)}_{ABC}$ given in Eq. (12) suggests that the non-classical correlations may exist between qubit C and subsystem AB because more than two non-orthogonal states of qubit C are correlated with states of AB. In other words, since $\rho^{(1)}_{ABC}$ cannot be written in the form

$$\sum_{j=1}^{2} \rho^{j}_{AB} \otimes \Pi^{j}_{C}, \tag{14}$$

where $\{\Pi^j_C\}$ form a complete set of orthogonal projectors on C, we know that there must be non-zero quantum discord in the $AB - C$ split of $\rho^{(1)}_{ABC}$.

The quantum discord in the $AB - C$ split when qubit C is measures is given by

$$\mathcal{D} = S\left(\rho^{(1)}_{C}\right) - S\left(\rho^{(1)}_{ABC}\right) + \min_{\{\Pi^j_C\}} \sum_j p_j S\left(\rho^{(1)}_{AB|\Pi^j_C}\right),$$

The eigenvalues of $\rho^{(1)}_{ABC}$ are $(1/3,\ 1/6,\ 1/6,\ 1/6,\ 1/6,\ 0,\ 0,\ 0,\ 0)$ and we can evaluate,

$$S\left(\rho^{(1)}_{ABC}\right) = \frac{2}{3} + \log 3,$$

where the logarithm is taken to base 2. The reduced state $\rho^{(1)}_{C} = \mathbb{1}/2$ and so

$$S\left(\rho^{(1)}_{C}\right) = 1.$$

In Eq. (13) we already have the post measurement state of AB given a projective measurement on C along the arbitrary state $|M_1\rangle$. The post measurement state of AB given a projective measurement on C along the state

$$|M_2\rangle = -\sin\theta|0\rangle + e^{-i\phi}\cos\theta|1\rangle, \qquad \langle M_1|M_2\rangle = 0$$

is

$$\rho^{(M_2)}_{AB} = \frac{1}{3}\Bigg(|00\rangle\langle 00| + \sin^2\theta|01\rangle\langle 01| + \cos^2\theta|10\rangle\langle 10| + |11\rangle\langle 11| \\ - \frac{1}{2}e^{-i\phi}\sin 2\theta|00\rangle\langle 11| - \frac{1}{2}e^{i\phi}\sin 2\theta|11\rangle\langle 00|\Bigg). \tag{15}$$

Fig. 2 The entropy of the post measurement state $\rho^{(1)}_{AB}$ as a function of the angle θ that specifies the measurement on C

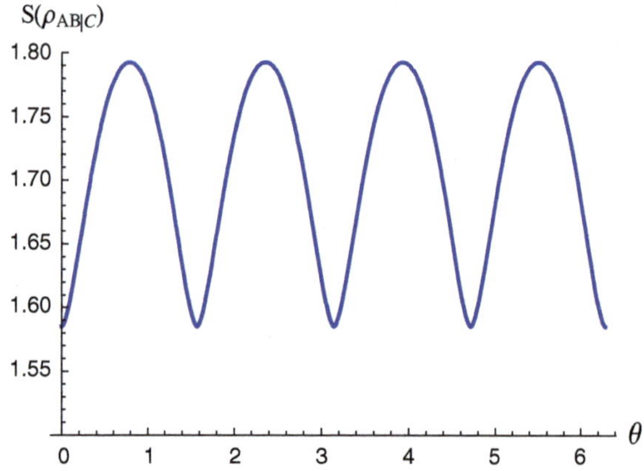

The eigenvalues of both post-measurement states in Eqs. (13) and (15) are

$$\left\{ \frac{\cos^2\theta}{3}, \frac{\sin^2\theta}{3}, \frac{1-\cos\theta\sin\theta}{3}, \frac{1+\cos\theta\sin\theta}{3} \right\}.$$

The probabilities of the two measurement outcomes when projective measurements along the orthogonal set $\{|M_1\rangle\langle M_1|, |M_2\rangle\langle M_2|\}$ are both equal to 1/2. A plot of $S(\rho^{(1)}_{AB|\Pi^j_C})$ is shown in Fig. 2 as a function of θ We see that the minima of the conditional entropy occurs at $\theta = 0, \pi/2, \pi, \ldots$. At $\theta = 0$, the eigenvalues of the post measurement states are $(1/3, 1/3, 1/3, 0)$. Accordingly,

$$\min_{\{\Pi^j_C\}} \sum_j p_j S\left(\rho^{(1)}_{AB|\Pi^j_C}\right) = \log 3.$$

So we have

$$\mathcal{D} = 1 - \frac{2}{3} - \log 3 + \log 3 = \frac{1}{3} \text{ ebits}.$$

The initial state $\rho^{(0)}_{ABC}$ is a zero discord state across the $AB - C$ split since it can be written in the form given in Eq. (14). The first CNOT operation generates a non-zero discord state with exactly 1/3 ebits worth of discord. It is interesting that the entanglement that can be extracted from the final state $\rho^{(3)}_{ABC}$ is also exactly 1/3 ebits. When the entanglement is characterized in terms of ebits and quantum discord is used to quantify the non-classical correlations in the intermediate state we see that the inequality in (8) is saturated for the specific entanglement distribution example protocol that was proposed in [20].

5 Open Quantum Dynamics

Open quantum dynamics [24, 25] may not quite qualify as quantum information processing but the importance of understanding open dynamics for development and implementation of information processing protocols is quite obvious. Apart from identifying, understanding and removing the deleterious effects of an uncontrollable environment quantum systems of interest, exploring information dynamics in open systems also provides insights into the possible role for the environment in enabling a computation particularly in the paradigm of mixed state quantum information processing.

Finite time dynamics of a quantum system evolving in contact with a (quantum) environment has traditionally been modelled in terms of completely positive, trace and hermiticity preserving maps:

$$\rho_S(t) = \mathrm{tr}_E[U\rho_{SE}(0)U^\dagger] = \Phi\rho_S(0),$$

where ρ_S is the state of the system and ρ_{SE} is the joint state of the system and the environment. The necessity of the map being completely positive is clear when we consider composite systems. The most widely accepted argument for demanding complete positivity is to introduce an arbitrary third system called the witness [26]. To start with, assume that we have a map Φ describing the evolution of the system. A witness system which is both blind (no interaction with the system) and dead (no independent evolution) is introduced and the evolution of $S + W$ is looked at. As the witness is blind and dead, it is clear that its evolution is given by the map $\mathbb{1}_W$. Since the mere presence of the witness cannot make the evolution of the system unphysical, it is reasonable to expect $\Phi \otimes \mathbb{1}_W$ to be positive for any choice of W. It turns out that for this condition to be satisfied, Φ should be completely positive.

A completely positive map is obtained when the open dynamics can be modelled as an interaction with an environment that together with the system is initially in a factorised (product) state with the environment being discarded after the interaction [5, 27, 28]. But from an experimental point of view, such isolation of the system is difficult and it is only natural to think about the kind of maps describing the evolution when the system and environment are initially correlated. It turns out that the reduced evolution of the system alone in these cases may not be described by completely positive maps.

In this case, the dynamics can be properly defined only on a subset of initial system states. The trivial extension to all possible states may not be physically realizable and may not even be positive let alone completely positive [29, 30]. Further, it has been shown that, in the presence of initial system environment correlations, not completely positive maps are perfectly good in describing the dynamics. It is true that these maps result in non physical evolutions for certain system states, but the point is that these are exactly those states which are excluded from being the possible initial states compatible with the specified initial correlations with the environment. In other words, the map succeeds in preserving the positivity of the states in the compatibility domain [31, 32]. Also, it is interesting to ask, if not completely positive maps are good candidates to describe open quantum dynamics as well, then in general, what kind of initial co relations between the system and environment guarantee completely positive reduced dynamics? A complete answer to this question is still being searched for.

In a recent notable attempt that did answer several questions regarding complete positivity or the lack of it for open dynamics using an information theoretic framework, the complete positivity of open dynamics is related to the quantum data processing inequality [33]. The approach in [33] adopted to comprehensively describe conditions on initial correlations that lead to completely positive reduced evolutions involves considering a tripartite scenario involving the reference (R), the system (S) and the environment (E) and going back to the bipartite system by using 'steering', that is by steering states on the system by measurements on the reference. It turns out that whether or not the initial tripartite states are Markovian in the sense that they satisfy the quantum data processing inequality has a lot to do with how the reduced dynamics of the system manifests.

Checking whether a given family of bipartite system-environment states gives rise to completely positive reduced dynamics is equivalent to checking whether said family can be obtained by steering from a tripartite Markov state of S, E and R, with the steering done on R by implementing all possible measurements on the state of R. Also, as long as the condition of Markovianity of the initial tripartite state is met (i.e. $I(R : E|S) = 0$, where I is the quantum mutual information between R and E conditioned on S), this approach does not place any restriction on the initial system-environment correlations. The accessible subset of system states on which the map gives completely positive dynamics in the case of initial system-environment correlations can in this approach be identified as the subset of steerable states of the system from the initial tripartite case. Furthermore, if the steerable set S is to coincide with the whole set of system states S, the initial factorisation condition is naturally recovered.

The information theoretic approach to open quantum dynamics in [33] captures the idea that more information cannot normally be extracted from the system S regarding the reference system R by letting S interact with an environment E first and then measuring S rather than looking at S directly. SE states that satisfy this criteria lead to completely positive dynamics on the subset of states of S that are compatible with these SE states (states of S that can be obtained as partial traces over the allowed SE states). Going beyond the discussion in [33], one can start with the idea that in general open dynamics may be positivity preserving only on a subset of system states and explore the consequences of nonClassical correlations between the three protagonists, namely the system, environment and reference [34]. Of particular interest is the case where there are nonClassical correlations between R and E which can potentially lead to violations of the quantum data processing inequality and tripartite Markovian state condition. Initial investigations suggest that nonClassical correlations in the RE subsystem, as expected, lead to not completely positive evolution even when the SE initial state is taken to be separable and with zero discord.

6 Discussion

Coming back to the picture of a quantum information processor that uses a register of qubits for input and output and an additional register of ancillary qubits for the computation with both input and output as human-readable bit strings, the initial state of the ancilla may be chosen so that the starting state of the quantum computer has no nonclassical correlation across any bipartite or multipartite division. The computational process is then a sequence of discrete steps that takes the all the qubits from the initial, zero-discord, state to the final, zero-discord, state. If we run a deterministic or probabilistic classical algorithm on the quantum information processor instead, then at each step the state of the quantum computer has no nonclassical correlation across any bipartite division. This means that one has to go from the initial to the final state jumping from one concordant state to another without ever leaving this set which is known to be of measure zero [13]. A true quantum computation however, lets one take alternate paths involving intermediate states with nonclassical correlations which might result in reaching the desired output state in significantly - perhaps even exponentially - fewer computational steps [35].

Is it possible that a quantum computer that works by acting on mixed quantum states may in some situations be harnessing the computational power of a much larger part of the universe that just the computer itself. In probabilistic classical computation, the computer samples from a given probability distribution defined on the input bits as part of the computational algorithm. However if the dimensionality of the distribution grows exponentially with the input size (number of bits, n, as the input) then the sampling is not efficient. One particular case of interest in the context of the DQC1 model of computation is the sampling problem when each bit string of length n is equally probable. Each one of the 2^n possible n bit strings have to be generated at least once to sample judiciously from the uniform distribution. The completely mixed state of n qubits that is the input into one of the registers of the DQC1 model achieves the uniform sampling in a single shot since the mixed state can be considered to be an equally weighted mixture of a complete set of 2^n basis states of the Hilbert space of n qubits. The fully mixed state of n qubits can be purified by expanding on to the Hilbert space of at least n more qubits ($2n$ in total). Can the ability of the mixed state to sample efficiently from the uniform distribution on n-bit strings be attributed to the computational power of the larger environment of at least n more qubits that it is a part of?

The open questions above are just a few of the interesting avenues to be pursued in exploring the role of nonClassical correlations in quantum information processing tasks of various kinds. While quantum discord in particular may have received quite a lot of attention over the past decade, the discussion should generalise to the broader notion of nonClassical correlations and exploration of several fundamental and applied problems remain. The experimental challenges and other applied aspects that would aid us in leveraging these correlations as a practical resource have not been touched upon in the above. Other contributions to this volume would certainly be addressing these as well.

Acknowledgements The technical content of this lecture note is based on work done - some of it published, some not - with several collaborators; particularly so with Animesh Datta, Carlton M Caves, Kavan Modi and Cesar Rodriguez-Rosario. The author's role here was to primarily put some of the key ideas and results together into a coherent narrative.

References

1. D. Bruß, J. Math. Phys. **43**, 4237 (2002)
2. R. Horodecki, P. Horodecki, M. Horodecki, K. Horodecki, Rev. Mod. Phys. **81**, 865 (2009)
3. R. Jozsa, N. Linden, Proc. R. Soc. A Math. Phys. Eng. Sci. **459**, 2011 (2003)
4. G. Vidal, Phys. Rev. Lett. **91**, 147902 (2003)
5. M.A. Nielsen, I.L. Chuang, *Quantum Computation and Quantum Information* (Cambridge University Press, Cambridge, 2000)
6. E. Knill, R. Laflamme, Phys. Rev. Lett. **81**, 5672 (1998)
7. A. Datta, S.T. Flammia, C.M. Caves, Phys. Rev. A **72**, 042316 (2005)
8. A. Datta, A. Shaji, C.M. Caves, Phys. Rev. Lett. **100**, 050502 (2008)
9. H. Ollivier, W.H. Zurek, Phys. Rev. Lett. **88**, 017901 (2001)
10. L. Henderson, V. Vedral, J. Phys. A Math. Gen. **34**, 6899 (2001)
11. J. Emerson, Y.S. Weinstein, M. Saraceno, S. Lloyd, D.G. Cory, Science **302**, 2098 (2003)
12. K. Modi, A. Brodutch, H. Cable, T. Paterek, V. Vedral, Rev. Mod. Phys. **84**, 1655 (2012)
13. B. Dakic, V. Vedral, C. Brukner, Phys. Rev. Lett. **105**, 190502 (2010)
14. B. Eastin, *Simulating Concordant Computations* (2010). arXiv:quant-ph/1006.4402
15. H. Cable, D.E. Browne, New J. Phys. **17**, 113049 (2015)
16. S. Luo, Phys. Rev. A **77**, 022301 (2008)
17. S. Wu, U.V. Poulsen, K. Mølmer, Phys. Rev. A **80**, 032319 (2009)
18. W.H. Zurek, Phys. Rev. A **67**, 012320 (2003)
19. M.D. Lang, C.M. Caves, A. Shaji, Int. J. Quanum Inf. **09**, 1553 (2011)
20. T.S. Cubitt, F. Verstraete, W. Dür, J.I. Cirac, Phys. Rev. Lett. **91**, 037902 (2003)
21. T.K. Chuan, J. Maillard, K. Modi, T. Paterek, M. Paternostro, M. Piani, arXiv:1203.1268 (2012)
22. A. Streltsov, H. Kampermann, D. Bruß, arXiv:1203.1264 (2012)
23. A. Datta, *Studies on the Role of Entanglement in Mixed-state Quantum Computation*, Ph.D. thesis, University of New Mexico, 2008
24. H.P. Breuer, F. Petruccione, *The Theory of Open Quantum Systems* (Oxford University Press, New York, 2002)
25. E.B. Davies, *Quantum Theory of Open Systems* (Academic Press, New York, 1976)
26. B. Schumacher, Phys. Rev. A **54**, 2614 (1996)
27. K. Kraus, *States, Effects and Operations: Fundamental Notions of Quantum Theory*, ed. by A. Bohm, J. D. Dollard, W.H. Wooters. Lecture Notes in Physics, vol. 190 (Springer, New York, 1983)
28. W.F. Stinespring, Proc. Am. Math. Soc. **6**, 211 (1955)
29. P. Pechukas, Phys. Rev. Lett. **73**, 1060 (1994)
30. P. Pechukas, Phys. Rev. Lett. **75**, 3021 (1995)
31. T.F. Jordan, A. Shaji, E.C.G. Sudarshan, Phys. Rev. A **70**, 52110 (2004)
32. A. Shaji, E.C.G. Sudarshan, Phys. Lett. A. **341**, 48 (2005)
33. F. Buscemi, Phys. Rev. Lett. **113**, 140502 (2014)
34. C.A. Rodriguez, K. Modi, A. meng Kuah, A. Shaji, E.C.G. Sudarshan, J. Phys. A Math. Theor. **41**, 205301 (2008)
35. A. Datta, A. Shaji, Int. J. Quantum Inf. **09**, 1787 (2011)

Part III
Quantum Correlations in Open Quantum Dynamics

The Local Detection Method: Dynamical Detection of Quantum Discord with Local Operations

Manuel Gessner, Heinz-Peter Breuer and Andreas Buchleitner

Abstract Quantum discord in a bipartite system can be dynamically revealed and quantified through purely local operations on one of the two subsystems. To achieve this, the local detection method harnesses the influence of initial correlations on the reduced dynamics of an interacting bipartite system. This article's aim is to provide an accessible introduction to this method and to review recent theoretical and experimental progress.

1 Introduction

Experimental achievements in the last decades have established the precise quantum control of individual quantum systems [1, 2]. Furthermore, recent efforts are focussed on the assembly and monitoring of interacting quantum systems, with various applications in the context of quantum information [3–6]. The efficient characterization of the underlying quantum states in high-dimensional state spaces, however, remains a challenge due to the large number of parameters.

One possible strategy for the analysis of systems whose size, complexity or structure is beyond the reach of a detailed microscopic examination is therefore to restrict access to a small, easily controllable subsystem [7, 8]. By interaction with the remaining system, the locally observable quantities of the subsystem may be able to convey information about the global properties of the interacting system.

M. Gessner (✉)
QSTAR, INO-CNR and LENS, Largo Enrico Fermi 2, 50125 Firenze, Italy
e-mail: manuel.gessner@ino.it

M. Gessner
Istituto Nazionale di Ricerca Metrologica (INRiM), 10135 Torino, Italy

H.-P. Breuer · A. Buchleitner
Physikalisches Institut, Albert-Ludwigs-Universität Freiburg,
Hermann-Herder-Straße 3, 79104 Freiburg, Germany

A. Buchleitner
Keble College, University of Oxford, Oxford OX1 3PG, UK

F.F. Fanchini et al. (eds.), *Lectures on General Quantum Correlations and their Applications*, Quantum Science and Technology,
DOI 10.1007/978-3-319-53412-1_14

While in general it is not always clear whether sufficient information about a possibly complex surounding system can be obtained from the few variables of the accessible subsystem, such an approach has proven to be suitable for probing the presence of correlations between the probe and its environment in a variety of situations [8, 9]. In the present article, we review recent progress in the local detection method [10, 11]—an interaction-assisted method, able to reveal quantum discord of the global system through the dynamics of a local subsystem. The method can be implemented when access is restricted to a controllable subsystem, and it has been tested in various different experimental settings. With the help of the examples reviewed in this article we discuss under which physical circumstances a successful local detection based on this method can generally be expected.

The concept of quantum discord can be intuitively understood in terms of local measurements of a bipartite quantum system. Measurements usually induce disturbances of the quantum system under observation [12, 13]. An exception to this textbook rule is found if the system is initially prepared in an eigenstate of the measured observable. More generally, if observable M and quantum state ρ commute, i.e., $[M, \rho] = 0$, a non-selective measurement of M will leave the quantum state ρ unchanged [12, 13]. Such a measurement projects the system into the eigenstate $|\varphi_m\rangle$ with probability $p_m = \langle\varphi_m|\rho|\varphi_m\rangle$, where we assume a non-degenerate observable with spectral decomposition $M = \sum_m \lambda_m |\varphi_m\rangle\langle\varphi_m|$. The state at the outcome of the projective measurement is consequently given as

$$\begin{aligned}\Phi(\rho) &= \sum_m p_m |\varphi_m\rangle\langle\varphi_m| \\ &= \sum_m |\varphi_m\rangle\langle\varphi_m|\rho|\varphi_m\rangle\langle\varphi_m|. \end{aligned} \tag{1}$$

In fact, we find that $\Phi(\rho) = \rho$ if and only if $[M, \rho] = 0$. Hence, for any given quantum state ρ, we can construct a family of observables M which can be measured non-selectively without disturbance. This family is comprised of all observables with the same eigenvectors as ρ, assuming no degeneracies.

Let us now consider the case of a bipartite quantum system, described by a tensor product of Hilbert spaces $\mathcal{H} = \mathcal{H}_A \otimes \mathcal{H}_B$. Under which circumstances is it possible to construct *local* observables whose non-selective measurement does not disturb the *total* quantum state ρ? The post-measurement state of a non-selective measurement of a *local* observable $M_A \otimes \mathbb{I}_B = \sum_m \lambda_m |\varphi_m\rangle\langle\varphi_m| \otimes \mathbb{I}_B$ is given by

$$(\Phi \otimes \mathbb{I})\rho = \sum_m (|\varphi_m\rangle\langle\varphi_m| \otimes \mathbb{I}_B)\rho(|\varphi_m\rangle\langle\varphi_m| \otimes \mathbb{I}_B), \tag{2}$$

with $\mathbb{I}_B$ the identity on $\mathcal{H}_B$. Again considering only non-degenerate observables, one finds that $(\Phi \otimes \mathbb{I})\rho = \rho$ is indeed equivalent to $[M_A \otimes \mathbb{I}_B, \rho] = 0$. The above question can thus be reformulated as: Which quantum states ρ commute with at least one local, non-degenerate observable? Obviously, if systems A and B are completely uncorrelated, i.e., if the total quantum state factorizes as $\rho = \rho_A \otimes \rho_B$, then we can conclude that a family of local observables, e.g. in system A, can always be constructed from the eigenvectors of ρ_A. The presence of correlations between the two systems, however, changes the situation.

Only a certain set of quantum states admit the existence of a non-degenerate observable M_A, such that $[M_A \otimes \mathbb{I}_B, \rho] = 0$. This family is known as the states of zero discord. They can always be written as [14–16]

$$\rho_{zd} = \sum_m p_m |\varphi_m\rangle\langle\varphi_m| \otimes \rho_B^m, \tag{3}$$

where p_m is a probability distribution and ρ_B^m are density operators on system B. It is important to note that the $|\varphi_m\rangle$ form an orthonormal basis of $\mathcal{H}_A$ since they are the eigenvectors of the Hermitian operator M_A. This distinguishes states of zero discord from separable states with the general form [17]

$$\rho_{sep} = \sum_i p_i \rho_A^i \otimes \rho_B^i, \tag{4}$$

where the states ρ_A^i are arbitrary and need neither be pure nor orthogonal. Furthermore, unlike entanglement, discord is an asymmetric property, requiring specification of the system which is measured. Throughout this article this will always be system A.

Quantum discord therefore characterizes the presence or absence of a local observable which commutes with the full quantum state. Nonzero discord can only be observed in correlated (i.e. not factorizing) quantum states, however, even some separable states exhibit discord. For pure states, the concepts of discord and entanglement coincide. Hence, in general, discord is a concept closely connected to correlations but does not itself measure correlations. In particular, a local operation on system A may change the orthogonality properties of the $|\varphi_m\rangle$ in Eq. (3) [18–20] and thereby create discord without creating correlations [21, 22].

The inability to commute with any local observable renders quantum states of nonzero discord furthermore suitable for certain technological tasks [16]. For instance, a phase shift φ, imprinted by a local unitary transformation $e^{-iM_A\varphi} \otimes \mathbb{I}_B$ can be estimated with high precision [23] only if the initial quantum state ρ is strongly affected by this transformation [24]. Conversely, if $M_A \otimes \mathbb{I}_B$ happens to commute with ρ, the state is completely invariant under this transformation and, consequently, an estimation of the phase shift φ is impossible. While states of zero discord are insensitive to the action of certain local operators, this can be excluded for all states of nonzero discord, since no local operator commutes with these states [24].

Among other applications [16, 25], discord was further shown to be useful for the distribution [26–31] and activation [32–34] of entanglement. To exploit these phenomena experimentally, one needs to first find convenient ways to generate sufficiently robust discordant quantum states [19, 20, 22, 35–37]. Second, methods to recognize the presence of discord, and perhaps to even quantify discord in experimentally relevant situations are required [16, 38–40]. The local detection method [10, 11], to be introduced in the next section, is a dynamical method which allows to detect and quantify discord in a bipartite system with limited experimental requirements.

2 The Local Detection Method

The efficient detection of properties such as entanglement or discord is a challenging task [41–43], which usually requires measurements of correlated observables. A popular approach for low-dimensional systems is to first obtain full information about the quantum state, and then to calculate a suitable quantifier based on the measured entries of the density matrix [22, 44]. Not only require the results of tomographic reconstructions of quantum states careful statistical analysis [45], their experimental realization soon becomes prohibitively expensive when high-dimensional or multipartite systems are of interest. In these cases, a complete characterization of the full quantum state can no longer be of interest. Alternatively, the measurement of carefully designed observables (such as entanglement witnesses) may reveal the presence of entanglement [42, 46–49] or discord [38–40, 50–52] without knowledge of the full quantum state. Yet, such procedures are often restricted to Hilbert spaces of a certain dimension and structure (e.g. qubit systems or two-mode systems of continuous variables), cf. [53]. Their implementation furthermore often requires a high degree of control over the full quantum system (or even multiple copies thereof [54–56]) which is difficult to achieve with increasing system size.

We may also encounter situations in which the experimenter may not even have full access to the complete quantum state of systems A and B but instead may be limited to measurements and operations on the subsystem A. This limitation may be due to a fundamental inability to access the second system, when, for instance party B is spatially separated from the experimenter at A or describes degrees of freedom that cannot be measured experimentally. It may also be a deliberate choice such as to restrict the dimension of the quantum system which is to be controlled. In either case one may consider the subsystem B an ancilla system or environment to system A. Since the reduced dynamics of system A may be strongly influenced by correlations with system B, a dynamical witness for discord may be observable, even by restricting to local measurements of system A.

2.1 *Witnessing Discord with Local Operations*

To introduce the basic idea of the local detection method [10, 11] let us recall that states of zero discord are characterized by their invariance under non-selective measurements, i.e., a state ρ has zero discord if and only if there exists a complete set of one-dimensional orthogonal projectors $\mathbf{\Pi} = \{\Pi_1, \Pi_2, \ldots,\}$, such that $\rho = (\Phi_{\Pi} \otimes \mathbb{I})\rho$, where

$$(\Phi_{\boldsymbol{\Pi}} \otimes \mathbb{I})\rho = \sum_i (\Pi_i \otimes \mathbb{I}_B)\rho(\Pi_i \otimes \mathbb{I}_B). \tag{5}$$

This operation (5) is a purely local operation on system A and its implementation does not require any knowledge of system B. It furthermore describes complete dephasing in the basis $\boldsymbol{\Pi}$ and is therefore called a *local dephasing operation* [10].

Let us assume the state ρ has zero discord. A local dephasing operation that leaves the state invariant thus exists, but in which particular local basis $\boldsymbol{\Pi}$? Let us first express the full quantum state ρ in terms of families of completely arbitrary local operator bases as $\rho = \sum_\alpha A_\alpha \otimes B_\alpha$. To answer the above question, we study the reduced density matrix of system A, which is obtained by performing the partial trace over B, after the dephasing operation. We obtain

$$\begin{aligned}\mathrm{Tr}_B\{(\Phi_{\boldsymbol{\Pi}} \otimes \mathbb{I}_B)\rho\} &= \sum_i \sum_\alpha \Pi_i A_\alpha \Pi_i \mathrm{Tr}\{B_\alpha\} \\ &= \sum_i p_i \Pi_i,\end{aligned} \tag{6}$$

where $p_i = \sum_\alpha \mathrm{Tr}\{\Pi_i A_\alpha\}\mathrm{Tr}\{B_\alpha\} = \mathrm{Tr}\{(\Pi_i \otimes \mathbb{I}_B)\rho\}$. The invariance property $\rho_A = \mathrm{Tr}_B\{\rho\} = \mathrm{Tr}_B\{(\Phi_{\boldsymbol{\Pi}} \otimes \mathbb{I})\rho\} = \sum_i p_i \Pi_i$ shows that the basis $\boldsymbol{\Pi}$ under which a local dephasing operation has no effect on the total quantum state must coincide with the eigenbasis of the reduced density matrix ρ_A [10]. For now, we do not consider the case of degeneracies, which may complicate the situation [11].

Hence, one may test for the presence of discord by realizing a local dephasing operation in the eigenbasis of ρ_A. The full quantum state is invariant under this operation if and only if it contains no discord. However, a possible change of the full quantum state under the local dephasing operation cannot be directly observed in system A: Even if the state contains discord, i.e., it changes under the local dephasing, the resulting reduced density matrix $\rho'_A = \mathrm{Tr}_B \rho'$ with $\rho' = (\Phi_{\boldsymbol{\Pi}} \otimes \mathbb{I}_B)\rho$ will always coincide with ρ_A [10]. This can be seen by realizing that the result of Eq. (6) was derived without making any assumption about the full quantum state ρ. We will therefore always observe that $\rho_A = \rho'_A$, regardless of whether $\rho = \rho'$ holds or not.

Moreover, since the local dephasing operation does not act on system B, one further finds that also $\rho'_B = \rho_B$ must always hold [11]. In fact, we can conclude that if any difference between the original state ρ and the locally dephased reference state ρ' exists, it must be contained in the correlations between the two subsystems—their respective local descriptions are always unchanged by the local dephasing. Does this mean that it is impossible to observe $\rho \neq \rho'$ (which would constitute a witness for discord) by purely local measurements of any of the two systems? A solution can be found by considering the dynamics of this bipartite system. In fact, a change of the correlation properties of the initial state can have a strong observable impact on the reduced dynamics of one of the subsystems. This is especially well known in the theory of open quantum systems [57], where the influence of initial system-environment correlations poses a considerable theoretical challenge [58–60].

Here, however, it can be exploited to reveal a change of the correlations between the subsystems to the local dynamics. Thus, even if ρ and ρ' are indistinguishable to measurements of system A at some initial time $t = 0$, they may become locally distinguishable after the two systems have been interacting for a time $t > 0$.

Let us assume that systems A and B are subject to some interaction. For simplicity, we consider a unitary evolution of the composite system [61], such that the evolution in subsystem A, given the initial state ρ, is governed by

$$\rho_A(t) = \mathrm{Tr}_B\{U(t)\rho U^\dagger(t)\}. \tag{7}$$

If the state ρ was subject to local dephasing before the time evolution, the state of system A at time t instead reads

$$\rho'_A(t) = \mathrm{Tr}_B\{U(t)\rho' U^\dagger(t)\}. \tag{8}$$

We had already noted that $\rho_A(0) = \rho'_A(0)$, regardless of the properties of ρ. However, if we observe

$$\rho_A(t) \neq \rho'_A(t) \tag{9}$$

at a later time $t > 0$ we can safely conclude that $\rho \neq \rho'$ which implies the presence of discord in the state ρ [10, 11]. All of the necessary steps, i.e.,

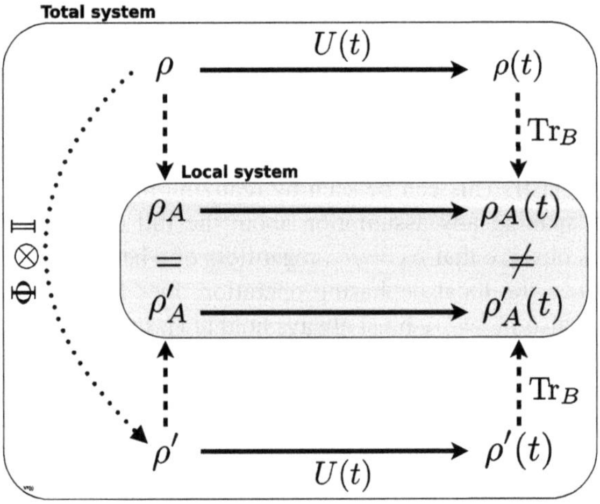

Fig. 1 Schematic representation of the local detection method. A local dephasing operation $\Phi \otimes \mathbb{I}$ produces a reference state ρ', which differs from the original state ρ only in lacking discord. The reduced dynamics of the accessible system A can be strongly influenced by the removal of discord. Hence, at a later time t, the system may evolve differently when the initial state ρ is replaced by ρ'. If $\rho_A(t) \neq \rho'_A(t)$ is indeed observed, it represents a witness for discord of ρ and the distance among the local states further serves as a quantitative measure of discord. Adapted from [62]

- Finding the eigenbasis of ρ_A
- Observing the time evolution of $\rho_A(t)$
- Local dephasing of the initial state in the eigenbasis of ρ_A
- Observing the time evolution of $\rho'_A(t)$

can be carried out with strictly local access only to system A, whereas no control or even knowledge of system B is required. The local detection protocol is illustrated in the diagram in Fig. 1.

2.2 *Quantifying Discord with Local Operations*

Detecting the mere presence of discord by following the protocol outlined above does not provide any quantitative information about the discord of the state ρ. Considering that discord is a rather ubiquitous phenomenon [63], it is also relevant to estimate how strongly discordant a given initial state is. Certain quantifiers furthermore allow for an operational interpretation and therefore directly quantify how well a certain quantum information task can be carried out [24, 25, 27, 28, 32, 33].

A straight-forward way to quantify discord emerges from the local dephasing operation. From the discussion above, we know that ρ' differs from ρ if and only if ρ contains discord. A simple quantifier of discord is thus given by the *dephasing disturbance* [8, 64], expressed by the trace distance

$$D(\rho) = \|\rho - \rho'\|, \tag{10}$$

where $\|X\| = 1/2\mathrm{Tr}\sqrt{X^\dagger X}$ [65]. The trace distance has several appealing properties, most notably for our purposes is its contractivity under trace-preserving and positive operations [66]. The unitary time evolution and the partial trace operation are both positive operations (they map positive operators, such as the density operator, to positive operators) [65]. Thus, using the contractivity property we find that the locally observable distance between the reduced density matrices $\rho_A(t)$ and $\rho'_A(t)$ provides a lower bound for the dephasing disturbance [11]:

$$\begin{aligned} d(t) = \|\rho_A(t) - \rho'_A(t)\| &= \|\mathrm{Tr}_B\{U(t)(\rho - \rho')U^\dagger(t)\}\| \\ &\leq \|U(t)(\rho - \rho')U^\dagger(t)\| \\ &= \|\rho - \rho'\|. \end{aligned} \tag{11}$$

The above inequality holds for all $t \geq 0$, hence one may optimize the locally accessible lower bound by observing the time evolution of the local system for as long as possible and then taking the maximum distance [62]

$$d_{\max} = \max_t \|\rho_S(t) - \rho'_S(t)\| \leq \|\rho - \rho'\|. \tag{12}$$

The above definition assumes that the local dephasing operation is unique, which requires that the eigenbasis of ρ_A is unambiguous. This, however, is not the case if degeneracies are present in the spectrum of ρ_A. In these cases the dephasing disturbance does not produce a suitable quantifier of discord [39, 67]. This problem can be circumvented by including a minimization over all possible dephasing bases. Recalling the $\mathbf{\Pi}$-dependent definition (5) of a general dephasing operation, we introduce the *minimal dephasing disturbance* [8, 68]

$$D_{\min}(\rho) = \min_{\mathbf{\Pi}} \|\rho - (\Phi_{\mathbf{\Pi}} \otimes \mathbb{I})\rho\|. \tag{13}$$

This measure in fact quantifies the minimal amount of entanglement which can be activated from discord in a measurement (*minimal entanglement potential*) [32–34, 69] when the accessible subsystem is a qubit, i.e., $\mathcal{H}_A = \mathbb{C}^2$.

A locally accessible bound for the *minimal dephasing disturbance* can be obtained by dephasing over different local bases instead of just the eigenbasis of ρ_A [68]. We introduce

$$\rho_A^{\mathbf{\Pi}}(t) = \mathrm{Tr}_B\{U(t)(\Phi_{\mathbf{\Pi}} \otimes \mathbb{I})\rho U^\dagger(t)\} \tag{14}$$

and the corresponding local trace distance

$$d_{\mathbf{\Pi}}(t) = \|\rho_A(t) - \rho_A^{\mathbf{\Pi}}(t)\|. \tag{15}$$

A more rigorous bound for discord than Eq. (12) is then obtained as [68]

$$d_{\min}(\rho) = \max_t \min_{\mathbf{\Pi}} d_{\mathbf{\Pi}}(t) \leq D_{\min}(\rho). \tag{16}$$

To measure the above quantity, one first records the time evolution of $\rho_S^{\mathbf{\Pi}}(t)$ for different dephasing bases $\mathbf{\Pi}$. At each time t, the minimum of all $d_{\mathbf{\Pi}}(t)$ is obtained, the minimum being taken over all $\mathbf{\Pi}$. Then, within the set of minima one finds the maximum value over all times t to obtain the strongest available local witness. Ideally, the optimization over $\mathbf{\Pi}$ should be carried out over all possible bases, which is experimentally impossible. In a realistic situation a systematic sampling over a sufficiently closely spaced grid of basis vectors can yield a good estimate with reasonable overhead, see, e.g., [34].

2.3 Pure States: Locally Accessible Lower Bound for Negativity

Let us consider the simple case of a pure state with a controllable qubit subspace, $\mathcal{H}_A = \mathbb{C}^2$. As mentioned before, the concept of discord reduces for pure states to entanglement—in the absence of classical mixing, this is the only form of correlation that can be present. In this case, the dephasing disturbance (10) can be evaluated analytically and yields [8, 68]

$$D(\rho) = \mathcal{N}(\rho), \tag{17}$$

where $\mathcal{N}$ denotes the *negativity* [70],

$$\mathcal{N}(\rho) = \frac{\|\rho^{\Gamma}\| - 1}{2}, \tag{18}$$

and ρ^{Γ} is the partial transpose of ρ.

On the other hand, the minimal dephasing disturbance (13) coincides with the minimal entanglement potential, which for pure states also reduces to the negativity [71]. Hence, in the above scenario, the dephasing disturbance (10) coincides with the minimal dephasing disturbance (13), and the minimum is achieved by dephasing in the eigenbasis of ρ_A [68]. Moreover, the local distance (12) yields a locally accessible lower bound for the negativity.

2.4 Efficacy of the Local Detection Method

Discord can ultimately be traced back to those two-body coherences that are present in ρ but are no longer found in ρ', i.e., after local dephasing. Those coherences are not detectable in either of the two subsystems. Therefore, the performance of the local detection method depends crucially on the interacting dynamics between the two subsystems. Its role is to map these initially hidden two-body coherences to measurable elements of the reduced density matrix of system A at a later time.

Certainly some dynamical processes will work better than others in detecting these correlations. For instance, if no interaction between the two systems were present, i.e., $U(t) = U_A(t) \otimes U_B(t)$, this task could never be achieved [11]. In the course of this article we will observe a number of different time evolutions and thereby explore the limitations of the local detection method based on these examples. Let us already mention that early estimates for the efficacy of the method have been obtained based on a formulation in terms of the Hilbert–Schmidt distance, which allows for analytical evaluation of measure theoretic averages over the unitary group [72]. It was shown that [10]

$$\int d\mu(U)\|\mathrm{Tr}_B\{U(\rho - \rho')U^{\dagger}\}\|_2^2 = \frac{d_A^2 d_B - d_B}{d_A^2 d_B^2 - 1}\|\rho - \rho'\|_2^2, \tag{19}$$

with the Hilbert–Schmidt norm $\|X\|_2^2 = \mathrm{Tr}X^{\dagger}X$, d_A and d_B being the dimensions of the Hilbert spaces $\mathcal{H}_A$ and $\mathcal{H}_B$, respectively, and $d\mu$ representing the Haar measure on the unitary group. These group-theoretic methods further allow for analytical evaluation of the variance corresponding to the above average [11], as well as time-dependent averages with respect to more realistic random matrix ensembles [72]. These results show that the locally observable signal, obtained from a generic dynamical system, is directly proportional to the discord of the initial state. Hence a generic unitary evolution is expected to reveal the quantum discord based on the local detection method. For further details on the Haar-measure integration techniques and additional numerical and analytical studies of the local detection method in this context, we refer to Refs. [10, 11, 72, 73]. Note, however, that the Hilbert–Schmidt distance is not contractive under positive maps. This can lead to unphysical behavior of Hilbert-Schmidt based quantifiers for discord and related quantities [74, 75]. For this reason, it is generally recommended to use the trace distance instead [76].

We also observe that the proportionality factor on the right-hand side of Eq. (19) shrinks to zero as d_B increases. This might suggest the conclusion that the signal becomes undetectably small when the observable system is coupled to a truly infinite environment. The result (19) however makes statements about generic evolutions which are well represented by the average over all unitaries. Generally speaking, systems that lead to such a dynamics are typically strongly chaotic, and deviations from the average result (19) are certainly expected. In common situations one might as well encounter highly non-generic evolutions, in particular in quantum optical systems. Later in this article we will discuss a successful experimental detection of system-environment discord by means of the local detection method where the accessible system couples to an infinite-dimensional, Markovian (memoryless) environment [77].

3 Experiments

The local detection method has been used to reveal discord in experiments with photons [77, 78] and trapped ions [62]. In all experimental applications reported so far, the controllable subspace was two-dimensional, whereas correlations were detected with ancilla systems ranging from two-dimensional systems to continuous variables.

3.1 Trapped-Ion Experiment

An experimental realization of the local detection method was first reported in [62]. A single trapped ion is used to simulate both a two-dimensional, locally accessible quantum system by means of its electronic degree of freedom, and a bosonic ancilla system, comprised of the same ion's motional degree of freedom. Since the ion is confined in a harmonic trapping potential, the ancilla system is described by a quantum harmonic oscillator [79, 80]. When the ion is driven by a laser whose detuning from the ion's resonance transition coincides with the frequency of the harmonic motion, an interaction between the two degrees of freedom, described to a good approximation by the anti-Jaynes-Cummings Hamiltonian

$$H = \frac{\hbar\Omega\eta}{2}(\sigma_+ a^\dagger + \sigma_- a). \tag{20}$$

is evoked [4, 8, 79–82]. Here, a and $a^\dagger$ are the bosonic creation and annihilation operators of the harmonic oscillator mode, σ_+, σ_- denote the ladder operators for the electronic qubit system, Ω denotes the Rabi frequency and η is the Lamb-Dicke parameter [80]. In the above expression we assumed, for ease of notation, the Lamb-Dicke regime, i.e., $\eta\sqrt{\langle(a+a^\dagger)^2\rangle} \ll 1$; the analysis of the trapped-ion experiment, however, can be extended beyond this regime and beyond ideal experimental conditions [62]; for a detailed account see [8]. The Hamiltonian (20) induces a coherent coupling between the states $|g, n\rangle$ and $|e, n+1\rangle$, where $|g\rangle$ and $|e\rangle$ describe the electronic ground- and excited states, respectively, and $|n\rangle$ is a Fock state of the harmonic motion. An ion initially prepared in the state $|g, n\rangle$ hence undergoes a Rabi oscillation of the form

$$|\Psi(t)\rangle = U(t)|g, n\rangle = \cos\left(\frac{\Omega_n}{2}t\right)|g, n\rangle + \sin\left(\frac{\Omega_n}{2}t\right)|e, n+1\rangle, \tag{21}$$

where $U(t) = \exp(-iHt/\hbar)$ and $\Omega_n = \sqrt{n+1}\eta\Omega$.

Initially, the system is prepared by optical pumping of the electronic level to the ground state and laser cooling of the motional degree of freedom, leading to the product state

$$\rho_0 = |g\rangle\langle g| \otimes \sum_{n=0}^{\infty} p_n |n\rangle\langle n|. \tag{22}$$

The thermal populations $p_n = \bar{n}^n/(\bar{n}+1)^{n+1}$ can be given in terms of the mean phonon number $\bar{n}$; see, e.g., [80]. When this initial state is exposed to the laser interaction for a duration t_0 (*state preparation*), it evolves as

$$\begin{aligned}
\rho(t_0) &= U(t_0)\rho_0 U^\dagger(t_0) \\
&= \sum_{n=0}^{\infty} p_n \Bigg[\cos\left(\frac{\Omega_n}{2}t_0\right)^2 |g, n\rangle\langle g, n| + \sin\left(\frac{\Omega_n}{2}t_0\right)\cos\left(\frac{\Omega_n}{2}t_0\right)|g, n\rangle\langle e, n+1| \\
&\quad + \sin\left(\frac{\Omega_n}{2}t_0\right)\cos\left(\frac{\Omega_n}{2}t_0\right)|e, n+1\rangle\langle g, n| + \sin\left(\frac{\Omega_n}{2}t_0\right)^2 |e, n+1\rangle\langle e, n+1|\Bigg].
\end{aligned} \tag{23}$$

The goal is now to detect the presence of discord between the electronic and motional degrees of freedom in the state $\rho(t_0)$. To this end, the local detection method is employed, which allows us to limit experimental access to the electronic degree of freedom. The first task is to obtain the eigenbasis of the reduced density matrix, which determines the basis for the local dephasing operation. By tracing over the motional degrees of freedom (system B), we obtain the quantum state of the qubit (system A),

$$\rho_A(t_0) = \mathrm{Tr}_B\rho(t_0) = \sum_{n=0}^{\infty} p_n \left[\cos\left(\frac{\Omega_n}{2}t_0\right)^2 |g\rangle\langle g| + \sin\left(\frac{\Omega_n}{2}t_0\right)^2 |e\rangle\langle e|\right], \quad (24)$$

which is diagonal in the basis $\{|g\rangle, |e\rangle\}$ at all times.

To detect the discord of the state at time t_0, the dynamical evolution is interrupted and a local dephasing is performed by projecting onto the local subspaces, spanned by $|g\rangle$ and $|e\rangle$, respectively. Experimentally this is achieved by inducing a weak ac-Stark shift on the ground state for a well-controlled period of time. To this end, another laser which off-resonantly addresses a transition between the ground state and another short-lived excited state is used. This adds a relative phase shift to any superposition that involves the states $|e\rangle$ and $|g\rangle$. By performing an average over a suitably chosen family of phase shifts, the relative phase relation between the states $|e\rangle$ and $|g\rangle$ can be completely removed, thereby effectively realizing the local dephasing operation [8, 62]. This technique can be combined with local, coherent laser manipulations to achieve dephasing in an arbitrary basis [8, 62].

The total state after local dephasing is then given as

$$\begin{aligned}\rho'(t_0) &= (\Phi\otimes\mathbb{I})\rho(t_0)\\ &= \sum_{i\in\{e,g\}} (|i\rangle\langle i|\otimes\mathbb{I}_B)\rho(t_0)(|i\rangle\langle i|\otimes\mathbb{I}_B)\\ &= \sum_{n=0}^{\infty} p_n \left[\cos\left(\frac{\Omega_n}{2}t_0\right)^2 |g,n\rangle\langle g,n| + \sin\left(\frac{\Omega_n}{2}t_0\right)^2 |e,n+1\rangle\langle e,n+1|\right].\end{aligned} \quad (25)$$

By construction, $\rho(t_0)$ and $\rho'(t_0)$ only differ in as much as $\rho(t_0)$ contains discord while $\rho'(t_0)$ does not, thus, comparison with Eq. (23) now allows us to precisely identify those terms that produce the discord in $\rho(t_0)$. As anticipated, these are the two-body coherences $|e,n\rangle\langle g,n+1|$ (and its adjoint counterpart). Since these matrix elements are indeed off-diagonal in both of the sub-systems, any local measurement of the qubit or the ions' motion will be unable to detect their presence. One readily confirms that $\rho'(t_0)$ has the same reduced density matrices as $\rho(t_0)$. Before we proceed to study the signature of discord in the subsequent qubit dynamics, we evaluate the dephasing disturbance, which reads [62],

$$D(\rho(t_0)) = \sum_{n=0}^{\infty} p_n \left|\sin\left(\frac{\Omega_n}{2}t_0\right)\cos\left(\frac{\Omega_n}{2}t_0\right)\right|. \quad (26)$$

As expected, this quantifies precisely the magnitude of the above-mentioned two-body coherences.

Despite being hidden from local measurements, the discord contained in the state $\rho(t_0)$ can be detected at a later time by observing deviating evolutions of the reduced density matrices $\rho_A(t_0 + t_1)$ and $\rho'_A(t_0 + t_1)$. The state is again subjected to the laser interaction, for a duration t_1 (*detection*). The evolution of the unperturbed state was given in Eq. (24), whereas the dephased state evolves as

$$\begin{aligned}\rho'_A(t_1+t_0) &= \mathrm{Tr}_B\{U(t_1)\rho'(t_0)U^\dagger(t_1)\} \\ &= \sum_{n=0}^{\infty} p_n \left[\left(\cos\left(\frac{\Omega_n}{2}t_0\right)^2 \cos\left(\frac{\Omega_n}{2}t_1\right)^2 + \sin\left(\frac{\Omega_n}{2}t_0\right)^2 \sin\left(\frac{\Omega_n}{2}t_1\right)^2\right)|g\rangle\langle g| \right. \\ &\quad \left. + \left(\cos\left(\frac{\Omega_n}{2}t_0\right)^2 \sin\left(\frac{\Omega_n}{2}t_1\right)^2 + \sin\left(\frac{\Omega_n}{2}t_0\right)^2 \cos\left(\frac{\Omega_n}{2}t_1\right)^2\right)|e\rangle\langle e|\right].\end{aligned} \tag{27}$$

The difference between the two evolutions can be observed by measuring the excited-state population. In the trapped-ion experiment, this is realized by a highly efficient fluorescence readout method [79, 80]. We observe the difference

$$\begin{aligned}d_e(t_0,t_1) &= \langle e|\rho_A(t_1+t_0) - \rho'_A(t_1+t_0)|e\rangle \\ &= \frac{1}{2}\sum_{n=0}^{\infty} p_n \sin(\Omega_n t_0)\sin(\Omega_n t_1),\end{aligned} \tag{28}$$

where we have used the identity $2\sin\alpha\cos\alpha = \sin 2\alpha$. The fact that both states $\rho_A(t_0)$ and $\rho'_A(t_0)$ are diagonal in the basis $\{|g\rangle, |e\rangle\}$ allows us to determine the trace distance directly from the excited-state deviations as

$$d(t_0,t_1) = \|\rho_A(t_1+t_0) - \rho'_A(t_1+t_0)\| = |d_e(t_0,t_1)|. \tag{29}$$

This quantity is locally measurable in subsystem A and provides a lower bound to the dephasing disturbance (26), a global property of the full quantum state. Whenever a nonzero deviation (29) is observed, we can conclude that $\rho(t_0)$ contained nonzero discord, and the magnitude of the local trace distance further allows for a quantification of the initial discord.

The experimental protocol and the measured local witness for discord is shown in Fig. 2. The experiment was performed for two different initial temperatures of the ion's motion. The theoretical description used for the plot includes the effects of experimental imperfections, such as small detunings and fluctuating parameters [8, 62]. For both environmental temperatures, a strong signature of the initial discord is observed.

Finally, the tightest bound to the dephasing disturbance (26) is obtained by the largest deviation $d_{\max}$, as introduced in Eq. (12). This quantity is plotted for different t_0 in Fig. 3. Comparison with the predictions show that the locally recovered signatures

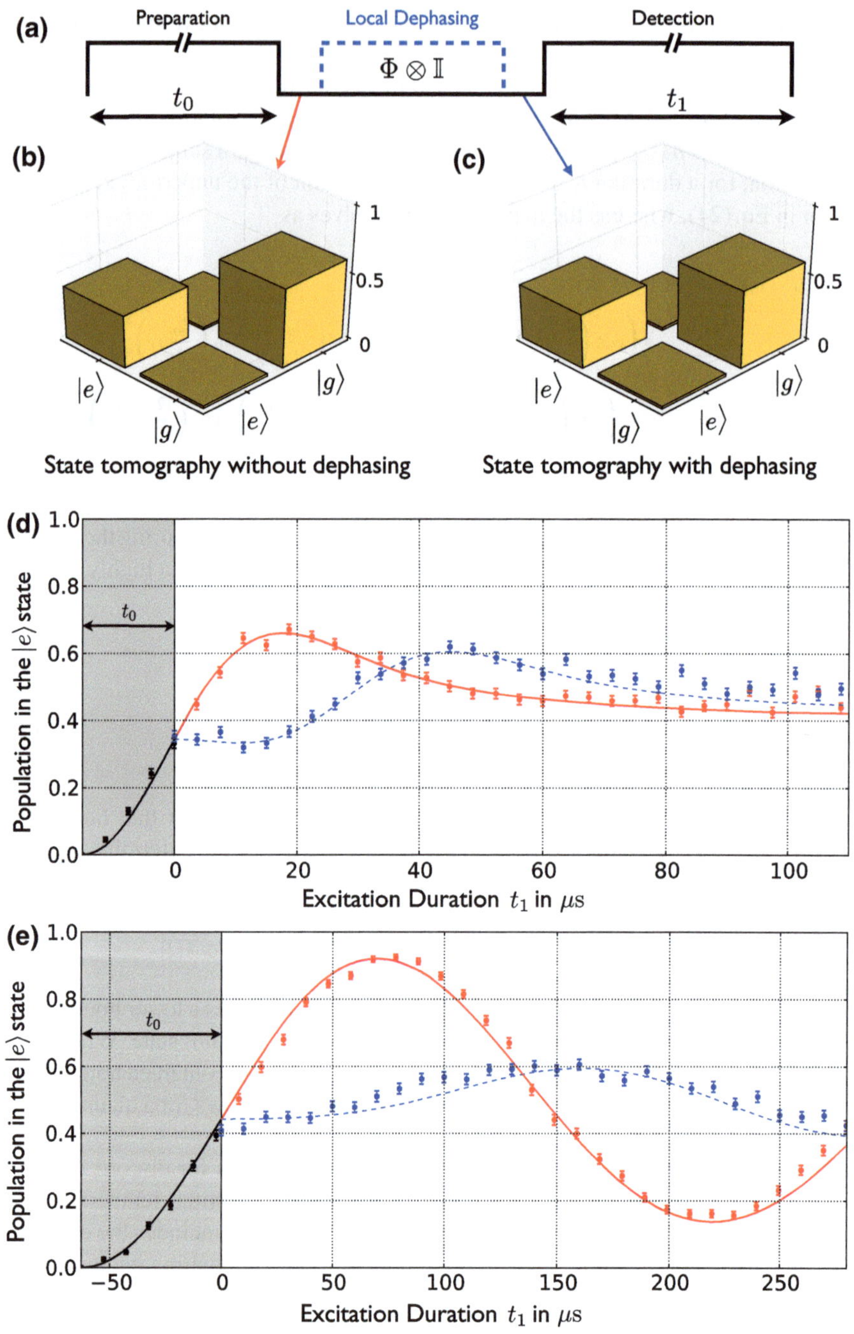
(a)
Preparation
Local Dephasing
Detection
$\Phi \otimes \mathbb{I}$
t_0
t_1
(b)
(c)
State tomography without dephasing
State tomography with dephasing
(d)
Population in the $|e\rangle$ state
Excitation Duration t_1 in μs
(e)
Population in the $|e\rangle$ state
Excitation Duration t_1 in μs

◀ **Fig. 2** The dynamics of the electronic qubit after state preparation is observed with and without dephasing (**a**). Local state tomography (**b**) identifies the local basis for the local dephasing operation $\Phi \otimes \mathbb{I}$. Another tomography after dephasing confirms that the local state is initially unchanged (**c**). The ensuing dynamics (**d**, **e**) is, however, strongly influenced by the removal of discord through local dephasing, and any observed difference between the *red* (unperturbed dynamics) and *blue* (dynamics after dephasing) data points indicates the presence of discord in $\rho(t_0)$. The average phonon numbers are $\bar{n} = 5.9$ in (**d**) and $\bar{n} = 0.2$ in (**e**). Figure taken from [62]

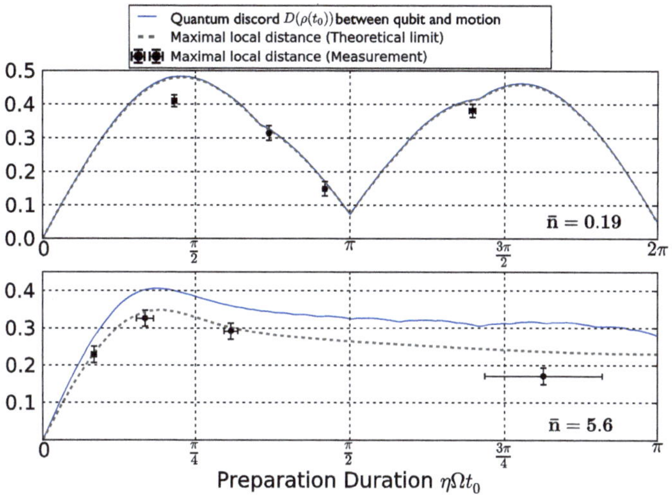

Fig. 3 To obtain the tightest possible bound, the maximum deviation between the local quantum states is taken. In the case of the low-temperature environment the obtained local witness almost saturates the actual distance between the global states. In the higher-temperature case, such a tight estimation is not possible, as shown by the theoretical prediction. Figure taken from [8, 62]

of the initial discord are as large as theoretically possible, and almost saturate the inequality (12) in the case of the low-temperature initial state.

The signal of the higher-temperature state is not as pronounced as the one obtained from the low-temperature distribution. The question arises whether the local signal would vanish completely if the temperature was increased even further, as one might expect if the usability of the local detection method was limited to effectively finite-dimensional environments. This, is however not the case [8]. A simple estimate of the signal for higher temperatures can be obtained by fixing the preparation and detection pulse durations to the value $\Omega_0 t_0 = \Omega_0 t_1 = \pi/2$. This leads to the maximum local signal of 1/2 if the initial state of motion has temperature zero.

From Eq. (29), we obtain

$$d(t_0, t_1)|_{t_0=t_1=\pi/(2\Omega_0)} = \frac{1}{2}\sum_{n=0}^{\infty} \frac{\bar{n}^n}{(\bar{n}+1)^{n+1}} \sin^2\left(\frac{\pi\sqrt{n+1}}{2}\right). \tag{30}$$

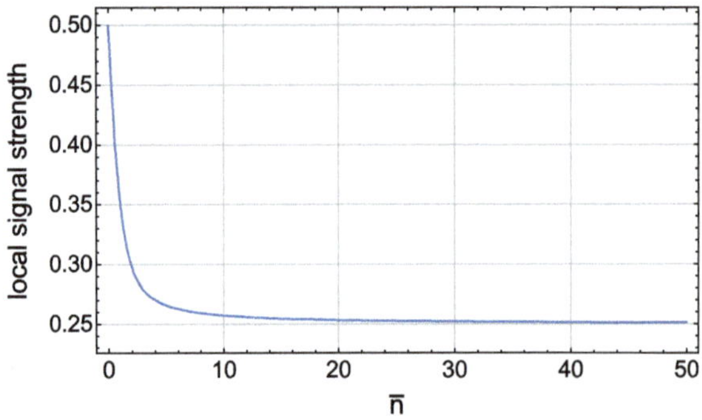

Fig. 4 Expected local signal of discord as a function of the average thermal phonon number $\bar{n}$, when preparation and detection time are both chosen as $t_0 = t_1 = \pi/(2\Omega_0)$. The prediction (30) shows that the local signal remains at a finite value for phonon distributions of much higher temperatures than those realized ($\bar{n} = 0.2$ and $\bar{n} = 5.9$) in the experiment [62]. Figure taken from [8]

The signal is shown in Fig. 4 as a function of $\bar{n}$. After an initial drop, the signal remains close to the finite value of 1/4 for much higher average phonon numbers than those tested in the experiment. This shows that even for a high-temperature thermal distribution, the local detection method is able to reveal the qubit-motion discord dynamically under an evolution governed by the Hamiltonian (20). We remark that the derivation presented in this section was based on the Lamb-Dicke limit. For sufficiently small values of η, the expression (30) still represents a valid approximation for the exact expression even for large values of $\bar{n}$. In particular, values up to $\bar{n} \approx 50$ can be adequately described as long as $\eta \lesssim 0.05$, which applies to the parameter reported in the experiment [62]. An exact expression for the effective Rabi frequency Ω_n, beyond the Lamb-Dicke limit, can be given in terms of the generalized Laguerre polynomials $L_n^{(\alpha)}(x)$ as $\Omega_n = \eta e^{-\eta^2}(n+1)^{-1/2} L_n^{(1)}(\eta^2)$ [8, 79, 80]. Numerical simulations with the exact expression produce nonzero values of the local signal even when $\bar{n}$ and η attain values outside of the Lamb-Dicke limit. For very high values of $\bar{n}$, however, the validity of the effective description of the laser-ion interaction through Eq. (20) reaches its limits, since the fast-moving ion can no longer be laser-addressed with sufficient precision; hence the truly infinite limit $\bar{n} \to \infty$ cannot be tested with this ansatz.

3.2 *Photonic Experiment with Continuous-Variable Ancilla*

The first optical realization of the local detection method was reported in [77]. The accessible system here is represented by a photon's polarization degrees of freedom, which interact with the same photon's frequency degree of freedom when passing through a birefringent material. In contrast to the trapped-ion experiment, the ancilla

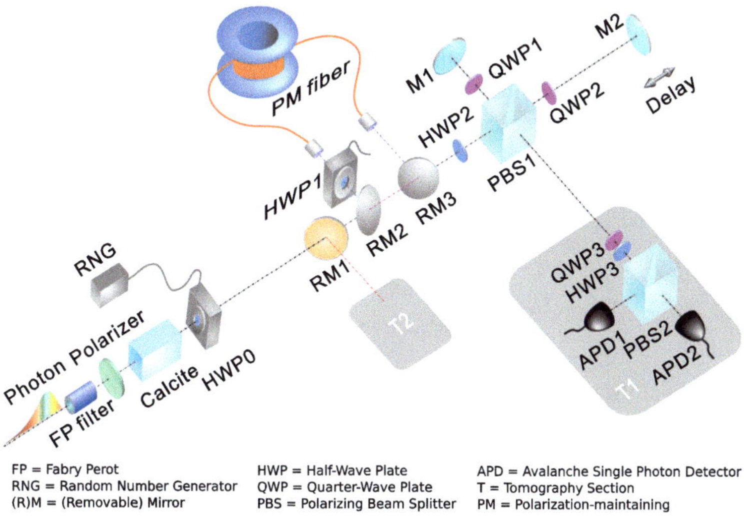

Fig. 5 Experimental setup for local detection with single photons. State preparation is realized by a Fabry-Perot (FP) filter, a polarizer and a calcite crystal, followed by a half-wave plate (HWP0) with a random orientation to scramble the local basis. A removable mirror (RM1) can be inserted to send the photon into the local state tomography unit (T2) which is used to obtain the full quantum state of polarization. When the mirrors RM1-3 are removed, the state is sent into the Michelson interferometer to generate an interacting dynamics between polarization and frequency, by adjusting the position of mirror M2. This evolution is followed by another tomography section T1 to measure the local dynamics. To reveal discord, local dephasing is realized by placing RM2 and RM3 to send the photon through a long polarization-maintaining (PM) fiber which removes all discord, when the local eigenbasis has been mapped onto its principal axes by means of HWP1. This may affect the ensuing polarization dynamics observed in T1 which would constitute a witness for discord. Figure adapted from [77]

system is no longer described by a single harmonic oscillator mode, but instead by a continuum of modes.

The experimental setup is summarized in Fig. 5. Initially single photons are created in the quantum state

$$\rho_{pi} = \sum_{\omega} \Delta\omega G(\omega) \left(\frac{1}{2} |H, \omega\rangle\langle H, \omega| + \beta e^{i\varphi} |H, \omega\rangle\langle V, \omega| + \beta e^{-i\varphi} |V, \omega\rangle\langle H, \omega| + \frac{1}{2} |V, \omega\rangle\langle V, \omega| \right), \tag{31}$$

where the mixed frequency distribution is described by a Lorentzian line shape,

$$G(\omega) = \frac{1}{\pi} \frac{\delta\omega}{\delta\omega^2 + (\omega - \omega_0)^2}. \tag{32}$$

Here, we have discretized the frequency space by introducing a small frequency interval $\Delta\omega$; later on we will consider the continuum limit $\Delta\omega \to 0$. A basis for the polarization state is defined by the states $\{|H\rangle, |V\rangle\}$, describing horizontal and vertical polarization, respectively. When passing through a birefringent material, states with a specific polarization direction travel with a modified velocity, and, due to a different dwell time inside the material, experience a modified phase shift. Formally, we find

$$U_c(t) : \begin{cases} |H, \omega\rangle \to |H, \omega\rangle \\ |V, \omega\rangle \to e^{-i\omega t}|V, \omega\rangle \end{cases}, \tag{33}$$

where the dwell time is given as $t = \Delta n_c L/c$, with the speed of light c, the length L of the crystal and, the birefringence Δn_c describing the difference between the refractive indices for the two polarization directions. This evolution produces the correlated states

$$\begin{aligned} \rho &= U_{\text{cal}}(t)\rho_{pi}U_{\text{cal}}^{\dagger}(t) \\ &= \sum_{\omega} \Delta\omega G(\omega)\left(\frac{1}{2}|H, \omega\rangle\langle H, \omega| + \beta e^{i(\omega t+\varphi)}|H, \omega\rangle\langle V, \omega|\right. \\ &\qquad \left. + \beta e^{-i(\omega t+\varphi)}|V, \omega\rangle\langle H, \omega| + \frac{1}{2}|V, \omega\rangle\langle V, \omega|\right). \end{aligned} \tag{34}$$

In the experiment, the initial phase φ was chosen such that $\varphi = -\omega_0 t$.

To reveal the discord of ρ with the local detection method, one first determines the reduced state of the accessible system, which in this case is the qubit state

$$\rho_A = \begin{pmatrix} 1/2 & \beta C(t) \\ \beta C(t) & 1/2 \end{pmatrix}, \tag{35}$$

with the real-valued function

$$C(t) = \sum_{\omega} \Delta\omega G(\omega) e^{i(\omega-\omega_0)t}. \tag{36}$$

Experimentally, this state is obtained by inserting the removable mirror RM1, and using the tomography section T2, as is pictured in Fig. 5. In contrast to the trapped-ion experiment, where the local eigenbasis was always given by the computational basis, here, the eigenvectors are given by $|\pm\rangle = \frac{1}{\sqrt{2}}(|H\rangle \pm |V\rangle)$. Note that the local eigenbasis is first hidden by a random local unitary basis rotation of the qubit, controlled by the half-wave plate HWP0 in Fig. 5. While this step renders the experimental detection of discord more challenging, it does not affect the theoretical treatment, since it can be accounted for by an effective redefinition of the local basis, which does not alter the correlation properties.

To realize a dephasing of the qubit experimentally, the eigenstates of the photon are mapped onto the principal axes of a long polarization-maintaining fiber by means of HWP1 after removing the mirror RM1, and placing the mirrors RM2 and RM3; see Fig. 5. The small birefringence of the fiber leads to an effective dephasing after a sufficiently long interaction time, such that the locally dephased reference state

$$\begin{aligned}\rho' &= (\Phi \otimes \mathbb{I})\rho \\ &= \sum_{i\in\{+,-\}} (|i\rangle\langle i| \otimes \mathbb{I}_B)\rho(|i\rangle\langle i| \otimes \mathbb{I}_B) \\ &= \frac{1}{2}\sum_{\omega} \Delta\omega G(\omega)\Big[|H,\omega\rangle\langle H,\omega| + |V,\omega\rangle\langle V,\omega| \\ &\qquad + \beta(e^{i(\omega-\omega_0)t} + e^{-i(\omega-\omega_0)t})|H,\omega\rangle\langle V,\omega| \\ &\qquad + \beta(e^{-i(\omega-\omega_0)t} + e^{i(\omega-\omega_0)t})|V,\omega\rangle\langle H,\omega|\Big],\end{aligned} \tag{37}$$

is created [77].

One may again confirm that the reduced density matrices describing polarization and frequency degrees of freedom coincide for ρ and ρ' [8]. The dephasing disturbance (10) is evaluated in the continuum limit $\Delta\omega \to 0$, i.e., $\sum_{\omega} \Delta\omega \to \int d\omega$, and yields [8, 77]

$$D(\rho) = \frac{\beta}{2}\int d\omega G(\omega)\left|e^{i(\omega-\omega_0)t} - e^{-i(\omega-\omega_0)t}\right|. \tag{38}$$

For the dynamical detection of the discord (38), the Michelson delay setup, consisting of HWP2 with tunable angle $\eta/2$, and a polarizing beam splitter (PBS1) is used. Theoretically the realized dynamics is equivalent to the one described in the birefringent material, since, again one of the two polarization states acquires a modified phase shift due to a different dwell time. The dwell time $\tau = 2x/c$ in the Michelson setup is determined by the delay x of the mirror M2.

The resulting local trace distance can be shown to be independent of η, and reads in the continuum limit [77],

$$\begin{aligned}d(\tau) &= \frac{\beta}{2}\left|\int d\omega G(\omega)\left(e^{i(\omega-\omega_0)t} - e^{-i(\omega-\omega_0)t}\right)e^{i\omega\tau}\right| \\ &= \frac{\beta}{2}\left|e^{-\delta\omega|t+\tau|} - e^{-\delta\omega|t-\tau|}\right|.\end{aligned} \tag{39}$$

Its maximum value

$$\max_{\tau} d(\tau) = \frac{\beta}{2}(1 - e^{-2\delta\omega t}), \tag{40}$$

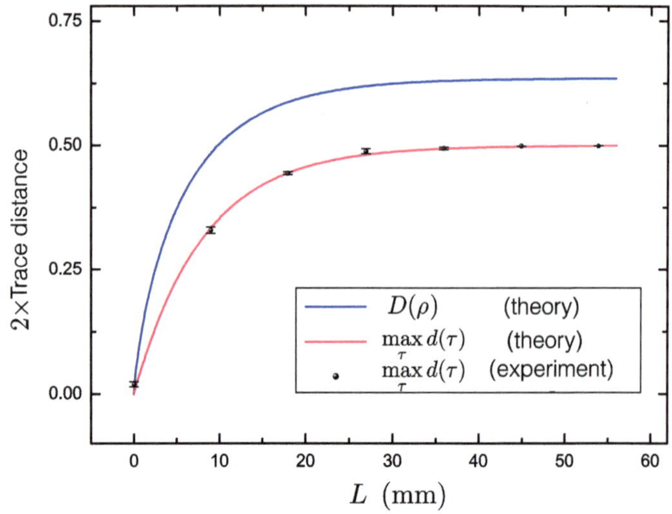

Fig. 6 The maximum local trace distance (40), $\max_\tau d(\tau)$, provides a lower bound for the total trace distance (38), $D(\rho)$, for all states (34). While the full amount of the correlations cannot be revealed locally, the theoretical limit is reached with high precision. Figure adapted from [77]

produces a bound to the dephasing disturbance (38), as depicted in Fig. 6 for different values of t. The figure shows a strong dynamical signal of the discord, as well as excellent agreement between experiment and theory.

The dynamics of the qubit system, evoked through interaction with its frequency degrees of freedom, could be described as pure dephasing. No excitations are exchanged between the polarization and frequency degrees of freedom. Instead the presence of correlations between these two subsystems effectively leads to the decay of coherences in the qubit system. This dynamics if furthermore completely irreversible, and can thus be considered as Markovian, i.e., memoryless open-system evolution [83, 84], with the time scale of the decay being determined by the width $\delta\omega$ of the initial frequency distribution [57]. Hence, in the case of a purely dephasing coupling between system and environment, the reported experiment demonstrates the applicability of the local detection method to reveal initial system-environment discord even if the environment is completely memoryless [8].

3.3 Photonic Experiment with Discrete-Variable Ancilla

In another photonic experiment, discord between the polarization and momentum degrees of freedom of a photon, created in the process of parametric downconversion, was detected using the local detection method [78]. The momentum space is restricted here to two possible channels, denoted by $|0\rangle$ and $|1\rangle$. Hence, the ancilla system is effectively described by a discrete two-dimensional state space. This distinguishes the setup from the two experiments discussed before, where the ancilla was described by single- [62] and multi-mode [77] harmonic oscillators, respectively.

The state of one of the two photons which are created during parametric down-conversion is, after suitable manipulations by a double-slit and a tunable half-wave plate, described by

$$\rho = \lambda |H\rangle\langle H| \otimes |0\rangle\langle 0| + (1-\lambda)|\theta\rangle\langle\theta| \otimes |1\rangle\langle 1|, \tag{41}$$

with $|\theta\rangle = \cos\theta|H\rangle + \sin\theta|V\rangle$. This state contains no discord only if the angle θ is chosen such that $|\theta\rangle = |H\rangle$ or $|\theta\rangle = |V\rangle$. For all other values the state contains discord since $|\theta\rangle$ and $|H\rangle$ are neither orthogonal nor parallel. As a side remark, we note that this type of discordant state can be (and was in fact) generated via a local operation from a zero-discord state, which is reflected by its low correlation rank [21].

The accessible system is given, again, by the polarization degree of freedom, and the associated qubit state is obtained through full state tomography. The local dephasing operation is then implemented using suitably adjusted polarizers. For the dynamical detection of discord a unitary evolution is realized: By inserting a relative phase shift between the two polarization states in only one of the two momentum channels, an effective interaction between the polarization and the momentum degrees of freedom is mediated. This effectively leads to pure dephasing of the polarization state and is close in spirit to the dynamics described in the previous experiment. Using these ingredients and following the local detection protocol, the discord of the initial states (41) was successfully revealed in the experiment.

When $|\theta\rangle = |V\rangle$, however, the initial state does not contain discord. Nevertheless it is still classically correlated since it cannot be written as a factorizing product state. The deviation from a product state can also be dynamically revealed by observing an increase of the trace distance above its initial value, when comparing the evolution of the unperturbed state with the evolution after an arbitrary local operation [9, 84]; for earlier experiments see [85–87]. In the experiment, this is realized whenever the local detection method did not produce a witness for discord. In this case, a local unitary operation, implemented through suitably placed half-wave plates, produces a reference state, such that the presence of correlations in the initial state was detected by means of an increase of the trace distance above its initial value, except when the initial state is indeed factorized [78]. This way, the resulting two-step protocol, comprised of a combination of the local detection method [10, 11] and the trace distance witness for initial correlations [9], is able, on the one hand, to detect discord, and on the other hand, to reveal classical correlations in the absence of discord [78].

4 Theoretical Studies

Aside from providing an experimentally convenient method for the detection of discord, the local detection method may also be helpful to gain insight into the impact of correlations and discord through theoretical studies of interacting systems. Such studies may further provide a useful characterization of the local detection method itself, by indicating the conditions under which the presence of discord can be successfully revealed through the local dynamics.

4.1 Dynamical Single-Spin Signature of a Quantum Phase Transition

A special situation arises when the state $|\Psi\rangle$ to which the local detection method is applied, is at the same time an eigenstate of the Hamiltonian which governs the interacting time evolution of the system [68]. In this case, the dynamics without dephasing is trivially constant and any dynamics that arises after the local dephasing operation constitutes a witness for discord. Additionally, the local trace distance provides a lower bound to the bipartite entanglement of the system (the bipartition is defined by the subspace on which the dephasing was implemented and the rest of the system). Such a situation is particularly interesting when $|\Psi\rangle$ is chosen as the ground state of a many-body system [68], since its quantum correlations can disclose information about the existence of a quantum phase transition [88, 89].

The local detection method has been applied in a theoretical study to reveal the quantum correlations of the ground state, as well as quantum discord of finite-temperature thermal equilibrium states in the context of a quantum phase transition [68]. The properties of the ground state are often regarded as a principal indicator of critical phenomena, since quantum phase transitions are defined as abrupt qualitative changes of the ground state as a function of some external control parameter [90]. By using the local detection method to reveal entanglement properties of the ground state to the dynamics of a single spin, we connect these ground-state properties to the entire excitation spectrum, which is relevant for the dynamical evolution of the spin.

The system studied here is the Ising model with variable interaction range [8, 91–97]

$$H_\alpha = -\sum_{\substack{i,j=1\\(i<j)}}^{N} \frac{J_0}{|i-j|^\alpha}\sigma_x^{(i)}\sigma_x^{(j)} - B\sum_{i=1}^{N}\sigma_y^{(i)}, \tag{42}$$

where $\sigma_k^{(i)}$ describe the Pauli matrices for spin i with $k \in \{x, y, z\}$, J_0 determines the strength of the spin-spin interaction, which stands in competition with the transverse external field of strength B. The ground state provides an intuitive understanding of the quantum phase transition: For small values of B the contribution of the internal spin-spin interaction dominates and the relative orientation of the spins is chosen such that the associated potential energy is minimized, thereby describing a ferromagnetic state for $J_0 > 0$. When B is increased above a critical value, which depends on α and

J_0, the spins tend to align along the direction of the external fields and the system describes a paramagnet.

4.1.1 Local Bound for the Ground-State Negativity

To apply the local detection method to the ground state of the above system, we consider the left-most spin as the controllable subsystem and consider the ensemble of all other spins as the inaccessible ancilla. This reduces the state space from an exponentially large dimension of 2^N to the easily manageable size of a qubit.

The initial state $|\Psi_0\rangle$, being the ground state of H_α, does not evolve in time. However, by applying the local dephasing operation to the state $|\Psi_0\rangle$ excited states are incoherently populated, which does no longer necessarily result in a time-invariant state. Any time evolution is, according to the local detection method, a witness for discord, and, since the state is pure, in this case also a witness for entanglement. As described in Sect. 2.3, local measurements of the single spin dynamics can be used to obtain a lower bound for the negativity (a simple entanglement measure) of $|\Psi_0\rangle$.

The local eigenbasis of each individual spin is for symmetry reasons (Z_2-symmetry: invariance of H under a π-rotation around the y-axis) always given by the y-axis [8, 68]. The local dephasing is therefore always described by

$$\rho_\Phi = (\Phi \otimes \mathbb{I})|\Psi_0\rangle\langle\Psi_0| = \sum_{\varphi\in\{\uparrow_y,\downarrow_y\}} (|\varphi\rangle\langle\varphi| \otimes \mathbb{I}_B)\,|\Psi_0\rangle\langle\Psi_0|\,(|\varphi\rangle\langle\varphi| \otimes \mathbb{I}_B)\,, \quad (43)$$

where $|\uparrow_y\rangle$ and $|\downarrow_y\rangle$ describe the eigenstates of $\sigma_y^{(1)}$ and, here, $\mathbb{I}_B$ is the identity operator on all remaining spins. Dephasing in this basis further yields the minimal trace distance (13), as was shown in Sect. 2.3, and the dephasing disturbance thus coincides with the negativity [68],

$$D(|\Psi_0\rangle\langle\Psi_0|) = \||\Psi_0\rangle\langle\Psi_0| - \rho_\Phi\| = \mathcal{N}(|\Psi_0\rangle\langle\Psi_0|). \quad (44)$$

The local evolution of the controllable spin is governed by

$$\rho_A(t) = \mathrm{Tr}_B\{U(t)(\Phi \otimes \mathbb{I})|\Psi_0\rangle\langle\Psi_0|U^\dagger(t)\}, \quad (45)$$

where $U(t) = e^{-iH_\alpha t/\hbar}$. At all times t, the local trace distance

$$d(t) = \|\rho_A(t) - \rho_A(0)\| \quad (46)$$

yields a lower bound for the negativity $\mathcal{N}(|\Psi_0\rangle\langle\Psi_0|)$. This quantity is fully determined by the magnetization $m_y(t) = \mathrm{Tr}\{\rho_A(t)\sigma_y^{(1)}\}$ along the y direction as [68]

$$d(t) = \frac{1}{2}|m_y(t) - m_y(0)|. \quad (47)$$

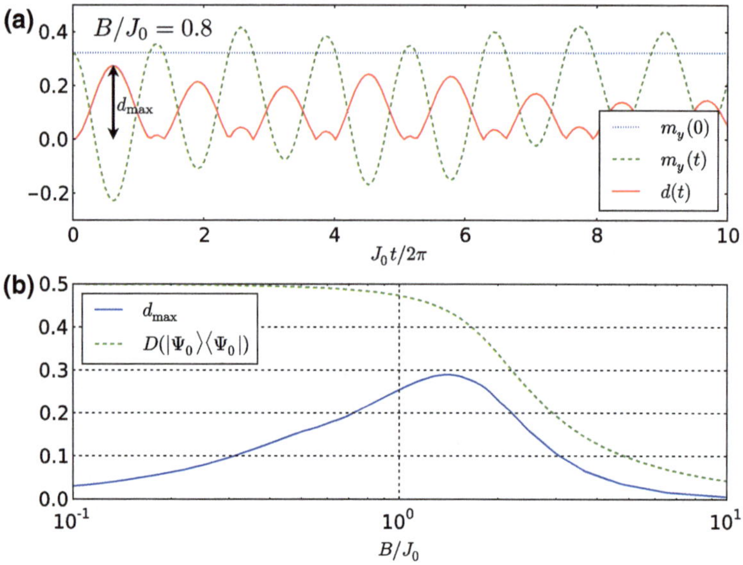

Fig. 7 Local dephasing of the ground state induces a dynamical evolution (47) of the observed spin, thereby detecting and quantifying the ground-state entanglement between the measured spin and the rest of the chain, as quantified by the negativity (44). Further optimization of the local signal by maximization over all observed times t discloses a finite-size precursor of the quantum phase transition in form of a pronounced peak in the vicinity of the critical point. Figure taken from [68]; M. Gessner et al., "Observing a quantum phase transition by measuring a single spin", Europhysics Letters, vol. **107**, issue 4, 2014, available at http://iopscience.iop.org/article/10.1209/0295-5075/107/40005

Again, we may take the time-maximum $d_{\max}$ of all local distances, Eq. (12), here coinciding with Eq. (16), to obtain the strongest available lower bound on $\mathcal{N}$, as plotted in Fig. 7.

A peak of the local signal around $B \simeq J_0$, indicates the quantum phase transition and hints at the position of the critical field. For larger values of B both the total ground-state entanglement and the local signal decrease. For small values of B, entanglement is present, but not dynamically revealed. This can be understood through an analysis of the dephasing-induced excitations of the state ρ_Φ [8].

4.1.2 Distribution of Dephasing-Induced Excitations

The excitation spectrum of ρ_Φ is determined by the matrix elements of ρ_Φ in a basis of eigenstates $|\Psi_j\rangle$ of H_α:

$$c_j = \langle\Psi_j|\rho_\Phi|\Psi_j\rangle. \tag{48}$$

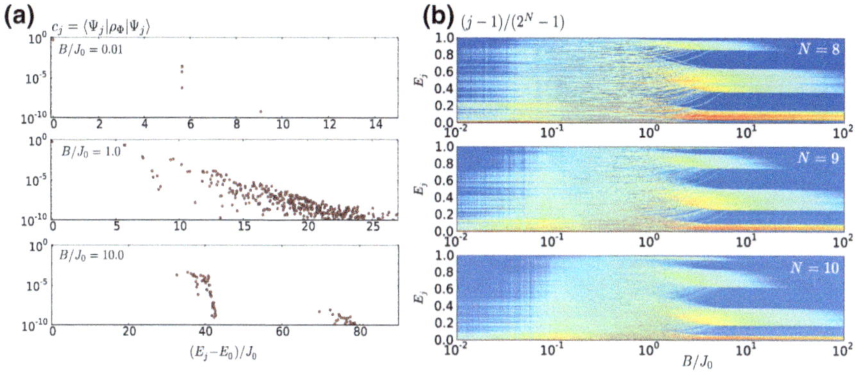

Fig. 8 Dephasing-induced excitations from the ground state for $\alpha = 1$ and $N = 10$ (*left panels*), as quantified by the overlap c_j with energy eigenstates, Eq. (48). For small values of B/J_0 only few states above the ground state are populated, whereas around $B \simeq J_0$ a broad excitation spectrum is observed. For large values of B/J_0 regular bands, characteristic of the paramagnetic spectrum, are observed. The *right panel* shows the (renormalized) index of the excited states along y, where the color code is logarithmically scaled and normalized to 100 steps between the respective minimum and maximum values of c_j, increasing from *blue* to *red*. We observe quick convergence of the images with increasing N towards an almost homogeneous excited-state distribution around the critical point of the quantum phase transition. Half of the paramagnetic bands are not populated for symmetry reasons (see text and Fig. 9). Figure adapted from [8]

Figure 8 shows the distribution of the dephasing-induced excitations c_j on a logarithmic scale for different values of B/J_0, $J_0 > 0$. For small values of B/J_0 hardly any significant excited-state populations are created due to the local dephasing. In contrast, the intermediate regime $B/J_0 \approx 1$ is characterized by a broadly distributed excitation spectrum. This shows, on the one hand, that the energy spectrum is widely spread, and on the other hand, that states of all energies are coupled to the ground state by means of the local dephasing. These features can be understood as the consequence of quantum chaotic structures [98] that emerge close to the critical point in this model [8]. For such dynamics, the local detection method is expected to be highly efficient since large parts of the state space are explored in the course of the dynamics, basically regardless of the initial condition [98], recall also Sect. 2.4. For this reason, a complex dynamical evolution has a higher chance of successfully mapping the initial two-body coherences, responsible for discord, into the locally accessible subsystem.

For large values of B/J_0 the energy depends linearly on B, with a slope proportional to the number of spins that orient along the y-direction, and thus the characteristic band structure of paramagnetic systems emerges around $E/J_0 = -NB, -(N-2)B, -(N-4)B, \ldots$. The finite width of these bands is due to nonzero values of J_0. However, this implies that besides those dephasing-induced populations at $(E - E_0)/J_0 = 40, 80$ which are observed in Fig. 8 there are other bands around $(E - E_0)/J_0 = 20, 60$ which are not reached by the local dephasing. The reason for this is again found in the Z_2-symmetry of the Hamiltonian: The ground

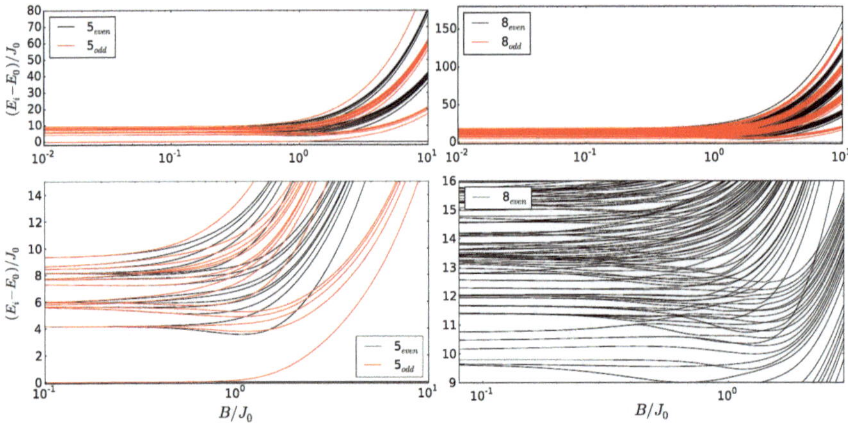

Fig. 9 Spectra of the spin chain Hamiltonian (42) for $\alpha = 1$, $J_0 > 0$ and $N = 5$ (*left*) and $N = 8$ (*right*), respectively. Spectral lines corresponding to the two parity subspaces are distinguished by color, as indicated in the legends. As is seen from the *top panels*, the paramagnetic energy bands comprise only states of a definite parity subspace, causing the local dephasing operation to leave half of them unpopulated. We further find broadly distributed spectra around $B \simeq J_0$, even if restricting to only one of the two subspaces. Figure taken from [8]

state, being member of the even-parity subspace, cannot be mapped onto states of the odd-parity subspace by local dephasing (43), since this operation commutes with the parity operator [8]. Hence, when applying local dephasing to the ground state, we remain in the parity subspace of the ground state.

Two ingredients are necessary for a successful mapping of the two-body coherences of $|\Psi_0\rangle$ to $\rho'_A(t_0)$: So far we have discussed the crucial aspect of populating a family of excited states such that $\rho'(t) = U(t)\rho_\Phi U^\dagger(t)$ experiences a suitable dynamics. However, this condition is not sufficient, since the partial trace operation may not disclose this dynamics to the observable subsystem. To see whether the state $\rho'(t)$ actually shows richer dynamics than $\rho'_A(t)$ in the case of $B \ll J_0$, we may consult the global time-autocorrelation function

$$C(t) = \frac{1}{\mathcal{P}(\rho_\Phi)} \mathrm{Tr}\{\rho_\Phi U(t)\rho_\Phi U^\dagger(t)\}, \tag{49}$$

which is normalized by the purity $\mathcal{P}(\rho_\Phi) = \mathrm{Tr}\rho_\Phi^2$ such that $C(0) = 1$. The time evolution of $\rho'(t)$ depends on the coherences of ρ_Φ in the energy eigenbasis, as seen from the expression

$$C(t) = \frac{1}{\mathcal{P}(\rho_\Phi)} \sum_{ij} |\langle\Psi_j|\rho_\Phi|\Psi_i\rangle|^2 e^{-i(E_i-E_j)t/\hbar}. \tag{50}$$

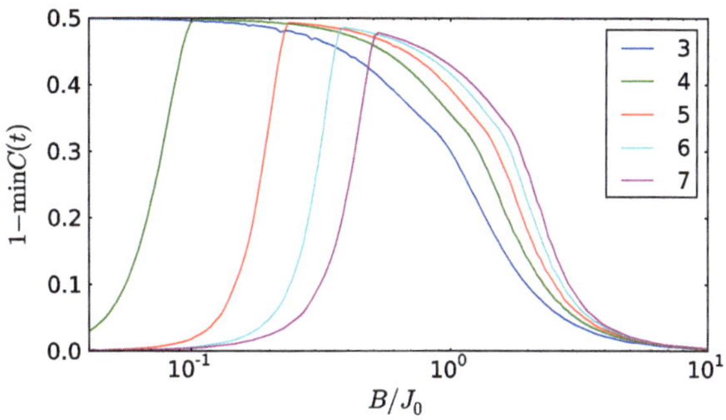

Fig. 10 Deviations of the dephased state ρ' from the original state ρ are quantified using the global time-autocorrelation function $C(t)$. Plotting the minimum value (over all t) as a function of B/J_0 confirms that no evolution takes place when B is very large or very small. The plots display different values of N (see legend) with $J_0 > 0$ and $\alpha = 1$. Hence, the local signal, as shown in Fig. 7 captures the qualitative behavior of the global dynamics, and little information is lost through the partial trace operation. Figure taken from [8]

However, as Fig. 10 shows, the fact that no dynamical witness for entanglement is obtained around $B \ll J_0$ is not due to the local observation of the quantum system. In this parameter regime, hardly any dynamical evolution of $\rho'(t)$ can be observed [8].

4.1.3 Thermal States: Local Bound for the Minimum Entanglement Potential

Below the critical point, the energy gap between the ground state and the first excited state decreases rapidly with increasing N. Rather than a preparation of the pure ground state, one would, in realistic conditions, therefore expect to find a thermal state,

$$\rho_\beta = \frac{e^{-\beta H}}{\mathrm{Tr} e^{-\beta H}}, \tag{51}$$

with inverse temperature $\beta = 1/kT$. When the energy gap, which also decreases with decreasing B for fixed N, becomes smaller than the thermal energy kT, then the two neighboring states are mixed incoherently. As a consequence, all quantum correlations which are present in $|\Psi_0\rangle$ are removed in ρ_β, which can be shown to have zero discord [8]. Thus, the minimal dephasing disturbance $D_{\min}$, as introduced in Eq. (13), approaches zero when B is decreased below a temperature-dependent value. Far away from any degeneracy it reduces, as expected, to the negativity of the energetically lower-lying state, see Fig. 11.

To obtain a local bound on $D_{\min}$, one uses the optimized local witness, described in Eq. (16). As seen in Fig. 11, this quantity is less sensitive to the mixing process

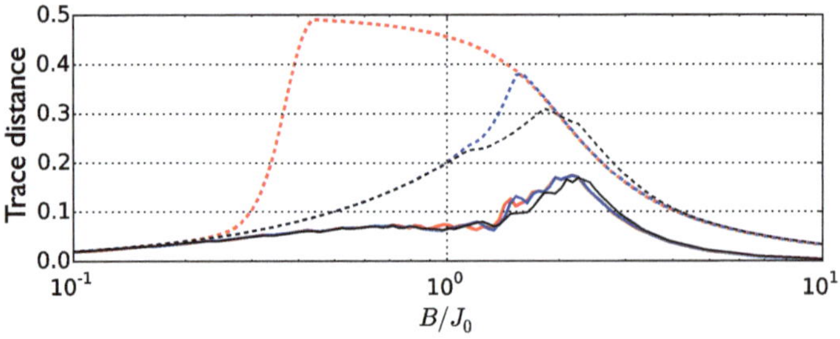

Fig. 11 The minimum dephasing disturbance $D_{\min}$ (*dashed lines*), Eq. (13), decreases dramatically when the thermal energy kT exceeds the energy gap between the two states of lowest energy is smaller than the thermal energy kT. The local signal $d_{\min}$ (*continuous lines*), Eq. (16), is more robust to increasing temperatures. Parameters are $N = 7$ with $kT = 10^{-5}$ (*thick*, *red lines*), $kT = 0.1$ (*medium*, *blue*), and $kT = 1$ (*black*, *thin*). Figure adapted from [68]; M. Gessner et al., "Observing a quantum phase transition by measuring a single spin", Europhysics Letters, vol. **107**, issue 4, 2014, available at http://iopscience.iop.org/article/10.1209/0295-5075/107/40005

than the total correlations, and a maximum signal can still be observed around the critical point.

4.2 Atom-Photon Correlations During Spontaneous Emission

Let us finally discuss an example of a dynamical system where the local detection method is unable to reveal initial entanglement in the subdynamics. In the spontaneous emission processes, atom and field start and end in factorized conditions, while the intermediate state contains atom-field entanglement. In a theoretical study reported in [8] the local detection method was applied to such an entangled intermediate state using the atomic two-level system as the accessible subsystem, while the spontaneous emission process into the electromagnetic field modes provides the interacting dynamics.

The situation differs conceptually from the photonic experiment reported in Sect. 3.2, since the interaction here exchanges energy instead of only imprinting a relative phase. Moreover, due to the irreversibility of the process, the energy which is transmitted from the atom to the photons is irretrievably lost. In the trapped ion experiment, reported in Sect. 3.1, energy was also exchanged between the two systems, but the ancilla system was described by a single mode instead of a continuum of modes, with the possibility of feedback from the environment to the controllable system.

The spontaneous emission process can be described by a unitary evolution $U(t) = e^{-iHt/\hbar}$, generated by the Hamiltonian $H = H_0 + V$ with [99, 100]

$$H_0 = E_e|e\rangle\langle e| + E_g|g\rangle\langle g| + \sum_{\mathbf{k}} \hbar\omega_{\mathbf{k}} a_{\mathbf{k}}^{\dagger} a_{\mathbf{k}} \tag{52}$$

and

$$V = \sum_{\mathbf{k}} (g_{\mathbf{k}} a_{\mathbf{k}}^{\dagger} |g\rangle\langle e| + g_{\mathbf{k}}^{*} a_{\mathbf{k}} |e\rangle\langle g|). \tag{53}$$

The modes of the electromagnetic field are labeled by $\mathbf{k}$ and $|\mathbf{k}\rangle$ denotes a one-photon state created by the bosonic operators $a_{\mathbf{k}}^{\dagger}$. The atom-field coupling strength is determined by the constants $g_{\mathbf{k}}$.

The atom is initially prepared in the excited state, while the field starts out in the vaccum. The initial state $|e, 0\rangle$ evolves as

$$|\Psi(t_0)\rangle = u_{00}(t_0)|e, 0\rangle + \sum_{\mathbf{k}} u_{\mathbf{k}0}(t_0)|g, \mathbf{k}\rangle, \tag{54}$$

with the matrix elements $u_{00}(t) = \langle e, 0|U(t)|e, 0\rangle$ and $u_{\mathbf{k}0}(t) = u_{0\mathbf{k}}^{*}(-t) = \langle g, \mathbf{k}|U(t)|e, 0\rangle$. The local eigenbasis is readily found to be $\{|e\rangle, |g\rangle\}$, and local dephasing transforms the above entangled state into the classically correlated reference state,

$$\rho'(t_0) = (\Phi \otimes \mathbb{I})|\Psi(t_0)\rangle\langle\Psi(t_0)|. \tag{55}$$

The difference between the two states,

$$\rho(t_0) - \rho'(t_0) = \sum_{\mathbf{k}} \left(u_{00}(t_0) u_{\mathbf{k}0}^{*}(t_0)|e, 0\rangle\langle g, \mathbf{k}| + u_{00}^{*}(t_0) u_{\mathbf{k}0}(t_0)|g, \mathbf{k}\rangle\langle e, 0| \right), \tag{56}$$

quantifies the atom-field negativity at time t_0. To determine the negativity explicitly, the matrix elements of the unitary time evolution are evaluated with the resolvent method [100] and the continuum limit is performed; for details see [8]. One finds [8]

$$\mathcal{N}(|\Psi(t_0)\rangle\langle\Psi(t_0)|) = c\sqrt{e^{-\Gamma t_0}(1 - e^{-\Gamma t_0})}, \tag{57}$$

where c is a constant, independent of t_0, and Γ is the spontaneous emission rate. This confirms the presence of atom-field entanglement in the intermediate states of the emission process.

The ensuing difference in the evolutions of the atomic system from t_0 to t_1 is described by

$$\begin{aligned} \rho_A(t_1, t_0) - \rho_A'(t_1, t_0) &= \mathrm{Tr}_B\{U(t_1 - t_0)(\rho(t_0) - \rho'(t_0))U^{\dagger}(t_1 - t_0)\} \\ &= 2\mathrm{Re}\left[\sum_{\mathbf{k}} u_{00}(t_0) u_{\mathbf{k}0}^{*}(t_0) u_{00}(t_1 - t_0) u_{0\mathbf{k}}^{*}(t_1 - t_0) \right] \sigma_z. \end{aligned} \tag{58}$$

Using the same techniques as before to evaluate this quantity, one finds it to be zero for all values of t_0 and t_1 [8]. This shows that the atomic evolution is insensitive to a replacement of all quantum correlations by classical correlations at any intermediate time t_0. Hence, the atom-field correlations that are created during the spontaneous emission process cannot be detected using the same dynamical evolution. This is expected to change when modifications of the uniform exponentially decaying evolution, e.g., through higher-order corrections [100] or a structured environment [57, 84, 101, 102] are introduced.

5 Conclusions

In conclusion, the quantum discord of an interacting bipartite system can be probed with manageable overhead by using the local detection method. To do this, control can be limited to only one of the two correlated subsystems while the second system might even be completely unknown and inaccessible. The protocol requires the realization of local state tomography of the accessible system and a local dephasing operation, which may be realized by a non-selective local measurement. Due to the destructive nature of the measurement process, the protocol requires multiple copies of the initial state, as is common practice in quantum mechanical experiments.

By limiting access to one of the two subsystems, only a small Hilbert space of much lower dimension than the full quantum system, needs to be controlled. This permits the detection of discord in high-dimensional and infinite-dimensional systems, where full tomographic methods and the measurement of witness operators can no longer be realized. The method is further applicable in an open-system scenario, where a controllable quantum system couples to an environment which is generally difficult to access [57].

The efficacy of the method depends strongly on the dynamical behavior of the interacting system. A non-vanishing local dynamical signature of the initial discord is expected to be found generically for systems with complex, e.g., chaotic dynamics, exploring large parts of the state space. The case studies summarized in this article also show that for regular quantum optical model systems, the local detection method can be implemented successfully. In the context of system-environment dynamics, the photonic experiment reported in [77] demonstrates that non-Markovian effects [84] are not needed to achieve this: The evolution of the controllable subsystem was described by irreversible pure dephasing but a strong signature of the initial discord was recorded.

Conversely, the theoretical case study of the spontaneous emission process showed that a local dynamical signature of the atom-field entanglement cannot be recorded [8]. In this extreme case the dynamics is no longer of a purely dephasing nature, but instead, excitations are decaying irretrievably from the controllable system into an environment. Thus, the examples discussed in this article suggest that, for dissipative dynamics, the local detection method relies on the presence of structure in the environment, such that excitations are indeed being exchanged both ways between the subsystems, as was the case in the trapped-ion experiment [62].

Further uses of the local detection method lie in the analysis of large interacting many-body systems by means of a small "quantum probe" [7, 8, 68, 87, 103, 104]. This was illustrated in the context of a quantum phase transition, where a strong dynamical signal of ground-state entanglement and thermal discord was observed in the vicinity of the critical point, with the measurements being limited to a single spin [68].

Acknowledgements MG thanks the German Academic Scholarship Foundation (Studienstiftung des deutschen Volkes) for support during the work on his PhD thesis. HPB & AB acknowledge support by the EU Collaborative project QuProCS (Grant Agreement 641277). AB thanks Dieter Jaksch and his group, as well as Keble College, for the hospitality he enjoyed during a research visit to Oxford.

References

1. D.J. Wineland, Rev. Mod. Phys. **85**, 1103 (2013)
2. S. Haroche, Rev. Mod. Phys. **85**, 1083 (2013)
3. I. Bloch, J. Dalibard, W. Zwerger, Rev. Mod. Phys. **80**, 885 (2008)
4. H. Häffner, C. Roos, R. Blatt, Phys. Rep. **469**, 155 (2008)
5. C. Schneider, D. Porras, T. Schaetz, Rep. Prog. Phys. **75**, 024401 (2012)
6. R. Blatt, C.F. Roos, Nat. Phys. **8**, 277 (2012)
7. P. Haikka, S. Maniscalco, Open Syst. Inf. Dyn. **21**, 1440005 (2014)
8. M. Gessner, *Dynamics and Characterization of Composite Quantum Systems*, (Springer International Publishing 2017)
9. E.-M. Laine, J. Piilo, H.-P. Breuer, Europhys. Lett. **92**, 60010 (2010)
10. M. Gessner, H.-P. Breuer, Phys. Rev. Lett. **107**, 180402 (2011)
11. M. Gessner, H.-P. Breuer, Phys. Rev. A **87**, 042107 (2013)
12. J. von Neumann, *Mathematical Foundations of Quantum Mechanics* (Princeton University Press, Princeton, 1955)
13. G. Lüders, Annalen der Physik **8**, 322 (1951)
14. H. Ollivier, W.H. Zurek, Phys. Rev. Lett. **88**, 017901 (2001)
15. L. Henderson, V. Vedral, J. Phys. A: Math. General **34**, 6899 (2001)
16. K. Modi, A. Brodutch, H. Cable, T. Paterek, V. Vedral, Rev. Mod. Phys. **84**, 1655 (2012)
17. R.F. Werner, Phys. Rev. A **40**, 4277 (1989)
18. B. Dakić, V. Vedral, C. Brukner, Phys. Rev. Lett. **105**, 190502 (2010)
19. F. Ciccarello, V. Giovannetti, Phys. Rev. A **85**, 010102 (2012)
20. A. Streltsov, H. Kampermann, D. Bruß, Phys. Rev. Lett. **107**, 170502 (2011)
21. M. Gessner, E.-M. Laine, H.-P. Breuer, J. Piilo, Phys. Rev. A **85**, 052122 (2012)
22. B.P. Lanyon, P. Jurcevic, C. Hempel, M. Gessner, V. Vedral, R. Blatt, C.F. Roos, Phys. Rev. Lett. **111**, 100504 (2013)
23. C.W. Helstrom, *Quantum Detection and Estimation Theory*. Mathematics in Science and Engineering, vol. 123 (Academic Press, New York, 1976)
24. D. Girolami, A.M. Souza, V. Giovannetti, T. Tufarelli, J.G. Filgueiras, R.S. Sarthour, D.O. Soares-Pinto, I.S. Oliveira, G. Adesso, Phys. Rev. Lett. **112**, 210401 (2014)
25. D. Girolami, T. Tufarelli, G. Adesso, Phys. Rev. Lett. **110**, 240402 (2013)
26. T.S. Cubitt, F. Verstraete, W. Dür, J.I. Cirac, Phys. Rev. Lett. **91**, 037902 (2003)
27. A. Streltsov, H. Kampermann, D. Bruß, Phys. Rev. Lett. **108**, 250501 (2012)

28. T.K. Chuan, J. Maillard, K. Modi, T. Paterek, M. Paternostro, M. Piani, Phys. Rev. Lett. **109**, 070501 (2012)
29. A. Fedrizzi, M. Zuppardo, G.G. Gillett, M.A. Broome, M.P. Almeida, M. Paternostro, A.G. White, T. Paterek, Phys. Rev. Lett. **111**, 230504 (2013)
30. C.E. Vollmer, D. Schulze, T. Eberle, V. Händchen, J. Fiurášek, R. Schnabel, Phys. Rev. Lett. **111**, 230505 (2013)
31. C. Peuntinger, V. Chille, L. Mišta, N. Korolkova, M. Förtsch, J. Korger, C. Marquardt, G. Leuchs, Phys. Rev. Lett. **111**, 230506 (2013)
32. A. Streltsov, H. Kampermann, D. Bruß, Phys. Rev. Lett. **106**, 160401 (2011)
33. M. Piani, S. Gharibian, G. Adesso, J. Calsamiglia, P. Horodecki, A. Winter, Phys. Rev. Lett. **106**, 220403 (2011)
34. G. Adesso, V. D'Ambrosio, E. Nagali, M. Piani, F. Sciarrino, Phys. Rev. Lett. **112**, 140501 (2014)
35. E.G. Carnio, A. Buchleitner, M. Gessner, Phys. Rev. Lett. **115**, 010404 (2015)
36. A. Orieux, M.A. Ciampini, P. Mataloni, D. Bruß, M. Rossi, C. Macchiavello, Phys. Rev. Lett. **115**, 160503 (2015)
37. E.G. Carnio, A. Buchleitner, M. Gessner, New J. Phys. **18** 073010, (2016)
38. C. Zhang, S. Yu, Q. Chen, C.H. Oh, Phys. Rev. A **84**, 032122 (2011)
39. D. Girolami, M. Paternostro, G. Adesso, J. Phys. A: Math. Theor. **44**, 352002 (2011)
40. S. Rahimi-Keshari, C.M. Caves, T.C. Ralph, Phys. Rev. A **87**, 012119 (2013)
41. F. Mintert, A.R. Carvalho, M. Kuś, A. Buchleitner, Phys. Rep. **415**, 207 (2005)
42. O. Gühne, G. Tóth, Phys. Rep. **474**, 1 (2009)
43. R. Horodecki, P. Horodecki, M. Horodecki, K. Horodecki, Rev. Mod. Phys. **81**, 865 (2009)
44. H. Häffner, W. Hänsel, C.F. Roos, J. Benhelm, D. Chek-al kar, M. Chwalla, T. Korber, U.D. Rapol, M. Riebe, P.O. Schmidt, C. Becher, O. Guhne, W. Dür, R. Blatt. Nature **438**, 643 (2005)
45. C. Schwemmer, L. Knips, D. Richart, H. Weinfurter, T. Moroder, M. Kleinmann, O. Gühne, Phys. Rev. Lett. **114**, 080403 (2015)
46. L.-M. Duan, G. Giedke, J.I. Cirac, P. Zoller, Phys. Rev. Lett. **84**, 2722 (2000)
47. A. Sørensen, L.M. Duan, J.I. Cirac, P. Zoller, Nature **409**, 63 (2001)
48. V. Giovannetti, S. Mancini, D. Vitali, P. Tombesi, Phys. Rev. A **67**, 022320 (2003)
49. L. Pezzé, A. Smerzi, Phys. Rev. Lett. **102**, 100401 (2009)
50. I.A. Silva, D. Girolami, R. Auccaise, R.S. Sarthour, I.S. Oliveira, T.J. Bonagamba, E.R. deAzevedo, D.O. Soares-Pinto, G. Adesso, Phys. Rev. Lett. **110**, 140501 (2013)
51. D. Wecker, B. Bauer, B.K. Clark, M.B. Hastings, M. Troyer, Phy. Rev. A **90**, 022305 (2014)
52. S. Hosseini, S. Rahimi-Keshari, J.Y. Haw, S.M. Assad, H.M. Chrzanowski, J. Janousek, T. Symul, T.C. Ralph, P.K. Lam, J. Phys. B: At. Mol. Opt. Phys. **47**, 025503 (2014)
53. M. Gessner, L. Pezzè, A. Smerzi, Phys. Rev. A **94**, 020101(R) (2016)
54. S.P. Walborn, P.H. Souto, Ribeiro, L. Davidovich, F. Mintert, A. Buchleitner, Nature **440**, 1022 (2006)
55. C. Schmid, N. Kiesel, W. Wieczorek, H. Weinfurter, F. Mintert, A. Buchleitner, Phys. Rev. Lett. **101**, 260505 (2008)
56. R. Islam, R. Ma, P.M. Preiss, M. Eric Tai, A. Lukin, M. Rispoli, M. Greiner, Nature **528**, 77 (2015)
57. H.-P. Breuer, F. Petruccione, *The Theory of Open Quantum Systems* (Oxford University Press, Oxford, 2002)
58. P. Pechukas, Phys. Rev. Lett. **73**, 1060 (1994)
59. R. Alicki, Phys. Rev. Lett. **75**, 3020 (1995)
60. P. Pechukas, Phys. Rev. Lett. **75**, 3021 (1995)
61. This assumption is not essential for the local detection method [11]
62. M. Gessner, M. Ramm, T. Pruttivarasin, A. Buchleitner, H.-P. Breuer, H. Häffner, Nat. Phys. **10**, 105 (2014)
63. A. Ferraro, L. Aolita, D. Cavalcanti, F.M. Cucchietti, A. Acín, Phys. Rev. A **81**, 052318 (2010)
64. S. Luo, Phys. Rev. A **77**, 022301 (2008)

65. M.A. Nielsen, I.L. Chuang, *Quantum Computation and Quantum Information* (Cambridge University Press, New York, 2000)
66. M.B. Ruskai, Rev. Math. Phys. **6**, 1147 (1994)
67. S. Campbell, L. Mazzola, M. Paternostro, Int. J. Quantum Inf. **9**, 1685 (2011)
68. M. Gessner, M. Ramm, H. Häffner, A. Buchleitner, H.-P. Breuer, Europhys. Lett. **107**, 40005 (2014)
69. T. Nakano, M. Piani, G. Adesso, Phys. Rev. A **88**, 012117 (2013)
70. G. Vidal, R.F. Werner, Phys. Rev. A **65**, 032314 (2002)
71. M. Piani, G. Adesso, Phys. Rev. A **85**, 040301 (2012)
72. M. Gessner, H.-P. Breuer, Phys. Rev. E **87**, 042128 (2013)
73. M. Gessner, *Initial Correlations in Open Quantum Systems* Diplomarbeit, Albert-Ludwigs-Universität Freiburg (2011)
74. M. Ozawa, Phys. Lett. A **268**, 158 (2000)
75. M. Piani, Phys. Rev. A **86**, 034101 (2012)
76. F.M. Paula, T.R. de Oliveira, M.S. Sarandy, Phys. Rev. A **87**, 064101 (2013)
77. J.-S. Tang, Y.-T. Wang, G. Chen, Y. Zou, C.-F. Li, G.-C. Guo, Y. Yu, M.-F. Li, G.-W. Zha, H.-Q. Ni, Z.-C. Niu, M. Gessner, H.-P. Breuer, Optica **2**, 1014 (2015)
78. S. Cialdi, A. Smirne, M.G.A. Paris, S. Olivares, B. Vacchini, Phys. Rev. A **90**, 050301 (2014)
79. D.J. Wineland, C. Monroe, W.M. Itano, D. Leibfried, B.E. King, D.M. Meekhof, J. Res. Natl. Inst. Stand. Technol. **103**, 259 (1998)
80. D. Leibfried, R. Blatt, C. Monroe, D. Wineland, Rev. Mod. Phys. **75**, 281 (2003)
81. D.J. Wineland, J.J. Bollinger, W.M. Itano, F.L. Moore, D.J. Heinzen, Phys. Rev. A **46**, R6797 (1992)
82. C.A. Blockley, D.F. Walls, H. Risken, Europhys. Lett. **17**, 509 (1992)
83. B.-H. Liu, L. Li, Y.-F. Huang, C.-F. Li, G.-C. Guo, E.-M. Laine, H.-P. Breuer, J. Piilo, Nat. Phys. **7**, 931 (2011)
84. H.-P. Breuer, E.-M. Laine, J. Piilo, B. Vacchini, Rev. Mod. Phys. **88**, 021002 (2016)
85. C.-F. Li, J.-S. Tang, Y.-L. Li, G.-C. Guo, Phys. Rev. A **83**, 064102 (2011)
86. A. Smirne, D. Brivio, S. Cialdi, B. Vacchini, M.G.A. Paris, Phys. Rev. A **84**, 032112 (2011)
87. A. Smirne, S. Cialdi, G. Anelli, M.G.A. Paris, B. Vacchini, Phys. Rev. A **88**, 012108 (2013)
88. A. Osterloh, L. Amico, G. Falci, R. Fazio, Nature **416**, 608 (2002)
89. T.J. Osborne, M.A. Nielsen, Phys. Rev. A **66**, 032110 (2002)
90. S. Sachdev, *Quantum Phase Transitions*, 1st edn. (Cambridge University Press, Cambridge, 1999)
91. F. Dyson, Commun. Math. Phys. **12**, 91 (1969)
92. D. Porras, J.I. Cirac, Phys. Rev. Lett. **92**, 207901 (2004)
93. A. Friedenauer, H. Schmitz, J.T. Glueckert, D. Porras, T. Schaetz, Nat. Phys. **4**, 757 (2008)
94. J.W. Britton, B.C. Sawyer, A.C. Keith, C.C.J. Wang, J.K. Freericks, H. Uys, M.J. Biercuk, J.J. Bollinger, Nature **484**, 489 (2012)
95. P. Jurcevic, B.P. Lanyon, P. Hauke, C. Hempel, P. Zoller, R. Blatt, C.F. Roos, Nature **511**, 202 (2014)
96. P. Richerme, Z.-X. Gong, A. Lee, C. Senko, J. Smith, M. Foss-Feig, S. Michalakis, A.V. Gorshkov, C. Monroe, Nature **511**, 198 (2014)
97. M. Gessner, V.M. Bastidas, T. Brandes, A. Buchleitner, Phys. Rev. B **93**, 155153 (2016)
98. F. Haake, *Quantum Signatures of Chaos* (Springer, Berlin, 2001)
99. V. Weisskopf, E.P. Wigner, Zeitschrift für Physik **63**, 54 (1930)
100. C. Cohen-Tannoudji, J. Dupont-Roc, G. Grynberg, *Atom-Photon Interactions* (Wiley, Weinheim, 1992)
101. R.R. Puri, G.S. Agarwal, Phys. Rev. A **33**, 3610 (1986)
102. P. Lambropoulos, G.M. Nikolopoulos, T.R. Nielsen, S. Bay, Rep. Prog. Phys. **63**, 455 (2000)
103. M. Gessner, F. Schlawin, H. Häffner, S. Mukamel, A. Buchleitner, New J. Phys. **16**, 092001 (2014)
104. M.A.C. Rossi, M.G.A. Paris, Phys. Rev. A **92**, 010302 (2015)

The Sudden Change Phenomenon of Quantum Discord

Lucas C. Céleri and Jonas Maziero

Abstract Even if the parameters determining a system's state are varied smoothly, the behavior of quantum correlations alike to quantum discord, and of its classical counterparts, can be very peculiar, with the appearance of non-analyticities in its rate of change. Here we review this sudden change phenomenon (SCP) discussing some important points related to it: Its uncovering, interpretations, and experimental verifications, its use in the context of the emergence of the pointer basis in a quantum measurement process, its appearance and universality under Markovian and non-Markovian dynamics, its theoretical and experimental investigation in some other physical scenarios, and the related phenomenon of double sudden change of trace distance discord. Several open questions are identified, and we envisage that in answering them we will gain significant further insight about the relation between the SCP and the symmetry-geometric aspects of the quantum state space.

Keywords Quantum discord · Open systems · Decoherence · Sudden change

1 Introduction

In principle, it is always possible to find a local observable whose measurement does not disturb a system in a classical-incoherent state. However, the presence of quantum coherence in composite-correlated systems makes local interrogation without

L.C. Céleri (✉)
Instituto de Física, Universidade Federal de Goiás, Goiânia, Goiás 74001-970, Brazil
e-mail: lucas@chibebe.org

J. Maziero
Departamento de Física, Centro de Ciências Naturais e Exatas, Universidade Federal de Santa Maria, Avenida Roraima 1000, Santa Maria, RS 97105-900, Brazil
e-mail: jonas.maziero@ufsm.br

J. Maziero
Instituto de Física, Facultad de Ingeniería, Universidad de la República, J. Herrera y Reissig 565, 11300 Montevideo, Uruguay

F.F. Fanchini et al. (eds.), *Lectures on General Quantum Correlations and their Applications*, Quantum Science and Technology,
DOI 10.1007/978-3-319-53412-1_15

disturbance inconceivable [1]. Quantum discord (QD) is the generic name we give to correlations that have a quantum character. Actually, the minimum "distance" between the states of a multipartite system after and before a local measurement is quantified by discord-like functions [2]. This kind of correlation is more general than quantum entanglement and has been shown to be important not only for the fundamentals of physics [3–8] but also as a resource for quantum information science (QIS) [9–11] applications. A few examples of QIS "protocols" for which there are strong evidences that QD is key for fueling the quantum advantage are noisy computation [12], state merging [13, 14], assisted state discrimination [15], cryptography [16], energy transport [17], quantum illumination [18], remote state preparation [19], data hidding [20], and metrology [21, 22].

Although many useful theoretical developments are usually made in QIS by considering first idealized isolated systems, subsequently their realistic open dynamics must be taken into account [9–11]. Quantum features such as entanglement and discord are particularly fragile to the influence of the environment. This kind of interaction usually smears the quantumness of the system of interest, inducing it to behave in a more classical manner. Such type of process is detrimental for applications of these quantities in QIS and generally goes under the name of decoherence. Motivated by that observation, much effort have been made in order to describe how useful quantum features are affected by several kinds of decoherent dynamics. In particular, the dynamic behavior of quantum discord has been shown to be very peculiar [23], with sudden discontinuities in its derivative being identified and analyzed in diverse physical scenarios.

This sudden change phenomenon (SCP) of classical and quantum correlations has been the theme of numerous works in the last few years, with considerable developments being put forward. Its geometric description and interpretation were addressed [24–29]. It has been regarded in the context of classical and quantum phase transitions [30–35] and in the dynamics of critical systems [36–38]. Considering Markovian [39–45] and non-Markovian [46] decoherent dynamics, the establishment of system-environment correlations was investigated in Refs. [47–51], while the protection of quantum correlations was addressed in Refs. [52–56]. Moreover, investigations of systems in relativistic motion [57–59] or subjects to classical, chaotic, thermal, and other kinds of noise environments were also reported [60–71]. Studies considering interesting physical systems like quantum dots [72, 73], atoms interacting with separated or common cavities [74–76] or interacting via plasmonic waveguides [77, 78] appeared in the literature. Finally, some interesting physical interpretations, as e.g. related to the emergence of the pointer basis in a quantum measurement process [79], or the worst-case-scenario fidelity of quantum teleportation [80] and the quantum speed limits [81] shed new light on the subject. Here we shall make a survey of some important points related to these developments intending to open up the path for reaching a better physical and information-theoretic understanding of the SCP of classical and quantum correlations, bringing out thus its possible role in physics and in QIS.

2 Quantum Mechanics, Open System Dynamics and Quantum Correlations

In this section, we present some important concepts of quantum mechanics and the quantification of correlations. We start, in Sect. 2.1, by recalling the postulates of quantum mechanics, with particular emphasis on the description of open quantum dynamics via the Kraus' operator-sum representation. In Sect. 2.2 we present the primary concepts involved in quantum, classical, and total correlation quantification and describe some of the correlation measures considered in this article. Our exposition shall be limited to discrete systems divided in two parts, but the main ideas can be straightforwardly translated to multipartite systems.

2.1 Quantum Mechanics of Closed and Open Systems

In standard quantum mechanics [9], the states of closed systems are described by normalized vectors on a projective Hilbert space $\mathcal{H}$. In Schrödinger's picture, these vectors evolve unitarily (and therefore reversibly): $|\psi_t\rangle = U|\psi\rangle$, with U being a linear operator such that $UU^\dagger = U^\dagger U = \mathbb{I}$ and $\mathbb{I}|\psi\rangle = |\psi\rangle$ for all $|\psi\rangle \in \mathcal{H}$. The time evolution operator U satisfies Schrödinger's equation: $i\hbar\partial_t U = HU$, where H is the system Hamiltonian. Observables are described by Hermitian operators, $O = O^\dagger$. For a system prepared in state $|\psi\rangle$, measurements of O at a time t yield one of its real eigenvalues o_i with probability given by Born's rule: $p_i = |\langle o_i|\psi_t\rangle|^2$ ($O = \sum_i o_i|o_i\rangle\langle o_i|$ is the spectral decomposition of O). Immediate subsequent measurements of O always produce the same result. Because of this repeatability of measurements, one says that the system's state "jumps" to the observable's corresponding eigenvector $|o_i\rangle$, which will be the system's state immediately after the measurement. Thus, in quantum mechanics the measurement process will irreversibly change the state of the system unless it is prepared in one of the eigenstates of the observable being measured.

All systems in Nature generally interact with their surrounds, thus breaking local unitarity. Considering that quantum states lies at the core of quantum information science, this fact has been particularly important in the development of this theory, whose main focus is the manipulation of the information stored in the quantum degrees of freedom [9]. In this open system scenario, we may use the standard quantum description for the whole universe, but more general mathematical tools are needed when we focus on the description of some particular subsystem. In this case, because of the correlations generated between this system and the rest of the universe (the environment), the state of the first must be described by a convex mixture of state vectors, a density operator ρ, which is a positive-semidefinite, $\rho \geq 0$, linear operator with unit trace, $\mathrm{Tr}(\rho) = 1$.

In what follows we briefly describe the dynamics of open quantum systems in terms of the so called operator sum representation [9]. Let $\mathcal{D}(\mathcal{H})$ be the space of density operators, $\mathcal{H}_S \otimes \mathcal{H}_E$ be the whole system Hilbert space, and $\rho = \rho^S \otimes |E_0\rangle\langle E_0|$ be the initial system-environment state. $\rho^S \in \mathcal{D}(\mathcal{H}_S)$ is the system initial state and $\{|E_i\rangle\}$ is an orthonormal basis for $\mathcal{H}_E$. We can take the environment as being described by a pure state since we can always treat its purification without changing physical conclusions.

Then, assuming that the entire system evolves by means of a unitary operation, $\rho_t = U\rho U^\dagger$, and considering the partial trace over the environment degrees of freedom we obtain

$$\rho_t^S = \mathrm{Tr}_E(\rho_t) = \sum_l K_l \rho^S K_l^\dagger, \tag{1}$$

which is the operator sum representation of the evolved system state [82–84]. The set $\{K_i\}$ are the Kraus operators, whose matrix elements in the basis $|S_i\rangle \in \mathcal{H}_S$ are defined as

$$\langle S_m|K_l|S_n\rangle \equiv \langle S_m E_l|U|S_n E_0\rangle. \tag{2}$$

In the last equation and hereafter we shall use the notation $|\phi\psi\rangle := |\phi\rangle \otimes |\psi\rangle$. One can easily verify that $\sum_l K_l^\dagger K_l = \mathbb{I}_S$, what implies that the evolution (1) is trace-preserving, i.e. $\mathrm{Tr}(\rho_t^S) = 1$ for all times. Equation (1) is the most general quantum evolution, known as a completely-positive and trace-preserving map.

2.2 Quantum Discord Quantifiers

Let's consider a bipartite system with Hilbert space $\mathcal{H}_a \otimes \mathcal{H}_b$. When can the correlations between the parties a and b be regarded as being classical? One way to answer this question is by recalling that coherent superpositions are a fundamental character of quantum systems [85]; and that the measurement of an observable, for a system prepared in a state other than its eigenstates, will involve wave function collapse and disturbance. So, it is reasonable to say that if the state of a bipartite system is invariant, not disturbed, under the composition of (non-selective) local projective measurement maps (LPMMs), i.e., if $\exists\, \Pi_a, \Pi_b \mid \Pi_a \circ \Pi_b(\rho) = \rho$, then the correlations between its constituent parts are of a classical nature. In the above equations, a LPMM applied to sub-system a is defined as

$$\Pi_a(\rho) := \sum_j \Pi_j^a \otimes \mathbb{I}_b \rho \Pi_j^a \otimes \mathbb{I}_b, \tag{3}$$

with $\sum_{j=1}^{\dim \mathcal{H}_a} \Pi_j^a = \mathbb{I}_a$ and $\Pi_j^a \Pi_k^a = \delta_{jk}\Pi_j^a$; and the analogous definition follows for $\Pi_b(\rho) := \sum_j \mathbb{I}_a \otimes \Pi_j^b \rho \mathbb{I}_a \otimes \Pi_j^b$.

One can easily verify that, for the so dubbed classical-classical states,

$$\rho_{cc} = \sum_{j,k} p_{jk} \Pi_j^a \otimes \Pi_k^b, \tag{4}$$

where p_{jk} is an arbitrary probability distribution ($p_{jk} \geq 0$ and $\sum_{j,k} p_{jk} = 1$), we have $\Pi_a \circ \Pi_b(\rho_{cc}) = \rho_{cc}$. For a system prepared in such a state, there is no quantum uncertainty associated with measurements of the local observables $A = \sum_j a_j \Pi_j^a$ and $B = \sum_k b_k \Pi_k^b$. Following these lines, the classical-quantum states (invariance under a LPMM on a: $\Pi_a(\rho_{cq}) = \rho_{cq}$),

$$\rho_{cq} = \sum_j p_j \Pi_j^a \otimes \rho_j^b, \tag{5}$$

and the quantum-classical states (not disturbed by a LPMM on b: $\Pi_b(\rho_{qc}) = \rho_{qc}$),

$$\rho_{qc} = \sum_j p_j \rho_j^a \otimes \Pi_j^b, \tag{6}$$

are also important for the theory and applications of quantum discord quantifiers. In the last two equations, p_j is a probability distribution and $\rho_j^s \in \mathcal{D}(\mathcal{H}_s)$ are density operators for $s = a, b$. We observe that in addition to the nondisturbability-based characterization above, the classicality of the correlations between the subsystems a and b, prepared in either one of these three classes of states, can be given an operational interpretation in terms of the possibility of locally broadcast them [86, 87].

Now that we have defined what may be considered as being the classical states, i.e., those states not possessing quantum correlations, in order to quantify the amount of quantumness in the correlations of a generic bipartite density operator ρ, we can use the distance or distinguishability between ρ and the regarded classical states. But using any classical state would be ambiguous; thus we utilize the classical states which "seems more" like ρ. That is to say, we define

$$D_d(\rho) := \min_{\rho_{cc}} d(\rho, \rho_{cc}) \equiv \min_{p_{jk}, \Pi_j^a, \Pi_j^b} d(\rho, \sum_{j,k} p_{jk} \Pi_j^a \otimes \Pi_k^b), \tag{7}$$

$$D_d^a(\rho) := \min_{\rho_{cq}} d(\rho, \rho_{cq}) \equiv \min_{p_j, \Pi_j^a, \rho_j^b} d(\rho, \sum_j p_j \Pi_j^a \otimes \rho_j^b), \tag{8}$$

$$D_d^b(\rho) := \min_{\rho_{qc}} d(\rho, \rho_{qc}) \equiv \min_{p_j, \rho_j^a, \Pi_j^b} d(\rho, \sum_j p_j \rho_j^a \otimes \Pi_j^b), \tag{9}$$

where d is a distance or distinguishability measure for density operators.

It is worthwhile noticing that there is a related, but somewhat more operational way to define discord quantifiers. In this approach one starts by recognizing that a LPMM transforms ρ into a classical state. Thus, QD is defined as the distance between a state ρ and the closest classical state obtained by applying a LPMM to it, i.e.,

$$\mathcal{D}_d(\rho) = \min_{\Pi_a, \Pi_b} d(\rho, \Pi_a \circ \Pi_b(\rho)), \tag{10}$$

$$\mathcal{D}_d^a(\rho) = \min_{\Pi_a} d(\rho, \Pi_a(\rho)), \tag{11}$$

$$\mathcal{D}_d^b(\rho) = \min_{\Pi_b} d(\rho, \Pi_b(\rho)). \tag{12}$$

These quantifiers capture the essence of quantum discord, which is equal to the minimum amount of correlation that is inevitably erased by a LPMM. It is noticeable that there is no extremization over probability distributions or over local states in these last three expressions. This is so because, in this case, they are determined by ρ and by the LPMM. It is interesting observing that we would have $\mathcal{D}_d \equiv D_d$ if we set $p_{jk} = \mathrm{Tr}(\Pi_j^a \otimes \Pi_k^b \rho)$ in the equation for D_d. Besides, $\Pi_a(\rho) \equiv \rho_{cq}$ if we make $p_j = \mathrm{Tr}(\Pi_j^a \otimes \mathbb{I}_b \rho)$ and $\rho_j^b = \mathrm{Tr}_a(p_j^{-1} \Pi_j^a \otimes \mathbb{I}_b \rho \Pi_j^a \otimes \mathbb{I}_b)$ in Eq. (8) and $\Pi_b(\rho) \equiv \rho_{qc}$ if we use $p_j = \mathrm{Tr}(\mathbb{I}_a \otimes \Pi_j^b \rho)$ and $\rho_j^a = \mathrm{Tr}_b(p_j^{-1} \mathbb{I}_a \otimes \Pi_j^b \rho \mathbb{I}_a \otimes \Pi_j^b)$ in Eq. (9). But now let us address an important issue about these two types of QD quantifiers, one of then that, to the best of our knowledge, has not been addressed in literature. It is true, for instance, that any state in the class ρ_{cc} can in principle be produced. Nevertheless, it is also well known that the probability distribution induced by local measurements (PDILM) on a system prepared in a certain state ρ cannot, in general, simulate all PDILM. One famous example of this fact is found in the Bell nonlocality scenario [88]. Therefore, in addition to make the optimization problem more involved, the measures in Eqs. (7)–(9) may lead to misleading results due to the application of classical states possible "unrelated" to ρ, i.e., that cannot be obtained from ρ via LPMMs.

After a LPMM is applied, a and b can still be correlated if, e.g., $p_{jk} \neq p_j^a p_k^b$, with p_j^a and p_k^b being probability distributions. This remaining correlation may be said to have a classical nature. But then, before the LPMM, the system posses two kinds of correlation. And it would be nothing but natural assuming that both types of correlation, the classical correlation (CC) and the quantum discord, add up to give the total correlation (TC) in ρ. Surprisingly, the panorama of the theory for measures of CCs and of TCs is even less satisfactorily developed than for QD. Next we regard some possible approaches that may be followed for the quantification of TC and of CC in bipartite states. A good starting point for that is the definition of uncorrelated states. If the sub-systems a and b are prepared independently, respectively, in states ϖ_a and ϖ_b, then their joint density operator would be the product state $\varpi_a \otimes \varpi_b$. Similarly to what was done for QD, these states can be used to define a measure of TC

$$I_d(\rho) := \min_{\varpi_a, \varpi_b} d(\rho, \varpi_a \otimes \varpi_b). \tag{13}$$

We use I_d instead of T_d because of the historic and present importance mutual information (MI) has for TC quantification. Actually, the MI, which is defined as

$$I(\rho) := S(\rho_a) + S(\rho_b) - S(\rho) \tag{14}$$

with

$$S(x) := -\mathrm{Tr}(x \log_2 x) \tag{15}$$

being the von Neumann's entropy, is the only TC measure already given an operational meaning [89].

It is intuitively expected that the closest product state of ρ is obtained from the tensor product of its reductions. And for the distinguishability measure named quantum relative entropy [90],

$$d_{re}(\rho, \xi) := \mathrm{Tr}(\rho(\log_2 \rho - \log_2 \xi)), \tag{16}$$

this is indeed the case [91, 92], i.e.,

$$I_{re}(\rho) := \min_{\varpi_a, \varpi_b} d_{re}(\rho, \varpi_a \otimes \varpi_b) = d_{re}(\rho, \rho_a \otimes \rho_b) \equiv I(\rho). \tag{17}$$

So, this TC is equal to mutual information. It is rather curious that e.g. for the trace distance,

$$d_{tr}(x, y) := ||x - y||_{tr}, \tag{18}$$

where

$$||x||_{tr} := \mathrm{Tr}\sqrt{x^{\dagger}x} \tag{19}$$

is the trace norm of x, the state $\rho_a \otimes \rho_b$ is not in general the uncorrelated state most similar to ρ [93]. This issue certainly deserves additional-thoroughly investigation. This fact also motivates using a more operational definition also for TC. One possibility would be defining

$$\mathcal{I}_d(\rho) := \min_{\Omega} d(\rho, \Omega(\rho)), \tag{20}$$

where

$$\Omega(\rho) := \textstyle\sum_j p_j(U_j^a \otimes U_j^b)\rho(U_j^a \otimes U_j^b)^{\dagger} \tag{21}$$

is a randomizing (decoupling) map leading any ρ into product states and U_j^s are appropriately chosen unitary operators acting on $\mathcal{H}_s$. Although being more difficult to calculate, the total correlation in Eq. (20) may be more suitable because in general (some property of) ρ may restrict the possible product states that can be produced by Ω acting on it.

What about classical correlation quantifiers? One possibility is arguing for additivity for correlations and simply define

$$C_d(\rho) := I_d(\rho) - D_d(\rho). \tag{22}$$

Or one can use the classical state minimizing the equation for e.g. D_d, and define the CC of ρ as the TC of this state. We could continue presenting several other ways in which CC may be defined. Actually, this observation is in its own an indication that these definitions are not, in general, operationally satisfying. The only CC quantifier with a well defined information theoretic interpretation that we know of was inspired by the Holevo bound and was proposed by Henderson and Vedral in Ref. [94]. But before presenting their CC measure, let us consider the following equivalent distinguishability-based quantifier

$$C^a_{hv}(\rho) := \max_{\Pi_a} I_{re}(\Pi_a(\rho)) \equiv \max_{\Pi_a} I(\Pi_a(\rho)) = \max_{\Pi^a_j}[S(\sum_j p_j \Pi^a_j) + S(\sum_j p_j \rho^b_j) - S(\sum_j p_j \Pi^a_j \otimes \rho^b_j)]. \quad (23)$$

As the states $\Pi^a_j \otimes \rho^b_j$ have support in orthogonal subspaces of $\mathcal{H}_a \otimes \mathcal{H}_b$, one can verify that

$$S(\sum_j p_j \Pi^a_j \otimes \rho^b_j) = H(p_j) + \sum_j p_j S(\Pi^a_j \otimes \rho^b_j) = S(\sum_j p_j \Pi^a_j) + \sum_j p_j S(\rho^b_j), \quad (24)$$

where $H(p_j) := -\sum_j p_j \log_2 p_j$ is Shannon's entropy. With this we get

$$C^a_{hv}(\rho) = \max_{\Pi^a_j}[S(\sum_j p_j \rho^b_j) - \sum_j p_j S(\rho^b_j)], \quad (25)$$

which is the Holevo quantity maximized over LPMMs on particle a, which are used here to acquire information about the subsystem b. Now, we recall that any *local quantum operation* can be written as

$$\$_a(\rho) = \sum_j K^a_j \otimes \mathbb{I}_b \rho K^{a\dagger}_j \otimes \mathbb{I}_b, \quad (26)$$

with $\sum_j K^{a\dagger}_j K^a_j = \mathbb{I}_a$. As expected, $\$_a$ does not change the state of b:

$$\begin{aligned}\mathrm{Tr}_a(\$_a(\rho)) &= \sum_l \langle a_l| \otimes \mathbb{I}_b \sum_j K^a_j \otimes \mathbb{I}_b \rho K^{a\dagger}_j \otimes \mathbb{I}_b |a_l\rangle \otimes \mathbb{I}_b \\ &= \sum_{l,j} \langle a_l| K^a_j (\sum_m |a_m\rangle\langle a_m|) \otimes \mathbb{I}_b \rho K^{a\dagger}_j |a_l\rangle \otimes \mathbb{I}_b \\ &= \sum_{l,j,m} \langle a_m| \otimes \mathbb{I}_b \rho K^{a\dagger}_j |a_l\rangle\langle a_l| K^a_j |a_m\rangle \otimes \mathbb{I}_b \\ &= \sum_m \langle a_m| \otimes \mathbb{I}_b \rho |a_m\rangle \otimes \mathbb{I}_b = \mathrm{Tr}_a(\rho) = \rho_b. \end{aligned} \quad (27)$$

As Π_a is a particular kind of quantum operation, then $\rho_b = \mathrm{Tr}_a(\rho) = \mathrm{Tr}_a(\Pi_a(\rho)) = \sum_j p_j \rho^b_j$ and [94]

$$C^a_{hv}(\rho) = S(\rho_b) - \min_{\Pi^a_j} \sum_j p_j S(\rho^b_j). \quad (28)$$

This expression leads to a nice entropic interpretation for this CC. The state of b is the mixture $\sum_j p_j \rho^b_j$ before and after Π_a is applied to a. However, because of the correlations between a and b, the information we get when a collapses to Π^a_j forces b to be in one of the states ρ^b_j. And, on average, this decreases our uncertainty about the state of b.

The notion of quantum discord as a quantum correlation quantifier was first introduced by Ollivier and Zurek in Ref. [95] and is directly related to the Henderson–Vedral CC:

$$D^a_{oz}(\rho) := I(\rho) - C^a_{hv}(\rho). \quad (29)$$

The motivation for the name "quantum discord" comes from the fact that classically the two definitions for correlation, measured by mutual information, are equivalent, but in the quantum realm we can have $I(\rho) \neq C^a_{hv}(\rho)$.

As we have seen in this sub-section, there are several motivations one can follow to define a quantum discord quantifier. Of course, the same holds for classical and total correlations. And even within a certain "kind" of QD quantifier, we can employ, for instance, several dissimilarity measures, which will imply in multiple QD functions. Actually, many functions involving ρ and e.g. $\Pi_a(\rho)$ can be used to defined QD quantifiers. For some examples of such quantities, see Refs. [1, 96–112] and references therein. This scenario has led to a variety of QDQs and has motivated the discussion about which properties they should enjoy [113, 114]. But, as this is not the focus of this work, any other QD quantifier, and the related issues, will be introduced as needed in our subsequent analysis of the sudden change phenomenon.

3 Uncovering of the SCP and Its Experimental Verification

The sudden change phenomenon was discovered in 2009 by us, R.M. Serra and V. Vedral, and was first reported in Ref. [23]. We considered the simple situation with two qubits prepared initially in a Bell-diagonal state and let them to interact with local-independent environments. For the sake of definiteness, we will consider the so called Pauli channels [10, 11]. These are the phase-damping channel, whose Kraus operators are

$$K_0^{pd} = \sqrt{1-p}\,\mathbb{I},\; K_1^{pd} = \sqrt{p}\,|0\rangle\langle 0| \text{ and } K_2^{pd} = \sqrt{p}\,|1\rangle\langle 1|, \tag{30}$$

the bit flip, described by the operators

$$K_0^{bf} = \sqrt{1-p}\,\mathbb{I} \text{ and } K_1^{bf} = \sqrt{p}\,\sigma_1, \tag{31}$$

phase flip,

$$K_0^{pf} = \sqrt{1-p}\,\mathbb{I} \text{ and } K_1^{pf} = \sqrt{p}\,\sigma_3 \tag{32}$$

and, finally, the bit-phase flip

$$K_0^{bpf} = \sqrt{1-p}\,\mathbb{I} \text{ and } K_1^{bpf} = \sqrt{p}\,\sigma_2. \tag{33}$$

p is the parametrized time, usually written as $p = 1 - e^{-\gamma t}$, with γ being the relaxation rate associated with the environment. These quantum channels have the property that they preserve the Bell-diagonal form of the evolved state, i.e.,

$$\rho_p^{bd} = \textstyle\sum_{i,j} K_i^{qc} \otimes K_j^{qc} \rho_0^{ab} K_i^{qc^\dagger} \otimes K_j^{qc^\dagger} = 2^{-2}\left(\mathbb{I}^{ab} + \Xi_p^{qc} \cdot \Upsilon\right). \tag{34}$$

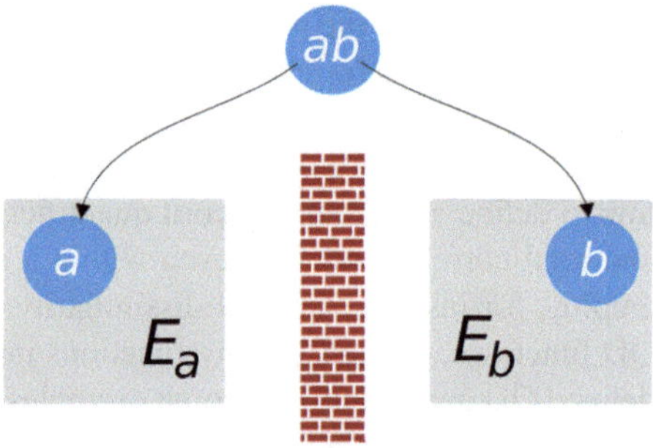

Fig. 1 Illustration of the situation in which two systems are initially prepared in a quantum-correlated state and afterwards let to interact with local-independent environments

In the last equation $\Upsilon = (\Upsilon_1, \Upsilon_2, \Upsilon_3) := (\sigma_1 \otimes \sigma_1, \sigma_2 \otimes \sigma_2, \sigma_3 \otimes \sigma_3)$ and the evolved correlation vectors $\Xi_p^{qc} = (c_1^{(p)}, c_2^{(p)}, c_3^{(p)})$, with $c_j^{(p)} = \mathrm{Tr}(\rho_p^{ab}\Upsilon_j)$, are given by (Fig. 1)

$$\Xi_p^{pd} = \left(c_1(1-p)^2, c_2(1-p)^2, c_3\right), \tag{35}$$

$$\Xi_p^{bf} = \left(c_1, c_2(1-2p)^2, c_3(1-2p)^2\right), \tag{36}$$

$$\Xi_p^{pf} = \left(c_1(1-2p)^2, c_2(1-2p)^2, c_3\right), \tag{37}$$

$$\Xi_p^{bpf} = \left(c_1(1-2p)^2, c_2, c_3(1-2p)^2\right). \tag{38}$$

For the Bell-diagonal class of states, the Ollivier–Zurek quantum discord (symmetric or asymmetric) can be computed analytically and reads [115–117]:

$$D_{oz}(\rho_p^{bd}) = I(\rho_p^{bd}) - C_{hv}(\rho_p^{bd}) = \sum\nolimits_{i,j=0}^{1} \lambda_{ij} \log_2(4\lambda_{ij}) \\ - 2^{-1}\sum\nolimits_{i=0}^{1} \left(1 + (-1)^i c_p\right) \log_2 \left(1 + (-1)^i c_p\right), \tag{39}$$

where $\lambda_{ij} = 2^{-2}\left(1 + (-1)^i c_1^{(p)} - (-1)^{i+j} c_2^{(p)} + (-1)^j c_3^{(p)}\right)$ and

$$c_p = \max(|c_1^{(p)}|, |c_2^{(p)}|, |c_3^{(p)}|). \tag{40}$$

For the sake of understanding the *mathematical origin* of the SCP from these equations, we first call the attention for the fact that the components of Ξ_p^{qc} change differently with time, with their decaying rates depending on the kind of environment the system is interacting with. So, fixed an initial state Ξ_0^{qc} and the quantum channel, with exception of those Ξ_0^{qc} with the constant component null or greater than all the others, at the time dubbed sudden change time, p_{sc}, the index of the coefficient with greater modulus shall become different. And this can happens only if the change rate of $c_{p<p_{sc}}$ is different from that of $c_{p>p_{sc}}$ (see the example in Fig. 2). Well, and in this scenario it is this change of decay rate of c_p that leads to the sudden change

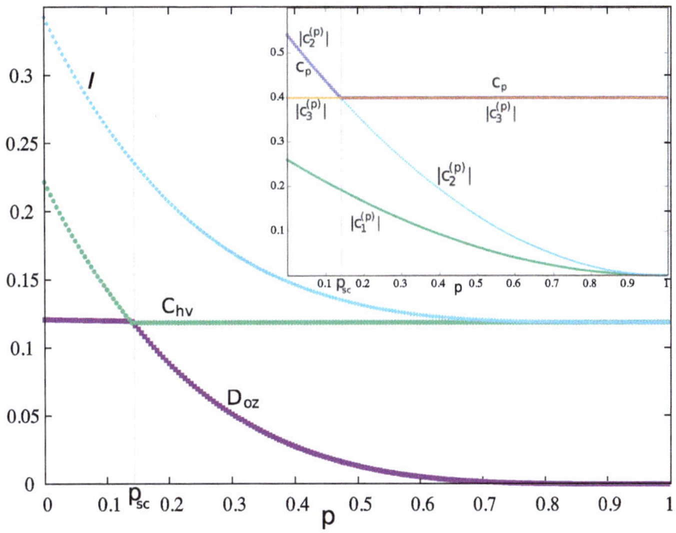

Fig. 2 Example of sudden change phenomenon taking place when an initial Bell-diagonal state with correlation vector $\Xi_0 = (-0.26, 0.54, 0.40)$ undergoes time evolution under local phase damping channels. Here, the sudden change time is $p_{sc} \approx 0.14$. The *inset* shows the time evolution of the components of the correlation vector and c_p. We see that $|c_3^{(p)}| \lessgtr |c_2^{(p)}|$ for $p \lessgtr p_{sc}$. So, as the rate of change of these two components is different, so is the decaying rate of the classical, C_{hv}, and quantum, D_{oz}, correlation in the two quite distinct dynamical behavior regimes. For the environments and correlation measures regarded in this section, the classical correlation is constant from p_{sc} thereon and there is only one possible value for p_{sc}, what shall be relevant in the discussion about the emergence of the pointer basis included in Sect. 4. Because the mutual information $I = C_{hv} + D_{oz}$ is a monotonically decaying function of p, for $p > p_{sc}$ the decaying rate of the quantum discord must be equal to that of the MI: $\partial_p D_{oz} = \partial_p I$. This fact indicates that if $\partial_p C_{hv} = \partial_p I$ for $p < p_{sc}$, then we shall have what has been called the freezing of quantum discord or the classical decoherence regime

phenomenon of the classical correlation, C_{hv}, and of quantum discord, D_{oz}. Note that once the quantum mutual information, I, does not depend on c_p, it decays monotonically with time, without abrupt changes.

For all channels mentioned in this section, the SCP happens when $|\kappa| = \max(|\kappa'|, |\kappa''|)$, with κ being the constant component of the correlation vector and κ' and κ'' being the components of Ξ_0^{qc} that shall be affected by the interaction with the environment. Thus, in the case of the phase damping channel, the *sudden change time* is

$$p_{sc} = 1 - \sqrt{\frac{|\kappa|}{\max(|\kappa'|, |\kappa''|)}} = 1 - \sqrt{\frac{|c_3|}{\max(|c_1|, |c_2|)}}, \tag{41}$$

and for the other three channels $p_{sc} = 2^{-1}(1 - \sqrt{|\kappa|/\max(|\kappa'|, |\kappa''|)})$. It is worthwhile noticing also that if $|\kappa| \neq 1$ and $\kappa \neq 0$, there always exists a quantum channel and an initial state for which $p_{sc} \in (0, 1)$; thus the SCP is seen to be universal.

Now, in order to comprehend the *physical origin* of the SCP, we start by observing that a positive value of QD is obligatorily linked to a minimum amount of correlation between the subsystems that has to be destroyed by the non-selective measurement of local-compatible observables. Although, to our knowledge, there is no proof for general states, in the case of Bell-diagonal states we can verify that at the sudden change time the less disturbing observable is also altered. We anticipate that the resource theory of asymmetry [118] can shed more light on the interpretation and possible applications of the SCP in the general scenario. Determining formally the relationship between the SCP and the resource theory of asymmetry is an important open problem.

Soon after been reported in Ref. [23], the SCP was verified experimentally by Xu et al., as described in Ref. [119]. They used as qubits the polarization degrees of freedom of two photons generated via parametric down conversion. One of the qubits was then subject to a simulated-controlled dephasing environment. Moreover, the authors also verified the transition between the classical and quantum decoherence regimes, that have been theoretically noticed in Ref. [120]. The dynamics shown in Fig. 2 is one example of such a transition, where for $p < p_{sc}$ ($p > p_{sc}$) only classical (quantum) correlations are affected by the interaction with the environment. In 2011, we and our co-workers observed experimentally these dynamical behaviors for the real dephasing and dissipative environments encountered in liquid state nuclear magnetic resonance, where the two qubits were encoded in the ^{1}H and ^{13}C spin-1/2 nuclei [121]. A detailed description of related theory and experiments can be found in Refs. [122–124].

4 The SCP and the Emergence of the Pointer Basis

In quantum mechanics, for any observable we can think of, coherent superpositions of its eigenstates are the rule, not the exception. And, as QM lists several of the known rules of Nature, an immediate question we are urged to ask is: Why it is so hard seeing or maintaining physical systems in states that are coherent with relation to certain observables? Decoherence [125], einselection [126], and quantum Darwinism [127] are theories that give fine answers to this question. An important conceptual hint for these theories can be traced back to 1935, when Schrödinger exposed the idea that when two particles get entangled, they cannot be described "by endowing each of them with a representative of its own" [128]. Decoherence theory recognizes that in the system-environment dynamics, in Zeh's words, "Any sufficiently effective interaction will induce correlations … defining in general a large value of entropy" of the system [125]. This increase in disorder is generally associated with the diminishing of quantum coherence. By its turn, environment-induced superselection, or einselection [126], explains why some observables, the so called pointer observables, are the chosen ones for classical reality. It is the commutativity of the system-environment interaction Hamiltonian with the pointer observables that leads to the robustness of its eigenstates; and to the disappearance of its superpositions. Quantum Darwinism identifies the special role information has in this scenario. In addition to be able to "survive" under the influence of the environment, the pointer states are the ones that most efficiently spread its information by getting correlated with many parts of the surroundings. This leads to the classical feature of the classical world: redundancy, i.e., several observers can have non-disturbing access to the same information [127].

The next natural question to be asked is about how much time it takes for the quantum-to-classical transition to occur. That is to say, we want to understand when the pointer basis emerges. This question is usually addressed in the following scenario. We consider a quantum system S and obtain information about it via another quantum system M, the measurement apparatus. To do that, a pre-measurement evolution must take place to correlate S and M, reflecting the possible observable values (or states) of S into the evolved states of M [129]. But S and M are, in general, not isolate; and the interaction with the environment E correlates all the three entities, leading to increased mixing of the joint state of S and M. We recall here that the pointer basis is said to have emerged if, in principle, we can perform a non-selective von Neumann measurement in this basis without disturbing the system being measured. The issue now is if such measurement must refer to S or to M. Starting with the first choice, the figure of merit that have been usually used to quantify the time for the emergence of classicality is called the decoherence time, or decoherence half-life. This quantity may represent the time taken for S to lose its coherence with relation to the pointer basis [130–133]. It is here that the *SCP enters the scene*. As reported in Ref. [79], once in practice we look at M to obtain information about S, it is the disappearance of the coherence of M in the pointer basis that marks the transition to classicality. Such disappearance can be identified using p_{sc}, the time in which a sudden change in the classical correlation C_{hv}^{M} between S and M occurs, if C_{hv}^{M} remains constant for later times [79, 134, 135]. It is worthwhile mentioning that for non-Markovian environments, a more complex and interesting situation with the possibility for more than one pointer observable and metastable pointer bases was reported in Ref. [136].

5 Dynamics of Correlations

As described earlier, the SCP was predicted by studying the non-unitary evolution of the system. In other words, considering that the system of interest interacts with uncontrolled degrees of freedom, collectively called environment. This interaction generates correlations between system and environment, causing an irreversible loss of information from the system. Since there is no system that can be regarded as truly isolated, the study of the behavior of the correlations (inside the system) under the action of decoherence channels is of major importance both from theoretical and practical issues.

The usual description of the dynamics of open system is given by the so called master equation, a first order differential equation for the reduced density operator describing the system, derived by means of second order perturbation theory [137].

This equation, which must have a well defined Lindblad structure [138], relies on two main approximations: Weak coupling (between system and environment) and short correlation times (compared with the typical decoherence time of the system). When such approximations are not satisfied, the dynamical description of the system must change accordingly, and non-Markovian effects can appear. In classical mechanics, non-Markovian effects are identified with memory. However, in quantum mechanics, this definition is not so direct and care must be taken. We do not intend to discuss such issue here and we refer the reader to reference [46] for a recent review. In what follows we consider some results concerning the Markovian and the non-Markovian dynamical evolutions of correlations, in the context of the SCP.

5.1 Markovian Dynamics of Correlations

There is a vast literature on the subject of quantum correlations under Markovian dynamics. Here we just comment on a few results, referring the reader to more technical, and complete, references.

Shortly after the publication of Ref. [23], the related phenomena of quantum and classical decoherence was predicted [120]. Considering energy-conserving dissipative maps (the ones described in Sect. 3), a class of states were identified such that before the sudden change point only the classical correlations are affected by decoherence and, after this critical point, only the quantum correlations are affected. These two regimes were then called classical and quantum decoherence [120]. An elegant geometric interpretation of the sudden change behavior, for the simple case of Bell-diagonal states, was provided in Ref. [24], while the conditions for the correlations to stay constant, the so called freezing phenomena, were put forward in Ref. [139], considering the phase damping channel.

Following these lines, in Ref. [140] it was proved that virtually all *bona fide* measures of quantum correlations present the freezing effect under the same dynamical conditions. The authors considered the case of Bell-diagonal states under the action of non-dissipative environments. A geometric interpretation was also provided.

Another interesting study was reported in Ref. [141]. Considering quantum correlations measured by quantum discord and quantum work deficit, the multipartite case, under local noise, was addressed. Among the results of the paper, a complementarity relation between the freezing time and the value of quantum correlation was provided. Moreover, a freezing index was introduced and its usefulness as a witness for quantum phase transition was discussed.

Instead of correlations shared by distinct parts of the system, in Ref. [142] the authors addressed the dynamical evolution of the correlations between the system and the environment. Specifically, they considered a two-qubit system under the action of two independent channels (the Pauli and the amplitude damping maps). They found that decoherence may occur without entanglement between system and environment and also that the initial non-classical correlations, presented in the system, completely disappears, under certain conditions, being not transferred to the environments [142].

A new interpretation of the SCP was introduced employing the idea of complementary correlations [143]. Considering two complementary observables acting on each one of the subsystems of a bipartite system, it was shown that the sum of the local correlations between such observables is a good measure of the quantum correlations shared by the composite system. The general conclusion is that the mixedness of the initial bipartite state is not enough for the SCP, but the state also needs to present certain asymmetry with respect to local complementary observables [143]. Moreover, they also proved that a pure state will never present the SCP and that the freezing phenomena is not a general property of all the Bell-diagonal states. It is important to mention that these results were obtained using the quantum discord as a measure for quantum correlations.

An interesting connection between the SCP and quantum teleportation was reported in Ref. [144]. The transition point between the classical and quantum decoherence was associated with a transition point appearing in the fidelity of the teleported state, signaling a change in the class of states that are harder to teleport.

5.2 *Non-Markovian Dynamics of Correlations*

Due to memory effects, non-Markovian evolution may presents a much richer dynamics than Markovian ones. In this subsection we briefly describe some results in this field regarding the SCP.

The case of two qubits interacting with two independent non-Markovian environments was considered in Ref. [145], where the dynamics of entanglement was compared with that one for quantum discord. The authors verified that while entanglement can present a sudden death (entanglement disappears for all times after the critical one), quantum discord can only vanish at some specific times. In Ref. [146] the authors addressed a similar problem, but now they studied the case of a common reservoir, and the SCP was again observed. An important result of this last work is the indication that the SCP is a characteristic feature of the evolution, for general initial conditions.

Regarding the freezing phenomenon, in Ref. [147] the case of two qubits under the action of local colored-noise dephasing channels was considered. As a main result, it was observed that, depending on the geometry of the initial state, the freezing phenomenon and the appearance of multiples SC was observed. Similar results were experimentally observed in Ref. [148].

In Ref. [149] the non-Markovian dynamics of two geometric measures of quantum correlations were compared with that one for the quantum discord considering the class of Bell-diagonal states. Although all the three considered measures share a common sudden change point, one of the geometric measures does not present the freezing phenomenon. This scenario was then extended to include the treatment of the correlations between the system and the environment in Ref. [150], where the appearance of the SC was studied as function of the system-environment coupling.

Relying on trace distance, several measures of quantum correlations were defined in [93]. The main result of this paper is the observation that the freezing behavior (and thus, the SCP) occurs for a larger class of states under non-Markovian dynamics than the Bell-diagonal states.

The geometric quantum discord was employed in the development of a witness for the SCP for both Markovian and non-Markovian dynamics in Ref. [151]. The dependence of the freezing effect on the choice of the correlation measure were also analyzed in this work.

Reference [152] proved that all geometric based measures of quantum correlations under the action of independent quantum channels exhibit the freezing phenomenon, and thus also the SC, for the Bell-diagonal class of two qubit states.

An interesting paper reported a study of several measures of quantum correlations, including entanglement, steering and the generalized discord, i.e. a definition of quantum correlations just like quantum discord, but based on the Tsallis q-entropy [153]. In this work the authors discovered a hierarchy among all the quantum resources and were able to identify a chronology of deaths and births (sudden changes) under non-Markovian channels.

All of the above studies considered the case in which the environments are all identical. In Ref. [154] the problem of a two qubit system (considering Bell-diagonal states) under the action of distinct environments were addressed, considering both Markovian and non-Markovian dynamics. The rules governing the time evolution of the classical and quantum correlations, including the sudden change points, were established.

Studying the information flow for qubits under an Ohmic environment, in Ref. [155] it was discovered a class of initial states for which quantum discord is forever frozen and the time-invariant discord was linked with non-Markovianity.

6 The Double SCP of Trace Distance Discord

In this section we discuss the double sudden change phenomenon of trace distance discord (D_{tr}), a puzzling effect with two sudden changes (SCs) for D_{tr} and only one SC for the associated classical correlation C_{tr} [134, 156]. But before doing that, let us mention that the possible existence of multiple sudden changes points for classical and quantum correlation is known since the first studies regarding their dynamics under non-Markovian global environments (see e.g. Ref. [157]). Also, as noticed in Ref. [154], two sudden change times can be obtained even with a Bell-diagonal state subject to local Markovian environments, if the qubits are acted on by suitable nonidentical surroundings. As an illustrative example, let us consider the dynamics of two qubits with the "qubit a" and "qubit b" let to evolve under the action of the bit flip (with parametrized time p) and phase flip (with parametrized time $q := rp$) channels, respectively (see Sect. 3). Under these conditions the evolved Bell-diagonal state is determined by the correlation vector:

$$\Xi_{p,q}^{bf\otimes pf} = ((1-2q)c_1, (1-2q)(1-2p)c_2, (1-2p)c_3)\,. \tag{42}$$

Let's assume also that the probability for the bit flip error is not greater than that for the phase flip: $q < p \therefore r \in (0, 1)$. We shall have thus that the second component of $\Xi_{p,q}^{bf\otimes pf}$ decays faster than the third one, which by its turn decays faster than the first one. Therefore, in this setting, any initial state with

$$|c_2| > |c_3| > |c_1| \tag{43}$$

will lead to two sudden changes for C_{hv} and D_{oz} if the crossing $|c_2^{(p,q)}| = |c_3^{(p,q)}|$ happens before than the intersection $|c_2^{(p,q)}| = |c_1^{(p,q)}|$ does. This leads to the following additional requirement for the existence of a double sudden change:

$$r > \frac{|c_2| - |c_3|}{|c_2| - |c_1|}. \tag{44}$$

Once all these conditions are fulfilled and $p \in [0, 2^{-1}]$ is adopted, we will see the SCP taking place at the times

$$p_{sc}^{(1)} = \frac{|c_2| - |c_3|}{2r|c_2|} \text{ and } p_{sc}^{(2)} = \frac{|c_3| - |c_1|}{2(|c_3| - r|c_1|)}, \tag{45}$$

as exemplified in Fig. 3. We remark that similar conditions can be obtained in an analogous manner for other combinations of local channels. Besides, we observe that if $r = 1 \therefore q = p$ then $p_{sc}^{(2)} = 2^{-1}$ and there is only one SCP.

Now we shall address the double sudden change phenomenon (DSCP) per se. In order to do that we will make use of the analytical formulas [135, 158] for the trace distance correlations of two-qubit X states:

$$\rho^x = \begin{bmatrix} \rho_{11} & 0 & 0 & |\rho_{14}|e^{i\phi_{14}} \\ 0 & \rho_{22} & |\rho_{23}|e^{i\phi_{23}} & 0 \\ 0 & |\rho_{23}|e^{-i\phi_{23}} & \rho_{33} & 0 \\ |\rho_{14}|e^{-i\phi_{14}} & 0 & 0 & \rho_{44} \end{bmatrix}. \tag{46}$$

This class of state can be transformed into the standard form

$$\rho^{xr} = \begin{bmatrix} \rho_{11} & 0 & 0 & |\rho_{14}| \\ 0 & \rho_{22} & |\rho_{23}| & 0 \\ 0 & |\rho_{23}| & \rho_{33} & 0 \\ |\rho_{14}| & 0 & 0 & \rho_{44} \end{bmatrix} \tag{47}$$

by applying the following local unitary (LU) transformation:

$$U = U_a \otimes U_b = e^{-i(\phi_{14}+\phi_{23})\sigma_3/4} \otimes e^{-i(\phi_{14}-\phi_{23})\sigma_3/4}. \tag{48}$$

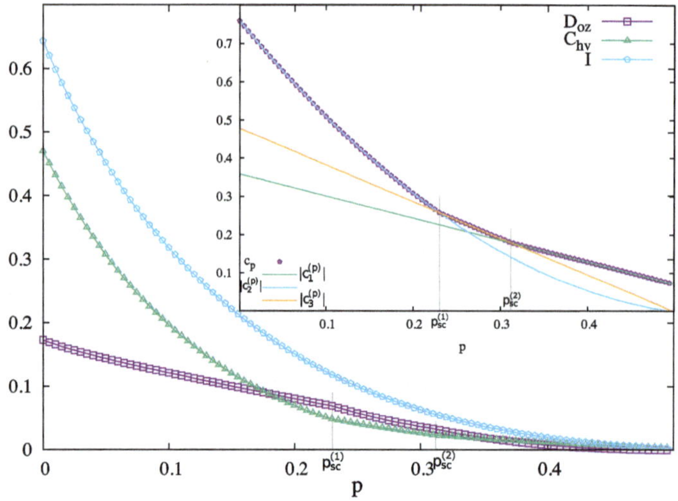

Fig. 3 Double sudden changes of Henderson–Vedral classical correlation and Ollivier–Zurek quantum discord for two qubits prepared in a Bell-diagonal state with $\Xi_0 = (0.36, -0.76, 0.48)$ and with the qubits a and b undergoing, respectively, bit and phase flip channels. In the inset are shown the evolved components of the correlation vector and the maximal one. Here we used $r = 0.8$ and the sudden change times are seen to be $p_{sc}^{(1)} \approx 0.23$ and $p_{sc}^{(2)} \approx 0.31$, as marked by the *gray lines*

As a basic property required for correlation quantifiers is invariance under local unitaries, we can study the correlations of ρ^x using its LU-equivalent version ρ^{xr}. When dealing with distance-based correlation measures, it is frequently useful utilizing the Bloch representation for the density operator; which in this case is

$$\rho^{xr} = 2^{-2}\left(\mathbb{I}^{ab} + a_3\sigma_3 \otimes \mathbb{I}^b + b_3\mathbb{I}^a \otimes \sigma_3 + \sum_{j=1}^{3} c_{jj}\sigma_j \otimes \sigma_j\right), \tag{49}$$

where the non-null elements of the Bloch vectors and correlation matrix are

$$a_3 = \mathrm{Tr}(\rho^{xr}\sigma_3 \otimes \mathbb{I}^b) = 2(\rho_{11} + \rho_{22}) - 1, \tag{50}$$
$$b_3 = \mathrm{Tr}(\rho^{xr}\mathbb{I}^a \otimes \sigma_3) = 2(\rho_{11} + \rho_{33}) - 1, \tag{51}$$
$$c_{11} = \mathrm{Tr}(\rho^{xr}\sigma_1 \otimes \sigma_1) = 2(|\rho_{23}| + |\rho_{14}|), \tag{52}$$
$$c_{22} = \mathrm{Tr}(\rho^{xr}\sigma_2 \otimes \sigma_2) = 2(|\rho_{23}| - |\rho_{14}|), \tag{53}$$
$$c_{33} = \mathrm{Tr}(\rho^{xr}\sigma_3 \otimes \sigma_3) = 1 - 2(\rho_{22} + \rho_{33}). \tag{54}$$

We observe that the effect of the LU above when transforming ρ^x into ρ^{xr} is: c_{11} : $2(\mathrm{Re}\rho_{23} + \mathrm{Re}\rho_{14}) \to 2(|\rho_{23}| + |\rho_{14}|)$, $c_{22} : 2(\mathrm{Im}\rho_{23} - \mathrm{Im}\rho_{14}) \to 0$, $c_{22} : 2(\mathrm{Re}\rho_{23} - \mathrm{Re}\rho_{14}) \to 2(|\rho_{23}| - |\rho_{14}|)$, and $c_{21} : -2(\mathrm{Im}\rho_{23} + \mathrm{Im}\rho_{14}) \to 0$.

Regarding the analytical formulas for the correlations, the trace distance discord of X states was obtained in Ref. [158] and reads

$$D_{tr}^{a}(\rho^{x}) := \min_{\rho_{cq}} d_{tr}(\rho, \rho_{cq}) \tag{55}$$

$$= \sqrt{\frac{\max\left(c_{33}^{2}, a_{3}^{2}+\min\left(c_{11}^{2}, c_{22}^{2}\right)\right)\max\left(c_{11}^{2}, c_{22}^{2}\right)-\min\left(c_{33}^{2}, \max\left(c_{11}^{2}, c_{22}^{2}\right)\right)\min\left(c_{11}^{2}, c_{22}^{2}\right)}{\max\left(c_{33}^{2}, a_{3}^{2}+\min\left(c_{11}^{2}, c_{22}^{2}\right)\right)-\min\left(c_{33}^{2}, \max\left(c_{11}^{2}, c_{22}^{2}\right)\right)+\max\left(c_{11}^{2}, c_{22}^{2}\right)-\min\left(c_{11}^{2}, c_{22}^{2}\right)}}. \tag{56}$$

Closed formulas were obtained in Ref. [135] for the trace distance classical and total correlations:

$$C_{tr}^{a}(\rho^{x}) := \max_{\Pi_a} d_{tr}(\Pi_a(\rho^{x}), \Pi_a(\rho_a^{x}\otimes\rho_b^{x})) = \kappa_{+}, \tag{57}$$

$$I_{tr}(\rho^{x}) := d_{tr}(\rho^{x}, \rho_a^{x}\otimes\rho_b^{x}) = \frac{1}{4}\sum_{j,k=0}^{1}|c_{11}+(-1)^{j}c_{22}+(-1)^{k}(c_{33}-a_3b_3)|$$

$$= \frac{1}{2}\left(\kappa_{+}+\max(\kappa_{+}, \kappa_{0}+\kappa_{-})\right). \tag{58}$$

Above, κ_{+}, κ_{0} and κ_{-} are, respectively, the maximum, intermediate, and minimum value among the elements in the list $\{|c_{11}|, |c_{22}|, |c_{33}-a_3b_3|\}$, Π_a is a complete non-selective measurement map, and $\rho_{a(b)}^{x} = \mathrm{Tr}_{b(a)}(\rho^{x})$ are the reduced density matrices. Its is interesting noticing here that for X states this definition of trace distance classical correlation is symmetric, $C_{tr}^{a}(\rho^{x}) = C_{tr}^{b}(\rho^{x})$, while the associated discord measure is generally asymmetric [117], $D_{tr}^{a}(\rho^{x}) \neq D_{tr}^{b}(\rho^{x})$, if $a_3 \neq b_3$.

Next we consider the dynamics of ρ^{xr} under local phase damping channels, which maintain the X form for ρ^{x}, and for its LU equivalent ρ_p^{xr}, with the only changes:

$$c_{11} \to c_{11}(1-p)^2 \text{ and } c_{22} \to c_{22}(1-p)^2; \tag{59}$$

all the other parameters remain constant with time. For easily seeing the mathematical origin of the DSCP, let us look at the special case of Bell-diagonal states ($a_3 = b_3 = 0$). For these states $D_{tr}(\rho_p^{bd})$ and $C_{tr}(\rho_p^{bd})$ are given, respectively, by the intermediate and maximum values among those in the list

$$(|c_{11}^{(p)}|, |c_{22}^{(p)}|, |c_{33}^{(p)}|) = (|c_{11}|(1-p)^2, |c_{22}|(1-p)^2, |c_{33}|). \tag{60}$$

Thus, the analysis of the sudden changes of C_{tr} is equivalent to that for C_{hv} (see Sect. 3). On the other side, we readily see that for a Bell-diagonal initial state with

$$|c_{11}| > |c_{22}| > |c_{33}| \tag{61}$$

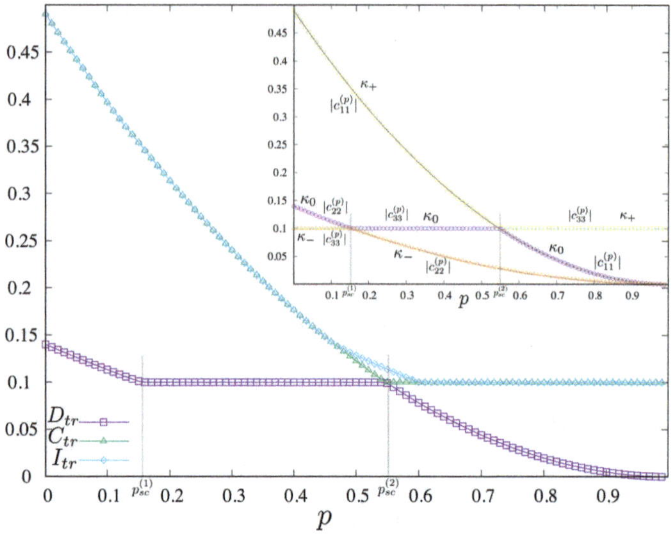

Fig. 4 Double sudden change phenomenon of trace distance quantum discord for Bell-diagonal states (with $(c_{11}, c_{22}, c_{33}) = (0.49, -0.14, -0.10)$) evolving under the action of local independent phase damping channels. As shown by the gray lines, there are two sudden changes of D_{tr}, at $p_{sc}^{(1)} \approx 0.16$ and at $p_{sc}^{(2)} \approx 0.55$, while the decaying rate of the classical correlation C_{tr} changes abruptly only at $p_{sc}^{(2)}$. It is noteworthy that the total correlation is (nearly) equal to the classical correlation for most values of p; a fact that may motivate the questioning about the suitability of these correlations definitions. Moreover, we identify the existence of abrupt changes also for I_{tr} at $p_{\pm} = 1 - \sqrt{|c_{33}|/(|c_{11}| \pm |c_{22}|)}$ (for this initial state $p_- \approx 0.47$ and $p_+ \approx 0.60$). As the definition of I_{tr} entails not extremizations at all, this effect has to be induced by the trace distance itself

at the times

$$p_{sc}^{(1)} = 1 - \sqrt{\frac{|c_{33}|}{|c_{22}|}} \text{ and } p_{sc}^{(2)} = 1 - \sqrt{\frac{|c_{33}|}{|c_{11}|}} \tag{62}$$

the intermediate value changes, and so does the decay rate of D_{tr}. In summary, the number of SCs of D_{tr} is greater than the number of abrupt changes of C_{tr}. This is the main feature of the DSCP, which is exemplified in Fig. 4. We notice that the analysis for X states can be made in a similar fashion, just by replacing c_{33} with $c_{33} - a_3 b_3$.

As can be seen in Fig. 4, $D_{tr}(\rho_p^{bd}) = |c_{33}|$ is constant, or freezed, for $p \in (p_{sc}^{(1)}, p_{sc}^{(2)})$. Given the recognized utility of quantum discord as a resource for some tasks in quantum information science, its robustness to noise is a most welcome feature; one that we should try to maintain for how much time as we can. In the case of Bell-diagonal states described above, we can begin doing that by choosing $|c_{22}| = |c_{33}|$, what implies in $p_{sc}^{(1)} = 0$. The next step is trying to get $p_{sc}^{(2)}$ as close to one as possible. But we perceive that the positivity of ρ^{bd} restricts the possible

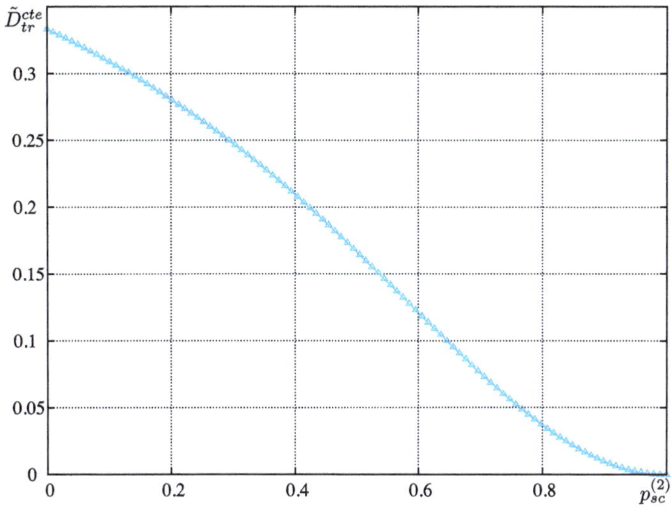

Fig. 5 Upper bound for the constant value of trace distance discord as a function of the second sudden change time. This bound holds for any Bell-diagonal state with $|c_{22}| = |c_{33}|$ and evolving under local-independent phase damping channels

values of c_{33} and c_{11}; and this will lead to the following tradeoff between $p^{(2)}_{sc}$ and the constant value of D_{tr}:

$$D^{cte}_{tr} \leq \frac{(1 - p^{(2)}_{sc})^2}{1 + 2(1 - p^{(2)}_{sc})^2} =: \tilde{D}^{cte}_{tr}.$$

Therefore, for instance, $p^{(2)}_{sc} = 2^0 \Rightarrow D^{cte}_{tr} = 0$, $p^{(2)}_{sc} = 2^{-1} \Rightarrow D^{cte}_{tr} \leq 0.16$, and $p^{(2)}_{sc} = 2^{-2} \Rightarrow D^{cte}_{tr} \leq 0.26$. The upper bound $\tilde{D}^{cte}_{tr}$ for the constant value of TDD is shown in Fig. 5 for $p^{(2)}_{sc} \in [0, 1]$.

The double sudden change phenomenon, which has been considered to be a genuine quantum feature [134], is a very interesting effect. Notwithstanding, as discussed in this section, there are several open questions motivated by the analysis of this phenomenon. Some examples of such questions are: Can we give a physical interpretation to the DSCP? Isn't the DSCP simply a byproduct of the distance measure utilized? Shouldn't we further analyze which properties a distance measure should have to be used in correlations quantification [159]? For which other distance or distinguishability measures this kind of effect can be observed [160, 161]? We think that in addressing these questions, we can deep our understanding about quantum systems and its correlations.

7 Final Remarks, Open Questions, and Perspectives

The sudden change phenomenon of quantum discord is an interesting and elusive effect. Despite the considerable amount of work dedicated or related to it have been constructing some knowledge, we are yet faced with more open questions than answers. Some of the main issues we leave here are:

Symmetry is a central concept in physics and is central e.g. for conservations laws. Can the resource theory of asymmetry bring to light the possible relevance of the SCP in physics?

Quantum discord has been shown to be a surprisingly general "order parameter" for phase transitions. As the sudden modification of its rate of change with the physical parameter determining the system state (phase) seems to play a key role here, it would be fruitful to pursue more formal connections between the SCP and symmetry (breaking) in phase transitions.

It is now clear that composite-discordant quantum states are the resource making possible the efficient implementation of several tasks in quantum information science. And, in our ever going fight against some noise interactions, the complete robustness of QD to environment influences is a most welcome feature it possesses. On the other side, one can also identify the existence of tradeoffs between the constant value of QD and the amount of time it can stay unchanged. It is important thus to analyze what kind of limitation this last effect can have on the possible practical applications of quantum discord.

Also, from the overall analysis we made here, on one side we can conclude that, as different discord functions are related to different physical or operational quantities, then the interpretation of the SCP could also be measure dependent. On the other hand, if future investigations happen to give a universal physical meaning to the SCP, then this fact could be considered in order to infer further conditions discord functions should have to satisfy to be faithfully used for the sake of quantum correlations quantification.

Acknowledgements We acknowledge financial support from the Brazilian funding agencies CNPq (Grants No. 401230/2014-7, 445516/2014-3, 305086/2013-8, 441875/2014-9 and 303496/2014-2) and CAPES (Grant No. 6531/2014-08), the Brazilian National Institute of Science and Technology of Quantum Information (INCT/IQ). JM gratefully acknowledges the hospitality of the Physics Institute and Laser Spectroscopy Group at the Universidad de la República, Uruguay.

References

1. D. Girolami, T. Tufarelli, G. Adesso, Characterizing nonclassical correlations via local quantum uncertainty. Phys. Rev. Lett. **110**, 240402 (2013)
2. K. Modi, A. Brodutch, H. Cable, T. Paterek, V. Vedral, The classical-quantum boundary for correlations: discord and related measures. Rev. Mod. Phys. **84**, 1655 (2012)
3. A. Shabani, D.A. Lidar, Vanishing quantum discord is necessary and sufficient for completely positive maps. Phys. Rev. Lett. **102**, 100402 (2009)

4. H.M. Wiseman, Quantum discord is Bohr's notion of non-mechanical disturbance introduced to counter the Einstein–Podolsky–Rosen argument. Ann. Phys. **338**, 361 (2013)
5. C.-J. Shan, W.-W. Cheng, J.-B. Liu, Y.-S. Cheng, T.-K. Liu, Scaling of geometric quantum discord close to a topological phase transition. Sci. Rep. **4**, 4473 (2014)
6. G.-M. Zeng, L.-A. Wu, H.-J. Xing, Symmetry restoration and quantumness reestablishment. Sci. Rep. **4**, 6377 (2014)
7. J. Li, T. Yu, H.-Q. Lin, J.Q. You, Probing the non-locality of Majorana fermions via quantum correlations. Sci. Rep. **4**, 4930 (2014)
8. J. Martin, V. Vennin, Quantum discord of cosmic inflation: Can we show that CMB anisotropies are of quantum-mechanical origin? Phys. Rev. D **93**, 023505 (2016)
9. M.A. Nielsen, I.L. Chuang, *Quantum Computation and Quantum Information* (Cambridge University Press, New York, 2000)
10. J. Preskill, Quantum Information and Computation. http://theory.caltech.edu/people/preskill/ph229/
11. M.M. Wilde, *Quantum Information Theory* (Cambridge University Press, Cambridge, 2013)
12. A. Datta, A. Shaji, C.M. Caves, Quantum discord and the power of one qubit. Phys. Rev. Lett. **100**, 050502 (2008)
13. V. Madhok, A. Datta, Interpreting quantum discord through quantum state merging. Phys. Rev. A **83**, 032323 (2011)
14. D. Cavalcanti, L. Aolita, S. Boixo, K. Modi, M. Piani, A. Winter, Operational interpretations of quantum discord. Phys. Rev. A **83**, 032324 (2011)
15. L. Roa, J.C. Retamal, M. Alid-Vaccarezza, Dissonance is required for assisted optimal state discrimination. Phys. Rev. Lett. **107**, 080401 (2011)
16. S. Pirandola, Quantum discord as a resource for quantum cryptography. Sci. Rep. **4**, 6956 (2014)
17. S. Lloyd, V. Chiloyan, Y. Hu, S. Huberman, Z.-W. Liu, G. Chen, No Energy Transport Without Discord (2015). https://arxiv.org/abs/1510.05035
18. C. Weedbrook, S. Pirandola, J. Thompson, V. Vedral, M. Gu, How discord underlies the noise resilience of quantum illumination. New J. Phys. **18**, 043027 (2016)
19. B. Dakić, Y.O. Lipp, X. Ma, M. Ringbauer, S. Kropatschek, S. Barz, T. Paterek, V. Vedral, A. Zeilinger, Č. Brukner, P. Walther, Quantum discord as resource for remote state preparation. Nat. Phys. **8**, 666 (2012)
20. M. Piani, V. Narasimhachar, J. Calsamiglia, Quantumness of correlations, quantumness of ensembles and quantum data hiding. New J. Phys. **16**, 113001 (2014)
21. D. Girolami, A.M. Souza, V. Giovannetti, T. Tufarelli, J.G. Filgueiras, R.S. Sarthour, D.O. Soares-Pinto, I.S. Oliveira, G. Adesso, Quantum discord determines the interferometric power of quantum states. Phys. Rev. Lett. **112**, 210401 (2014)
22. N. Friis, M. Skotiniotis, I. Fuentes, W. Dür, Heisenberg scaling in Gaussian quantum metrology. Phys. Rev. A **92**, 022106 (2015)
23. J. Maziero, L.C. Céleri, R.M. Serra, V. Vedral, Classical and quantum correlations under decoherence. Phys. Rev. A **80**, 044102 (2009)
24. M.D. Lang, C.M. Caves, Quantum discord and the geometry of Bell-diagonal states. Phys. Rev. Lett. **105**, 150501 (2010)
25. B. Li, Z.-X. Wang, S.-M. Fei, Quantum discord and geometry for a class of two-qubit states. Phys. Rev. A **83**, 022321 (2011)
26. M. Shi, F. Jiang, C. Sun, J. Du, Geometric picture of quantum discord for two-qubit quantum states. New J. Phys. **13**, 073016 (2011)
27. Y. Yao, H.-W. Li, Z.-Q. Yin, Z.-F. Han, Geometric interpretation of the geometric discord. Phys. Lett. A **376**, 358 (2012)
28. C. Liu, Y.-L. Dong, S.-Q. Zhu, Geometric discord for non-X states. Chin. Phys. B **23**, 060307 (2014)
29. Z. Huang, D. Qiu, P. Mateus, Geometry and dynamics of one-norm geometric quantum discord. Quantum Inf. Process. **15**, 301 (2016)

30. M.S. Sarandy, Classical correlation and quantum discord in critical systems. Phys. Rev. A **80**, 022108 (2009)
31. T. Werlang, C. Trippe, G.A.P. Ribeiro, G. Rigolin, Quantum correlations in spin chains at finite temperatures and quantum phase transitions. Phys. Rev. Lett. **105**, 095702 (2010)
32. A.K. Pal, I. Bose, Quantum discord in the ground and thermal states of spin clusters. J. Phys. B: At. Mol. Opt. Phys. **44**, 045101 (2011)
33. M.S. Sarandy, T.R. de Oliveira, L. Amico, Quantum discord in the ground state of spin chains. Int. J. Mod. Phys. B **27**, 1345030 (2013)
34. S. Campbell, J. Richens, N. Lo, Gullo and T. Busch. Criticality, factorization, and long-range correlations in the anisotropic XY model. Phys. Rev. A **88**, 062305 (2013)
35. G. Karpat, B. Çakmak, F.F. Fanchini, Quantum coherence and uncertainty in the anisotropic XY chain. Phys. Rev. B **90**, 104431 (2014)
36. Z. Xi, X.-M. Lu, Z. Sun, Y. Li, Dynamics of quantum discord in a quantum critical environment. J. Phys. B: At. Mol. Opt. Phys. **44**, 215501 (2011)
37. Y.-C. Li, H.-Q. Lin, J.-B. Xu, Dynamics of correlations and scaling behaviours in a spin-chain environment. EPL **100**, 20002 (2012)
38. L.-J. Tian, C.-Y. Zhang, L.-G. Qin, Sudden transition in quantum discord dynamics: role of three-site interaction. Chin. Phys. Lett. **30**, 050303 (2013)
39. X.-M. Lu, Z. Xi, Z. Sun, X. Wang, Geometric measure of quantum discord under decoherence. Quantum Inf. Comp. **10**, 0994 (2010)
40. L.-X. Jia, B. Li, R.-H. Yue, H. Fan, Sudden change of quantum discord under single qubit noise. Int. J. Quantum Inf. **11**, 1350048 (2013)
41. F.-J. Jiang, H.-J. Lu, X.-H. Yan, M.-J. Shi, A symmetric geometric measure and the dynamics of quantum discord. Chin. Phys. B **22**, 040303 (2013)
42. J.-L. Guo, H. Li, G.-L. Long, Decoherent dynamics of quantum correlations in qubit–qutrit systems. Quantum Inf. Process. **12**, 3421 (2013)
43. L. Qiu, G. Tang, X. Yang, Z. Xun, B. Ye, A. Wang, Sudden change of quantum discord in qutrit-qutrit system under depolarising noise. Int. J. Theor. Phys. **53**, 2769 (2014)
44. B.-L. Ye, Y.-K. Wang, S.-M. Fei, One-way quantum deficit and decoherence for two-qubit X states. Int. J. Theor. Phys. **55**, 2237 (2016)
45. J.-D. Shi, D. Wang, Y.-C. Ma, L. Ye, Revival and robustness of Bures distance discord under decoherence channels. Phys. Lett. A **380**, 843 (2016)
46. H.P. Breuer, E.-M. Laine, J. Piilo, B. Vacchini, Colloquium: Non-Markovian dynamics in open quantum systems. Rev. Mod. Phys. **88**, 021002 (2016)
47. Z.-X. Man, Y.-J. Xia, N.B. An, The transfer dynamics of quantum correlation between systems and reservoirs. J. Phys. B: At. Mol. Opt. Phys. **44**, 095504 (2011)
48. F. Han, The dynamics of quantum correlation and its transfer in dissipative systems. Int. J. Theor. Phys. **50**, 1785 (2011)
49. X.-X. Zhang, F.-L. Li, Controlling transfer of quantum correlations among bi-partitions of a composite quantum system by combining different noisy environments. Chin. Phys. B **20**, 110302 (2011)
50. P. Huang, J. Zhu, X.-X. Qi, G.-Q. He, G.-H. Zeng, Different dynamics of classical and quantum correlations under decoherence. Quantum Inf. Process. **11**, 1845 (2012)
51. Z.-D. Hu, J. Wang, Y. Zhang, Y.-Q. Zhang, Dynamics of nonclassical correlations with an initial correlation. J. Phys. Soc. Jpn. **83**, 114004 (2014)
52. H.-S. Xu, J.-B. Xu, Protecting quantum correlations of two qubits in independent non-Markovian environments by bang-bang pulses. J. Opt. Soc. Am. B **29**, 2074 (2012)
53. F.F. Fanchini, E.F. de Lima, L.K. Castelano, Shielding quantum discord through continuous dynamical decoupling. Phys. Rev. A **86**, 052310 (2012)
54. C. Addis, G. Karpat, S. Maniscalco, Time-invariant discord in dynamically decoupled systems. Phys. Rev. A **92**, 062109 (2015)
55. H. Song, Y. Pan, Z. Xi, Dynamical control of quantum correlations in a common environment. Int. J. Quantum Inf. **11**, 1350012 (2013)

56. M.-L. Hu, H. Fan, Robustness of quantum correlations against decoherence. Ann. Phys. **327**, 851 (2012)
57. L.C. Céleri, A.G.S. Landulfo, R.M. Serra, G.E.A. Matsas, Sudden change in quantum and classical correlations and the Unruh effect. Phys. Rev. A **81**, 062130 (2010)
58. Z. Tian, J. Jing, How the Unruh effect affects transition between classical and quantum decoherences. Phys. Lett. B **707**, 264 (2012)
59. M. Ramzan, Decoherence dynamics of geometric measure of quantum discord and measurement induced nonlocality for noninertial observers at finite temperature. Quantum Inf. Process. **12**, 2721 (2013)
60. Y.Y. Xu, W.L. Yang, M. Feng, Dissipative dynamics of quantum discord under quantum chaotic environment. EPL **92**, 10005 (2010)
61. G. Karpat, Z. Gedik, Correlation dynamics of qubit–qutrit systems in a classical dephasing environment. Phys. Lett. A **375**, 4166 (2011)
62. J.-Q. Li, J.-Q. Liang, Quantum and classical correlations in a classical dephasing environment. Phys. Lett. A **375**, 1496 (2011)
63. C. Wang, C. Li, L. Nie, X. Li, J. Li, Classical correlation, quantum discord and entanglement for two-qubit system subject to heat bath. Opt. Commun. **284**, 2393 (2011)
64. L. Xu, J.B. Yuan, Q.S. Tan, L. Zhou, L.M. Kuang, Dynamics of quantum discord for two correlated qubits in two independent reservoirs at finite temperature. Eur. Phys. J. D **64**, 565 (2011)
65. Z.X. Man, Y.J. Xia, N.B. An, Quantum dissonance induced by a thermal field and its dynamics in dissipative systems. Eur. Phys. J. D **64**, 521 (2011)
66. X.Q. Yan, Z.L. Yue, Dynamics of quantum and classical correlations of a two-atom system in thermal reservoirs. Chaos Solitons Fractals **57**, 117 (2013)
67. B.-Y. Yang, M.-F. Fang, Y.-N. Guo, Dissipative dynamics of quantum discord of two strongly driven qubits. Int. J. Theor. Phys. **53**, 921 (2014)
68. M.-L. Hu, D.-P. Tian, Preservation of the geometric quantum discord in noisy environments. Ann. Phys. **343**, 132 (2014)
69. J.-Q. Li, J. Liu, J.-Q. Liang, Environment-induced quantum correlations in a driven two-qubit system. Phys. Scr. **85**, 065008 (2012)
70. J.-B. Yuan, L.-M. Kuang, J.-Q. Liao, Amplification of quantum discord between two uncoupled qubits in a common environment by phase decoherence. J. Phys. B: At. Mol. Opt. Phys. **43**, 165503 (2010)
71. X.-P. Liao, J.-S. Fang, M.-F. Fang, B. Liu, Z. Huang, Entanglement and quantum discord dynamics of two atoms in a broadband squeezed vacuum bath. Int. J. Theor. Phys. **52**, 1729 (2013)
72. K. Berrada, Investigation of quantum and classical correlations in a quantum dot system under decoherence. Laser Phys. **23**, 095201 (2013)
73. P. Mazurek, K. Roszak, P. Horodecki, The decay of quantum correlations between quantum dot spin qubits and the characteristics of its magnetic field dependence. EPL **107**, 67004 (2014)
74. Q.-L. He, J.-B. Xu, Sudden transition and sudden change of quantum discord in dissipative cavity quantum electrodynamics system. J. Opt. Soc. Am. B **30**, 251 (2013)
75. V. Eremeev, N. Ciobanu, M. Orszag, Thermal effects on sudden changes and freezing of correlations between remote atoms in a cavity quantum electrodynamics network. Opt. Lett. **39**, 2668 (2014)
76. J.S. Sales, W.B. Cardoso, A.T. Avelar, N.G. de Almeida, Dynamics of nonclassical correlations via local quantum uncertainty for atom and field interacting into a lossy cavity QED. Phys. A **443**, 399 (2016)
77. Q.-L. He, J.-B. Xu, D.-X. Yao, Mediating and inducing quantum correlation between two separated qubits by one-dimensional plasmonic waveguide. Quantum Inf. Process. **12**, 3023 (2013)
78. N. Iliopoulos, A.F. Terzis, V. Yannopapas, E. Paspalakis, Two-qubit correlations via a periodic plasmonic nanostructure. Ann. Phys. **365**, 38 (2016)

79. M.F. Cornelio, O. Jiménez Farías, F.F. Fanchini, I. Frerot, G.H. Aguilar, M.O. Hor-Meyll, M.C. de Oliveira, S.P. Walborn, A.O. Caldeira, P.H. Souto Ribeiro, Emergence of the pointer basis through the dynamics of correlations. Phys. Rev. Lett. **109**, 190402 (2012)
80. K. Roszak, Ł. Cywiński, The relation between the quantum discord and quantum teleportation: the physical interpretation of the transition point between different quantum discord decay regimes. EPL **112**, 10002 (2015)
81. W. Han, K.-X. Jiang, Y.-J. Zhang, Y.-J. Xia, Quantum speed limits for Bell-diagonal states. Chin. Phys. B **24**, 120304 (2015)
82. K. Kraus, General state changes in quantum theory. Ann. Phys. **64**, 311 (1971)
83. J. Maziero, The Kraus representation for the dynamics of open quantum systems. Rev. Bras. Ensino Fis. **38**, e2307 (2016)
84. J. Maziero, Computing partial traces and reduced density matrices. Int. J. Mod. Phys. C **28**, 1750005 (2016)
85. R.P. Feynman, R.B. Leighton, M. Sands, *The Feynman Lectures on Physics*, vol. 3 (Addison-Wesley Publishing Company, Massachusetts, 1965)
86. M. Piani, P. Horodecki, R. Horodecki, No-local-broadcasting theorem for multipartite quantum correlations. Phys. Rev. Lett. **100**, 090502 (2008)
87. S. Luo, W. Sun, Decomposition of bipartite states with applications to quantum no-broadcasting theorems. Phys. Rev. A **82**, 012338 (2010)
88. N. Brunner, D. Cavalcanti, S. Pironio, V. Scarani, S. Wehner, Bell nonlocality. Rev. Mod. Phys. **86**, 419 (2014)
89. B. Groisman, S. Popescu, A. Winter, Quantum, classical, and total amount of correlations in a quantum state. Phys. Rev. A **72**, 032317 (2005)
90. V. Vedral, The role of relative entropy in quantum information theory. Rev. Mod. Phys. **74**, 197 (2002)
91. K. Modi, T. Paterek, W. Son, V. Vedral, M. Williamson, Unified view of quantum and classical correlations. Phys. Rev. Lett. **104**, 080501 (2010)
92. J. Maziero, Distribution of mutual information in multipartite states. Braz. J. Phys. **44**, 194 (2014)
93. B. Aaronson, R. Lo Franco, G. Compagno, G. Adesso, Hierarchy and dynamics of trace distance correlations. New J. Phys. **15**, 093022 (2013)
94. L. Henderson, V. Vedral, Classical, quantum and total correlations. J. Phys. A: Math. Gen. **34**, 6899 (2001)
95. H. Ollivier, W.H. Zurek, Quantum discord: A measure of the quantumness of correlations. Phys. Rev. Lett. **88**, 017901 (2001)
96. J. Oppenheim, M. Horodecki, P. Horodecki, R. Horodecki, Thermodynamical approach to quantifying quantum correlations. Phys. Rev. Lett. **89**, 180402 (2002)
97. S. Luo, Using measurement-induced disturbance to characterize correlations as classical or quantum. Phys. Rev. A **77**, 022301 (2008)
98. A. Brodutch, D.R. Terno, Quantum discord and local demons. Phys. Rev. A **81**, 062103 (2010)
99. B. Dakić, V. Vedral, Č. Brukner, Necessary and sufficient condition for nonzero quantum discord. Phys. Rev. Lett. **105**, 190502 (2010)
100. C.C. Rulli, M.S. Sarandy, Global quantum discord in multipartite systems. Phys. Rev. A **84**, 042109 (2011)
101. M.D. Lang, C.M. Caves, A. Shaji, Entropic measures of nonclassical correlations. Int. J. Quantum Inf. **09**, 1553 (2011)
102. F.M. Paula, T.R. de Oliveira, M.S. Sarandy, Geometric quantum discord through the Schatten 1-norm. Phys. Rev. A **87**, 064101 (2013)
103. T. Nakano, M. Piani, G. Adesso, Negativity of quantumness and its interpretations. Phys. Rev. A **88**, 012117 (2013)
104. D. Spehner, M. Orszag, Geometric quantum discord with Bures distance. New J. Phys. **15**, 103001 (2013)
105. A. Farace, A. De Pasquale, L. Rigovacca, V. Giovannetti, Discriminating strength: a bona fide measure of non- classical correlations. New J. Phys. **16**, 073010 (2014)

106. W. Roga, S.M. Giampaolo, F. Illuminati, Discord of response. J. Phys. A: Math. Theor. **47**, 365301 (2014)
107. U. Singh, A.K. Pati, Super quantum discord with weak measurements. Ann. Phys. **343**, 141 (2014)
108. S.J. Akhtarshenas, H. Mohammadi, S. Karimi, Z. Azmi, Computable measure of quantum correlation. Quantum Inf. Process. **14**, 247 (2015)
109. H. Cao, Z.-Q. Wu, L.-Y. Hu, X.-X. Xu, J.-H. Huang, An easy measure of quantum correlation. Quantum Inf. Process. **14**, 4103 (2015)
110. A.L.O. Bilobran, R.M. Angelo, A measure of physical reality. Europhys. Lett. **112**, 40005 (2015)
111. K.P. Seshadreesan, M. Berta, M.M. Wilde, Rényi squashed entanglement, discord, and relative entropy differences. J. Phys. A: Math. Theor. **48**, 395303 (2015)
112. L. Li, Q.-W. Wang, S.-Q. Shen, M. Li, Geometric measure of quantum discord with weak measurements. Quantum Inf. Process. **15**, 291 (2016)
113. A. Brodutch, K. Modi, Criteria for measures of quantum correlations. Quantum Inf. Comp. **12**, 0721 (2012)
114. F.M. Paula, A. Saguia, T.R. de Oliveira, M.S. Sarandy, Overcoming ambiguities in classical and quantum correlation measures. EPL **108**, 10003 (2014)
115. S. Luo, Quantum discord for two-qubit systems. Phys. Rev. A **77**, 042303 (2008)
116. S. Luo, Q. Zhang, Observable correlations in two-qubit states. J. Stat. Phys. **136**, 165 (2009)
117. J. Maziero, L.C. Céleri, R.M. Serra, Symmetry Aspects of Quantum Discord (2010). https://arxiv.org/abs/1004.2082
118. I. Marvian, R.W. Spekkens, Extending Noether's theorem by quantifying the asymmetry of quantum states. Nat. Commun. **5**, 3821 (2014)
119. J.-S. Xu, X.-Y. Xu, C.-F. Li, C.-J. Zhang, X.-B. Zou, G.-C. Guo, Experimental investigation of classical and quantum correlations under decoherence. Nat. Commun. **1**, 7 (2010)
120. L. Mazzola, J. Piilo, S. Maniscalco, Sudden transition between classical and quantum decoherence. Phys. Rev. Lett. **104**, 200401 (2010)
121. R. Auccaise, L.C. Céleri, D.O. Soares-Pinto, E.R. deAzevedo, J. Maziero, A.M. Souza, T.J. Bonagamba, R.S. Sarthour, I.S. Oliveira, R.M. Serra, Environment-induced sudden transition in quantum discord dynamics. Phys. Rev. Lett. **107**, 140403 (2011)
122. D.O. Soares-Pinto, M.H.Y. Moussa, J. Maziero, E.R. deAzevedo, T.J. Bonagamba, R.M. Serra, L.C. Céleri, Equivalence between Redfield- and master-equation approaches for a time-dependent quantum system and coherence control. Phys. Rev. A **83**, 062336 (2011)
123. D.O. Soares-Pinto, R. Auccaise, J. Maziero, A. Gavini-Viana, R.M. Serra, L.C. Céleri, On the quantumness of correlations in nuclear magnetic resonance. Phil. Trans. R. Soc. A **370**, 4821 (2012)
124. J. Maziero, R. Auccaise, L.C. Céleri, D.O. Soares-Pinto, E.R. deAzevedo, T. Bonagamba, R. Sarthour, I. Oliveira, R. Serra, Quantum discord in nuclear magnetic resonance systems at room temperature. Braz. J. Phys. **43**, 86 (2013)
125. H.D. Zeh, On the interpretation of measurement in quantum theory. Found. Phys. **1**, 69 (1970)
126. W.H. Zurek, Pointer basis of quantum apparatus: Into what mixture does the wave packet collapse? Phys. Rev. D **24**, 1516 (1981)
127. W.H. Zurek, Quantum Darwinism. Nat. Phys. **5**, 181 (2009)
128. E. Schrödinger, Discussion of probability relations between separated systems. Math. Proc. Camb. Philos. Soc. **31**, 555 (1935)
129. J. von Neumann, *Mathematical Foundations of Quantum Mechanics* (Princeton University Press, Princeton, 1955)
130. W.H. Zurek, Reduction of the wave packet: How long does it take?, in *Frontiers of Nonequilibrium Statistical Physics*, ed. by P. Meystre, M.O. Scully (Plenum, New York, 1984)
131. C. Monroe, D.M. Meekhof, B.E. King, D.J. Wineland, A "Schrödinger Cat" superposition state of an atom. Science **272**, 1131 (1996)
132. A.C.S. Costa, R.M. Angelo, M.W. Beims, Monogamy and backflow of mutual information in non-Markovian thermal baths. Phys. Rev. A **90**, 012322 (2014)

133. M. Fuchs, J. Schliemann, B. Trauzettel, Ultra long spin decoherence times in graphene quantum dots with a small number of nuclear spins. Phys. Rev. B **88**, 245441 (2013)
134. F.M. Paula, I.A. Silva, J.D. Montealegre, A.M. Souza, E.R. deAzevedo, R.S. Sarthour, A. Saguia, I.S. Oliveira, D.O. Soares-Pinto, G. Adesso, M.S. Sarandy, Observation of environment-induced double sudden transitions in geometric quantum correlations. Phys. Rev. Lett. **111**, 250401 (2013)
135. P.C. Obando, F.M. Paula, M.S. Sarandy, Trace-distance correlations for X states and emergence of the pointer basis in Markovian and non-Markovian regimes. Phys. Rev. A **92**, 032307 (2015)
136. F. Lastra, C.E. López, S.A. Reyes, S. Wallentowitz, Emergence of metastable pointer states basis in non-Markovian quantum dynamics. Phys. Rev. A **90**, 062103 (2014)
137. H.-P. Breuer, F. Petruccine, *The theory of open quantum systems* (Oxford University Press, Oxford, 2012)
138. G. Lindblad, Commun. Math. Phys. **48**, 119 (1976)
139. B. You, L.-X. Cen, Necessary and sufficient conditions for the freezing phenomena of quantum discord under phase damping. Phys. Rev. A **86**, 012102 (2012)
140. B. Aaronson, R. Lo Franco, G. Adesso, Comparative investigation of the freezing phenomena for quantum correlations under nondissipative decoherence. Phys. Rev. A **88**, 012120 (2013)
141. T. Chanda, A.K. Pal, A. Biswas, A. Sen(De), U. Sen, Freezing of quantum correlations under local decoherence. Phys. Rev. A **91**, 062119 (2015)
142. J. Maziero, T. Werlang, F.F. Fanchini, L.C. Céleri, R.M. Serra, System-reservoir dynamics of quantum and classical correlations. Phys. Rev. A **81**, 022116 (2010)
143. P. Deb, M. Banik, Role of complementary correlations in the evolution of classical and quantum correlations under Markovian decoherence. J. Phys. A: Math. Theor. **48**, 185303 (2015)
144. K. Roszak, L. Cywiński, The relation between the quantum discord and quantum teleportation: the physical interpretation of the transition point between different quantum discord decay regimes. EPL **112**, 10002 (2015)
145. B. Wang, Z.-Y. Xu, Z.-Q. Chen, M. Feng, Non-Markovian effect on the quantum discord. Phys. Rev. A **81**, 014101 (2010)
146. F.F. Fanchini, T. Werlang, C.A. Brasil, L.G.E. Arruda, A.O. Caldeira, Non-Markovian dynamics of quantum discord. Phys. Rev. A **81**, 052107 (2010)
147. L. Mazzola, J. Piilo, S. Maniscalco, Frozen discord in non-Markovian depolarizing channels. Int. J. Quant. Inf. **9**, 981 (2011)
148. J.-S. Xu, C.-F. Li, C.-J. Zhang, X.-Y. Xu, Y.-S. Zhang, G.-C. Guo, Experimental investigation of the non-Markovian dynamics of classical and quantum correlations. Phys. Rev. A **82**, 042328 (2010)
149. Z.Y. Xu, W.L. Yang, X. Xiao, M. Feng, Comparison of different measures for quantum discord under non-Markovian noise. J. Phys. A: Math. Theor. **44**, 395304 (2011)
150. L. Chuan-Feng, W. Hao-Tian, Y. Hong-Yuan, G. Rong-Chun, G. Guang-Can, Non-Markovian dynamics of quantum and classical correlations in the presence of system-bath coherence. Chin. Phys. Lett. **28**, 120302 (2011)
151. C.-S. Yu, B. Li, H. Fan, The witness of sudden change of geometric quantum correlation. Quantum Inf. Comput. **14**, 0454 (2013)
152. M. Cianciaruso, T.R. Bromley, W. Roga, R. Lo Franco, G. Adesso, Universal freezing of quantum correlations within the geometric approach. Sci. Rep. **5**, 10177 (2015)
153. A.C.S. Costa, M.W. Beims, R.M. Angelo, Generalized discord, entanglement, Einstein-Podolsky-Rosen steering, and Bell nonlocality in two-qubit systems under (non-)Markovian channels: Hierarchy of quantum resources and chronology of deaths and births (2016). https://arxiv.org/abs/1311.5702v2
154. B.-C. Ren, H.-R. Wei, F.-G. Deng, Correlation dynamics of a two-qubit system in a Bell-diagonal state under non-identical local noises. Quantum Inf. Process. **13**, 1175 (2014)
155. P. Haikka, T.H. Johnson, S. Maniscalco, Non-Markovianity of local dephasing channels and time-invariant discord. Phys. Rev. A **87**, 010103(R) (2013)

156. J.D. Montealegre, F.M. Paula, A. Saguia, M.S. Sarandy, One-norm geometric quantum discord under decoherence. Phys. Rev. A **87**, 042115 (2013)
157. F.F. Fanchini, T. Werlang, C.A. Brasil, L.G.E. Arruda, A.O. Caldeira, Non-Markovian dynamics of quantum discord. Phys. Rev. A **81**, 052107 (2010)
158. F. Ciccarello, T. Tufarelli, V. Giovannetti, Toward computability of trace distance discord. New J. Phys. **16**, 013038 (2014)
159. J. Maziero, Non-monotonicity of trace distance under tensor products. Braz. J. Phys. **45**, 560 (2015)
160. H. Yuan, L.-F. Wei, Geometric measure of quantum discord under decoherence and the relevant factorization law. Int. J. Theor. Phys. **52**, 987 (2013)
161. V. Eremeev, N. Ciobanu, M. Orszag, Thermal effects on the sudden changes and freezing of correlations between remote atoms in cavity QED network. Opt. Lett. **39**, 2668 (2014)

Frozen and Invariant Quantum Discord Under Local Dephasing Noise

Göktuğ Karpat, Carole Addis and Sabrina Maniscalco

1 Introduction

In nature, there exist various different types of correlations among the constituents of composite physical systems. While macroscopic systems only form correlations of classical nature, quantum mechanics allows for the existence of curious correlations devoid of a classical analogue, such as quantum entanglement. The idea of entanglement is as old as the quantum theory itself. In fact, Schrödinger himself believed that entanglement is not one but rather the characteristic trait of quantum mechanics. Besides its foundational importance for quantum mechanics, entanglement has attracted renewed interest in the quantum physics community during the past few decades. The reason lies in the fact that the concept of entanglement has emerged as the main resource of quantum information science [1–3], that is, it has been shown to be fundamental to the applications of quantum information processing, quantum computation and quantum cryptography.

G. Karpat (✉)
Faculdade de Ciências, UNESP - Universidade Estadual Paulista, Bauru, SP 17033-360, Brazil
e-mail: gkarpat@fc.unesp.br

G. Karpat
Faculty of Arts and Sciences, Department of Physics, Izmir University of Economics, 353300 Izmir, Turkey

C. Addis
SUPA, EPS/Physics, Heriot-Watt University, Edinburgh EH14 4AS, UK

S. Maniscalco
Department of Physics and Astronomy, Turku Center for Quantum Physics, University of Turku, 20014 Turku, Finland

S. Maniscalco
Department of Applied Physics, School of Science, Centre for Quantum Engineering, Aalto University, P.O. Box 11000, 00076 Espoo, Finland

F.F. Fanchini et al. (eds.), *Lectures on General Quantum Correlations and their Applications*, Quantum Science and Technology,
DOI 10.1007/978-3-319-53412-1_16

Despite the central role of entanglement in quantum information science, recent investigations have proved that it might not be the only kind of quantum correlation serving as a resource for quantum information tasks. For instance, it has been demonstrated that some quantum systems in separable states can perform more efficiently than their classical counterparts in certain applications [4]. As a consequence, numerous different correlation measures have been introduced in the recent literature to be able to characterize and quantify the non-classical correlations in quantum systems [5–11], which cannot be captured by entanglement measures. Among them, quantum discord, originally proposed by Ollivier and Zurek [5] and independently by Henderson and Vedral [6] as a measure of the quantumness of correlations, has received remarkable attention. This is mainly due to its usefulness as a resource in various quantum protocols such as the distribution of entanglement [9–14], quantum locking [15], entanglement irreversibility [16] and many others [17, 18].

Realistic quantum mechanical systems unavoidably interact with their surrounding environments. The effects of this inevitable interaction between the principal system of interest and its environment can be understood within the framework of open quantum systems theory [19–22]. Indeed, such interactions are typically detrimental for the crucial quantum traits present in the principal system as environment induced decoherence quickly destroys them in general. Therefore, protecting the precious quantum resources against the effects of the environment is of great importance, and also a major challenge for the realization of quantum computing devices.

One of the most striking outcomes of the decoherence process is the total loss of entanglement in composite systems in a finite time interval, also known as the sudden death of entanglement [23, 24]. On the other hand, quantum discord has been shown to be robust against sudden death [25–28], that is, where entanglement vanishes in finite time, quantum discord disappears only asymptotically. Another remarkable result related to quantum discord is the existence of a sharp transition between the loss of classical and quantum correlations during the time evolution of the open system [29, 30]. In particular, this phenomenon implies that, under a suitable local decoherence setting, quantum correlations as quantified by quantum discord remain frozen at its initial value for a certain time interval, unaffected by the destructive effects of the environment, before they finally start to decay. Moreover, considering non-Markovian open system models, where the role of memory effects in the dynamics is no longer neglected, several intervals of frozen quantum discord throughout the dynamics have been observed [31]. In fact, assuming a particular sort of local non-Markovian environment model with coherence trapping, it has also been demonstrated that quantum discord can be forever frozen during the whole time evolution thus becoming time invariant [32]. Although it has also been recently shown that entanglement might also become time invariant under suitable conditions, this requires the existence of global environmental interactions and decoherence free subspaces [33, 34].

In this chapter, we intend to explore and review some remarkable dynamical properties of quantum discord under various different open quantum system models. Specifically, our discussion will include several concepts connected to the phenomena of time invariant and frozen quantum discord. Furthermore, we will elaborate on how these two phenomena are related to both the non-Markovian features of the open system dynamics and the usage of dynamical decoupling protocols.

2 Quantum Discord

Correlations of purely quantum nature and more general than entanglement can be quantified by quantum discord [35], which is defined as the discrepancy between the two natural yet non-identical quantum generalizations of classically equivalent expressions for mutual information. Assuming that one has two classical random variables denoted by A and B, the first classical expression for the mutual information is given as follows

$$I(A:B) = H(A) + H(B) - H(A, B), \tag{1}$$

where the Shannon entropy $H(X) = -\sum_x p_x \log_2 p_x$ measures the amount of information one can obtain on average after learning the outcome of the measurement of the random variable $X \in (A, B)$, and the probability of obtaining the outcome labelled by x is denoted as p_x. Here, $H(A, B) = -\sum_{x,y} p_{x,y} \log_2 p_{x,y}$ is the joint entropy of the random variables A and B, which quantifies the total amount of ignorance about the pair, and $p_{x,y}$ is their joint probability distribution. On the other hand, the second definition of the mutual information can be given as

$$J(A:B) = H(A) - H(A|B), \tag{2}$$

where $H(A|B)$ is the entropy of the random variable A conditioned on the outcome of B. Considering that $H(A|B) = H(A, B) - H(B)$ in classical information theory, then it is not difficult to observe that both definitions of the classical mutual information are equivalent.

A rather straightforward quantum analogue of the classical mutual information, the so-called quantum mutual information, can be deduced by replacing the probability distributions related to the random variables with density operators. In particular, while the reduced density matrices $\rho_A = \mathrm{Tr}_B\{\rho_{AB}\}$ and $\rho_B = \mathrm{Tr}_A\{\rho_{AB}\}$ take the place of the probability distributions for the random variables A and B respectively, ρ_{AB} replaces the joint probability distribution for the A, B pair. Moreover, the Shannon entropy is replaced by the von Neumann entropy $S(\rho) = -\mathrm{Tr}(\rho \log \rho)$. With these considerations, one can define the quantum mutual information in the following way

$$\mathcal{I}(\rho_{AB}) = S(\rho_A) + S(\rho_B) - S(\rho_{AB}), \tag{3}$$

which, although first named by Cerf and Adami [36], is known to have already been considered by Belavkin and Stratonovich [37] many years ago. Today, it is widely agreed in the literature that the quantum mutual information is the information-theoretic quantifier of total correlations in a bipartite quantum state. This is also supported by several operational interpretations. For instance, Groisman et al. have argued that quantum mutual information can be defined via the amount of noise that is required to completely destroy the correlations present in the system [38]. Another supportive example has been provided by Schumacher and Westmoreland, where they have shown that if two separated parties share a correlated quantum system to be used as the key for a "one-time pad cryptographic system", the maximum amount of information that one party can send securely to the other is equal to the quantum mutual information of the shared correlated state [39].

The other quantum analogue of the classical mutual information is based on the concept of conditional entropy. By definition, classical conditional entropy $H(A|B)$ depends on the outcome of the random variable B. However, as is well known in quantum theory, measurements generally do disturb quantum systems. In other words, performing measurements on the system B affects our knowledge of the system A, and how much the system A is disturbed by a measurement of B depends on the type of measurement performed, i.e., on the measurement basis. Here, we assume that the measurements on the system B are of von Neumann type (projective measurements), i.e., they can be described by a complete set of orthonormal projectors $\{\Pi_k^B\}$ corresponding to the outcome k. Then, the quantum generalization of the classical mutual information in terms of the conditional entropy now reads

$$\mathcal{J}(\rho_{AB}) = S(\rho_A) - \sum_k p_k S\left(\rho_{A|k}\rho_{A|k}\right), \tag{4}$$

where $\rho_{A|k} = Tr_B(\Pi_k^B \rho_{AB} \Pi_k^B)/p_k$ is the remaining state of the subsystem A after obtaining the outcome k with probability $p_k = Tr_{AB}(\Pi_k^B \rho_{AB} \Pi_k^B)$ in the subsystem B. Supposing that the reduced state of the subsystem to be measured here is a two-level quantum system, one can construct the local von Neumann measurement operators $\{\Pi_1^B, \Pi_2^B\}$ in the following way:

$$\Pi_1^B = \frac{1}{2}\left(I_2^B + \sum_{j=1}^{3} n_j \sigma_j^B\right), \qquad \Pi_2^B = \frac{1}{2}\left(I_2^B - \sum_{j=1}^{3} n_j \sigma_j^B\right), \tag{5}$$

where $\sigma_j (j = 1, 2, 3)$ are the Pauli spin operators and $n = (\sin\theta\cos\phi, \sin\theta\sin\phi, \cos\theta)^T$ is a unit vector on the Bloch sphere with $\theta \in [0, \pi)$ and $\phi \in [0, 2\pi)$. It should be quite clear now that the two classically equivalent definitions of the mutual information do not agree in the quantum realm in general, as $\mathcal{J}(\rho_{AB})$ is basis dependent and reduces to $\mathcal{I}(\rho_{AB})$ only under special conditions.

A reasonable assumption is that the total amount of correlations in a bipartite quantum system can be divided into two parts and written as the sum of quantum ($\mathcal{Q}$) and classical ($\mathcal{C}$) correlations,

$$\mathcal{I}(\rho_{AB}) = \mathcal{Q}(\rho_{AB}) + \mathcal{C}(\rho_{AB}). \tag{6}$$

It has been recently suggested that the classical correlations in a bipartite quantum system can be quantified by the maximization of $\mathcal{J}(\rho_{AB})$ over all projective measurements on the subsystem B, that is, $\mathcal{C}(\rho_{AB}) = \max_{\Pi_k^B}\{\mathcal{J}(\rho_{AB})\}$ [6]. This quantity captures the maximum amount of information that can be obtained about the subsystem A by performing local measurements on the subsystem B, i.e., the locally accessible information. Consequently, genuine quantum correlations can be measured by subtracting the classical part of the correlations from the total amount,

$$\mathcal{Q}(\rho_{AB}) = \mathcal{I}(\rho_{AB}) - \max_{\Pi_k^B}\{\mathcal{J}(\rho_{AB})\},$$
$$= \mathcal{I}(\rho_{AB}) - \max_{\Pi_k^B}\left\{S(\rho_A) - \sum_k p_k S\left(\rho_{A|k}\rho_{A|k}\right)\right\}, \tag{7}$$

which is the well known quantum discord [5], also known as the locally inaccessible information. Quantum discord quantifies the genuine quantum correlations in a bipartite quantum system. It is important to emphasize that quantum discord is more general than entanglement in the sense that it is possible for some separable mixed states to have non-zero quantum discord and thus non-classical correlations. For instance, despite the fact that the state $\rho = 1/2(|0\rangle\langle 0|_A \otimes |-\rangle\langle -|_B + |+\rangle\langle +|_A \otimes |1\rangle\langle 1|_B)$ with $|\pm\rangle = (|0\rangle \pm |-\rangle)/\sqrt{2}$ has zero entanglement, it still cannot be described by classical means, i.e., by a classical bivariate probability distribution. The reason lies in the non-orthogonality of the reduced states of the subsystems A and B, which guarantees the impossibility of locally distinguishing the states of each subsystem. Indeed, in order to evaluate classical correlations $\mathcal{J}(\rho_{AB})$ and thus quantum discord $\mathcal{Q}(\rho_{AB})$, one might more generally perform the optimization over all possible positive operator valued measures (POVMs) instead of the set of projective measurements. Nevertheless, projective measurements are most widely considered in the literature since, even for this simpler case, there exists no available generic analytical expression for quantum discord and analytical results have been obtained only in few restricted cases, such as Bell-diagonal [40] or X shaped states of two qubits [41]. On the other hand, it has also been shown that projective measurements are almost sufficient for calculating the quantum discord of two qubits, and they are always optimal for the case of rank-2 states [42].

Let us lastly mention some of the important properties of quantum discord. Being a measure for correlations, it is expectedly non-negative, which is due to the concavity of the conditional entropy [43]. One can also notice that quantum discord is not a symmetric quantity in general, meaning that its value depends on whether the measurement is performed on subsystem A or subsystem B, which is a result of the asymmetry of the conditional entropy. It is invariant under local change of basis, that is, invariant under local unitary transformations. Furthermore, for pure states, quantum discord becomes a measure of entanglement being reduced to entanglement entropy, and it vanishes if and only if the considered state is classical-quantum [5]. Finally, even though there are now numerous discord-like or related quantum correlation measures introduced in the literature [11], we will mainly focus on the original entropic discord in this chapter.

3 Open System Dynamics of Quantum Discord

In this section, we will explore some remarkable dynamical features of quantum discord for bipartite systems under local dephasing noise. Particularly, we will discuss the phenomena of frozen and time-invariant quantum correlations in the case of both Markovian and non-Markovian interactions between the open quantum system and its environment. We will also review the role of dynamical decoupling protocols in protecting quantum correlations from environment induced decoherence.

3.1 *Frozen Quantum Discord*

3.1.1 Decoherence of Classical and Quantum Correlations

In the following, we will consider a bipartite system of two non-interacting qubits evolving in time under the action of a locally acting dephasing (or equivalently phase-flip) channel. One can start from a Markovian time-local master equation to obtain the dynamics of the bipartite open system. For each of the two-level systems, the Markovian master equation is given by

$$\dot{\rho}_{A(B)} = \frac{\gamma}{2}[\sigma_z^{A(B)} \rho_{A(B)} \sigma_z^{A(B)} - \rho_{A(B)}], \tag{8}$$

where $\sigma_z^{A(B)}$ is the usual Pauli spin operator in the z-direction acting on the subsystem $A(B)$. Using the solution of the above master equation, the time evolution of the open system in such a decoherence scenario can be expressed with the help of the operator-sum representation as follows

$$\rho_{AB}(t) = \sum_{i,j=1}^{2} (M_i^A \otimes M_j^B) \rho_{AB}(0) (M_i^A \otimes M_j^B)^\dagger, \tag{9}$$

where the Kraus operators M_1 and M_2 defining the dynamical dephasing map are given as

$$M_1^{A(B)}(t) = \begin{pmatrix} \sqrt{1-p(t)/2} & 0 \\ 0 & \sqrt{1-p(t)/2} \end{pmatrix}, \quad M_2^{A(B)}(t) = \begin{pmatrix} \sqrt{p(t)/2} & 0 \\ 0 & -\sqrt{p(t)/2} \end{pmatrix}, \tag{10}$$

and the explicit time dependence of the dephasing factor is $p(t) = 1 - \exp(-\gamma t)$ with γ being the dephasing rate. Note that here the local noise channels act on both qubits in the same way with identical dephasing factors. In Ref. [29], the dynamics of quantum discord and classical correlations have been discussed for a system of two qubits under such a local dephasing setting.

For this purpose, one can consider the local dephasing dynamics of the Bell-diagonal states, i.e., the class of two-qubit states with maximally mixed reduced density operators,

$$\rho_{AB} = \frac{1}{4}\left(I_{AB} + \sum_{i=1}^{3} c_i \sigma_i^A \otimes \sigma_i^B\right), \tag{11}$$

where I_{AB} is the 4×4 identity matrix and c_j are real numbers such that $0 \leq |c_j| \leq 1$. This class of states are indeed nothing but the convex combination of the four Bell states given as

$$\rho_{AB} = \lambda_\phi^+ \left|\phi^+\right\rangle \left\langle\phi^+\right| + \lambda_\psi^+ \left|\psi^+\right\rangle \left\langle\psi^+\right| + \lambda_\psi^- \left|\psi^-\right\rangle \left\langle\psi^-\right| + \lambda_\phi^+ \left|\phi^+\right\rangle \left\langle\phi^+\right| \tag{12}$$

where the non-negative eigenvalues of the bipartite density matrix ρ_{AB} read

$$\lambda_\psi^\pm = [1 \pm c_1 \mp c_2 + c_3]/4, \qquad \lambda_\phi^\pm = [1 \pm c_1 \pm c_2 - c_3]/4, \tag{13}$$

and $|\psi^\pm\rangle = (|00\rangle \pm |11\rangle)/\sqrt{2}$, $|\phi^\pm\rangle = (|01\rangle \pm |10\rangle)/\sqrt{2}$ are the four maximally entangled Bell states. From a geometrical perspective, the class of Bell-diagonal states can be thought to form a tetrahedron with the four maximally entangled Bell states sitting in the extreme points. In this case, dynamics of the mutual information $\mathcal{I}(\rho_{AB}(t))$, the classical correlations $\mathcal{C}(\rho_{AB}(t))$ and the quantum correlations measured by quantum discord $\mathcal{Q}(\rho_{AB}(t))$ are given by [29]

$$\mathcal{I}(\rho_{AB}(t)) = 2 + \sum_{kl} \lambda_k^l(t) \log_2 \lambda_k^l(t), \tag{14}$$

$$\mathcal{C}(\rho_{AB}(t)) = \sum_{j=1}^{2} \frac{1 + (-1)^j \mathcal{X}(t)}{2} \log_2[1 + (-1)^j \mathcal{X}(t)] \tag{15}$$

$$\mathcal{Q}(\rho_{AB}(t)) = 2 + \sum_{kl} \lambda_k^l(t) \log_2 \lambda_k^l(t) - \sum_{j=1}^{2} \frac{1 + (-1)^j \mathcal{X}(t)}{2} \log_2[1 + (-1)^j \mathcal{X}(t)], \tag{16}$$

where $\mathcal{X}(t) = \max\{|c_1(t)|, |c_2(t)|, |c_3(t)|\}$, $k = \phi, \psi$, and $l = \pm$. We should stress that, since the Bell diagonal states preserve their form under the local dephasing map, time dependence of the eigenvalues λ_k^l, and thus the coefficients c_i, can be easily obtained from the time evolved density matrix $\rho_{AB}(t)$ as $c_1(t) = c_1(0)\exp(-2\gamma t)$, $c_2(t) = c_2(0)\exp(-2\gamma t)$, and $c_3(t) = c_3(0) = c_3$.

The authors of Ref. [29] have identified three fundamentally different types of dynamical behavior for the correlations depending on the region of parameters c_i of the initial Bell-diagonal state. In the first region where $|c_3| > |c_1|, |c_2|$, classical correlations $\mathcal{C}(\rho_{AB}(t))$ stay constant independently of the dephasing parameter $p(t)$ and quantum discord $\mathcal{Q}(\rho_{AB}(t))$ decays in a monotonic fashion. In the

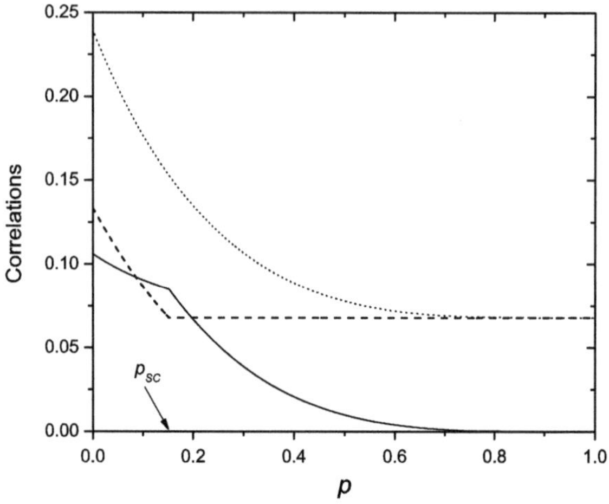

Fig. 1 Classical correlations $\mathcal{C}(\rho_{AB}(t))$ (*dashed line*), quantum correlations $\mathcal{Q}(\rho_{AB}(t))$ (*solid line*), and total correlations $\mathcal{I}(\rho_{AB}(t))$ (*dotted line*) versus the parametrized time $p(t)$ under local dephasing noise. The chosen initial Bell diagonal state parameters are $c_1 = 0.06$, $c_2 = 0.42$, and $c_3 = 0.30$, and the sudden change occurs at the point $p_{sc} = 0.15$. Note that the quantum discord $\mathcal{Q}(\rho_{AB}(t))$ is greater than the classical correlations $\mathcal{C}(\rho_{AB}(t))$ for $0.09 \leq p \leq 0.20$

second region in which $|c_3| = 0$, we have a monotonic decay for both classical and quantum correlations, which is not particularly interesting. Lastly, and most interestingly, in the region of parameters where $|c_1| > |c_2|, |c_3|$ or $|c_2| > |c_1|, |c_3|$, after decaying monotonically until a specific parametrized time p_{SC}, classical correlations $\mathcal{C}(\rho_{AB}(t))$ experience a sudden change in their dynamics and remain constant afterwards. At the same specific time p_{sc}, quantum discord exhibits an abrupt change in its decay trend, and then continues to decay monotonically over time. An example of this scenario is displayed in Fig. 1. We note that such a sudden change in the dynamics had not been observed before in the literature for the other known quantum correlation measures. In addition, another interesting point we can observe here is that the early conjecture that $\mathcal{C} \geq \mathcal{Q}$ for any quantum state [44] is shown to be invalid, as can be clearly seen in Fig. 1.

Finally, the result that the classical correlations might become unaffected from the detrimental effects of the environment induced decoherence for a certain noise channel can be in principle used to simplify the evaluation of classical correlations and thus quantum discord, by removing the typically difficult optimization procedure in their definitions. More specifically, if one knows the dynamical map (assuming that such a map actually exists), for a given initial bipartite quantum state which preserves the classical correlations in the system, quantum discord will be given by the difference between the mutual information $\mathcal{I}(\rho_{AB})$ and the completely decohered mutual information $\mathcal{I}(\rho_{AB}|_{p=1})$, that is,

$$\mathcal{Q}(\rho_{AB}) = \mathcal{I}(\rho_{AB}) - \mathcal{I}(\rho_{AB}|_{p=1}), \tag{17}$$

due to the simple fact that $\mathcal{I}(\rho_{AB}) = \mathcal{Q}(\rho_{AB}) + \mathcal{C}(\rho_{AB})$ and also $C(\rho_{AB}) = \mathcal{I}(\rho_{AB}|_{p=1})$ in this case.

3.1.2 Sudden Transition Between Classical and Quantum Decoherence

Following the discussion of frozen and time-invariant behavior of classical correlations under non-dissipative dephasing noise, we will now turn our attention to the possibility of observing frozen quantum correlations as measured by quantum discord. While it was first thought that quantum discord decays and vanishes asymptotically under Markovian noise, the authors of Ref. [30] remarkably identified a class of two-qubit initial states for which quantum discord preserves its initial value and therefore becomes frozen for a finite time interval until a critical time point is reached. This striking result, valid for dephasing noise, was the first evidence that precious quantum features in open quantum systems can remain intact naturally even in the presence of decoherence. Clearly, the existence of the phenomenon of frozen quantum discord might be highly important for quantum information protocols that rely on it as a resource.

Similarly to the previous subsection, we consider a system of two non-interacting qubits in Bell-diagonal states under a locally acting Markovian dephasing noise. Consequently, the formulas for the classical correlations $\mathcal{C}(\rho_{AB}(t))$ and the quantum discord $\mathcal{Q}(\rho_{AB}(t))$ are once again given by Eqs. (15) and (16), respectively. However, one can alternatively focus on a special class of initial states with parameters $c_1(0) = \pm 1$, $c_2(0) = \mp c_3(0)$ and $|c_3| < 1$, which can be written as

$$\rho_{AB} = \frac{(1+c_3)}{2} \left|\Psi^{\pm}\right\rangle \left\langle\Psi^{\pm}\right| + \frac{(1-c_3)}{2} \left|\Phi^{\pm}\right\rangle \left\langle\Phi^{\pm}\right| \tag{18}$$

For this class of initial Bell-diagonal states, it is straightforward to see from Eq. (14) that the quantum mutual information $\mathcal{I}[\rho_{AB}(t)]$ takes the following form:

$$\mathcal{I}(\rho_{AB}(t)) = \sum_{j=1}^{2} \frac{1+(-1)^j c_3}{2} \log_2[1+(-1)^j c_3] + \sum_{j=1}^{2} \frac{1+(-1)^j c_1(t)}{2} \log_2[1+(-1)^j c_1(t)]. \tag{19}$$

Recalling that $c_1(t) = \exp(-2\gamma t)$ and the form of the classical correlations $\mathcal{C}(\rho_{AB}(t))$ given in Eq. (15), it follows that, in case one has $t < \tilde{t} = -\ln(|c_3|)/(2\gamma)$, the second term on the right-hand side of Eq. (19) is equal to the classical correlations $\mathcal{C}(\rho_{AB}(t))$ due to the fact that $|c_1(t)| > |c_2(t)|$ and $|c_3(t)| = |c_3|$. Hence, the quantum discord is given by the first term on the right-hand side, which is independent of time. In other words, quantum correlations $\mathcal{Q}(\rho_{AB}(t))$ as quantified by quantum discord, become frozen until a critical time $\tilde{t}$ is reached. Furthermore, one can control the time interval in which quantum discord is constant in time by adjusting the value of $|c_3|$. However, a trade-off exists, that is, the longer the time interval of frozen quantum correlations, the smaller the initial value of quantum discord one can obtain. Figure 2 shows the dynamical behavior of the quantum discord, the classical correlations and

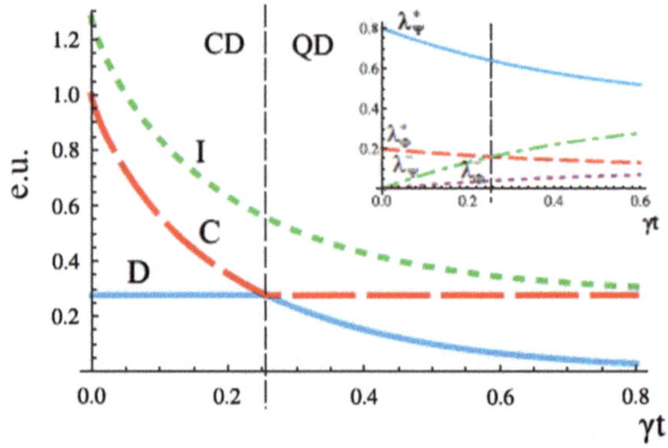

Fig. 2 Dynamics of mutual information (*green dotted line*), classical correlations (*red dashed line*) and quantum discord (*blue solid line*) as a function of t for $c_1(0) = 1$, $c_2(0) = -c_3$ and $c_3 = 0.6$. In the inset, we plot the eigenvalues, λ_ψ^+ (*blue solid line*), λ_ψ^- (*green dash dotted line*), λ_ϕ^+ (*red dashed line*) and λ_ϕ^- (*violet dotted line*) as a function of γt for the same parameters

the mutual information for the initial Bell-diagonal state with parameters $c_1(0) = 1$, $c_2(0) = -c_3$ and $c_3 = 0.6$. Here, one can observe that until the critical time $\tilde{t}$ only the classical correlations decay and quantum discord remains frozen. After this critical instant, where a sudden change occurs, classical correlations freeze and quantum discord begins to decay asymptotically. This is recognized as the phenomenon of sudden transition between classical and quantum decoherence [30].

One can further investigate the roots of this curious sudden transition between classical and quantum decoherence from a geometrical point of view. To this aim, it is customary to adopt a quantity to measure the distance between quantum states. Here, following the approach of Ref. [7], we choose the relative entropy defined by $S(\rho_1||\rho_2) = -\mathrm{Tr}(\rho_1 \log \rho_2) - S(\rho_1)$ as an entropic distance measure, although it cannot be technically considered as a distance. An alternative definition of the quantum discord can then be given as the distance of the considered state from the closest classical state ρ_{cl}. It turns out that the closest classical state to our system of choice given in Eq. (12), i.e. Bell-diagonal states, during the open system dynamics will be given by [7]

$$\rho_{\mathrm{cl}}(t) = \frac{q(t)}{2} \sum_{i=1,2} |\Psi_i\rangle\langle\Psi_i| + \frac{1-q(t)}{2} \sum_{i=3,4} |\Psi_i\rangle\langle\Psi_i| \tag{20}$$

where $q(t) = \lambda_1(t) + \lambda_2(t)$, and $\lambda_1(t)$ and $\lambda_2(t)$ are the two greatest time dependent eigenvalues of ρ_{AB} given by Eq. (13) with $|\Psi_i\rangle$ being the corresponding Bell states. The inset of Fig. 2 shows the time evolution of the eigenvalues $\lambda_\psi^\pm$ and $\lambda_\phi^\pm$, which can be considered as weights of the four Bell states forming the Bell-diagonal state. One can see that there is a transition time $\tilde{t}$ at which the weight of $|\phi_+\rangle$ becomes identical to the weight of $|\psi_-\rangle$. This results in a sudden change in the second greatest eigenvalue. Thus, while the closest classical state for $t < \tilde{t}$ is given by

$$\rho_{\rm cl}(t<\tilde{t})=\frac{1+e^{-2\gamma t}}{4}(|\Psi^{+}\rangle\langle\Psi^{+}|+|\Phi^{+}\rangle\langle\Phi^{+}|)+\frac{1-e^{-2\gamma t}}{4}(|\Phi^{-}\rangle\langle\Phi^{-}|+|\Psi^{-}\rangle\langle\Psi^{-}|), \tag{21}$$

in case of $t > \tilde{t}$, the closest classical state will be given by

$$\rho_{\rm cl}(t>\tilde{t})=\frac{1+c_3}{4}(|\Psi^{+}\rangle\langle\Psi^{+}|+|\Psi^{-}\rangle\langle\Psi^{-}|)+\frac{1-c_3}{4}(|\Phi^{-}\rangle\langle\Phi^{-}|+|\Phi^{+}\rangle\langle\Phi^{+}|). \tag{22}$$

Note that,

$$D(\rho_{AB}(t)||\rho_{\rm cl}(t))=-{\rm Tr}(\rho_{AB}(t)\log_2\rho_{\rm cl}(t))+{\rm Tr}(\rho_{AB}(t)\log_2\rho_{AB}(t)), \tag{23}$$

from which it is straightforward to show that $D(\rho_{AB}||\rho_{\rm cl})=\mathcal{D}(\rho_{AB})$, proving the equivalence of the original and relative entropy based definitions of quantum discord, for the whole class of Bell-diagonal states. Based on this result, we see that the closest classical state to the considered one before the critical time at $\tilde{t}$ remains unchanged even though the state evolves in time. On the other hand, after the critical time is reached, the closest state starts to change over time and thus quantum discord exhibits a monotonically decaying behavior.

Moreover, to compare the dynamics of quantum discord and entanglement, it is instructive to study the relative entropy of entanglement E, which is defined as the distance to the closest separable state ρ_S. In this case, entanglement for the Bell-diagonal states can be calculated as

$$E=1+\lambda_1\log_2\lambda_1+(1-\lambda_1)\log_2(1-\lambda_1), \tag{24}$$

where λ_1 is the greatest of the eigenvalues given by Eq. (13). Thus, entanglement decays in a monotonic way and vanishes when $t\geq t_s=-\ln[(1-|c_3|)/(1+|c_3|)]/2\gamma$. More importantly, for $t_s<\tilde{t}$, entanglement vanishes even before quantum discord begins to decay, which occurs for $0<|c_3|<\sqrt{2}-1$. Figure 3 displays an

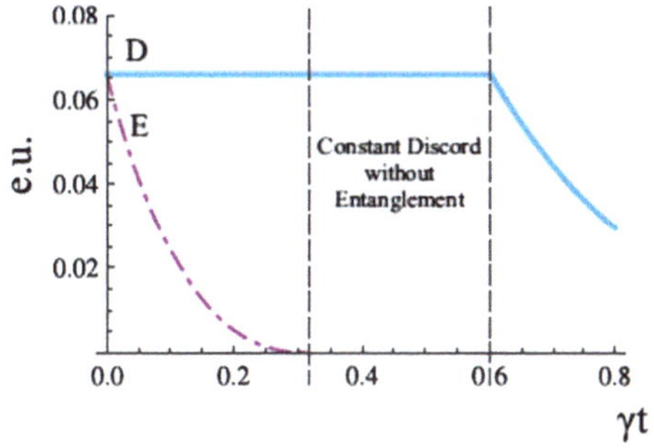

Fig. 3 Dynamics of the entanglement (*violet dashed-dotted line*) and the quantum discord (*blue solid line*) as a function of γt for the initial state $c_1(0)=1$, $c_2(0)=-c_3$ and $c_3=0.6$

example of this situation, where one has constant quantum discord without entanglement, for the initial state with $c_1(0) = 1$, $c_2(0) = -c_3$ and $c_3 = 0.6$.

After its first discovery, the phenomenon of frozen quantum discord has been explored in the literature in greater detail. For instance, necessary and sufficient conditions giving rise to the freezing of quantum discord have been explored for the Bell-diagonal states under local dephasing noise [45] and for certain multipartite states under other non-dissipative local noise channels [46]. Note that although the transition from constant to decaying quantum discord has been shown to occur abruptly for specific two-qubit states, it has been suggested that this transition might be sudden only for an idealized zero-measure subset of states within the whole set of two qubit states [47]. The investigation of the dynamics of quantum discord in open quantum systems has not been limited to the case of two level systems. In particular, frozen quantum discord has been also observed in hybrid qubit-qutrit states under local dephasing noise [48]. Furthermore, due to the difficulties in the analytical evaluation of the original entropic quantum discord, numerous different geometric versions have been introduced to measure the distance of the considered state to the closest classical state according to a chosen metric. It has also been demonstrated that sudden change and freezing behavior of quantum discord can also be observed for geometric discord under non-dissipative local noise [49]. It is even possible to have double sudden transitions for a particular version of geometric discord, i.e., the trace-norm discord [50, 51]. Later on, the phenomenon of frozen correlations has been shown to be a common feature of all bona fide measures of quantum correlations [52]. Lastly, it has been proven that all geometric quantifiers of quantum correlations, having the properties of invariance under transposition, convexity, and contractivity under quantum channels, might lead to the freezing phenomenon under suitable local non-dissipative noise [53].

On the other hand, the phenomena of sudden change and freezing of quantum discord have not remained as a purely theoretical construct. There have been several experiments demonstrating these peculiar effects with different physical systems. The first of these investigations has been performed by Xu et al. with an all optical experimental setup, where they have simulated the effects of a one-sided phase damping channel on a particular Bell-diagonal state [54]. In their experiment, the open system is represented by a pair of photons that are entangled in their polarization degrees of freedom and generated by a spontaneous parametric downconversion process. The dephasing effect is produced when photons pass through a quartz plate of adjustable thickness which causes their polarization degrees of freedom to couple to their frequency degrees of freedom, acting as the environment of the photon. They have observed a clear evidence of the occurrence of the phenomenon of frozen quantum discord and also experimentally proved that quantum correlations can be greater than classical correlations disproving the earlier conjecture. Another important experiment on the open system dynamics of quantum discord has been performed by Auccaise et al. using a room temperature nuclear magnetic resonance setup, where the environment induced sudden change takes place during the relaxation of two nuclear spins to the Gibbs state [55]. In their work, they have presented an evidence of both the sudden change and freezing of discord.

3.1.3 Frozen Discord in Non-Markovian Dephasing Channels

In this section, we will extend the investigation of the dynamics of classical and quantum correlations to the case where the bipartite open system is under the effect of non-Markovian dephasing local noise channels. We will elaborate on the non-trivial consequences of the memory effects, emerging due to the non-Markovian nature of the noise, for the phenomenon of frozen quantum discord. To this end, we will consider a well established pure dephasing model where white noise producing Markovian evolution is replaced by colored noise giving rise to non-Markovian time evolution [56].

Let us now introduce the open quantum system model which describes the independent dephasing of two non-interacting qubits under local identical dephasing noise channels. The time evolution of each of the qubits can be given by a master equation of the form

$$\dot{\rho} = K\mathcal{L}\rho, \tag{25}$$

where ρ is the density matrix of open quantum system, $\mathcal{L}$ is a Lindblad superoperator describing the time evolution of the open quantum system, and K is a time-dependent integral operator acting on the system as $K\phi = \int_0^t k(t-t')\phi(t')dt'$. The function $k(t-t')$ is a memory kernel that determines the type of memory in the dynamics. Such a master equation emerges if one considers the time evolution of a two-level system interacting with an environment having the properties of random telegraph signal noise. To give a physical example, this type of a model might be used to describe a spin in the presence of a constant intensity magnetic field that changes its sign randomly in time.

One can start to analyze such a system by writing a time-dependent Hamiltonian as

$$H(t) = \hbar \sum_{k=1}^{3} \Gamma_k(t)\sigma_k, \tag{26}$$

where $\hbar$ is the Planck's constant and $\Gamma_k(t)$ are independent random variables respecting the statistics of a random telegraph signal which can be expressed as $\Gamma_k(t) = a_k n_k(t)$. Here, $n_k(t)$ has a Poisson distribution with a mean equal to $t/2\tau_k$ and a_k is a coin-flip random variable having the values $\pm a_k$. With the help of the von Neumann equation $\dot{\rho} = -(i/\hbar)[H,\rho]$, it is possible to find a solution for the density matrix of the two-level system given by

$$\rho(t) = \rho(0) - i\int_0^t \sum_k \Gamma_k(s)[\sigma_k, \rho(s)]ds. \tag{27}$$

If one substitutes Eq. (27) back into the von Neumann equation and performs a stochastic average, one obtains a master equation of the form

$$\dot{\rho}(t) = -\int_0^t \sum_k e^{-(t-t')/\tau_k} a_k^2 [\sigma_k, [\sigma_k, \rho(t')]] dt', \tag{28}$$

where the memory kernel originates from the correlation functions of the random telegraph signals given as $\langle \Gamma_j(t)\Gamma_k(t')\rangle = a_k^2 \exp(-|t-t'|/\tau_k)\delta_{jk}$.

In Ref. [56], Daffer et al. have studied the requirements of completely positive time evolution for this type of a master equation. They have found that the dynamics generated by Eq. (28) is completely positive when two of the a_k are zero, which corresponds to the situation where the colored noise only acts in a single direction. In particular, when the condition $a_3 = a$ and $a_1 = a_2 = 0$ is satisfied, the resulting time evolution experienced by the open quantum system is that of a colored noise dephasing channel with non-Markovian features. In this case, the Kraus operators that describes the dynamics of the each two-level system are given by

$$M_1 = \sqrt{[1+\Lambda(\nu)]/2} I_2, \tag{29}$$

$$M_2 = \sqrt{[1-\Lambda(\nu)]/2} \sigma_3. \tag{30}$$

with $\Lambda(\nu) = e^{-\nu}[\cos(\mu\nu) + \sin(\mu\nu)/\mu]$, $\mu = \sqrt{(4a\tau)^2 - 1}$ and $\nu = t/2\tau$ is the dimensionless time. Since we will consider the dynamics of two-qubits independently interacting with identical colored dephasing environments, we can obtain the dynamical map using Eq. (9).

We will once again study the problem for initial Bell diagonal states of the form given in Eq. (12). As a result of the dynamics considered here, the Bell diagonal states preserve their general form, and the real three parameters defining them evolve in time as follows

$$c_1(\nu) = c_1(0)\Lambda(\nu)^2, \qquad c_2(\nu) = c_2(0)\Lambda(\nu)^2, \qquad c_3(\nu) = c_3(0), \tag{31}$$

Similarly to the case of Markovian dephasing noise discussed in the previous subsection, one can identify three regions of distinct dynamical features depending on the relations among the initial state parameters c_1, c_2 and c_3 [31].

In the first region of parameters where $|c_3(0)| \geq |c_1(0)|, |c_2(0)|$, both the quantum mutual information and the quantum discord display damped oscillations, eventually vanishing asymptotically. There exist no sudden changes in either quantum or classical correlations throughout the time evolution of the open system. On the other hand, classical correlations do not feel the effect of the noise and remain constant at all times, which can be easily seen from Eq. (15) as $\chi(t)$ is always equal to $|c_3(0)|$ in the course of the dynamics. The second regime includes the time evolution of the initial Bell diagonal states with $c_3 = 0$. Here, classical, quantum and also total correlations asymptotically tend to zero which means that final state of the open system is a product state, possessing no correlations of either classical or quantum nature. The last and the most interesting region is characterized by the initial states having $|c_3(0)| < |c_1(0)|$ and/or $|c_2(0)|$. Since $\chi(t)$ is initially equal to $\max\{|c_1(t)|, |c_2(t)|\}$, classical correlations begin to decay at first until a certain critical time. At this time point, they become suddenly frozen due to the fact that $|c_3(0)|$, which is constant during the dynamics, becomes greater than $\max\{|c_1(t)|, |c_2(t)|\}$. Recalling in the non-Markovian case that, $c_1(t)$ and $c_2(t)$ are oscillating functions of time, $|c_1(t)|$ or $|c_2(t)|$ might grow bigger than $|c_3(0)|$ once again during the dynamics, in which case classical correlations would start oscillating until they become abruptly constant. In fact, in this last region, we can distinguish two different dynamical behaviors for quantum discord, namely, sudden change dynamics with or without frozen behavior. An example of the first case is shown in Fig. 4a, where one can observe that quantum and classical correlations display multiple abrupt changes in their decay rates due to the non-Markovian memory effects in the open system dynamics. Indeed, such a behavior can be seen for the whole subclass of Bell diagonal states with $c_{1(2)}(0) = k$, $c_{2(1)}(0) = -c_3 k$ and $c_3(0) = c_3$ with k real and $|k| > |c_3|$. On the other hand, an example of the sudden change without frozen discord behavior is shown in Fig. 4b. Here, both classical correlations and quantum discord decay until a sudden change point is reached, where classical correlations become constant while quantum discord abruptly changes its decay rate without getting frozen. As time passes, this behavior is repeated with intervals of constant classical correlations. In order the understand the underlying reason behind the multiple periods of constant discord and sudden changes, one can conduct a geometrical analysis in terms of relative entropy based discord, similarly to what has been done in the previous subsection. Recall that original quantum discord and relative entropy based geometric discord turn out to be equivalent for the Bell diagonal states. Then, in this case, it is possible to track the trajectory of the state under investigation and its closest classical state during the dynamics to understand the geometrical origins of sudden changes and freezing behavior.

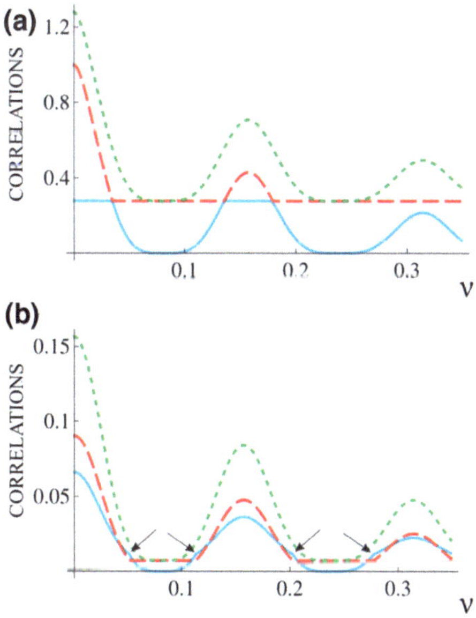

Fig. 4 Dynamics of the quantum mutual information (*green dotted line*), classical correlations (*red dashed line*) and quantum discord (*blue solid line*) as a function of dimensionless time $\nu = t/2\tau$ with $\tau = 5$ and $a = 1$. **a** Frozen discord with sudden multiple transitions where initial state parameters are $c_1(0) = 1$, $c_2(0) = 0.6$ and $c_3 = 0.6$. **b** Sudden change without frozen behavior with parameters $c_1(0) = 0.35$, $c_2(0) = 0.3$ and $c_3 = 0.1$. The arrows show the sudden change points

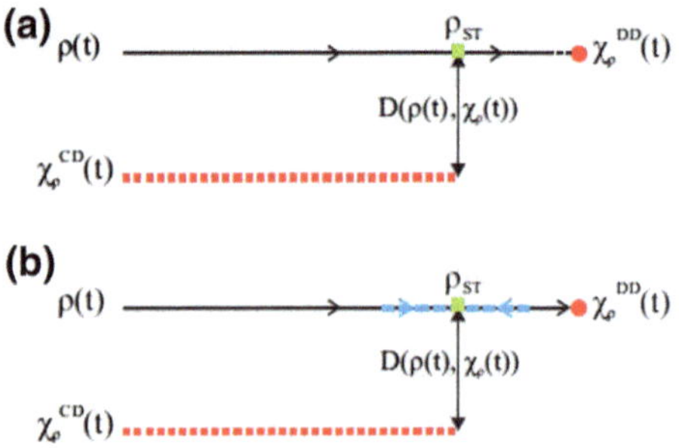

Fig. 5 Sketch of the trajectories of the state of the system under investigation and its closest classical state in the Hilbert space in the **a** Markovian, **b** non-Markovian case

Let us first have a look at the Markovian dephasing case illustrated in Fig. 5a, where there are no oscillations in the dynamics of correlations. Whereas the solid black line represents the trajectory of the considered state $\rho(t)$, the dotted red line depicts the path traced by the closest classical state $\chi_{\rho}^{CD}(t)$. As can be seen from the figure, the trajectory of the $\rho(t)$ is parallel to the one followed by $\rho^{CD}(t)$, that is, the system has frozen discord until the sudden change point is reached. The green square shows the sudden change point, where the state of the system ρ_{ST} has two closest classical states in equal distance, namely $\chi_{\rho}^{CD}(t)$ and $\chi_{\rho}^{DD}(t)$ in correspondence with the eigenvalue crossing point $\lambda_2 = \lambda_3$. As the open quantum system keeps evolving in time, it continues to travel along the black line. Therefore, the state under investigation $\rho(t)$ gets closer to its closest classical state $\chi_{\rho}^{DD}(t)$ asymptotically resulting in decaying quantum discord.

The consequences of the non-Markovian memory effects for the dynamics of the open system can be seen in Fig. 5b. Note that here the meaning of symbols and trajectories, and the structure of the set of closest classical states remain the same as in Fig. 5a. The crucial difference is that the state of the system under investigation oscillates around the sudden transition point throughout the dynamics, due to the non-Markovian features of the noise channel. When the considered state $\rho(t)$ first passes through the transition point as it travels from left to right, it reaches the closest classical state $\chi_{\rho}^{DD}(t)$ in a finite time interval. After this point, non-Markovian memory effects force the state $\rho(t)$ to travel back along part of its previous trajectory, which is shown in Fig. 5b by the thick dashed blue line partially overlapping with the black one. As it travels back from right to left on the blue line, it once again enters the frozen discord region after crossing the sudden change point. Finally, the direction of path of the state $\rho(t)$ flips once more and it enters the decaying quantum discord region to remain there. Xu et al. observed the dynamics of the correlations in a non-Markovian dephasing environment with an all-optical setup [57] and found the same features as described here.

3.2 *Time-Invariant Discord*

3.2.1 Time-Invariant Discord and Non-Markovian Open Quantum Systems

The simple but exact characterisation of local pure dephasing non-Markovian noise in qubit systems has lead to a plethora of studies surrounding the quantification and usefulness of non-Markovianity [58–66]. In this subsection, we examine this exact model in order to reveal the origin of time invariant discord. In more detail, the sudden transition between classical and quantum decoherence, inevitable in Markovian systems, does not necessarily occur in non-Markovian systems. Without this transition, quantum correlations may persist at all times while other dynamical quantities evolve.

We consider the following microscopic Hamiltonian describing the local interaction of a qubit and a bosonic reservoir in units of $\hbar$ [58–60],

$$H = \omega_0 \sigma_z + \sum_k \omega_k a_k^\dagger a_k + \sum_k \sigma_z (g_k a_k + g^* a_k^\dagger), \tag{32}$$

with ω_0 the qubit frequency, ω_k the frequencies of the reservoir modes, $a_k(a_k^\dagger)$ the annihilation (creation) operators and g_k the coupling constant between each reservoir mode and the qubit. In the continuum limit $\sum_k |g_k|^2 \to \int d\omega J(\omega)\delta(\omega_k - \omega)$ where $J(\omega)$ is the reservoir spectral density. The time local master equation for the qubit is given by,

$$\dot{\rho} = \gamma(t)[\sigma_z \rho \sigma_z - \rho]/2. \tag{33}$$

If the environment is initially in a thermal state at T temperature, the time-dependent dephasing rate takes the form,

$$\gamma(t) = \int d\omega J(\omega) \coth[\hbar\omega/2k_B T] \sin(\omega t)/\omega \tag{34}$$

resulting in the decay of the density matrix off-diagonal elements, $\rho_{ij}(t) = e^{-\Gamma(t)}$ $\rho_{ij}(0), i \neq j$, with dephasing factor $\Gamma(t) = \int_0^t \gamma(t')dt'$ given by

$$\begin{aligned} \Gamma(t) &= \int_0^\infty d\omega J(\omega) \coth[\hbar\omega/2k_B T][1 - \cos(\omega t)]/\omega^2 \\ &= \int_0^\infty d\omega g(\omega, T)[1 - \cos(\omega t)], \end{aligned} \tag{35}$$

where the Ohmic spectral densities are given by

$$J(\omega) = \frac{\omega^s}{\omega_c^{s-1}} e^{-\omega/\omega_c} \tag{36}$$

with ω_c the reservoir cutoff frequency. By changing the s parameter, one goes from sub-Ohmic reservoirs ($s < 1$) to Ohmic ($s = 1$) and super-Ohmic ($s > 1$) reservoirs, respectively. Such engineering of the Ohmicity of the spectrum is possible, e.g., when simulating the dephasing model in trapped ultracold atoms, as described in Ref. [62]. A closed analytical expression for the time-dependent dephasing rate can be found in both the zero T and high T limit. In the former case, one obtains,

$$\gamma_0(t,s) = [1 + (\omega_c t)^2]^{-s/2} \Gamma_E[s] \sin[s \arctan(\omega_c t)], \tag{37}$$

with $\Gamma_E[x]$ the Euler gamma function. For high T instead we have,

$$\gamma_{\mathrm{HT}}(t,s) = 2k_B T \gamma_0(t, s-1)/\omega_c. \tag{38}$$

Mathematically, the effect of the qubit on its environment is described by a displacement operator acting on each environment mode, with the associated phase conditional on the state of the qubit. The two qubit states excite each mode with opposing phases, leading to a decrease in the overlap between the states of the mode in each case. Destructive interference between excitations of a mode at different times reverses decoherence leading to recoherence at the frequency of the mode; it is the balance between these two effects for different modes, captured by Eq. (35) that determines whether the dynamics is non-Markovian. From Eq. (35), a simple link between the onset of non-Markovianity and the form of the reservoir spectrum can be established. As the cosine transform of a convex function is monotonically increasing, a sufficient condition for Markovianity is that $g(\omega, T)$ is convex or, equivalently, the non-convexity of $g(\omega, T)$ is a necessary condition for non-Markovianity. Physically, a convex $g(\omega, T)$ means that any recoherence is always outweighed by more decoherence from lower frequency modes.

For Ohmic class spectra, a simple connection between the general form of the spectrum and memory effects in the reduced system can be established from the form of the decay rate. Memory effects originate from non-divisible maps, corresponding to dissipators with decay rates which take temporarily negative values. It is straightforward to realize from Eq. (37) that for zero temperature, the dephasing rate takes negative values if and only if $s_{\mathrm{crit}} > 2$. Equally, from Eq. (38), the onset of non-Markovianity for high temperatures is $s_{\mathrm{crit}} > 3$. Hence, for the pure dephasing model, non-Markovianity only occurs for super-Ohmic environments. For intermediate temperatures, s_{crit} increases monotonically until $s_{\mathrm{crit}} = 3$ at infinite temperatures. Moreover, it can be shown that for s_{crit} and for all temperatures, the function $g(\omega, T)$ changes from a convex to non-convex function of ω, implying that the condition on the non-convexity of the spectrum is necessary and sufficient for non-Markovianity at all T.

For a system of two qubits in Bell-diagonal states experiencing local dephasing, the dynamics of the mutual information $\mathcal{I}(\rho_{AB}(t))$, the classical correlations $\mathcal{C}(\rho_{AB}(t))$ and the quantum correlations measured by quantum discord $\mathcal{Q}(\rho_{AB}(t))$ are given by Eqs. (14)–(16) with $c_1(t) = c_2(t) = \exp(-2\Gamma(t))$, $c_3(t) = c > 0$ and $\chi(t) = \max\{e^{-2\Gamma(t)}, c\}$. Hence, one immediately sees that when $e^{-2\Gamma(t)} > c$, classi-

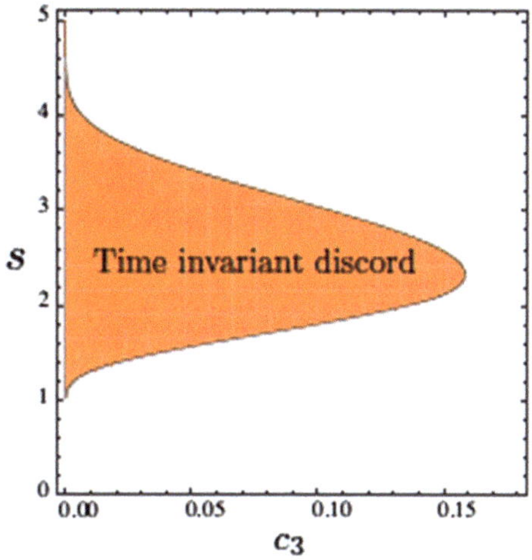

Fig. 6 The shaded region marks the range of parameters s and c for which the discord is frozen forever for $T = 0$. Outside this region one will always observe a transition from classical to quantum decoherence

cal correlations decay while discord remains constant. On the other hand, if a finite transition time $\tilde{t}$ such that,

$$e^{-2\Gamma(\tilde{t})} = c \tag{39}$$

exists, then for $t > \tilde{t}$ the discord starts decaying and the classical correlations remain constant. Contrary to the Markovian dephasing model, the transition time $\tilde{t}$ now crucially depends not only on initial state of the two-qubit system through the parameter c but also on parameters describing the structure of the reservoir spectral density, specifically s and the reservoir temperature T, through $\Gamma(t)$.

In Fig. 6, the values of s and c for which the condition in Eq. (39) is satisfied are illustrated, for $c = 0.1$ and $T = 0$. For a certain range of the parameter s, Eq. (39) has a solution and accordingly the system has a sudden transition from classical to quantum decoherence at time $\tilde{t}$. However, there exist a range of values of s for which Eq. (39) has no solution and no transition time $\tilde{t}$ exists leading to decaying classical correlations, while discord remains frozen forever. The two different cases are illustrated in Fig. 7 where classical correlations and discord for the Ohmic case $s = 1$ and $s = 2.5$ are plotted, respectively. By looking at the asymptotic long time limit of Eq. (39) we can define the s and c parameter space for which time-invariant discord exists (see Fig. 6). The value of frozen discord is,

$$\mathcal{Q} = (1 + c)\log_2(1 + c)/2 + (1 - c)\log_2(1 - c)/2. \tag{40}$$

Therefore, frozen discord takes significant values only for small values of c, approximately corresponding to the Ohmic range $2 \leq s \leq 3$ when the dynamics is non-Markovian. Increasing the temperature rapidly destroys the time invariant discord phenomenon and in the high temperature limit, values of s and c for which this effect occurs can not be found.

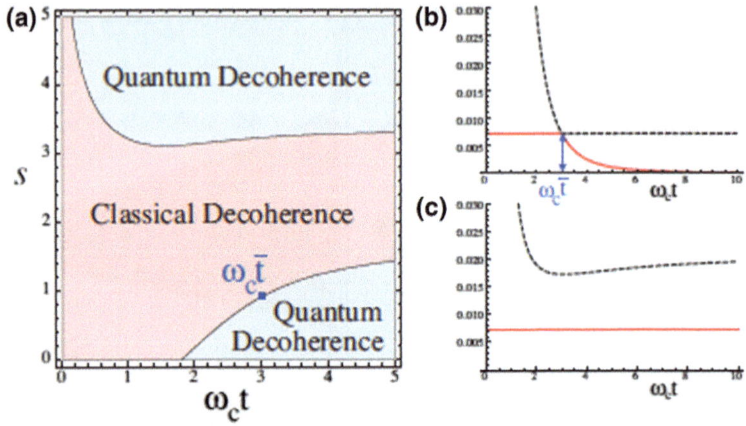

Fig. 7 **a** Landscape of correlation dynamics in the $s-t$ plane, for $c=0.1$ and $T=0$. *Blue* areas denote parameters (t,s) corresponding to classical decoherence, *red* areas to quantum decoherence and the intersection between the two, marking the values of s and $\tilde{t}$ satisfying Eq. (39) defines the transition time $\tilde{t}$ as a function of the reservoir spectrum parameter s. The two insets show discord (*solid red line*) and classical correlations (*dashed black line*) for two specific choices of s; **b** for $s=1$ the system has a sudden transition from quantum to classical decoherence while for **c** $s=2.5$ the discord is frozen forever. The *blue dot* in **a** and **b** points the transition time $\tilde{t}$ for $s=1$

The occurrence of time invariant discord in two qubit dynamics can be connected to the form of the reservoir spectrum and non-Markovianity in the single qubit dynamics. It is straightforward to show that the time invariant discord phenomenon can occur only for reservoir spectra leading to a bounded value of $\Gamma(t)$. This ensures the existence of values of c such that $e^{-2\Gamma(t)} > c$ for all t implying that Eq. (39) is never satisfied. On the other hand, an asymptotic divergence of $\Gamma(t)$ allows for the existence of a transition time $\tilde{t}$. Such a divergence, and therefore absence of time invariant discord rests on the divergence of $\omega g(\omega, T)$ when $\omega \to 0$ occurring for $s \leq 1(2)$ at zero (finite) temperature. Futhermore, convexity and thus Markovianity is ensured if $g(\omega, T)$ diverges at low frequencies, occurring for $s \leq s_{\text{crit}} = 2(3)$. Hence, time invariant discord and non-Markovianity are intimately related and ultimately rely on the eventual dominance of recoherence over decoherence, thus both require the suppression of coupling to low frequency modes, embodied by the low frequency dependences of $J(\omega)$ and $g(\omega, T)$.

3.2.2 Time-Invariant Discord in Dynamically Decoupled Systems

We now continue our study of the exact purely dephasing system in order to reveal the connection between time invariant discord and optimal control in the form of dynamical decoupling. In more detail, we now compare reservoir engineering techniques based on a modification of the reservoir spectral density through the Ohmicity parameter in order to change the Markovian character of the dynamics with reservoir engineering exploiting dynamical decoupling (which in turn, can also be seen as effective filtering of the spectral density). Dynamical decoupling (DD) techniques for open quantum systems are considered one of the most successful control protocols to suppress decoherence in qubit systems [67, 70]. Inspired by the spin-echo effect, dynamical decoupling involves the application of a sequence of external pulses to the system which induce unitary rotations in order to counter the harmful effects of the environment [68–74]. Specifically, "bang bang" periodic dynamical decoupling (PDD) schemes have been shown to prolong coherence times and restore decaying correlations in quantum systems which are undergoing decoherence [69].

We now recall the exact dynamics obtained in Ref. [75] which address the purely dephasing qubit behavior in the presence of an arbitrary sequence of instantaneous bang-bang pulses. Each pulse is modelled as an instantaneous π-rotation around the x-axis. An arbitrary storage time is considered, t, during which a total number of N pulses are applied at instants $\{t_1, ...t_n, ...t_f\}$, with $0 < t_1 < t_2 < \cdots < t_f < t$. As shown by Uhrig [76, 77], the controlled coherence function $\Gamma(t)$ can be worked out as,

$$\Gamma(t) = \begin{cases} \Gamma_0(t) & t \leq t_1 \\ \Gamma_n(t) & t_n < t \leq t_{n+1}, 0 < n < N \\ \Gamma_N(t) & t_f < t \end{cases}, \tag{41}$$

where the exact representation of the controlled decoherence function $\Gamma_n(t)$ for $1 \leq n \leq N$, can be written in the following form:

$$\Gamma_n(t) = \int_0^{\infty} \frac{J(\omega)}{2\omega^2} |y_n(\omega t)|^2 d\omega, \quad n \geq 0, \tag{42}$$

Further, from Eq. (42), $|y_0(\omega t)|^2 = |1 - e^{i\omega t}|^2$, and

$$y_n(z) = 1 + (-1)^{n+1} e^{iz} + 2\sum_{m=1}^{n} (-1)^m e^{iz\delta_m}, \quad z > 0. \tag{43}$$

Here, it is understood that the nth pulse occurs at time $t_n - \delta_n t$ and $0 < \delta_1 < \cdots < \delta_n < \cdots < \delta_s < 1$. In order to express the controlled decoherence function $\Gamma_n(t)$ in terms of its uncontrolled counterpart $\Gamma_0(t)$, we simply relate $|y_1(\omega t)|^2$ to $|y_0(\omega t)|^2$ to write,

$$\Gamma_1(t) = -\Gamma_0(t) + 2\Gamma_0(t_1) + 2\Gamma_0(t - t_1). \tag{44}$$

Upon iteration, and relating again, $|y_n(\omega t)|^2$ to $|y_0(\omega t)|^2$, we find the decoherence rate for n pulses,

$$\Gamma_n(t) = 2\sum_{m=1}^{n} (-1)^{m+1} \Gamma_0(t_m)$$

$$+ 4\sum_{m=2}^{n}\sum_{j<m}\Gamma_0(t_m - t_j)(-1)^{m-1+j}$$
$$+ 2\sum_{m=1}^{n}(-1)^{m+n}\Gamma_0(t - t_m) + (-1)^n\Gamma_0(t). \tag{45}$$

It is well known that the performance of dynamical decoupling techniques crucially depends on the temporal separation of the pulses. Moreover, the performance can be linked to the timescale of the environment correlation function, highlighting the important role played by spectral properties of the noise causing decoherence and introducing errors [78]. Following this, in Ref. [79], an explicit connection between the direction of information flow and dynamical decoupling is established. The influence of dynamical decoupling on the direction of information flow can be seen directly from the following relation, connecting $\gamma_n(t_n)$, i.e., $\dot{\Gamma}_n(t_n)$ at the moment t_n when the system is pulsed and the corresponding quantity at the previous instant:

$$\gamma_n(t_n) = -\gamma_{n-1}(t_n), \tag{46}$$

where for $1 \leq n \leq N$,

$$\gamma_n(t) = 2\sum_{m=1}^{n}(-1)^{m+n}\gamma_0(t - t_m) + (-1)^n\gamma_0(t). \tag{47}$$

From this expression, it is immediate to associate the change of sign of the decay rate at time t_n to a reversal of information flow occurring at the instant the system is pulsed. As a direct consequence, a Markovian open system dynamics will always be transformed to a non-Markovian dynamics. Hence, from a reservoir engineering perspective, in order for each pulse to result in revivals in dynamic quantities, one should choose environments with Ohmicity parameters $s < 2$, corresponding to unperturbed Markovian dynamics.

It is well established that the use of either DD techniques or non-Markovian effects prolongs the preservation of both entanglement and discord in the presence of environmental noise [80–85]. Unfortunately, as realized from Eq. (47), simultaneous use of non-Markovian reservoir engineering and DD protocols is counterproductive for avoiding decoherence. In more detail, if a pulse occurs during an interval of information back flow, information flow is reversed, inducing rapid decay in dynamical quantities such as coherence. We now consider the preservation of quantum and classical correlations and the possibility of creating time invariant discord through PDD techniques in relation to the initial conditions (unperturbed dynamics). For a more detailed insight, we consider two different scenarios of local dephasing noise in the presence of DD. Specifically, we first consider the scenario where both qubits A and B interact locally with identical dephasing environments, where in this case, both qubits are subject to pulsing. Secondly, we consider the case where only one of

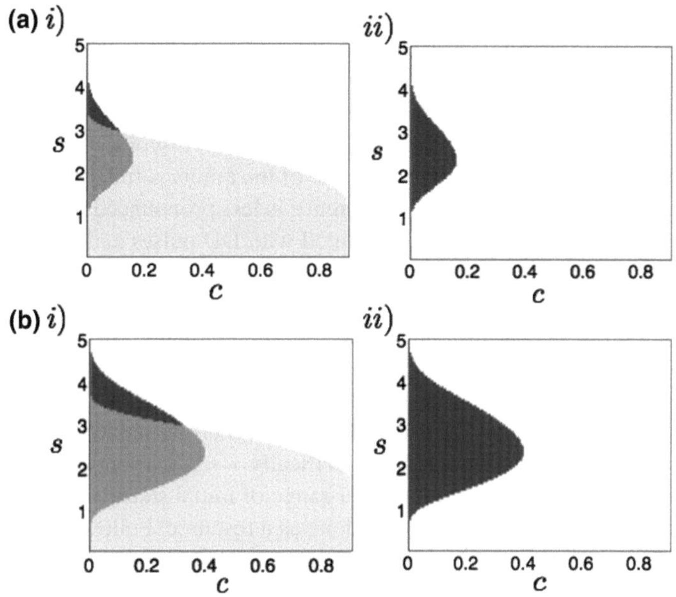

Fig. 8 The *black* region shows the range of s and c parameters for which the discord is frozen forever for a free system evolving without pulsing. The *grey* region shows the range of s and c for dynamical decoupled systems with small interval spacing $\Delta t = 0.3\omega_c^{-1}$ (i) and long interval spacing $\Delta t = 3\omega_c^{-1}$ (ii). The plots in **a** are for the case where both qubits are affected by noise, as the plots in **b** are for single qubit noise. Outside these regions, one will always observe a transition from classical to quantum decoherence and thus no time-invariant discord. The final pulse is applied at $t_f = N_{\max}\Delta t \leq 25\omega_c^{-1}$ where $N_{\max}$ is the maximum number of pulses that can be applied within the time interval $0 \leq t \leq 25\omega_c^{-1}$. Quantities plotted are dimensionless

the qubits interacts with the dephasing environment, while the other qubit is entirely protected from decoherence (the expressions for classical and quantum correlations hold in this case provided we exchange $e^{-\Gamma(t)}$ with $e^{-2\Gamma(t)}$). We note that only the qubit experiencing noise is subject to DD. Without the possibility of analytically or numerically defining continuously pulsed dynamics in the asymptotic time limit, we focus on discord which remains "time-invariant" within a chosen experimental time interval rather than forever.

We now compare the regions of s and c for which time-invariant discord exists for both the pulsed and unperturbed case. For a short pulse $\Delta t = 0.3\omega_c^{-1}$, we see from Fig. 8a, b (i), for two-sided and one-sided noise respectively, time invariant discord is created for a wider range of parameters compared to the unperturbed case. Specifically, time-invariant discord is created for sub-Ohmic values of s, (i.e., $s < 1$) for up to very high values of c, only when the system is subject to DD. Moreover, increasingly significant values of time invariant discord, corresponding to increasing values of c, occur for $s < 2$. These conclusions are independent of the specific pulse interval chosen provided that $\Delta t < \bar{t}$ where $\bar{t}$ defines the first time instant information back flow occurs for the unperturbed dynamics. For $s \lesssim 1$, coherence is maintained close to unity with no degradation shown to occur within computable times. Hence, one can conjecture that time invariant discord will be created in the asymptotic long time limit ($t_f \to \infty$). On the other hand, for $s > 1$, as t_f increases, the region of time-invariant discord will decrease as the coherence decays to increasingly small values. Physically, it is self-evident that the regions of invariant discord, in the absence and presence of control, become larger when one of the qubits is fully protected against noise. We point out however that the difference is less pronounced between the one-sided and two-sided noise for discord created with DD pulses and more significant in the case of unpulsed non-Markovian dynamics.

As the pulse interval Δt increases, the overlap between the regions of uncontrolled and controlled time-invariant discord becomes smaller. In Fig. 8a (ii), b (ii), the destructive influence of DD on time-invariant discord for large pulse intervals is immediately evident. Indeed, time-invariant discord is completely destroyed for both one-sided and two-sided noise. Hence, we find that short pulse interval DD schemes paired with Markovian environments (specifically $s < 1$), are optimal for the creation of time-invariant discord for a larger range of initial states when compared to strategies relying on non-Markovianity alone as a resource. Following this, one can say that relying only on non-Markovianity as a resource of time-invariant discord becomes more preferable in general as the pulse interval Δt increases.

3.3 Conclusions

In this chapter, we have presented a comprehensive exploration of the phenomena of frozen and time invariant quantum discord for bipartite quantum systems evolving under several different open quantum system models. We have discussed the suitable conditions for the initial states and the properties of the considered environmental models so that the behaviors of frozen or time invariant discord can be consistently observed. We have also considered non-Markovian open system models and shown how the effects of the memory in the dynamics affect the time intervals during which quantum discord becomes frozen. Finally, we have elaborated on the effect of pulsed dynamical decoupling techniques on the preservation of quantum discord at all times during the dynamics.

Acknowledgements S.M. acknowledges the Horizon 2020 EU collaborative project QuProCS (Grant Agreement 641277), the Academy of Finland (Project no. 287750) and the Magnus Ehrnrooth Foundation. C.A. acknowledges financial support from the EPSRC (UK) via the Doctoral Training Centre in Condensed Matter Physics. G.K. is grateful to Sao Paulo Research Foundation (FAPESP) for the fellowship given under grant number 2012/18558-5.

References

1. B. Gilles, Quantum cryptography: public-key distribution and coin tossing, in *Proceedings of IEEE International Conference on Computers, Systems and Signal Processing 1984* (IEEE Computer Society, 1984), pp. 175–179
2. M.A. Nielsen, I.L. Chuang, *Quantum Computation and Quantum Information* (Cambridge University Press, Cambridge, 2000)
3. P.W. Shor, Algorithms for quantum computation: discrete logarithms and factoring, in *Proceedings of the 35th Annual Symposium on Foundations of Computer Science* (IEEE Computer Society Press, Los Alamitos, CA, 1994), pp. 124D134
4. B.P. Lanyon, M. Barbieri, M.P. Almeida, A.G. White, Experimental quantum computing without entanglement. Phys. Rev. Lett. **101**, 200501 (2008)
5. H. Ollivier, W.H. Zurek, Quantum discord: a measure of the quantumness of correlations. Phys. Rev. Lett. **88**, 017901 (2001)
6. L. Henderson, V. Vedral, Classical, quantum and total correlations. J. Phys. A **34**, 6899 (2001)
7. K. Modi, T. Paterek, W. Son, V. Vedral, M. Williamson, Unified view of quantum and classical correlations. Phys. Rev. Lett. **104**, 080501 (2010)
8. B. Dakic, V. Vedral, C. Brukner, Necessary and sufficient condition for non-zero quantum discord. Phys. Rev. Lett. **105**, 190502 (2010)
9. V. Madhok, A. Datta, Interpreting quantum discord through quantum state merging. Phys. Rev. A **83**, 032323 (2011)
10. A. Brodutch, D.R. Terno, Quantum discord, local operations, and Maxwell's demons. Phys. Rev. A **81**, 062103 (2010)
11. K. Modi, A. Brodutch, H. Cable, T. Paterek, V. Vedral, The classical-quantum boundary for correlations: discord and related measures. Rev. Mod. Phys. **84**, 1655 (2012)
12. D. Cavalcanti, L. Aolita, S. Boixo, K. Modi, M. Piani, A. Winter, Operational interpretations of quantum discord. Phys. Rev. A. **83**, 032324 (2011)
13. A. Streltsov, H. Kampermann, D. Bruss, Quantum cost for sending entanglement. Phys. Rev. Lett. **108**, 250501 (2012)
14. T.K. Chuan, H. Maillard, K. Modo, T. Paterek, M. Paternostro, M. Pianai, Quantum discord bounds the amount of distributed entanglement. Phys. Rev. Lett. **109**, 070501 (2012)
15. S. Boixo, L. Aolita, D. Cavalcanti, K. Modi, A. Winter, Quantum locking of classical correlations and quantum discord of classical-quantum states. Int. J Quant. Inf. **9**, 1643 (2011)
16. M.F. Cornélio, M.C. de Oliveira, F.F. Fanchni, Entanglement irreversibility from quantum discord and quantum deficit. Phys. Rev. Lett. **107**, 020502 (2011)
17. A. Streltsov, H. Kampermann, D. Bruss, Linking quantum discord to entanglement in a measurement. Phys. Rev. Lett. **106**, 160401 (2011)
18. Werlang, C. Trippe, G.A.P. Ribeiro, G. Rigolin, Quantum correlations in spin chains at finite temperatures and quantum phase transitions. Phys. Rev. Lett. **105**, 095702 (2010)
19. H.-P. Breuer, F. Petruccione, *The Theory of Open Quantum Systems* (Oxford University, Press, 2007)
20. U. Weiss, *Quantum Dissipative Systems*, 3rd edn. (World Scientific, Singapore, 2008)
21. C.W. Gardiner, P. Zoller, *Quantum Noise* (Springer, Berlin, 2010)
22. M.A. Schlosshauer, *Decoherence and the Quantum-To-Classical Transition* (Springer, Berlin, 2007)
23. T. Yu, J.H. Eberly, Finite-time disentanglement via spontaneous emission. Phys. Rev. Lett. **93**, 140404 (2004)
24. T. Yu, J.H. Eberly, Quantum open system theory: bipartite aspects. Phys. Rev. Lett. **97**, 140403 (2006)
25. B. Wang, Z.Y. Xu, Z.Q. Chen, M. Feng, Non-Markovian effect on the quantum discord. Phys. Rev. A **81**, 014101 (2010)
26. A. Auyuanct, L. Davidovich, Quantum correlations as precursors of entanglement. Phys. Rev. A **82**, 032112 (2010)

27. Z.Y. Sun, L. Li, K.L. Yao, G.H. Du, J.W. Liu, B. Luo, N. Li, H.N. Li, Quantum discord in matrix product systems. Phys. Rev. A **82**, 032310 (2010)
28. T. Werlang, S. Souza, F.F. Fanchini, C.J. Villas Boas, Robustness of quantum discord to sudden death. Phys. Rev. A **80**, 024103 (2009)
29. J. Maziero, L.C. Celeri, R.M. Serra, V. Vedral, Classical and quantum correlations under decoherence. Phys. Rev. A **80**, 044102 (2009)
30. L. Mazzola, J. Piilo, S. Maniscalco, Sudden transition between classical and quantum decoherence. Phys. Rev. Lett. **104**(20), 200401 (2010)
31. L. Mazzola, J. Piilo, S. Maniscalco, Frozen discord in non-Markovian dephasing channels. Int. J. Quantum Inf. **9**(03), 981 (2011)
32. P. Haikka, T.H. Johnson, S. Maniscalco, Non-Markovianity of local dephasing channels and time-invariant discord. Phys. Rev. A **87**(1), 010103 (2013)
33. G. Karpat, Z. Gedik, Correlation dynamics of qubit-qutrit systems in a classical dephasing environment. Phys. Lett. A **375**, 4166 (2011)
34. E.G. Carnio, A. Buchleitner, M. Gessner, Robust asymptotic entanglement under multipartite collective dephasing. Phys. Rev. Lett. **115**, 010404 (2015)
35. K. Modi, A. Brodutch, H. Cable, T. Paterek, V. Vedral, The classical-quantum boundary for correlations: discord and related measures. Rev. Mod. Phys. **84**, 1655 (2012)
36. N.J. Cerf, C. Adami, Quantum information theory of entanglement and measurement. Physica **D120**, 62–81 (1998)
37. V.P. Belavkin, R.L. Stratonovich, Optimization of quantum information processing maximizing mutual information. Radio Eng. Electron. Phys. **19**(9), 1349 (1973). [trans. from Radiotekhnika i Electronika, 1973, 19, 9, 1839. 844]
38. B. Groisman, S. Popescu, A. Winter, Quantum, classical, and total amount of correlations in a quantum state. Phys. Rev. A **72**, 032317 (2005)
39. B. Schumacher, M.D. Westmoreland, Quantum mutual information and the one-time pad. Phys. Rev. A **74**, 042305 (2006)
40. S. Luo, Quantum discord for two-qubit systems. Phys. Rev. A **77**, 042303 (2008)
41. M. Ali, A.R.P. Rau, G. Alber, Quantum discord for two-qubit X states. Phys. Rev. A **81**, 042105 (2010)
42. F. Galve, G.L. Giorgi, R. Zambrini, Orthogonal measurements are almost sufficient for quantum discord of two qubits. EPL **96**, 40005 (2011)
43. A. Wehrl, General properties of entropy. Rev. Mod. Phys. **50**, 221 (1978)
44. G. Lindblad, Quantum entropy and quantum measurement, in *Quantum Aspects of Optical Communications*, vol. 378, Lecture Notes in Physics, ed. by C. Bendjaballah, et al. (Springer, Heidelberg, 1991), pp. 71–80
45. B. You, L.-X. Cen, Necessary and sufficient conditions for the freezing phenomena of quantum discord under phase damping. Phys. Rev. A **86**, 012102 (2012)
46. T. Chanda, A.K. Pal, A. Biswas, A. Sen (De), U. Sen, Freezing of quantum correlations under local decoherence. Phys. Rev. A **91**, 062119 (2015)
47. J.P.G. Pinto, G. Karpat, F.F. Fanchini, Sudden change of quantum discord for a system of two qubits. Phys. Rev. A **88**, 034304 (2013)
48. G. Karpat, Z. Gedik, Correlation dynamics of qubit-qutrit systems in a classical dephasing environment. Phys. Lett. A **375**, 4166 (2011)
49. B. Bellomo, R. Lo Franco, G. Compagno, Dynamics of geometric and entropic quantifiers of correlations in open quantum systems. Phys. Rev. A **86**, 012312 (2012)
50. J.D. Montealegre, F.M. Paula, A. Saguia, M.S. Sarandy, One-norm geometric quantum discord under decoherence. Phys. Rev. A **87**, 042115 (2013)
51. F.M. Paula, I.A. Silva, J.D. Montealegre, A.M. Souza, E.R. deAzevedo, R.S. Sarthour, A. Saguia, I.S. Oliveira, D.O. Soares-Pinto, G. Adesso, M.S. Sarandy, Observation of environment-induced double sudden transitions in geometric quantum correlations. Phys. Rev. Lett. **111**, 250401 (2013)
52. B. Aaronson, R. Lo Franco, G. Adesso, Comparative investigation of the freezing phenomena for quantum correlations under nondissipative decoherence. Phys. Rev. A **88**, 012120 (2013)

53. M. Cianciaruso, T.R. Bromley, W. Roga, R. Lo Franco, G. Adesso, Universal freezing of quantum correlations within the geometric approach. Sci. Rep. **5**, 10177 (2015)
54. J.-S. Xu, X.-Y. Xu, C.-F. Li, C.-J. Zhang, X.-B. Zou, G.-C. Guo, Experimental investigation of classical and quantum correlations under decoherence. Nature Commun. **1**, 7 (2010)
55. R. Auccaise, L.C. Celeri, D.O. Soares-Pinto, E.R. deAzevedo, J. Maziero, A.M. Souza, T.J. Bonagamba, R.S. Sarthour, I.S. Oliveira, R.M. Serra, Environment-induced sudden transition in quantum discord dynamics. Phys. Rev. Lett. **107**, 140403 (2011)
56. S. Daffer, K. Wdkiewicz, J.D. Cresser, J.K. McIver, Depolarizing channel as a completely positive map with memory. Phys. Rev. A **70**, 010304(R) (2004)
57. J.-S. Xu, C.-F. Li, C.-J. Zhang, X.-Y. Xu, Y.-S. Zhang, G.-C. Guo, Experimental investigation of the non-Markovian dynamics of classical and quantum correlations. Phys. Rev. A **82**, 042328 (2010)
58. J. Luczka, Spin in contact with thermostat: exact reduced dynamics. Physica A **167**, 919 (1990)
59. G.M. Palma, K.-A. Suominen, A.K. Ekert, Quantum computers and dissipation. Proc. Roy. Soc. Lond. A **452**, 567 (1996)
60. J.H. Reina, L. Quiroga, N.F. Johnson, Decoherence of quantum registers. Phys. Rev. A **65**, 032326 (2002)
61. M.A. Cirone, G. De Chiara, G.M. Palma, A. Recati, Collective decoherence of cold atoms coupled to a Bose–Einstein condensate. New J. Phys. **11**, 103055 (2009)
62. P. Haikka, S. McEndoo, G. De Chiara, G.M. Palma, S. Maniscalco, Quantifying, characterizing, and controlling information flow in ultracold atomic gases. Phys. Rev. A **84**, 031602 (2011)
63. A.W. Chin, S.F. Huelga, M.B. Plenio, Quantum metrology in non-Markovian environments. Phys. Rev. Lett. **109**, 233601 (2012)
64. G.R. Fleming, S.F. Huelga, M.B. Plenio, Focus on quantum effects and noise in biomolecules. New J. Phys. **13**, 115002 (2011)
65. C. Uchiyama, M. Aihara, Multipulse control of decoherence. Phys. Rev. A **66**, 032313 (2002)
66. C. Addis, F. Ciccarello, M. Cascio, G.M. Palma, S. Maniscalco, Dynamical decoupling efficiency versus quantum non-Markovianity. New J. Phys. **17**, 123004 (2015)
67. L. Viola, E. Knil, S. Lloyd, Dynamical decoupling of open quantum systems. Phys. Rev. Lett. **82**, 2417 (1998)
68. T.E. Hodgson, L. Viola, I. D'Amico, Towards optimised suppression of dephasing in systems subject to pulse timing constraints. Phys. Rev. A **81**, 062321 (2010)
69. T.E. Hodgson, L. Viola, I. DAmico, Decoherence-protected storage of exciton qubits through ultrafast multipulse control. Phys. Rev. B **78**, 165311 (2008)
70. L. Viola, S. Lloyd, Dynamical suppression of decoherence in two-state quantum systems. Phys. Rev. A **58**, 2733 (1998)
71. M.J. Biercuk et al., Optimised dynamical decoupling in a model quantum memory. Nature **458**, 996–1000 (2009)
72. E.L. Hahn, Spin echoes. Phys. Rev. **80**, 580 (1950)
73. K. Khodjasteh, D.A. Lidar, Fault-tolerant quantum dynamical decoupling. Phys. Rev. Lett. **95**, 180501 (2005)
74. K. Khodjasteh, D.A. Lidar, Performance of deterministic dynamical decoupling schemes: concatenated and periodic pulse sequences. Phys. Rev. A **75**, 062310 (2007)
75. T.E. Hodgson, L. Viola, I. DAmico, Towards optimised suppression of dephasing in systems subject to pulse timing constraints. Phys. Rev. A **81**, 062321 (2010)
76. G.S. Uhrig, Keeping a quantum bit alive by optimised π-pulse sequences. Phys. Rev. Lett. **98**, 100504 (2007)
77. G.S. Uhrig, Exact results on dynamical decoupling by π pulses in quantum information processes. New J. Phys. **10**, 083024 (2008)
78. C. Arenz, R. Hillier, M. Fraas, D. Burgarth, Distinguishing decoherence from alternative quantum theories by dynamical decoupling. Phys. Rev. A **92**, 022102 (2015)
79. C. Addis, F. Ciccarello, M. Cascio, G.M. Palma, S. Maniscalco, Dynamical decoupling efficiency versus quantum non-Markovianity. New J. Phys. **17**, 123004 (2015)

80. R. Lo Franco, A. D'Arrigo, G. Falci, G. Compagno, E. Paladino, Preserving entanglement and nonlocality in solid-state qubits by dynamical decoupling. Phys. Rev. B **90**, 054304 (2014)
81. S. Singha Roy, T.S. Mahesh, G.S. Agarwal, Storing entanglement of nuclear spins via Uhrig dynamical decoupling. Phys. Rev. A **83**, 062326 (2011)
82. F.F. Fanchini, E.F. de Lima, L.K. Castelano, Shielding quantum discord through continuous dynamical decoupling. Phys. Rev. A **86**, 052310 (2012)
83. A. Rosario et al., On the relationship between non-Markovianity and entanglement protection. J. Phys. B At. Mol. Opt. Phys. **45**, 095501 (2012)
84. B. Bellomo, R. Lo Franco, G. Compagno, Non-Markovian effects on the dynamics of entanglement. Phys. Rev. Lett. **99**, 160502 (2007)
85. J.-S. Zhang, A.-X. Chen, Controlling sudden transitions of bipartite quantum correlations under dephasing via dynamical decoupling. J. Phys. B At. Mol. Opt. Phys. **47**, 21 (2014)

Overview on the Phenomenon of Two-Qubit Entanglement Revivals in Classical Environments

Rosario Lo Franco and Giuseppe Compagno

Abstract The occurrence of revivals of quantum entanglement between separated open quantum systems has been shown not only for dissipative non-Markovian quantum environments but also for classical environments in absence of back-action. While the phenomenon is well understood in the first case, the possibility to retrieve entanglement when the composite quantum system is subject to local classical noise has generated a debate regarding its interpretation. This dynamical property of open quantum systems assumes an important role in quantum information theory from both fundamental and practical perspectives. Hybrid quantum-classical systems are in fact promising candidates to investigate the interplay among quantum and classical features and to look for possible control strategies of a quantum system by means of a classical device. Here we present an overview on this topic, reporting the most recent theoretical and experimental results about the revivals of entanglement between two qubits locally interacting with classical environments. We also review and discuss the interpretations provided so far to explain this phenomenon, suggesting that they can be cast under a unified viewpoint.

1 Introduction

Quantum correlations, such as entanglement, nonlocality, steering and discord, among parts of composite systems are at the core of quantum theory and have also been acquiring a paramount importance as a resource for quantum information processes [1–20]. Realistic systems are open and interact with the surrounding environment which usually has the effect to eventually destroy the quantum features

R. Lo Franco (✉)
Dipartimento di Energia, Ingegneria dell'Informazione e Modelli Matematici,
Università di Palermo, Viale delle Scienze, Edificio 9, 90128 Palermo, Italy
e-mail: rosario.lofranco@unipa.it

R. Lo Franco · G. Compagno
Dipartimento di Fisica e Chimica, Università di Palermo, via Archirafi 36,
90123 Palermo, Italy

F.F. Fanchini et al. (eds.), *Lectures on General Quantum Correlations and their Applications*, Quantum Science and Technology,
DOI 10.1007/978-3-319-53412-1_17

of the system, even at a finite time, thus compromising their exploitation [7–10, 21–27]. Such a fate for quantum properties especially manifests within the configuration of independent qubits each one locally embedded in its own environment [8], which is the one required for implementing quantum communication and information protocols with distant individually addressable particles [18, 19]. Efforts are thus necessary to design efficient and feasible procedures to protect quantum correlations against detrimental noise.

Under this perspective, in contrast to Markovian (memoryless) environments, suitable engineered environments capable to maintain quantum memory effects have been employed [5–9, 28–51]. Such non-Markovian environments exhibit the general property to be necessary for revivals of quantum correlations to occur, irrespective of the fact whether they have either a quantum nature (e.g., a bosonic or fermionic environment) [7–9, 28, 29, 50, 52, 53] or a classical nature (e.g., stochastic noise, random field, phase noisy laser) [54–74]. The possibility to have revivals of quantum correlations allow an extension of their exploitation time for some specific protocol. In order to make the revival phenomenon in open quantum systems easily reproducible and effective, it is of basic interest to understand its underlying mechanisms, particularly in light of the fact that it may happen under noise conditions originating from fundamentally different surrounding environments.

Revivals of entanglement between independent qubits after a finite time of complete disappearance have been first shown in the presence of non-Markovian dissipative quantum environments [28, 29]. Although the emergence of entanglement revivals under these conditions may appear strange at a first rapid look, it has been successively explained in terms of periodic entanglement exchanges among the qubits and the quantum constituents of the environment, because of the back-action of the local quantum environments on the qubits themselves allowed by the memory effects (see Fig. 1a) [28, 75–79]. On the other hand, the possibility to retrieve entanglement once it is destroyed between distant qubits locally subject to classical environments seems particularly counterintuitive, especially when such environments do not back react on the quantum system and are not able to store or share any quantum correlations. The first theoretical observations of entanglement revivals without environment back-action, for instance under random telegraph noise for solid state qubits [54–56, 80, 81], put in evidence the importance of the phenomenon yet leaving open its interpretation. Closing this issue is not only relevant from a fundamental point of view regarding the classical-quantum border, but it also provides insights for the classical control of quantum systems with potential applications in future quantum technology requiring classical interfaces to operate [82–85]. These considerations justify the wide interest in studying the evolution of quantum coherence and correlations in hybrid quantum-classical systems during the recent years [47, 54–74, 86–90].

A convenient approach to understand the mechanisms underlying entanglement revivals in classical environments without back-action is to study simple feasible systems in order to minimize undesired side effects and make the role of classical noise prominent. The simplest possible open quantum system which fulfills this requirement is that depicted in Fig. 1b, made of two initially entangled qubits, one of which (qubit A) is isolated, evolving according to its free Hamiltonian, while

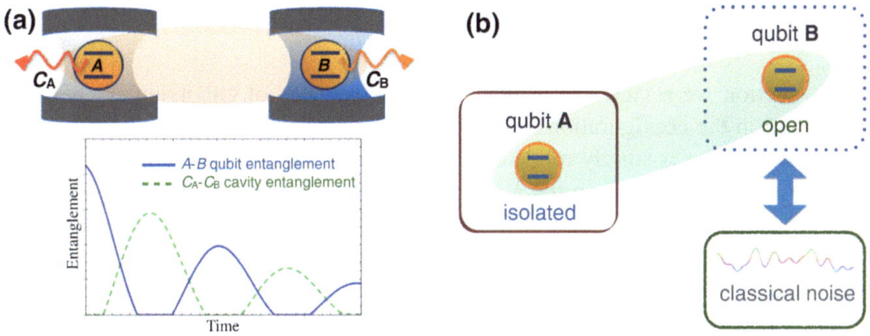

Fig. 1 **Illustrations of the basic systems. a** Two separated initially entangled qubits A and B locally interact with their own quantum environment represented by a cavity with high quality factor. The plot qualitatively show that the initial two-qubit entanglement spontaneously revive after being periodically transferred back and forth to the two cavities, thanks to the memory effects of the leaky cavities under non-Markovian conditions (see also Ref. [77]). **b** A classical noise acts on the qubit B, whereas qubit A is isolated. The two qubits are initially entangled. The two-qubit entanglement evolution under this configuration is qualitatively analogous to the one with both qubits locally interacting with their own environments

the second one (qubit B) interacts with a classical noise and thus evolves under the action of a non-unitary dynamical map. The two-qubit system and the classical environment are initially decoupled. Such a situation, which is paradigmatic for decoherence problems, is also known as the "spectator configuration" [37]. This chapter is devoted to review the main theoretical results about the topic within this configuration, the experimental observations and the interpretations supplied so far. Moreover, the physical aspect that gathers the various interpretations under a unified framework is here provided.

In particular, the chapter is organized as follows. In Sect. 2 we first discuss entanglement revivals in the case of classical environment modeled by a random external field [57, 58], which represents the first attempt to provide a simple model for deepening and understanding the phenomenon; we then also report two other models typical of the solid state where the classical noise is a low-frequency noise [68] and a random telegraph noise (RTN) [55]. In Sect. 3 we describe two quantum optics experiments which reproduce the models with random external field [59] and low-frequency noise [69], respectively, and verify the existence of entanglement revivals. In Sect. 4 we report the three known interpretations of the phenomenon of entanglement revivals in classical environments [58, 59, 68], putting them under a unified view. We give our conclusions and outlook on the topic in Sect. 5.

2 Theoretical Predictions

In this section we review the results about the revivals of entanglement between two qubits in the configuration of Fig. 1b. We particularly focus on the case when the classical noise is simply modeled by a random external field and, successively, consider also the cases when the noise is the typical one encountered by superconducting qubits in the solid state such as longitudinal low-frequency noise and RTN. The two qubits are considered identical, that is with the same transition frequency $\omega_{0A} = \omega_{0B} = \omega_0$, and separated. The total Hamiltonian is thus in general given by $\mathcal{H}_{\text{tot}} = \mathcal{H}_A + \mathcal{H}_B$, where $\mathcal{H}_A = -(\omega_0/2)\sigma_z$ is the free Hamiltonian of qubit A where $\sigma_z = |0\rangle\langle 0| - |1\rangle\langle 1|$ is the third Pauli matrix. We shall see that the first two types of noise are capable to make entanglement revive spontaneously during the evolution, while the third one needs a local operation to obtain the desired entanglement revival.

2.1 Random External Field

As mentioned in the introduction, when the local environment is a quantum non-Markovian one (for instance, a bosonic reservoir of photons inside a high-quality factor cavity [28, 29]), the discovery that the entanglement between two separated noninteracting qubits can reappear during the evolution after complete vanishing has been interpreted by repeated bipartite entanglement exchanges among the quantum parts of the global system [8, 68, 77]. In fact, the initial two-qubit entanglement is redistributed between the two quantum reservoirs and between a qubit and the other qubit's reservoir and, thanks to the memory effects, returns to the two qubits with a partial loss [77–79]. Revivals of two-qubit entanglement were successively predicted also for local non-Markovian classical environments which do not back-react and cannot share quantum excitations, such as random telegraph noises [54–56] and phase noisy lasers [64], that poses a very simple question: in this case, where do the initial quantum correlations go when they disappear?

In order to answer this question, the best strategy appears that of finding a simple yet paradigmatic model which allows a straightforward treatment of the phenomenon. Starting from the phase noisy laser, which is a classical field with a randomly fluctuating phase [64, 91, 92], the natural simplification is that to consider a field with a random phase assuming only two possible values [57, 58]. While the first study with such a random external field employs a model where both qubits locally interact with the noise [57], here we review the simplest case where a qubit is isolated (Fig. 1b) [58], whose all-optical experimental simulation has been realized [59] and shall be discussed in Sect. 3.

As illustrated in Fig. 2a, the environment is a classical field (laser) with a random phase φ, which can be $\varphi_\pm = \pm\frac{\pi}{2}$ with probability $p_\pm = \frac{1}{2}$ [57, 58]. The dynamical map which gives the evolved state of the two-qubit system is

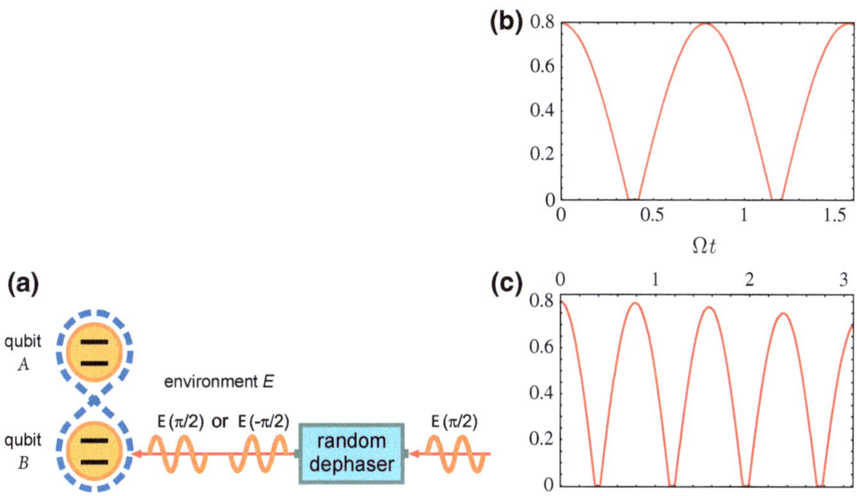

Fig. 2 Revivals under random external field. a A random external classical field acts on the qubit B. The random dephaser either shifts of π the phase of the input field with probability $1/2$ or leaves it unchanged with probability $1/2$ (figure from Ref. [58]). **b** Concurrence $C(t)$ of $\rho_{AB}(t)$ versus Ωt for initial conditions $x = z = 1$ and $y = 0.9$ in the case of periodic dynamics ($\sigma \to 0$). **c** Concurrence $C(t)$ of $\rho_{AB}(t)$ for the same initial conditions in the case of decoherent dynamics ($\sigma = 0.1\,\Omega$)

$$\rho_{AB}^{\Omega}(t) = \frac{1}{2}\sum_{\varphi=\varphi_\pm}\Big(\mathbb{1}_A \otimes U_{\varphi,\Omega}(t)\Big)\rho_{AB}(0)\Big(\mathbb{1}_A \otimes U^{\dagger}_{\varphi,\Omega}(t)\Big), \tag{1}$$

where $\mathbb{1}_A$ is the identity matrix in the Hilbert space of the qubit A and

$$U_{\varphi,\Omega}(t) = \begin{pmatrix} \cos(\Omega t/2) & \mathrm{e}^{-\mathrm{i}\varphi}\sin(\Omega t/2) \\ -\mathrm{e}^{\mathrm{i}\varphi}\sin(\Omega t/2) & \cos(\Omega t/2) \end{pmatrix}, \tag{2}$$

is the unitary matrix in the basis $\{|0\rangle, |1\rangle\}$ of the time evolution operator associated to the interaction between qubit B and a classical electric field $\mathbf{E}$ with phase φ. This interaction is described, in the rotating frame at the qubit-field frequency and within the rotating wave approximation, by the Hamiltonian [57, 59]

$$\mathcal{H}_\varphi = \mathrm{i}\hbar(\Omega/2)(\sigma_+ e^{-\mathrm{i}\varphi} - \sigma_- e^{\mathrm{i}\varphi}), \tag{3}$$

where Ω is the qubit-field coupling constant (Rabi frequency) proportional to the field amplitude and $\sigma_+ = |1\rangle\langle 0|, \sigma_- = |0\rangle\langle 1|$ are the qubit raising, lowering operators. The non-Markovian dynamical map of Eq. (1) is a completely positive trace preserving map representing a unital channel Λ_t (that is, $\Lambda_t \mathbb{1} = \mathbb{1}$) of the class of random unitaries [3, 60, 91, 93]. A useful feature of this dynamical map is that, if the two-qubit initial state belongs to the class of Bell-diagonal states, which are mixtures of the four Bell states, the evolved state will remain inside this class during the evolution [94–98].

In realistic situations a signal inhomogeneous broadening can be present [59] whose effect is a Gaussian distribution in the field amplitude and thus in the Rabi oscillation frequency Ω_g, which must be traced out in order to get the evolved two-qubit state $\rho_{AB}(t)$. In this case one has

$$\rho_{AB}(t) = \int_{-\infty}^{\infty} d\Omega_g \, G(\Omega_g) \, \rho_{AB}^{\Omega_g}(t), \quad G(\Omega_g) = \frac{1}{\sigma\sqrt{\pi}} e^{-\frac{(\Omega_g-\Omega)^2}{4\sigma^2}}, \tag{4}$$

where Ω is the Rabi frequency without dissipation (the central Rabi frequency) and σ the standard deviation (the Rabi frequency width). The effect of the noise on the random field is transferred to the intrinsic evolution of the quantum system.

To investigate the dynamics originating from different initial conditions, a convenient two-qubit initial state is [58]

$$\rho_{AB}^{0}(x, y, z) = y|x_+\rangle\langle x_+| + (1-y)|z_-\rangle\langle z_-|, \tag{5}$$

where

$$|x_+\rangle = x|2_+\rangle + \sqrt{1-x^2}|1_+\rangle, \quad |z_-\rangle = z|2_-\rangle + \sqrt{1-z^2}|1_-\rangle, \tag{6}$$

and $|1_\pm\rangle = (|01\rangle \pm |10\rangle)/\sqrt{2}$, $|2_\pm\rangle = (|00\rangle \pm |11\rangle)/\sqrt{2}$ are the one-excitation and two-excitation Bell (maximally entangled) states. Such an initial state allows both a linear combination (quantum coherence) between Bell states of different kinds and a statistical mixture of them. Here we limit to the case of an initial Bell diagonal state and utilize the concurrence C [99] to quantify the two-qubit entanglement. For convenience, we recall that the concurrence is defined as $C(\rho_{AB}) = \max\{0, \sqrt{\chi_1} - \sqrt{\chi_2} - \sqrt{\chi_3} - \sqrt{\chi_4}\}$, where χ_j's are the eigenvalues in decreasing order of the matrix $\rho_{AB}(\sigma_y \otimes \sigma_y)\rho_{AB}^*(\sigma_y \otimes \sigma_y)$ with σ_y denoting the second Pauli matrix and ρ_{AB}^* corresponding to the complex conjugate of the two-qubit density matrix ρ_{AB} in the canonical basis $\{|00\rangle, |01\rangle, |10\rangle, |11\rangle\}$.

In Fig. 2b–c the dynamics of entanglement is plotted starting from an initial Bell-diagonal state $\rho_{AB}^{0}(1, 0.9, 1)$, which has a concurrence $C = 0.8$ and is the same initial state considered in the experiment of Ref. [59]. Figure 2b shows the periodic evolution corresponding to the case of fixed qubit-field coupling (that is, fixed Rabi frequency, $\sigma = 0$) [57]; Fig. 2c instead represents the case when the Gaussian distribution of the Rabi frequency of Eq. (4) is considered and entanglement peaks decay with a decoherence time proportional to σ^{-1}. The periodic dynamics can be meant as the dynamics of the system at times much shorter than the (Gaussian-induced) decoherence time. Revivals and dark periods of entanglement spontaneously shows up in both cases. The interpretations related to this model shall be discussed in Sect. 4.

2.2 Local Pulse Under Low-Frequency Noise

Here we briefly review the model of a pure-dephasing classical noise, that gathers basic characteristics of many nanodevices under low-frequency noise [100–103], with a local pulse applied at a certain time of the evolution [68]. The Hamiltonian which rules the dynamics of the open qubit B is given by ($\hbar = 1$)

$$\mathcal{H}_B(t) = [-\Omega_A\sigma_z + \varepsilon(t)\sigma_z + \mathcal{V}(t)\sigma_x]/2, \tag{7}$$

where $\varepsilon(t)$ is a stochastic process and $\mathcal{V}(t)$ an external control field. This $\mathcal{V}(t)$ represents an echo π-pulse at a given time $\bar{t}$ with evolution operator $\mathrm{e}^{-\mathrm{i}\sigma_x\pi/2} = -\mathrm{i}\sigma_x$, short enough to neglect the effect of noise during its application. The stochastic process has an exponential autocorrelation function $\langle\varepsilon(t)\varepsilon(0)\rangle = \sigma^2\mathrm{e}^{-t/\tau}$, with noise correlation time τ. For simplicity, the stochastic process $\varepsilon(t)$ is chosen slow enough to be approximatively static $\varepsilon(t) \approx \varepsilon$ during the evolution time t, which means $\tau \to \infty$. The parameter ε is a Gaussian random variable with zero expectation value and standard deviation σ. This static noise produces an effect analogous to inhomogeneous broadening in nuclear magnetic resonance (NMR) [104].

Taking the two qubits initially in any of the four Bell states, indicated with $|\Psi_0\rangle$, applying the evolution operator due to the Hamiltonian of Eq. (7) and tracing out the static noise degrees of freedom, one finds that the two-qubit system evolves in a mixed state $\rho(t) = \int d\varepsilon p(\varepsilon)|\Psi_\varepsilon(t)\rangle\langle\Psi_\varepsilon(t)|$, where $|\Psi_\varepsilon(t)\rangle = \hat{T}\mathrm{e}^{-\mathrm{i}\int_0^t \mathcal{H}_A(t')dt'} \otimes \hat{T}\mathrm{e}^{-\mathrm{i}\int_0^t \mathcal{H}_B(t')dt'}|\Psi_0\rangle$ and $p(\varepsilon)$ is the Gaussian probability density function of ε. The corresponding time-dependent concurrence is given by [68]

$$C(\rho(t)) = \begin{cases} \mathrm{e}^{-\frac{1}{2}\sigma^2 t^2}, & 0 \le t \le \bar{t}, \\ \mathrm{e}^{-\frac{1}{2}\sigma^2 (t-2\bar{t})^2}, & \bar{t} < t \le 2\bar{t}. \end{cases} \tag{8}$$

In Fig. 3 the entanglement of formation $E_f(\rho(t))$ is plotted, which is monotonically related to the concurrence by [99]

$$E_f(\rho(t)) = \mathrm{h}\Big(\frac{1 + \sqrt{1 - C(\rho(t))^2}}{2}\Big), \tag{9}$$

where $\mathrm{h}(x) = -x\log_2 x - (1-x)\log_2(1-x)$. It is displayed that, if no pulse is applied, $E_f(\rho(t))$ decays and tends to zero at times $\sigma t \gg 1$. Differently, the action of a local pulse at $t = \bar{t}$ makes $E_f(\rho(t))$ revive and reach its initial maximum value $E_f(\rho(2\bar{t})) = E_f^{\max} = 1$. This value coincides with the average entanglement $E_{\mathrm{av}}(\mathcal{A}(t)) = \int p(\varepsilon)E_f(|\Psi_\varepsilon(t)\rangle\langle\Psi_\varepsilon(t)|)d\varepsilon = 1$ of the evolved physical ensemble $\mathcal{A} = \{p(\varepsilon)d\varepsilon,\ |\Psi_\varepsilon(t)\rangle\}$ [68]. Notice that in this situation the entanglement revival is not spontaneously found during the evolution but created by means of the local pulse, which makes the dynamical map indivisible and thus non-Markovian [68, 105–107]. A discussion about the interpretation of this result shall be reported in Sect. 4.

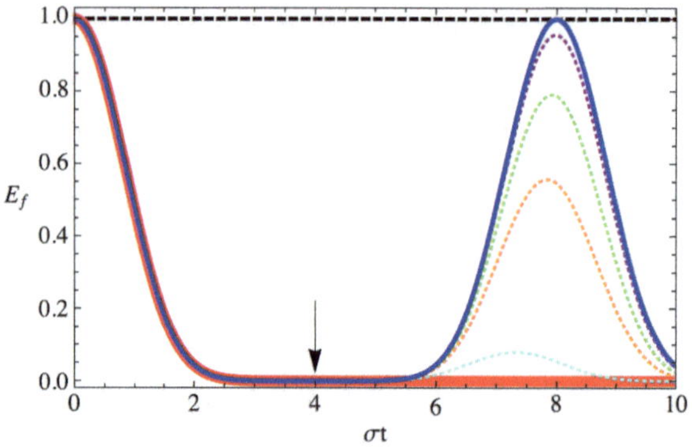

Fig. 3 Revival by local operation under low-frequency noise. Entanglement of formation $E_f(\rho(t))$ as a function of the dimensionless time σt. The *thick red line* gives the free evolution under static noise while the *thin blue solid line* represents the evolution when an echo pulse is applied at time $\sigma\bar{t} = 4$ (individuated by the *black arrow*). The *black dashed line* is the average entanglement of the system $E_{\text{av}} = 1$. *Dotted lines* represent $E_f(\rho(t))$ for a non-static $\varepsilon(t)$ with increasing values of $\sigma\tau$ from *bottom* to *top*. Total entanglement recovery is obtained only in the limit of static noise ($\tau/\bar{t} \to \infty$). Figure from Ref. [68]

2.3 Random Telegraph Noise

The system we review here consists in a pair of independent superconducting qubits, A and B, where qubit B interacts with a bistable impurity (fluctuating charge) which produces pure dephasing RTN [55]. The Hamiltonian of qubit B is ($\hbar = 1$) [108]

$$\mathcal{H}_B = -(\omega_0/2)\sigma_z - (v/2)\xi(t)\sigma_z, \tag{10}$$

where $\xi(t)$ establishes the RTN switching at a rate γ between ± 1 and v is the qubit-RTN coupling constant. The ratio $g = v/\gamma$ is the characteristic parameter that rules the crossover between a Markovian noise for weakly coupled impurities ($g < 1$) and a non-Markovian noise for strong coupled impurities ($g > 1$) [108]. The exact evolution of single-qubit coherence $q(t) \equiv \rho_{01}(t)/\rho_{01}(0)$ is known [55, 108] and in turn allows to obtain the evolved two-qubit density matrix by a standard procedure based on the independence of the two qubits and their own environments [28, 29].

With the qubits initially prepared in the extended Werner-like (EWL) states [29]

$$\rho_1 = r|1_a\rangle\langle 1_a| + \frac{1-r}{4}\mathbb{1}_4, \quad \rho_2 = r|2_a\rangle\langle 2_a| + \frac{1-r}{4}\mathbb{1}_4, \tag{11}$$

where $|1_a\rangle = a|01\rangle + b|10\rangle$, $|2_a\rangle = a|00\rangle + b|11\rangle$ with $|a|^2 + |b|^2 = 1$ and $\mathbb{1}_4$ is the two-qubit identity matrix, one can follow the entanglement dynamics by the concurrence $C = C(t)$. The density matrix of EWL states has an X form [29] and this structure is maintained during the pure dephasing evolution. We recall that entangled states of superconducting qubits with purity ≈ 0.87 and fidelity to Bell states ≈ 0.90

Fig. 4 Revivals under random telegraph noise. Concurrence as a function of the dimensionless time γt for values of g equal to 0.5 (*solid black line*), 1.1 (*red dashed line*), 2 (*green dot-dashed line*), 5 (*blue dotted line*). Initial state parameters are $r = r_{\exp} = 0.91$, $a = b = 1/\sqrt{2}$. Figure from Ref. [55]

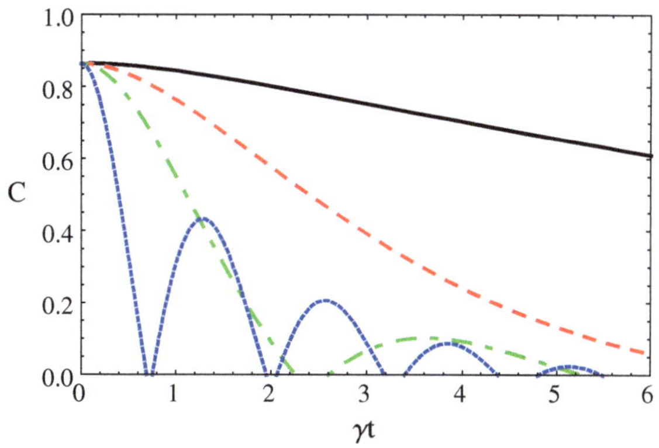

have been experimentally generated [109] and can be approximately described by EWL states with a purity parameter $r_{\exp} \approx 0.91$.

The concurrences at time t for both the two initial states of Eq. (11) are equal to $C(t) = \max\{0, 2K(t)\}$, where $K(t) = r|a|\sqrt{1-|a|^2}|q(t)| - (1-r)/4$, $q(t)$ being the single-qubit coherence. The plots of Fig. 4 display that a sequence of revivals of entanglement occur provided that the coupling parameter g reaches sufficiently high values which enable a non-Markovian dynamics for the system, with the frequency of revivals increasing as g increases. Therefore, a noise of completely classical nature as the RTN causes two-qubit entanglement to reappear after dark periods once the non-Markovian feature of the noise has been activated by a sufficiently strong qubit-impurity coupling [54–56].

3 Experimental Observations

This section deals with the description of two all-optical experiments that, simulating two of the models reported above, confirm that quantum entanglement can either spontaneously revive [59] or be recovered by local operations [69] in classical environments.

3.1 Experimental Entanglement Revivals Under Random External Field

The simplicity of the theoretical model of Fig. 2a has the advantage of making it realizable by a neat experimental setup which avoids any side effect that can influence the expected dynamics and complicate its interpretation. In particular, an all-optical experiment was reported that simulates this model, with the random external field mimicked by quantum degrees of freedom of the optical devices, and allows observation and control of entanglement revivals without system-environment back-action [59]. The experimental setup is shown in Fig. 5. The bipartite quantum system is made of two polarized photons, each one representing a qubit with basis states $|H\rangle$ (horizontal polarization) and $|V\rangle$ (vertical polarization). We omit the very technical details of the devices employed in the setup (available in Ref. [59]) while focusing on the general aspects of the experiment which determine the realization of the target model.

The *preparation part* of the setup generates a pair of polarization entangled photons in a desired Bell-diagonal state ρ_{ab}^{in}. The photon in mode a (the isolated qubit) is directly sent to the state tomography part while the photon in mode b goes to the environment part and finally to state tomography part.

The *environment part* of the setup simulates the random external field on qubit b by exploiting a beam-splitter that creates two photon paths (reflected p_+ and transmitted p_-), corresponding to the effect of the field with either phase, plus the measurement process that does not distinguish the two paths $\text{p}_\pm$ in a classically probabilistic fashion, creating a statistical mixture of them with equal probabilities (1/2). The two photonic paths are designed such as to induce, apart from an unimportant global phase factor, the unitary transformations

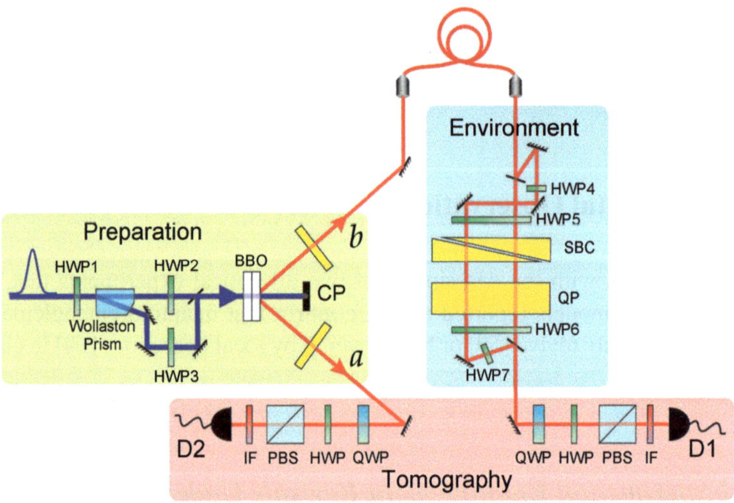

Fig. 5 Experimental setup simulating the random external field. The all-optical setup which realizes the model of Fig. 2a is made of three main parts. (i) *Preparation*. This part is devoted to initialize the two-photon state in the desired Bell-diagonal state. The photon in mode a (corresponding to the isolated qubit A) is directly sent to the measurement apparatus. (ii) *Environment*. The photon in mode b (representing the open qubit B) reaches the environment part, with two possible probabilistic paths representing the two unitaries (random phases). (iii) *Tomography*. This part performs suitable measures of the photon polarizations which allow the construction of the output (evolved) density matrix. Figure from Ref. [59]

$$|H\rangle \xrightarrow{\mathrm{p}_\pm} \cos(\phi/2)|H\rangle \pm i\sin(\phi/2)|V\rangle,$$
$$|V\rangle \xrightarrow{\mathrm{p}_\pm} \pm i\sin(\phi/2)|H\rangle + \cos(\phi/2)|V\rangle, \tag{12}$$

where ϕ is the phase difference between $|H\rangle$ and $|V\rangle$ introduced by the Soleil-Babinet compensator (SBC) and the quartz plates (QPs). This phase difference is defined as $\phi = \omega\tau$, where ω is the photon frequency and $\tau \equiv L\Delta n/c$ is the time taken by the photon to cross the optical element (SBC or QP), L being the thickness of the optical element, c the vacuum speed of light, Δn the difference between the refraction indices of H and V polarizations. It is immediate to see that the two paths $\mathrm{p}_\pm$ of Eq. (12) define unitaries $U_{\mathrm{p}_\pm}(\phi)$ on the basis states $\{|H\rangle, |V\rangle\}$ of b which act exactly as the two time evolution operators $U_{\varphi_\mp}(t)$ of Eq. (2) on the qubit B, respectively, with the connections $|0\rangle \leftrightarrow |H\rangle$, $|1\rangle \leftrightarrow |V\rangle$ and $\phi = \omega\tau \leftrightarrow \Omega t$. The overall output state thus becomes

$$\Lambda\rho_{ab}^{\mathrm{in}} = \frac{1}{2}\sum_{p=\mathrm{p}_\pm}(\mathbb{1}_a \otimes U_p)\rho_{ab}^{\mathrm{in}}(\mathbb{1}_a \otimes U_p^\dagger), \tag{13}$$

which reproduces the two-qubit evolved density matrix of Eq. (1). In the experiment the photon has an intrinsic Gaussian frequency distribution $f(\omega) = (2/\sigma\sqrt{\pi})\exp[-4(\omega-\omega_0)^2/\sigma^2]$ [21, 110], where ω_0 is the center frequency and σ the frequency width (standard deviation). This frequency degree of freedom is a decoherence source analogous to that due to the Rabi frequency distribution in the model of Fig. 2a described above. The experimental evolved state ρ_{ab}^{out} is then determined by tracing out the photon frequency stochastic variable from $\Lambda\rho_{ab}^{\mathrm{in}}$, giving rise to an evolved state analogous to that of Eq. (4).

The *tomography part* finally performs standard quantum state tomography for constructing the output (evolved) density matrix ρ_{ab}^{out} of the two photons at many values of the experimental time τ.

The two-photon system is initialized in the Bell-diagonal state $\rho_{ab}^{\mathrm{in}}=\rho_{AB}^{0}(1, 0.9, 1)$ of Eq. (5). The entanglement evolution is followed by resorting to the concurrence $C(\rho_{ab}^{\mathrm{out}})$, the experimental points being acquired from the reconstructed output density matrices by state tomography. The results for the coherent evolution, where only the SBC is used, are plotted as a function of the relative phase $\phi = \omega_0\tau$ in Fig. 6a. Entanglement exhibits dark periods (around points $\pi/2$ and $3\pi/2$ in a 2π period) and revivals. In Fig. 6b the experimental results are displayed for the decoherent evolution, where both SBC and QPs are used. In particular, the main panel shows the envelope dynamics of correlations as a function of the quartz plate length L, given by the maximum amplitudes of revivals. The maxima of entanglement revivals monotonously decrease and totally vanishes at $L \approx 258\lambda_0$. The inset of Fig. 6b is a plot of the theoretical curves exhibiting these decaying revivals. The coherent evolution is part of the decoherent evolution, as highlighted by the red box in Fig. 6b.

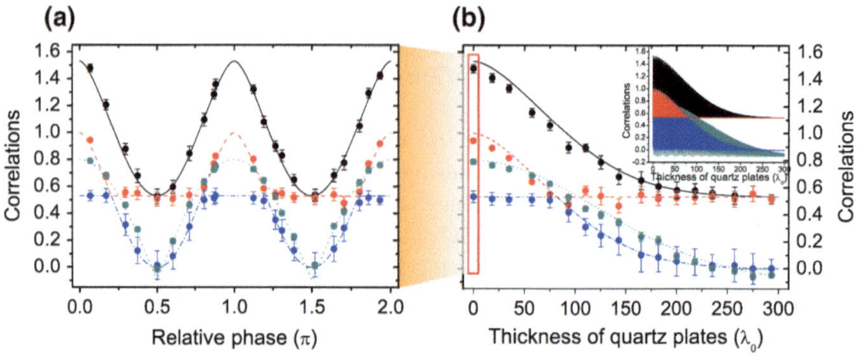

Fig. 6 Experimental observations for the random external field. a Theoretical (*cyan line*) and experimental (*cyan points*) two-qubit entanglement as a function of the relative phase in the coherent evolution. Only the SBC is used in the setup and the relative phase (in units of π) is $\phi = \omega_0\tau$. Revivals of entanglement are clearly visible. **b** Theoretical (*cyan line*) and experimental (*cyan points*) envelope of entanglement dynamics as a function of the quartz plate length L (in units of $\lambda_0 = 800$ nm, that is the center photon wavelength). The coherent evolution with revivals in panel **a** is a part of the decoherent evolution, evidenced by the *red box* in panel **b**. The *inset* shows the theoretical dynamics of the various correlations. In both panels, the *black*, *red* and *blue lines* (points) represent, respectively, the theoretical (experimental) total correlations, classical correlations and quantum discord, whose description is out of the scopes of this review. The initial state ρ_{ab}^{in} is the Bell-diagonal state $\rho_{AB}^0(1, 0.9, 1)$ of Eq. (5). Figure from Ref. [59]

3.2 Experimental Entanglement Revival by Local Operations Under Low-Frequency Noise

An all-optical experiment [69] has been reported which reproduces the theoretical model of a longitudinal low-frequency noise with a local echo pulse, described above and ruled by the Hamiltonian of Eq. (7). We review here the main aspects and results of this experiment.

The experimental setup is illustrated in Fig. 7a. The qubit information is encoded, as usual, in the horizontal H and vertical V polarizations of two photons A and B propagating in the freespace. The two-photon system is initially prepared in the (maximally entangled) Bell state $|\Psi^-\rangle = (|H_A V_B\rangle - |V_A H_B\rangle)/\sqrt{2}$ by standard parametric down-conversion. Photon A, representing the isolated qubit, is directly sent to the measurement device, whilst photon B interacts with a classical environment described by a stochastic process $x(t)$, playing the role of the variable $\epsilon(t)$ in the Hamiltonian of Eq. (7).

The designed noisy channel acting on B induces pure dephasing at times t_k by means of a sequence of four liquid crystals retarders (LC_k), each one introducing a phase $x_k \equiv x(t_k)$ between the photon polarization components, that is $\alpha|H_B\rangle + \beta|V_B\rangle \to \alpha|H_B\rangle + e^{ix_k}\beta|V_B\rangle$. This procedure realizes the desired interaction Hamiltonian $H_B(t) = x(t)\delta(t - t_k)\sigma_z/2$, where $\sigma_z = |H\rangle\langle H| - |V\rangle\langle V|$ and $\delta(t)$ is the Dirac delta function. The induced phase $x_k \in [0, \pi]$ can be arbitrarily adjusted

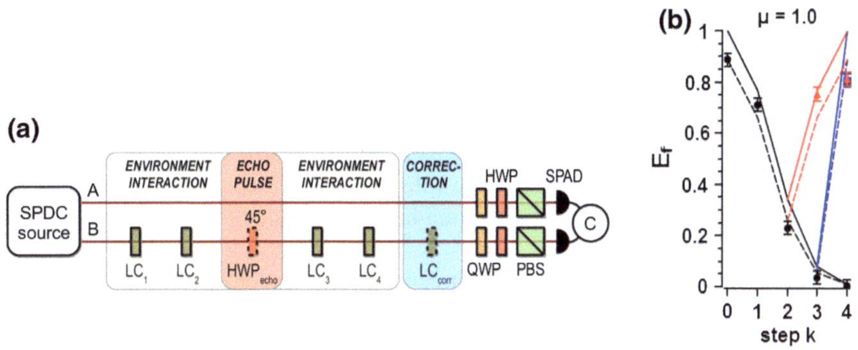

Fig. 7 Experimental setup and observations for the low-frequency noise with local operation. **a** The experimental apparatus initially prepares qubits A and B in the Bell state $|\Psi^-\rangle$ by standard parametric down conversion (SPDC). While qubit A directly goes to measurement part, qubit B stroboscopically interacts with the environment through four random phases induced by liquid crystal (LC) retarders. The noise induced by the environment is compensated by an echo-pulse unitary $U_{\text{echo}} = \sigma_x$ produced by an half-wave plate (HWP). Other elements in the measurement part are quarter-wave plate (QWP), polarizing beam-splitter (PBS), single photon avalanche photodiode (SPAD) and coincidence counting electronics (C). The "correction" LC_{corr} represents a rephasing unitary which is able to compensate the dephasing noise when the latter is known (a situation that is not treated here). **b** Entanglement of formation E_f measured at each step k ($k = 1, 2, 3, 4$) for $\mu = 1$ (static noise). Points and lines represent the experimental data and the theoretical calculations, respectively. *Dashed lines* are simulations for a state with a fidelity $F = 0.96$ to a Bell state. *Black* and *red colors* correspond, respectively, to the uncontrolled and pulsed dynamics (the *blue* one is the controlled dynamics). Figures from Ref. [69]

by the voltage applied to each LC_k. The stochastic process is simulated by generating an ensemble of N random phase sequences $\{x_1, x_2, x_3, x_4\}$, where each phase x_k is a Gaussian random variable with same variance σ^2 and (normalized) autocorrelation $\mu \equiv \langle x_k x_{k+1}\rangle/\sigma^2$ ($\mu \in [0, 1]$). The local echo pulse on photon B is then produced by means of a half-wave plate (HW) at 45° between LC_2 and LC_3 (see Fig. 7a) which realizes a local bit-flip operation $U_{\text{echo}} = \sigma_x$, flipping the polarization of photon B ($\sigma_x|H\rangle = |V\rangle$ and viceversa). The dynamics of the two-photon system, which must be averaged with respect to all the phase sequences in order to trace out the noise degrees of freedom, is finally determined by mixing together the tomographic measurement data obtained for each realization of the N random phase sequences [69]. Here, we focus on the case of static noise which is produced for $\mu = 1$ (see discussion after Eq. (7)).

The evolution of the entanglement of formation of the two photons is shown in Fig. 7b. In absence of local control, entanglement monotonously decays as evidenced by black points and lines. Differently, entanglement is recovered when a local pulse is applied, as displayed by red points and lines. An entanglement echo is thus realized in the system dynamics as predicted by the theoretical model [68].

4 Interpretations of the Phenomenon

So far, three interpretations for the phenomenon of revivals of entanglement in classical environments have been proposed that can be summarized as follows:

(i) the classical environment plays a role as a control mechanism which keeps a classical record for what operation has been applied to the quantum system [57, 59];
(ii) there is an interchange between threepartite correlations and two-qubit entanglement due to system-environment information flows [58];
(iii) the quantum system contains hidden entanglement [68], that is the amount of quantum correlations not revealed by the density matrix description of the system which is recoverable by local operations [68, 69, 80].

In this section we review these interpretations, providing the physical aspect that put them under a unified point of view.

4.1 Quantum-Classical State and Classical Environment as a Controller

The model with random external field described in Sect. 2 can be conveniently described by means of a quantum-classical state, the quantum part played by the two qubits A, B and the classical part by the environment E[57, 59]. For the sake of clearness, we consider here the case when decoherence due to the Gaussian distribution of the Rabi frequency is negligible (analogous argumentations hold even if decoherence is present [58]). Since the classical environment can only be in a time-invariant maximal mixture of its basis states, the overall initial state can be written as

$$\rho_{ABE}(0) = \rho_{AB}(0) \otimes \rho_E = \rho_{AB}(0) \otimes \frac{1}{2} \sum_{\varphi=\varphi_\pm} |\varphi\rangle\langle\varphi|, \tag{14}$$

where $|\varphi_+\rangle$ ($|\varphi_-\rangle$) corresponds to the state of the field with phase $\varphi = \frac{\pi}{2}$ ($\varphi = -\frac{\pi}{2}$). It is then possible to define a unitary evolution $U_{BE}(t)$ acting on the bipartition B-E

$$U_{BE}(t) = \sum_{\varphi=\varphi_\pm} U_{\varphi,\Omega}(t) \otimes |\varphi\rangle\langle\varphi|, \tag{15}$$

where $U_{\varphi,\Omega}(t)$ is the unitary operator of Eq. (2). By the introduction of $U_{BE}(t)$, the evolved state of the threepartite system ABE is obtained by

$$\rho_{ABE}(t) = [\mathbb{1}_A \otimes U_{BE}(t)]\rho_{ABE}(0)[\mathbb{1}_A \otimes U^\dagger_{BE}(t)]. \tag{16}$$

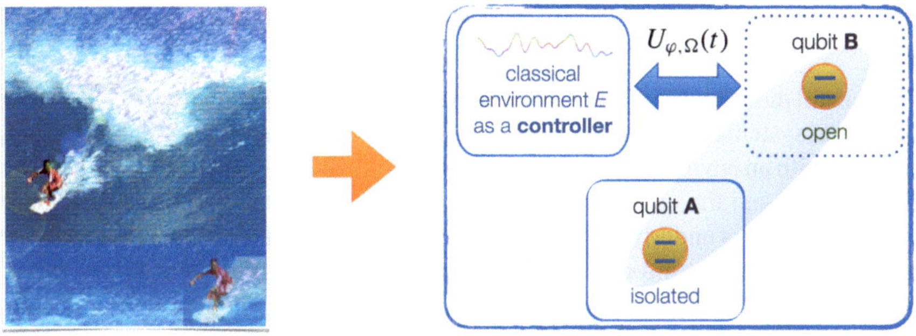

Fig. 8 Classical environment as a control system. On the *left*, a wave rules the dynamics of two coordinated surfers where only one of them is on the wave which remains unaffected by the surfer's motion. This "classical world" situation may supply a pictorial description of a classical environment without back-action whose states control which unitary $U_{\varphi,\Omega}(t)$ is applied to its qubit thus determining the dynamics of the two initially correlated qubits

The two-qubit evolved state $\rho_{AB}(t)$ of Eq. (4) is then straightforwardly determined by tracing out the environmental degrees of freedom ($|\varphi_+\rangle$, $|\varphi_-\rangle$ in this case) from $\rho_{ABE}(t)$. The dynamics of the open system is non-Markovian [59, 60], as witnessed by well-known measures of non-Markovianity [105, 111]. During the evolution due to $\mathbb{1}_A \otimes U_{BE}$, the states of the classical environment remain invariant, the qubit B does not influence the environment E and the qubit-environment back-action is thus absent [57, 59]. Moreover, a classical environment cannot store any quantum correlations on its own. The bipartition B-E evolves under the local unitary operation U_{BE} so that the quantum correlations between B-E, including entanglement, are invariant. If one traces out the isolated qubit A from $\rho_{ABE}(t)$, it is easy to see that the qubit B and its environment E never become quantum correlated. For instance, for an initial A-B Bell-diagonal state, like that considered in the model and in the experiment described above, the reduced state of B-E during the evolution is the uncorrelated state $(\mathbb{1}_B/2) \otimes (\mathbb{1}_E/2)$. Therefore, the qubit-environment correlations do not enter the phenomenon of entanglement revivals.

The introduction of the unitary evolution $U_{BE}(t)$ of Eq. (15) has a crucial role in suggesting an interpretation of these revivals by means of the role of the classical environment as a *controller* for which unitary operation is acting on the system, as pictorially shown in Fig. 8. By memory effects, being the dynamics non-Markovian, the environment E keeps a classical record of what unitary operation has been applied to the qubit B and this occurs even without back-action. The information about the quantum system held by the environment E is therefore due to what action E performs on the system itself. At times when the environment loses this classical information (statistical mixing of the two different unitary operations $U_{\varphi_\pm,\Omega}(t)$, e.g., at $\Omega t = \pi/2$), entanglement disappears; at times when this information is recovered (both unitaries $U_{\varphi_\pm,\Omega}(t)$ act as the same operation, e.g., at $\Omega t = \pi$ they are equal to σ_x), entanglement revives [57, 59].

4.2 Tripartite Correlations and Information Flows

The general model of Fig. 1 can be viewed from the standard decoherence paradigm of a quantum system (qubit A) entangled with a measurement apparatus (qubit B) which interacts with an environment (E) [112], and studying the information fluxes between the system A and the environment E [113, 114]. This fact allows the investigation of the mechanisms underlying the revivals of two-qubit entanglement by approaching the problem from an information-theoretic point of view. For the case of random external field of Fig. 2a, by suitably tracing out the degrees of freedom of the unwanted subsystem in the evolved threepartite state $\rho_{ABE}(t)$ of Eq. (16), it is straightforward to obtain the evolved reduced density matrices of the various components of the global system. Relations between the two-qubit entanglement and the genuine threepartite correlations of the system can be thus found, together with the flows of information among the different parties which provide physical grounds of this relationship [58].

A suitable measure of genuine tripartite correlations of the system $\{A, B, E\}$ is [115, 116]

$$\tau(\rho_{ABE}) = \min\left\{I(\rho_{AB,E}), I(\rho_{AE,B}), I(\rho_{BE,A})\right\}, \tag{17}$$

where $I(\rho_{ij,k}) = S(\rho_{ij}) + S(\rho_k) - S(\rho_{ijk})$ is the quantum mutual information across any possible bipartition ij-k of the tripartite system $\{A, B, E\}$ and $S(\rho) = -\mathrm{Tr}\rho\ln\rho$ is the von Neumann entropy of the quantum state ρ. We recall that genuine tripartite correlations are those which cannot be described as bipartite correlations within any bipartition of a threepartite system [117]. The above measure τ takes into account both classical and quantum correlations of the hybrid quantum-classical system. To evidence dynamical relations between two-qubit entanglement and tripartite correlations, the entanglement is quantified, as usual, by the concurrence $C(\rho_{AB})$ of the two-qubit reduced state.

The evolutions of $C(t) = C(\rho_{AB}(t))$ and $\tau(t) = \tau(\rho_{ABE}(t))$, starting from the Bell-diagonal state $\rho^0_{AB}(1, 0.9, 1)$ considered both in the theoretical model of Sect. 2.1 and in the experiment of Sect. 3.1, are plotted in Fig. 9a–b for a direct comparison. The qubit-field interaction reduces the entanglement while correlating the environment with the two-qubit system. Since B and E always remain uncorrelated during the dynamics (as discussed in the previous subsection), correlations in the overall system can only turn into genuine tripartite correlations, as seen from Fig. 9a–b. When entanglement decreases, genuine tripartite correlations increase, C and τ showing a time behavior in phase opposition such that the maxima of τ coincide with the minima of C and viceversa.[1]

The study of the information fluxes within the system $\{A, B, E\}$ can be conducted by exploiting a relation involving mutual informations I, genuine tripartite correlations τ and von Neumann entropies, which is given by [118]

[1] It is worth to mention here that, if a coherence between Bell states is introduced in the initial two-qubit state, for instance for a state as $\rho_2 = \rho^0_{AB}(0.6, 0.8, 0.3)$ of Eq. (5), freezing of genuine tripartite correlations occurs for finite time periods, showing a plateau in correspondence of the plateau of zero entanglement [58]. The discussion of this behavior is out of the scopes of this chapter.

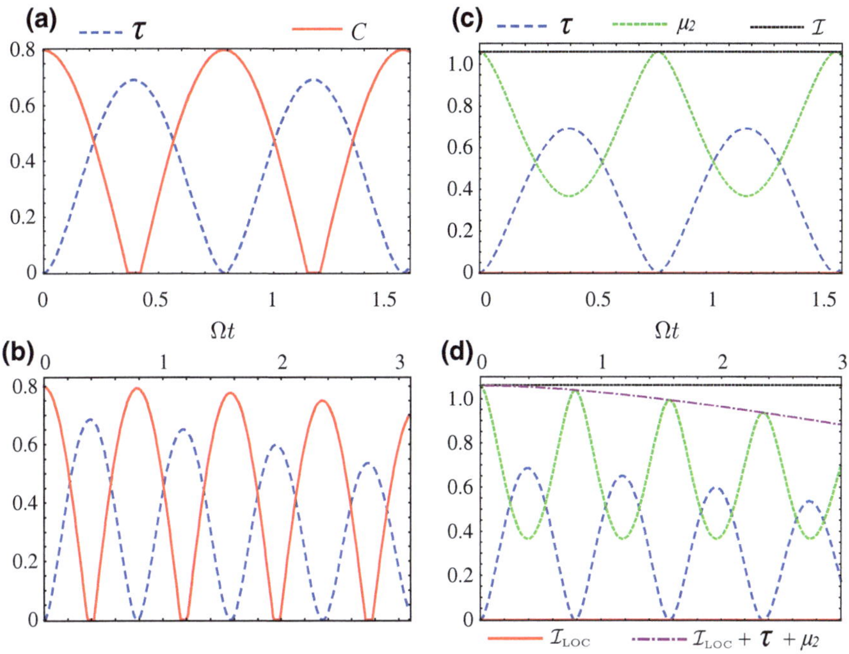

Fig. 9 **Correlation dynamics and information flows.** **a** Total tripartite correlations $\tau(\rho_{ABE}(t))$ (*blue dashed line*) and concurrence $C(\rho_{AB}(t))$ (*red solid line*) versus Ωt for the initial Bell-diagonal state $\rho^0_{AB}(1, 0.9, 1)$ in the case of periodic dynamics ($\sigma \to 0$, no Gaussian distribution of the Rabi frequency). **b** $\tau(\rho_{ABE}(t))$ (*blue dashed line*) and $C(\rho_{AB}(t))$ (*red solid line*) for the same initial conditions under decoherent dynamics (Gaussian distribution of the Rabi frequency with $\sigma = 0.1\,\Omega$). **c** Genuine tripartite correlations τ (*dashed blue line*), total state information $\mathcal{I}$ (*dotted black line*), maximal bipartite correlations μ_2 (*green dashed line*) and local state information $\mathcal{I}_{\rm LOC}$ (*red solid line*) versus Ωt for the initial Bell-diagonal state $\rho^0_{AB}(1, 0.9, 1)$ in the case of periodic dynamics ($\sigma \to 0$). **d** The same quantities of panel **c** plotted in the case of decoherent evolution ($\sigma = 0.1\,\Omega$). Figures from Ref. [58]

$$\mathcal{I} = \mathcal{I}_{\rm LOC} + \tau + \mu_2, \tag{18}$$

where $\mathcal{I} = \mathcal{I}(\rho_{ABE}) = \ln d - S(\rho_{ABE})$ is the state information of the total tripartite state ρ_{ABE} in the Hilbert space of dimension $d = 2^3 = 8$, $\mathcal{I}_{\rm LOC} = \mathcal{I}(\rho_A) + \mathcal{I}(\rho_B) + \mathcal{I}(\rho_E)$ is the total state information locally stored in each part, with $\mathcal{I}(\rho_i) = \ln d_i - S(\rho_i)$ ($i = A, B, E$; $d_i = 2$), and $\mu_2 = \max\{I(\rho_{ij})\}$ is the maximal mutual information over any possible bipartite reduced state ρ_{ij}. According to the expression of Eq. (18), local information, tripartite and bipartite correlations thus constitute three containers where the system can store its total information. In Fig. 9c–d the dynamics of all the quantities involved in Eq. (18) is plotted, when the two qubits are initially in the Bell-diagonal state $\rho^0_{AB}(1, 0.9, 1)$. While in the case of periodic dynamics (closed system) $\mathcal{I} \equiv \mathcal{I}(0)$ is constant, in the decoherent (Gaussian-induced) dynamics the total state information $\mathcal{I}(t) = \tau + \mathcal{I}_{\rm LOC} + \mu_2$ decays, as shown in panel Fig. 9d.

A particular feature of the dynamics is that the local state information $\mathcal{I}_{\mathrm{LOC}}$ is constantly zero. The information regarding the total state is always stored in bipartite and (or) tripartite correlations. Precisely, the information is periodically transferred back and forth between bipartite and tripartite correlations.

This behavior is physically understandable by looking at the meaning of the involved quantities within the quantum-classical system under consideration. Since the reduced state of the classical environment is a time invariant maximally mixed state $\rho_E = \frac{1}{2}\sum_{\varphi=\varphi_\pm}|\varphi\rangle\langle\varphi|$, it gives $\mathcal{I}(\rho_E) = 0$. The red line in Fig. 9c–d therefore describes the local information $\mathcal{I}_{\mathrm{LOC}}$ due to the two qubits. In this particular case, $\mathcal{I}_{\mathrm{LOC}} = \mathcal{I}(\rho_A) + \mathcal{I}(\rho_B)$ is also zero because $\rho_{AB}(t)$ remains a Bell-diagonal state, having maximally mixed marginals for definition and thus $\mathcal{I}(\rho_A) = \mathcal{I}(\rho_B) = 0$ at any time. Genuine tripartite correlations τ, involving all the three parties of the system, represent the information shared among qubit A, qubit B and environment E. As a consequence, $\mathcal{I}_{\mathrm{LOC}}$ and τ are two different and non-mixable forms of information stored in the system. A convenient qualitative behavior of the possible fluxes can be provided in a nutshell as [58]

$$\mathcal{I}_{\mathrm{LOC}} \nrightarrow \tau, \quad \mathcal{I}_{\mathrm{LOC}} \rightleftarrows \mu_2 \rightleftarrows \tau, \tag{19}$$

which also indicates that local information and bipartite correlations can transform into each other as well as bipartite correlations and genuine tripartite correlations can. Under this scheme, it is clear how in Fig. 9c–d, where $\mathcal{I}_{\mathrm{LOC}}$ is zero, all the information is stored in correlations which, periodically, change from the bipartite to the tripartite kind, enabling entanglement revivals.

4.3 Hidden Entanglement and Lack of Classical Information

Another viewpoint about the occurrence of entanglement revivals under local classical noise is based on the local operation and classical communication (LOCC) principle that, if quantum entanglement is restored by means of a local operation after its disappearance, there must be entanglement hidden in the system which does not emerge in the density matrix description of the quantum system [1]. It is thus useful to introduce the concept of "hidden entanglement" [68].

Let us take a bipartite system defined by an ensemble of states $\mathcal{A} = \{(p_i, |\psi_i\rangle)\}$, for which the statistical distribution of the bipartite pure states $\{|\psi_i\rangle\}$ occurring with probabilities $\{p_i\}$ is known, giving the density matrix $\rho = \sum_i p_i|\psi_i\rangle\langle\psi_i|$. The hidden entanglement of the ensemble is defined as [68]

$$E_h(\mathcal{A}) \equiv E_{\mathrm{av}}(\mathcal{A}) - E(\rho) = \sum_i p_i E(|\psi_i\rangle\langle\psi_i|) - E\Big(\sum_i p_i|\psi_i\rangle\langle\psi_i|\Big), \tag{20}$$

where $E_{\mathrm{av}}(\mathcal{A}) = \sum_i p_i E(|\psi_i\rangle\langle\psi_i|)$ is the average entanglement of the ensemble [119–121] and $E(\rho)$ is a convex quantifier of the entanglement of the state (e.g., entanglement of formation) [4]. The convexity of $E(\rho)$ ensures that $E_h \geq 0$. Notice that in the case of a continuous-variable ensemble of the system (see, for instance, the model of low-frequency noise of Sect. 2.2), the sums become integrals. The hidden entanglement E_h represents the amount of entanglement being unexploitable due to the lack of knowledge of which state of the mixture one is handling. When this classical information is supplied, such amount of entanglement can be recovered with the only aid of local operations. Notice that E_h strictly depends on the particular quantum ensemble description of the state of the system. Its role is thus principally relevant in those dynamical situations which, starting from a pure state of the system, univocally determine a physical decomposition in terms of an ensemble of evolved pure states. Interestingly, this situation is always verified when the system is subject to classical noise, which can be treated as an ensemble of local unitaries (random unitaries) applied to the quantum system [68].

For the case of random external field of Sect. 2.1 with fixed Rabi frequency it is then easy to see that, once fixed the initial two-qubit state $\rho_{AB}(0)$, the physical ensemble is univocally given by

$$\begin{aligned} \mathcal{A}(t) = \Bigg\{ & \left(\frac{1}{2}, (\mathbb{1}_A \otimes U_{\varphi_+,\Omega}(t))\rho_{AB}(0)(\mathbb{1}_A \otimes U^\dagger_{\varphi_+,\Omega}(t))\right), \\ & \left(\frac{1}{2}, (\mathbb{1}_A \otimes U_{\varphi_-,\Omega}(t))\rho_{AB}(0)(\mathbb{1}_A \otimes U^\dagger_{\varphi_-,\Omega}(t))\right)\Bigg\}, \end{aligned} \tag{21}$$

which implies $E_{\rm av}(\mathcal{A}(t)) = E(\rho_{AB}(0))$ at any times, since the amount of entanglement is invariant under local unitary operations (note that for an initial Bell state, one would have $E_{\rm av}(\mathcal{A}(t)) = E(\rho_{AB}(0)) = 1$). Therefore, at times $\bar{t}$ when the entanglement of $\rho_{AB}(\bar{t})$ is zero ($E_f(\rho(\bar{t})) = C(\rho(\bar{t})) = 0$), one has a nonzero hidden entanglement $E_h = E_{\rm av}(\mathcal{A}(\bar{t})) = E(\rho_{AB}(0))$. The ensemble description points out that this hidden entanglement is due to the lack of knowledge about which local operation is acting on the system. At times t^* when this lack of knowledge has no effect, as happens when the two unitaries act as the same operation, entanglement revives reaching its initial value, with $E_f(\rho(t^*)) = E(\rho_{AB}(0))$ and $E_h(\mathcal{A}(t^*)) = 0$ (see the argumentations at the end of above Sect. 4.1).

For the model with local pulse under low-frequency noise of Sect. 2.2, where the two-qubit system starts from a Bell state, each realization of ε gives a pure maximally entangled state forming the ensemble $\mathcal{A} = \{p(\varepsilon)d\varepsilon,\ |\Psi_\varepsilon(t)\rangle\}$. The average entanglement is $E_{\rm av}(\mathcal{A}(t)) = 1$ at any time. Entanglement decay is due to the lack of classical knowledge on the system A-B, namely on the random frequency ε. When the pulse is applied (at $t = \bar{t}$), $E_h \approx 1$ and $E_f \approx 0$. Entanglement is not destroyed during the evolution but hidden. After the pulse, this lack of classical knowledge is gradually reduced until $E_h = 0$ and the entanglement reaches its initial value $E_f = 1$ (at $t = 2\bar{t}$). The classical information needed to recover entanglement is therefore acquired by means of the local echo pulse.

4.4 Unifying Aspect of the Interpretations

The three mechanisms discussed above which explain the phenomenon of entanglement revivals in classical environment have all a necessary common root: the system dynamics is non-Markovian as defined by the presence of backflows of (classical) information from the environment to the system [111, 122].

By collecting the main aspects of the interpretations provided so far, the following qualitative considerations can be done:

- the classical environment keeps memory of which unitary is acting on the qubit thanks to the occurrence of backflows of classical information;
- the periodic transformation of genuine tripartite correlations into two-qubit entanglement is activated by system-environment information fluxes;
- local control leads to a partial coherent exchange of information between system and the environment, as also highlighted in the context of discrete qubit dynamics [106], thus allowing the recovery of the hidden entanglement.

These considerations can be cast under a general unified view by showing that non-Markovianity of the system dynamics defined by information backflows is the required condition for entanglement revivals to occur in the presence of classical environments.

As previously said, local classical noise can be suitably described as an ensemble of local unitaries [68] which make the corresponding dynamical map of the system unital [3, 93]. Under the spectator configuration adopted here typical of the decoherence paradigm (an isolated qubit plus an open qubit interacting with its local environment), it is straightforward to prove that two-qubit entanglement revivals necessarily enable information backflows from the environment to the system and viceversa. In fact, the occurrence of an entanglement non-monotonic evolution within this configuration is just the ground aspect for the non-Markovianity quantifier based on the indivisibility of the dynamical map [105] which coincides, for unital maps [40, 60, 123], with the quantifier based on distinguishability of quantum states as measured by trace distance [111]. The latter is then interpreted in terms of information backflows from the environment to the system, where this information can be either quantum (for the case of dissipative quantum environments) or classical (for the case of nondissipative and classical environments) [40, 111, 113, 114]. Being the dynamical map associated to a classical environment without back-action a unital channel, one finally has the equivalence

$$\textit{Backflows of Classical Information} \Leftrightarrow \textit{Entanglement Revivals in Classical Environments.} \tag{22}$$

Hence, if there is classical information flowing back from the classical environment to the bipartite quantum system in absence of backaction, then entanglement revivals occur; viceversa, if bipartite quantum entanglement revives during the system evolution under a local interaction with a classical environment which does not back react, then system-environment backflows of classical information occur.

5 Conclusion

In this chapter we have presented an overview about some of the main theoretical and experimental results presenting the phenomenon of revivals of quantum entanglement between two qubits where one qubit only is locally interacting with a classical environment, the other qubit being isolated. This configuration is the simplest one to study the effects of the classical environment on system dynamics and its role in restoring entanglement initially present in the two-qubit system. This has been employed by many theoretical studies considering classical noise made, for instance, of a random external field, pure-dephasing low-frequency noise and random telegraph noise, which are the ones we have explicitly presented here (see Sect. 2). Major emphasis has been given to the case of random external field characterized by two random phases, since it constitutes the first instance where a tentative interpretation of the phenomenon of entanglement revivals in classical environments without back-action has been provided [57, 58].

We have then discussed two all-optical experiments reproducing, respectively, the model with a two-phase random external field [59] and the model with dephasing low-frequency noise where a local pulse is applied to make entanglement revive [69]. Both the experiments confirm the theoretical predictions, presenting direct observations of entanglement revivals (spontaneous or inducted by a local operation) in a classical environment.

We have also reviewed the three interpretations provided so far for the phenomenon treated in the chapter, all of them supplying responses to the question: where does quantum entanglement go before reappearing during the system dynamics in absence of back-action? This question stands at the basis of the comprehension of the physical mechanisms allowing entanglement revivals under this condition. We notice that for any nondissipative environment, either quantum or classical, back-action is absent and the entanglement revivals should be thus interpreted by the same mechanisms: for instance, this is the case of unital quantum channels such as bit flip, bit-phase flip and phase flip [96–98, 124]. The three interpretations respectively rely on three different concepts, which can be summed up as follows: (i) classical environment as a controller keeping a record for what unitary operation acts on the qubit [57, 59]; (ii) interchange between threepartite correlations and two-qubit entanglement [58]; (iii) hidden entanglement existing in the system which is recoverable by a local operation [68]. We have finally shown that these explanations of the phenomenon can be collected under a unified physical aspect, namely the presence of non-Markovianity as defined by the occurrence of backflows of classical information from the classical environment without backaction to the quantum system. In general, all the studies developed so far suggest that information backflows (quantum or classical) are the essential requisite to obtain revivals of quantum features, as also pointed out in Ref. [70], independently of the quantum or classical nature of the environment.

The reviewed results and the argumentations here reported supply a wide insight on the mechanisms underlying the recovery of entanglement in hybrid quantum-classical systems. Such a knowledge can be useful to motivate and boost further studies on the manipulation of hybrid systems for quantum technology [82].

Acknowledgements R.L.F. and G.C. acknowledge Felipe Fanchini, Diogo O. Soares-Pinto and Gerardo Adesso for giving them the possibility to contribute to the present book: *Lectures on general quantum correlations and their applications*.

References

1. M.B. Plenio, S. Virmani, Quantum Inf. Comput. **7**, 1 (2007)
2. G. Benenti, G. Casati, G. Strini, *Principles of Quantum Computation and Information* (World Scientific, Singapore, 2007)
3. R. Alicki, K. Lendi, *Quantum Dynamical Semigroups and Applications*. Lecture Notes in Physics, vol. 717 (Springer, Berlin, 2007)
4. R. Horodecki, P. Horodecki, M. Horodecki, K. Horodecki, Rev. Mod. Phys. **81**, 865 (2009)
5. H.-P. Breuer, F. Petruccione, *The Theory of Open Quantum Systems* (Oxford University Press, Oxford, New York, 2002)
6. Á. Rivas, S.F. Huelga, M.B. Plenio, Rep. Prog. Phys. **77**, 094001 (2014)
7. T. Yu, J.H. Eberly, Science **323**, 598 (2009)
8. R. Lo Franco, B. Bellomo, S. Maniscalco, G. Compagno, Int. J. Mod. Phys. B **27**, 1345053 (2013)
9. L. Aolita, F. de Melo, L. Davidovich, Rep. Prog. Phys. **78**, 042001 (2015)
10. K. Modi, A. Brodutch, H. Cable, T. Paterek, V. Vedral, Rev. Mod. Phys. **84**, 1655 (2012)
11. G. Adesso, T.R. Bromley, M. Cianciaruso, arXiv:1605.00806 (2016)
12. R. Lo Franco, G. Compagno, Sci. Rep. **6**, 20603 (2016)
13. R. Lo Franco, G. Compagno, A. Messina, A. Napoli, Open Sys. Inf. Dyn. **13**, 463 (2006)
14. R. Lo Franco, G. Compagno, A. Messina, A. Napoli, Int. J. Quantum Inf. **7**, 155 (2009)
15. R. Lo Franco, G. Compagno, A. Messina, A. Napoli, Phys. Lett. A **374**, 2235 (2010)
16. L. Amico, R. Fazio, A. Osterloh, V. Vedral, Rev. Mod. Phys. **80**, 517576 (2008)
17. M.C. Tichy, F. Mintert, A. Buchleitner, J. Phys. B At. Mol. Opt. Phys. **44**, 192001 (2011)
18. T.D. Ladd, F. Jelezko, R. Laflamme, Y. Nakamura, C. Monroe, J.L. O'Brien, Nature **464**, 45 (2010)
19. N. Gisin, R. Thew, Nat. Photonics **1**, 165 (2007)
20. V. Vedral, Nat. Phys. **10**, 256 (2014)
21. J.-S. Xu et al., Nat. Commun. **1**, 7 (2010a)
22. T. Werlang, S. Souza, F.F. Fanchini, C.J. Villas, Boas. Phys. Rev. A **80**, 024103 (2009)
23. A. Salles, F. de Melo, M.P. Almeida, M. Hor-Meyll, S.P. Walborn, P.H.S. Ribeiro, L. Davidovich, Phys. Rev. A **78**, 022322 (2008)
24. B. Bellomo, G. Compagno, A. D'Arrigo, G. Falci, R. Lo Franco, E. Paladino, Int. J. Quantum Inf. **9**, 63 (2011)
25. B. Bellomo, G. Compagno, A. D'Arrigo, G. Falci, R. Lo, Franco, E. Paladino. Phys. Rev. A **81**, 062309 (2010)
26. J. Laurat, K.S. Choi, H. Deng, C.W. Chou, H.J. Kimble, Phys. Rev. Lett. **99**, 180504 (2007)
27. M.P. Almeida et al., Science **316**, 579 (2007)
28. B. Bellomo, R. Lo Franco, G. Compagno. Phys. Rev. Lett. **99**, 160502 (2007)
29. B. Bellomo, R. Lo Franco, G. Compagno, Phys. Rev. A **77**, 032342 (2008a)
30. K.M.F. Romero, R. Lo Franco, Phys. Scr. **86**, 065004 (2012)
31. R. Lo Franco, Quantum Inf. Process. **15**, 2393 (2016)
32. Z.-X. Man, Y.-J. Xia, R. Lo Franco, Phys. Rev. A **92**, 012315 (2015a)
33. Z.-X. Man, Y.-J. Xia, R. Lo Franco, Sci. Rep. **5**, 13843 (2015b)

34. B. Bellomo, R. Lo Franco, S. Maniscalco, G. Compagno, Phys. Rev. A **78**, 060302(R) (2008b)
35. B. Bellomo, R. Lo Franco, G. Compagno, Adv. Sci. Lett. **2**, 459 (2009)
36. P. Haikka, T.H. Johnson, S. Maniscalco, Phys. Rev. A **87**, 010103(R) (2013)
37. C. González-Gutiérrez, R. Román-Ancheyta, D. Espitia, R. Lo Franco, Int. J. Quantum Inform. (in press). Preprint at arXiv:1604.04671 (2016)
38. R. Lo Franco, New J. Phys. **17**, 081004 (2015)
39. F. Brito, T. Werlang, New J. Phys. **17**, 072001 (2015)
40. B. Bylicka, D. Chruściński, S. Maniscalco, Sci. Rep. **4**, 5720 (2014)
41. C. Addis, G. Brebner, P. Haikka, S. Maniscalco, Phys. Rev. A **89**, 024101 (2014)
42. L. Mazzola, S. Maniscalco, J. Piilo, K.-A. Suominen, B.M. Garraway, Phys. Rev. A **80**, 012104 (2009)
43. J. Tan, T.H. Kyaw, Y. Yeo, Phys. Rev. A **81**, 062119 (2010)
44. Z.X. Man, Y.J. Xia, N.B. An, New J. Phys. **12**, 033020 (2010)
45. Z.-X. Man, N.B. An, Y.-J. Xia, Phys. Rev. A **90**, 062104 (2014)
46. Z.-X. Man, N.B. An, Y.-J. Xia, Opt. Express **23**, 5763 (2015c)
47. C. Benedetti, F. Buscemi, P. Bordone, M.G.A. Paris, Phys. Rev. A **93**, 042313 (2016)
48. B. Bellomo, R. Lo Franco, S. Maniscalco, G. Compagno, Phys. Scr. **T140**, 014014 (2010)
49. B. Bellomo, R. Lo Franco, G. Compagno, Phys. Rev. A **78**, 062309 (2008)
50. B. Bellomo, G. Compagno, R. Lo Franco, A. Ridolfo, S. Savasta, Phys. Scr. **T143**, 014004 (2011)
51. B. Bellomo, G. Compagno, R. Lo Franco, A. Ridolfo, S. Savasta, Int. J. Quantum Inform. **9**, 1665 (2011)
52. J.-S. Xu et al., Phys. Rev. Lett. **104**, 100502 (2010b)
53. F.F. Fanchini, T. Werlang, C.A. Brasil, L.G.E. Arruda, A.O. Caldeira, Phys. Rev. A **81**, 052107 (2010)
54. D. Zhou, A. Lang, R. Joynt, Quantum Inf. Process. **9**, 727 (2010)
55. R. Lo Franco, A. D'Arrigo, G. Falci, G. Compagno, E. Paladino, Phys. Scr. **T147**, 014019 (2012a)
56. P. Bordone, F. Buscemi, C. Benedetti, Fluct. Noise Lett. **11**, 1242003 (2012)
57. R. Lo Franco, B. Bellomo, E. Andersson, G. Compagno, Phys. Rev. A **85**, 032318 (2012b)
58. B. Leggio, R. Lo Franco, D.O. Soares-Pinto, P. Horodecki, G. Compagno, Phys. Rev. A **92**, 032311 (2015)
59. J.-S. Xu, K. Sun, C.-F. Li, X.-Y. Xu, G.-C. Guo, E. Andersson, R. Lo Franco, G. Compagno, Nat. Commun. **4**, 2851 (2013)
60. M. Mannone, R. Lo Franco, G. Compagno, Phys. Scr. **T153**, 014047 (2013)
61. F. Altintas, A. Kurt, R. Eryigit, Phys. Lett. A **377**, 53 (2012)
62. C. Benedetti, F. Buscemi, P. Bordone, M.G.A. Paris, Phys. Rev. A **87**, 052328 (2013)
63. J.H. Wilson, B.M. Fregoso, V.M. Galitski, Phys. Rev. B **85**, 174304 (2012)
64. B. Bellomo, R. Lo Franco, E. Andersson, J.D. Cresser, G. Compagno, Phys. Scr. **T147**, 014004 (2012)
65. J. Trapani, M. Bina, S. Maniscalco, M.G.A. Paris, Phys. Rev. A **91**, 022113 (2015)
66. A. D'Arrigo, R. Lo Franco, G. Benenti, E. Paladino, G. Falci, Phys. Scr. **T153**, 014014 (2013)
67. A. D'Arrigo, R. Lo Franco, G. Benenti, E. Paladino, G. Falci, Int. J. Quantum Inform. **12**, 1461005 (2014)
68. A. D'Arrigo, R. Lo Franco, G. Benenti, E. Paladino, G. Falci, Ann. Phys. **350**, 211 (2014)
69. A. Orieux, G. Ferranti, A. D'Arrigo, R. Lo Franco, G. Benenti, E. Paladino, G. Falci, F. Sciarrino, P. Mataloni, Sci. Rep. **5**, 8575 (2015)
70. J. Trapani, M.G.A. Paris, Phys. Rev. A **93**, 042119 (2016)
71. C. Benedetti, F. Buscemi, P. Bordone, M.G.A. Paris, Int. J. Quantum Inf. **10**, 1241005 (2012)
72. C. Benedetti, M.G.A. Paris, F. Buscemi, P. Bordone, *22nd International Conference on Noise and Fluctuations (ICNF)* pp. 1–4 (2013)
73. F. Buscemi, P. Bordone, Phys. Rev. A **87**, 042310 (2013)
74. M. Rossi, C. Benedetti, M.G.A. Paris, Int. J. Quantum Inf. **12**, 1560003 (2014)
75. B.-H. Liu et al., Nat. Phys. **7**, 931 (2011)

76. C.E. López, G. Romero, F. Lastra, E. Solano, J.C. Retamal, Phys. Rev. Lett. **101**, 080503 (2008)
77. C.E. López, G. Romero, J.C. Retamal, Phys. Rev. A **81**, 062114 (2010)
78. A. Chiuri, C. Greganti, L. Mazzola, M. Paternostro, P. Mataloni, Sci. Rep. **2**, 968 (2012)
79. J.H. Reina, C.E. Susa, F.F. Fanchini, Sci. Rep. **4**, 7443 (2014)
80. R. Lo Franco, A. D'Arrigo, G. Falci, G. Compagno, E. Paladino, Phys. Rev. B **90**, 054304 (2014)
81. R.L. Franco, A. D'Arrigo, G. Falci, G. Compagno, E. Paladino, Phys. Scr. **T153**, 014043 (2013)
82. G. Kurizki, P. Bertet, Y. Kubo, K. Mlmer, D. Petrosyan, P. Rabl, J. Schmiedmayer, PNAS **112**, 3866 (2015)
83. G.J. Milburn, Phil. Trans. R. Soc. A **370**, 4469 (2012)
84. D. Pavlovic, Proc. Sym. App. Math. **71**, 233 (2012)
85. C. Altafini, F. Ticozzi, I.E.E.E. Trans, Autom. Control **57**, 1898 (2012)
86. A. Beggi, F. Buscemi, P. Bordone, Quantum Inf. Process. 1–33 (2016). doi:10.1007/s11128-016-1334-8
87. C. Benedetti, F. Buscemi, P. Bordone, M.G.A. Paris, Phys. Rev. A **89**, 032114 (2014)
88. C. Benedetti, M.G.A. Paris, Phys. Lett. A **378**, 2495 (2014a)
89. C. Benedetti, M.G.A. Paris, Int. J. Quantum Inf. **12**, 1461004 (2014b)
90. D. Calvani, A. Cuccoli, N.I. Gidopoulos, P. Verrucchi, PNAS **110**, 6748 (2013)
91. E. Andersson, J.D. Cresser, M.J. Hall, J. Mod. Opt. **54**, 1695 (2007)
92. J.D. Cresser, C. Facer, Opt. Commun. **283**, 773 (2010)
93. F.A. Wudarski, P. Nalezyty, G. Sarbicki, D. Chruscinski, Phys. Rev. A **91**, 042105 (2015)
94. S. Luo, Phys. Rev. A **77**, 042303 (2008)
95. B. Bellomo, R. Lo Franco, G. Compagno, Phys. Rev. A **86**, 012312 (2012)
96. B. Aaronson, R. Lo Franco, G. Adesso. Phys. Rev. A **88**, 012120 (2013)
97. B. Aaronson, R. Lo Franco, G. Compagno, G. Adesso, New J. Phys. **15**, 093022 (2013)
98. M. Cianciaruso, T.R. Bromley, W. Roga, R. Lo Franco, G. Adesso, Sci. Rep. **5**, 10177 (2015)
99. W.K. Wootters, Phys. Rev. Lett. **80**, 2245 (1998)
100. G. Falci, A. D'Arrigo, A. Mastellone, E. Paladino, Phys. Rev. Lett. **94**, 167002 (2005a)
101. G. Ithier, E. Collin, P. Joyez, P.J. Meeson, D. Vion, D. Esteve, F. Chiarello, A. Shnirman, Y. Makhlin, J. Schriefl et al., Phys. Rev. B **72**, 134519 (2005)
102. J. Bylander et al., Nat. Phys. **7**, 565 (2011)
103. F. Chiarello, E. Paladino, M.G. Castellano, C. Cosmelli, A. D'Arrigo, G. Torrioli, G. Falci, New J. Phys. **14**, 023031 (2012)
104. L.M.K. Vandersypen, I.L. Chuang, Rev. Mod. Phys. **76**, 1037 (2005)
105. Á. Rivas, S.F. Huelga, M.B. Plenio, Phys. Rev. Lett. **105**, 050403 (2010)
106. J. Sun, Y.-N. Sun, C.-F. Li, G.-C. Guo, K. Luoma, J. Piilo, Sci. Bull. **61**, 1031 (2016)
107. C. Addis, F. Ciccarello, M. Cascio, G.M. Palma, S. Maniscalco, New J. Phys. **17**, 123004 (2015)
108. G. Falci, A. D'Arrigo, A. Mastellone, E. Paladino, Phys. Rev. Lett. **94**, 167002 (2005b)
109. L. DiCarlo et al., Nature **460**, 240 (2009)
110. J.-S. Xu et al., Phys. Rev. Lett. **103**, 240502 (2009)
111. H.-P. Breuer, E.-M. Laine, J. Piilo, Phys. Rev. Lett. **103**, 210401 (2009)
112. W.H. Zurek, Rev. Mod. Phys. **75**, 715 (2003)
113. F.F. Fanchini, G. Karpat, B. Cakmak, L.K. Castelano, G.H. Aguilar, O.J. Farías, S.P. Walborn, P.H. Souto Ribeiro, M.C. de Oliveira, Phys. Rev. Lett. **112**, 210402 (2014)
114. S. Haseli, G. Karpat, S. Salimi, A.S. Khorashad, F.F. Fanchini, B. Cakmak, G.H. Aguilar, S.P. Walborn, P.H. Souto Ribeiro. Phys. Rev. A **90**, 052118 (2014)
115. G.L. Giorgi, B. Bellomo, F. Galve, R. Zambrini, Phys. Rev. Lett. **107**, 190501 (2011)
116. J. Maziero, F.M. Zimmer, Phys. Rev. A **86**, 042121 (2012)
117. C.H. Bennett, A. Grudka, M. Horodecki, P. Horodecki, R. Horodecki, Phys. Rev. A **83**, 012312 (2011)
118. A.C.S. Costa, R.M. Angelo, M.W. Beims, Phys. Rev. A **90**, 012322 (2014)

119. C.H. Bennett, D.P. DiVincenzo, J.A. Smolin, W.K. Wootters, Phys. Rev. A **54**, 3824 (1996)
120. O. Cohen, Phys. Rev. Lett. **80**, 2493 (1998)
121. A.R.R. Carvalho, M. Busse, O. Brodier, C. Viviescas, A. Buchleitner, Phys. Rev. Lett. **98**, 190501 (2007)
122. H.-P. Breuer, E.-M. Laine, J. Piilo, B. Vacchini, Rev. Mod. Phys. **88**, 021002 (2016)
123. S. Haseli, S. Salimi, A.S. Khorashad, Quant. Inf. Process. **14**, 3581 (2015)
124. M. Ali, Phys. Lett. A **378**, 2048 (2014)

Quantum Correlations and Synchronization Measures

Fernando Galve, Gian Luca Giorgi and Roberta Zambrini

Abstract The phenomenon of spontaneous synchronization is universal and only recently advances have been made in the quantum domain. Being synchronization a kind of temporal correlation among systems, it is interesting to understand its connection with other measures of quantum correlations. We review here what is known in the field, putting emphasis on measures and indicators of synchronization which have been proposed in the literature, and comparing their validity for different dynamical systems, highlighting when they give similar insights and when they seem to fail.

1 Introduction

Synchronization phenomena [1, 2] and quantum correlations [3–5] have been studied for a long time by two different communities, and only recently their relation started to be explored. The common ingredient for the emergence of both features is the mutual interaction between the components of a system, and in the quantum regime the potential relation between synchronization and the presence of mutual information, discord, entanglement or other correlations has been recently explored.

The phenomenon of spontaneous or mutual synchronization refers to the ability of two or more systems, that would display different dynamics when separate, to evolve coherently when coupled. In the case of oscillatory dynamics this corresponds to achieving oscillation at a common frequency. This concept has been further refined and generalized in chaotic systems to encompass several scenarios such as, for instance, lag synchronization, generalized synchronization, or phase synchronization [6]. In general, the definition of classical synchronization itself refers

F. Galve (✉) · G. Luca Giorgi · R. Zambrini (✉)
IFISC (UIB-CSIC), Instituto de Física Interdisciplinar y Sistemas Complejos,
Palma de Mallorca, Spain
e-mail: fernando@ifisc.uib-csic.es

R. Zambrini
roberta@ifisc.uib csic.es

F.F. Fanchini et al. (eds.), *Lectures on General Quantum Correlations and their Applications*, Quantum Science and Technology,
DOI 10.1007/978-3-319-53412-1_18

to some similarity in the time evolutions, i.e. some temporal *correlation* between the local dynamical variables of the involved systems. Therefore, this is a definition associated to classical trajectories. The counterpart, and eventually generalization, in the quantum regime can follow different approaches.

The first works on the subject of quantum synchronization were actually dealing with entrainment, where an external driving acts as a pacemaker, in systems such as spin-boson with modulated driving [7], driven resonator with one [8] or two superconducting qubits [9], and more recently driven quantum Van der Pol oscillators [10, 11]. In the case of entrainment, or forced synchronization, the driver is a strong external field, generally classical, and is not influenced by the interaction with the system. Different is the case of mutual synchronization that refers to the emergence of synchronization as a collective phenomenon, leading to a coherent dynamics out of different coupled units, in the absence of a driver.

Mutual synchronization has been recently predicted for spins interacting with a common bath [12] and for the average positions of quantum optomechanical systems [13, 14]. The first analysis in the quantum regime dealt with harmonic networks [15, 16] looking at quantum noise synchronization and showing its counterpart in relation with classical and quantum correlations, showing the same trend for mutual information and quantum discord. Synchronization in the dynamics of second-order quadratures (squeezing) was considered there with an exact approach and role of local, global and independent baths was elucidated. In Ref. [17] it was instead considered the question about the limits to perfect synchronization imposed by quantum fluctuations and uncertainty relations. The phenomenon is characterized with a synchronization error and is discussed in the context of coupled optomechanical devices. Also in Ref. [10] there is a discussion of coherently coupled Van der Pol oscillators characterizing synchronization through their phase-locking in phase space.

When extending the concepts of synchronization into the quantum regime, a first question is about what defines this phenomenon, as actually in the classical regimes it refers to the dynamics of phase space trajectories and, in general, classical variables. This chapter reviews this question showing different approaches as well as the specific peculiarities reported so far for quantum synchronization with respect to the classical case. In the following section we give a brief overview of the platforms where quantum synchronization is under study and then review the characterizations and measures proposed for this phenomenon. We then discuss some general questions and possible future directions.

2 Synchronization in Quantum Systems

The phenomenon of synchronization is paradigmatic in a large variety of biological, physical, and chemical systems, operating in the classical regime, as reviewed for instance in [1, 2, 6, 18, 19]. Some fascinating examples are fireflies flashing at once Fig. 1 cardiac myocytes acting as pacemakers, or the swaying motion of the millennium bridge due to the crowd walk in synchrony. The first reported observation of classical synchronization was described as "sympathy of two clocks" and dates

Fig. 1 A historical perspective of synchronization. *Left panel* Original drawing of Huygens of two synchronizing pendulum clocks attached to a common support. *Middle panel* Swarms of fireflies illuminate the undergrowth in a forest (photo by Kei Nomiyama/Barcroft Media). *Right panel* Micromechanical oscillator arrays coupled through light (figure taken from Ref. [21])

back to XVII century, when Huygens observed pendula hanging on a wall in a boat (see extracts and references in [1]). A reproduction of Huygens' original drawing is presented in Fig. 1. An equivalent popular experiment is with metronomes on the same bar placed on two cylinders (cans) free to roll [20]. The extension to chaotic systems has also been a wide field of research, establishing the possibility to observe this phenomenon in spite of the high sensitivity to small differences in the initial conditions or device parameters [6].

When moving to microscopic systems, synchronization phenomena are expected to take place, and several recent works address quantum synchronization in nanomechanical devices, harmonic oscillators and spin systems.

2.1 *Nanomechanical Devices*

Optomechanical devices exploit radiation pressure to couple coherently light and matter motional degrees of freedom allowing to explore different aspects of synchronization in a flexible, hybrid and highly sensitive platform, where operation in the quantum regime has been achieved [22]. Spontaneous synchronization of optomechanical devices has been predicted theoretically considering mechanical coupling [13] as well as coupling through a common optical mode [14], focusing on the average positions and laying the base for the study of quantum signatures of synchronization in these devices. Phase-coherent mechanical oscillations have been shown in regular optomechanical crystals considering the effects of quantum noise [23].

Reported experiments with microdisks [24] and arrays [21] and nanomechanical resonators interacting through an optical racetrack [25] display synchronization of the average (classical) motional degree of freedom. In Fig. 1 the device used in Ref. [21] is reproduced. The possibility to lock distant optomechanical oscillators has also been explored in the classical regime [26, 27]. Recently it was also reported the experimental realization of spontaneous synchronization among micro- [28] and nano-electromechanical [29] autonomous oscillators. Quantum signatures of synchronization phenomena in experiments on optomechanical devices have not yet been reported.

2.2 Linear and Non-linear Oscillators

Among theoretical models, both linear and non-linear oscillators have been considered in the quantum regime. Van der Pol oscillators have been investigated in the quantum regime [10, 11, 30, 31] and in comparison with the classical one [1]. These models exhibit self-sustained oscillations and spontaneous synchronization due to coherent [10, 32] and dissipative coupling [30, 31], as well as phase locking [10] and frequency entrainment [11] in presence of external drive. The realization of Van der Pol oscillators in physical platforms operating in the quantum regime has been suggested in trapped ions [10] and optomechanical oscillators [31].

Self-sustained oscillators, like Van der Pol oscillators are a well-known platform for studying synchronization. However, due to their non-linearity the analysis in the quantum regime can be only addressed in limited cases and under various approximations, such as truncation of the Wigner function, linearization around stable states [33] or in the limit of infinite non-linear couplings favoring few low-energy Fock states [10]. An *exact* analysis can be performed in linear systems like harmonic networks; these have been considered in order to identify the conditions for quantum synchronization beyond approximations and to clarify the role of losses, comparing diffusive and reactive couplings in Refs. [15, 16, 34]. The analysis of networks in squeezed vacuum [16] shows that under dissipation in separate equivalent baths (a common assumption), independently on the strength of the coupling the oscillators will not be able to synchronize, while in any other more complex dissipation scenarios (common bath, local bath, etc.), the presence of one less damped normal mode of the system allows for a transient or asymptotic synchronization. By accessing few oscillators' parameters this synchronization in the squeezing dynamics can be tuned in the network or in clusters.

2.3 Spin Models

As mentioned before, quantum spin synchronization was first discussed under the perspective of entrainment to an external driving force, either in the general spin-boson framework [7] or considering superconducting two-level systems [8, 9]. An experimental observation was recently reported considering a damped current-biased Josephson junction [35].

Studies of spontaneous synchronization of spins within abstract theoretical models were performed in Refs. [12, 36] considering spin-boson dissipation. In Ref. [12], it was first observed that synchronization is induced by the coherent exchange of bath excitations between the two spins, while in Ref. [36] it was shown that pure dephasing is not able to generate synchronization. The formation of Chimera states was discussed in [37] considering an extended spin chain described by a non-Hermitian Hamiltonian. In Ref. [32], the authors analyzed the behavior of two qubits placed inside two coupled cavities where only the first one is driven by a laser. The steady-state synchronization of ensembles of dissipative, driven two-level atoms collectively coupled to a cavity mode, was studied in [38] and, under more general conditions in [39], where the authors also provided a direct analogy with the synchronization of classical phase oscillators. Following the experimental results of Ref. [40], where a self-rephasing mechanism was observed on the ground state of magnetically trapped ultracold atoms, synchronization within a full quantum model for the case of two non-dissipative, interacting macro-spins, was studied in Ref. [41].

Finally, a platform to probe synchronization was introduced in Ref. [42]. There, the authors considered two cold ions in microtraps and studied the synchronization between their motional degrees of freedom. The presence of synchronization was witnessed by the correlations developed by the electronic, discrete, degrees of freedom of the two ions.

2.4 Applications

Synchronization is clearly a resource in biological systems [1, 2]. The synchronized flashing of fireflies is a strategy so that the female can identify her species-specific flashing signal (Fig. 1). Synchronization of neuronal activity by phase locking of self-generated network oscillations dynamics is one of the coordinating mechanisms of the brain, and abnormalities in this process are at the basis of several diseases and dysfunctions, like epilepsy.

The achievement of a coherent dynamics out of different components (also due to experimental imperfections) is clearly a resource also in physical systems. An interesting application, for instance, is in cryptographic protocols based on chaotic carriers of signals [43]. In general, synchronization allows for enhancing of frequency stability, coherence and power output. Therefore applications are envisaged for precise frequency sources, time-keeping, and sensing [21, 29] and can be taken also to the domain of quantum technologies.

An application of a synchronization transition was recently proposed as an effective tool to probe the dynamics of a quantum system dissipating through a thermal bath [44]. Indeed, coupling the system to an external, detuned object (which plays the role of the probe) a transition between in-phase and anti-phase system-probe synchronization is observed as a function of the detuning and of the spectral density of the bath. Clearly, this transition can be observed monitoring the dynamics of the probe alone. Then, measuring the critical detuning at which the transition takes place amounts to getting information about the whole dissipative process.

3 Measures of Mutual Quantum Synchronization

Synchronization refers to some coherence in the temporal dynamics of coupled systems and several measures are known in the classical realm [1, 2, 6]. Synchronization in presence of driving, entrainment, is typically encoded in the phase locking of the slave system with respect to the drive: the detuning between the slave oscillation and the driving frequency is typically plotted as a function of the frequency of the driver to identify the region of locking (zero detuning) and when this region's width is considered for different driving strengths one gets the synchronization region known as Arnold tongue [1].

In autonomous systems, synchronization can arise as a mutual phenomenon, the final dynamics coming from the interaction between components. The equivalent of an Arnold tongue appears by considering the relative coupling and detuning of the system components. Despite its intuitive conceptual definition, the quantification of synchronization in the quantum realm is a challenging problem where both temporal and quantum correlations come into play. Two or more objects, irrespective of their quantum or classical nature, do spontaneously synchronize if they adjust their own local dynamics to a common pace determined by their mutual interaction. Then, a good synchronization measure is expected to be able to capture this adjustment of rhythms, that can only be detected monitoring the behavior of all local units. Synchronization can be inferred observing how similar the local density matrices are, according to some meaningful criteria, or considering local observables and looking at their correlations in time. Furthermore, it is sometimes possible to deduce the behavior of local observables inspecting overall quantities, like, for instance, emission spectra.

A broad plethora of studies of synchronization in classical systems in the last three decades provides useful hints about possible approaches when moving into the quantum regime. The main difference with respect to classical systems is clearly that synchronization there generally refers to time trajectories and limit cycles in the phase space not present in the quantum approach. On the other hand, classical synchronization has been already generalized in presence of noise and of chaos [6] where it is identified by assessing the 'similarity' of local dynamical evolutions quantified by several indicators. Indeed a number of these indicators can be taken into the quantum regime providing insightful approaches to quantum synchronization, as it is the case of the Pearson function and synchronization error introduced below.

These considerations do not exclude that manifestations of synchronization can be found also in global indicators, including mutual information and correlations, and that can be associated to genuine quantum properties of the whole state. Overall, when addressing the question of the identification of genuine quantum synchronization phenomena, two main approaches can be distinguished: one is to define synchronization in local observables and look for the presence of quantum correlations triggered by this phenomenon; an alternative approach is to define synchronization itself as a form of quantum correlation. In the following we give an overview of different physical cases where synchronization is expected to come out, discussing the interplay between local indicators and collective ones.

3.1 Pearson Factor

The Pearson's correlation coefficient is a widely used measure of the degree of linear dependence between two variables. Calling X and Y the variables, it is defined as

$$\mathcal{C}_{X,Y} = \frac{E[XY] - E[X]E[Y]}{\sqrt{E[X^2] - E[X]^2}\sqrt{E[Y^2] - E[Y]^2}}, \tag{1}$$

where $E[.]$ is an average value. As a consequence of the definition, $\mathcal{C}_{X,Y}$ gives a value between $+1$ and 1, where 1 indicates total positive correlation, 0 is the absence of correlation, and 1 is total negative correlation. Considering two functions $f(t)$ and $g(t)$ evolving in time, the Pearson's coefficient $\mathcal{C}_{f(t),g(t)}(t)$ can be calculated over a sliding window of length Δt replacing the expectation values with time averages: in this case

$$E[f](t, \Delta t) \equiv \overline{f(t, \Delta t)} = \frac{1}{\Delta t}\int_t^{t+\Delta t} f(t). \tag{2}$$

Given two time-dependent variables A_1 and A_2 the Pearson synchronization measure reads

$$\mathcal{C}_{A_1,A_2}(t|\Delta t) = \frac{\int_t^{t+\Delta t}(A_1 - \bar{A}_1)(A_2 - \bar{A}_2)dt}{\sqrt{\int_t^{t+\Delta t}(A_1 - \bar{A}_1)^2 dt \int_t^{t+\Delta t}(A_2 - \bar{A}_2)^2 dt}}, \tag{3}$$

where

$$\bar{A}_i = \frac{1}{\Delta t}\int_t^{t+\Delta t} A_i dt. \tag{4}$$

By definition, this measure quantifies the temporal correlation between two classical trajectories and has been widely used in classical synchronization problems [1, 6].

In the quantum framework the trajectories A_i can be the expectation values of quantum operators, as moments at different orders of local observables, like $\langle \hat{N}_i \rangle$, $\langle \hat{x}_i^2 \rangle$, $\langle \hat{x}_i^4 \rangle$, σ_i^x.... This measure was first adopted in the framework of quantum synchronization in Ref. [15], using the quantum-mechanical expectation values of the second moments of the positions and momenta of two linearly coupled harmonic oscillators dissipating in a common environment. In this way the synchronization in the dynamics of second-order quadratures was captured. The same quantification of synchronization was also carried out in the case of an extended network of linear harmonic oscillators [16].

In the case of dissipating spin pairs, the Pearson's coefficient was adopted in Refs. [36, 44] to quantify the degree of synchronization between $\langle \sigma_1^x \rangle$ and $\langle \sigma_2^x \rangle$. A slightly different version of it, especially tailored to detect phase synchronization, was also analyzed in Ref. [32]. A qualitative analysis of the similarity between the time evolution of local averages of spin operators, even though without any explicit reference to the Pearson's measure, was also invoked by Orth et al. [12].

In general, this measure can be applied to any quantum problem when looking at temporal dynamics of local observables. The main advantages are (i) that it depends on the quantum signatures of the system (e.g. quantum noise, when going beyond first order moments) and (ii) that this measure has absolute reference values: reaching the maximum (minimum) value $C_{X,Y} = 1(-1)$ for perfect (anti-)synchronization.

3.2 Synchronization Error

The average distance between classical trajectories has been largely used to study synchronization of chaotic systems, see e.g. the example of a pair of bidirectionally coupled Lorenz systems in Ref. [6] of two coupled chaotic systems. This synchronization error was first considered in [17] to study quantum synchronization of coupled optomechanical oscillators attaining limit cycles. For continuous variable (CV) systems the synchronization error reads

$$\mathcal{S}_c(t) = \left\langle q_-(t)^2 + p_-(t)^2 \right\rangle^{-1}, \tag{5}$$

where $q_- = (q_1 - q_2)/\sqrt{2}$ is the difference in position, and the same for momentum, of the objects of interest. At variance with the classical case where the average distance can go to zero, in the quantum domain this measure is bounded. It achieves a maximum value when the two quantum objects are synchronized, and is upper-bounded by the uncertainty principle

$$\mathcal{S}_c(t) \leq \frac{1}{2\sqrt{\langle q_-(t)^2 \rangle \langle p_-(t)^2 \rangle}} \leq 1. \tag{6}$$

A poor value of this quantity can come from two possible origins: either the mean value (first moment) of q_- and p_- is big, or because the variances of these operators are big. In order to neglect the first cause, it is also interesting to define a modified measure with

$$q_-(t) \rightarrow q_-(t) - \langle q_-(t) \rangle \ , \ p_-(t) \rightarrow p_-(t) - \langle p_-(t) \rangle \,, \tag{7}$$

which is preferable if we want to study purely quantum effects.

While synchronization error in classical systems is generally addressed between the time series of two deterministic variables, as for example $q_1(t)$ and $q_2(t)$, the intrinsic probabilistic nature in the quantum domain enlarges this scenario. As for the Pearson factor, when comparing two operators $\hat{q}_1(t)$ and $\hat{q}_2(t)$ (hats are omitted elsewhere), the corresponding first moments can behave independently of the second moments or moments of higher order. This is particularly the case for Brownian oscillators initialized in vacuum squeezed states. The intention of the authors in [17] is to be able to compare the two operators by introducing a quadratic error measure, as reported in optomechanical settings (see Sect. 4.2), gauging well (as can be seen from comparison to other synchronization indicators) both the synchronization of first moments and second moments. The relation of this measure with the synchronization of local dynamics is in general an open question.

In a similar spirit a measure of phase synchronization is also introduced in [17], by writing the operator $a_j(t) := [q_j(t) + ip_j(t)]/\sqrt{2}$ of the jth system in the following way

$$a_j(t)) = [r_j(t) + a'_j(t)]e^{i\phi_j(t)}, \tag{8}$$

where r_j and ϕ_j are the amplitude and phase of the expectation value of $a_j(t)$: $\langle a_j(t)\rangle = r_j(t)e^{i\phi_j(t)}$. Now the Hermitian and anti-Hermitian parts of $a'_j(t) := [q'_j(t) + ip'_j(t)]/\sqrt{2}$ can be interpreted as amplitude and phase fluctuations, and we can say that whenever $\langle a_1(t)\rangle$ and $\langle a_2(t)\rangle$ are phase locked we can define a phase shift with respect to this locking by the operator $p'_-(t) = [p'_1(t) - p'_2(t)]/\sqrt{2}$. Hence, a measure of phase synchronization is

$$\mathcal{S}_p(t) = \frac{1}{2}\left\langle p'_-(t)^2\right\rangle^{-1}, \tag{9}$$

which in contrast to $\mathcal{S}_c$ can be arbitrarily large [17]. The authors point out though that $\mathcal{S}_p \le 1$ whenever two CV quantum systems can be represented by a positive P function (quantum optics notion of classicality), whereas the opposite would require collective squeezing.

3.3 Mutual Information and Other Information-Based Correlations

Entropic measures are often used in different contexts to quantify the correlation between sub-parts. In many cases, these quantities have been compared to other classical synchronization measures. Here we briefly review the case where they have been proposed in relation to quantum synchronization.

Classical mutual information, associated to time series of system observables, has been used as a measure of classical synchronization [6]. The quantum mutual information (MI) of a whole density matrix ρ_{AB} is defined as

$$I(\rho) = S(\rho_A) + S(\rho_B) - S(\rho_{AB}), \tag{10}$$

where ρ_A (ρ_B) is the reduced density matrix obtained tracing the subpart B (A) and where S stands for the von Neumann entropy: $S(\rho) = -\mathrm{Tr}\{\rho \log \rho\}$. MI was proposed in Ref. [32] as synchronization witness. The authors considered two different models showing synchronization, that is, two coupled Van der Pol oscillators and two qubits inside optical cavities in the presence of driving. It was shown that the steady-state MI had the same qualitative behavior of, respectively, the complete synchronization measure of Eq. (5), and relative phase between the two qubits (measured with an indicator close to Pearson's parameter) during the transient. Comparisons between mutual information and synchronization had already been performed in Refs. [15, 16, 36] showing that in harmonic systems [15, 16] MI is more robust in the synchronization regime, while for spins coupled through the environment [36] it is not distinctive signature of synchronization.

In Refs. [15, 16, 32, 36], synchronization was also compared to quantum discord, the part of mutual information quantifying nonclassical correlations [45, 46], leading to similar results. Given a bipartite system AB, it is defined as the difference between $I(\rho)$ and the classical part of correlations $\mathcal{J}(\rho)_{\{\Pi_j^B\}} = S(\rho_A) - S(A|\{\Pi_j^B\})$, where the conditional entropy is $S(A|\{\Pi_j^B\}) = \sum_i p_i S(\varrho_{A|\Pi_i^B})$, $p_i = \mathrm{Tr}_{AB}(\Pi_i^B \varrho)$ and where $\varrho_{A|\Pi_i^B} = \Pi_i^B \varrho \Pi_i^B / p_i$ is the density matrix after an optimal, complete projective measurement ($\{\Pi_j^B\}$) has been performed on B.

Generalized versions of MI can be obtained using the Rényi entropy $S_\alpha(\rho) = (1-\alpha)^{-1} \log \mathrm{Tr}\{\rho^\alpha\}$ (which reduces to S in the limit of $\alpha \to 1$). The Rényi-2 mutual information ($I_2(\rho) = S_2(\rho_A) + S_2(\rho_B) - S_2(\rho_{AB})$) was used by Bastidas et al. to detect chimera-type synchronization in a quantum network of coupled Van der Pol oscillators [47]. Chimera states describe the coexistence of synchronized and unsynchronized components [48].

Entanglement has also been considered in the context of synchronization [15]. Lee and coworkers, studying the case of two dissipatively coupled Van der Pol oscillators, argued that the steady-state exhibits an entanglement tongue, the quantum analogue of the Arnold tongue [30]. Entanglement, after a truncation of the total Hilbert space of the two oscillators, was quantified using the concurrence E [49]. The concurrence between a pair of qubits, whose density matrix is ρ, is defined as $E = \max\{0, \lambda_1 - \lambda_2 - \lambda_3 - \lambda_4\}$, where λ_r is the square root of the rth eigenvalue of $R = \rho\tilde{\rho}$ in descending order. Here, we have introduced $\tilde{\rho} = (\sigma_y \otimes \sigma_y)\rho^*(\sigma_y \otimes \sigma_y)$, where ρ^* is the complex conjugate of ρ.

The linear entropy of the sub-part i $S(\rho_i) = 1 - \mathrm{Tr}\{\rho_i^2\}$ can be used as entanglement quantifier provided that the whole state is pure. It was put in relation with synchronization of chimera states in Ref. [37] in the case of a closed spin chain.

3.4 Correlations of Observables

In Ref. [50] the synchronization between coupled non-linear cavities (a and b) was addressed considering normalized intensity correlations $g_2(a,b) = \langle n_a n_b\rangle / (\langle n_a\rangle\langle n_b\rangle)$ between the cavities (first and second harmonic) modes. The average $\langle\cdots\rangle$ was temporal in the classic limit (neglecting quantum noise and considering classical trajectories), and it was in this limit that synchronization was addressed. In the quantum regime, the expectation value over the quantum (steady) state was considered so that g_2 is a measure of intensity correlations (capturing bunching/antibunching effects between the coupled systems). The transition between classical and quantum regime was described but addressing synchronization only in the classical regime and comparing it with steady state correlations when moving into the quantum regime.

The average of the collective operator

$$Z = \langle(\sigma_1^+\sigma_2^- + \sigma_2^+\sigma_1^-)\rangle \tag{11}$$

was used in Ref. [39] to detect the presence of phase locking between two (ensembles of) spins and then the synchronization between them. There, it was shown that decay rates of these correlations encode information about the spectral content of the emitted radiation, which, in turn can be directly calculated using the two-time correlation function $Z = \langle(\sigma_1^+(\tau)\sigma_2^-(0) + \sigma_2^+(\tau)\sigma_1^-(0))\rangle$, as already done in Ref. [38].

The value of spin-spin correlations $\langle\sigma_1^\alpha\sigma_2^\alpha\rangle - \langle\sigma_1^\alpha\rangle\langle\sigma_2^\alpha\rangle$ $(\alpha = x, y, z)$ was also used by Hush et al. [42] as a sufficient criterion to assess the synchronization between the motion of two trapped ions. In such set-up, the spins represent the electronic degrees of freedom of the ions, and the value of their correlations was shown to be related to the relative phase distribution of the density matrix of the two motional degrees of freedom.

Looking at quantum correlations between observables to assess quantum synchronization is a natural strategy to identify non-classical signatures but it does not always capture the emergence of similarity of time evolutions between different sub-systems. This approach is actually often considered looking at stationary states loosing any relation with the classical counterpart of synchronization.

3.5 *Kuramoto Models*

The Kuramoto model [51] was introduced by Yoshiki Kuramoto to study the synchronous behavior of a large set of coupled oscillators, which appears naturally in the context of chemical/biological systems. The equations of motion for the phase variable of the N oscillators take the form:

$$\dot{\theta}_i = \omega_i + \xi_i + \frac{K}{N}\sum_j \sin(\theta_i - \theta_j), \tag{12}$$

where ω_i are oscillator frequencies, ξ_i noise terms, and K quantifies the coupling. An order parameter is defined assuming mean-field coupling

$$re^{i\psi} = \frac{1}{N}\sum_i e^{i\theta_i}, \tag{13}$$

where r represents the phase-coherence of the population of oscillators. Moving to a rotating frame where $\psi = 0$, the equations of motion become (without noise)

$$\dot{\theta}_i = \omega_i - K\,r\,\sin(\theta_i), \tag{14}$$

which models a particle in a washboard potential. This means that whenever $|\omega_i| < Kr$ the phase is trapped and we have synchronization, otherwise the phase slips down the washboard. If the distribution of frequencies $g(\omega)$ is unimodal and centered around Ω, the phase transition to complete synchronization occurs for the critical coupling value $K_c = 2/\pi g(\Omega)$.

This model and similar ones have spurred a vast amount of research reviewed in [52]. Here we note that whenever a system can be reduced to a set of equations similar to the Kuramoto model, an analogous argument can be made to check whether there is phase locking or not. This is the case for coupled optomechanical oscillators as reported in Ref. [13]. The authors are able to reduce the mean field dynamics of the optomechanical array with all-to-all couplings (after carefully eliminating the amplitudes from the dynamics) to a Kuramoto-type model, which for only two units simplifies to

$$\delta\dot{\theta} = -\delta\Omega - C\cos(\delta\theta) - K\,\sin(2\delta\theta), \tag{15}$$

with $\delta\theta = \theta_2 - \theta_1$ and $\delta\Omega$ the frequency difference. Once here, pure analysis of the parameters in the equation directly point to either phase locking or phase slip.

Notably, translation of the Kuramoto model to the semiclassical domain was recently achieved [53] and it was shown that the picture of the washboard potential is a good intuitive guide, where now quantum tunneling can allow the phase to tunnel through maxima of the (otherwise phase-locking inducing) potential. This leads to the necessity of a higher critical coupling constant K_c in order to pin down and lock the phases in the model.

3.6 Other Approaches

Quantum synchronization has also been addressed considering phase space representations. In Ref. [23] the collective phase-coherence among the components of an optomechanical array was characterized through an order parameter, Eq. (13), and looking at the transition of the Wigner representation from an angular-symmetric distribution to a coherent displaced state with time dependent phase. Under the assumption of infinite non-linearity, Van der Pol oscillators in presence of driving were considered in [10, 31] in strongly non-classical regimes, where the dynamics is governed by few Fock states: the authors discuss the similarity of the Wigner distribution (W) with limit cycle appearing in the classical regime and the breaking of rotational invariance of W to characterize phase locking. The marginals of the relative phase are also a signature of phase locking as considered in Ref. [42]. Interestingly, phase state tomography has been recently reported in order to characterize phase diffusion and locking for an on-fiber optomechanical cavity operating into the classical regime [54]. Still, when writing this chapter, there are not experimental results on synchronization reported in the quantum regime.

4 Synchronization of Oscillators

4.1 Linear Networks of Brownian Oscillators

Coupled harmonic oscillators dissipating in a bosonic environment represent the simplest setup where synchronization of quantum systems can occur [55]. Their intrinsic linearity precludes the possibility of synchronous limit cycles, but not the presence of a long transient where synchronization is present and also of steady oscillating states protected from dissipation. They have been studied for example in [15, 16, 34], where it was shown that eigenmodes in the system of coupled oscillators can display very different dissipation timescales for some parameter ranges. If one of the eigenmodes' dissipative rate is much smaller than that of the rest, this eigenmode dominates the dynamics of the coupled oscillators, and hence they become synchronized at the eigenfrequency of that mode. This synchronization is temporary but can be very long if such rate is small [15]. For certain situations of high symmetry, one of the eigenmodes of the coupled oscillators system can be isolated from the environment and then synchronization can last forever [16]. Situations of higher symmetry can also occur where more than one eigenmode is not dissipating, whereby synchronization can not happen [34]. More elaborate situations have been studied, for example in harmonic networks [16] it has been shown that the full network or only a motif inside it can be synchronized by tuning one of the frequencies.

One of the main conclusions in this type of systems is that dissipation induces quantum synchronization through diffusive coupling: be it for some parameter region or another, synchronization can always be achieved, with the sole exception of the separate baths case, where each coupled oscillator is attached to its own independent heat bath. It is easy to demonstrate mathematically that in this case, also eigenmodes are dissipating into equivalent independent heat baths, and therefore their dissipative rates are of equal size: no eigenmode then survives longer than the rest and thus synchronization is avoided. Different dissipation mechanisms that can arise in extended environments [56] lead to specific forms of diffusive couplings that can induce synchronization.

The Hamiltonian that describes this dynamics is

$$H = \sum_j \frac{p_j^2}{2m_j} + m_j \frac{\omega_j^2}{2} x_j^2 + \sum_{i \neq j} \lambda_{ij} (x_i - x_j)^2, \tag{16}$$

whereas dissipation is introduced by coupling each oscillator to its own environment -eventually with different dissipation strengths-, to a common one -with a homogeneous coupling or not-, or other combinations. For ease of exposition here we will present the commonly denominated 'common bath' case with system-bath interaction

$$H_{diss.} = \sum_j x_j \sum_k c_k Q_k, \tag{17}$$

where all system components couple equally to the same environment. In the case of two units with equal (unit) mass and different frequency the system reads

$$H = \frac{p_1^2}{2m} + \frac{p_2^2}{2m} + \frac{\omega_1^2}{2}x_1^2 + \frac{\omega_2^2}{2}x_2^2 + \lambda x_1 x_2, \tag{18}$$

where we have already absorbed quadratic contributions from the spring coupling into the oscillators frequencies. The oscillators in the bath are uncoupled among them and have frequencies ω_j, which together with the coupling constants c_k define in the continuum limit what is usually called the spectral density $J(\omega) = \sum_k \frac{c_k^2}{\omega_k}\delta(\omega - \omega_k)$. This function encodes basically all information about the dissipation characteristics and is usually taken to be 'Ohmic' [that is, $J(\omega) \sim \omega$] which gives a damping of the oscillators which is just proportional to their velocity. The linearity of the dynamics means that if we use an initial state which is Gaussian (all its information can be described from first and second moments), it will remain so at all times. Examples of Gaussian states [57, 58] widely used and relevant are displaced, squeezed and thermal states.

Synchronization can for example be displayed by the time evolution of second moments of quantum states, such as vacuum squeezed states, with the first moments being zero at all times. An example of this can be seen in Fig. 2, where the Pearson indicator is drawn for different detunings and coupling strength [15]. The shape is reminiscent of an Arnold tongue in classical physics, and has been also later observed with Van der Pol oscillators [32, 50]. The fact that first moments are zero while second moments are oscillating and synchronize does not seem to fit well with the synchronization error indicator, which in fact grows in the regions where the Pearson indicator clearly points to worse values.

Furthermore, information-based correlations are preserved through the common eigenmode that does not dissipate, and thus survive longer in the synchronized case. However, correlations cannot assess synchronization if they were not already present at some initial time and thus are neither good synchronization measures for this

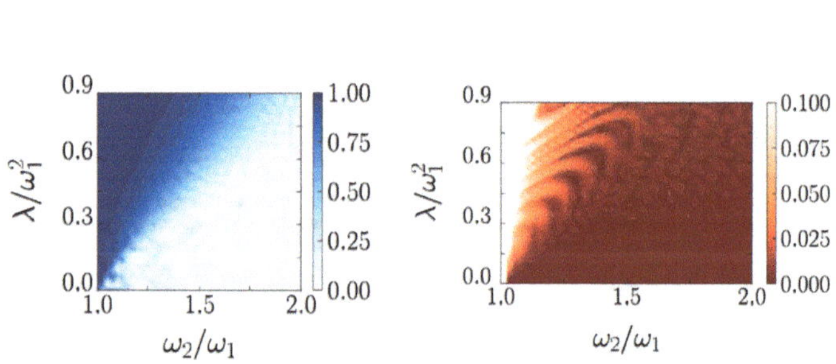

Fig. 2 *Left* Pearson indicator of synchronization for different detunings and coupling strengths in the two coupled oscillators model dissipating into a common bath. They are initialized to squeezed vacuum states with zero first order moments. Synchronization is measured in the quadratures of each oscillators, i.e. $C_{\langle x_1^2(t)\rangle\langle x_2^2(t)\rangle}$ is plotted. *Right* Quantum discord at the same value of time. A similar, although narrower, "tongue" is observed

example. We note that the considered system focuses on the effect of dissipative coupling in harmonic arrays, and this analysis could be applied -for instance- to the noisy precursor of an optomechanical system below the oscillation threshold.

4.2 Self-sustained Optomechanical Oscillators

An important example of synchronization dynamics are optomechanical systems: capable of displaying limit cycles of constant amplitude, they provide an intuitive connection to what is known in the classical realm where the Kuramoto model is paradigmatic. Their ability to synchronize was first proven in [13], where it was shown that mechanically coupled optomechanical oscillators above their dynamical Hopf bifurcation can be described with a Kuramoto-type model; phase and anti-phase synchronization are displayed. The analysis in the case of a common bosonic mode [14] and experimental demonstration of the classical synchronization of such system [24] followed shortly. Quantum synchronization has been later reported: above a threshold mechanical coupling between optomechanical units, a regime of quantum phase-coherent mechanical oscillations arises [23]. The measure of quantum synchronization analogue to synchronization error was considered for optomechanical oscillators in [17].

Optomechanical systems [22] comprise a mechanical mode and a confined optical mode which is typically driven by a laser field. These modes are coupled nonlinearly through radiation pressure, provided by a movable mirror or a structure which can be deformed by the action of light such as a dielectric medium. The combination of an external pump and a nonlinear coupling provides stable limit cycles upon which synchronization can occur. The usual form of the Hamiltonian describing such dynamics is

$$H = \Delta a^{\dagger} a + \omega b^{\dagger} b + g a^{\dagger} a (b + b^{\dagger}) + i E (a - a^{\dagger}) \tag{19}$$

with $a(a^{\dagger})$ the annihilation (creation) operators of light in the cavity, $b(b^{\dagger})$ those of the mechanical mode and E is the intensity of the laser input. The Hamiltonian is written in the frame rotating with the laser frequency ω_L, which is detuned by $\Delta = \omega_C - \omega_L$ with respect to the cavity frequency ω_C. Both modes are further coupled to noise sources with strength κ, γ respectively, providing dissipation. Below a given threshold intensity of the laser, the amplitude of light and mechanical modes simply decay to the value enforced by the noise sources. However, above that intensity, multistability and limit cycles of increasing complexity can be observed. The final ingredient to observe synchronization is coupling, via the mechanical or optical degrees of freedom, several optomechanical systems. For illustrative purposes we choose here mechanical coupling of the type $H_{int} = \mu(b_1 b_2^{\dagger} + b_1^{\dagger} b_2)$. We will follow here the example of dynamics given in [17] to compare different measures of synchronization and correlations. After a transient, limit cycles are achieved for each degree of freedom and we can linearize them around their stable orbits with the usual ansatz: $\hat{a}_j = A_j(t) + \delta\hat{a}_j$ and $\hat{b}_j = B_j(t) + \delta\hat{b}_j$, with capital letters representing the

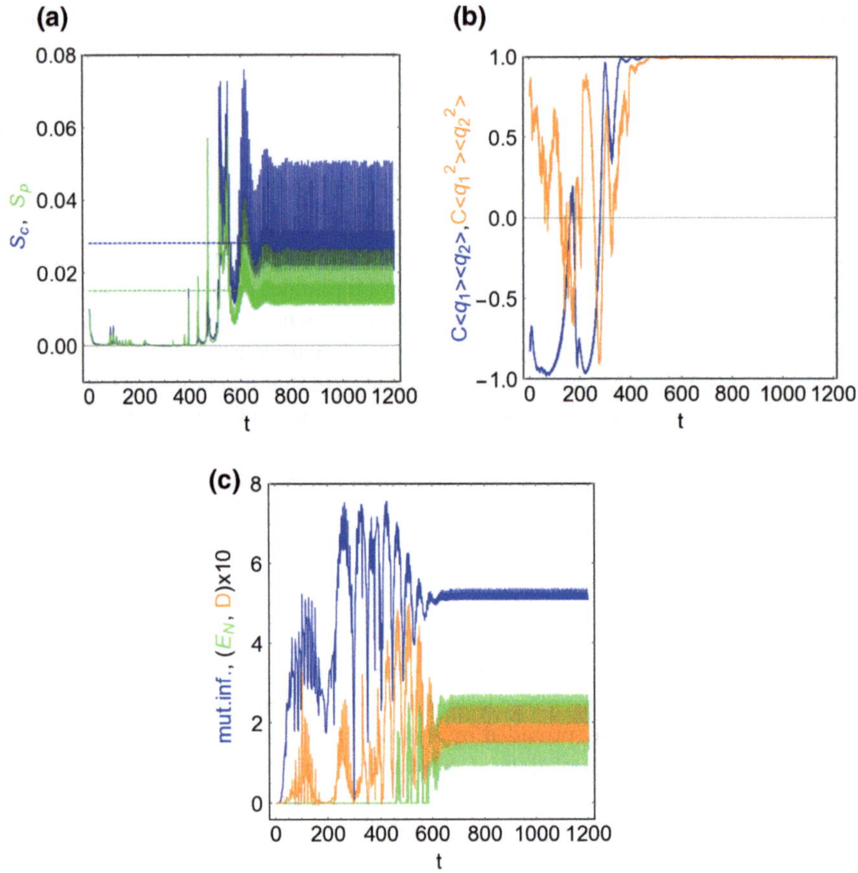

Fig. 3 Comparison of several synchronization measures: **a** error synchronization S_c, S_p, **b** Pearson indicator for first and second momenta $C_{\langle q_1\rangle\langle q_2\rangle}$, $C_{\langle q_1^2\rangle\langle q_2^2\rangle}$, and **c** correlations: mutual information, logarithmic negativity and quantum discord. We set parameters $\omega_1 = 1$, $\omega_2 = 1.005$, $g = 0.005$, $\mu = 0.02$, detunings $\Delta_j = \omega_j$, $\kappa = 0.15$, $\gamma = 0.005$ and laser input $E = 320\kappa = 48$, as in Ref. [17]. The initial condition for first moments are $\langle q_1(0)\rangle = 100$ and $\langle q_2(0)\rangle = -100$, and all other first moments zero. The second order moments at $t = 0$ are 100 times their vacuum value. Changing these initial conditions does not change qualitatively the results

limit cycles as a classical variable, and $\delta\cdot$ the linear displacements (fluctuations) with respect to them. The ansatz is then used to neglect nonlinear terms in the dynamics. Finally, the fluctuations can be arranged in the form of a covariance matrix whose time evolution can be integrated, yielding both the dynamical content and the quantum/classical information content.

Coupled optomechanical oscillators represent a good platform combining sufficiently complex dynamics and the possibility to assess different informational measures, so that proposed indicators of synchronization can be compared, as seen in Fig. 3. We consider two optomechanical systems like in Ref. [17]. The qualitative

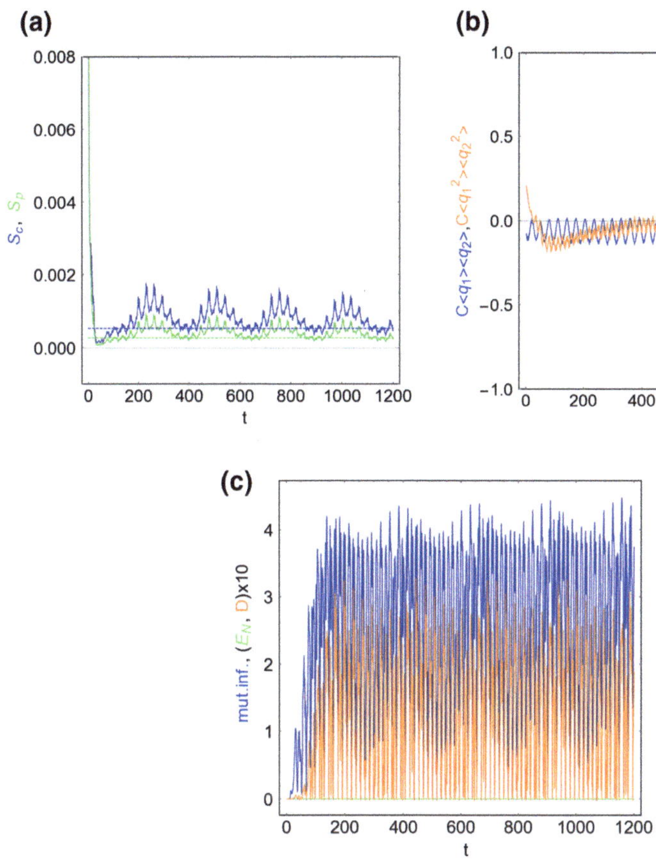

Fig. 4 Same parameters as Fig. 3, but with higher detuning $\omega_2 = 1.2$, so there is no synchronization. Notice that the scale in **a** is ten-fold lower than in Fig. 3a

behaviour is similar for all indicators: an initial build up and a final stage of full synchronization by a stationary value of all quantities.

As mentioned before, perfect synchronization can be identified looking at the absolute value of the indicator only for the Pearson coefficient (Sect. 3.1), which really shows perfect synchronization (value $\sim$1) for first and second moments of the mechanical observables (Fig. 3b, time $\gtrsim$500). The synchronization error indicators have a rather low value, Fig. 3a, compared to their maximum attainable 1 (see Sect. 3.2), and we have checked that for different initial conditions of the mechanical first moments, it can be increased, with very similar qualitative behaviour. The possibility to reach its maximum bound Eq. (6) in such system has not been reported, although for fixed initial conditions it might be informative to assess the quality of synchronization when changing Hamiltonian parameters (as when comparing Figs. 3 and 4). As a note, while these different indicators can signal stable synchronization for $t > 600$, the Pearson coefficient spots it sooner. Comparing Figs. 3c and 4c we can conclude too that any measure based on classical/quantum correlations (of information-theoretic character) is not of too much value if we compare absolute values. Introducing initial squeezing in the mechanical modes changes the time profile of all correlations, but as expected not their final stable values.

Finally, what is important to note is that the stability in time of the indicators is a necessary condition for signaling synchronization, and once that is achieved, comparison of absolute values might yield some extra information, although only the Pearson indicator is a bona fide absolute measure in this sense.

5 Synchronization of Interacting Spins

When considering precessing spins, the behavior of local spectra can be used to extract direct information about the presence of a synchronized dynamics. This method was adopted by Orth and coworkers in Ref. [12], where they considered two interacting spins dissipating through a common environment, and observed that there is a regime in the parameter space of the system where only a single frequency appears in the spectrum of both the local observables. This single-line spectrum is not the only possible manifestation of synchronization. It is indeed possible, like in the infinite-dimension Hilbert space cases discussed above, that the collective dynamics favours the suppression of some spectral lines, while one of them has a very long lifetime, leaving the system synchronized during the transient decay leading to steady state [36, 44]. In these cases, the Pearson's measure can be adopted to observe the dynamical setting-up of synchronization. It can be defined considering the expectation values of any arbitrary operator for each spin A_k, $k = 1, 2$, decomposed in the single-spin basis $\{\sigma_x^k, \sigma_y^k, \sigma_z^k, I_d^k\}$:

$$A_k = a_x^k \sigma_x^k + a_y^k \sigma_y^k + a_z^k \sigma_z^k + a_d I_d^k. \tag{20}$$

Actually, in many cases, it is enough to consider one spin direction. For instance, in Ref. [36], the z components of the two spins were synchronized from the beginning and the interesting part of the dynamics concerned the $x - y$ plane.

In the following, we are going to compare the Pearson's measure of quantum synchronization with correlation indicators in a model of two detuned spins interacting through an Ising-like coupling:

$$H_S = \frac{\omega_1}{2}\sigma_1^z + \frac{\omega_2}{2}\sigma_2^z + \lambda \sigma_1^x \sigma_2^x. \tag{21}$$

Let us assume that the spins experience a dissipative dynamics induced by the presence of a thermal environment weakly coupled to the system through

$$H_I = \sum_k g_k (a_k^\dagger + a_k)(A\sigma_1^x + \sigma_2^x). \tag{22}$$

Here, the annihilation (creation) operators a_k ($a_k^\dagger$) act on the bath degrees of freedom and the coefficient A determines the ratio between the strength of the two spin-bath couplings. For the sake of clarity, in the following we will only discuss the two extreme cases $A = 0$ (local environment) and $A = 1$ (common environment). In order to derive a master equation describing the dynamics of the two spins, it is necessary to know the diagonal form of H_S, which can be obtained applying the standard Jordan-Wigner transformation, mapping spins into spinless fermions, defined as $\sigma_1^z = 1 - 2c_1^\dagger c_1$, $\sigma_2^z = 1 - 2c_2^\dagger c_2$, $\sigma_1^x = c_1^\dagger + c_1$, $\sigma_2^x = (1 - 2c_1^\dagger c_1)(c_2^\dagger + c_2)$ [59]. We have

$$H_S = E_1(\eta_1^\dagger \eta_1 - 1/2) + E_2(\eta_2^\dagger \eta_2 - 1/2), \tag{23}$$

with $E_1 = \frac{1}{2}\left(\sqrt{4\lambda^2 + \omega_+^2} + \sqrt{4\lambda^2 + \omega_-^2}\right)$ and $E_2 = \frac{1}{2}\left(\sqrt{4\lambda^2 + \omega_+^2} - \sqrt{4\lambda^2 + \omega_-^2}\right)$ where $\omega_\pm = \omega_1 \pm \omega_2$. The quasi-particle fermion operators are obtained combining the Bogoliubov transformation $c_1 = \cos\theta_+ \xi_1 + \sin\theta_+ \xi_2^\dagger$, $c_2 = \cos\theta_+ \xi_2 - \sin\theta_+ \xi_1^\dagger$ together with the rotation $\xi_1 = \cos\theta_- \eta_1^\dagger + \sin\theta_- \eta_2^\dagger$, $\xi_2 = \cos\theta_- \eta_2^\dagger - \sin\theta_- \eta_1^\dagger$.

The spectral density $J(\omega) = \sum_k g_k^2 \delta(\omega - \Omega_k)$ is assumed to follow, apart from a high-frequency cut-off, the Ohmic power law $J(\omega) \sim \omega$. Assuming weak dissipation, the qubit pair dynamics can be studied in the Born–Markov and secular approximations [60] with Lindblad master equation $\dot{\rho}(t) = -i[H_S + H_{LS}, \rho(t)] + \mathcal{D}[\rho(t)]$, where the Lamb shift H_{LS} commutes with H_S and where $\mathcal{D}[\rho(t)]$, which takes into account dissipation, is the sum of four terms, each of them associated to one of the four transition frequencies $\pm E_i$ ($i = 1, 2$):

$$\mathcal{D}(\rho) = \sum_{i=1}^{2} \tilde{\gamma}_i^+ \mathcal{L}[\eta_i](\rho) + \sum_{i=1}^{2} \tilde{\gamma}_i^- \mathcal{L}[\eta_i^\dagger](\rho), \tag{24}$$

Here, the Lindblad superoperators are defined as $\mathcal{L}[\hat{X}](\rho) = \hat{X}\rho\hat{X}^\dagger - \{\hat{\rho}, \hat{X}^\dagger\hat{X}\}/2$. The exact value of the decay rates $\gamma_i^\pm$ will depend on the nature of the system-bath coupling. In the case of local dissipation $A = 0$, their specific form can be found in Ref. [44]. As already discussed, synchronization takes place if there is substantial separation between the two largest γ's determining the dynamics. In this case, local degrees of freedom undergo quasi-monochromatic oscillations and their relative phases get locked.

5.1 *Spin Synchronization*

As discussed in Ref. [44], the case of a local environment $A = 0$ shows an "anomalous" synchronization pattern. Indeed, unlike classical manifestations of synchronization and quantum synchronization induced by a common environment (Refs. [15, 16, 36]), it is greatly enhanced in the strong detuning $\Delta = |\omega_1 - \omega_2|$ regime, while

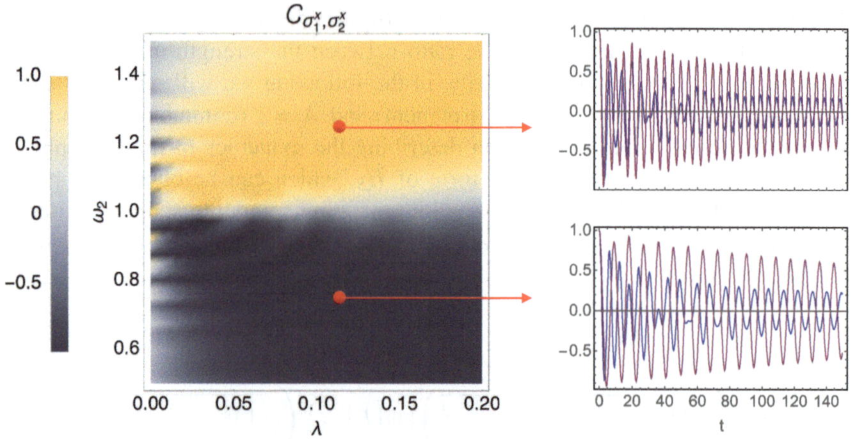

Fig. 5 *Left panel* synchronization diagram as a function of ω_2 and λ for a local bath. $C_{\sigma_1^x,\sigma_2^x}$ has been calculated at time $t = 75$ (in units of ω_1) using a time window of $\tau = 10$. As explained in the text, the local bath ($A = 0$) is assumed to be Ohmic with cut-off frequency $\omega_c = 20$, and its temperature is $T = 0$. The initial state is $|\psi(0)\rangle = (|\uparrow\rangle + |\downarrow\rangle)(|\uparrow\rangle + |\downarrow\rangle)/2$. *Right panels* $\sigma_1^x(t)$ (*blue*) and $\sigma_2^x(t)$ (*red*) assuming $\omega_2 = 1.25$ and $\lambda = 0.11$ (*top*) and $\omega_2 = 0.75$ and $\lambda = 0.11$ (*bottom*)

the direct spin-spin coupling λ has a partially detrimental effect. Turning attention to the case $A = 1$, the presence of a common bath has the tendency to facilitate spontaneous synchronization. In order to observe it, it is fundamental for the two spins to have an interaction strong enough as to compensate the detuning. Furthermore, in the local-bath case, depending on the Hamiltonian parameters, synchronization can appear both in phase and in anti-phase. This feature is suppressed in the presence of a common environment, where only anti-synchronization can be observed. This is due to the different interplay between the γ's in the two scenarios.

The anomalous synchronization emerging from a local environment is shown in the $\{\Delta - \lambda\}$ diagram of Fig. 5. The local observable used to calculate the Pearson's parameter $\mathcal{C}$ are, respectively, σ_1^x and σ_2^x, even if the calculation could be extended to generic local operators without qualitative changes in the results. In the left panel, we show the synchronization diagram, showing the transition from phase to anti-phase, while in the right panels we plot the trajectories of the two local observables in the two distinct regimes.

On the other hand, the behavior of $\mathcal{C}$ for a common bath is displayed in Fig. 6 and it is very much similar to the characteristic Arnold tongues emerging in classical synchronization problems. In this case, anti-synchronization emerges provided that the spin-spin coupling is not too small with respect to the detuning. As a singular behavior, around $\Delta = 0$, the system shows "trivial" synchronization, given that the two spins become indistinguishable.

The complementarity and the qualitative difference of the synchronization diagrams emerging in the two cases under study will be used in the following of this

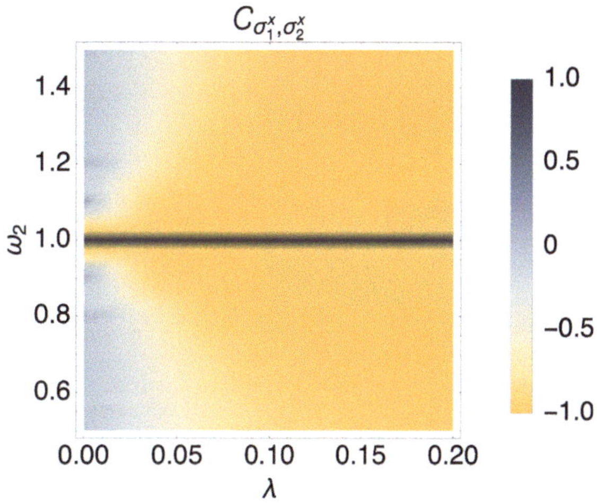

Fig. 6 $C_{\sigma_1^x,\sigma_2^x}(t = 75, \tau = 10)$ as a function of ω_2 and λ in the presence of a common bath ($A = 1$). All other conditions as in Fig. 5

section to compare the Pearson's measure to correlation quantifiers, namely, spin-spin correlations $\langle \sigma_i^+ \sigma_j^- \rangle$, mutual information, and entanglement.

5.2 *Spin Correlations*

As noticed in Ref. [39], $Z = \langle \sigma_1^+ \sigma_2^- + \sigma_2^+ \sigma_1^- \rangle$ plays the role of a phase locking indicator when applied to interacting spins, and then can be used as a synchronization measure (see Sect. 3.4). At a first sight, C and Z evolve independently, as the sets of equations of motion of their respective matrix elements are not coupled to each other. Actually, the constraints, to which a physical density matrix is subject to, make the behavior of the two indicators very close to each other. This aspect is discussed in great detail in Ref. [61], in the case of independent spins, where the interplay between spontaneous synchronization and superradiance is studied. In fact, in order for spontaneous synchronization to emerge, the whole system needs to support a long-lasting collective mode, which unavoidably displays spin-spin correlations.

These qualitative considerations are confirmed in the cases we are discussing here: in both models (of local and global dissipation), low-quality synchronization C is always accompanied by a value for Z close to zero. On the other hand, within the synchronized regions, Z is significantly enhanced. Furthermore, the sign of Z is reminiscent of the phase–anti-phase form of synchronization. These results are shown in Fig. 7, where, in order to wash out faster oscillations and regularize the picture, the time integral of Z ($Z_I = \int_{t=0}^{100} Z(t')dt'$) is plotted as a function of ω_2 and λ. Both cases of local ($A = 0$) and global ($A = 1$) dissipation are shown. For a local bath, Fig. 7a, the change from positive to negative values for the spin-spin correlation parameter takes place in the same region where C passes from synchronization to anti-synchronization (compare with Fig. 5a, b). In view of the previous considerations,

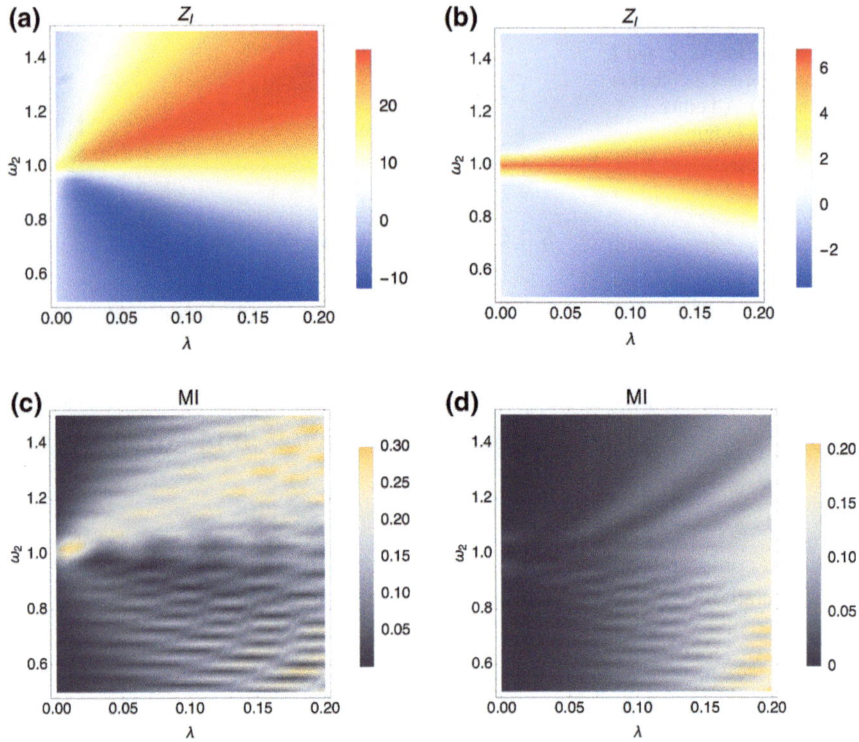

Fig. 7 Z_I (**a**, **b**) and mutual information MI (**c**, **d**) as a function of ω_2 and λ for a local bath $A = 0$ (**a**, **c**) and for a common bath $A = 1$ (**b**, **d**). All parameters as in Fig. 5

this change is suppressed for $A = 1$, Fig. 7b. It is worth noticing that the "anomalous" synchronization peak around $\omega_1 = \omega_2$ displayed by $\mathcal{C}$ (Fig. 6) is broadened by Z_I (Fig. 7b), whose value appears smoother at critical changes.

5.3 Mutual Information and Entanglement

Mutual information (MI) as a quantum synchronization witness in spin systems was proposed by Ameri et al. in a work dealing with a system of two qubits placed in two coupled cavities where only the first one is driven by a laser, while the second one is populated by the photons leaking from the first cavity [32]. In that example, it was shown that the steady-state mutual information was reminiscent of the synchronized oscillations of local operators in the pre-steady-state regime.

In the following we show that, in the models we are investigating, MI does not play the witnessing role suggested in Ref. [32]. As a first observation, we notice that we deal with systems decaying towards an equilibrium state (the Gibbs state) that only depends on the Hamiltonian parameters, while the synchronization diagram depends critically on the properties of the environment. To make this point clearer, both cases we are considering here admit the same equilibrium state, whilst the two synchronization diagrams are radically different. One may ask if some information about synchronization appears in the dynamical behavior of MI instead of its asymptotic value. For this reason, we considered the time at which synchronization starts to be solid ($t = 80$ in Fig. 7c, d) and calculated MI for the two models (at longer times MI would rapidly converge to zero everywhere). The two behaviors indicate a very weak connection between MI and spontaneous synchronization. Starting from a factorized state, the coupling λ immediately produces a quite robust amount of MI (depending on the strength of λ) between the two spins. Then, the presence of dissipation makes this correlation disappear, but it seems that the way MI fades away is not connected with the building up of a synchronized dynamics. A very similar argument can be applied to entanglement, that can be quantified using the concurrence E [49]. The dynamical behavior of E (not shown) displays qualitative features very close to the ones of MI.

Besides the specific models studied here, the argument can be made more general considering the case of a purely dephasing dynamics. On the one hand, as discussed in Ref. [36], such a process is not able to induce any synchronization, as no time scale separation takes place. On the other hand, entanglement and mutual information can converge to a finite value under the same circumstances. Therefore quantum correlations in the steady state, in general, do not witness a previous syncronization during relaxation.

6 Discussion and Conclusions

Present research on quantum synchronization has just started to unveil the distinctive features of this phenomenon. Several factors, known to influence it in the classical regime [1] are under study, including, for instance, non-linearity, dissipation, noise, forcing, mutual or directional coupling between inhomogeneous components, or time delay. Experiments reporting distinctive signatures of quantum synchronization are expected to flourish in the next years.

The question about what is essential of quantum synchronization with respect to the classical one is intimately related to the interplay between temporal and quantum correlations. From the previous analysis we can establish few relevant criteria to approach the phenomenon of quantum synchronization as described by different measures and to assess usefulness and meaningfulness in each specific context:

- **Absolute reference value**: In order to be able to assess the amount of synchronization in different regimes, it is important for a measure to be bounded and to have a definite value associated to the perfect emergence of full synchronization. The Pearson's parameter Eq. (3) reaches values very close to the maximum attainable $|\mathcal{C}_{A_1,A_2}(t|\Delta t)| \simeq 1$ whenever good synchronization emerges. Similarly, the synchronization error Eq. (5) is bounded in the quantum case, whereas the classical one is not. Still, reported values are rather modest in the case of coupled optomechanical oscillators [17] and saturation of this bound that would correspond to the best quantum synchronization has not yet been reported.

- **Time dynamics dependence**: The concept of synchronization is relative to the time evolution of system's observable or variables and a measure of synchronization should reflect it covering a time window of the system dynamics, like in temporal averages for instance, or being robust during evolution. This is the case for several measures in different ways: some are based on time averages (e.g. Pearson's parameter (3)), others maintain distinct higher values during synchronization (e.g. synchronization error (5)), and others assess the time stability of the process (as Lyapunov exponents [62]). In general, the problem when looking at instantaneous (quantum) correlations, MI etc. is that they can be instantaneously huge even when there is no synchronization, as shown in Fig. 4. On the other hand, looking at asymptotic values is not always leading to an insightful synchronization condition.
- **Local versus non-local** Among the reported measures of synchronization, some refer to local observables of the synchronized systems (like Pearson factors, or local phases in Kuramoto models) while other refer to quantum correlations present in the composed system (in this sense being 'non-local'). The possibility to associate a genuine quantum correlation to synchronization is clearly appealing to distinguish it from classical synchronization, as for instance with the synchronization error [17]. On the other hand, this can give rise to spurious definitions of quantum synchronization, actually not related at all with this dynamical phenomenon. This question is still open and few further considerations are given below.

In the attempt to identify a measure for a genuine quantum synchronization, different indicators have been proposed that actually do not refer to observables but to the full quantum state, as discussed in Sect. 3. Invoking generic quantum correlation as a measure of synchronization is in general not convincing. As an example, a bipartite Bell-state is strongly correlated under any possible definition of correlation, but in general this has nothing to do with synchronization. As a matter of fact, any local unitary would leave it unchanged, while altering the dynamics of the components of the system can alter dynamical synchronization. Even if non-local correlations are not necessarily associated to specific synchronization phenomena, it is important to remind that quantum correlations and dynamical synchronization can occur under the same conditions in some systems [15–17, 30]. Still, quantum correlations that capture specific signatures of synchronization are not generally established. Looking at the examples we have treated here we can also draw some conclusions.

Optomechanical self-sustained oscillators can achieve synchronization and several parameters like the error and Pearson indicators give a similar insight to mutual information or other correlations (Figs. 3 and 4). Synchronization error however displays a small value far for the maximum bound not providing an absolute indication for the synchronization degree (Fig. 3a). Furthermore, correlations signal synchronization not by their value, but by having a final stable nonzero value, in contrast to a highly oscillating one in the case of no synchronization.

In the case of linear dissipative oscillators the situation worsens. The Pearson factor gives a valuable guide to look for synchronization, while all the other indicators fail: synchronization error does not provide a good estimate of the behaviour of synchronization with respect to the system's parameters and other information measures are strongly initial state dependent. Still, robust quantum correlations (such as discord) can witness synchronization, as both emerge under the same circumstances in this case. At some level, all of the measures can be used to yield some insight, however they require some craft efforts as compared to Pearson coefficients.

The literature about quantum synchronization in spins is much more limited with respect to the case of harmonic oscillators or optomechanical systems. Furthermore, in many cases, synchronization has been assessed using ad-hoc witness measures more than quantifiers. This chapter represents the first attempt to compare such quantities. As a result, we observed consistent indication of synchronization between Pearson's and spin-spin Z indicators, due to a strong interplay between phase-locking dynamics and the dynamics of the local observables. In contrast, mutual information and entanglement fail to give any useful information.

Finally, all the previous measures of synchronization could be modified to account for more general forms of synchronization. In all the discussed cases, synchronization is either in-phase or anti-phase. It is worth remarking that, in general, delayed synchronization can also emerge and synchronization indicators need to be improved to catch this effect. This can be easily done, for instance in the case of the Pearson's parameter, allowing one of the two sliding windows to open at a time different from the other one, that is equivalent to delay the time of one of the observable expectation value

$$\mathcal{C}_{A_1(t),A_2(t+\tau)}(t|\Delta t). \tag{25}$$

Similarly this could be done for all indicators based on local observables. Another way of improving synchronization indicators consists in correcting possible relative amplitude mismatch effects, similarly to the conditional variance factor appearing in the context of the EPR correlations [63].

To conclude we would like to stress that the field is still in its early stages and this work is the first attempt to assess meaning and utility of different synchronization measures as well as their possible dependence to the specific features of the system under study. Up to now, no experimental results in the quantum domain are at hand. Therefore there is plenty of room for improvement and surprises, both regarding the theoretical framework and possible practical applications of potential use as quantum technologies.

Acknowledgements Funding from EU project QuProCS (Grant Agreement No. 641277), MINECO and FEDER/AEI (NOMAQ FIS2014-60343-P and QuStruct FIS2015-66860-P), and "Vicerectorat d'Investigació i Postgrau" of the UIB are acknowledged.

References

1. A. Pikovsky, M. Rosenblum, J. Kurths, *Synchronization: A Universal Concept in Nonlinear Sciences* (Cambridge edition, 2001)
2. S.H. Strogatz, *Nonlinear Dynamics and Chaos: With Applications to Physics, Biology, Chemistry, and Engineering* (Westview p edition, 2001)
3. R. Horodecki, P. Horodecki, M. Horodecki, K. Horodecki, Quantum entanglement. Rev. Mod. Phys. **81**, 865–942 (2009)
4. K. Modi, A. Brodutch, H. Cable, T. Paterek, V. Vedral, The classical-quantum boundary for correlations: discord and related measures. Rev. Mod. Phys. **84**, 1655–1707 (2012)
5. G. Adesso, T.R. Bromley, M. Cianciaruso, Measures and applications of quantum correlations. J. Phys. A: Math. Theor. **49**, 473001 (2016)
6. S. Boccaletti, J. Kurths, G. Osipov, D.L. Valladares, C.S. Zhou, The synchronization of chaotic systems. Phys. Rep. **366**, 1–101 (2002)
7. I. Goychuk, J. Casado-Pascual, M. Morillo, J. Lehmann, P. Hänggi, Quantum stochastic synchronization. Phys. Rev. Lett. **97**(21), 210601 (2006)
8. O. Zhirov, D. Shepelyansky, Synchronization and bistability of a qubit coupled to a driven dissipative oscillator. Phys. Rev. Lett. **100**(1), 014101 (2008)
9. O. Zhirov, D. Shepelyansky, Quantum synchronization and entanglement of two qubits coupled to a driven dissipative resonator. Phys. Rev. B **80**(1), 014519 (2009)
10. T.E. Lee, H.R. Sadeghpour, Quantum synchronization of quantum van der Pol oscillators with trapped Ions. Phys. Rev. Lett. **111**, 234101 (2013)
11. S. Walter, A. Nunnenkamp, C. Bruder, Quantum synchronization of a driven self-sustained oscillator. Phys. Rev. Lett. **112**(9), 094102 (2014)
12. P.P. Orth, D. Roosen, W. Hofstetter, K. Le Hur, Dynamics, synchronization, and quantum phase transitions of two dissipative spins. Phys. Rev. B **82**, 144423 (2010)
13. G. Heinrich, M. Ludwig, J. Qian, B. Kubala, F. Marquardt, Collective dynamics in optomechanical arrays. Phys. Rev. Lett. **107**(4), 043603 (2011)
14. C.A. Holmes, C.P. Meaney, G.J. Milburn, Synchronization of many nanomechanical resonators coupled via a common cavity field. Phys. Rev. E **85**, 066203 (2012)
15. G.L. Giorgi, F. Galve, G. Manzano, P. Colet, R. Zambrini, Quantum correlations and mutual synchronization. Phys. Rev. A **85**, 052101 (2012)
16. G. Manzano, F. Galve, G.L. Giorgi, E. Hernández-García, R. Zambrini, Synchronization, quantum correlations and entanglement in oscillator networks. Sci. Rep. **3**, 1439 (2013)
17. A. Mari, A. Farace, N. Didier, V. Giovannetti, R. Fazio, Measures of quantum synchronization in continuous variable systems. Phys. Rev. Lett. **111**(10), 103605 (2013)
18. S.C. Manrubia, A.S. Mikhailov, D.H. Zanette, *Emergence of Dynamical Order. Synchronization Phenomena in Complex Systems* (World Scientific Publishing Co., Singapore, 2004). Lecture no edition
19. A. Arenas, A. Diaz-Guilera, J. Kurths, Y. Moreno, C. Zhou, Synchronization in complex networks. Phys. Rep. **469**(3), 93–153 (2008)
20. J. Pantaleone, Synchronization of metronomes. Am. J. Phys. **70**(10), 992 (2002)
21. M. Zhang, S. Shah, J. Cardenas, M. Lipson, Synchronization and phase noise reduction in micromechanical oscillator arrays coupled through light. Phys. Rev. Lett. **115**, 163902 (2015)
22. M. Aspelmeyer, T.J. Kippenberg, F. Marquardt, Cavity optomechanics. Rev. Mod. Phys. **86**, 1391 (2014)

23. M. Ludwig, F. Marquardt, Quantum many-body dynamics in optomechanical arrays. Phys. Rev. Lett. **111**(7), 073603 (2013)
24. M. Zhang, G.S. Wiederhecker, S. Manipatruni, A. Barnard, P. McEuen, M. Lipson, Synchronization of micromechanical oscillators using light. Phys. Rev. Lett. **109**(23), 233906 (2012)
25. M. Bagheri, M. Poot, L. Fan, F. Marquardt, H.X. Tang, Photonic cavity synchronization of nanomechanical oscillators. Phys. Rev. Lett. **111**, 213902 (2013)
26. S.Y. Shah, M. Zhang, R. Rand, M. Lipson, Master-slave locking of optomechanical oscillators over a long distance. Phys. Rev. Lett. **114**, 113602 (2015)
27. W. Li, F. Zhang, C. Li, H. Song, Quantum synchronization in a star-type cavity QED network. Commun. Nonlinear Sci. Numer. Simul. **42**, 121–131 (2017)
28. D.K. Agrawal, J. Woodhouse, A.A. Seshia, Observation of locked phase dynamics and enhanced frequency stability in synchronized micromechanical oscillators. Phys. Rev. Lett. **111**, 084101 (2013)
29. M.H. Matheny, M. Grau, L.G. Villanueva, R.B. Karabalin, M.C. Cross, M.L. Roukes, Phase synchronization of two anharmonic nanomechanical oscillators. Phys. Rev. Lett. **112**, 014101 (2014)
30. T.E. Lee, C.-K. Chan, S. Wang, Entanglement tongue and quantum synchronization of disordered oscillators. Phys. Rev. E **89**(2), 022913 (2014)
31. S. Walter, A. Nunnenkamp, C. Bruder, Quantum synchronization of two Van der Pol oscillators. Annalen der Physik **527**(1–2), 131–138 (2015)
32. V. Ameri, M. Eghbali-Arani, A. Mari, A. Farace, F. Kheirandish, V. Giovannetti, R. Fazio, Mutual information as an order parameter for quantum synchronization. Phys. Rev. A **91**(1), 012301 (2015)
33. H. Carmichael, *An Open Systems Approach to Quantum Optics: Lectures Presented at the Université Libre de Bruxelles, October 28 to November 4, 1991*, Lecture Notes in Physics Monographs (Springer, Berlin, 2009)
34. G. Manzano, F. Galve, R. Zambrini, Avoiding dissipation in a system of three quantum harmonic oscillators. Phys. Rev. A **87**(3), 032114 (2013)
35. G.M. Xue, M. Gong, H.K. Xu, W.Y. Liu, H. Deng, Y. Tian, H.F. Yu, Y. Yu, D.N. Zheng, S.P. Zhao, S. Han, Observation of quantum stochastic synchronization in a dissipative quantum system. Phys. Rev. B **90**, 224505 (2014)
36. G.L. Giorgi, F. Plastina, G. Francica, R. Zambrini, Spontaneous synchronization and quantum correlation dynamics of open spin systems. Phys. Rev. A **88**(4), 042115 (2013)
37. D. Viennot, L. Aubourg, Quantum chimera states. Phys. Lett. A. **380**(5–6), 678–683 (2016)
38. X. Minghui, D.A. Tieri, E.C. Fine, J.K. Thompson, M.J. Holland, Synchronization of two ensembles of atoms. Phys. Rev. Lett. **113**, 154101 (2014)
39. B. Zhu, J. Schachenmayer, M. Xu, F. Herrera, J.G. Restrepo, M.J. Holland, A.M. Rey, Synchronization of interacting quantum dipoles. New J. Phys. **17**, 083063 (2015)
40. C. Deutsch, F. Ramirez-Martinez, C. Lacroûte, F. Reinhard, T. Schneider, J.N. Fuchs, F. Piéchon, F. Laloë, J. Reichel, P. Rosenbusch, Spin self-rephasing and very long coherence times in a trapped atomic ensemble. Phys. Rev. Lett. **105**(2), 020401 (2010)
41. Y. Liu, F. Piéchon, J.N. Fuchs, Quantum loss of synchronization in the dynamics of two spins. EPL (Europhys. Lett.) **103**(1), 17007 (2013)
42. M.R. Hush, W. Li, S. Genway, I. Lesanovsky, A.D. Armour, Spin correlations as a probe of quantum synchronization in trapped-ion phonon lasers. Phys. Rev. A **91**, 061401(R) (2015)
43. A. Argyris, D. Syvridis, L. Larger, V. Annovazzi-Lodi, P. Colet, I. Fischer, J. García-Ojalvo, C.R. Mirasso, L. Pesquera, K. Alan Shore, Chaos-based communications at high bit rates using commercial fibre-optic links. Nature **438**(7066), 343–346 (2005)
44. G.L. Giorgi, F. Galve, R. Zambrini, Probing the spectral density of a dissipative qubit via quantum synchronization. Phys. Rev. A **94**, 052121 (2016)
45. H. Ollivier, W.H. Zurek, Quantum discord: a measure of the quantumness of correlations. Phys. Rev. Lett. **88**, 017901 (2001)
46. L. Henderson, V. Vedral, Classical, quantum and total correlations. J. Phys. A: Math. Gen. **34**(35), 6899 (2001)

47. V.M. Bastidas, I. Omelchenko, A. Zakharova, E. Schöll, T. Brandes, Quantum signatures of chimera states. Phys. Rev. E **92**, 062924 (2015)
48. A.E. Motter, Nonlinear dynamics: spontaneous synchrony breaking. Nat. Phys. **6**(3), 164–165 (2010)
49. W.K. Wootters, Entanglement of formation of an arbitrary state of two qubits. Phys. Rev. Lett. **80**, 2245–2248 (1998)
50. T.E. Lee, M.C. Cross, Quantum-classical transition of correlations of two coupled cavities. Phys. Rev. A **88**, 013834 (2013)
51. Y. Kuramoto, *International Symposium on Mathematical Problems in Theoretical Physics*, vol. 39 (Springer, New York, 1975)
52. J.a. Acebrón, L.L. Bonilla, C.J. Pérez Vicente, F. Ritort, R. Spigler, The Kuramoto model: a simple paradigm for synchronization phenomena. Rev. Mod. Phys. **77**(1), 137–185 (2005)
53. I.H. de Mendoza, L.a. Pachón, J. Gómez-Gardeñes, D. Zueco, Synchronization in a semiclassical Kuramoto model. Phys. Rev. E **90**(5), 052904 (2014)
54. K. Shlomi, D. Yuvaraj, I. Baskin, O. Suchoi, R. Winik, E. Buks, Synchronization in an optomechanical cavity. Phys. Rev. E **91**, 032910 (2015)
55. C. Benedetti, F. Galve, A. Mandarino, M.G.A. Paris, R. Zambrini, Minimal model for spontaneous synchronization. Phys. Rev. A **94**, 052118 (2016)
56. F. Galve, A. Mandarino, M.G.A. Paris, C. Benedetti, R. Zambrini, Microscopic description for the emergence of collective dissipation in extended quantum systems. Sci. Reps. **7**, 42050 (2017)
57. A. Ferraro, S. Olivares, M.G.a. Paris, Gaussian states in continuous variable quantum information. (Bibliopolis, Napoli 2005; ISBN 88-7088-483-X)
58. G. Adesso, F. Illuminati, Entanglement in continuous-variable systems: recent advances and current perspectives. J. Phys. A **40**, 7821 (2007)
59. E. Lieb, T. Schultz, D. Mattis, Two soluble models of an antiferromagnetic chain. Ann. Phys. **16**(3), 407–466 (1961)
60. H.P. Breuer, F. Petruccione, *The Theory of Open Quantum Systems*. (OUP Oxford, Oxford, 2007)
61. B. Bellomo, G.L. Giorgi, G.M. Palma, R. Zambrini, Quantum synchronization as a local signature of super and subradiance. arXiv:1612.07134
62. W. Li, C. Li, H. Song, Criterion of quantum synchronization and controllable quantum synchronization based on an optomechanical system. J. Phys. B: At. Mol. Opt. Phys. **48**(3), 035503 (2015)
63. P.D. Drummond, M.D. Reid, Correlations in nondegenerate parametric oscillation. II. below threshold results. Phys. Rev. A **41**(7), 3930–3949 (1990)

How Does Interference Fall?

Patrick J. Orlando, Felix A. Pollock and Kavan Modi

Abstract We study how single- and double-slit interference patterns fall in the presence of gravity. First, we demonstrate that universality of free fall still holds in this case, i.e., interference patterns fall just like classical objects. Next, we explore lowest order relativistic effects in the Newtonian regime by employing a recent quantum formalism which treats mass as an operator. This leads to interactions between non-degenerate internal degrees of freedom (like spin in an external magnetic field) and external degrees of freedom (like position). Based on these effects, we present an unusual phenomenon, in which a falling double slit interference pattern periodically decoheres and recoheres. The oscillations in the visibility of this interference occur due to correlations built up between spin and position. Finally, we connect the interference visibility revivals with non-Markovian quantum dynamics.

Since the days of Galileo and Newton, it has been known that acceleration under the influence of gravity is independent of an object's mass [1, 2]. This peculiarity has led to the proposition of various gravitational equivalence principles which, if broken, represent a departure from our current understanding of the theory of gravity. Einstein's theory of general relativity is fundamentally classical, describing gravity on large length scales in terms of curvature of the underlying spacetime metric. Although it is possible to formulate quantum field theories on a static curved metric, it remains unclear how existing theory should be modified to describe gravity on the quantum mechanical scale [3]. Whilst the work we present here does not attempt

P.J. Orlando (✉) · F.A. Pollock · K. Modi
School of Physics and Astronomy, Monash University,
Melbourne, Victoria 3800, Australia
e-mail: patrick.james.orlando@gmail.com

F.A. Pollock
e-mail: felix.pollock@monash.edu

K. Modi
e-mail: kavan.modi@monash.edu

F.F. Fanchini et al. (eds.), *Lectures on General Quantum Correlations and their Applications*, Quantum Science and Technology,
DOI 10.1007/978-3-319-53412-1_19

to quantise gravity, it demonstrates that there is much insight to be gained from exploring non-relativistic quantum mechanics in weak-field gravity.

In the weak-field limit, a Newtonian description of gravity provides a satisfactory approximation and is, advantageously, compatible with the Hamiltonian formulation of quantum mechanics; however, its disadvantage lies in the concealment of relativistic effects, such as gravitational time dilation and the gravitational redshift of photons. Fortunately, one need not utilise the complete machinery of general relativity to take these effects into account. In fact, lowest order relativistic effects can be introduced by simply considering the mass contributions of different energy states, as given by the mass-energy relation $E = mc^2$ of special relativity [4].

This is true even in the case of internal energy and becomes particularly interesting for quantum systems, whose internal energy can exist in superposition. Recent work by Zych and Brukner [5] treats this by promoting mass to an operator, the purpose of which is to account for the effective mass of quantised internal energy. In addition to introducing lowest order relativistic effects, this construction provides a new quantum mechanical generalisation of the Einstein equivalence principle to superpositions of energy eigenstates.

The role that Newtonian gravity plays in quantum theory was perhaps best highlighted by the famous experiment of Colella, Overhauser and Werner (COW), who demonstrated interference of cold neutrons due to a relative phase acquired due to the difference in gravitational potential between two arms of an interferometer. We include details of the COW experiment in Appendix A.

More recently, the theory of ultra-cold atom condensates has provided a way to test gravitational equivalence principles with quantum systems, by using optically trapped atomic gases as an integrated interferometer [6–10]. The short de Broglie wavelength of an atom makes atomic interferometers highly sensitive, whilst the macroscopic nature of the condensate allows for a high degree of control. Proposals for tests on board the international space station have been put forward which, if performed, are expected to surpass the best classical tests by a factor of 100 [11]. Finally, tests of the uniquely quantum mechanical equivalence principle for superpositions have also been proposed [12].

In this article, we study how single- and double-slit interference patterns fall due to gravity. Initially, we ignore the lowest order relativistic effects introduced by internal degrees of freedom and find (unsurprisingly) that the interference patterns fall just as classical objects do; in other words, the universality of free fall holds for spatially delocalised quantum systems.

We then pedagogically introduce the mass operator and use it to explore non-Newtonian effects on quantum systems with quantised internal energy. One such system is a particle with intrinsic spin incident on a double slit in a gravitation field. We demonstrate that when placed in a uniform magnetic field, the internal energy results in periodic decoherence and re-coherence of the double-slit pattern. This result is an example of decoherence due to gravitational time dilation presented by Pikovski et al. [13] and other related works [14, 15].

The decoherence occurs due to the buildup of correlations between the spin and position degrees of the particle. We identify the oscillations in the visibility of the interference fringes as a signature of non-Markovian quantum dynamics [16], and demonstrate explicitly how memory effects play a role in the evolution of these fringes. This illustrates that the tools of open quantum systems theory can help us clearly understand Newtonian gravity in a quantum mechanical context.

1 Dropping a Quantum Interference Pattern

General relativity arose from the concept that gravitational effects are a result of the underlying spacetime geometry. Whilst three fundamental forces of nature: electromagnetism, the strong force and the weak force; all depend on the internal properties of matter, gravity, in the Newtonian regime, depends only on the mass of the particle. Further, its dependence on the mass is such that the dynamics are completely independent of the particle itself. This is often attributed to Galileo in a famous thought experiment, devised to refute Aristotle's claim that the gravitational acceleration of a body is proportional to its mass. His very elegant thought experiment, described in Fig. 1, led to the conclusion that all objects must fall at the same rate, regardless of their mass. This is known as the *universality of free fall*, and has profound consequences for theories of gravity.

In this section we study how quantum interference patterns fall due to gravity. We imagine that massive quantum particles (say neutrons) are ejected towards a single or double slit. Once the particle passes through the slit, it falls freely under the influence of gravity, while simultaneously interfering with itself.

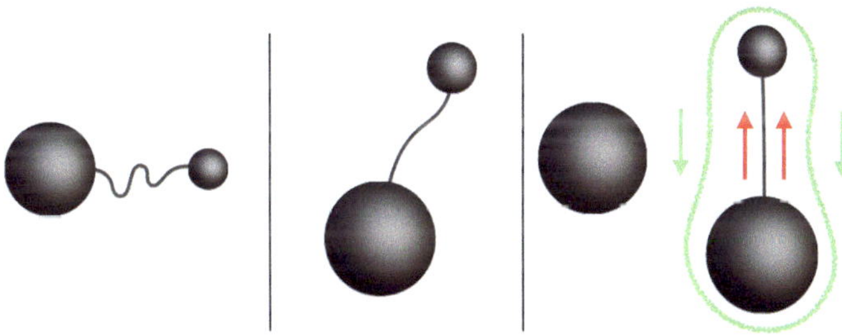

Fig. 1 Galileo's thought experiment. Galileo considered three spheres composed of the same material. Two of the spheres had identical mass, whilst the third sphere was much lighter. He then imagined attaching a rope between the small mass and one of the larger masses, and wondered what would happen if all three were simultaneously dropped from the leaning tower of Pisa. According to Aristotle, the small mass should fall slower than the large mass, pulling the rope taught and impeding the acceleration of the larger mass. One would then expect to see the solitary large mass hit the ground before the attached pair. However, one could also consider the pair of attached masses as a single body, whose mass exceeds that of the large mass alone. In this case, the attached pair of masses would be expected to hit the ground before the solitary large mass. This results in a logical contradiction, from which the only escape is to conclude that both the small and large masses fall with the same acceleration

Here, we are concerned with the possibility of interesting gravitational effects appearing in a single slit diffraction or double slit interference experiment. In accordance with the Einstein equivalence principle, we have come to expect that all objects should fall identically under the influence of gravity, and this by no means excludes quantum particles exhibiting their wave-like nature. However, this does not discount the possibility of COW-like phases [17] skewing the wavefunction at the screen to give *apparent* violations.

This leads us to our first result, which is to explore how the phase generated by the gravitational potential results in an interference pattern that appears to fall like a classical object. It also provides the foundation for a more sophisticated problem explored in a later section. From a conceptual point of view, this is an interesting scenario to investigate, especially when one considers the path integral formulation of quantum mechanics.

In simple terms, the Feynman propagator is the Green's function for the Schrödinger equation, the solution resulting from the initial spatial wavefunction being a dirac-delta distribution. It represents the amplitude for a particle at position x and time t to be found at cx' a later time t'. Once the propagator is known, the evolution for any initial wavefunction can be found by convolution with the propagator. The one dimensional propagator is often expressed as

$$\langle x'|U(t'-t)|x\rangle = K_0(x',t';x,t) = \int \mathcal{D}(x(t)) \exp\left[\frac{i}{\hbar}\int_t^{t'} L(x(s),\dot{x}(s),s)\,ds\right], \tag{1}$$

where $U(\delta t) = \exp(-i\hat{H}\delta t/\hbar)$ is the time evolution operator, $x(t)$ is a parametrised path in space, $\mathcal{D}(x(t))$ is the Feynman measure over all possible paths and $L(x,\dot{x},t)$ is the Lagrangian describing the system.

The path-integral formulation of quantum mechanics is conceptually very appealing, since it can be interpreted as a statement about how quantum mechanical objects may deviate from the laws of classical dynamics. In fact, even in the presence of a gravitational field, there is a non-zero amplitude which corresponds to the quantum system not falling at all: $\langle x,t'|x,t\rangle > 0$ for some $t' > t$. Thus, from a foundational point of view, we would like to use the path integral approach to examine the way in which the gravitational potential affects a single-particle, double-slit interference pattern.

A short outline of the derivation is shown here, with full details available in Appendix B. The Lagrangian for a particle in a Newtonian gravitational potential is $L = \frac{1}{2}m\dot{x}^2 - mgx$. With reference to the propagator defined in Eq. (1), we parametrise the path $x(t)$ in terms of deviations $\delta x(t)$ from the classical trajectory, $x_c(t)$, between the two points. This gives $x(t) = x_c(t) + \delta x(t)$, with $\delta x(t) = \delta x(t') = 0$. This parametrisation leaves the Feynman measure unchanged, as a sum over all paths is equivalent to a sum over all deviations from a specific path. We are then left with two terms: a phase dependent on the action of the classical trajectory and a Feynman integral over the deviations that has a form identical to that of a free particle. We substitute the integral with the free particle propagator, but acknowledge that, since this is sum over deviations, we must set $x = x' = 0$. The propagator for a particle in a Newtonian gravitation potential is then

$$K_g(x', t'; x, t) = \frac{\exp\left[\frac{i}{\hbar} S\left[x_c(t)\right]\right]}{\sqrt{2\pi i \hbar (t' - t)/m}}, \tag{2}$$

where $S\left[x_c(t)\right]$ is the functional that gives action associated with the classical trajectory between the points. We can express it as a function of (x, t, x', t') by solving the equations of motion for the boundary conditions $x_c(t) = x$ and $x_c(t') = x'$. With complete details in Appendix B.1, the general form for the classical action is given by,

$$S[x_c(t)] = \frac{m}{2}\left\{\frac{(x' - x)^2}{t' - t} - g(x + x')(t' - t) - \frac{g^2}{12}(t' - t)^3\right\}. \tag{3}$$

1.1 *Single and Double Slit Interference*

We now consider applying this propagator to the problem at hand. Let's begin by assuming that the slits are long enough to ignore diffraction effects in the y-direction (perpendicular to the gravitational field – which is in the negative x-direction – but in the plane of the screen), this allows us to effectively reduce the problem to two dimensions. Consider a source of particles at the origin (0, 0) and let a double slit be located at distance D from the source in the z-direction. Each slit has width $2a$ with centre located at $x = \pm b$. The screen is then a further distance L away from the slits. The two-dimensional propagator required for this problem is given by a free particle propagator in the z-direction, multiplied by the gravitational propagator for the x direction, as calculated in Eq. (2). This propagator allows us to ask the question: *If a particle initially starts at position* $\vec{r} = (0, 0)$, *what is the probability of finding it at position* $\vec{r}\,' = (x, D + L)$ *on the screen?* This distribution in x will be the the two slit interference pattern that we seek.

When computing this amplitude, we consider a semi-classical approach. We assume that the 'trajectory' of the neutron can be separated into two parts: (a) the path from the source to the slits, followed by (b) the path from the slits to the screen. Quantum mechanically, the particles need not pass through the slits and there even exists the possibility of them passing through the slits multiple times before hitting the screen. That being said, the probabilities associated with these events are negligible under certain conditions: The semi-classical approach is valid, provided that the majority of the particle's momentum is in the z direction, such that the wavelength is approximately the z-direction wavelength, $\lambda \approx \frac{2\pi\hbar}{mv_z}$. We assume that this wavelength is much smaller than the relevant z-direction length scales, D and L, in conjunction with the assumption that these are much larger than the relevant x direction length scales, b and a. Within this regime, the problem reduces to a single dimension. After a rather tedious calculation (included in Appendix B.2 for completeness), the wavefunction at the screen due to a single slit centred at $x = b$, the instant the particle hits it in the semi-classical approximation ($\tau = L/v_z$), is given by

$$\psi^{(1)}(x) = \frac{e^{i\phi(x)}}{i2\sqrt{\eta a}} \left\{ C[\sigma_+(x)] - C[\sigma_-(x)] + iS[\sigma_+(x)] - iS[\sigma_-(x)] \right\}, \quad (4)$$

where $C[u] \equiv \int_0^u \cos\left(\frac{\pi}{2}x^2\right) dx$ is the Fresnel cosine function, $S[u] \equiv \int_0^u \sin\left(\frac{\pi}{2}x^2\right) dx$ is the Fresnel sine function and $\eta = 1 + \frac{L}{D}$. Above

$$\sigma_\pm(x) = \sqrt{\frac{2}{\lambda L}\eta} \left\{ (b \pm a) - \frac{x}{\eta} - \frac{1}{2} g \frac{m^2\lambda^2}{h^2} DL \right\} \quad \text{and} \quad (5)$$

$$\phi(x) = \pi \left\{ \frac{x^2}{\lambda(D+L)} - mgx\frac{\lambda(D+L)}{h^2} - \frac{g^2}{12}\frac{m^4\lambda^3}{h^4}(D+L)(D-L)^2 \right\}. \quad (6)$$

If b is set to zero, then this gives single slit diffraction. Extension to double slit or even N-slit interference is given by taking a normalised superposition of the wavefunctions corresponding to the different slit positions.

The square of this wavefunction will give the observed probability distribution for the position at which the particle hits the screen; this is plotted for a single slit in Fig. 2. The pattern clearly appears to shift towards the negative x direction as the screen is moved further from the slit. In general, this is far easier to identify in single

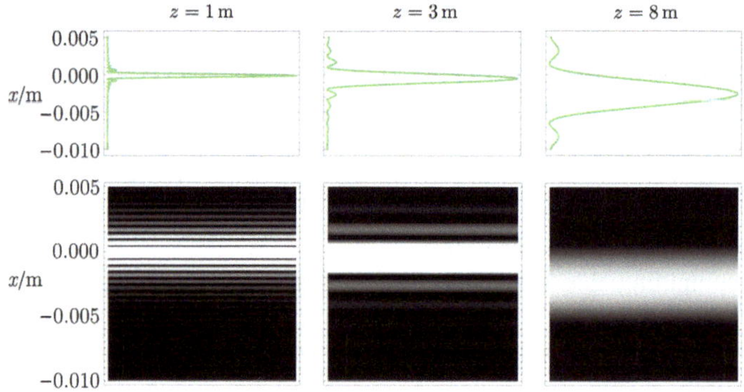

Fig. 2 Single Slit Diffraction in a Gravitational field. In the *top row*, the magnitude squared of the wavefunction in Eq. (4) $|\psi^{(1)}(x)|^2$ is plotted for source to screen distance $D = 2$ m and slit to screen distances $z = \{1\,\text{m}, 3\,\text{m}, 8\,\text{m}\}$. The second row shows the same information as a two-dimensional probability density on the screen. The particle was chosen to be a neutron with wavelength $\lambda \sim 10^{-9}$ m, and the gravitational field strength $g = 9.8\,\text{ms}^{-2}$. In addition to the typical spreading of the pattern, we observe an apparent translation of the pattern, which we can interpret as falling

slit diffraction, as the spreading of the pattern is less noticeable than in the double slit case.

The location of the central maximum is indicative of the position at which a classical point particle would arrive. If we consider a single particle incident on the slit, then there exists a possibility that it will be detected above the central maximum. We could interpret this as the particle having fallen less than expected classically. Similarly we could detect a particle below the central maximum, indicating that it fell faster than expected classically. Although it would be appealing to label this as violation of the equivalence principle, to do so would be incorrect. The easiest way to verify this is to transform to an accelerated coordinate system, taking us to the freely falling frame, in which gravitational effects should completely vanish.

When calculating the propagator in Eq. (2), we made use of the general form for the classical trajectory, which satisfies the Euler-Lagrange equations of motion. Using the solution, we find that the classical parabolic trajectory an object takes from the source at $(x = 0, t = 0)$ to the slit at $(x = 0, t = T)$ requires an initial upward speed $v_x(t = 0) = \frac{1}{2}gT$ and final vertical speed of $v_x(t = T) = -\frac{1}{2}gT$. The position of the object τ seconds later would then be,

$$x_c(T + \tau) = -\frac{g\tau}{2}g(\tau + T) = -\frac{gm^2\lambda^2}{2h^2}\left(L^2 + LD\right). \tag{7}$$

We would expect that, by performing the coordinate transformation $x = \xi + x_c(T + \tau)$, the pattern should become identical to the case where $g = 0$, in accordance with the equivalence principle. Since the x dependence in the wavefunction $\psi^{(1)}$ appears only through the function $\sigma_\pm(x)$, we can work directly with the expression given in Eq. (5),

$$\sigma_\pm(\xi) = \sqrt{\frac{2}{\lambda L}}\eta\left\{(b \pm a) - \frac{\xi - \frac{1}{2}g\frac{m^2\lambda^2 D^2}{h^2}\left(L^2 + LD\right)}{\eta} - \frac{1}{2}g\frac{m^2\lambda^2}{h^2}DL\right\} \tag{8}$$

$$= \sqrt{\frac{2}{\lambda L}}\eta\left\{(b \pm a) - \frac{\xi}{\eta} + \frac{1}{2}g\frac{m^2\lambda^2 D^2}{h^2}L\left[\frac{D\,(L + D)}{L + D} - D\right]\right\} \tag{9}$$

$$= \sqrt{\frac{2}{\lambda L}}\eta\left\{(b \pm a) - \frac{\xi}{\eta}\right\}. \tag{10}$$

The result above shows that the gravitational effect on the interference can be eliminated by transforming to an accelerated coordinate system. It is now clear that there are no equivalence principle violations; if we detect a particle away from the central maximum, it is interpreted as the usual deviations of a quantum particle from its classically expected trajectory.

We don't need to calculate the double slit pattern to identify the absence of a COW phase. Since the multi-slit wavefunction is simply a superposition of single slit wavefunctions, the coordinate transform above extends to the general case; the only effect gravity will have is a translation of the entire pattern. The reason for this is that in the COW experiment, the Mach–Zehnder interferometer constrains the path of the particle to an approximate binary. Whilst confined to these paths, the relative Aharonov–Bohm-like phase is accumulated. In the case of single-slit diffraction, there is no path confinement, and even for multiple slits, where there are discrete variations between paths, there is no relative phase accumulated; this is because the slits are effectively infinitely thin in our scenario.

Therefore, it would appear that there are no peculiar quantum effects that appear in a freely falling interference pattern, beyond what one would expect in the absence of gravity. The quantum mechanical deviations from the classically expected trajectory represent a departure from the laws of classical physics, and, although the deviations might seem to constitute a violation of the universality of free fall, the effects are completely consistent with quantum behaviour as viewed from an accelerating coordinate system. In other words, an interference pattern falls like a classical object.

2 Effects of Internal Degrees of Freedom

In this section, we examine some of the gravitational effects that appear at leading relativistic order for particles with internal degrees of freedom. These effects, which can be seen as arising from relative time dilation of different internal levels, were first investigated in detail by Zych et al. [14, 15] and Pikovski et al. [13], and were further discussed by Zych and Brukner in the context of the equivalence principle [5].

2.1 *The Hamiltonian Formulation*

According to the Einstein equivalence principle, all internal energy acts as a mass from the perspective of both general and special relativity. That is, the mass terms appearing in the kinetic and potential energy of a system in a gravitational field should depend on the internal energy state. When the internal state corresponds to a dynamically varying degree of freedom, with its own Hamiltonian H^{int}, then all terms involving the mass should couple it to the external degree of freedom. In other words, the mass is promoted to an operator on the internal degree of freedom:

$$m \to M = m\,\mathbb{1}^{\text{int}} + \frac{H^{\text{int}}}{c^2}. \tag{11}$$

The full Hamiltonian for a particle in a uniform gravitational field, including the newly defined mass operator is then (to leading relativistic order) [5]

$$H = Mc^2 + \frac{P^2}{2M} + Mgx \tag{12}$$

$$= \left(m\mathbb{1}^{\text{int}} + \frac{H^{\text{int}}}{c^2}\right)c^2 + \frac{P^2}{2\left(m\mathbb{1}^{\text{int}} + \frac{H^{\text{int}}}{c^2}\right)} + \left(m\mathbb{1}^{\text{int}} + \frac{H^{\text{int}}}{c^2}\right)gx \tag{13}$$

$$= mc^2 + H^{\text{int}} + \frac{P^2}{2m} + mgx + \frac{1}{mc^2}\left\{-\frac{P^2}{2m}H^{\text{int}} + gxH^{\text{int}}\right\} + \mathcal{O}\left(c^{-4}\right), \tag{14}$$

where, in the last line, we have expanded the M^{-1} in a Taylor expansion. This is valid, provided that the largest eigenvalue of the internal Hamiltonian, denoted by $\|H^{\text{int}}\|$, satisfies $\|H^{\text{int}}\|/mc^2 \ll 1$, i.e., the internal energy is small compared to the rest mass. The additional terms introduced by the mass operator give lowest order relativistic effects. The first effect is introduced by the coupling of the internal energy to the kinetic energy operator, which represents lowest order special relativistic time dilation. The other interaction term, coupling the internal energy to the Newtonian potential, represents lowest order gravitational time dilation effects.

We can verify this by looking at the evolution of the internal degree of freedom. Provided that the internal evolution is not trivial, i.e., that it is not in an eigenstate of the internal Hamiltonian, it can be considered operationally as a clock [13]. If we denote q to be an observable of the internal degree of freedom, then the evolution given in the Heisenberg picture is, as described in Ref. [5],

$$\dot{q} = \frac{1}{i\hbar}[q, H] = \frac{1}{i\hbar}\left\{[q, H^{\text{int}}]\mathbb{1}^{\text{ext}} - [q, H^{\text{int}}]\frac{P^2}{2m^2c^2} + [q, H^{\text{int}}]\frac{gx}{c^2}\right\} \tag{15}$$

$$= \dot{q}_{\text{loc}}\left(1 - \frac{P^2}{2m^2c^2} + \frac{gx}{c^2}\right). \tag{16}$$

Here $\dot{q}_{\text{loc}}$, is the normal rate of internal evolution as given in the system's rest frame. Recalling that the rate of change of proper time, in the non-relativistic, weak-field limit, is $d\tau = (1 - \frac{v^2}{2c^2} - \frac{\phi(x)}{c^2})dt$, we can easily identify these additional terms as a result of lowest order time dilation. For semi-classical evolution of the external degrees of freedom, the evolution of the internal degree of freedom is affected in a manner that is consistent with our understanding of relativistic effects. Interestingly this equation is valid not just for semi-classical systems, but also for non-local systems or systems with momentum that is not well defined. In these cases however, we cannot apply any of our classical intuition [13].

This result can be interpreted in the following way; general relativity provides a description for the evolution of clocks which are attached to observers evolving according to the laws of classical mechanics. On the other hand, the mass operator, has in a sense, allowed us to describe the evolution of a clock attached to an observer who evolves according to the laws of quantum mechanics.

Though this intuition can be applied to the internal evolution, we will present results that show this is not true when observing the external evolution. The evolution of the position degree of freedom is given by,

$$\dot{x} = \frac{1}{i\hbar}[x, H] = \frac{1}{i\hbar}\frac{[x, P^2]}{2m\left(1 + \frac{H^{\text{int}}}{mc^2}\right)} = \frac{P}{m}\left(1 - \frac{H^{\text{int}}}{mc^2}\right) + \mathcal{O}\left(m^{-2}c^{-4}\right). \tag{17}$$

Again, if we consider a semi-classical wavepacket, and take the expectation value of the equation above, we find that the velocity of this wavepacket depends on the state of the internal degree of freedom. In particular, a particle in an excited state will have a slower expected velocity than one in its ground state. If the particle is prepared in a superposition state of internal energy, then its position at a later time will be entangled with the internal degree of freedom. Thus, the mass operator introduces spatial decoherence, which even appears in the case of a free particle [13].

2.2 The Path Integral Formulation

To examine these effects further, we will investigate the mass operator from the perspective of the path integral formalism. We motivate the work here with the question: *How does the mass operator affect the falling interference presented in the last section?* In order for there to be any effect, the particle must have some non-degenerate internal energy levels. We will restrict ourselves to the simple case discussed in Ref. [12], where the particle is spin-$\frac{1}{2}$ with a Zeeman splitting induced by an external magnetic field.[1] Before we can answer the above question, we need to find the form for the propagator in this scenario.

We begin with the newly defined Lagrangian for this problem,

$$L(x, \dot{x}) = \frac{1}{2}M\dot{x}^2 - Mgx - Mc^2, \tag{18}$$

For a particle with magnetic moment μ in a uniform magnetic field of strength B, the mass operator is given by

$$M = \begin{bmatrix} m - \frac{\mu B}{2c^2} & 0 \\ 0 & m + \frac{\mu B}{2c^2} \end{bmatrix}. \tag{19}$$

From this point, we can construct the Feynman propagator, in accordance with Eq. (1).

$$K^{\chi',\chi}(x', t'; x, t) = \langle x', \chi' | U(t' - t) | x, \chi \rangle \tag{20}$$

This is still a matrix element of the time-evolution operator; however, the evolution operator now contains an index for spin, accounting for the two-dimensional internal Hilbert space. This also naturally leads to a matrix representation:

[1] We will, however, still consider the particle to be neutral, so there is no coupling to the electromagnetic field beyond its spin interaction.

$$K^{\chi',\chi}(x',t';x,t) = \begin{bmatrix} K^{\uparrow\uparrow}(x',t';x,t) & K^{\uparrow\downarrow}(x',t';x,t) \\ K^{\downarrow\uparrow}(x',t';x,t) & K^{\downarrow\downarrow}(x',t';x,t) \end{bmatrix}. \tag{21}$$

Since the interaction terms appearing in Eq. (18) all commute with the mass operator in Eq. (19), the propagator can be greatly simplified, as it is then diagonal in the internal energy eigenbasis:

$$\mathbf{K}(x',t';x,t) = \begin{bmatrix} K^{m_-}(x',t';x,t) & 0 \\ 0 & K^{m_+}(x',t';x,t) \end{bmatrix}, \tag{22}$$

where $K^{m_\pm}(x',t';x,t)$ is the propagator for a particle of mass $m_\pm = m \pm \mu B/(2c^2)$; from the perspective of the propagator, the different internal energy states just appear as modified masses. Typically, the rest mass energy is excluded from the Lagrangian; it has no effect on the dynamics, and merely leads to an unmeasurable global phase $\exp(-imc^2t/\hbar)$. When promoting mass to an operator, a relative phase of $\exp(-i\mu Bt/\hbar)$ is introduced between the two internal states, which could in principle have a measurable effect.

We wish to use this propagator to calculate the wavefunctions for particles which initially have spin in the superposition $|\chi_0\rangle = \alpha|\uparrow\rangle + \beta|\downarrow\rangle$, with $|\alpha|^2 + |\beta|^2 = 1$. However, if we are only interested in the interference pattern observed on the screen, and do not measure the spin state of the particle, then we need to trace out the spin information. We review how to do this in Appendix C. We find that the spatial probability distribution for an initial state $|\chi_0\rangle \otimes |\psi_0\rangle$ is then

$$\langle x\rangle = |\alpha|^2 \left|\int dx\, K^{m_-}(x,t;x,0)\psi_0(x)\right|^2 + |\beta|^2 \left|\int dx\, K^{m_+}(x,t;x,0)\psi_0(x)\right|^2, \tag{23}$$

which is a convex sum of the contributions coming from each internal state. This is immediately identifiable as decoherence, which is consistent with our interpretation of Eq. (17). Additionally, this demonstrates that, when tracing out the spin degree of freedom, the phase introduced by the rest mass operator becomes irrelevant.

The form of Eq. (23) allows for easy calculation of the decohered spatial distribution, which we will illustrate with an example. Take the propagator to be that of a free particle and choose the initial wavefunction to be a Gaussian wavepacket with momentum p. This wavefunction is given by

$$\psi_0(x) = (\pi\sigma)^{-\frac{1}{4}} \exp\left[-\frac{x^2}{2\sigma^2} + \frac{ipx}{\hbar}\right]. \tag{24}$$

After convolving this with the free space propagator $K_0^{m_\pm}(x,t,x_0,0)$, we have

$$\psi_{m_\pm}(x,t) = \frac{\exp\left[i\phi - 2\frac{(z-p/m_\pm t)^2}{\sigma^2(1+\gamma_{m_\pm}^2)}\right]}{(\pi\sigma^2)^{1/4}\sqrt{i-\gamma_{m_\pm}}}, \tag{25}$$

where $\gamma_{m_\pm} = \frac{\hbar t}{m_\pm \sigma^2}$ and ϕ is an irrelevant phase factor. We notice that the mean of this Gaussian moves with speed $p/m_\pm$. If the particle is prepared in a superposition state of internal energy then Eq. (23) states that the probability distribution will be given by

$$P(x,t) = |\alpha|^2 \left|\psi_{m_-}(x,t)\right|^2 + |\beta|^2 \left|\psi_{m_+}(x,t)\right|^2. \tag{26}$$

In other words, the spatial distribution is given by a mixture of Gaussian wavepackets propagating with different speeds.

Given the initial state $|\Psi\rangle = (\alpha\,|\uparrow\rangle + \beta\,|\downarrow\rangle) \otimes |\psi_0\rangle$, the coupling introduced by the mass operator evolves the state to $|\Psi(t)\rangle = \alpha\,|\uparrow\rangle \otimes \left|\psi_{m_-}(t)\right\rangle + \beta\,|\downarrow\rangle \otimes \left|\psi_{m_+}(t)\right\rangle$. If a detector is placed at a distance from the source far enough for the Gaussian distributions described by $\langle x|\psi_{m_-}(t)\rangle$ and $\langle x|\psi_{m_+}(t)\rangle$ to become distinct, then the arrival time of the particle will be bimodal. Again, if the position degree of freedom is considered to be a 'clock' – its non-trivial evolution permits this – then this may be considered to be a special relativistic time dilation effect [13].

3 Gravitational Decoherence in Double Slit Interference

We have now developed the tools to explore falling double slit interference with an internal degree of freedom. Again, we consider a particle incident on slits of width $2a$ centred at $x = \pm b$. Our earlier calculation in the first section used the semi-classical approximation for the z-direction, to replace arrival times T and τ with the classically expected times, D/v and L/v. We have just shown, however, that the arrival time of the particle will no longer be well defined when the internal degree of freedom plays a dynamical role.

We also saw, in the previous section, that the effect of using the gravitational propagator was equivalent to performing a mass-independent coordinate transformation. This means that, for a wavepacket with zero initial average momentum, the mass operator has no effect on the position expectation value under the influence of a gravitational potential.[2] This leads to an interesting effect if we consider a two dimensional Gaussian wavepacket with zero average momentum in the x-direction (in the direction of the gravitational field) and a non-zero average momentum in the z-direction. At some fixed distance along z from the particle's initial location, the difference in expected arrival times will mean that, depending on the internal state, gravity will have displaced the wavepacket for different amounts of time. As a result, the higher energy state will fall further than the lower energy state, causing gravity to act, in some sense, like an asymmetric Stern–Gerlach device. This effect will be very small, as it depends on the magnitude of $\|H^{\text{int}}\|/(mc^2)$, but can be sensitively detected by introducing an interference pattern along x.

We calculate the pattern produced by the double slit by returning to a two dimensional propagator, and simplifying the problem to a Gaussian particle distribution

[2] It will however affect the spreading of the wavepacket and therefore the variance in the position.

incident on the slits which is then detected at a screen L metres away. In this case, the wavefunction just beyond the slits, for a particle of mass m, is given by

$$\psi^{(1)}(x,z,t) = \frac{\int_{b-a}^{b+a}\int_{-\infty}^{\infty} dx_0 dz_0 K_0(z,t;z_0,t) K_g(x,t;x_0,0)\psi_0(x_0,z_0)}{\int_{b-a}^{b+a} dx\psi(x_0,0)}. \tag{27}$$

This avoids the semi-classical approximations made in the previous calculation, but leads to a time-dependent wavefunction. However, we are only interested in the spatial distribution observed at the screen, with no measurement performed regarding the time of arrival. The simplest way to account for this is to average the distribution over some length of time so that

$$\bar{\psi}(x, z = L) = \frac{1}{2\Delta t}\int_{-\Delta t}^{\Delta t} |\psi(x, L, t)|^2 dt, \tag{28}$$

where Δt will have some relationship with the spatial spread of the Gaussian packet in the z direction, such that the majority of the probability lies within $\pm\Delta t$. Figure 3 shows a plot of the resulting two-slit interference, calculated for a neutron in equal superposition of its internal energy states. The energy splitting is $\Delta E \approx 10^{-14}$ J, corresponding to a magnetic field on the order of 10^{12} T. Even with this infeasibly large energy splitting, the decoherence effect occurs over tens of metres. If a more reasonable value for the energy splitting is used, then spreading of the wavepacket delocalises the particle before the decoherence is even detectable.

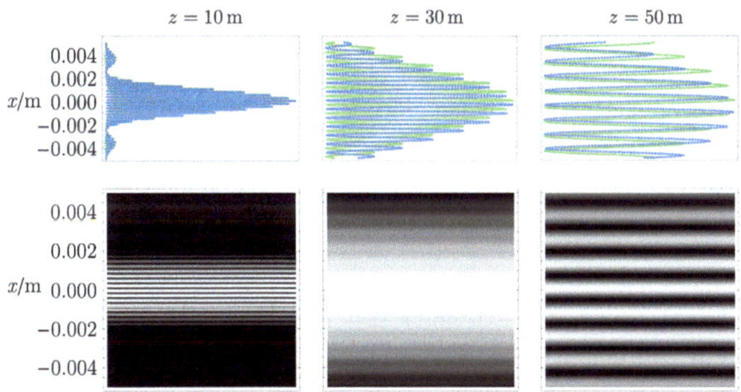

Fig. 3 Decoherence of Double Slit Interference in a Gravitational Field. The internal energy splitting leads to different expected arrival times for the wavepacket at the screen. The pattern corresponding to the higher energy spin state will fall further than its counterpart, resulting in periodic reductions in the visibility of the interference. The intensity at the screen for distances of 10, 30 and 50 m is shown for a neutron with wavelength $\lambda \sim 10^{-8}$ m and internal energy splitting of $\Delta E = 10^{-14}$ J. The *top row* shows the time-averaged spatial probability distribution for the up (*green*, solid) and down (*blue*, *dotted*) spin components, while the *bottom row* plots the spin-averaged probability distribution as it would be observed on the screen. All x positions are relative to the position of a classical particle with mass m_-

This decoherence is suggestive of the effects described in Ref. [14]; where Zych et al. demonstrate that the interference pattern in a Mach–Zehnder interferometer decoheres as a result of proper time. The work we present here is a free space interference analogue, which similarly exhibits periodic decoherence effects. This gives strength to the argument in Ref. [13] that the effects on external evolution, introduced by the mass operator, are complementary to the interpretation of time dilation, occurring for evolution of the internal degrees of freedom. To put it more elegantly, embedding an operational clock in a system which behaves quantum mechanically results in an evolution which destroys this quantum nature.

4 Coherence, Correlations, and Non-Markovian Dynamics

The decoherence of the interference fringes discussed above can be better understood in terms of correlations between internal and external degrees of freedom. It also turns out that spin coherence is necessary for the generation of non-classical correlations between the internal and external degrees of freedom. We discuss each of these ideas successively below, beginning with coherence theory, before providing a further interpretation in terms of non-Markovian open dynamics.

Incoherent operations. The total Hamiltonian in Eq. (14) is diagonal in the spin basis and can therefore be expressed as $H = |\uparrow\rangle\langle\uparrow| \otimes H_{m_-} + |\downarrow\rangle\langle\downarrow| \otimes H_{m_+}$. Therefore, the unitary operator for the joint dynamics has the form of a controlled-unitary on the external degree of freedom:

$$U = |\uparrow\rangle\langle\uparrow| \otimes U_{m_-} + |\downarrow\rangle\langle\downarrow| \otimes U_{m_+}, \tag{29}$$

which has the potential to generate correlations between the internal and external degrees of freedom. However, when considering the reduced dynamics of the spin alone, evolution between two points in time is described by an incoherent operation (one which maps incoherent states to incoherent states) in the energy eigenbasis [18], i.e.,

$$U\,|s\,\psi(0)\rangle = |s\,\psi_{m_r}(t)\rangle, \tag{30}$$

where $s \in \{\downarrow, \uparrow\}$, $r = +$ if $s = \downarrow$, and $r = -$ if $s = \uparrow$. More specifically, U is incoherent since it will map a mixed state of the form $\sigma_{IC} = q\,|\uparrow\rangle\langle\uparrow| + (1-q)\,|\downarrow\rangle\langle\downarrow|$ to itself:

$$\mathrm{tr}_{ext}[U\sigma_{IC} \otimes \rho U^\dagger] = \sigma_{IC}, \tag{31}$$

where ρ is any state of the external degree and IC stands for incoherent.

On the other hand, due to the non-vanishing commutator $[P^2, x] \neq 0$, the dynamics of the external degree of freedom alone is not described by an incoherent operation in the position basis. That is, $U\,|s\,x\rangle = |s\,\psi_{m_r}(t)\rangle$, and coherence of the wavefunction can increase. For the same reason, both conditional unitary operations U_{m_-} and U_{m_+} will also look like incoherent operations from the perspective of the spin.

Entanglement and discord. In recent years, researchers studying coherence theory have shown that incoherent operations can lead to generation of entanglement and quantum discord when the initial spin state possesses coherence. The generation of entanglement is easily checked by taking the spin to initially be in the pure state $\alpha\left|\downarrow\right\rangle + \beta\left|\uparrow\right\rangle$ for $\alpha, \beta \neq 0$ and the position state to be $|\psi\rangle$:

$$U\left(\alpha\left|\downarrow\psi(0)\right\rangle + \beta\left|\uparrow\psi(0)\right\rangle\right) = \alpha\left|\downarrow\psi_{m_+}(t)\right\rangle + \beta\left|\uparrow\psi_{m_-}(t)\right\rangle, \tag{32}$$

which is an entangled state, since the marginal states are not pure.

It is known that any coherence can be turned into entanglement via some incoherent operation and a pure ancilla [19]. However, in our setup we are limited to a specific incoherent operation, which may not be able to generate entanglement for all coherent initial states. Consider the case when the initial spin state is the mixed state

$$\sigma_{CO} = w\left|\uparrow\right\rangle\!\left\langle\uparrow\right| + (1-w)\left|+\right\rangle\!\left\langle+\right|, \tag{33}$$

where $|+\rangle = (|\uparrow\rangle + |\downarrow\rangle)/\sqrt{2}$ and CO stands for coherent. For $0 \leq w \leq 1$, the time-evolved state will have quantum correlations, but for some values of w will have no entanglement; the future state will be fully separable for $w = 1$ and entangled for $w = 0$. Therefore, there must be a critical value for $w = w_c$ where the transition from entangled state to separable state occurs. In the regime where the state is separable, it will necessarily have quantum discord [20–22] as measured by the internal or external degree of freedom.

In fact, the only time quantum discord vanishes for $t > 0$ is when the initial spin state has the form $\sigma_{IC} = q\left|\uparrow\right\rangle\!\left\langle\uparrow\right| + (1-q)\left|\downarrow\right\rangle\!\left\langle\downarrow\right|$. Let us further suppose that the initial external state is given by a density matrix ρ. After evolving for some time t, the system will be in state

$$U\sigma_{IC}\otimes\rho U^{\dagger} = q\left|\uparrow\right\rangle\!\left\langle\uparrow\right|\otimes\rho_{m_-}(t) + (1-q)\left|\downarrow\right\rangle\!\left\langle\downarrow\right|\otimes\rho_{m_+}(t). \tag{34}$$

This clearly becomes a classical mixture of the states $\rho_{m_-}(t)$ and $\rho_{m_+}(t)$ with weighting w when the internal degree of freedom is measured (whichever measurement basis is chosen). That is, the unitary operation in Eq. (29), being an incoherent operation on the internal degree of freedom, will not generate any non-classical correlations when the initial spin state is a classical mixture of energy eigenstates.

On the other hand, with the exception of pathological cases where the initial wavefunction does not have support everywhere, the two spin-conditioned external states will never be exactly orthogonal, i.e., $\mathrm{tr}[\rho_{m_+}(t)\rho_{m_-}(t)] \neq 0$. This means that the spin state after a measurement on the external degree of freedom will, in general, depend on the choice of measurement basis; in other words, there are non-classical correlations (discord) in one direction.

Let us now consider the case where both initial states can be arbitrary mixed states. Then the time evolved states have form

$$U\begin{pmatrix} a & b \\ b^* & 1-a \end{pmatrix} \otimes \rho U^\dagger = \begin{pmatrix} a\,\rho_{m_-} & b\,U_{m_-}\,\rho\,U_{m_+}^\dagger \\ b^* U_{m_+}\,\rho\,U_{m_-}^\dagger & (1-a)\rho_{m_+}. \end{pmatrix} \tag{35}$$

If we can make (non-unitary) operations on the spin degree of freedom, such as projections, we will see different interference patterns corresponding to different outcomes. By making strong measurements on the spin by, e.g., introducing a Stern–Gerlach apparatus, the correlations could be used to steer the interference pattern.

For example, to see how correlations affect the reduced dynamics, consider the middle column of Fig. 3. If the magnetic field, and hence the effective coupling, was turned off for $z > 30\,\mathrm{m}$, the interference pattern would subsequently evolve unitarily with a single mass-m propagator. It would continue to fall as if it were a classical object, and the fringe visibility would never return. This is because the spin degree of freedom, which 'remembers' the original two-slit pattern, is no longer interacting with the position degree of freedom. However, if the spin components were filtered out at a later time using a Stern–Gerlach apparatus, the visibility could be recovered in full; the spin acts as a memory, hiding information about the particle's earlier trajectory. In other words, the periodic re-coherence of the spatial wavefunction is indicative of non-Markovian behaviour, which we will now discuss further.

Reduced non-Markovian dynamics. In order to see the effects of these correlations from another perspective, suppose we only look at the position of the particle. From the perspective of an observer who cannot measure the internal degree of freedom, the evolution of the particle appears to be open, with the spin acting as an environment. We immediately see that the same features are seen whether the initial spin state possesses coherence or not. The external state is obtained by tracing over the spin in Eq. (35) to get

$$\rho_{ext} = a\rho_{m_-} + (1-a)\rho_{m_+}. \tag{36}$$

That is, the observed interference pattern is indistinguishable and independent of b. When $b = 0$ the internal and external degrees of freedom become classically correlated, and both entanglement and discord are vanishing. Conversely, when $b \neq 0$ discord (and possibly entanglement) will be present.

Whenever the past state of a system directly affects its future evolution, a process is called non-Markovian. While there have been several mathematical definitions of 'non-Markovianity' proposed for quantum processes [23, 24] (with variable levels of descriptive success), the operational meaning of the term is clear cut [16]: If the causal continuity of a system's evolution is broken at some time t by, for example, making a measurement and re-preparing the system in a fixed state $|\phi\rangle$, independently of the measurement outcome, then the process is non-Markovian if the system's density operator $\rho_\tau(x, x')$ at a later time τ depends on the measurement outcome k or on the system's history h prior to t. Formally,

$$\rho_\tau(x, x' \,|\, |\phi\rangle\,, k, h) \neq \rho_\tau(x, x' \,|\, |\phi\rangle\,, k', h') \Rightarrow \text{Non-Markovian.} \tag{37}$$

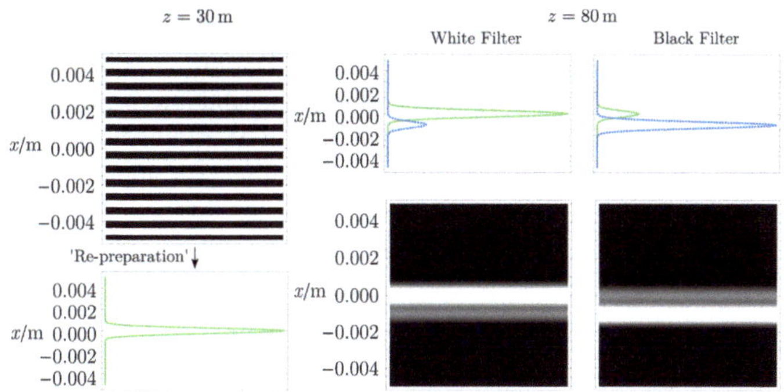

Fig. 4 Non-Markovian Behaviour of an Interfering Particle with Spin. A grating filter is introduced at $z = 30$ m, allowing particles to pass through either the white or *black* regions in the *top-left panel*. The particle is then subsequently re-prepared in a Gaussian state, whose probability density is shown in the *bottom-left panel*; this state is the same for either choice of filter. The plots on the *right* show the probability distribution to find the particle at different positions on a screen placed at $z = 80$ m for the two choices of filter; *top curves* show spin-up (*green*, *solid*) and spin-down (*blue*, *dotted*) projections, the lower plots show the spin-averaged position distribution. Distinguishability of the two cases indicates non-Markovian behaviour. All x positions are relative to the corresponding position of a classical particle with mass m_-

This kind of behaviour implies that there is some sort of memory transmitting information from the past across the causal break. We will see that this is the case for the falling particle described earlier in this section.

In order to introduce a causal break in the evolution, we will put the spatial filter shown in the left column of Fig. 4 at $z = 30$ m. This can be set to either allow particles through the white region (which has the greatest overlap with the spin-up wavefunction) or the black region (which has the greatest overlap with the spin-down wavefunction). After the filter, the particle is rapidly (effectively instantaneously) collimated into a Gaussian state along x, which does not depend on whether the black or white filter is chosen.

Since the overlap of the white (black) filter function $f_{w(b)}(x)$ with the spin-up and spin-down wavefunctions at $z = 30$ m is different, the subsequent spin state will be conditioned on the choice of filter. The post-filter spin density operator is given by

$$\rho_{w(b)} = |\alpha|^2 \left| \int \mathrm{d}x\; f_{w(b)}(x)\psi_{m_-}(x) \right| |\!\uparrow\rangle\langle\uparrow\!| + |\beta|^2 \left| \int \mathrm{d}x'\; f_{w(b)}(x)\psi_{m_+}(x) \right| |\!\downarrow\rangle\langle\downarrow\!|$$
$$+ \alpha\beta^* \int \mathrm{d}x\mathrm{d}x'\; f_{w(b)}(x)\psi_{m_-}(x) f_{w(b)}(x')\psi^*_{m_+}(x')\, |\!\uparrow\rangle\langle\downarrow\!| + \text{h.c.}, \qquad (38)$$

where $\psi_{m_\pm}(x)$ is the time-averaged wavefunction for the relevant spin branch at the z position the filter is applied. For $\alpha = \beta = 1/\sqrt{2}$, the post-filter probabilities for the spin-up and down states are $\sim\frac{4}{5}$ and $\sim\frac{1}{5}$ respectively for the white filter, and *vice versa* for the black filter.

The right hand side of Fig. 4 shows the probability distribution further from the slits after each of the filters is applied (note that the two spin components have already begun to separate again). Since the two conditional distributions are clearly different, the dynamics of the spatial distribution must be non-Markovian; the only way the post re-preparation evolution can depend on which filter was applied is through the spin state, which is acting as a memory.

5 Conclusion

The universality of free fall is a pervasive phenomenon, and one which has inspired more fundamental gravitational equivalence principles. This includes Einstein's famous equivalence between mass and energy which, ultimately, forms part of the foundation for our current understanding of gravity. Here, we have explored how a self-interfering quantum particle falls under the influence of Newtonian gravity. We have shown that the universality of free fall holds even in this case, as the interference pattern itself fall just like a classical particle.

We have also considered interference of falling neutrons in the presence of a strong magnetic field. The presence of the magnetic field leads to splitting of the internal energy of the neutrons which, according to the Einstein equivalence principle, should make spin-down neutrons more massive than spin-up neutrons. Moreover, if a neutron is prepared in a spin-superposition state (with respect to the internal energy eigenbasis), this seemingly leads to the violation of a super-selection rule, i.e., superposition of masses. However, we use the mass operator formalism [5] to show that, if the energy splitting of the internal spin contributes to the mass of the neutron, then the visibility of the interference pattern periodically decreases and increases.

Our results indicate that these decoherence effects are a consequence of an operational clock embedded within a quantum mechanical rest frame. That is, the internal degree of freedom keeps track of the time the particle spends being in different mass states. Finally, we have shown that this accounting of the internal energy (mass) state can be understood as non-Markovian dynamics for the position degree of freedom, with the spin acting as a memory. We show the non-Markovian behaviour by operational methods using the notion of causal break introduced in Ref. [16]. In particular, we have given an operational recipe to witness the non-Markovian memory by solely acting on the external degree of freedom.

In Ref. [13] it is argued that gravity may be the culprit for quantum decoherence. This mechanism does not depart from how we think of decoherence in open systems theory more generally. This view is fundamentally different from that posited by the proponents of collapse theory who claim that gravity leads to fundamentally irreversible dynamics, cf. Ref. [25]. Thankfully, one can differentiate between the two hypotheses by checking whether the decoherence can be reversed [26], which we do here demonstrating that coherence-information loss due to gravity can be recovered.

Acknowledgements We thank Robert Mann for insightful discussions and German Valencia for pointing out errors in an earlier version of this work.

Appendices

A The COW Experiment

The Colella-Overhauser-Werner experiment provided the first evidence of a gravitational effect that is purely quantum mechanical [17]. In this experiment, Colella et al. used a silicon crystal interferometer to split a beam of neutrons, placing one of the beam paths in a higher gravitational potential (see Fig. 5). The difference in the gravitational potential between each arm results in a relative phase shift, which, when recombined, can be measured as modulated intensity. On the length scale of the interferometer, the gravitational field is approximately constant. This allows the relative phase difference between the beams to be calculated using the Wentzel-Kramers-Brillouin (WKB) approximation; that is to integrate the potential difference between the classical trajectories over time [27]. The two vertical paths of the interferometer contribute phases which cancel out, leaving only the horizontal paths. The phase shift is found to be

$$\Delta\Phi = \frac{2\pi m_I m_G g A \lambda}{h^2} \sin\phi. \tag{39}$$

A phase shift of this form would be predicted for a quantum mechanical particle in the presence of any scalar potential; in this case, it is the Newtonian gravitational potential. A full description of this effect requires only regular quantum mechanics and Newtonian theory, needing no metric description of gravity, but being unexplainable by classical Newtonian gravity alone. It represents the first evidence of gravity

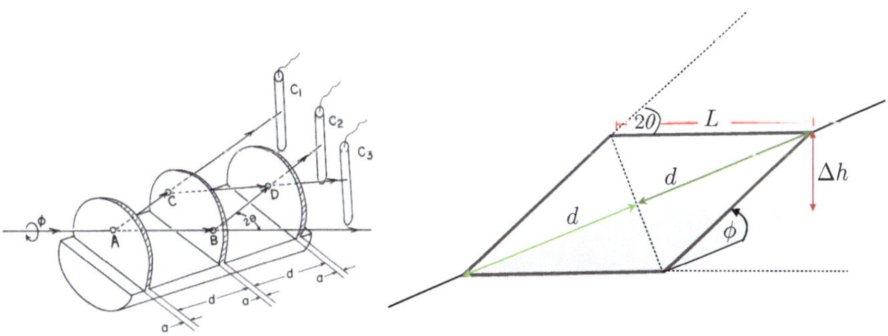

Fig. 5 COW Interferometer | *Left* Schematic of the apparatus used in the COW experiment, taken from Ref. [17]. The interferometer is rotated about the axis of the first Bragg angle of diffraction. *Right* Simplified diagram used to derive the induced phase shift

interacting in a truly quantum mechanical way. However, from the perspective of quantum theory, this effect is well understood as a scalar Aharanov–Bohm effect and manifests similarly for charged particles in electric potentials [14, 28].

B Single Slit Diffraction in a Newtonian Gravitational Potential

B.1 Derivation of the Propagator

First consider the Lagrangian for a free particle $L = \frac{1}{2}m\dot{x}^2$. The Feynman propagator is given by

$$\langle x', t'|x, t\rangle = K_0(x', t'; x, t) = \int \mathcal{D}(x(t)) \exp\left[\frac{i}{\hbar}\int_t^{t'} \frac{1}{2}m\dot{x}(s)^2 ds\right] \tag{40}$$

$$= \frac{\exp\left[\frac{im}{2\hbar}\frac{(x'-x)^2}{t'-t}\right]}{\sqrt{2\pi i\hbar(t'-t)/m}}. \tag{41}$$

This result will be needed when we consider the propagator for a particle in a linear gravitational potential. In this case the Lagrangian is given by $L = \frac{1}{2}m\dot{x}^2 - mgx$, which gives the Feynman propagator

$$K_g(x', t'; x, t) = \int \mathcal{D}(x(t)) \exp\left[\frac{i}{\hbar}\int_t^{t'} ds \left\{\frac{1}{2}m\dot{x}(s)^2 - mgx(s)\right\}\right]. \tag{42}$$

To simplify this calculation we express the path $x(t)$ in terms of deviations from the classical trajectory $x_c(t)$ which satisfies the Euler-Lagrange equations of motion. The Feynman measure which sums over all possible paths then becomes a sum over all possible deviations from the classical path. The action expressed in terms of this new parametrisation is

$$S[x_c(t) + \delta x(t)] = \int_t^{t'} dt \left\{\frac{1}{2}m\Big(\dot{x} + \delta\dot{x}\Big)^2 - mg\Big(x_c + \delta x\Big)\right\} \tag{43}$$

$$= \int_t^{t'} dt \left\{\frac{1}{2}m\dot{x}_c^2 - mgx_c + m\dot{x}_c\delta\dot{x} + \frac{1}{2}m(\delta\dot{x})^2 - mg\delta x\right\}, \tag{44}$$

where $\frac{1}{2}m\dot{x}_c^2 - mgx_c = s[x_c(t)]$, which is evidently the extremised action given by the classical trajectory. The term containing $\dot{x}_c\delta\dot{x}$ can be integrated by parts, realising that the deviations are zero at the endpoints of the path:

$$S[x_c(t) + \delta x(t)] = S[x_c(t)] + \int_t^{t'} \frac{1}{2} m(\delta \dot{x})^2 \, dt + [\ddot{x} \delta x]_t^{t'} - \int_t^{t'} m \delta x \, (\ddot{x}_c + g) \, dt. \tag{45}$$

The last two terms vanish ($\ddot{x}_c = -g$) and substituting the rest into Eq. (42) and factoring out the classical action we arrive at

$$K_g(x', t'; x, t) = \exp\left[\frac{i}{\hbar} S\left[x_c(t)\right]\right] \int \mathcal{D}\left(\delta x(t)\right) \exp\left[\frac{i}{\hbar} \int_t^{t'} \frac{1}{2} m \delta \dot{x}(s)^2 ds\right]. \tag{46}$$

The remaining Feynman integral over the deviations is recognised as the free particle propagator in Eq. (40), but with the subtle difference being that $x = x' = 0$. Substituting the integral with the expression from Eq. (41) the propagator becomes

$$K_g(x', t'; x, t) = \frac{\exp\left[\frac{i}{\hbar} S\left[x_c(t)\right]\right]}{\sqrt{2\pi i \hbar (t' - t)/m}}. \tag{47}$$

Now, all that remains is to find the explicit form for the classical action. We begin with the classical equation of motion, $\ddot{x}_c = -g$, and solve to find the general solution,

$$x_c(t) = -\frac{1}{2} g t^2 + a t + b. \tag{48}$$

We now impose the boundary conditions $x(t) = x$ and $x(t') = x'$ and solve for the constants a and b:

$$x = -\frac{1}{2} g t^2 + a t + b \quad \text{and} \quad x' = -\frac{1}{2} g t'^2 + a t' + b. \tag{49}$$

Solving for a and b gives

$$a = \frac{x' - x}{t' - t} - \frac{g(t^2 - t'^2)}{2(t' - t)} = \frac{x' - x}{t' - t} + \frac{1}{2} g (t + t'), \tag{50}$$

and

$$b = \frac{1}{2} \left(x' + x + \frac{1}{2} g (t^2 + t'^2) - a(t + t') \right) \tag{51}$$

$$= \frac{1}{2} \left(x' + x - (x' - x) \frac{t + t'}{t' - t} + \frac{1}{2} g (t^2 + t'^2) - \frac{1}{2} g (t + t')^2 \right) \tag{52}$$

$$= \frac{1}{2}\left(\frac{(x'+x)(t'-t)-(x'-x)(t+t')}{t'-t} - gtt'\right) \tag{53}$$

$$= \frac{1}{2}\left(\frac{2(xt'-x't)}{t'-t} - gtt'\right) = \frac{xt'-x't}{t'-t} - \frac{1}{2}gtt'. \tag{54}$$

Thus, the action of the path taken from (x, t) to (x', t') is

$$S[x_c(t)] = \int_t^{t'} dt \left\{\frac{1}{2}m(-gt+a)^2 - mg(-\frac{1}{2}gt^2+at+b)\right\} \tag{55}$$

$$= \frac{m}{2}\int_t^{t'} dt \left\{2g^2t^2 - 2gat + a^2 - 2gb\right\} \tag{56}$$

$$= \frac{m}{2}\left(a^2(t'-t) - 2g\left(a(t'-t)+b\right)(t'-t) + \frac{2g^2}{3}(t'-t')^3\right) \tag{57}$$

$$= \frac{m}{2}\left\{\frac{(x'-x)^2}{t'-t} - g(x+x')(t'-t) - \frac{g^2}{12}(t'-t)^3\right\}. \tag{58}$$

Finally, substituting this into Eq. (47), we arrive at the complete expression for the propagator for a particle in a gravitational potential,

$$K_g(x', t'; x, t) = \frac{\exp\left[\frac{im}{2\hbar}\left\{\frac{(x'-x)^2}{t'-t} - g(x+x')(t'-t) - \frac{g^2}{12}(t'-t)^3\right\}\right]}{\sqrt{2\pi i\hbar(t'-t)/m}}. \tag{59}$$

B.2 Calculating the Single Slit Wavefunction

We now consider applying this propagator to the problem at hand. Let's begin by assuming that the slits are long enough to ignore diffraction effects in the y direction. Consider a source of particles at the origin $(0, 0)$ and let a double slit be located at distance $z = D$ metres from the source. Each slit has width $2a$ with centre located at $x = \pm b$. The screen is then a further L metres away from the slits. The two dimensional propagator required for this problem is given by a free particle propagator in the z-direction multiplied by the gravitational propagator for the x direction as calculated in Eq. (2). This propagator allows us to ask the question of *If a particle initially starts at position* $\vec{r} = (0, 0)$, *what is the probability of finding the it at position* $\vec{r}\,' = (x, D+L)$ *on the screen?* This distribution in x will be the the two slit interference pattern that we seek. When computing this amplitude we consider a semi-classical approach. We assume that the 'trajectory' of the neutron can be separated into two parts: (a) the path from the source to the slits, followed by (b) the path from the slits to the screen. Quantum mechanically the particles need not pass through the slits and there even exists the possibility of them passing through the slits multiple times before hitting the screen. That being said the probabilities associated with these events are negligible.

The semi-classical approach is valid provided that the majority of the particle's momentum is in the z direction, such that the wavelength is approximately the z-direction wavelength, $\lambda \approx \frac{2\pi\hbar}{mv_z}$. We assume that this wavelength is much smaller than the z-direction scale lengths D and L in conjunction with the assumption that these are much larger than the x direction scale lengths. This allows us to consider the particles motion in the z-direction as approximately classical and allows for the motion to be partitioned about the slits. The specific propagator $K_g^{(1)}(x, T+\tau; 0, 0)$, for the process of starting at point $(x = 0, z = 0)$ at time $t = 0$, passing through position $(\omega \in [b-a, b+a], D)$ at time T and then arriving at position $(x, D+L)$ at time $T+\tau$ will simply be a product of propagator for each independent component of the path, integrated over the slit distribution $\Omega(\omega)$,

$$\Omega(\omega) = \begin{cases} 1 \; b-a < \omega < b+a \\ 0 \qquad \text{otherwise} \end{cases}, \tag{60}$$

$$\begin{aligned} K_g^{(1)}(x, T+\tau; 0, 0) = & K_0(D, T; 0, 0) K_0(D+L, T+\tau; D, T) \\ & \times \int_{b-a}^{b+a} K_g(\omega, T; 0, 0) K_g(x, T+\tau; \omega, T) d\omega. \end{aligned} \tag{61}$$

Evidently for any particular choice of D and L, the two z propagators will only give global phase which is identical for all x. This global phase has no measurable effects, allowing us to discard the z propagators. This integral is performed in below giving the result

$$\begin{aligned} K_g^{(1)}(x, T+\tau; 0, 0) = & \frac{e^{i\phi(x)}}{i\sqrt{2\lambda(D+L)}} \\ & \times \left\{ C[\sigma_+(x)] - C[\sigma_-(x)] + iS[\sigma_+(x)] - iS[\sigma_-(x)] \right\}, \end{aligned} \tag{62}$$

where $C[u] \equiv \int_0^u \cos\left(\frac{\pi}{2}x^2\right) dx$ is the Fresnel cosine function, $S[u] = \int_0^u \sin\left(\frac{\pi}{2}x^2\right) dx$ is the Fresnel sine function and $\eta = 1 + \frac{L}{D}$ and

$$\sigma_\pm(x) = \sqrt{\frac{2}{\lambda L}\eta} \left\{ (b \pm a) - \frac{x}{\eta} - \frac{1}{2} g \frac{m^2\lambda^2}{h^2} DL \right\}, \tag{63}$$

$$\phi(x) = \pi \left\{ \frac{x^2}{\lambda(D+L)} - mgx \frac{\lambda(D+L)}{h^2} - \frac{g^2}{12} \frac{m^4\lambda^3}{h^4} (D+L)(D-L)^2 \right\}. \tag{64}$$

The propagator obtains its name for good reason. An initial wavefunction $\psi_0(x)$ convoluted with the propagator will give the future state of the wavefunction for all time $\psi(x,t) = \int G(x,t;s,0)\psi_0(s)ds$. For the purposes of this calculation we can assume a point source of particles such that the initial spacial distribution of the particle is a δ-function. This however means that the wavefunction is the 'square root of a δ-function', which is not guaranteed to be defined. That aside, we can calculate the spatial distribution of the particle at the screen, but in order to have this distribution be normalised, we must account for the fact that a large portion of the wavefunction does not pass through the slit. Thus in actual fact the distribution at the screen is given by the conditional probability to be at position x and time $T+\tau$ given that it was at position $x' \in [-a, a]$ at time T. Fortunately as outlined below, the normalised wavefunction is simply the propagator in Eq. (62) multiplied by a factor $\sqrt{\frac{\lambda D}{2a}}$. Finally we arrive at the normalised wavefunction at the screen, for a particle passing through a single slit of width $2a$, centred at $x = b$,

$$\psi^{(1)}(x) = \frac{e^{i\phi(x)}}{i2\sqrt{\eta a}}\Big\{C[\sigma_+(x)] - C[\sigma_-(x)] + iS[\sigma_+(x)] - iS[\sigma_-(x)]\Big\}, \tag{65}$$

with $\phi(x)$ and $\sigma_\pm(x)$ defined in Eqs. (6) and (5), and $\eta = 1 + L/D$. The square of this wavefunction will give the probability distribution for the particle at the slit, which is plotted in Fig. 2 for various distances between slit and screen. The pattern clearly appears to shift towards the negative x direction as the screen is moved further from the slit.

Integrating over the Slit Profile

The propagator to arrive at x having passed through a single slit of width $2a$ with centre at $x = b$ is found by integrating over the slit distribution, $\Omega(\omega)$, which is 1 for $b - a < \omega < b + a$ and 0 otherwise:

$$K_g^{(1)}(x;a,b) = \int_{-\infty}^{\infty} A(x,\omega)\Omega(\omega)d\omega \tag{66}$$

$$= \int_{b-a}^{b+a} K_g(\omega,T;0,0)K_g(x,T+\tau;\omega,T)d\omega \tag{67}$$

$$= \int_{-\infty}^{\infty} d\omega\,\Omega(\omega)\sqrt{\frac{m}{2\pi i\hbar T}}\sqrt{\frac{m}{2\pi i\hbar\tau}}\exp\left[\frac{im}{2\hbar}\left\{\frac{\omega^2}{T} - g\omega T - \frac{g^2}{12}T^3\right\}\right] \tag{68}$$

$$\times \exp\left[\frac{im}{2\hbar}\left\{\frac{(x-\omega)^2}{\tau} - g(x+\omega)\tau - \frac{g^2}{12}\tau^3\right\}\right]. \tag{69}$$

Completing the square in ω,

$$\frac{\omega^2}{T} - g\omega T - \frac{g^2}{12}T^3 + \frac{(x-\omega)^2}{\tau} - g(x+\omega)\tau - \frac{g^2}{12}\tau^3 \tag{70}$$

$$= \frac{\omega^2}{T} + \frac{\omega^2}{\tau} - \frac{2x\omega}{\tau} - g\omega(T+\tau) + \frac{x^2}{\tau} - gx\tau + \frac{g^2}{4}(T^3+\tau^3) \tag{71}$$

$$= \omega^2\frac{T+\tau}{T\tau} - 2\omega\left(\frac{x}{\tau} + \frac{1}{2}g(T+\tau)\right) + \frac{x^2}{\tau} - gx\tau - \frac{g^2}{12}(T^3+\tau^3) \tag{72}$$

$$= \zeta\left(\omega - \frac{\kappa}{\zeta}\right)^2 - \frac{\kappa^2}{\zeta} + \frac{x^2}{\tau} - gx\tau - \frac{g^2}{12}(T^3+\tau^3), \tag{73}$$

where $\zeta = \frac{T+\tau}{T\tau}$ and $\kappa = \frac{x}{\tau} + \frac{1}{2}g(T+\tau)$. Returning to Eq. (69),

$$K_g^{(1)}(x;a,b) = e^{i\phi(x,T,\tau)}\sqrt{\frac{m}{2\pi i\hbar T}}\sqrt{\frac{m}{2\pi i\hbar\tau}}\int_{-\infty}^{\infty} d\omega \exp\left[\frac{im\zeta}{2\hbar}\left(\omega - \frac{\kappa}{\zeta}\right)^2\right], \tag{74}$$

where $\phi(x,T,\tau) = \frac{m}{2\hbar}\left(\frac{x^2}{\tau} - \frac{\kappa^2}{\zeta} - gx\tau - \frac{g^2}{12}(T^3+\tau^3)\right)$ is the phase produced by terms not dependent on ω. We make the substitution $v = \sqrt{\frac{m\zeta}{\pi\hbar}}(\omega - \frac{\kappa}{\zeta})$, and define new limits of integration $\sigma_\pm(x) = \sqrt{\frac{m\zeta}{\pi\hbar}}\left((b \pm a) - x\frac{T}{T+\tau} - \frac{1}{2}gT\tau\right)$:

$$K_g^{(1)}(x;a,b) = e^{i\phi(x,T,\tau)}\sqrt{\frac{2\pi i\hbar}{m\zeta}}\sqrt{\frac{m^2}{(2\pi i\hbar)^2 T\tau}}\int_{\sigma_-}^{\sigma_+} \exp\left[\frac{i\pi}{2}v^2\right] dv \tag{75}$$

$$= \frac{e^{i\phi(x,T,\tau)}}{\sqrt{(2i)^2\pi\hbar(T+\tau)/m}}\int_{\sigma_-}^{\sigma_+}\left\{\cos\left(\frac{i\pi}{2}v^2\right) + i\sin\left(\frac{i\pi}{2}v^2\right)\right\} dv \tag{76}$$

$$= \frac{e^{i\phi(x,T,\tau)}}{2i\sqrt{\pi\hbar(T+\tau)/m}}\left\{C[\sigma_+(x)] - C[\sigma_-(x)] + iS[\sigma_+(x)] - iS[\sigma_-(x)]\right\}, \tag{77}$$

where $C[u] \equiv \int_0^u \cos\left(\frac{\pi}{2}x^2\right) dx$ is the Fresnel cosine function, $S[u] \equiv \int_0^u \sin\left(\frac{\pi}{2}x^2\right) dx$ is the Fresnel sine function. Now to simplify $\phi(x,T,\tau)$ we first have

$$\kappa^2 = \frac{x^2}{\tau^2} + \frac{x}{\tau}g(T+\tau) - \frac{g^2}{12}(T+\tau)^2, \tag{78}$$

$$\frac{\kappa^2}{\zeta} = \frac{x^2}{\tau^2}\frac{T\tau}{T+\tau} + \frac{x}{\tau}g(T+\tau)\frac{T\tau}{T+\tau} - \frac{g^2}{12}(T+\tau)^2\frac{T\tau}{T+\tau} \tag{79}$$

$$= \frac{x^2 T}{\tau(T+\tau)} + gxT - \frac{g^2}{12}T\tau(T+\tau) \tag{80}$$

to get

$$\phi(x,T,\tau) = \frac{m}{2\hbar}\left\{\frac{x^2}{\tau} - \frac{\kappa^2}{\zeta} - gx\tau - \frac{g^2}{12}(T^3+\tau^3)\right\} \tag{81}$$

$$= \frac{m}{2}\left\{\frac{x^2(T+\tau) - x^2T}{\tau(T+\tau)} - gx(T+\tau) - \frac{g^2}{12}(T^3+\tau^3 - T\tau(T+\tau))\right\} \tag{82}$$

$$= \frac{m}{2}\left\{\frac{x^2}{T+\tau} - gx(T+\tau) - \frac{g^2}{12}(T+\tau)(T-\tau)^2\right\}. \tag{83}$$

We can make use of the approximation $v_z \gg v_x$ and that $\lambda \approx \frac{h}{mv_z}$ to find expressions $T = \frac{m\lambda D}{h}$ and $\tau = \frac{m\lambda L}{h}$. Thus, we have

$$T \pm \tau = \frac{m\lambda(D \pm L)}{h} \quad \text{and} \quad T\tau = \frac{m^2\lambda^2}{h^2}DL. \tag{84}$$

Using these we get

$$\frac{m\zeta}{\pi\hbar} = \frac{2}{\lambda}\left(\frac{1}{D} + \frac{1}{L}\right) \qquad \text{where} \qquad \zeta = \frac{T+\tau}{T\tau} = \frac{h}{m\lambda}\frac{D+L}{DL} = \frac{h}{m\lambda}\left(\frac{1}{D} + \frac{1}{L}\right). \tag{85}$$

Next, let

$$\eta = \frac{T}{T+\tau} = \frac{D}{D+L} = \frac{1}{1+L/D}, \tag{86}$$

allowing us to express ϕ and $\sigma_\pm$ in terms of L, D and λ:

$$\sigma_\pm(x) = \sqrt{\frac{m\zeta}{\pi\hbar}}\left((b \pm a) - x\frac{T}{T+\tau} - \frac{1}{2}gT\tau\right)$$

$$= \sqrt{\frac{2}{\lambda L}\eta}\left\{(b \pm a) - \frac{x}{\eta} - \frac{1}{2}g\frac{m^2\lambda^2}{h^2}DL\right\}, \tag{87}$$

$$\phi(x,T,\tau) = \frac{m}{2}\left\{\frac{x^2}{T+\tau} - gx(T+\tau) - \frac{g^2}{12}(T+\tau)(T-\tau)^2\right\}, \tag{88}$$

$$\phi(x) = \pi\left\{\frac{x^2}{\lambda(D+L)} - mgx\frac{\lambda(D+L)}{h^2} - \frac{g^2}{12}\frac{m^4\lambda^3}{h^4}(D+L)(D-L)^2\right\}. \tag{89}$$

This is the form of the propagator given in Eq. (62).

Normalisation of the Distribution at the Screen

As derived in the first section the propagator for the process of starting at position $\vec{r} = (0, 0)$, passing through the point $(x' \in [b-a, b+a], D)$ and finally being detected at position $\vec{r}\,' = (x, D+L)$ on the screen is not the same as the wavefunction at the screen. To obtain this we must first convolve the propagator with a initial wavefunction whose square magnitude is a δ-function giving the wavefunction as seen at the other side of the slit. This wavefunction however will not be normalised due to the fact that only a portion of the initially normalise wavefunction has been propagated beyond the slits. It can be renormalised however by scaling by the probability of passing through the slit. Unfortunately the 'square root of a δ-function' is not always well defined as is the case for operators acting on any distribution. We can however attempt to use a Gaussian with variance σ as the initial wavefunction, compute the quantity of interest and take the limit $\sigma \to 0$. Under suitable circumstances the limit will be defined giving the desired result.

We will begin with an initial wavefunction that is the square root of a Gaussian

$$\psi_\sigma(x) = g_\sigma(x) = \frac{1}{(2\pi\sigma^2)^{\frac{1}{4}}} e^{-\frac{x^2}{4\sigma^2}}. \tag{90}$$

However we notice that the square root of a Gaussian is simply another Gaussian of variance $\rho = \sigma\sqrt{2}$ multiplied by the factor $(8\pi\sigma^2)^{\frac{1}{4}}$. So the initial function can represented as

$$\psi_\rho(x) = (4\pi\rho^2)^{\frac{1}{4}} \frac{e^{-\frac{x^2}{2\rho^2}}}{\sqrt{2\pi\rho^2}} = (4\pi\rho^2)^{\frac{1}{4}} g_\rho(x). \tag{91}$$

Convolving this with the propagator $K_g^{(1)}(x, T+\tau; x_0, 0)$ will give the un-normalised wavefunction at the screen:

$$\psi_\rho(x, T+\tau) = \int_\infty^\infty dx_0 K_g^{(1)}(x, T+\tau; x_0, 0)\psi_\rho(x_0). \tag{92}$$

To renormalise this, we scale by the probability of the particle passing through the slit. The probability of the particle being in $x \in [b-a, b+a]$ at time T is

$$P(x \in [b-a, b+a]; T) = \int_{b-a}^{b+a} \left| \int_{-\infty}^{\infty} K_g(x', T; x_0, 0)\psi_\rho(x_0)dx_0 \right|^2 dx', \tag{93}$$

which gives that the renormalised wavefunction at the screen is

$$\psi_\rho{}'(x, T+\tau) = \frac{\psi(x, T+\tau)}{\sqrt{\int_{-a}^{a} | \int_{-\infty}^{\infty} K_g(x', T; x_0, 0)\psi_\rho(x_0)dx_0|^2 dx'}} \tag{94}$$

$$= \frac{\int_\infty^\infty K_g^{(1)}(x, T+\tau; x_0, 0)\psi_\rho(x_0)dx_0}{\sqrt{\int_{b-a}^{b+a} | \int_{-\infty}^{\infty} K_g(x', T; x_0, 0)\psi_\rho(x_0)dx_0|^2 dx'}} \tag{95}$$

$$= \frac{(4\pi\rho^2)^{\frac{1}{4}} \int_\infty^\infty K_g^{(1)}(x, T+\tau; x_0, 0)g_\rho(x_0)dx_0}{(4\pi\rho^2)^{\frac{1}{4}}\sqrt{\int_{b-a}^{b+a} | \int_{-\infty}^{\infty} K_g(x', T; x_0, 0)g_\rho(x_0)dx_0|^2 dx'}}. \tag{96}$$

Now the limit $\sigma \to 0$ can equivalently be taken as $\rho \to 0$. The Gaussians $g_\rho(x)$ then become delta functions $\delta(x)$:

$$\psi'(x, T+\tau) = \lim_{\rho \longrightarrow 0} \frac{\int_{\infty}^{\infty} K_g^{(1)}(x, T+\tau; x_0, 0) g_\rho(x_0) dx_0}{\sqrt{\int_{b-a}^{b+a} \mid \int_{-\infty}^{\infty} K_g(x', T; x_0, 0) g_\rho(x_0) dx_0 \mid^2 dx'}} \tag{97}$$

$$= \frac{K_g^{(1)}(x, T+\tau; 0, 0)}{\sqrt{\int_{b-a}^{b+a} |K_g(x', T; 0, 0)|^2 dx'}} = \frac{K_g^{(1)}(x, T+\tau; 0, 0)}{\sqrt{\int_{b-a}^{b+a} (2\pi\hbar T/m)^{-1} dx'}} \tag{98}$$

$$= K_g^{(1)}(x, T+\tau; 0, 0)\sqrt{\frac{\frac{h}{m}T}{2a}} = K_g^{(1)}(x, T+\tau; 0, 0)\sqrt{\frac{\frac{m\lambda D}{h}\frac{h}{m}}{2a}} \tag{99}$$

$$= K_g^{(1)}(x, T+\tau; 0, 0)\sqrt{\frac{\lambda D}{2a}}, \tag{100}$$

where $|K_g(x', T; 0, 0)|^2$ was taken from Eq. (59). Thus we see that the normalised wavefunction at the screen is given by multiplying the propagator by the factor $\sqrt{\frac{\lambda D}{2a}}$.

C Tracing Out Spin from a Matrix Propagator

This is best achieved using the density operator prescription. The pure density operator ρ for a quantum state $|\psi\rangle$ is $\rho = |\psi\rangle\langle\psi|$. For a state comprised of two subsystems, we can ignore the state of a subsystem by tracing it out. This is given by the operation $\mathrm{tr}_B[X_{AB}] = \sum_k \langle k|_B X_{AB} |k\rangle_B$, where X_{AB} is an operator on the composite system AB and $\{|k\rangle_B\}$ forms a complete basis for subsystem B.

Here, we would like to trace out the spin state. The initial density operator is $\rho_0 = |\chi_0\rangle\langle\chi_0| \otimes |\psi_0\rangle\langle\psi_0|$, where $\langle x|\psi_0\rangle = \psi_0(x)$, is the spatial distribution of the particle, and it is assumed that, initially, the spatial location of the particle is uncorrelated with the spin state. The state of the system at later time t is given by $\rho(t) = U(t)\rho_0 U^\dagger(t)$. We can represent the time evolution operator in terms of the propagator by making use of the resolution of the identity $\sum_{\{\chi,\chi'\}\in\{\downarrow,\uparrow\}} \int dx\, dx' \, |x', \chi'\rangle\langle x, \chi| = \mathbb{1}$:

$$U(t) = \sum_{\{\chi,\chi'\}\in\{\downarrow,\uparrow\}} \int dx\, dx' K_g^{\chi',\chi}(x', t; x, 0)\, |x', \chi'\rangle\langle x, \chi|, \tag{101}$$

which gives that,

$$\rho(t) = \sum \int dx\, dx' dy\, dy' K_g^{\chi',\chi}(x', t; x, 0) K_g^{*\phi',\phi}(y', t; y, 0) \left|x', \chi'\right\rangle \times \left\langle x, \chi\middle|\rho_0\middle|y, \phi\right\rangle \left\langle y', \phi'\right|, \tag{102}$$

where the sum is over all spin variables. We can then take the trace over the spin subspace to give,

$$\tilde{\rho}(t) = \mathrm{tr}_{\mathrm{spin}}[\rho(t)] = \sum \int dx\, dx' dy\, dy' K_g^{\chi',\chi}(x', t; x, 0) K_g^{*\phi',\phi}(y', t; y, 0) \times \left\langle x, \chi\middle|\rho_0\middle|y, \phi\right\rangle \langle\chi'|\phi'\rangle \left|x'\right\rangle\!\left\langle y'\right|. \tag{103}$$

The spatial probability distribution is then given by the expectation of the position operator $\langle\hat{x}\rangle = \mathrm{tr}[\tilde{\rho}(t)\hat{x}]$. Making use of the fact that $\langle\chi'|\phi'\rangle = \delta_{\chi',\phi'}$ and $\langle y'|x'\rangle = \delta(x' - y')$, we find

$$\langle\hat{x}\rangle = \sum \int dx\, dy K_g^{\chi',\chi}(x', t; x, 0) K_g^{*\chi',\phi}(x', t; y, 0) \left\langle x, \chi\middle|\rho_0\middle|y, \phi\right\rangle. \tag{104}$$

At this point, we work with the term $\left\langle x, \chi\middle|\rho_0\middle|y, \phi\right\rangle$

$$\left\langle x, \chi\middle|\rho_0\middle|y, \phi\right\rangle = \left\langle x, \chi\right|\Big(|\chi_0\rangle\langle\chi_0| \otimes |\psi_0\rangle\langle\psi_0| \Big)\left|y, \phi\right\rangle \tag{105}$$

$$= \langle\chi| \Big(\alpha\,|\uparrow\rangle + \beta\,|\downarrow\rangle\Big)\Big(\alpha^*\,\langle\uparrow| + \beta^*\,\langle\downarrow|\Big)\,|\phi\rangle\,\langle x|\psi_0\rangle\langle\psi_0|y\rangle \tag{106}$$

$$= \langle\chi| \Big(|\alpha|^2\,|\uparrow\rangle\langle\uparrow| + |\beta|^2\,|\downarrow\rangle\langle\downarrow| + \alpha^*\beta\,|\downarrow\rangle\langle\uparrow| + \beta * \alpha\,|\uparrow\rangle\langle\downarrow|\Big)\,|\phi\rangle \times \psi_0(x)\psi_0^*(y). \tag{107}$$

Since the matrix propagator in Eq. (22) is diagonal, we can immediately discard the $|\downarrow\rangle\langle\uparrow|$ and $|\uparrow\rangle\langle\downarrow|$ terms when substituting into the expression for the spatial distribution in Eq. (104),

$$\langle\hat{x}\rangle = \int dx\, dy |\alpha|^2 K_g^{\uparrow,\uparrow}(x', t; x, 0)\psi_0(x) \Big(K_g^{\uparrow,\uparrow}(x', t; y, 0)\psi_0(y)\Big)^* + |\beta|^2 K_g^{\downarrow,\downarrow}(x', t; x, 0)\psi_0(x) \Big(K_g^{\downarrow,\downarrow}(x', t; y, 0)\psi_0(y)\Big)^* \tag{108}$$

$$= |\alpha|^2 \left| \int dx K_g^{\uparrow,\uparrow}(x', t; x, 0)\psi_0(x)\right|^2 + |\beta|^2 \left|\int dx K_g^{\downarrow,\downarrow}(x', t; x, 0)\psi_0(x)\right|^2. \tag{109}$$

This result is simply a convex sum of the initial spatial distribution evolved by each propagator.

References

1. G. Galilei, *Dialogo sopra i due massimi sistemi del mondo (Dialogue Concerning the Two Chief World Systems, Ptolemaic and Copernican)* (University of California Press, California, 1953)
2. I. Newton, *Philosophiæ Naturalis Principia Mathematica (The Principia: Mathematical Principles of Natural Philosophy)*, vol. 1687 (University of California Press, California, 1999)
3. N. Birrell, P. Davies, *Quantum Fields in Curved Space*, Cambridge Monographs on Mathematical Physics (Cambridge University Press, Cambridge, 1984)
4. A. Einstein, Über den einfluß der schwerkraft auf die ausbreitung des lichtes. Ann. Phys. (Berlin) **35**, 898 (1911). [*The Collected Papers of Albert Einstein, Vol. 3: The Swiss Years: Writings, 1909-1911* (University of Chicago Press, 1995)]
5. M. Zych, C. Brukner, Quantum formulation of the Einstein equivalence principle (2015). arXiv:1502.00971
6. M. Inguscio, L. Fallani, *Atomic Physics: Precise Measurements and Ultracold Matter* (Oxford University Press, Oxford, 2013)
7. T. Berrada, S. van Frank, R. Bücker, T. Schumm, J.F. Schaff, J. Schmiedmayer, Integrated Mach–Zehnder interferometer for Bose–Einstein condensates. I, Nat. Commun. **4**, 1 (2013)
8. A. Peters, K.Y. Chung, S. Chu, Measurement of gravitational acceleration by dropping atoms. Nature **400**, 849–852 (1999)
9. A. Bonnin, N. Zahzam, Y. Bidel, A. Bresson, Characterization of a simultaneous dual-species atom interferometer for a quantum test of the weak equivalence principle. Phys. Rev. A **92**, 023626 (2015)
10. L. Zhou, S. Long, B. Tang, X. Chen, F. Gao, W. Peng, W. Duan, J. Zhong, Z. Xiong, J. Wang, Y. Zhang, M. Zhan, Test of equivalence principle at 10^{-8} level by a dual-species double-diffraction Raman atom interferometer. Phys. Rev. Lett. **115**, 013004 (2015)
11. J. Williams, S.-w. Chiow, H. Mueller, N. Yu, Quantum Test of the Equivalence Principle and Space-Time Aboard the International Space Station (2015). arXiv:1510.07780
12. P.J. Orlando, R. Mann, K. Modi, F.A. Pollock, A test of the equivalence principle(s) for quantum superpositions. Class. Quantum Grav. **33**, 19LT01 (2015)
13. I. Pikovski, M. Zych, F. Costa, C. Brukner, Universal decoherence due to gravitational time dilation. Nat. Phys. **11**, 668–672 (2015)
14. M. Zych, F. Costa, I. Pikovski, C. Brukner, Quantum interferometric visibility as a witness of general relativistic proper time. Nat. Commun. **2**, 505 (2011)
15. M. Zych, F. Costa, I. Pikovski, T.C. Ralph, C. Brukner, General relativistic effects in quantum interference of photons. Class. Quantum Grav. **29**, 224010 (2012)
16. F.A. Pollock, C. Rodríguez-Rosario, T. Frauenheim, M. Paternostro, K. Modi, Complete Framework for Efficient Characterisation of Non-Markovian Processes (2015). arXiv:1512.00589
17. R. Colella, A.W. Overhauser, S.A. Werner, Observation of gravitationally induced quantum interference. Phys. Rev. Lett. **34**, 1472–1474 (1975)
18. A. Streltsov, G. Adesso, M.B. Plenio, Quantum Coherence as a Resource (2016). arXiv:1609.02439
19. A. Streltsov, U. Singh, H.S. Dhar, M.N. Bera, G. Adesso, Measuring quantum coherence with entanglement. Phys. Rev. Lett. **115**, 020403 (2015)
20. L.C. Céleri, J. Maziero, R.M. Serra, Theoretical and experimental aspects of quantum discord and related measures. Int. J. Quantum Inf. **09**(07n08), 1837–1873 (2011)
21. K. Modi, A. Brodutch, H. Cable, T. Paterek, V. Vedral, The classical-quantum boundary for correlations: discord and related measures. Rev. Mod. Phys. **84**, 1655–1707 (2012)
22. G. Adesso, T.R. Bromley, M. Cianciaruso, Measures and Applications of Quantum Correlations (2016). arXiv:1605.00806
23. A. Rivas, S.F. Huelga, M.B. Plenio, Quantum non-Markovianity: characterization, quantification and detection. Rep. Prog. Phys. **77**, 094001 (2014)
24. H.-P. Breuer, Foundations and measures of quantum non-Markovianity. J. Phys. B: At. Mol. Opt. Phys. **45**(15), 154001 (2012)

25. M. Snadden, J. McGuirk, P. Bouyer, K. Haritos, M. Kasevich, Measurement of the Earth's gravity gradient with an atom interferometer-based gravity gradiometer. Phys. Rev. Lett. **81**, 971–974 (1998)
26. C. Arenz, R. Hillier, M. Fraas, D. Burgarth, Distinguishing decoherence from alternative quantum theories by dynamical decoupling. Phys. Rev. A **92**, 022102 (2015)
27. B. Hall, *Quantum Theory for Mathematicians*, Graduate Texts in Mathematics (Springer, New York, 2013)
28. B.E. Allman, W.T. Lee, O.I. Motrunich, S.A. Werner, Scalar Aharonov–Bohm effect with longitudinally polarized neutrons. Phys. Rev. A **60**, 4272–4284 (1999)

Part IV
Physical Realizations and Experimental Progress

Quantum Discord and Entropic Measures of Quantum Correlations: Optimization and Behavior in Finite XY Spin Chains

N. Canosa, M. Cerezo, N. Gigena and R. Rossignoli

Abstract We discuss a generalization of the conditional entropy and one-way information deficit in quantum systems, based on general entropic forms. The formalism allows to consider simple entropic forms for which a closed evaluation of the associated optimization problem in qudit-qubit systems is shown to become feasible, allowing to approximate that of the quantum discord. As application, we examine quantum correlations of spin pairs in the exact ground state of finite XY spin chains in a magnetic field through the quantum discord and information deficit. While these quantities show a similar behavior, their optimizing measurements exhibit significant differences, which can be understood and predicted through the previous approximations. The remarkable behavior of these quantities in the vicinity of transverse and non-transverse factorizing fields is also discussed.

1 Introduction

Non-classical correlations in mixed states of composite quantum systems have attracted strong attention in recent years [1, 2]. In pure states they can be identified with entanglement [3–6] and are essential for quantum teleportation [7] and for achieving exponential speed-up in pure state based quantum algorithms [8, 9]. However, in the case of mixed states it is now well known that separable states, defined in general as convex mixtures of product states [10], i.e. those which can then be created by local operations and classical communication [5, 10], may still exhibit non-classical features, such as a non-zero value of the quantum discord [11–14]. The latter is defined as the difference between two distinct quantum extensions of the classical mutual information or conditional entropy, becoming zero for classically correlated states and reducing to the entanglement entropy for pure states. A finite quantum discord is also present in the mixed state based quantum algorithm of Knill

N. Canosa · M. Cerezo · N. Gigena · R. Rossignoli (✉)
Departamento de Física-IFLP, Universidad Nacional de La Plata,
C.C. 67, 1900 La Plata, Argentina
e mail: rossigno@fisica.unlp.edu.ar

F.F. Fanchini et al. (eds.), *Lectures on General Quantum Correlations and their Applications*, Quantum Science and Technology,
DOI 10.1007/978-3-319-53412-1_20

and Laflamme [15], as shown in [16], which achieves an exponential speed-up over classical algorithms without substantial entanglement [17]. This fact triggered the interest not only in the quantum discord and its fundamental properties [18–23] but also in other related measures with similar features [1, 2], which include among others the one-way information deficit [1, 24–26], the geometric discord [27], the generalized entropic measures introduced in [28, 29] (which contain the previous ones as particular cases), the local quantum uncertainty [30, 31], the trace distance discord [32–34] and more recently coherence based measures [2, 35, 36]. Besides, various operational interpretations of the quantum discord and other related measures have been provided [1, 19, 26, 37–41]. It is worth mentioning, however, that most of these measures require the determination of an optimizing local measurement, which makes their evaluation difficult in a general situation (shown to be NP-complete [42]).

Interacting spin chains provide a useful scenario for studying the previous measures and their behavior in the vicinity of critical points [1, 43–53]. In general, ground states of interacting spin chains are strongly entangled states, implying that the state of a reduced spin pair or group of spins will typically be a mixed state. Hence, for these subsystems differences between discord type measures and entanglement will arise already at zero temperature.

In this chapter we first briefly review in Sect. 2 the quantum discord and the associated local measurement dependent conditional entropy on which it is based. We then discuss the consistent generalization of this entropy to general entropic forms [54, 55]. This extension enables in particular the consideration of simple forms which allow an analytic solution of the associated optimization problem, i.e., that of determining the local measurement leading to the lowest conditional entropy, for general mixed states of qudit-qubit systems [54, 55]. The solution is given in terms of an eigenvalue equation which admits a simple geometrical picture [54]. We then examine the generalized information deficit [28], based on general entropic forms, which contains the standard one-way information deficit [24–26] as a particular case, together with its exact minimization for simple quadratic entropic forms for general states of qudit-qubit systems [27, 29].

In Sect. 3 we will analyze the exact behavior of the quantum discord and the information deficit associated with spin pairs in the ground state of finite XY spin $1/2$ chains immersed in a magnetic field. We will show that while their behavior is quite similar, significant differences do arise in their corresponding minimizing measurements, which can be correctly predicted and understood by the approximations based on simple entropic forms. A remarkable effect in these chains is the possibility of exhibiting a completely separable exact ground state at a factorizing field. The existence of a factorizing field was first discussed in Ref. [56] and its properties together with the general conditions for its existence at transverse fields were analyzed in [57–66]. The transverse factorizing field actually corresponds to the last ground state parity transition [61, 64, 65], and accordingly, it will be shown that in finite chains the quantum discord and information deficit exhibit full range in its vicinity, with an appreciable finite limit value at this field. We will also discuss the behavior for a non-transverse field [67], which will differ from the previous one due to the broken spin parity symmetry. Conclusions are finally given in Sect. 4.

2 Formalism

We first describe the main features of the quantum discord and the generalized conditional entropy and information deficit, together with some analytic results for general states of qudit-qubit systems.

2.1 *Quantum Discord and Conditional Entropy*

Let us start with the well known quantum discord, introduced in [11, 12]. For a bipartite quantum system $A + B$ initially in a state ρ_{AB}, it can be defined as the minimum difference between two distinct quantum versions of the mutual information, or equivalently, of the conditional entropy:

$$D(A|B) = \underset{M_B}{\text{Min}}[I(A, B) - I(A, B_{M_B})] \tag{1}$$

$$= \underset{M_B}{\text{Min}}\, S(A|B_{M_B}) - S(A|B) \tag{2}$$

where the minimization is over all possible local POVM measurements [5] M_B on B, characterized by a set of operators $M_j = I_A \otimes M_{jB}$ satisfying $\sum_j M_j^\dagger M_j = I_A \otimes I_B$. Here $I(A, B) = S(\rho_A) - S(A|B)$ represents the quantum mutual information [68] before the measurement and $I(A, B_{M_B}) = S(\rho_A) - S(A|B_{M_B})$ a measurement dependent mutual information, with

$$S(A|B) = S(\rho_{AB}) - S(\rho_B) \tag{3}$$

$$S(A|B_{M_B}) = \sum_j p_j S(\rho_{A/j}) \tag{4}$$

the corresponding conditional entropies, where

$$\rho_{A/j} = p_j^{-1} \text{Tr}_B\, \rho_{AB} M_j^\dagger M_j \tag{5}$$

is the reduced state of A after outcome j at B, with $p_j = \text{Tr}\, \rho_{AB} M_j^\dagger M_j$ the probability of such outcome, and $S(\rho) = -\text{Tr}\, \rho \log_2 \rho$ the von Neumann entropy. In the case of complete local *projective* measurements $M_j = P_j = I_A \otimes P_{jB}$, with $P_{jB} \equiv |j_B\rangle\langle j_B|$ one-dimensional orthogonal projectors ($P_{jB} P_{kB} = \delta_{jk} P_{jB}$), then

$$S(A|B_{M_B}) = S(\rho'_{AB}) - S(\rho'_B) \tag{6}$$

where ρ'_{AB} is the joint state after the (unread) local measurement,

$$\rho'_{AB} = \sum_j P_j \rho_{AB} P_j = \sum_j p_j \rho_{A/j} \otimes P_{jB} \tag{7}$$

and $\rho'_B = \mathrm{Tr}_A \rho'_{AB} = \sum_j p_j P_{jB}$ the ensuing state of B.

As is well known, the mutual information $I(A, B)$ is a measure of all correlations between subsystems A and B, being non-negative and vanishing just for product states $\rho_{AB} = \rho_A \otimes \rho_B$ [68]. Equations (1)–(2) can then be regarded as the difference between all correlations present in the original state and the classical correlations that remain after the local measurement on B, measuring then the quantum correlations. Accordingly, $D(A|B)$ is always non-negative [11, 12], a property which stems from the concavity of the *conditional* von Neumann entropy $S(A|B)$ [68]. It vanishes just for semi-quantum states ρ_{AB}, which are already of the form (7) and which then remain invariant under the local measurement determined by the projectors P_{jB}. The quantum discord is then non-zero not only in entangled states but also in most separable mixed states, i.e., those not of the form (7) (and hence not diagonal in a conditional product basis $\{|i_A^j\rangle|j_B\rangle\}$). For pure states ($\rho^2_{AB} = \rho_{AB}$) it reduces to the entanglement entropy $S(\rho_A) = S(\rho_B)$ of the system, as $S(A|B_{M_B}) = 0$ for *any* measurement based on rank one projectors.

We remark that in the general case, the minimum in Eq. (2) is always reached for measurements based on rank one projectors $M_{jB} \propto P_{jB}$, not necessarily orthogonal [1, 54, 55], with a minimization based on standard projective measurements, Eq. (6), providing normally a good approximation. Nevertheless, the minimization in Eq. (2) is in general still difficult, being in fact an NP-complete problem [42].

2.2 Generalized Conditional Entropy After a Local Measurement

Due the previous difficulty, and in order to obtain a more clear picture of the optimization problem associated with the quantum discord, it is convenient to consider more simple entropic forms, which may enable an easier evaluation of the minimum conditional entropy. We then consider first the generalized conditional entropy [54, 55]

$$S_f(A|B_{M_B}) = \sum_j p_j S_f(\rho_{A/j}) \tag{8}$$

where $S_f(\rho) = \mathrm{Tr}\, f(\rho)$ is a generalized trace form entropy [68, 69]. Here $f : [0, 1] \to \mathbb{R}$ is a smooth strictly concave function satisfying $f(0) = f(1) = 0$, such that $S_f(\rho) \geq 0$, with $S_f(\rho) = 0$ just for pure states. Concavity of f implies, for $S_f(A) \equiv S_f(\rho_A)$ [54],

$$S_f(A) \geq S_f(A|B_{M_B})$$

so that the average conditional mixedness of A after measurement is never greater than the original mixedness, irrespective of the measure S_f used to quantify it. Moreover, the minimum of $S_f(A|M_B)$ is also always reached for rank one projectors $M_{jB} \propto P_{jB}$ [55], as in the von Neumann case. Hence, these properties remain valid for general concave functions f.

In particular, we may consider simple entropic forms, like the quadratic entropy

$$S_2(\rho) = 2[1 - \mathrm{Tr}\,\rho^2] \tag{9}$$

which follows from $f_2(\rho) = 2\rho(1-\rho)$ and is also known as linear entropy since it corresponds to the linear approximation $-\rho \ln \rho \approx \rho(1-\rho)$. It is a particular case of the Tsallis entropies [70] $S_q(\rho) = \frac{1-\mathrm{Tr}\,\rho^q}{1-2^{1-q}}$, obtained for $f_q(\rho) \propto \rho - \rho^q$, $q > 0$, which approach the von Neumann entropy for $q \to 1$ (we set $S_f(\rho) = 1$ for a maximally mixed single qubit state).

Equation (9) is just a linear function of the purity $\mathrm{Tr}\,\rho^2$ and does not require the explicit knowledge of the eigenvalues of ρ, thus enabling an easier evaluation, both theoretically and experimentally [71–73]. For instance, writing a general mixed state of a system with Hilbert space dimension d as

$$\rho = \frac{1}{d}(I + \boldsymbol{r} \cdot \boldsymbol{\sigma}) \tag{10}$$

where $\boldsymbol{\sigma} = (\sigma_1, \ldots, \sigma_{d^2-1})$ is an orthogonal basis for traceless operators in the system ($\mathrm{Tr}\,\sigma_\mu = 0$, $\mathrm{Tr}\,\sigma_\mu\sigma_\nu = d\delta_{\mu\nu}$), implying $\boldsymbol{r} = \mathrm{Tr}\,\rho\boldsymbol{\sigma} = \langle\boldsymbol{\sigma}\rangle$, we obtain the explicit expression

$$S_2(\rho) = \frac{2}{d}(d - 1 - |\boldsymbol{r}|^2)\,. \tag{11}$$

Equation (11) shows that $|\boldsymbol{r}|^2 \leq d-1$, with $|\boldsymbol{r}|^2 = d-1$ just for pure states.

2.3 The Qudit-Qubit Case

Let us now consider a composite system where A is a system with Hilbert space dimension d_A and B a single qubit. Denoting with $\boldsymbol{\sigma}_A$ an orthogonal basis for operators in A and $\boldsymbol{\sigma}_B \equiv \boldsymbol{\sigma}$ the Pauli matrices of B, a general state of this system can be written as [54]

$$\rho_{AB} = \rho_A \otimes \rho_B + \frac{1}{2d_A}\sum_{\mu,\nu} C_{\mu\nu}\sigma_{A\mu} \otimes \sigma_{B\nu} \tag{12}$$

where $\rho_A = \frac{1}{d_A}(I_A + \boldsymbol{r}_A \cdot \boldsymbol{\sigma}_A)$, $\rho_B = \frac{1}{2}(I_2 + \boldsymbol{r}_B \cdot \boldsymbol{\sigma})$ are the reduced states of A and B, with $\boldsymbol{r}_A = \langle \boldsymbol{\sigma}_A \rangle$, $\boldsymbol{r}_B = \langle \boldsymbol{\sigma} \rangle$, and

$$C_{\mu\nu} = \langle \sigma_{A\mu} \otimes \sigma_{B\nu} \rangle - \langle \sigma_{A\mu} \rangle \langle \sigma_{B\nu} \rangle \quad (13)$$

are the elements of the *correlation tensor*, represented by the $(d_A^2 - 1) \times 3$ matrix C.

We consider a local POVM measurement on the qubit B based on rank one operators $M_{kB} = \sqrt{q_k} P_{kB}$, where $P_{kB} = \frac{1}{2}(I_2 + \boldsymbol{k} \cdot \boldsymbol{\sigma})$, with $\boldsymbol{k}$ a unit vector ($|\boldsymbol{k}| = 1$), is the projector onto the pure qubit state with $\langle \boldsymbol{\sigma} \rangle = \boldsymbol{k}$, and $\sum_k q_k P_{kB} = I_2$. We may then express the ensuing conditional entropy (8) as

$$S_f(A|B_k) = \sum_k p_k S_f(\rho_{A/k}) \quad (14)$$

where

$$\rho_{A/k} = \rho_A + \frac{1}{d_A} \left(\frac{C\boldsymbol{k}}{1 + \boldsymbol{r}_B \boldsymbol{k}} \right) \cdot \boldsymbol{\sigma}_A \quad (15)$$

$$p_k = \frac{1}{2} q_k (1 + \boldsymbol{r}_B \cdot \boldsymbol{k}) \quad (16)$$

are, respectively, the conditional post measurement state of A after result $\boldsymbol{k}$ and the probability of obtaining this result. The vector $\boldsymbol{r}_{A/k}$ characterizing the post-measurement state of A is then

$$\boldsymbol{r}_{A/k} = \boldsymbol{r}_A + \frac{C\boldsymbol{k}}{1 + \boldsymbol{r}_B \cdot \boldsymbol{k}} . \quad (17)$$

For a standard projective spin measurement along direction $\boldsymbol{k}$ just vectors $\pm\boldsymbol{k}$ are to be considered in the previous sums, with $q_{\pm k} = 1$.

While in the general case the eigenvalues of $\rho_{A/k}$ are required for the evaluation of (14), for the quadratic entropy (9) a closed evaluation is directly feasible with Eq. (11). For a standard projective spin measurement along direction $\boldsymbol{k}$ we obtain [54]

$$S_2(A|B_k) = S_2(\rho_A) - \Delta S_2(A|B_k) , \quad (18)$$

$$\Delta S_2(A|B_k) = \frac{2}{d_A} \frac{|C\boldsymbol{k}|^2}{1 - (\boldsymbol{r}_B \cdot \boldsymbol{k})^2} = \frac{2}{d_A} \frac{\boldsymbol{k}^T C^T C \boldsymbol{k}}{\boldsymbol{k}^T N_B \boldsymbol{k}} , \quad (19)$$

where $C^T C$ and $N_B = I - \boldsymbol{r}_B \boldsymbol{r}_B^T$, are 3×3 positive semi-definite matrices. Equation (19) is non-negative and independent of $\boldsymbol{r}_A$, and represents the average conditional purity gain due to the measurement on B. Since Eq. (19) is a ratio of quadratic forms, the direction $\boldsymbol{k}$ which leads to the *maximum* entropy decrease, i.e. to the *minimum* conditional entropy, can be obtained by solving the generalized eigenvalue equation [54]

$$C^T C\boldsymbol{k} = \lambda N_B \boldsymbol{k}\,, \tag{20}$$

which implies $\mathrm{Det}[C^T C - \lambda N_B] = 0$, and selecting the eigenvector $\boldsymbol{k}$ associated with the largest eigenvalue $\lambda_{\max}$. This leads to $\Delta S_2(A|B_{\boldsymbol{k}}) \leq 2\lambda_{\max}/d_A\ \forall\ \boldsymbol{k}$, i.e.,

$$\underset{\boldsymbol{k}}{\mathrm{Min}}\, S_2(A|B_{\boldsymbol{k}}) = S_2(\rho_A) - \frac{2}{d_A}\lambda_{\max}\,. \tag{21}$$

An important remark is that generalized POVM measurements on qubit B *cannot decrease the projective minimum* (21) for this entropy [54].

It is then seen that the minimizing measurement is essentially determined by the correlation tensor (13), i.e., it is essentially a spin measurement along the direction of *maximum correlation*. We may also express (19) as the quadratic form $\Delta S_2(A|B_{\boldsymbol{k}}) = \frac{2}{d_A}\boldsymbol{k}_N^T C_N^T C_N \boldsymbol{k}_N$, where $C_N = C N_B^{-1/2}$ and $\boldsymbol{k}_N = N_B^{1/2}\boldsymbol{k}/|N_B^{1/2}\boldsymbol{k}|$, and write (20) as $C_N^T C_N \boldsymbol{k}_N = \lambda \boldsymbol{k}_N$, which shows that $\sqrt{\lambda_{\max}}$ is the *maximum singular value* of C_N. The counterpart at A of this equation is $C_N C_N^T \boldsymbol{k}_A = \lambda \boldsymbol{k}_A$, which has the same non-zero eigenvalues and provides a clear *geometric picture:* As $\boldsymbol{k}$ is varied in the Bloch sphere of qubit B, the set of post-measurement vectors (17) determining the post-measurement state of A form a three dimensional *correlation ellipsoid* on the $d_A^2 - 1$ dimensional space containing the vector $\boldsymbol{r}_{A/\boldsymbol{k}}$ [54] (see Fig. 1 for the two-qubit case) whose principal axes are precisely determined as the eigenvectors $\boldsymbol{k}_A$ of the previous equation. Therefore, the optimizing measurement of the quadratic entropy is that leading to $\delta\boldsymbol{r}_A = \boldsymbol{r}_A - \boldsymbol{r}_{A/\boldsymbol{k}} \propto C\boldsymbol{k}$ parallel to the major semi-axis of the correlation ellipsoid (see [54] for more details).

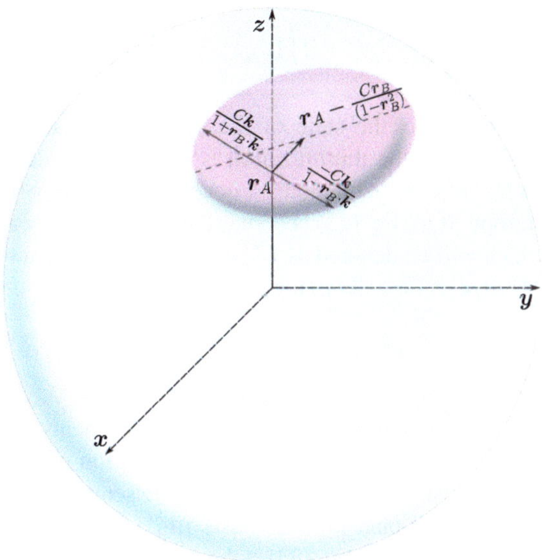

Fig. 1 The set of possible Bloch vectors $\boldsymbol{r}_{A/\boldsymbol{k}}$ of the post-measurement state of qubit A after a measurement on the qubit B form the correlation ellipsoid (from [54]). For a spin measurement along direction $\boldsymbol{k}$ at qubit B, the vectors $\boldsymbol{r}_{A/\pm\boldsymbol{k}}$ in A are the endpoints of a chord running through $\boldsymbol{r}_A$. The optimizing measurement determined by Eq. (20) leads to $\delta\boldsymbol{r}_A$ parallel to the major axis of this ellipsoid

While not strictly valid for other entropies, these results provide an approximate picture of the measurement minimizing the conditional entropy in these systems, which will typically lie close to that minimizing the quadratic entropy. In fact, all entropies $S_f(\rho)$ reduce essentially to the quadratic entropy if ρ is sufficiently close to maximum mixedness, as $S_f(\frac{I}{d}+\delta\rho)\approx S_f(\frac{I_d}{d})+\frac{1}{4}|f''(\frac{1}{d})|[(S_2(\frac{I}{d}+\delta\rho)-S_2(\frac{I_d}{d})]$ up to $O(\delta\rho^2)$. Moreover, for a sufficiently small correlation tensor, i.e. if $|\delta\boldsymbol{r}_A|=|\frac{C\boldsymbol{k}}{1\pm\boldsymbol{r}_B\boldsymbol{k}}|\ll 1\;\forall\,\boldsymbol{k}$, an expansion of the conditional entropy (8) up to second order in $\delta\rho_A$ leads to [54]

$$S_f(A|B_{\boldsymbol{k}})\approx S_f(\rho_A)-\frac{2}{d_A}\frac{\boldsymbol{k}^T C^T \Lambda_f(\rho_A) C\boldsymbol{k}}{\boldsymbol{k}^T N_B \boldsymbol{k}} \tag{22}$$

where $\Lambda_f(\rho_A)$ is a scaled $(d_A^2-1)\times(d_A^2-1)$ Hessian matrix [54], showing that in the present weakly correlated regime the effect of a general entropy is just to replace C by the "deformed" correlation tensor $C_f=\sqrt{\Lambda_f(\rho_A)}\,C$. Let us finally mention that the minimum generalized conditional entropy coincides with the associated generalized entanglement of formation between A and a third system C purifying the whole system [18, 55].

2.4 Generalized Information Deficit

As mentioned in the introduction, several other measures of quantum correlations with properties similar to those of the quantum discord have been considered. In particular, we have introduced in [28, 29] the *generalized information deficit*

$$I_f^B(\rho_{AB})=\underset{M_B}{\text{Min}}\;S_f(\rho'_{AB})-S_f(\rho_{AB})\,, \tag{23}$$

where ρ'_{AB} is the state of the system after an unread local measurement at B, Eq. (7), and the minimization is over all complete local projective measurements on B. Here $S_f(\rho)$ denotes a generalized entropy. In the case of the von Neumann entropy $S(\rho)$, Eq. (23) becomes the standard one-way information deficit [1, 24, 26], which will be denoted as I_1^B. It can be rewritten in terms of the relative entropy [68, 74] $S(\rho||\rho')=-\text{Tr}\,\rho(\log_2\rho'-\log_2\rho)$ as

$$I_1^B(\rho_{AB})=\underset{M_B}{\text{Min}}\,S(\rho'_{AB})-S(\rho_{AB})=\underset{M_B}{\text{Min}}\,S(\rho_{AB}||\rho'_{AB})\,. \tag{24}$$

Like the quantum discord, Eq. (23) (and hence (24)) is non-negative if $S_f(\rho)$ is Schur-concave [75], due to the majorization relation [28, 68] $\rho'_{AB}\prec\rho_{AB}$ satisfied by the post-measurement state (7). Essentially, the off-diagonal elements of ρ_{AB} in the conditional product basis $\{|i_A^j\rangle|j_B\rangle\}$ where ρ'_{AB} is diagonal are lost in the measurement, and Eq. (23) is then a measure of the minimum information loss under such measurement. It can also be considered as the minimum relative entropy of coherence [35] in this type of basis. And it is a measure of the minimum entanglement between the measurement device and the system generated by a complete local measurement [76], with (24) representing the minimum distillable entanglement [26, 39].

For strict concavity of S_f, Eq. (23) vanishes only if ρ_{AB} is already of the semi-quantum post-measurement form (7). And for pure states $\rho_{AB} = |\Psi_{AB}\rangle\langle\Psi_{AB}|$ it can be shown [28] that it reduces to the corresponding entanglement entropy:

$$I_f^B(|\Psi_{AB}\rangle) = S_f(\rho_A) = S_f(\rho_B) \tag{25}$$

with (24) becoming the standard entanglement entropy like the quantum discord. Nonetheless, unlike the latter (which in this case is minimized by a complete measurement in any local basis) the minimum of (23) and (24) for a pure state is always reached for a measurement in the basis of B which corresponds to the Schmidt decomposition of $|\Psi_{AB}\rangle$ (and hence diagonalizes ρ_B) [28], already indicating a different behavior of the minimizing measurement.

As in the case of the conditional entropy, the use of generalized entropies enables the possibility of using simple entropic forms like the quadratic entropy (9) or the Tsallis entropies, in which case Eq. (23) becomes $I_q^B(\rho_{AB}) = \text{Min}_{M_B} \frac{\text{Tr}\,(\rho_{AB}^q - \rho'^q_{AB})}{1-2^{1-q}}$. We may also consider the deficits based on the Renyi entropies [68] $S_{R_q}(\rho) = \frac{1}{1-q}\log_2 \text{Tr}\,\rho^q$, $q > 0$ (just an increasing function of S_q), which are given by [76]

$$I_{R_q}^B(\rho_{AB}) = \underset{M_B}{\text{Min}} \frac{1}{1-q} \log_2 \frac{\text{Tr}\,\rho'^q_{AB}}{\text{Tr}\,\rho^q_{AB}}\,. \tag{26}$$

They approach the von Neumann information deficit (24) for $q \to 1$ and likewise do not depend on the addition of an uncorrelated ancilla to A ($\rho_{AB} \to \rho_C \otimes \rho_{AB}$). Nonetheless, they are just increasing functions of I_q^B for fixed ρ_{AB} and the associated optimization problem is the same as that for I_q^B.

2.5 *Minimizing Measurement and Stationary Conditions*

The determination of the minimizing measurement M_B in (23) is, like in the case of the quantum discord, again a difficult problem in general. Complete projective measurements at B are determined by $d_B^2 - d_B$ real parameters if B has Hilbert space dimension d_B, growing then exponentially with the number of components of B. Nevertheless, it can be shown that the minimizing measurement should fulfill the stationary condition [29]

$$\text{Tr}_A[f'(\rho'_{AB}), \rho_{AB}] = 0\,, \tag{27}$$

which leads to $d_B(d_B-1)$ real equations [29, 77]. In the quantum discord (1), an additional term $-[f'(\rho'_B), \rho_B] = [\log_2 \rho'_B, \rho_B]$ is to be added in (27) for complete projective measurements [29].

Important differences between the measurements minimizing $I_f^B(\rho_{AB})$ and $D(A|B)$ may arise, as previously mentioned for the case of pure states. While for a general classically correlated state of the form (7) the minimum for *both* $D(A|B)$ and *all* $I_f^B(\rho_{AB})$ is attained for a measurement in the local basis defined by the projectors P_j^B (i.e., the pointer basis [11, 12]), in the particular case of product states $\rho_A \otimes \rho_B$, $D(A|B)$ (*but not* $I_f^B(\rho_{AB})$) becomes the same for *any* M_B, as for such states $S(A|M_B) = S(A)\ \forall\ M_B$. These differences will have important consequences in the results of the next section, leading to a quite different response of the minimizing measurement to the onset of quantum correlations. They reflect the fact that while in $I_f^B(\rho_{AB})$ one is looking for the *least disturbing local measurement*, such that ρ'_{AB} is as close as possible to ρ_{AB}, in $D(A|B)$ the search is for the measurement in B which makes the ensuing conditional entropy smallest, i.e., by which one can learn the most about A, which leads to those observables which are most correlated, as discussed before.

These differences become apparent in the case of the quadratic entropy (9), as an analytic evaluation of the associated deficit I_2^B for qudit-qubit systems becomes again feasible [27, 29]. In this case Eq. (23) becomes just a purity difference, $I_2^B(\rho_{AB}) = 2\text{Min}_{M_B}\text{Tr}\,(\rho_{AB}^2 - \rho'^2_{AB}) = 2\,\text{Min}_{\rho'_{AB}}\,||\rho_{AB} - \rho'_{AB}||^2$, where $||O||^2 = \text{Tr}\,O^\dagger O$ and the last minimization can be extended to any state of the general form (7). Through the last expression it is seen that it is then proportional to the geometric discord [1, 27], defined as the closest squared Hilbert-Schmidt distance between ρ_{AB} and a state of the form (7). For pure states I_2^B becomes the squared concurrence C_{AB}^2 [78], which for such states is just the quadratic entropy of any of the subsystems [79]. While as a measure it does not comply, due to the lack of additivity, with all the properties satisfied by the quantum discord or the von Neumann based information deficit, it has the advantage of enabling a simple analytic evaluation in qudit-qubit systems and admitting through its relation with the purity a more direct experimental access [71–73]. Moreover, the optimizing measurement will be the same measurement as that minimizing the associated Renyi deficit $I_{R_2}^B(\rho_{AB})$.

Writing again a general ρ_{AB} of a qudit-qubit system in the form (12), it can be shown that for a projective spin measurement at B along direction $\boldsymbol{k}$, the quadratic information loss becomes [27, 29]

$$I_2^B(\boldsymbol{k}) = \tfrac{1}{d_A}(||\boldsymbol{r}_B||^2 + ||J||^2 - \boldsymbol{k}^T M_2 \boldsymbol{k}) \tag{28}$$

where M_2 is the positive semi-definite matrix

$$M_2 = \boldsymbol{r}_B \boldsymbol{r}_B^T + J^T J\,, \tag{29}$$

with $J = C + \boldsymbol{r}_A \boldsymbol{r}_B^T$, i.e., $J_{\mu\nu} = \langle \sigma_{A\mu} \otimes \sigma_\nu \rangle$. Minimization of $I_2^B(\boldsymbol{k})$ leads then to the standard eigenvalue equation $M_2\boldsymbol{k} = \lambda \boldsymbol{k}$, implying $I_2^B(\rho_{AB}) = \frac{1}{d_A}(\text{tr}\,M_2 - \lambda_{\max})$, with $\lambda_{\max}$ the largest eigenvalue of M_2 and the minimizing $\boldsymbol{k}$ the associated eigenvector. Such direction will not necessarily coincide with that minimizing the quadratic conditional entropy, as the latter is determined essentially by the correlation tensor C while the present one by the tensor J and $\boldsymbol{r}_B$. While coinciding in some regimes (they become identical if $\boldsymbol{r}_B = \boldsymbol{0}$, i.e., ρ_B maximally mixed, in which case $J = C$), they can deviate considerably in others, as will be explicitly shown in the next section. In fact, a transition in the least disturbing measurement direction $\boldsymbol{k}$ from the main eigenvector of $J^T J$ to the direction of $\boldsymbol{r}_B$ can be expected as J decreases, which may not imply a concomitant change in the main eigenvector of (20). A closed expression for the minimum of $I_3^B(\rho_{AB})$ can also be obtained [29]. We finally note that for a general qubit-qubit state and entropy S_f, the stationary condition (27) becomes explicitly

$$(\alpha_1 \boldsymbol{r}_B + \alpha_2 J^T \boldsymbol{r}_A + \alpha_3 J^T J)\boldsymbol{k} = \lambda \boldsymbol{k} \tag{30}$$

which represents a non-linear eigenvalue equation since the coefficients α_i depend on $f'(\rho'_{AB})$ and hence on $\boldsymbol{k}$ [29]. Again, the prominent role of $J^T J$ is clearly evident.

3 Results in Spin Chains

We now consider the correlations of spin pairs in the ground state (GS) of finite spin $1/2$ arrays interacting through XY type Heisenberg couplings and immersed in a magnetic field $\boldsymbol{h}$. The Hamiltonian reads

$$H = -\sum_i \boldsymbol{h} \cdot \boldsymbol{S}_i - \tfrac{1}{2} \sum_{i \neq j, \mu = x, y} J_\mu^{ij} S_i^\mu S_j^\mu\,, \tag{31}$$

where i, j label the sites in the array and S_i^μ the spin components at site i.

In the transverse case $\boldsymbol{h} = (0, 0, h_z)$, the Hamiltonian commutes with the S_z *spin parity* $P_z = e^{i\pi \sum_i (S_i^z + 1/2)} = \prod_i (-2S_i^z)$, implying that the exact GS will have a definite parity if non-degenerate. In particular, in finite chains of N spins with first neighbor couplings and anisotropy $\chi = J_y/J_x \in (0, 1]$, the exact GS, which can be analytically obtained through the Jordan–Wigner fermionization [61, 80], will exhibit $N/2$ parity transitions as the field h_z increases from 0, where the lowest levels of each parity cross, the last one at the *transverse factorizing field* [61] $h_{zs} = \sqrt{J_y J_x} = J_x\sqrt{\chi}$. These transitions are reminiscent of the $N/2$ magnetization transitions of the XX case $\chi = 1$ [53, 81], where H commutes with the z component of the total spin $S^z = \sum_i S_i^z$.

At the factorizing field, the two crossing states generate a two dimensional GS subspace which is spanned, remarkably, by *completely separable* ground states $|\Theta\rangle = |\theta, \ldots, \theta\rangle$ and $|-\Theta\rangle = P_z|\Theta\rangle = |-\theta, \ldots, -\theta\rangle$ in the ferromagnetic case $J_x > 0$, where $|\theta\rangle = e^{-\imath\theta S_y}|\downarrow\rangle$ is the single spin state forming an angle θ with the $-z$ direction and $\cos\theta = h_{sz}/J_x = \sqrt{\chi}$. Hence, at this point the system possesses two

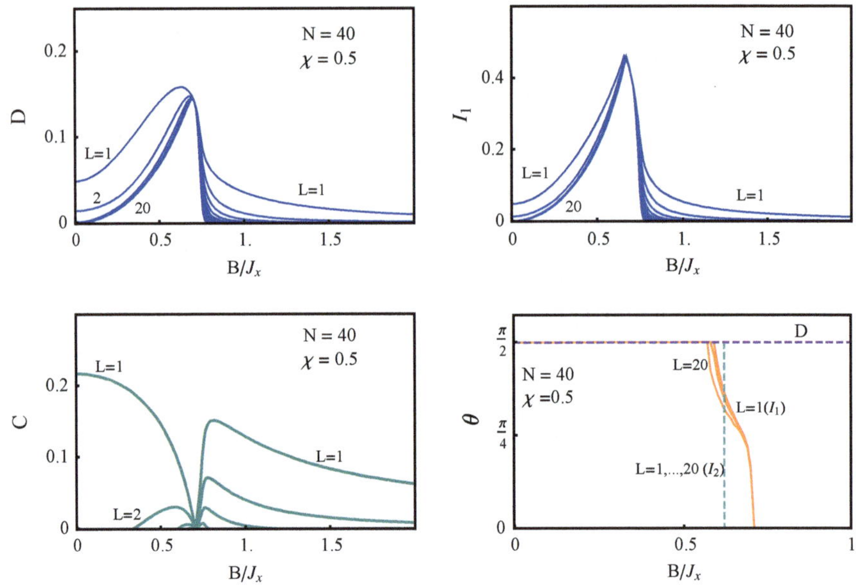

Fig. 2 The quantum discord (*top left*), the information deficit (*top right*), the concurrence (*bottom left*) and the angle θ determining the minimizing spin measurement of the first two quantities (*bottom right*) for reduced states of spin pairs in the exact ground state of a spin $1/2$ chain with first neighbor anisotropic XY coupling as a function of the transverse field ($B = h_z$). L indicates the separation between the spins of the pair ($L = 1$ denotes first neighbors) while N is the number of spins and $\chi = J_y/J_x$ the anisotropy. The angle θ is that formed between the measurement direction and the z axis in the xz plane, which is constant for D ($\theta = \pi/2$) but experiences an $x \to z$ transition in the information deficits I_1 (*solid line*) and I_2 (*dashed line*), which is sharp in the latter. All quantities reach full range at the factorizing field $B = J_x\sqrt{\chi}$, with common L-independent limits, which are negligible in the case of the concurrence but finite for the discord and information deficit

completely separable parity breaking degenerate ground states. Yet, the exact GS side-limits at this point are provided by the definite parity combinations [61]

$$|\Theta_\pm\rangle = \frac{|\Theta\rangle \pm |-\Theta\rangle}{\sqrt{2(1 \pm \langle -\Theta|\Theta\rangle)}} \tag{32}$$

approached for $h_z \to h_{zs}^\pm$, which are *entangled* states. They lead to *common reduced states* $\rho_{\theta\pm}$ *for any spin pair* $i \neq j$ [47, 61], becoming both identical with $\rho_\theta = (|\theta\theta\rangle\langle\theta\theta| + |-\theta-\theta\rangle\langle-\theta-\theta|)/2$ if the overlap $\langle-\Theta|\Theta\rangle = \cos^N\theta$ is neglected (it is negligible if N and θ are not too small). This is a separable mixed state, therefore leading to a zero concurrence (and hence zero entanglement of formation [78]) for any pair, as seen in the bottom left panel of Fig. 2. The concurrence actually approaches small common side limits $C_\pm = \frac{\chi^{n/2-1}(1-\chi)}{1\pm\chi^{n/2}}$ if the overlap is preserved [61, 64], not appreciable in the scale of Fig. 2.

However, ρ_θ is a discordant state for $\theta \in (0, \pi/2)$, leading to appreciable *finite limits* of the quantum discord D and the information deficit I_1 at the factorizing field, as seen in the top panels of Fig. 2. These limits can be analytically determined from the previous expression for ρ_θ [47, 76] and are independent of the separation L. Moreover, these quantities actually attain their maximum values in the vicinity of this point, remaining appreciable for all $h_z < h_{zs}$, since in this sector the reduced state of any pair in the exact GS will be essentially ρ_θ (with a field-dependent θ) plus smaller corrections. Let us note that in the cyclic chain considered, the reduced pair states ρ_{ij} depend just on the separation $L = |i - j|$, implying $D(i|j) = D(j|i) = D$ and $I_f^i(\rho_{ij}) = I_f^j(\rho_{ij}) = I_f \; \forall \, i \neq j$.

These results remain strictly valid for arbitrary range couplings with a common anisotropy $\chi = J_y^{ij}/J_x^{ij}$ in the ferromagnetic case $J_x^{ij} > 0$ [61], including dimer-type chains [65], since they also exhibit a factorizing field with the same factorized states. They hold as well in the antiferromagnetic case $J_x < 0$ for first neighbor couplings in a spin chain, since for a transverse field it can be mapped to the ferromagnetic case by a local rotation at even sites, which leads to $|\theta\theta\ldots\rangle \to |\theta, -\theta, \ldots\rangle$ in the factorized states and in ρ_θ. The same reduced state ρ_θ also follows from the mixture $\frac{1}{2}(|\Theta_+\rangle\langle\Theta_+| + |\Theta_-\rangle\langle\Theta_-|)$ if the overlap is neglected, which is the exact $T \to 0^+$ limit at h_{zs} of the thermal state $\propto \exp[-H/kT]$. We remark finally that in the thermodynamic limit $N \to \infty$, the lowest states for each parity become degenerate for $h_z < h_c = \frac{J_x + J_y}{2}$, so GS correlations actually depend on the choice of GS, the present results applying for the definite parity choice.

As seen in Fig. 2, although the quantum discord D and the information deficit I_1 exhibit a similar qualitative behavior, I_1 shows a more pronounced maximum in comparison with D. This feature reflects the transition in the orientation $\boldsymbol{k}$ of the local spin measurement minimizing I_1 as the field increases, which is absent in the quantum discord. This effect can be understood from the expressions (28)–(29) for the quadratic deficit I_2, which lead to a sharp $x \to z$ transition in the optimizing $\boldsymbol{k}$ for all separations as the maximum eigenvalue of M_2 shifts from that associated with $\boldsymbol{k} = \boldsymbol{e}_x$ to that for $\boldsymbol{k} = \boldsymbol{e}_z$ as the transverse field increases [29, 76]. In the case of I_1 such sharp transition is smoothed, as seen in the bottom right panel of Fig. 2, with θ covering all intermediate values in a narrow field interval centered at the I_2 measurement transition. In contrast, the quantum discord prefers a spin measurement (we consider here projective spin measurements) along the x axis for *all* transverse fields, for any separation L, following the strongest correlation [54, 55], which is along x for $|J_x| > |J_y|$. This is precisely the same measurement minimizing the quadratic conditional entropy, determined by Eqs. (19)–(21), since the largest eigenvalue in (20) of the contracted correlation matrix $C^T C$ corresponds to $\boldsymbol{k}$ along the x axis for the present anisotropic XY coupling $\forall \, h_z$ [54, 55].

The measurement transitions of the information deficit reflect, on the other hand, the qualitative change undergone by the reduced state of the pair (essentially by its dominant eigenstate) as the field increases [76]. The same $x \to z$ transition in I_1 is found in the XX case, where it reflects the transition in the dominant eigenstate of the reduced state of the pair from a Bell state $\frac{|\uparrow\downarrow\rangle + |\downarrow\uparrow\rangle}{\sqrt{2}}$ to the aligned state $|\downarrow\downarrow\rangle$ as the

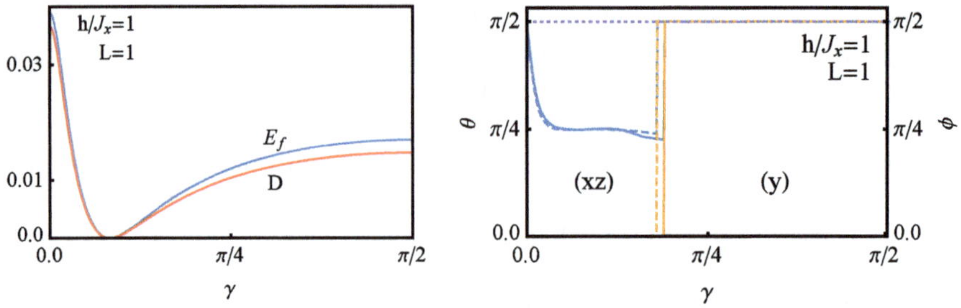

Fig. 3 The quantum discord D and the entanglement of formation E_f (*left panel*), and the angles θ, ϕ determining the minimizing spin measurement direction $\boldsymbol{k} = (\sin\theta\cos\phi, \sin\theta\sin\phi, \cos\theta)$ for D (*right panel*, *solid lines*), for a first neighbor pair in a XY spin chain with $\chi = 0.5$, and a non-transverse field $\boldsymbol{h}$ in the xz plane, as a function of the angle γ it forms with the z axis, for $|\boldsymbol{h}| = J_x$. Here both D and E_f vanish at the factorizing field due to the non-degeneracy of the factorized ground state. A transition from the xz plane to the y axis takes place in the minimizing projective measurement as γ increases, which can be predicted through the measurement optimizing the quadratic conditional entropy (*dashed lines* in *right panel*)

transverse field increases [53]. The main correlation in $C^T C$ stays, however, along the x axis. And in spin 1 systems, while the local optimizing measurements become more complex (they are not standard spin measurements), a similar transition pattern is observed in the measurement minimizing the information deficit [77].

In Fig. 3 we depict illustrative results for a non-transverse field $\boldsymbol{h} = (h_x, 0, h_z)$ in the xz plane, for an XY chain with coupling anisotropy $\chi = 0.5$ and small spin number $N = 8$. As recently shown [67], such chains also exhibit a non-transverse GS factorizing field in the xz plane, whose magnitude is given by

$$|\boldsymbol{h}_s| = \frac{h_{zs}\sin\theta}{\sin(\theta-\gamma)} \tag{33}$$

where $h_{zs} = J_x\cos\theta$ is the transverse factorizing field, with $\cos\theta = \sqrt{\chi}$ and $\gamma < \theta$ the angle formed by the field with the z axis. In contrast with the transverse case, such field is now associated with a *non-degenerate* separable GS $|\Theta\rangle = |\theta\theta\ldots\rangle$, as parity symmetry no longer holds. It is then seen that both D and the entanglement of formation E_f exactly vanish at $\boldsymbol{h}_s$, being now smaller than in the previous case since the GS no longer has parity symmetry. Moreover, for first neighbors the reduced pair state is much less mixed than before, and hence E_f and D have similar values, with E_f slightly larger than D, as also occurs for strong transverse fields [47]. For second and more distant neighbors, the behavior of D is qualitatively similar but becomes smaller (and larger than E_f). It should be remarked that E_f (and also D, I_f) continues to exhibit *long range* in the vicinity of the non-transverse factorizing field [67].

In addition, the quantum discord now also exhibits a measurement transition if the field is not too small, from the xz plane to the y axis ($\theta = \phi = \pi/2$) as the field rotates in the xz plane from the z axis to the x axis. This transition can be understood through the quadratic conditional entropy, as the maximum eigenvalue of the contracted correlation tensor $C^T C$ in (20) jumps from the xz block to the y block as the field is rotated, following the main correlation. As verified in the right panel, the measurement minimizing the quadratic conditional entropy lies very close to that minimizing the von Neumann based quantum discord. In contrast, even though the information deficit (not shown) still exhibits a behavior similar to that of D, the associated minimizing measurement tends to align with the field for strong $|\boldsymbol{h}|$, deviating again considerably from that minimizing the quantum discord.

4 Conclusions

We have first described a consistent extension to general concave entropic forms of the measurement dependent von Neumann conditional entropy for bipartite quantum systems. This extension, while providing a general characterization of the average information gain after such measurement, enables the use of simple entropic forms like the quadratic entropy, for which a closed evaluation of the minimizing measurement (leading to maximum purity gain) in terms of the correlation tensor becomes feasible for general states of qudit-qubit systems. Such solution admits a simple geometrical picture and allows to capture the main features of the projective measurement minimizing the quantum discord, which is then seen to follow essentially the direction of maximum correlation. In contrast, that minimizing the information deficit is essentially a least disturbing local measurement, and can then exhibit significant differences with the latter. The entropic generalization of the one way information deficit was also described, and for the quadratic entropy a closed evaluation for qudit-qubit states becomes again feasible, which allows to identify the previous differences.

When considered in spin pairs immersed in finite XY spin chains, both quantities, discord and information deficit, exhibit similar trends although with significant differences in the behavior of their optimizing measurements, which can be understood and predicted with the closed evaluations for the quadratic case. For transverse fields, these quantities exhibit appreciable values and long range for fields $h < h_c$ in the exact definite parity ground state, reaching full range and becoming independent of the pair separation in the vicinity of the factorizing field. A measurement transition takes place in the information deficit, which is absent in the quantum discord. In contrast, for non-transverse factorizing fields parity symmetry is broken and these quantities become smaller, strictly vanishing at factorization. Measurement transitions can occur in both quantities.

Acknowledgements The authors acknowledge support from CONICET (NG,NC,MC) and CIC (RR) of Argentina, and of CONICET grant PIP 11220150100732.

References

1. K. Modi et al., Rev. Mod. Phys. **84**, 1655 (2012)
2. G. Adesso, T.R. Bromley, M. Cianciaruso, J. Phys. A **49**, 473001 (2016)
3. B. Schumacher, Phys. Rev. A **51**, 2738 (1995)
4. C.H. Bennett, H.J. Bernstein, S. Popescu, B. Schumacher, Phys. Rev. A **53**, 2046 (1996)
5. M.A. Nielsen, I. Chuang, *Quantum Computation and Quantum Information* (Cambridge University, Press, 2000)
6. S. Haroche, J.M. Raimond, *Exploring the Quantum* (Oxford University Press, Oxford, 2007)
7. C.H. Bennett et al., Phys. Rev. Lett. **70**, 1895 (1993); Phys. Rev. Lett. **76**, 722 (1996)
8. R. Josza, N. Linden, Proc. R. Soc. **A459**, 2011 (2003)
9. G. Vidal, Phys. Rev. Lett. **91**, 147902 (2003)
10. R.F. Werner, Phys. Rev. A **40**, 4277 (1989)
11. H. Ollivier, W.H. Zurek, Phys. Rev. Lett. **88**, 017901 (2001)
12. L. Henderson, V. Vedral, J. Phys. A **34**, 6899 (2001)
13. V. Vedral, Phys. Rev. Lett. **90**, 050401 (2003)
14. W.H. Zurek, Phys. Rev. A **67**, 012320 (2003)
15. E. Knill, R. Laflamme, Phys. Rev. Lett. **81**, 5672 (1998)
16. A. Datta, A. Shaji, C.M. Caves, Phys. Rev. Lett. **100**, 050502 (2008)
17. A. Datta, S.T. Flammia, C.M. Caves, Phys. Rev. A **72**, 042316 (2005)
18. M. Koashi, A. Winter, Phys. Rev. A **69**, 022309 (2004)
19. A. Datta, S. Gharibian, Phys. Rev. A **79**, 042325 (2009)
20. A. Shabani, D.A. Lidar, Phys. Rev. Lett. **102**, 100402 (2009)
21. K. Modi et al., Phys. Rev. Lett. **104**, 080501 (2010)
22. A. Ferraro et al., Phys. Rev. A **81**, 052318 (2010)
23. F.F. Fanchini et al., Phys. Rev. A **84**, 012313 (2011)
24. M. Horodecki et al., Phys. Rev. A **71**, 062307 (2005)
25. J. Oppenheim et al., Phys. Rev. Lett. **89**, 180402 (2002)
26. A. Streltsov, H. Kampermann, D. Bruß, Phys. Rev. Lett. **106**, 160401 (2011)
27. B. Dakić, V. Vedral, C. Brukner, Phys. Rev. Lett. **105**, 190502 (2010)
28. R. Rossignoli, N. Canosa, L. Ciliberti, Phys. Rev. A **82**, 052342 (2010)
29. R. Rossignoli, N. Canosa, L. Ciliberti, Phys. Rev. A **84**, 052329 (2011)
30. D. Girolami, T. Tuffarelli, G. Adesso, Phys. Rev. Lett. **110**, 240402 (2013)
31. S. Luo, S. Fu, C.H. Oh, Phys. Rev. A **85**, 032117 (2012)
32. M. Paula, T.R. de Olivera, M.S. Sarandi, Phys. Rev. A **87**, 064101 (2013)
33. H. Hu, H. Fan, D.L. Zhou, W.M. Liu, Phys. Rev. A **87**, 032340 (2013)
34. T. Nakano, M. Piani, G. Adesso, Phys. Rev. A **88**, 012117 (2013)
35. T. Baumgratz, M. Cramer, M.B. Plenio, Phys. Rev. Lett. **113**, 140401 (2014)
36. I. Marvian, R.W. Spekkens, Phys. Rev. A **90**, 062110 (2014)
37. V. Madhok, A. Datta, Phys. Rev. A **83**, 032323 (2011)
38. D. Cavalcanti et al., Phys. Rev. A **83**, 032324 (2011)
39. M. Piani et al., Phys. Rev. Lett. **106**, 220403 (2011)
40. D. Girolami, G. Adesso, Phys. Rev. Lett. **108**, 150403 (2012)
41. T. Tufarelli et al., Phys. Rev. A **86**, 052326 (2012)
42. Y. Huang, New. J. Phys. **16**, 033027 (2014)
43. R. Dillenschneider, Phys. Rev. B **78**, 224413 (2008)
44. M.S. Sarandy, Phys. Rev. A **80**, 022108 (2009)
45. J. Maziero et al., Phys. Rev. A **82**, 012106 (2010)
46. T. Werlang, G. Rigolin, Phys. Rev. A **81**, 044101 (2010)
47. L. Ciliberti, R. Rossignoli, N. Canosa, Phys. Rev. A **82**, 042316 (2010)
48. T. Werlang et al., Phys. Rev. Lett. **105**, 095702 (2010)
49. T. Werlang, G.A.P. Ribeiro, G. Rigolin, Phys. Rev. A **83**, 062334 (2011)
50. B.Q. Liu et al., Phys. Rev. A **83**, 052112 (2011)
51. Y.C. Li, H.Q. Lin, Phys. Rev. A **83**, 052323 (2011)

52. N. Canosa, L. Ciliberti, R. Rossignoli, Int. J. Mod. Phys. B **27**, 1345033 (2012)
53. L. Ciliberti, N. Canosa, R. Rossignoli, Phys. Rev. A **88**, 012119 (2013)
54. N. Gigena, R. Rossignoli, Phys. Rev. A **90**, 042318 (2014)
55. N. Gigena, R. Rossignoli, J. Phys. A **47**, 015302 (2014)
56. J. Kurmann, H. Thomas, G. Müller, Physica A **112**, 235 (1982)
57. T. Roscilde et al., Phys. Rev. Lett. **93**, 167203 (2004)
58. T. Roscilde et al., Phys. Rev. Lett. **94**, 147208 (2005)
59. L. Amico et al., Phys. Rev. A **74**, 022322 (2006)
60. F. Baroni et al., J. Phys. A **40**, 9845 (2007)
61. R. Rossignoli, N. Canosa, J.M. Matera, Phys. Rev. A **77**, 052322 (2008)
62. S.M. Giampaolo, G. Adesso, F. Illuminati, Phys. Rev. Lett. **100**, 197201 (2008)
63. S.M. Giampaolo, G. Adesso, F. Illuminati, Phys. Rev. B **79**, 224434 (2009)
64. R. Rossignoli, N. Canosa, J.M. Matera, Phys. Rev. A **80**, 062325 (2009)
65. N. Canosa, R. Rossignoli, J.M. Matera, Phys. Rev. B **81**, 054415 (2010)
66. S. Campbell, J. Richens, N. Lo Gullo, T. Busch, Phys. Rev. A **88**, 062305 (2013)
67. M. Cerezo, R. Rossignoli, N. Canosa, Phys. Rev. B **92**, 224422 (2015)
68. H. Wehrl, Rev. Mod. Phys. **50**, 221 (1978)
69. N. Canosa, R. Rossignoli, Phys. Rev. Lett. **88**, 170401 (2002)
70. C. Tsallis, J. Stat. Phys. **52**, 479 (1988); C. Tsallis, *Introduction to Non-Extensive Statistical Mechanics* (Springer, New York, 2009)
71. R. Filip, Phys. Rev. A **65**, 062320 (2002)
72. H. Nakazato et al., Phys. Rev. A **85**, 042316 (2012)
73. T. Tanaka, G. Kimura, H. Nakazato, Phys. Rev. A **87**, 012303 (2013)
74. V. Vedral, Rev. Mod. Phys. **74**, 197 (2002)
75. R. Bhatia, *Matrix Analysis* (Springer, New York, USA, 1997)
76. N. Canosa, L. Ciliberti, R. Rossignoli, Entropy **17**, 1634 (2015)
77. R. Rossignoli, J.M. Matera, N. Canosa, Phys. Rev. A **86**, 022104 (2012)
78. S. Hill, W.K. Wootters, Phys. Rev. Lett. **78**, 5022 (1997); W.K. Wootters, Phys. Rev. Lett. **80**, 2245 (1998)
79. P. Rungta, C.M. Caves, Phys. Rev. A **67**, 012307 (2003); P. Rungta et al., Phys. Rev. A **64**, 042315 (2001)
80. E. Lieb, T. Schultz, D. Mattis, Ann. Phys. (NY) **16**, 407 (1961)
81. N. Canosa, R. Rossignoli, Phys. Rev. A **75**, 032350 (2007)

Experimental Investigation of the Dynamics of Quantum Discord in Optical Systems

Jin-Shi Xu, Chuan-Feng Li and Guang-Can Guo

One of the most remarkable properties in quantum systems is the existence of correlations without the classical counterparts. Entanglement, a special kind of nonclassical correlation, has been studied widely [1] and is found to be an useful resource in quantum communication and quantum computation [2]. However, entanglement is not the only kind of quantum correlation. There are other nonclassical correlations even existing in separable quantum states. Quantum discord [3, 4], as a measure of total quantum correlations encoded in a quantum system, has attracted great attentions [5]. Quantum discord without entanglement has been proposed to has the potential to implement quantum information tasks that can benefit from quantum advantages [5]. Several experimental works, such as the implementation of deterministic quantum computation with one qubit [6] and remote quantum state operation [7] have been reported.

On the other hand, since any quantum system is always unavoidably coupled to its surrounding environment, quantum correlations would be greatly affected during the evolution. Entanglement as a fragile quantum resource, has been shown to suffer from sudden death in its dynamic behavior, in which case entanglement disappears completely at a finite evolution time [8]. Quantum discord, which is shown to be more robust than entanglement against decoherence [9, 10], displays peculiar properties in the dynamical behavior in open quantum systems and they are shown to be of

J.-S. Xu · C.-F. Li (✉) · G.-C. Guo
CAS Key Laboratory of Quantum Information, University of Science and Technology of China, Hefei 230026, People's Republic of China
e-mail: cfli@ustc.edu.cn

J.-S. Xu · C.-F. Li · G.-C. Guo
Synergetic Innovation Center of Quantum Information and Quantum Physics, University of Science and Technology of China, Hefei 230026, People's Republic of China

F.F. Fanchini et al. (eds.), *Lectures on General Quantum Correlations and their Applications*, Quantum Science and Technology,
DOI 10.1007/978-3-319-53412-1_21

fundamental and practical importance [5]. Experimental investigation on the dynamic behavior of quantum discord therefore arises great interests. Optical systems provide a versatile and well-control platform, in which the distinguished dynamics behaviors of quantum discord can be conveniently investigated. In this chapter, we would focus on the optical investigation of the dynamics of quantum discord.

1 The Definition of Quantum Discord

Quantum discord represents the difference between the classical information theory and quantum information theory [3]. Consider a bipartite system with components A and B, the classical mutual information between them is characterized by [11]:

$$I(A:B) = H(A) + H(B) - H(A, B), \tag{1}$$

where $H(A)$ ($H(B)$) is the Shannon entropy representing the uncertainty of the outcomes of A (B). According to the Bayes rule, the conditional entropy of A given B can be represented as $H(A|B) = H(A, B) - H(B)$. As a result, the classical mutual information has the equivalent form

$$J(A:B) = H(A) - H(A|B). \tag{2}$$

In the quantum case, the entropy of the composite quantum state ρ_{AB} is characterized by the von Neumann entropy $S(\rho)$. The quantum mutual information can be directly extended from Eq. (1) and becomes

$$\mathcal{I}(\rho_{AB}) = S(\rho_A) + S(\rho_B) - S(\rho_{AB}), \tag{3}$$

where $\rho_A = \mathrm{Tr}_B\rho_{AB}$ ($\rho_B = \mathrm{Tr}_A\rho_{AB}$) is the reduced density matrix of the partition A (B). The extension of $J(A:B)$ to its quantum form is not so direct. Since the conditional entropy is greatly dependent on the measurement we chose on B. Considering a set of complete measurement $\{M_B^i\}$ on the subsystem B, the reduced state of subsystem A conditioned on the measurement labeled by i becomes $\rho_A^i = \frac{1}{n_i}\mathrm{Tr}_B[(\mathbb{1}_A \otimes M_B^i)\rho_{AB}(\mathbb{1}_A \otimes M_B^i)^\dagger]$ with probability $n_i = \mathrm{Tr}_{AB}[(\mathbb{1}_A \otimes M_B^i)\rho_{AB}(\mathbb{1}_A \otimes M_B^i)^\dagger]$ and $\mathbb{1}_A$ representing the identical operator on the subsystem A. If we define $S(A|B) = \sum_i n_i S(\rho_A^i)$, the quantum generalization of Eq. (2) becomes

$$J(\rho_{AB}) = S(\rho_A) - S(A|B). \tag{4}$$

$I(\rho_{AB})$ and $J(\rho_{AB})$ are normally unequal to each other. Quantum discord is then defined as [3]

$$\mathcal{Q}(\rho_{AB}) = \mathcal{I}(\rho_{AB}) - \mathcal{C}(\rho_{AB}), \tag{5}$$

where

$$\mathcal{C}(\rho_{AB}) = \max_{\{M_B^i\}}[J(\rho_{AB})] = \max_{\{M_B^i\}}[S(\rho_A) - S(A|B)], \tag{6}$$

represents the maximum information extractable from the measurement on B and is independently defined as the classical correlation of ρ_{AB}[4]. For bipartite quantum systems, the quantum mutual information $I(\rho_{AB})$ quantifies the total correlations [12, 13].

In order to calculate quantum discord, one usually need to perform the optimization process over all possible positive-operator-valued measurements (POVMs) [4]. However it would be difficult to do so in practical for even two-qubit systems. Many works have employed the form of quantum discord with the operators $\{M_B^i\}$ taking as the von Neumann measurements. Nevertheless, great efforts have been made for simplifying the process of optimization especially for two-qubit cases and it has been shown that projective measurements with rank-one are sufficient to maximize the classical correlation between two qubits [14]. Moreover, for some special two-qubit states, analytical results are obtained. The most common case is the Bell-diagonal states, which possess maximally-mixed marginal and can be written as

$$\rho_{AB} = \frac{1}{4}(\mathbb{1} + \Sigma_{i=1}^3 c_i \sigma_i \otimes \sigma_i), \tag{7}$$

where c_i are the three real parameters and σ_i are the three Pauli matrices. The reduced density matrix $\rho_A = \rho_B = \mathbb{1}/2$. If we choose the four Bell states $|\phi^\pm\rangle = \frac{1}{\sqrt{2}}(|00\rangle \pm |11\rangle)$ and $|\psi^\pm\rangle = \frac{1}{\sqrt{2}}(|01\rangle \pm |10\rangle)$ as the basis, the density matrix of Eq. (7) is written in the diagonal form. By optimizing the projective measurement on particle B, the classical correlation is given by [15]

$$\mathcal{C}(\rho_{AB}) = \frac{1-c}{2}\log_2(1-c) + \frac{1+c}{2}\log_2(1+c), \tag{8}$$

with $c = \max(|c_1|, |c_2|, |c_3|)$. The quantum discord is then given by

$$\mathcal{Q}(\rho_{AB}) = 2 + \Sigma_{i=1}^4 \lambda_i \log_2 \lambda_i - \mathcal{C}(\rho), \tag{9}$$

where λ_i are the four eigenvectors of the state ρ_{AB} with the values of $\{\lambda_1 = \frac{1}{4}(1 + c_1 - c_2 + c_3), \lambda_2 = \frac{1}{4}(1 + c_1 + c_2 - c_3), \lambda_3 = \frac{1}{4}(1 - c_1 + c_2 + c_3), \lambda_4 = \frac{1}{4}(1 - c_1 - c_2 - c_3)\}$ and the total correlation $\mathcal{I}(\rho_{AB}) = 2 + \Sigma_{i=1}^4 \lambda_i \log_2 \lambda_i$. The equivalence between the approaches by using orthogonal projectors and rank-one POVM to maximize the classical correlation of Bell-diagonal states has been demonstrated in Ref. [16].

There are other measures of quantum correlations in composite quantum systems, for example, the thermodynamic approach [17, 18], the measurement-induced disturbance [19], the relative entropy measurement [20] and the geometry measurement [21–23]. An explicit review on different nonclassical measurement can be found in

Ref. [5]. Usually these definitions of nonclassical correlation are not equal to each other. However, some of the measures are shown to be equivalent for the Bell-diagonal states [20]. In optical experiment, the Bell-diagonal states can be conveniently prepared by mixing different Bell states [24, 25]. We focus on such kind of states and demonstrate the particular dynamic behaviors of quantum discord in open quantum systems.

2 Dynamics of Quantum Discord in Dephasing Environment

Dephasing noise is of particular interests in quantum information processing. The interaction quantum map for the dephasing evolution of a two-level quantum system S with lower and upper states $|0\rangle_S$ and $|1\rangle_S$ can be represented as

$$\begin{aligned} |0\rangle_S \otimes |0\rangle_E &\rightarrow \sqrt{1-p}|0\rangle_S \otimes |0\rangle_E + \sqrt{p}|0\rangle_S \otimes |1\rangle_E, \\ |1\rangle_S \otimes |0\rangle_E &\rightarrow \sqrt{1-p}|1\rangle_S \otimes |0\rangle_E + \sqrt{p}|1\rangle_S \otimes |2\rangle_E. \end{aligned} \tag{10}$$

$|0\rangle_E$ is the initial state of the environment. During the evolution, the environment has the probability p to be scattered to its excited state of $|1\rangle_E$ ($|2\rangle_E$) when the system state is $|0\rangle_S$ ($|1\rangle_S$), and has the probability $1-p$ to remain unchanged. The system S remains in the initial state and its coherence decays exponentially. $p = 1-\exp(-\gamma t)$, where γ is the decay rate.

In optical systems, the coupling between the information carriers (photon polarization modes) and the degrees of freedom that are not involved in information representation (photon frequency modes) can occur on birefringent crystals, such as quartz plates. More explicitly, the evolution of $|H\rangle$ (horizontal polarization) and $|V\rangle$ (vertical polarization) in the birefringent crystal would become $|H\rangle \rightarrow e^{i\delta_H(\omega)}|H\rangle$ and $|V\rangle \rightarrow e^{i\delta_V(\omega)}|V\rangle$ (a global phase is omitted), with $\delta_H(\omega) = \omega L n_H/c$ and $\delta_V(\omega) = \omega L n_V/c$ depending on the polarization and frequency ω of the incident photon. Here L is the length of the crystal, c is the speed of the light in vacuum, and n_H and n_V are the refractive indexes of two polarizations in the birefringent crystals. Now, consider a photon frequency distribution with the complex amplitude $g(\omega)$, a general single photon state is initialized as $\int d\omega g(\omega)(\alpha|H\rangle + \beta|V\rangle) \otimes |\omega\rangle$, where α and β are the complex amplitudes of $|H\rangle$ and $|V\rangle$, respectively. Here we assume that the initial state is decoupled from the frequency degrees of freedom. For each frequency component ω passing through the crystal, the corresponding reduced density matrix of the qubit is (the information detecting process is frequency independent, i.e., by tracing over the frequency)

$$\begin{pmatrix} |\alpha|^2 & \alpha\beta^* e^{-i\delta(\omega)} \\ \alpha^*\beta e^{i\delta(\omega)} & |\beta|^2 \end{pmatrix}, \tag{11}$$

where $\delta(\omega) = L(n_V - n_H)/c\omega = \kappa\omega$. In general, the phases resulting from different frequency components will cause dephasing. For example, consider a Gaussian distribution $f(\omega) = |g(\omega)|^2 = \left(2/\tau\sqrt{\pi}\right)\exp(-4(\omega-\omega_0)^2/\tau^2)$ of the photon with the center frequency ω_0 and a distribution width τ. The value of the off-diagonal element of the final density matrix imposed by the decoherence parameter $\eta = \exp\left(-\frac{1}{16}\kappa^2\tau^2 + i\kappa\omega_0\right)$ decays exponentially [26] and a dephasing environment is simulated (such kind of environment also exists in the polarization-maintaining optical fibers [27]).

When one of two photons is sent to the depahsing environment, different from the exponentially decay of the coherence of a single qubit, photonic entanglement may completely disappear in a finite time, showing the phenomenon of entanglement sudden death [8], which has been experimentally demonstrated [28, 29]. The dynamical behavior of the more general quantum correlation-quantum discord in the dephasing environment would behaves quite different and has been experimentally investigate in an all optical system [30]. In this work, different kinds of two-photon Bell diagonal states are prepared by constructing unbalanced Mach-Zehnder interferometers. One of the two photons is passing through quartz plates to introduce dephasing. During the experiment, the minimum of conditional entropy $S(\rho_A^i)$ is scanned by performing different projective measurement on photon B ($|i\rangle = \cos\theta|H\rangle + \sin\theta|V\rangle$), as shown in Fig. 1a. Due to the symmetry feature of the evolved state, the minimum value of $S(A|B)$ in Eq. (6) is equal to the minimum value of $S(\rho_A^i)$ [32]. It is found that the minimum value of $S(\rho_A^i)$ is obtained by projecting the state of photon B to $1/\sqrt{2}(|H\rangle + |V\rangle)$ when the length of the quartz plate is smaller than $138\lambda_0$, and $|H\rangle$ ($|V\rangle$) when the length of quartz plates is larger than $138\lambda_0$ ($\lambda_0 = 0.78\,\mu\text{m}$). The final evolved state is reconstructed by the quantum state tomography process, in which $S(\rho_A)$, $S(\rho_B)$ and $S(\rho_{AB})$ can be obtained. Therefore, the values of quantum discord and classical correlation can then be calculated from Eqs. (5) and (6). During the evolution the sudden change in behavior in the decay rates of quantum discord and classical correlation are observed, as shown in Fig. 1b. When the length of quartz plates is smaller than $138\lambda_0$, quantum discord remains unchanged and classical correlation decays exponentially. With further increasing the length of quartz plates, quantum discord decays exponentially and classical correlation remains unchanged. The phenomenon is further explained as the sudden transition from classical to quantum decoherence regime [31]. By comparing the dynamics of quantum discord and quantum entanglement, we find that during the decoherence-free region of quantum discord (remain unchanged), entanglement (characterized by the entanglement of formation [33] and relative entropy of entanglement [34]) decays exponentially. Moreover, quantum discord still exists even when entanglement sudden death occurs. These observation confirm the prediction that quantum discord is more robust against decoherence than entanglement [9, 10]. We further show the dynamics of non-entanglement quantum correlation defined as the difference between quantum discord and the relative entropy of entanglement, which also suffers from sudden changes. By preparing a special input Bell diagonal state, quantum discord is observed to be larger than classical correlation as the system evolves,

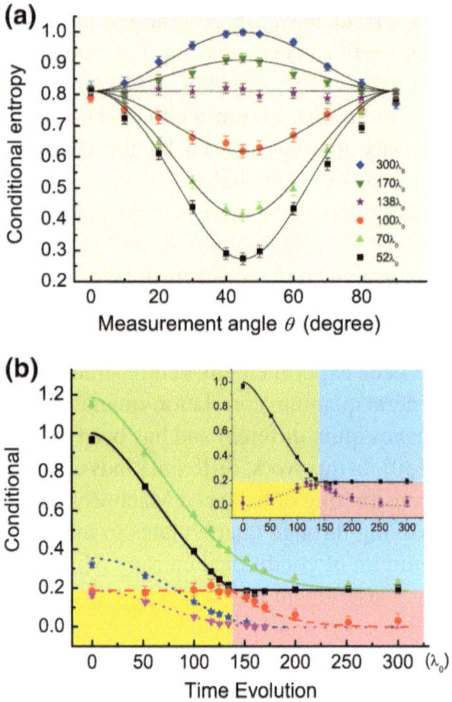

Fig. 1 The correlation dynamics of input state $\rho_{AB} = 0.75|\phi^-\rangle\langle\phi^-| + 0.25|\psi^-\rangle\langle\psi^-|$. **a** The values of conditional entropy $S(\rho_A^i)$, with photons evolving in different thicknesses of quartz plates, as a function of the measurement angle θ in mode B ($|i\rangle = \cos\theta|H\rangle + \sin\theta|V\rangle$). **b** The dynamics of correlations. *Green upward-pointing triangles*, *black squares*, *red dots*, *blue stars* and *magenta downward-pointing triangles* represent experimental results of mutual quantum information, classical correlation, quantum correlation, entanglement of formation and relative entropy of entanglement with the *green solid line*, *black solid line*, *red dashed line*, *blue dotted line* and *magenta dotted line* representing the corresponding theoretical predictions. Non-entanglement quantum correlation (*purple dots* with the *purple dotted line* representing the theoretical prediction) is further compared with classical correlation in the *inset*. The x axis represents the total thickness of quartz plates with $\lambda_0 = 0.78\mu m$ which corresponds to the time evolution of the photons. This figure is reproduced from Xu et al. Nat. Comm. 1, 7 (2010)

which disproves an early conjecture that classical correlation is always larger than quantum correlations [35].

Usually, the dissipative correlation evolution is essentially dependent on the type of noise that acts on the system. Markovian noise leads to the irreversible decay of quantum correlations, as shown above. Non-Markovian noise, on the other hand, would lead to the revival of quantum correlations. In order to simulate the non-Markovian environment in optical systems, we employ a Fabry-Perot cavity followed by quartz plates, which has been used in the investigation of non-Markovian evolution of quantum entanglement [28, 29]. The Fabry-Perot cavity behaves as an optical resonator. Optical wavelengths for which the cavity optical thickness is equal to an integer multiple of half wavelengths are resonant in the cavity and transmitted. Other wavelengths within the reflective band of the Fabry-Perot cavity are reflected. In the experiment, the Fabry-Perot cavity is a 0.2 mm thick quartz glass with coating films (reflectivity 90% at wavelengths around 780 nm) on both side. When a photon with a Gaussian frequency distribution is sent through the Fabry-Perot cavity, the frequency distribution is filtered to be the combination of three Gaussian frequency distributions and it can be written as $f(\omega) = \sum_{j=1}^{3} A_j(2/\sqrt{\pi}\tau_j)\exp[-4(\omega-\omega_j)^2/\tau_j^2]$. A_j is the relative amplitude for each Gaussian function distribution with the central frequency ω_j and frequency width τ_j. The photon is then sent through quartz plates. The decoherence parameter is calculated as $\eta = \sum_{i=1}^{3} A_j \exp[-\kappa^2\tau_j^2/16 + i\kappa\omega_j]$. During the dephasing process, the overall relative phase may refocus and the non-Markovian effect occurs, which leads to the revival of η and then the revival of quantum correlations.

The dynamics of quantum discord has been experimentally investigated in the non-Markovian environment described above [36]. In the experiment, one of the two correlated photons (denoted as B) is sent to the non-Markovian environment simulated by the Fabry-Perot cavity and quartz plates. The final evolved state can be transformed to a Bell diagonal form with local unitary operations, in which the relative entropy are used to measure the quantum correlation and classical correlation [20]. For Bell diagonal states, the entropy quantum discord and classical correlation only relate to the four eigenvalues of the final states, which can be directly calculated from the reconstructed density matrixes. During the experiment, the sudden transition from classical to quantum decoherence area was observed, which is similar to that in the Markovian evolution [30]. Due to the refocusing effect of the relative phase, quantum correlations revive from near zero. However, the non-Markovian effect is too small to revive classical correlation and it keeps constant even when the revival of quantum discord occurs. In order to obtain stronger non-Markovian effect, we can narrow the frequency width of the filtered wave packets. It is theoretically shown that when the frequency width are identically fitting to 0.2 nm (the value reads 0.85 nm in the experiment), quantum discord and classical correlation both revive. The sudden transition from quantum to classical revival regime can be further obtained. Usually, it would difficult to obtain narrower frequency width. We use a spin-echo approach to obtain the revival of classical correlation. A σ_x operation is implemented on the photon B followed by another quartz plates. The σ_x operation (realized by a half-wave plate with the angle setting to 45°) exchanges $|H\rangle$ and $|V\rangle$, i.e., $|H\rangle \rightarrow |V\rangle$ and $|V\rangle \rightarrow |H\rangle$. When the photon further passes through the same thickness of quartz plates, each of the information basis ($|H\rangle$ and $|V\rangle$) obtains the same phase as $e^{i(\delta_H(\omega)+\delta_V(\omega))}$ ($|H\rangle \rightarrow e^{i(\delta_H(\omega)+\delta_V(\omega))}|V\rangle$ and $|V\rangle \rightarrow e^{i(\delta_V(\omega)+\delta_H(\omega))}|H\rangle$), which can be treat as a global phase. The decoherence effect is compensated. Another σ_x is used to change the final state back to the initial state. All the correlations are revived to the initial values and correlation echoes are formed.

3 Revival of Quantum Discord in Classical Environment

In the above experiments, the frequency spectrum (unobserved degree of freedom) is initially decoupled from polarizations (observed degree of freedom). After the coupling in the birefringent crystals, the qubit states (polarization) entangled with environmental states (frequency). The revival of quantum correlations is explained to be the refocusing of relative phase in the non-Markovian quantum environment, which is due to the back-action from the environment to the qubits [36]. Recently, the revival of quantum correlations has been extended to the classical environment without back action on the quantum system [38–42]. In such case, the environment can not store or share the quantum correlations initially present in the quantum system.

Recently, the experimental recovery of quantum correlations in absence of system-environment back-action in an all-optical setup has been reported [43]. A simple theoretical model is proposed to exclude other side effects, in which the environment is considered to be a random external classical field, whose phase is either $\pi/2$ or $-\pi/2$ with equal probability $1/2$. The classical field acts only on one of the two qubits with the rotation operation corresponding to r_+ and r_-, respectively. In the optical experiment, the environment is simulated by separating the photon with a 50:50 beam-splitter and implementing half-wave plates and quartz plates in both paths, in which the operations of p_+ (reflected path) and p_- (transmitted path) can be realized. Since the measurement process traces over the path information, the measurement result would create a statistical mixture of them with equal probabilities 1/2. The unitary transformations of p_+ and p_- correspond to the random field operations of r_+ and r_-, respectively. The operations on the photon basis states $|H\rangle$ and $|V\rangle$ can be written as (a global phase factor is omitted)

$$\begin{aligned} |H\rangle &\xrightarrow{\mathrm{p}_\pm} \cos(\delta/2)|H\rangle \pm i\sin(\delta/2)|V\rangle, \\ |V\rangle &\xrightarrow{\mathrm{p}_\pm} \pm i\sin(\delta/2)|H\rangle + \cos(\delta/2)|V\rangle, \end{aligned} \tag{12}$$

where $\delta = \kappa\omega$ is the phase difference between $|H\rangle$ and $|V\rangle$ introduced by quartz plates. If the thickness of quartz plates is increased, there will be decoherence in the evolution in both paths due to the fact that the frequency is also traced over during the measurement process.

In the experiment, different kinds of Bell diagonal states are prepared as the initial states. After the evolution in the simulated classical environment, the final quantum states are reconstructed through quantum state tomography. The quantum discord and classical correlation defined in Eqs. (5) and (6) are obtained by numerically scan the projective operators $\{M_B^i\}$ represented by $\{\cos\theta|H\rangle + e^{i\varphi}\sin\theta|V\rangle, e^{-i\varphi}\sin\theta|H\rangle - \cos\theta|V\rangle\}$ on photon B with the scan precision of φ and θ setting to $\pi/100$. During the coherent evolution region (the length of quartz plates is small), the sudden changes in the dynamics of quantum and classical correlations occur several times, which is shown in Fig. 2. Quantum discord and classical correlations are observed to be robust against variations of the relative phase δ at certain evolution periods. When come

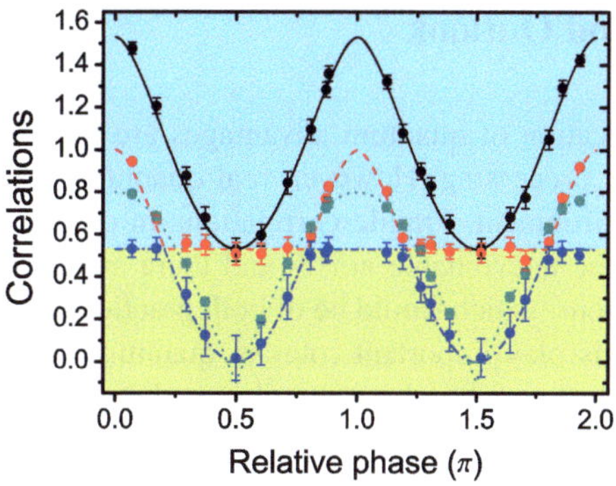

Fig. 2 Dynamics of correlations in the absence of system-environment back-action. The initial input state is $\rho_{AB} = 0.9|\phi^+\rangle\langle\phi^+| + 0.1|\phi^-\rangle\langle\phi^-|$. The *black solid line*, *dashed red line*, *dot-dashed blue line* and *dotted dark cyan line* represent, respectively, the theoretical predictions of total correlation, classical correlation, quantum discord and entanglement measured by concurrence [37]. The *black*, *red*, *blue* and *dark cyan dots* represent the corresponding experimental results. This figure is reproduced from Xu et al. Nat. Comm. 4, 2851 (2013)

to the decoherent evolution (the length of quartz plates is large), a sudden change in behaviour in the dynamics of the maximum values of classical correlation and quantum discord is also found. Although the decay behavior of the maximum values of correlations is similar to that in quantum Markovian environment [36], the internal noise mechanism is completely different. There are no revivals of correlations in the Markovian evolution in [36]. However, the correlations exhibit collapses and revivals during the evolution in the classical environment without back-action (as shown in Fig. 2). If the initial states are changed, the maximum values of quantum discord remain unchanged for a range of initial states and the sudden change behaviours of quantum and classical correlations are observed again. The revival without back-action are connect to the intrinsic non-Markovian of the system evolution, in wich the non-Markovianity based on the trance distance [44] are measured. The non-Markovianity can be identified by the positive rate of change of the trace distance [45]. It is found that the rate of change is positive when entanglement revivals occur. As a result, the revival of quantum correlations is a dynamical feature of open systems no matter what kind of environment is (classical or quantum).

The existence of non-Markovianity presents a necessary condition for the revival to occur. We further provide a possible interpretation of the observed phenomenon by treating the system-environment state as a quantum-classical state. The classical environment acts only on one of the two qubits B and it never correlates with the qubit B during the evolution. There is not the back-action by the environment on the qubit and the classical environment cannot store any quantum correlations on it own. However, the environment plays a control system (rotation operations) and keep records on what kinds of unitary operation (r_+ or r_-) is applied to qubit B. When such kind of classical information is lost, quantum correlations disappear. The recovery of the control classical information would lead to the recovery of quantum correlations.

4 Conclusion and Outlook

The origin and application of quantum advantages are the subjects in the heat of quantum information processing. However, real quantum systems are always surrounded by their environment. Fragile correlations in quantum systems would be easily destroyed by the unavoidable noises and there are also distinctive dynamic behaviors of correlations, which would be of both practical and fundamental importance. Optical systems play important roles in quantum information science [46, 47]. Due to well-control coupling between observed degree of freedom and unobserved degree of freedom, optical systems provide an ideal platform to investigate the dynamics behaviors of quantum systems in different kinds of environment (Markovian or non-Markovian, quantum or classical). Initial photon states can be prepared with high enough precision. After the well-control coupling and evolution, the final states can be reconstructed by the quantum state tomography. The dynamics of quantum discord and other correlations in quantum systems can then be deduced.

Recently, in order to understand the dynamical behaviors of correlations, the transference of correlation between the system and reservoir are further studied [48, 49]. Correlations between the system and reservoir can fundamentally influence the dynamics of system. It is shown that vanishing quantum discord of the initial system-environment state is the sufficient and necessary condition to induce a completely positive map of the quantum dynamical process [50]. Moreover, it is proved that a class of system-environment states called lazy states leads to the zero system entropy rate under any system-reservoir coupling [51]. We expect that the optical systems would play important roles in experimental investigation the system-environment dynamics.

The discussion of different kinds of correlations in the same context would provide a particular perspective on quantum discord. Quantum discord has been predicted to closely related to the generated entanglement [52, 53] and the experimental work has been reported [54]. Recently, the relation between quantum discord and quantum coherent has further been investigated [55]. From the experimental view, a more convenient method except for the quantum state tomography to investigate the dynamics of quantum discord is of great interest. Similar to the evolution equation of entanglement in noisy environment [28, 56, 57], the characterization of the dynamics of quantum discord may be further simplified.

References

1. R. Horodecki, P. Horodecki, M. Horodecki, K. Horodecki, Quantum entanglement. Rev. Mod. Phys. **81**, 865–942 (2009)
2. C.H. Bennett, D.P. DiVincenzo, Quantum information and computation. Nature **404**, 247–255 (2000)
3. H. Ollivier, W.H. Zurek, Quantum discord: a measure of the quantumness of correlations. Phys. Rev. Lett. **88**, 017901 (2001)

4. H. Henderson, V. Vedral, Classical, quantum and total correlations. J. Phys. A **43**, 6899–6905 (2001)
5. K. Modi, A. Brodutch, H. Cable, T. Paterek, V. Vedral, The classical-quantum boundary for correlations: discord and related measures. Rev. Mod. Phys. **84**, 1655–1707 (2012)
6. B.P. Lanyon, M. Barbieri, M.P. Almeida, A.G. White, Experimental quantum computing without entanglement. Phys. Rev. Lett. **101**, 200501 (2008)
7. B. Dakic et al., Quantum discord as resource for remote state preparation. Nature Phys. **8**, 666–670 (2012)
8. T. Yu, J.H. Eberly, Sudden death of entanglement. Science **323**, 598–601 (2009)
9. T. Werlang, S. Souza, F.F. Fanchini, C.J. Villas Boas, Robustness of quantum discord to sudden death. Phys. Rev. A **80**, 024103 (2009)
10. A. Ferraro, L. Aolita, D. Cavalcanti, F.M. Cucchietti, A. Acín, Almost all quantum states have nonclassical correlations. Phys. Rev. A **81**, 052318 (2010)
11. T.M. Cover, J.A. Thomas, *Elements of Information Theory* (Wiley, New York, 1991)
12. B. Groisman, S. Popescu, A. Winter, Quantum, classical, and total amount of correlations in a quantum state. Phys. Rev. A **72**, 032317 (2005)
13. B. Schumacher, M.D. Westmoreland, Quantum mutual information and the one-time pad. Phys. Rev. A **74**, 042305 (2006)
14. S. Hamieh, R. Kobes, H. Zaraket, Positive-operator-valued measure optimization of classical correlations. Phys. Rev. A **70**, 052325 (2004)
15. S. Luo, Quantum discord for two-qubit systems. Phys. Rev. A **77**, 042303 (2008)
16. M.D. Lang, C.M. Caves, Quantum discord and the geometry of Bell-diagonal states. Phys. Rev. Lett. **105**, 150501 (2010)
17. J. Oppenheim, M. Horodecki, P. Horodecki, R. Horodecki, Thermodynamical approach to quantifying quantum correlations. Phys. Rev. Lett. **89**, 180402 (2002)
18. M. Horodecki et al., Local versus nonlocal information in quantum-information theory: formalism and phenomena. Phys. Rev. A **71**, 062307 (2005)
19. S. Luo, Using measurement-induced disturbance to characterize correlations as classical or quantum. Phys. Rev. A **77**, 022301 (2008)
20. K. Modi, T. Paterek, W. Son, V. Vedral, M. Williamson, Unified view of quantum and classical correlations. Phys. Rev. Lett. **104**, 080501 (2010)
21. B. Dakić, V. Vedral, Č. Brukner, Necessary and sufficient condition for nonzero quantum discord. Phys. Rev. Lett. **105**, 190502 (2010)
22. S. Luo, S. Fu, Measurement-induced nonlocality. Phys. Rev. Lett. **106**, 120401 (2011)
23. M.-L. Hu, H. Fan, Dynamics of entropic measurement-induced nonlocality in structured reservoirs. Ann. Phys. **327**, 2343–2353 (2012)
24. G. Puentes, D. Voigt, A. Aiello, J.P. Woerdman, Tunable spatial decoherers for polarization-entangled photons. Opt. Lett. **31**, 2057–2059 (2006)
25. A. Aiello, G. Puentes, D. Voigt, A. Aiello, J.P. Woerdman, Maximally entangled mixed-state generation via local operations. Phys. Rev. A **75**, 062118 (2007)
26. A.J. Berglund, *Quantum Coherence and Control in One- and Two-Photon Optical Systems* (2000). arXiv:quant-ph/0010001
27. J.-S. Xu et al., Robust bidirectional links for photonic quantum networks. Sci. Adv. **2**, e1500672 (2016)
28. J.-S. Xu et al., Experimental characterization of entanglement dynamics in noisy channels. Phys. Rev. Lett. **103**, 240502 (2009)
29. J.-S. Xu et al., Experimental demonstration of photonic entanglement collapse and revival. Phys. Rev. Lett. **105**, 100502 (2010)
30. J.-S. Xu et al., Experimental investigation of classical and quantum correlations under decoherence. Nat. Commun. **1**, 7 (2010)
31. L. Mazzola, J. Piilo, S. Maniscalco, Sudden transition between classical and quantum decoherence. Phys. Rev. Lett. **104**, 200401 (2010)
32. J. Maziero, L.C. Céleri, R.M. Serra, V. Vedral, Classical and quantum correlations under decoherence. Phys. Rev. A **80**, 044102 (2009)

33. C.H. Bennett, D.P. DiVincenzo, J.A. Smolin, W.K. Wootters, Mixed-state entanglement and quantum error correction. Phys. Rev. A **54**, 3824–3851 (1996)
34. V. Vedral, B.M. Plenio, M.A. Rippin, P.L. Knight, Quantifying entanglement. Phys. Rev. Lett. **78**, 2275–2279 (1997)
35. G. Lindblad, in *Quantum Aspects, of Optical Communications-Lecture Notes in Physics*, ed. by C. Bendjaballah, et al. (Springer, Singapore, 1991), pp. 71–80
36. J.-S. Xu et al., Experimental investigation of the non-Markovian dynamics of classical and quantum correlations. Phys. Rev. A **82**, 042328 (2010)
37. W.K. Wootters, Entanglement of formation of an arbitrary state of two qubits. Phys. Rev. Lett. **80**, 2245–2248 (1998)
38. D. Zhou, A. Lang, R. Joynt, Disentanglement and decoherence from classical non-Markovian noise: random telegraph noise. Quantum Inf. Process **9**, 727–747 (2010)
39. R. Lo Franco, A. D'Arrigo, G. Falci, G. Compagno, E. Paladino, Entanglement dynamics in superconducting qubits affected by local bistable impurities. Phys. Scr. **T147**, 014019 (2012)
40. P. Bordone, F. Buscemi, C. Benedetti, Effect of Markov and non-Markov classical noise on entanglement dynamics. Fluct. Noise Lett. **T147**, 014019 (2012)
41. R. Lo Franco, B. Bellomo, E. Andersson, G. Compagno, Revival of quantum correlations without system-environment back-action. Phys. Rev. A **85**, 032318 (2012)
42. F. Altintas, A. Kurt, R. Eryigit, Classical memoryless noise-induced maximally discordant mixed separable steady states. Phys. Lett. A **377**, 53–59 (2012)
43. J.-S. Xu et al., Experimental recovery of qunatum correlations in absence of system-environment back-action. Nat. Commun. **4**, 2851 (2013)
44. B.-H. Liu et al., Experimental control of the transition from Markovian to non-Markovian dynamics of open quantum systems. Nat. Phys. **7**, 931–934 (2011)
45. M. Mannone, R. Lo Franco, G. Compagno, Comparison of non-Markovianity criteria in a qubit system under random external fields. Phys. Scr. **T153**, 014047 (2013)
46. N. Gisin, R. Thew, Quantum communication. Nat. Photon **1**, 165–171 (2007)
47. P. Kok et al., Linear optical quantum computing with photonnic qubits. Rev. Mod. Phys. **79**, 135–174 (2007)
48. J. Maziero, T. Werlang, F.F. Fanchini, L.C. Céleri, R.M. Serra, System-reservoir dynamics of quantum and classical correlations. Phys. Rev. A **81**, 022116 (2010)
49. R.-C. Ge, M. Gong, C.-F. Li, J.-S. Xu, G.-C. Guo, Quantum correlation and classical correlation dynamics in the spin-boson model. Phys. Rev. A **81**, 064103 (2010)
50. A. Shabani, D.A. Lidar, Vanishing quantum discord is necessary and sufficient for completely positive maps. Phys. Rev. Lett. **102**, 100402 (2009)
51. C.A. Rodríguez-Rosario, G. Kimura, H. Imai, A. Aspuru-Guzik, Sufficient and necessary condition for zero quantum entropy rates under any coupling to the environment. Phys. Rev. Lett. **106**, 050403 (2011)
52. M. Piani et al., All nonclassical correlations can be activated into distillable entanglement. Phys. Rev. Lett. **106**, 220403 (2011)
53. A. Streltsov, H. Kampermann, D. Bruß, Interpreting quantum discord through quantum state merging. Phys. Rev. Lett. **106**, 160401 (2011)
54. G. Adesso et al., Experiemntal entanglement activation from discord in a programmable quantum measurement. Phys. Rev. Lett. **112**, 140501 (2014)
55. J. Ma, B. Yadin, D. Girolami, V. Vedral, M. Gu, Converting coherence to quantum correlations. Phys. Rev. Lett. **116**, 160407 (2016)
56. T. Konrad et al., Evolution equation for quantum entanglement. Nat. Phys. **4**, 99 (2011)
57. O.J. Farías, C.L. Latune, S.P. Walborn, L. Davidovich, P.H.S. Ribeiro, Determining the dynamics of entanglement. Science **324**, 1414 (2009)

Experimental Investigation of Quantum Correlation in Solid-State Spin System

Jiangfeng Du, Fangzhou Jin and Xing Rong

Abstract Quantum systems exhibit diversified correlations which have no classical counterparts. It has been pointed out that quantum entanglement cannot describe all the nonclassicality in the correlations. The quantum discord, which can describe quantum correlations in separable states, has been demonstrated as important physical resources in quantum information processing. Since the solid-state spin systems are promising for applications of quantum computation and metrology, it is of practical significance to experimentally characterize the properties of quantum correlation in such systems. In this chapter, we review the experimental investigations of quantum correlations in solid-state spin system.

1 Introduction

Nowadays the significance of quantum correlations goes well beyond quantum entanglement, which has been widely investigated and believed to be the key resources of quantum information processing [1, 2]. However, the quantum entanglement does not provide a complete characterization of the nonclassicality: there exist quantum nonlocality without entanglement [3]. Moreover, quantum entanglement is not always necessary for quantum computation: we presented the experiment to perform a quantum algorithm for parity problem without using entanglement as early as 2001 [4], and recently a deterministic quantum computation with one qubit proposed in 1998 [5] has been realized experimentally [6] on the mixed separable states. It is likely the type of quantum correlation known as quantum discord [7, 8], rather than

J. Du (✉) · F. Jin · X. Rong
Hefei National Laboratory for Physical Sciences at the Microscale
and Department of Modern Physics, University of Science and Technology of China,
Hefei 230026, China
e-mail: qcmr@ustc.edu.cn

J. Du · X. Rong
Synergetic Innovation Center of Quantum Information and Quantum Physics,
University of Science and Technology of China, Hefei 230026, China

F.F. Fanchini et al. (eds.), *Lectures on General Quantum Correlations and their Applications*, Quantum Science and Technology,
DOI 10.1007/978-3-319-53412-1_22

quantum entanglement, that provides the enhancement for the computation [9], although there are different views [10, 11] on this. There are also many other aspects where the quantum discord has shown its significance, such as remote state preparation [12], encoding information [13], entanglement irreversibility [14], the link to distillable of entanglements [15, 16], entanglement distribution [17–19], and quantum metrology [20, 21] etc.

Apart from the wide applications, quantum discord has been exhibited peculiar characters in decoherent dynamics. For example, quantum discord has been claimed to have no sudden death [22] but has peculiar sudden change in its decay rates [23], and be robust in its initial period of decoherence but suffers a sudden change phenomenon [24]. Such behaviors has been experimentally investigated in various systems [25–28]. Quantum correlations also provide a powerful framework for the understanding of complex quantum systems. It has been shown that quantum discord, in contrast to entanglement, captures the critical points associated with quantum phase transitions (QPT) for *XXZ* and *XY* model, even at a finite temperature or in an external magnetic field [29–33]. Subsequently, there was experimental demonstration of the quantum discord for capturing the critical points associated with the behavior of the two-qubit *XXZ* Hamiltonian [34].

On the other hand, solid-state spin systems such as donor electron spins in phosphorus-doped silicon (Si:P) [35–37] and nitrogen-vacancy colour centres in diamond [38–40] are promising for applications of quantum computation and metrology. In Si:P ensemble system, there will be usually no quantum entanglement, but quantum discord could exist. This system can be served as bed for study the capability of using quantum discord in the absence of entanglement, which is in accord with the comments by Zurek on quantum discord [41] "There is no longer a question that discord works", and "the important thing now is to find out when discord without entanglement can be exploited most usefully". Thus experimental investigation of quantum correlations in such systems are of both fundamental and practical significance. In this chapter, we focus on two experimental work about quantum correlations in solid-state spin system [27, 34].

The chapter is organized as follows. In Sect. 2, we describe the definition and calculation of quantum correlations. In Sect. 3, we study the dynamics of quantum correlations in open solid systems, and protect the quantum correlations by dynamical decoupling (DD) technique. In Sect. 4, we use quantum discord for capturing the critical points of the two-qubit *XXZ* Hamiltonian, and we further describe the creation of quantum discord by the Unruh effect. The final discussions are presented in Sect. 5.

2 The Definition and Calculation of Quantum Correlation

We first briefly introduce the definition of quantum correlations. In a quantum system, a bipartite state *AB* described by the density matrix ρ_{AB}, with the subsystems ρ_A and ρ_B. The total correlations is quantified by the mutual information of the expression

$$I(\rho_{AB}) = S(\rho_A) + S(\rho_B) - S(\rho_{AB}), \tag{1}$$

where $S(\rho) = -Tr[\rho Log_2 \rho]$ is the von Neumann entropy. On the other hand, another form of mutual information which is equivalent to Eq. (1) in classical scenario, is expressed as

$$J(\rho_{AB}) = S(\rho_A) - S(\rho_{A|B}), \tag{2}$$

where $S(\rho_{A|B})$ is the conditional quantum entropy, which clearly depends on the observable we are measuring on B.

A general measurement on B is positive-operator-valued-measurement (POVM) $\{\Pi_k^B\}$. After the outcome corresponding to Π_k^B has been detected, the state of A is

$$\rho_{A|\Pi_k^B} = Tr_B[(\mathbb{1} \otimes \Pi_k^B)\rho_{AB}(\mathbb{1} \otimes \Pi_k^B)^\dagger]/p_k, \tag{3}$$

with probability $p_k = Tr[(\mathbb{1} \otimes \Pi_k^B)\rho_{AB}(\mathbb{1} \otimes \Pi_k^B)^\dagger]$. Then Eq. (2) can be written as

$$J(\rho_{AB})_{\{\Pi_k^B\}} = S(\rho_A) - S(\rho_{A|\{\Pi_k^B\}}), \tag{4}$$

where $S(\rho_{A|\{\Pi_k^B\}}) = \sum_k p_k S(\rho_{A|\Pi_k^B})$. This quantify represents the information gained about the system A as a result of the measurement $\{\Pi_k^B\}$. The maximal accessible information about A is the classical correlation [7, 8]

$$C_B(\rho_{AB}) = S(\rho_A) - \min_{\{\Pi_k^B\}} \sum_k p_k S(\rho_{A|\{\Pi_k^B\}}). \tag{5}$$

Then, the quantum part of the total correlation is defined as quantum discord

$$D_B(\rho_{AB}) = I(\rho_{AB}) - C_B(\rho_{AB}). \tag{6}$$

In the same way, D_A can also be calculated by optimizing over all projective measurements on A. Note that in general, discord is asymmetric, i.e., $D_B \neq D_A$. However, in the case of Bell-diagonal states, $D_B = D_A$.

Up to now, there isn't an analytical formula of quantum discord for general two-qubit states, except for the Bell-diagonal states [42] and two-qubit X states [43–45]. In an experiment, the density matrix is reconstructed by the state tomography, with almost all elements nonzero. The quantum discord of these reconstructed density matrices need to be numerically calculated according their definitions. The mutual information of ρ_{AB} is calculated with Eq. (1). The classical correlation and quantum discord are obtained with Eqs. (5) and (6) by optimizing over all one-qubit measurements on B. It has been proven [46] that the optimal measurement is always projective for two-qubit states, so it is sufficient to maximize over all the following projective measurements $\{\mathbb{1}\otimes \mid \Theta_k\rangle\langle\Theta_k \mid, k = \parallel, \perp\}$, where $\mid \Theta_\parallel\rangle = \cos\theta \mid 0\rangle + e^{i\phi}\sin\theta|1\rangle$ and $|\Theta_\perp\rangle = e^{-i\phi}\sin\theta|0\rangle - \cos\theta|1\rangle$ presents an arbitrary basis of B formed by two orthogonal states on the Bloch sphere, with

$0 \leq \theta \leq \frac{\pi}{2}$ and $0 \leq \phi \leq 2\pi$. Then the classical correlation and quantum discord can be obtained numerically.

3 Dynamics of Quantum Correlation in Open Solid Systems

Quantum systems inevitably interact with the environment, the evolution of the quantum discord in a noisy environment is certainly of interests. It is well-known that in a dissipative environment entanglement decays to zero in a finite time, which is called entanglement sudden death (ESD) [47]. In contrast, quantum discord has no sudden death but has peculiar sudden change in its decay rates. Meanwhile, as quantum discord is regarded as an important resource in many quantum tasks, it is essential to preserve the quantum correlation in a fragile quantum system, thus it is vital to overcome decoherence effect induced by the environment. One of the strategies is the dynamical decoupling (DD) technique [48, 49], which uses stroboscopic spin flips to reduce the average coupling to the environment to effectively zero. DD is a particularly promising strategy for combating decoherence [50, 51], since it can be naturally integrated with other desired functionalities, such as quantum gates.

Here we describe the experimental investigation of the dynamics of both the classical and quantum correlations in solids [27]. The phosphorous donors in silicon (P:Si) material with P concentration about $1 \times 10^{16}\,\text{cm}^{-3}$ is chosen for the study. Silicon is a particularly attractive material for hosting spin qubit for its low-orbit coupling and low natural abundance of nuclear-spin-bearing isotope [52]. The experiment was performed at temperature 8 K. The system consists of an electron spin $S = 1/2$ and a nuclear spin $I = 1/2$. It can be described by an isotropic spin Hamiltonian

$$H_{e,n} = \omega_e S_z - \omega_I I_z + 2\pi a \cdot \overrightarrow{\mathbf{S}} \cdot \overrightarrow{\mathbf{I}}, \tag{7}$$

where $\omega_e = g\beta_e B_0/\hbar$ and $\omega_I = g_I \beta_I B_0/\hbar$ characterize the Zeeman interactions for the electron and nuclear spins, respectively. $a = 117\,\text{MHz}$ is the strength of the isotropic hyperfine interaction, and $\overrightarrow{\mathbf{S}}$ ($\overrightarrow{\mathbf{I}}$) is the electron (nuclear) spin operator. The energy diagram of this system is plotted in Fig. 1a, where the four-level system can be manipulated by resonant the microwave (MW1, MW2) and radio-frequency (RF1, RF2) radiation. The dynamics of an open quantum system ρ_{AB} coupling to the environment in solids can be described by longitudinal and transverse relaxations of time constants T_1 and T_2, respectively. For the electron spin, the transverse relaxation time is $T_{2e} = 120\,\mu\text{s}$, the longitudinal (i.e., electron population) relaxation time is $T_{1e} = 5.6\,\text{ms}$ and the dephasing time $T_{2e}^* \approx 0.2\,\mu\text{s}$. For the nuclear spin, the dephasing time is determined as $T_{2n}^* = 24\,\mu\text{s}$ by the nuclear spin free induction decay (FID) experiment. The longitudinal relaxation time T_1 of both the electron and the nuclear spins are much larger than the transverse relaxation time T_2 and so can be neglected in our experiments. The decay of the off-diagonal elements depends on the time scale of dephasing time T_2^* of the electron and the nuclear spin. Since T_{2e}^* is almost two orders of magnitude smaller than T_{2n}^*, the decay of the secondary diagonal elements is dominated by T_{2e}^*. For the phosphorous donors in silicon with natural abundance of ^{29}Si, the hyperfine fields of the ^{29}Si nuclei cause random static shifts of the individual electron-spin resonant frequencies which satisfy a Gaussian distribution [53]. Averaging over the Gaussian-distributed resonant frequencies, the dephasing of the electron spin can be derived with the form $\sim \exp[-(t/T_{2e}^*)^2]$.

Here the Bell-diagonal state is considered for the initial state. It can be obtained from the thermal equilibrium state $\rho_0 = \frac{1}{4}\mathbb{1}_{4\times 4} - \varepsilon\sigma_z \otimes \mathbb{1}_{2\times 2}$, where $\varepsilon = g\beta_e B_0/8k_B T = 7.35 \times 10^{-3}$ (at temperature 8 K) is the ratio between the magnetic and thermal energies. Then with the dephasing of the electron spin, the time evolution of the total system is given by

$$\rho_{AB}(t) = \frac{1}{4}[\mathbb{1} + c_1(t)\sigma_x^A\sigma_x^B + c_2(t)\sigma_y^A\sigma_y^B + c_3(t)\sigma_z^A\sigma_z^B], \tag{8}$$

where $\sigma_i^{A(B)}$ $(i = x, y, z)$ are the Pauli matrices of the first (second) qubit, and the coefficients $c_1(t) = c_1(0)\exp[-(t/T_{2e}^*)^2]$, $c_2(t) = c_2(0)\exp[-(t/T_{2e}^*)^2]$ and $c_3(t) = c_3$.

The typical coefficients $c_i \sim 10^{-3}$ $(i = 1, 2, 3)$, therefore we focus on a class of reasonable and widely used states for which $c_1(0) = 0$, $|c_3(0)| < |c_2(0)| \ll 1$. By following Ref. [42], we obtain the analytical expressions for the mutual information, the classical correlation and the quantum discord. Expanding $I[\rho(t)]$ and $C[\rho(t)]$ in the Taylor series of $c_2(t)$ and c_3 and neglecting high-order terms, we obtain

$$I[\rho(t)] = \frac{1}{2\ln 2}\,[c_3^2 + c_2^2(t)], \tag{9}$$

$$C[\rho(t)] = \frac{1}{2\ln 2}\,c^2(t), \tag{10}$$

where $c(t) = \max\{c_2(t), c_3\}$. Hence, the quantum discord is calculated to be

$$D[\rho(t)] = \begin{cases} \frac{1}{2\ln 2}\,c_3^2 & \text{if } t \leqslant t_c, \\ \frac{1}{2\ln 2}\,c_2^2(t) & \text{if } t > t_c. \end{cases} \tag{11}$$

Here $t_c = \sqrt{-\ln[c_3/c_2(0)]}T_{2e}^*$ is obtained by setting $c_2(t_c) = c_3$. Consequently, in the noisy environment, the quantum discord is constant and the classical correlation decreases in the initial period $t \leqslant t_c$, while for $t > t_c$, the classical correlation does not change in time and only the quantum discord is reduced sharply.

Figure 1b shows the pulse sequence applied in our experiment. It consists of three steps: state preparation, evolution under the noisy environment, and the tomography. Firstly the quantum state is initialized to be a Bell-diagonal state, which satisfies the conditions $c_1(0) = 0$, $|c_3(0)| < |c_2(0)| \ll 1$. We present the result of the tomography of the initial state in Fig. 1c, d, where the real and imaginary parts of the deviation density matrix in unit of ε are presented. After the preparation step, the quantum system is left to evolve under the noisy environment. Then the resultant quantum states are reconstructed by state tomography [54].

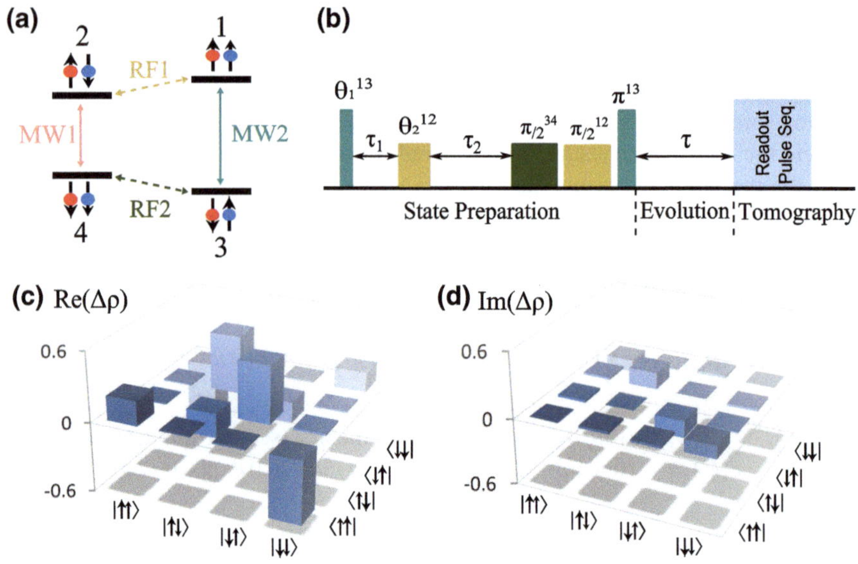

Fig. 1 **a** Energy level diagram for the P:Si system. There are four Zeeman product states which are labeled by states 1–4, respectively. ↑ and ↓ stand for the $\pm 1/2$ states of electron and nuclear spins. Electron spin resonance (ESR) and nuclear magnetic resonance (NMR) transitions are indicated by two-way arrows. **b** Diagram of the experimental pulse sequence, which includes three steps: state preparation, evolution under the noisy environment, and the tomography. **c**, **d** The real and imaginary parts of the reconstructed deviation density matrix $\Delta\rho = \rho - \mathbb{1}/4$, in unit of ε, respectively, where $\varepsilon \approx 7.35 \times 10^{-3}$. From Ref. [27]

The dynamics of the quantum and classical correlations depend on the decay of secondary diagonal elements according to Eqs. (10) and (11). The decay of ρ_{23} and ρ_{14} decay with function of $\sim \exp[-(t/T_{\text{decay}})^2]$, where $T_{\text{decay}} \approx 175$ ns. The values of mutual information (darkgreen circle), the quantum discord (red quadrate) and the classical correlation (black triangle), which are obtained by numerical calculating from the state tomography density matrix, are plotted in Fig. 2a. The theoretical predications plotted in Fig. 2a can successfully describe the experimental data. It clearly shows that the sudden transition from the classical to the quantum decoherence regime occurs at about 166 ns. In the initial period ($0 \le t \lesssim 166$ ns), the quantum discord remains constant but the classical correlation decreases, while for $t \gtrsim 166$ ns, the classical correlation does not change but the quantum correlation decreases dramatically.

Then the two-flip DD pulse sequence is applied to the electron spin to prolong both classical and quantum correlations. Dynamics decoupling is an effective method to preserve the coherence of a quantum system and it has been tested by experiments in both single qubit [50] and two-qubit [51] systems. Recently, ultralong coherence times of both electron and nuclear spins have been achieved by DD in high-purity silicon [55, 56]. However, the ability of DD to preserve the nonclassical correlations (quantum discord) between electron and nuclear spins remains elusive. Figure 2b

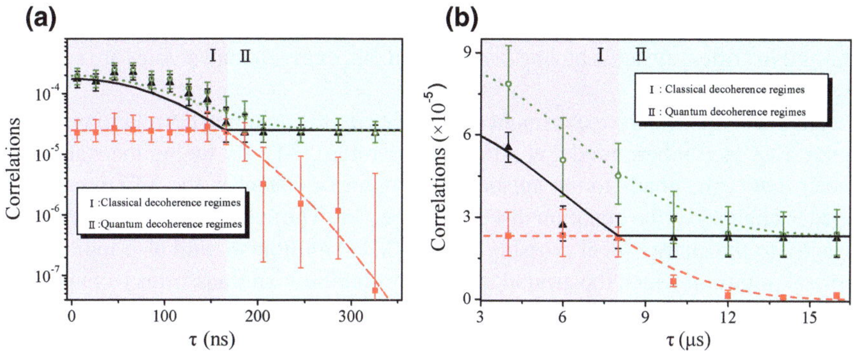

Fig. 2 **a** Dynamics of classical and quantum correlations. The logarithmic-scale values of mutual information (*darkgreen circle*), quantum discord (*red square*) and classical correlation (*black triangle*) are numerical computed with their original definitions (and so do the other experimental data in the chapter). The curves are the theoretical predication. **b** The evolution of classical and quantum correlations under the dynamical decoupling protection. The curves are drawn to follow the trend of the variation of correlations. I and II denote the classical and quantum decoherence regimes, respectively. Adapted from Ref. [27]

shows that the decay of classical and quantum correlations become much slower, and the period before the sudden transition of correlations is prolonged via DD by about 50 times to 8.0 μs.

The results show clearly a transition from classical decoherence regime to quantum decoherence regime at a critical point $t_c \approx 166$ ns, and t_c can be extended via a simple DD approach by about 50 times to 8.0 μs. Since quantum discord might be the reason for the power of quantum computation in some cases [6], and a resource for remote state preparation [12], together with the demonstration of an operation method to use quantum discord as a physical resource [13], the experimental demonstration of the existence of a non-decay region, the revival and prolonging of the quantum discord in a noisy solid-state system may have great potential applications in quantum information processing.

4 Quantum Correlation for Capture the Critical Points of the *XXZ* Model

QPT [58] is a purely quantum process occurring at absolute zero temperature when an external parameter of the Hamiltonian is varied. It is different from classical phase transition because they are strictly determined by the properties of the ground state. This phenomenon is one of the main interests in the field of condensed matter physics. On the other hand, correlations play an important role in critical systems, for instance, it has shown that the significant connection between entanglement and QPT in the *XY* spin chain [57]. Recently, quantum discord is shown to capture the critical points associated with QPT for *XXZ* and *XY* model, even at a finite temperature or in an external magnetic field [29–33]. With extensive theoretical studies about quantum discord in critical systems having been carried out, experimental studies are essential but very few.

Here we describe the experiment to characterize the quantum correlations in two-qubit *XXZ* Heisenberg model at a finite temperature [34]. By tuning the state parameter that corresponds to the anisotropic coupling constant in the *XXZ* model, the sudden changes of the quantum discord is observed. The sudden changes correspond exactly to the energy-level crossing of the *XXZ* Hamiltonian, and also indicate the critical points on which the ground state of the Hamiltonian transforms from product state to entangled one, or vice versa. When the number of the qubit on the *XXZ* chain tends to infinity (i.e., thermodynamic limit), the sudden changes of discord spotlight the critical points associated with QPT [31].

The Hamiltonian of a two-qubit *XXZ* Heisenberg chain is given by

$$H = \frac{J}{4}(\sigma_x^A\sigma_x^B + \sigma_y^A\sigma_y^B + \Delta\sigma_z^A\sigma_z^B), \tag{12}$$

where J is the coupling constant and Δ is the anisotropy parameter. In fact, the density matrix for this model at thermal equilibrium at temperature T is given by the canonical ensemble $\rho_{AB} = e^{-\beta H}/Z$, where $\beta = 1/k_B T$, and $Z = \mathrm{Tr}(e^{-\beta H})$ is the partition function. This is in form of Bell-diagonal states. The entanglement of formation (EoF) [59] and the quantum discord of the state ρ_{AB} can be easily obtained from Refs. [60] and [42], respectively. When $T = 0$, the state ρ_{AB} is the ground state of the Hamiltonian H. When the temperature T is appreciably larger than the maximum splitting of H in Eq. (12), the thermal-equilibrium state can be approximated as

$$\rho_{AB} = \frac{\mathbb{1}}{4} - \frac{\beta J}{16}(\sigma_x^A\sigma_x^B + \sigma_y^A\sigma_y^B + \Delta\sigma_z^A\sigma_z^B). \tag{13}$$

We depict the energy levels with respect to the anisotropic parameter Δ in Fig. 3a, where the units have been chosen such that $J = 1$ and k_B is unity [61]. It is shown that with the temperature increasing, the EoF becomes zero, while the quantum discord is always larger than zero [Fig. 3b]. Also, at finite temperature, the quantum discord changes suddenly at $\Delta = \pm 1$ which corresponds exactly to the energy-level crossing, but it is not the case for EoF, whose behavior is illustrated by dotted lines in Fig. 3b.

The P:Si material is chosen as a benchmark system as well. The energy diagram of this system is shown in Fig. 1a. The experiment consists of three steps: state initialization, evolution to the target states, and the tomography. After the target states corresponding to the different values of Δ are achieved, tomography technique [54] is then used to reconstruct the density matrix ρ_{AB}.

The values of quantum discord [triangles in Fig. 3c] are numerically calculated from the reconstructed density matrices according its definition. The experimental data agree with the theoretical prediction [line in the Fig. 3c]. It is clear that quantum discord experiences a sudden change at $\Delta = \pm 1$ where the energy level crossing occurs. It is worthy of pointing out that the entanglement, which is characterized by EoF, is measured to be zero in our experiments. Thus, we have presented a case

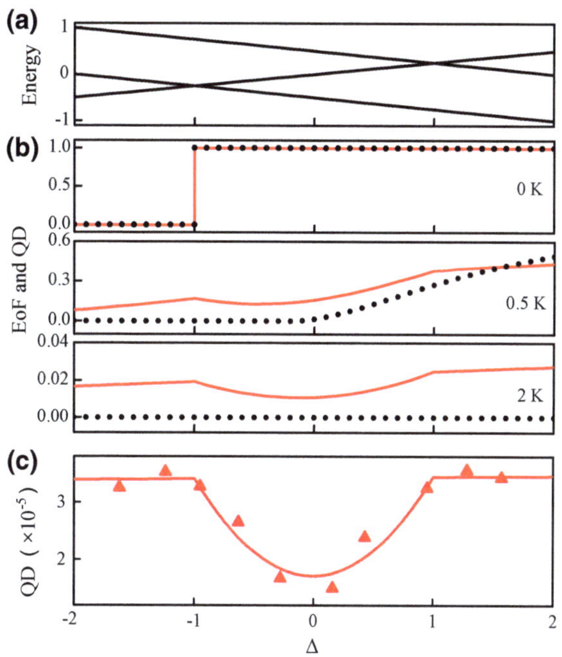

Fig. 3 **a** The energy levels of the Hamiltonian given by Eq. (12). **b** The EoF (*black dotted lines*) and quantum discord (*red solid lines*) of the state ρ_{AB}. When the temperature increases from 0 to 2 K, the EoF vanishes while the discord remains positive. The sudden changes of the discord correspond to the level crossings shown in **a**. **c** The behavior of quantum correlations when Δ varies. The *triangles* are experimental data of quantum discord derived from the reconstructed density matrices, which agree with the theoretical values (*red solid line*). The deviation between the experimental data and the theoretical prediction is due to the imperfection of the pulses. Adapted from Ref. [34]

where entanglement fails to capture the energy level crossing while the temperature is of a finite value. Note that there are theoretical studies which conclude that quantum discord can be utilized to capture the quantum phase transition even at finite temperatures [30, 31, 33]. The success in highlighting the sudden change of the ground state ($\Delta = -1$) in the two-qubit system employed in our experiments indicates that quantum correlation could be used to observe the quantum criticality at finite temperatures.

Here the two-qubit *XXZ* Heisenberg model has been taken as an example, and the abrupt change of its ground state has been unambiguously spotlighted by quantum correlations. The results reveal the capability of using quantum correlations to indicate the intrinsic change of the physical system. This experiment may serve as a preliminary meaningful step to observe quantum criticality at finite temperatures via quantum correlations.

Besides above peculiar characters, quantum discord has been shown playing an important role in relativistic quantum information [62–64]. It has been shown that quantum discord is observer-dependent and degrades with observer's acceleration for both bosonic [62] and fermionic systems [63] via the Unruh effect [65–67]. Interestingly, it was pointed out recently that quantum discord could be created by the Unruh effect even from the classically correlated states [68]. This remarkable phenomenon could be ascribed to the nonunital characteristics of the Unruh channel, for which quantum discord can be produced by the local nonunital channel [69–71].

Consider an inertial observer Alice (A) and a uniformly accelerated observer Bob (B) with a constant acceleration a. As shown in Fig. 4a, Alice moves in the Minkowski plane with coordinate (t, z). The setting of the uniform acceleration can be conveniently described by the Rindler coordinate (τ, ζ) with two disconnected regions I and II. One can describe the uniformly accelerated Bob to travel on a hyperbola constrained to region I, as Fig. 4a shown. Bob has no access to field modes in the causally disconnected region II. Therefore, he must trace over the inaccessible region II, which unavoidably leads to the detection of a mixed state. Under the single-mode approximation, the Minkowski vacuum state $|0\rangle_M$ and one-particle state $|1\rangle_M$ can be expressed as [63, 72]:

$$\begin{aligned} |0\rangle_M &= \cos r|0\rangle_{\mathrm{I}}|0\rangle_{\mathrm{II}} + \sin r|1\rangle_{\mathrm{I}}|1\rangle_{\mathrm{II}} \,, \\ |1\rangle_M &= |1\rangle_{\mathrm{I}}|0\rangle_{\mathrm{II}} \,. \end{aligned} \tag{14}$$

The dimensionless acceleration parameter r is defined by $\cos r = [\exp(\frac{-2\pi\Omega c}{a}) + 1]^{-1/2}$ with $r \in [0, \pi/4]$, $a \in [0, \infty)$, and Ω is the frequency of Unruh mode.

The Bell-diagonal states ρ_{AB} shared by Alice and Bob initially in an inertial frame are taken into account. Then the Minkowski particle states are expended into Rindler region I and II particle and antiparticle states. Since Bob is causally disconnected from region II, we trace over the mode in this region II. The quantum state shared by Alice and Bob can be expressed as

$$\begin{aligned} \rho_{AB_{\mathrm{I}}} = \frac{1}{4}\Big(&\mathbb{1} + c_1 \cos r\sigma_x^A \otimes \sigma_x^B + c_2 \cos r\sigma_y^A \otimes \sigma_y^B \\ &+ c_3 \cos^2 r\sigma_z^A \otimes \sigma_z^B - \sin^2 r\mathbb{1}^A \otimes \sigma_z^B\Big). \end{aligned} \tag{15}$$

This is an X-type state which has an analytical formula of quantum discord [44, 45]. For clarity, we focus on a special Bell-diagonal state with $c_1 = 1,\ c_2 = c_3 = 0$ which is a classically correlated state.

The behaviours of correlations are depicted in Fig. 4b. It is shown that both mutual information and classical correlation decay along with r, however, quantum discord can be created from this classically correlated state via the Unruh effect. Furthermore, this remarkable phenomenon has been experimentally simulated on a nuclear magnetic resonance quantum simulator [68]. The experimental results are shown in Fig. 4b as well. This result implies that, although the total correlations decay, the non-classical correlations can be generated under the Unruh effect. Note that the quantum entanglement in this case is always zero.

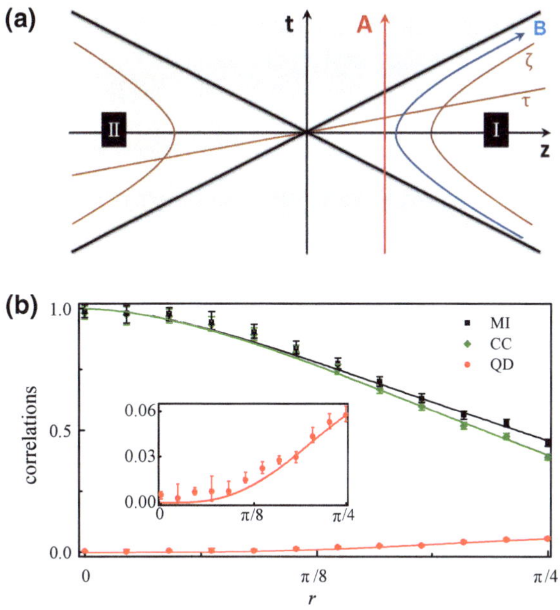

Fig. 4 **a** Alice (A) is an inertial observer and Bob (B) is a uniformly accelerated observer. The sets (t, z) and (τ, ζ) denote the Minkowski and Rindler coordinates, respectively. The *right* and *left* Rindler wedges are the regions with $|t| < z$ and $|t| < -z$, respectively. They are separated by the Rindler horizon so that they are causally disconnected from each other. **b** The behavior of mutual information (*black* - MI), classical correlation (*olive* - CC), and quantum discord (*red* - QD) under the Unruh effect. The discrete points are the experimental results with the nuclear spin quantum simulator. The curves are the corresponding theoretical predications. *Insets* the zoom-in plot of the sectional region. Adapted from Ref. [68]

5 Concluding Remarks

To summarize, quantum discord is a concept intended to describe the whole non-classical correlation in quantum systems. Although both quantum discord and entanglement are used to measure quantum correlation, it has been found that quantum discord can identify quantum correlation in separable states while entanglement cannot. Theoretical work has been carried out recently in various aspects, while experimental studies are few. As the promising applications of the solid-state spin systems, we believe that it is necessary to perform clear and reliable experiments to investigate experimentally the presence of quantum discord where entanglement vanishes in such systems.

This chapter has introduced mainly two experimental work about quantum correlations in solid-state spin system. One is the dynamics of quantum correlations in open solid systems and its protection by DD, another one is the utilization of quantum discord for capturing the critical points of the two-qubit *XXZ* Hamiltonian. These are just a sub branch of the characters and applications of quantum correlation. And, attentions should be paid to quantum correlations and related aspects, that studying the nature of correlations in the world around us might even help us catch a glimpse of the theory that comes to supersede quantum physics [2].

Acknowledgements The authors acknowledge financial support from the National Key Basic Research Program of China (Grant No.2013CB921800 and No.2014CB848700), the National Natural Science Foundation of China (Grant No.11227901, No.91021005, No.11375167, No.11374308, No.11104262, and No.11275183), and the Strategic Priority Research Program (B) of the CAS (Grant No. XDB01030400).

References

1. R. Horodecki, P. Horodecki, M. Horodecki, K. Horodecki, Rev. Mod. Phys. **81**, 865 (2009)
2. K. Modi, A. Brodutch, H. Cable, T. Paterek, V. Vedral, Rev. Mod. Phys. **84**, 1655 (2012)
3. C.H. Bennett, D.P. DiVincenzo, C.A. Fuchs, T. Mor, E. Rains, P.W. Shor, J.A. Smolin, W.K. Wootters, Phys. Rev. A **59**, 1070 (1999)
4. J. Du, M. Shi, X. Zhou, Y. Fan, B. Ye, R. Han, J. Wu, Phys. Rev. A **64**, 042306 (2001)
5. E. Knill, R. Laflamme, Phys. Rev. Lett. **81**, 5672 (1998)
6. B.P. Lanyon, M. Barbieri, M.P. Almeida, A.G. White, Phys. Rev. Lett. **101**, 200501 (2008)
7. H. Ollivier, W. Zurek, Phys. Rev. Lett. **88**, 017901 (2002)
8. L. Henderson, V. Vedral, J. Phys. A **34**, 6899 (2001)
9. A. Datta, A. Shaji, C.M. Caves, Phys. Rev. Lett. **100**, 050502 (2008)
10. A. Brodutch, D.R. Terno, Phys. Rev. A **83**, 010301 (2011)
11. F.F. Fanchini, M.F. Cornelio, M.C. de Oliveira, A.O. Caldeira, Phys. Rev. A **84**, 012313 (2011)
12. B. Dakić, Y.O. Lipp, X. Ma, M. Ringbauer, S. Kropatschek, S. Barz, T. Paterek, V. Vedral, A. Zeilinger, Č. Brukner, P. Walther, Nat. Phys. **8**, 666 (2012)
13. M. Gu, H.M. Chrzanowski, S.M. Assad, T. Symul, K. Modi, T.C. Ralph, V. Vedral, P.K. Lam, Nat. Phys. **8**, 671 (2012)
14. M.F. Cornelio, M.C. de Oliveira, F.F. Fanchini, Phys. Rev. Lett. **107**, 020502 (2011)
15. A. Streltsov, H. Kampermann, D. Bruß, Phys. Rev. Lett. **106**, 160401 (2011)
16. A. Streltsov, H. Kampermann, D. Bruß, Phys. Rev. Lett. **108**, 250501 (2012)
17. T.K. Chuan, J. Maillard, K. Modi, T. Paterek, M. Paternostro, M. Piani, Phys. Rev. Lett. **109**, 070501 (2012)
18. A. Fedrizzi, M. Zuppardo, G.G. Gillett, M.A. Broome, M.P. Almeida, M. Paternostro, A.G. White, T. Paterek, Phys. Rev. Lett. **111**, 230504 (2013)
19. C. Peuntinger, V. Chille, L. Mišta Jr., N. Korolkova, M. Förtsch, J. Korger, C. Marquardt, G. Leuch, Phys. Rev. Lett. **111**, 230506 (2013)
20. K. Modi, H. Cable, M. Williamson, V. Vedral, Phys. Rev. X **1**, 021022 (2011)
21. D. Girolami, A.M. Souza, V. Giovannetti, T. Tufarelli, J.G. Filgueiras, R.S. Sarthour, D.O. Soares-Pinto, I.S. Oliveira, G. Adesso, Phys. Rev. Lett. **112**, 210401 (2014)
22. T. Werlang, S. Souza, F.F. Fanchini, C.J. Villas, Boas. Phys. Rev. A **80**, 024103 (2009)
23. J. Maziero, L.C. Cleri, R.M. Serra, V. Vedral, Phys. Rev. A **80**, 044102 (2009)
24. L. Mazzola, J. Piilo, J. Maniscalco, Phys. Rev. Lett. **104**, 200401 (2010)
25. J.-S. Xu, X.-Y. Xu, C.-F. Li, C.-J. Zhang, X.-B. Zou, G.-C. Guo, Nat. Commun. **1**, 7 (2010)
26. R. Auccaise, L.C. Céleri, D.O. Soares-Pinto, E.R. deAzevedo, J. Maziero, A.M. Souza, T.J. Bonagamba, R.S. Sarthour, I.S. Oliveira, R.M. Serra, Phys. Rev. Lett. **107**, 140403 (2011)
27. X. Rong, F. Jin, Z. Wang, J. Geng, C. Ju, Y. Wang, R. Zhang, C. Duan, M. Shi, J. Du, Phys. Rev. B **88**, 054419 (2013)
28. F.M. Paula, I.A. Silva, J.D. Montealegre, A.M. Souza, E.R. deAzevedo, R.S. Sarthour, A. Saguia, I.S. Oliveira, D.O. Soares-Pinto, G. Adesso, M.S. Sarandy, Phys. Rev. Lett. **111**, 250401 (2013)

29. R. Dillenschneider, Phys. Rev. B **78**, 224413 (2008)
30. M.S. Sarandy, Phys. Rev. A **80**, 022108 (2009)
31. T. Werlang, C. Trippe, G.A.P. Ribeiro, G. Rigolin, Phys. Rev. Lett. **105**, 095702 (2010)
32. J. Maziero, H.C. Guzman, L.C. Cleri, M.S. Sarandy, R.M. Serra, Phys. Rev. A **82**, 012106 (2010)
33. T. Werlang, G.A.P. Ribeiro, G. Rigolin, Phys. Rev. A **83**, 062334 (2011)
34. X. Rong, Z. Wang, F. Jin, J. Geng, P. Feng, N. Xu, Y. Wang, C. Ju, M. Shi, J. Du, Phys. Rev. B **86**, 104425 (2012)
35. B.E. Kane, Nature **393**, 133 (1998)
36. A.M. Tyryshkin, J.J.L. Morton, S.C. Benjamin, A. Ardavan, G.A.D. Briggs, J.W. Ager, S.A. Lyon, J. Phys. Condens. Matter **18**, S783 (2006)
37. J.J.L. Morton, A.M. Tyryshkin, R.M. Brown, S. Shankar, B.W. Lovett, A. Ardavan, T. Schenkel, E.E. Haller, J.W. Ager, S.A. Lyon, Nature **455**, 1085 (2008)
38. A. Gruber, A. Dräbenstedt, C. Tietz, L. Fleury, J. Wrachtrup, C. von Borczyskowski, Science **276**, 2012 (1997)
39. F. Jelezko, T. Gaebel, I. Popa, A. Gruber, J. Wrachtrup, Phys. Rev. Lett. **92**, 076401 (2004)
40. F. Shi, Q. Zhang, P. Wang, H. Sun, J. Wang, X. Rong, M. Chen, C. Ju, F. Reinhard, H. Chen, J. Wrachtrup, J. Wang, J. Du, Science **347**, 1135 (2015)
41. Z. Merali, Nature **474**, 24–26 (2011)
42. S. Luo, Phys. Rev. A **77**, 042303 (2008)
43. M. Shi, F. Jiang, C. Sun, J. Du, New J. Phys. **13**, 073016 (2011)
44. M. Ali, A.R.P. Rau, G. Alber, Phys. Rev. A **81**, 042105 (2010)
45. Y. Huang, Phys. Rev. A **88**, 014302 (2013)
46. S. Hamieh, R. Kobes, H. Zaraket, Phys. Rev. A **70**, 052325 (2004)
47. T. Yu, J.H. Eberly, Science **323**, 598 (2009)
48. L. Viola, E. Knill, S. Lloyd, Phys. Rev. Lett. **82**, 2417 (1999)
49. G.S. Uhrig, Phys. Rev. Lett. **98**, 100504 (2007)
50. J. Du, X. Rong, N. Zhao, Y. Wang, J. Yang, R.B. Liu, Nature **461**, 1265 (2009)
51. Y. Wang, X. Rong, P. Feng, W. Xu, B. Chong, J.-H. Su, J. Gong, J. Du, Phys. Rev. Lett. **106**, 040501 (2011)
52. J.J.L. Morton, D.R. McCamey, M.A. Eriksson, S.A. Lyon, Nature **479**, 345 (2011)
53. E. Abe, A.M. Tyryshkin, S. Tojo, J.J.L. Morton, W.M. Witzel, A. Fujimoto, J.W. Ager, E.E. Haller, J. Isoya, S.A. Lyon, M.L.W. Thewalt, K.M. Itoh, Phys. Rev. B **82**, 121201 (2010)
54. L. Vandersypen, I. Chuang, Rev. Mod. Phys. **76**, 1037 (2004)
55. A.M. Tyryshkin, S. Tojo, J.J.L. Morton, H. Riemann, N.V. Abrosimov, P. Becker, H. Pohl, T. Schenkel, M.L.W. Thewalt, K.M. Itoh, S.A. Lyon, Nat. Mater. **11**, 143 (2012)
56. M. Steger, K. Saeedi, M.L.W. Thewalt, J.J.L. Morton, H. Riemann, N.V. Abrosimov, P. Becker, H.J. Pohl, Science **336**, 1280 (2012)
57. T.J. Osborne, M.A. Nielsen, Phys. Rev. A **66**, 032110 (2002)
58. S. Sachdev, *Quantum Phase Transitions* (Cambridge University Press, Cambridge, 1999)
59. C.H. Bennett, H.J. Bernstein, S. Popescu, B. Schumacher, Phys. Rev. A **53**, 2046 (1996)
60. W. Wootters, Phys. Rev. Lett. **80**, 2245 (1998)
61. X. Peng, J. Du, D. Suter, Phys. Rev. A **71**, 012307 (2005); X. Peng, S. Wu, J. Li, D. Suter, J. Du, Phys. Rev. Lett. **105**, 240405 (2010)
62. A. Datta, Phys. Rev. A **80**, 052304 (2009)
63. J. Wang, J. Deng, J. Jing, Phys. Rev. A **81**, 052120 (2010)
64. L.C. Céleri, A.G.S. Landulfo, R.M. Serra, G.E.A. Matsas, Phys. Rev. A **81**, 062130 (2010)
65. P.C.W. Davies, J. Phys. A **8**, 609 (1975)
66. W.G. Unruh, Phys. Rev. D **14**, 870 (1976)
67. L.C.B. Crispino, A. Higuchi, G.E.A. Matsas, Rev. Mod. Phys. **80**, 787 (2008)
68. F. Jin, H. Chen, X. Rong, H. Zhou, M. Shi, Q. Zhang, C. Ju, Y. Cai, S. Luo, X. Peng, J. Du, Sci. China-Phys. Mech. Astron. **59**, 630302 (2016)
69. A. Streltsov, H. Kampermann, D. Bruß, Phys. Rev. Lett. **107**, 170502 (2011)
70. F. Ciccarello, V. Giovannetti, Phys. Rev. A **85**, 010102 (2012)
71. B.P. Lanyon, P. Jurcevic, C. Hempel, M. Gessner, V. Vedral, R. Blatt, C.F. Roos, Phys. Rev. Lett. **111**, 100504 (2013)
72. P.M. Alsing, I. Fuentes-Schuller, R.B. Mann, T.E. Tessier, Phys. Rev. A **74**, 032326 (2006)

Quantum Correlations in NMR Systems

T.S. Mahesh, C.S. Sudheer Kumar and Udaysinh T. Bhosale

'Correlations cry out for explanation' - J.S. Bell in Speakable and Unspeakable in Quantum Mechanics, Cambridge university press (1989).

Abstract In conventional NMR experiments, the Zeeman energy gaps of the nuclear spin ensembles are much lower than their thermal energies, and accordingly exhibit tiny polarizations. Generally such low-purity quantum states are devoid of quantum entanglement. However, there exist certain nonclassical correlations which can be observed even in such systems. In this chapter, we discuss three such quantum correlations, namely, quantum contextuality, Leggett–Garg temporal correlations, and quantum discord. In each case, we provide a brief theoretical background and then describe some results from NMR experiments.

1 Introduction

Quantum physics is known for many nonintuitive phenomena including certain classically forbidden correlations. To study and understand these mysterious quantum correlations we require a suitable testbed. Nuclear Magnetic Resonance (NMR) [1, 2] of an ensemble of molecular nuclei in bulk liquids/solids form a convenient testbed even at room temperatures [3]. The weakly perturbed nuclear spins in such

T.S. Mahesh (✉) · C.S. Sudheer Kumar · U.T. Bhosale
Department of Physics and NMR Research Center, Indian Institute of Science Education and Research, Pune 411008, India
e-mail: mahesh.ts@iiserpune.ac.in

C.S. Sudheer Kumar
e-mail: sudheer.kumar@students.iiserpune.ac.in

U.T. Bhosale
e-mail: uday.bhosale@iiserpune.ac.in

F.F. Fanchini et al. (eds.), *Lectures on General Quantum Correlations and their Applications*, Quantum Science and Technology,
DOI 10.1007/978-3-319-53412-1_23

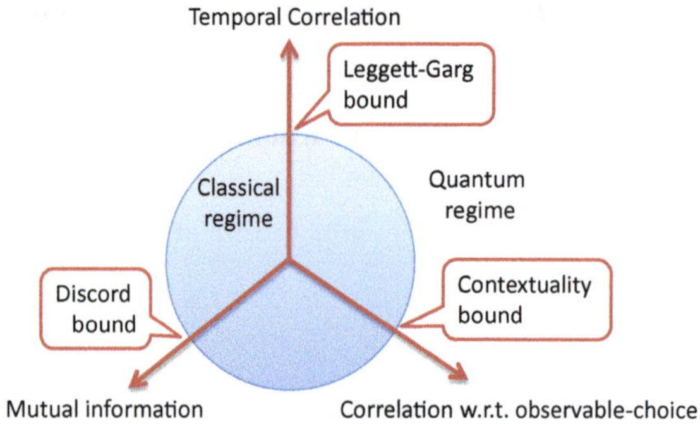

Fig. 1 Various correlations and corresponding bounds distinguishing quantum regime from classical

systems can store quantum superpositions for long durations ranging from seconds to minutes. In addition, excellent unitary controls via radio-frequency pulses allow precise manipulations of spin-dynamics. Even though one can not have local addressability of individual spins, and one works with the spin-ensemble as a whole, it is still possible to study many of the quantum correlations, namely contextuality, temporal correlation, discord etc (Fig. 1). The ensemble measurements are often sufficient, since many of the quantum correlations can be evaluated via expectation values. However, at room temperature there is little entanglement in conventional NMR systems [4]. In fact, this makes NMR a good candidate for studying quantum correlations without entanglement.

In the following sections we are going to review some NMR experiments investigating quantum contextuality, Leggett–Garg inequality, and quantum discord. For the sake of completeness, we have provided a brief theoretical background in each case.

2 Quantum Contextuality

As the name suggests, outcome of a quantum measurement in general depends on the context i.e., measurement-setting, arrangement, situation, circumstance, etc. Quantum contextuality (QC) states that the outcome of a measurement depends not only on the system and the observable being measured, but also on the context of the measurement, i.e., on other compatible observables which are measured along with [5–7]. QC signifies a mysterious nonclassical correlation between measurement outcomes corresponding to distinct observables. One consequence of QC is violation of Bell's inequality [8, 9], which has challenged the most cherished tenet of special theory of relativity, i.e., locality.

Peres explained quantum contextuality using a pair of electrons in a singlet state $(|01\rangle - |10\rangle)/\sqrt{2}$ [6]. Suppose we measure a Pauli observable $\sigma_{i\alpha}$, where $\alpha \in \{x, y, z\}$, on the ith particle, and obtain an outcome $\alpha_i = \pm 1$. For the singlet state, the result of measuring $\sigma_{1x}\sigma_{2x}$ is $x_1 x_2 = -1$ since $\langle \sigma_{1x}\sigma_{2x} \rangle = -1$. Similarly,

$y_1 y_2 = -1$. However, if one measures $\sigma_{1x}\sigma_{2y}$ followed by $\sigma_{1y}\sigma_{2x}$ one would obtain the outcome $x_1 y_2 y_1 x_2 = -1$ since $\langle \sigma_{1x}\sigma_{2y}\sigma_{1y}\sigma_{2x} \rangle = \langle \sigma_{1z}\sigma_{2z} \rangle = -1$, which is in contradiction with $x_1 x_2 = y_1 y_2 = -1$.

Later Mermin [10] generalized quantum contextuality to a state-independent scenario. Consider a pair of spin-1/2 particles and a set of nine Pauli-observables arranged in the following fashion:

$$\begin{array}{|c|c|c||c|} \hline \sigma_{1z} & \sigma_{2z} & \sigma_{1z}\sigma_{2z} & +\mathbb{1} \\ \hline \sigma_{2x} & \sigma_{1x} & \sigma_{1x}\sigma_{2x} & +\mathbb{1} \\ \hline \sigma_{1z}\sigma_{2x} & \sigma_{1x}\sigma_{2z} & \sigma_{1y}\sigma_{2y} & +\mathbb{1} \\ \hline \hline +\mathbb{1} & +\mathbb{1} & -\mathbb{1} & \\ \hline \end{array}. \quad (1)$$

Here the last column (row) lists the product along the row (column). In this arrangement, all the operators along any row, or any column, mutually commute and therefore they can be measured sequentially or simultaneously without any mutual disturbance. Whatever may be the state of the spin-pair, if one measures the three consecutive observables along any row one would obtain the outcome $+1$, the only eigenvalue of $\mathbb{1}$. Similarly, if one measures along first or second column one would obtain $+1$. On the other hand, choosing observables along the last column will lead to an outcome -1. However, no assignment of ± 1 values to individual measurements of all the nine observables can satisfy the above joint-measurement outcomes, indicating that such noncontextual preassignments of measurement outcomes is incompatible with quantum physics.

2.1 *Contextuality Studies Using NMR Systems*

The first demonstration of contextuality in NMR systems was reported by Moussa et al. [11]. Using a solid state NMR system, they evaluated the state independent inequality [8]

$$\beta = \langle \pi_{r_1} \rangle + \langle \pi_{r_2} \rangle + \langle \pi_{r_3} \rangle + \langle \pi_{c_1} \rangle + \langle \pi_{c_2} \rangle - \langle \pi_{c_3} \rangle \leq 4 \quad (2)$$

where $\langle \pi_{r_i} \rangle$ are the expectation values obtained when all the observables along the ith row of matrix in (1) are measured. Similarly $\langle \pi_{c_j} \rangle$ is the expectation value for measurements along the jth column. Exploiting the state independent property, they initialized the system in the maximally mixed state and obtained the value $\beta = 5.2 \pm 0.1$. While the result is in agreement with the quantum bound which is $\beta \leq 6$, it strongly violates the inequality in (2).

Later, Xi Kong et al. demonstrated QC by a single three level system in a NV center setup [12]. More recently, Dogra et al. demonstrated QC using a qutrit (spin-1) NMR system with a quadrupolar moment, oriented in a liquid crystalline environment. Using a set of 8 traceless observables (Gell-Mann matrices) and an inequality derived based on a noncontextual hidden variable (NCHV) model, they observed a clear violation of the NCHV inequality [13].

Contextuality via Psuedo Spin Mapping

Su et al. [14] have theoretically studied QC of eigenstates of one dimensional quantum harmonic oscillator (1D-QHO) by introducing two sets of pseudo-spin operators,

$$\mathbf{\Gamma} = (\Gamma_x, \Gamma_y, \Gamma_z), \;\; \mathbf{\Gamma}' = (\Gamma'_x, \Gamma'_y, \Gamma'_z)$$

with components,

$$\Gamma_x = \sigma_x \otimes \mathbb{1}_2, \Gamma_y = \sigma_z \otimes \sigma_y, \Gamma_z = -\sigma_y \otimes \sigma_y, \Gamma'_x = \sigma_x \otimes \sigma_z, \Gamma'_y = \mathbb{1}_2 \otimes \sigma_y, \Gamma'_z = -\sigma_x \otimes \sigma_x, \tag{3}$$

where $\mathbb{1}_2$ is 2×2 identity matrix. Defining the dichotomic unitary observables,

$$A = \Gamma_x, \quad B = \Gamma'_x \cos\beta + \Gamma'_z \sin\beta, \quad C = \Gamma_z, \quad D = \Gamma'_x \cos\eta + \Gamma'_z \sin\eta, \tag{4}$$

they setup Bell-Clauser-Horne-Shimony-Holt (Bell-CHSH) inequality [15],

$$\mathbf{I} = \langle AB\rangle + \langle BC\rangle + \langle CD\rangle - \langle AD\rangle \leq 2. \tag{5}$$

However, the quantum bound was shown to be $\mathbf{I}_Q \leq 2\sqrt{2}$, clearly violating the above Bell-CHSH inequality and thus exhibiting QC of QHO.

Katiyar et al. carried out an NMR investigation of this inequality by mapping the QHO eigenstates to the spin-states of a 2-qubit system (with an additional ancilla qubit) [16]. Using the Moussa protocol [11] (described in the next section) to extract the joint-expectation values in the inequality (5), they obtained $\mathbf{I}_Q \approx 2.4 \pm 0.1$. Although decoherence limited the experimental value to below the quantum bound ($\mathbf{I}_Q \leq 2.82$), it is clearly above the classical bound ($\mathbf{I} \leq 2$) and therefore establishes QC of 1D-QHO.

Thus, we observe that even when a system is in a separable state, measuring nonlocal observables leads to violation of Bell-CHSH inequality [17].

3 Temporal Correlations

Bell's inequalities (BI) are concerned with how two systems (each with a dimension of at least 2) are correlated over space, where as the Leggett–Garg inequality (LGI) is concerned with the correlation of a single system (with a dimension of at least 2), with itself at different time instants. While the former deals with context of the measurement, the latter deals with a temporal context.

LGI is based on the following two assumptions:

1. Macroscopic realism (MR): A macroscopic system, with two or more macroscopically distinct states available to it, exists in one of these states at any given point of time.
2. Noninvasive measurability (NM): It is possible to determine the state of the system with arbitrarily small perturbation to its future dynamics [18, 19].

Although the original motivation of Leggett and Garg was to test the existence of quantumness even at a macroscopic level, most of the violations of LGI reported so far are on microscopic systems [19]. The LGI violations in such systems were either due to invasive measurement or the system being in microscopic superpositions. Even though LGI violation in a macroscopic system such as a superconducting qubit has been reported [20], the existence of macroscopically distinct states in such a system is not clear [19]. Other experimental works on LGI include Nitrogen-Vacancy centers [21, 22], photonic systems [23], electron interferometers [24], superconducting qubit [25], and more recently in neutrino oscillations [26]. Recent theoretical extensions of LGI include its entropic formulation [27] and LGI in a large ensemble of qubits [28]. The violation of the former was recently observed using NMR experiments by Katiyar et al. [29]. LGI is also studied for a system of qubits coupled to a thermal environment [30]. For more details reader can refer to the review [19]. LGI violation in a 3-level NMR system has also been reported recently [31].

In the following we provide a brief theoretical as well as experimental review of LGI in the context of NMR.

3.1 Leggett–Garg String

Consider a system (the 'target') evolving under some Hamiltonian. Let $\mathbb{Q}$ be a dichotomic observable with eigenvalues $Q = \pm 1$, and let $Q(t_i)$ denotes the measurement outcome at time t_i. Repeating these measurements a large number of times we obtain the two-time correlation coefficient (TTCC) C_{ij} for each pair:

$$C_{ij} = \lim_{N\to\infty} \frac{1}{N} \sum_{r=1}^{N} Q_r\left(t_i\right) \cdot Q_r\left(t_j\right) = \langle Q\left(t_i\right) \cdot Q\left(t_j\right)\rangle, \tag{6}$$

where r is the trial number. Finally, the values of these coefficients are to be substituted in the n-measurement LG string given by:

$$K_n = C_{12} + C_{23} + C_{34} + \cdots + C_{(n-1)n} - C_{1n}. \tag{7}$$

Each TTCC C_{ij} is bounded by a maximum value of $+1$, corresponding to a perfect correlation, and a minimum value of -1, corresponding to a perfect anti-correlation. $C_{ij} = 0$ indicates no correlation. Thus, the upper bound for K_n consistent with *macrorealism* comes out to be $(n-2)$, while the lower bound is n for *odd* n,

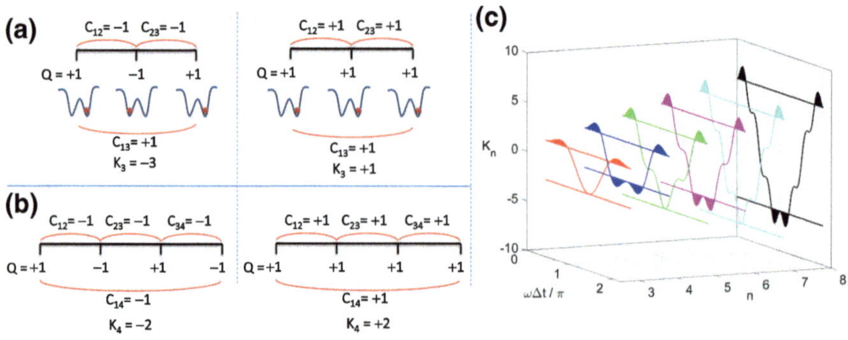

Fig. 2 Extreme values of TTCCs for a classical particle in a double-well potential for the cases of **a** three-time measurement and **b** four-time measurement. The left and right columns illustrate minimum and maximum values of K_n-strings respectively. **c** K_n versus n and $\omega\Delta t/\pi$ for a single qubit. The filled regions indicate LGI violations

and $-(n-2)$ for *even* n (see Fig. 2a and b). With these considerations LGI reads $-n \le K_n \le (n-2)$ for odd n, and $-(n-2) \le K_n \le (n-2)$ for even n.

In the following, we consider the case of a single qubit, namely a spin-1/2 nucleus precessing in an external static magnetic field.

3.2 Violation of LGI with a Single Qubit

A spin-1/2 nucleus precessing in an external magnetic field along z-axis has the following Hamiltonian: $\frac{1}{2}\omega\sigma_z$, where ω is the Larmor frequency. Let σ_x be the dichotomic observable [32]. Starting from the definition of TTCCs, we obtain for an arbitrary initial state ρ_0 [33, 34],

$$C_{ij} = \langle \sigma_x(t_i)\, \sigma_x(t_j) \rangle = \cos\left\{\omega(t_j - t_i)\right\}. \tag{8}$$

Dividing the total duration from t_1 to t_n into $(n-1)$ parts each of length Δt, we can express the LG string consistent with Eq. (8) as

$$K_n = (n-1)\cos\{\omega\Delta t\} - \cos\{(n-1)\omega\Delta t\}. \tag{9}$$

Figure 2c illustrates K_n curves for $n = 3$–8 and for $\omega\Delta t \in [0, 2\pi]$. The classical bounds in each case are shown by horizontal lines. As indicated by the filled areas, LGI is violated for each value of n at specific regions of $\omega\Delta t$. Quantum bounds of K_3 are -3 and $+1.5$ and that for K_4 are $-2\sqrt{2}$ and $+2\sqrt{2}$, and so on. In the following we discuss an experimental protocol for evaluating the LG strings.

3.3 Moussa Protocol

As described before, one needs to extract TTCCs in a way as noninvasive as possible. One way to achieve this is by using an ancilla qubit and employing Moussa protocol (Fig. 3). It involves preparing the ancilla in $|+\rangle$ state (an eigenstate of σ_x; or a pseudopure state $(1-\epsilon)\mathbb{1}/2 + \epsilon|+\rangle\langle+|$) followed by a pair of CNOT gates separated by the delay $t_j - t_i$. Finally σ_x observable of the ancilla qubit is measured in the form of transverse magnetization which reveals the corresponding TTCC [11]:

$$\langle\sigma_x\rangle_{\text{ancilla}} = \text{Tr}[\rho_s \sigma_x(t_i)\sigma_x(t_j)] = C_{ij}, \tag{10}$$

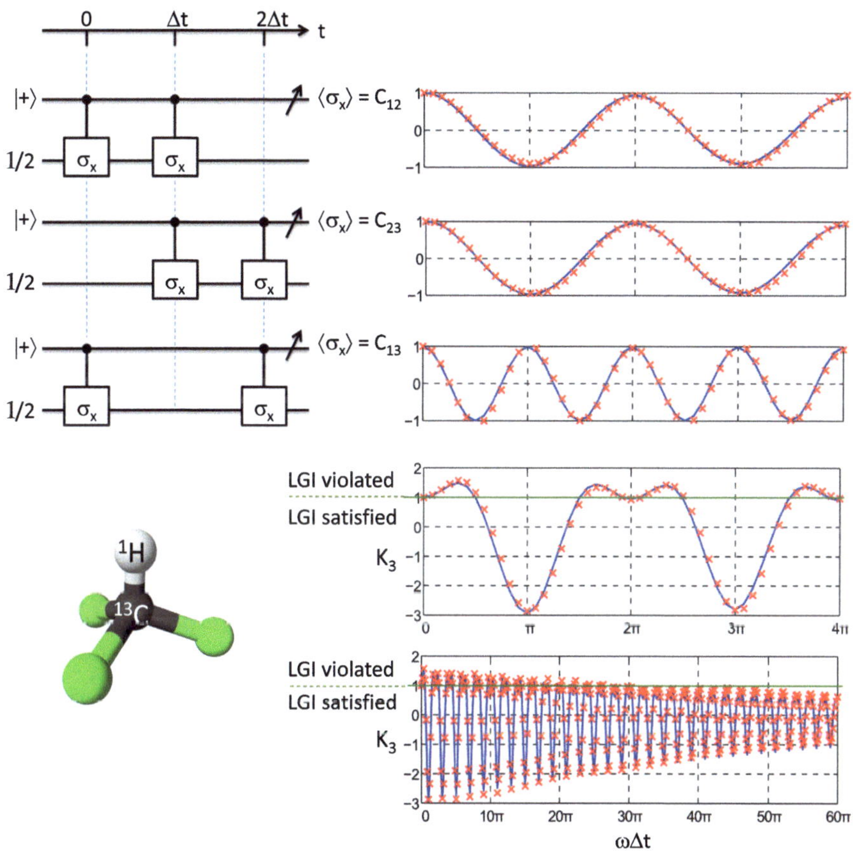

Fig. 3 Moussa circuits (*left*) to extract TTCCs for the three-measurement case and the experimental results (crosses in the *right*) of C_{ij} and K_3 obtained with ^{1}H (ancilla) and ^{13}C (system) spins of chloroform (molecular structure shown in *bottom-left*). Both short-time and long-time behavior of K_3 are shown. Here smooth curves are drawn with theoretical expression (Eq. 8) along with an appropriate decay factor. Parts of this figure are adapted from [32]

where $\rho_s = \mathbb{1}/2$ is the initial state of the system qubit.

The Moussa circuits are easy to implement using a two-qubit NMR system [32, 35]. Athalye et al. [32] have used ^{13}C and ^{1}H spins of ^{13}C-Chloroform as system and ancilla qubits respectively and found a clear violation of LGI by more than 10 standard deviations at short time scales. However, with longer time scales, the TTCCs decayed resulting in a gradual reduction in the violation, and ultimately satisfying the LGI bounds.

More recently, Knee et al. [36] have used ideal negative result measurements (INRM) to extract TTCCs noninvasively. The method involves two sets of experiments - one with CNOT and the other with anti-CNOT. In the former, the system qubit is unaltered if the ancilla (control-qubit) is in state $|0\rangle$, while in the latter, the system is unaltered if the ancilla is in state $|1\rangle$. Postselecting the subspaces wherein the system is unaltered is considered to be more noninvasive [36]. Using nuclear and electronic spins as system and ancilla, Knee et al. demonstrated LGI violation with INRM [36].

3.4 *Entropic Leggett–Garg Inequality (ELGI)*

In 2013, Usha Devi et al. [27] have formulated the entropic Leggett–Garg inequality in which they place bounds on amount of information associated with a noninvasive measurement of a macroscopic system. The amount of information stored in a classical observable $\mathbb{Q}(t_i)$ at time t_i is given by the Shannon entropy,

$$H(\mathbb{Q}(t_i)) = -\sum_{Q(t_i)} P(Q(t_i)) \log_2 P(Q(t_i)), \tag{11}$$

where $P(Q(t_i))$ is the probability of the measurement outcome $Q(t_i)$ at time t_i. The conditional entropy $H(\mathbb{Q}(t_j)|\mathbb{Q}(t_i))$ is related to the joint-entropy

$$H(\mathbb{Q}(t_j), \mathbb{Q}(t_i)) = -\sum_{Q(t_i),Q(t_j)} P(Q(t_i), Q(t_j)) \log_2 P(Q(t_i), Q(t_j)) \tag{12}$$

by Bayes' theorem, i.e.,

$$H(\mathbb{Q}(t_j)|\mathbb{Q}(t_i)) = H(\mathbb{Q}(t_i), \mathbb{Q}(t_j)) - H(\mathbb{Q}(t_i)). \tag{13}$$

For n measurements performed at equal intervals Δt, we denote $h(\Delta t) = H(\mathbb{Q}(\Delta t)|\mathbb{Q}(0)) = H(\mathbb{Q}(2\Delta t)|\mathbb{Q}(\Delta t)) = \cdots$, and $h((n-1)\Delta t) = H(\mathbb{Q}((n-1)\Delta t)|\mathbb{Q}(0))$. By setting up a quantity called information deficit

$$\mathcal{D}_n = \frac{(n-1)h(\Delta t) - h((n-1)\Delta t)}{\log_2(2s+1)}, \tag{14}$$

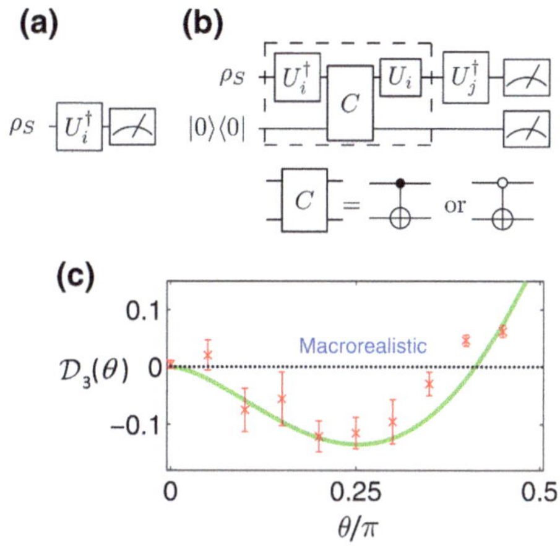

Fig. 4 **a**, **b** The quantum circuits for extracting single-time and joint probabilities. Here $U_i^\dagger$ denotes the back-evolution of the system in the computational basis which is equivalent to having the dynamical observable $\mathbb{Q}(t_i)$. **c** Experimental information deficit (crosses with errorbars) compared to theoretical values (solid curve) for a spin-1/2 particle. The dashed line indicates the macrorealistic bound. Here $\theta = (n-1)\omega\Delta t$. Parts of this figure are adapted from [29]

where $2s + 1$ is the number of distinct states (where s is spin number), Usha Devi et al. proved that $\mathcal{D}_n \geq 0$ for classical systems.

The experimental violation of ELGI was first demonstrated by Katiyar et al. [29] again using ^{13}C-Chloroform as the two-qubit register. The single-time probability $P(Q(t_i))$ and the joint probabilities $P(Q(t_i), Q(t_j))$ are extracted using the circuits shown in Fig. 4a and b respectively. Note that an ancilla spin is used to extract joint probabilities with the help of INRM procedure applied to the first measurement. The results displayed in Fig. 4c, indicate a clear violation of ELGI by four standard deviations.

4 Quantum Discord

In the early days of quantum information and quantum computation it was shown that entanglement is the key resource to perform various tasks [37]. However, it was later realized that quantum correlations beyond entanglement are also useful for quantum information processing [38–41]. It was shown theoretically [4, 42] as well as experimentally [43] that some tasks can be made efficient even with separable states, but with non-zero quantum correlations. Thus, quantifying the quantum correlation becomes important, and it can be achieved by using measures such as discord [44, 45] and geometric discord [46–48]. For more details on the topic of quantum correlations one may refer to the reviews in [49–54].

Discord has also been studied in the ground state of certain spin chains particularly close to quantum phase transitions [55]. Signatures of chaos in the dynamics of quantum discord are found using the model of the quantum kicked top [56]. Quantum critical behavior in the anisotropic XY spin chain is studied using geometric discord [57].

It is believed that discord is a resource behind the efficiency of the DQC-1 model [38, 42, 58–60]. Quantum advantage with no entanglement but with non-zero quantum discord has been demonstrated in single-photon states [61]. Quantum discord has also been estimated in optical systems using mixed states [43] and in an anti ferromagnetic Heisenberg compound [62].

Non-zero quantum discord in NMR systems has been observed by many researchers [63–66]. For various theoretical and experimental aspects of quantum discord and related measures reader can refer to review [67]. Investigations on the evolution of quantum discord under decoherence [68] and under decoherence-suppression sequences [66] have also been reported. In the following we briefly describe some aspects related to discord and geometric discord.

4.1 Discord and Mutual Information

Mutual Information $I(A:B)$ is defined as the amount of information that is common to both the subsystems A and B of a bipartite system, and is given in terms of Shannon entropy

$$I(A:B) = H(A) + H(B) - H(A,B). \tag{15}$$

It can be seen that mutual information is symmetric, i.e., $I(A:B) = I(B:A)$. Another classically equivalent expression based on Bayes rule can be obtained from Eq. (15) as follows:

$$J(A:B) = H(A) - H(A|B) = H(A) - \sum_i p_i^b H(A|b=i)\,. \tag{16}$$

These expressions can be intuitively understood using Fig. 5.

In the quantum information theory, the von Neumann entropy gives the information content of a density matrix and is defined as

$$\mathcal{H}(\rho) = -\sum_x \lambda_x \log_2 \lambda_x, \tag{17}$$

where λ_x's are the eigenvalues of the density matrix ρ. Although the two expressions of mutual information given in Eqs. (15) and (16) are equivalent in classical information theory this is not the case in quantum information theory. The reason for this difference is that the expression for mutual information given by Eq. (16) involves measurements and its value depends on the measurement outcomes. Measurements

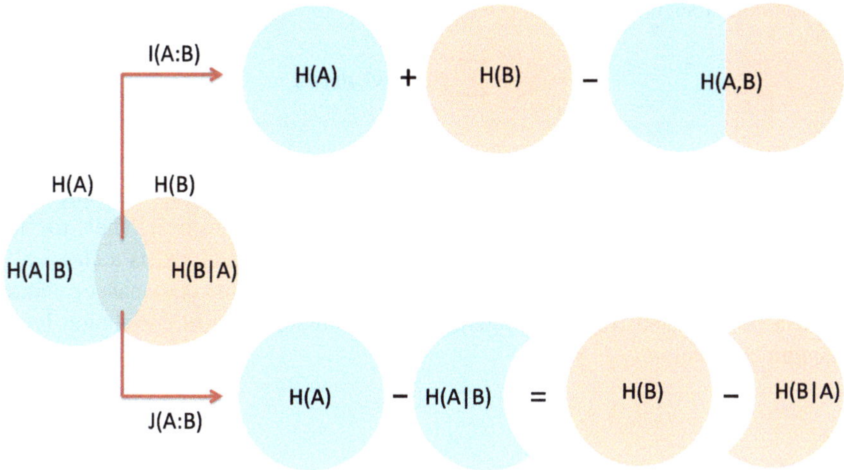

Fig. 5 Venn diagram representing total information $H(A, B)$, individual informations ($H(A)$, $H(B)$), the conditional information ($H(A|B)$, $H(B|A)$), and the mutual information $I(A : B) = J(A : B)$ in classical information theory

in quantum theory depends on the basis used and it changes the final state of the system. Henderson and Vedral [45] have proved that the total classical correlation can be obtained as the maximum value of

$$\mathcal{J}(A : B) = \mathcal{H}(B) - \mathcal{H}(B|A) = \mathcal{H}(B) - \sum_i p_i^a \mathcal{H}(B|a = i), \tag{18}$$

where the maximization is performed over all possible orthonormal measurement bases $\{\Pi_i^a\}$ for A. The quantum mutual information $\mathcal{I}(A : B)$ is defined in a way analogous to that of the classical mutual information, i.e.,

$$\mathcal{I}(A : B) = \mathcal{H}(A) + \mathcal{H}(B) - \mathcal{H}(A, B). \tag{19}$$

Therefore, the non-classical correlations can be quantified as the difference

$$D(B|A) = \mathcal{I}(A : B) - \max_{\{\Pi_i^a\}} \mathcal{J}(A : B). \tag{20}$$

Ollivier and Zurek had called this difference as '*discord*' [44]. Zero-discord states or "classical" states are the ones in which the maximal amount of information about a subsystem can be obtained without disturbing its correlations with the rest of the system.

It should be noted that discord is not a symmetric function in general, i.e. $D(B|A)$ and $D(A|B)$ can differ. Datta [69] has proved that a given state ρ_{AB} satisfies $D(B|A) = 0$ if and only if there exists a complete set of orthonormal measurement operators on A such that

$$\rho_{AB} = \sum_i p_i^a \Pi_i^a \otimes \rho_{B|a=i}. \tag{21}$$

When the first part of a general bipartite system is measured, the resulting density matrix is of the form given by Eq. (21). Since the final state after measurements is a classical state, one can extract the classical correlations from it. Thus, for any quantum state and every orthonormal measurement basis, there exists a classically correlated state. Maximization of $\mathcal{J}(A:B)$ gives the maximum classical correlation that can be extracted from the system, and the remaining extra correlation is the quantum correlation.

4.2 Evaluation of Discord

Given a density matrix ρ_{AB}, one can easily construct the reduced density matrices ρ_A and ρ_B of the individual subsystems. Then the total correlation $\mathcal{I}(A:B)$ can be found using the quantum mutual information Eq. (19). Maximization of $\mathcal{J}(A:B)$ to evaluate discord is nontrivial. The brute force method is to maximize $\mathcal{J}(A:B)$ over as many orthonormal measurement bases as possible, taking into account all constraints and symmetries. Strictly speaking, this method gives a lower bound on $\mathcal{J}(A:B)$ since the maximization may not be perfect.

While a closed analytic formula for discord does not exist for a general quantum state, analytical results are available for certain special classes of states [70]. For example, Chen et al. have described analytical evaluation of discord for two qubit X-states under specific circumstances [71–75]. Luo has given an analytical formula for discord of the Bell-diagonal states which are a subset of the X-states [76], and are defined as the states which are diagonal in the Bell basis

$$|\psi^{\pm}\rangle = \frac{1}{\sqrt{2}}(|01\rangle \pm |10\rangle)\,, \quad |\phi^{\pm}\rangle = \frac{1}{\sqrt{2}}(|00\rangle \pm |11\rangle). \tag{22}$$

The generic structure of a Bell-diagonal state is $\rho_{BD} = \lambda_1|\psi^-\rangle\langle\psi^-| + \lambda_2|\phi^-\rangle\langle\phi^-| + \lambda_3|\phi^+\rangle\langle\phi^+| + \lambda_4|\psi^+\rangle\langle\psi^+|$. This state is separable iff its spectrum lies in $[0, 1/2]$ [77].

Using only local unitary operations (so that the correlations remain unaltered), all Bell-diagonal states can be transformed to the form given by

$$\rho_{BD} = \frac{1}{4}\Big(\mathbb{1} + \sum_{j=1}^{3} r_j \sigma_j \otimes \sigma_j\Big), \tag{23}$$

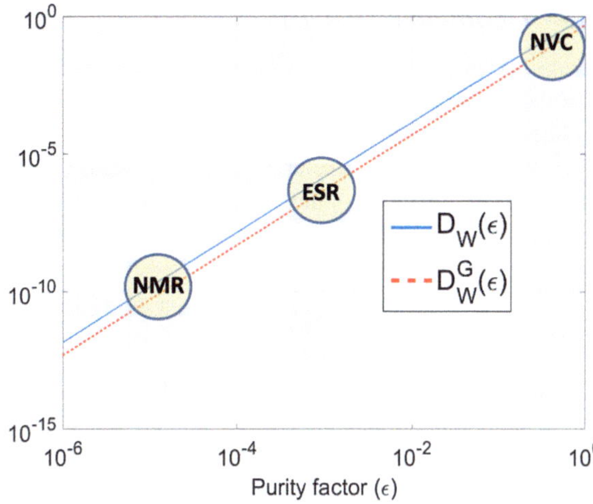

Fig. 6 Discord (D_W) and geometric discord (D_W^G) of Werner state as a function of its purity factor ϵ. Typical ranges of purity and discord values for some spin-based architectures such as NMR, low-field ESR, and optically polarized electronic spin of nitrogen-vacancy center (NVC) are indicated

where the real numbers r_j are constrained such that all eigenvalues of ρ_{BD} remain in [0, 1]. The symmetric form of ρ_{BD} also implies that it has symmetric discord, i.e., $D_{BD}(B|A) = D_{BD}(A|B)$. Thus, the analytical formula for discord in this case is, using Eq. (20),

$$D_{BD}(B|A) = 2 + \sum_{i=1}^{4} \lambda_i \log_2 \lambda_i - \left(\frac{1-r}{2}\right) \log_2(1-r) - \left(\frac{1+r}{2}\right) \log_2(1+r), \tag{24}$$

where $r = \max\{|r_1|, |r_2|, |r_3|\}$.

A special Bell-diagonal state, i.e., when $\lambda_1 = (1+3\epsilon)/4$ and $\lambda_2 = \lambda_3 = \lambda_4 = (1-\epsilon)/4$, is known as the Werner state

$$\rho_W(\epsilon) = \frac{1-\epsilon}{4}\mathbb{1} + \epsilon|\psi^-\rangle\langle\psi^-|. \tag{25}$$

It has entanglement iff $1/3 \leq \epsilon \leq 1$. In this case $r_j = -\epsilon$ for $j = 1, 2, 3$ and $r = \epsilon$. The discord using Eq. (24) is then given by

$$D_W(\epsilon) = \frac{1}{4}\log_2 \frac{(1-\epsilon)(1+3\epsilon)}{(1+\epsilon)^2} + \frac{\epsilon}{4}\log_2 \frac{(1+3\epsilon)^3}{(1-\epsilon)(1+\epsilon)^2} = \frac{\epsilon^2}{\ln 2} + O(\epsilon^3). \tag{26}$$

This expression is plotted in Fig. 6.

4.3 Geometric Discord

Geometric discord is a form of Discord that is relatively easier to compute [46, 47]. In the following, we discuss the case of two-qubit geometric discord [46, 78]. For every quantum state there exist a set of postmeasurement classical states (Ω_0), and the geometric discord is defined as the distance between the quantum state (ρ) and the nearest classical state (χ),

$$D^G(B|A) = \min_{\chi \in \Omega_0} \|\rho - \chi\|^2, \tag{27}$$

where $\|\rho - \chi\|^2 = \mathrm{Tr}[(\rho - \chi)^2]$ is the Hilbert–Schmidt quadratic norm. Obviously, $D^G(B|A)$ is invariant under local unitary transformations. Explicit and tight lower bound on the geometric discord for an arbitrary $A_{m\times m} \otimes B_{n\times n}$ state of a bipartite quantum system is available [47, 79]. Protocols to determine lower bounds on geometric discord without tomography have also been discovered recently [79, 80].

Following the formalism of Dakic et al. [46] analytical expression for the geometric discord for two-qubit states was obtained in [77]. The two-qubit density matrix in the Bloch representation is

$$\rho = \frac{1}{4}\left(\mathbb{1} \otimes \mathbb{1} + \sum_{i=1}^{3} x_i \sigma_i \otimes \mathbb{1} + \sum_{i=1}^{3} y_i \mathbb{1} \otimes \sigma_i + \sum_{i,j=1}^{3} T_{ij} \sigma_i \otimes \sigma_j\right), \tag{28}$$

where x_i and y_i represent the Bloch vectors for the two qubits, and $T_{ij} = \mathrm{Tr}[(\rho(\sigma_i \otimes \sigma_j))]$ are the components of the correlation matrix. The geometric discord for such a state is

$$D^G(B|A) = \frac{1}{4}\left(\|x\|^2 + \|T\|^2 - \eta_{\max}\right), \tag{29}$$

where $\|T\|^2 = \mathrm{Tr}[T^\dagger T]$, and $\eta_{\max}$ is the largest eigenvalue of the matrix $\vec{x}\vec{x}^\dagger + TT^\dagger$. Explicit form of $\eta_{\max}$ and a remarkable tight lower bound on geometric discord are given in [77].

Using the transformed form of Bell-diagonal states as given in Eq. (23) it can be seen that $x_i = y_i = 0$ and T is a diagonal matrix with elements $T_{ii} = r_i$. Then the geometric discord is given as

$$D^G_{BD} = \frac{1}{4}\left(\sum_{i=1}^{3} r_i^2 - \max(r_1^2, r_2^2, r_3^2)\right). \tag{30}$$

For the Werner state $r_i = -\epsilon$. Then $\|T\|^2 = 3\epsilon^2$ and all eigenvalues of $TT^\dagger$ are ϵ^2, yielding

$$D^G_W(\epsilon) = \frac{1}{4}\left(3\epsilon^2 - \epsilon^2\right) = \frac{\epsilon^2}{2}. \tag{31}$$

This expression is plotted versus the purity ϵ in Fig. 6. Comparison with Eq. (26) reveals that discord and geometric discord are proportional for low-purity Werner states. Also, the numerical difference between $D_W(\epsilon)$ and $2D_W^G(\epsilon)$ does not exceed 0.027 for all $\epsilon \in [0, 1]$. An analytical formula for symmetric geometric discord for two-qubit systems is given in [81] and geometric discord for qubit–qudit systems is given in [82].

4.4 NMR Studies of Quantum Discord

Katiyar et al. [66] have studied discord and its evolution in certain NMR systems. After preparing the pseudopure state $\rho_0 = (1-\epsilon)\mathbb{1}/2 + \epsilon|00\rangle\langle 00|$ they applied the pulse sequence shown in Fig. 7a. The initial state ρ_0 is transformed into a Werner state when θ is set to an odd integral multiple of $\pi/2$. Katiyar et al. measured quantum discord using extensive measurement method described earlier. Figure 7b displays discord as a function of θ. One can notice that discord is zero for the initial state ρ_0, grows with θ and reaches a maximum value at the Werner state. This experiment demonstrates the existence of small, but non-zero, nonclassical correlations in NMR systems even at room temperatures.

Maziero et al. studied the behavior of quantum discord under decoherence using an NMR testbed [68]. They observed a sudden change in the behavior of classical and quantum correlations at a particular instant of time and found distinct time intervals where classical and quantum correlations are robust against decoherence. Yurishchev [83] has analytically and numerically studied NMR dynamics of quantum discord in

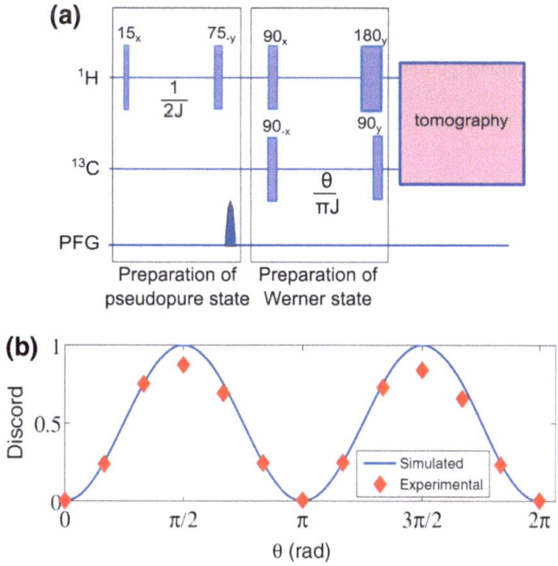

Fig. 7 **a** NMR Pulse-sequence used by Katiyar et al. to prepare Werner state and measure discord and **b** experimental and simulated discord as a function of the nonlocal rotation θ. Parts of this figure are adapted from [66]

gas molecules (with spin) confined in a closed nanopore. Kuznetsova and Zenchuk [84] have theoretically studied quantum discord in a pair of spin-1/2 particles (dimer) governed by the standard multiple quantum NMR Hamiltonian and shown the relation between discord and the intensity of the second-order multiple quantum coherence in NMR systems.

5 Summary

In this chapter, we have briefly discussed three types of quantum correlations, namely quantum contextuality, Leggett–Garg temporal correlations, and quantum discord. In each case, we have surveyed a few NMR experiments.

Exploiting the state independent nature of quantum contextuality, Moussa et al. [11] demonstrated that even a content-less maximally mixed-state ($\mathbb{1}/4$) violates certain noncontextual hidden variable inequalities when subjected to quantum measurements of certain observables. Similarly, the violation of Leggett–Garg inequalities can be observed even in a two-level quantum system (while quantum contextuality is exhibited by a quantum system with at least three levels). Hence, as demonstrated by Athalye et al. [32] the violation of LGI is observable even in a spin-1/2 NMR system at room temperature. Moreover, Maziero et al. [68] and Katiyar et al. [29] showed the existence of nonzero discord in NMR systems.

NMR has wide-ranging applications from spectroscopy to imaging, and quantum information testbed is the latest of them. Although NMR offers excellent control operations and long coherence times, highly mixed nature of spin-ensembles at room temperatures allows only separable quantum states. In the absence of entanglement, does it have any resource for quantum information studies? This question was answered in terms of above nonclassical correlations.

Acknowledgements TSM acknowledges support from DST/SJF/PSA-03/2012-13 and CSIR 03(1345)/16/EMR-II. UTB acknowledges support from DST-SERB-NPDF (File Number PDF/2015/000506).

References

1. A. Abragam, *The Principles of Nuclear Magnetism* (Oxford at the Clarendon press, Oxford, 1961)
2. M.H. Levitt, *Spin Dynamics-Basics of NMR* (Wiley, Chichester, 2008)
3. D.G. Cory, A.F. Fahmy, T.F. Havel, Proc. Natl. Acad. Sci. **94**, 1634 (1997)
4. S.L. Braunstein, C.M. Caves, R. Jozsa, N. Linden, S. Popescu, R. Schack, Phys. Rev. Lett. **83**, 1054 (1999)
5. S. Kochen, E.P. Specker, J. Math. Mech. **17**, 59 (1967)
6. A. Peres, Phys. Lett. A **151**, 107 (1990)
7. A. Peres, *Quantum Theory: Concepts and Methods* (Kluwer Academic Publishers, Dordrecht, 1995)

8. A. Cabello, Phys. Rev. Lett. **101**, 210401 (2008)
9. B. Hensen, H. Bernien, A. Drau, A. Reiserer, N. Kalb, M. Blok, J. Ruitenberg, R.F.L. Vermeulen, R. Schouten, C. Abelln, W. Amaya, V. Pruneri, M. Mitchell, M. Markham, D. Twitchen, D. Elkouss, S. Wehner, T. Taminiau, R. Hanson, Nature **526**, 682686 (2015)
10. N.D. Mermin, Phys. Rev. Lett. **65**, 3373 (1990)
11. O. Moussa, C.A. Ryan, D.G. Cory, R. Laflamme, Phys. Rev. Lett. **104**, 160501 (2010)
12. X. Kong, M. Shi, F. Shi, P. Wang, P. Huang, Q. Zhang, C. Ju, C. Duan, S. Yu, J. Du, arXiv:1210.0961 [quant-ph] (2012)
13. S. Dogra, K. Dorai, Arvind, Phys. Lett. A **380**, 1941 (2016)
14. H.-Y. Su, J.-L. Chen, C. Wu, S. Yu, C.H. Oh, Phys. Rev. A **85**, 052126 (2012)
15. M.A. Nielsen, I.L. Chuang, *Quantum Computation and Quantum Information p115–6* (Cambridge University Press, Cambridge, 2010)
16. H. Katiyar, C.S. Sudheer Kumar, T.S. Mahesh, EPL **113**, 20003 (2016)
17. O. Gühne, M. Kleinmann, A. Cabello, J.-A. Larsson, G. Kirchmair, F. Zähringer, R. Gerritsma, C.F. Roos, Phys. Rev. A **81**, 022121 (2010)
18. A.J. Leggett, A. Garg, Phys. Rev. Lett. **54**, 857 (1985)
19. C. Emary, N. Lambert, F. Nori, Rep. Prog. Phys. **77**, 016001 (2014)
20. A. Palacios-Laloy, F. Mallet, F. Nguyen, P. Bertet, D. Vion, D. Esteve, A.N. Korotkov, Nat. Phys. **6**, 442 (2010)
21. G. Waldherr, P. Neumann, S.F. Huelga, F. Jelezko, J. Wrachtrup, Phys. Rev. Lett. **107**, 090401 (2011)
22. R.E. George, L.M. Robledo, O.J.E. Maroney, M.S. Blok, H. Bernien, M.L. Markham, D.J. Twitchen, J.J.L. Morton, G.A.D. Briggs, R. Hanson, Proc. Natl. Acad. Sci. **110**, 3777 (2013)
23. J. Dressel, C.J. Broadbent, J.C. Howell, A.N. Jordan, Phys. Rev. Lett. **106**, 040402 (2011)
24. C. Emary, N. Lambert, F. Nori, Phys. Rev. B **86**, 235447 (2012)
25. J.P. Groen, D. Ristè, L. Tornberg, J. Cramer, P.C. de Groot, T. Picot, G. Johansson, L. DiCarlo, Phys. Rev. Lett. **111**, 090506 (2013)
26. J.A. Formaggio, D.I. Kaiser, M.M. Murskyj, T.E. Weiss, Phys. Rev. Lett. **117**, 050402 (2016)
27. A.R.U. Devi, H.S. Karthik, Sudha, A.K. Rajagopal, Phys. Rev. A **87**, 052103 (2013)
28. N. Lambert, K. Debnath, A.F. Kockum, G.C. Knee, W.J. Munro, F. Nori, Phys. Rev. A **94**, 012105 (2016)
29. H. Katiyar, A. Shukla, K.R.K. Rao, T.S. Mahesh, Phys. Rev. A **87**, 052102 (2013)
30. M. Łobejko, J. Łuczka, J. Dajka, Phys. Rev. A **91**, 042113 (2015)
31. H. Katiyar, A. Brodutch, D. Lu, R. Laflamme, arXiv:1606.07151v1 [quant-ph] (2016)
32. V. Athalye, S.S. Roy, T.S. Mahesh, Phys. Rev. Lett. **107**, 130402 (2011)
33. J. Kofler, C. Brukner, Phys. Rev. Lett. **99**, 180403 (2007)
34. T. Fritz, New J. Phys. **12**, 083055 (2010)
35. A.M. Souza, I.S. Oliveira, R.S. Sarthour, New J. Phys. **13**, 053023 (2011)
36. G.C. Knee, S. Simmons, E.M. Gauger, J.J.L. Morton, H. Riemann, N.V. Abrosimov, P. Becker, H.-J. Pohl, K.M. Itoh, M.L.W. Thewalt, G.A.D. Briggs, S.C. Benjamin, Nat. Commun. **3**, 606 (2012)
37. R. Horodecki, P. Horodecki, M. Horodecki, K. Horodecki, Rev. Mod. Phys. **81**, 865 (2009)
38. E. Knill, R. Laflamme, Phys. Rev. Lett. **81**, 5672 (1998)
39. C.H. Bennett, D.P. DiVincenzo, C.A. Fuchs, T. Mor, E. Rains, P.W. Shor, J.A. Smolin, W.K. Wootters, Phys. Rev. A **59**, 1070 (1999)
40. M. Horodecki, P. Horodecki, R. Horodecki, J. Oppenheim, A. Sen(De), U. Sen, B. Synak-Radtke, Phys. Rev. A **71**, 062307 (2005)
41. J. Niset, N.J. Cerf, Phys. Rev. A **74**, 052103 (2006)
42. D.A. Meyer, Phys. Rev. Lett. **85**, 2014 (2000)
43. B.P. Lanyon, M. Barbieri, M.P. Almeida, A.G. White, Phys. Rev. Lett. **101**, 200501 (2008)
44. H. Ollivier, W.H. Zurek, Phys. Rev. Lett. **88**, 017901 (2001)
45. L. Henderson, V. Vedral, J. Phys. A Math. Gen. **34**, 6899 (2001)
46. B. Dakić, V. Vedral, C. Brukner, Phys. Rev. Lett. **105**, 190502 (2010)
47. S. Luo, S. Fu, Phys. Rev. A **82**, 034302 (2010)

48. P. Giorda, M.G.A. Paris, Phys. Rev. Lett. **105**, 020503 (2010)
49. K. Modi, A. Brodutch, H. Cable, T. Paterek, V. Vedral, Rev. Mod. Phys. **84**, 1655 (2012)
50. J.-S. Zhang, A.-X. Chen, Quant. Phys. Lett. **1**, 69 (2012)
51. M. Horodecki, J. Oppenheim, Int. J. Mod. Phys. B **27**, 1345019 (2013)
52. K. Modi, Open Syst. Inf. Dyn. **21**, 1440006 (2014)
53. A. Streltsov, *Quantum Correlations Beyond Entanglement and their Role in Quantum Information Theory* (Springer International Publishing, Heidelberg, 2015)
54. G. Adesso, T.R. Bromley, M. Cianciaruso, arXiv:1605.00806 [quant-ph] (2016)
55. M.S. Sarandy, T.R. de Oliveira, L. Amico, Int. J. Mod. Phys. B **27**, 1345030 (2013)
56. V. Madhok, V. Gupta, D.-A. Trottier, S. Ghose, Phys. Rev. E **91**, 032906 (2015)
57. W.W. Cheng, C.J. Shan, Y.B. Sheng, S.M. Gong, L.Y. Zhao, B.Y. Zheng, Phys. E **44**, 1320 (2012)
58. A. Datta, A. Shaji, C.M. Caves, Phys. Rev. Lett. **100**, 050502 (2008)
59. G. Passante, O. Moussa, C.A. Ryan, R. Laflamme, Phys. Rev. Lett. **103**, 250501 (2009)
60. J. Maziero, L.C. Céleri, R.M. Serra, V. Vedral, Phys. Rev. A **80**, 044102 (2009)
61. A. Maldonado-Trapp, P. Solano, A. Hu, C.W. Clark, arXiv:1604.07351 [quant-ph] (2016)
62. H. Singh, T. Chakraborty, P.K. Panigrahi, C. Mitra, Quant. Inf. Process. **14**, 951 (2015)
63. D.O. Soares-Pinto, L.C. Céleri, R. Auccaise, F.F. Fanchini, E. R. deAzevedo, J. Maziero, T. J. Bonagamba, and R. M. Serra. Phys. Rev. A **81**, 062118 (2010)
64. G. Passante, O. Moussa, D.A. Trottier, R. Laflamme, Phys. Rev. A **84**, 044302 (2011)
65. R. Auccaise, J. Maziero, L.C. Céleri, D.O. Soares-Pinto, E.R. deAzevedo, T.J. Bonagamba, R.S. Sarthour, I.S. Oliveira, R.M. Serra, Phys. Rev. Lett. **107**, 070501 (2011)
66. H. Katiyar, S.S. Roy, T.S. Mahesh, A. Patel, Phys. Rev. A **86**, 012309 (2012)
67. L.C. Celeri, J. Maziero, R.M. Serra, Int. J. Quant. Inform. **09**, 1837 (2011)
68. J. Maziero, R. Auccaise, L.C. Céleri, D.O. Soares-Pinto, E.R. deAzevedo, T.J. Bonagamba, R.S. Sarthour, I.S. Oliveira, R.M. Serra, Braz. J. Phys. **43**, 86 (2013)
69. A. Datta, Ph.D. thesis, The University of New Mexico, arXiv:0807.4490; arXiv:1003.5256
70. D. Girolami, G. Adesso, Phys. Rev. A **83**, 052108 (2011)
71. T. Yu, J.H. Eberly, Quantum Inf. Comput. **7**, 459 (2007)
72. A.R.P. Rau, J. Phys. A Math. Theor. **42**, 412002 (2009)
73. M. Ali, A.R.P. Rau, G. Alber, Phys. Rev. A **81**, 042105 (2010)
74. F.F. Fanchini, T. Werlang, C.A. Brasil, L.G.E. Arruda, A.O. Caldeira, Phys. Rev. A **81**, 052107 (2010)
75. Q. Chen, C. Zhang, S. Yu, X.X. Yi, C.H. Oh, Phys. Rev. A **84**, 042313 (2011)
76. S. Luo, Phys. Rev. A **77**, 042303 (2008)
77. R. Horodecki, M. Horodecki, Phys. Rev. A **54**, 1838 (1996)
78. Z. Huang, D. Qiu, Quantum Inf. Process. **15**, 1979 (2016)
79. A.S.M. Hassan, B. Lari, P.S. Joag, Phys. Rev. A **85**, 024302 (2012)
80. S. Rana, P. Parashar, Phys. Rev. A **85**, 024102 (2012)
81. J. Feng-Jian, L. Hai-Jiang, Y. Xin-Hu, S. Ming-Jun, Chin. Phys. B **22**, 040303 (2013)
82. S. Vinjanampathy, A.R.P. Rau, J. Phys. A Math. Theor. **45**, 095303 (2001)
83. M.A. Yurishchev, J. Exp. Theor. Phys. **119**, 828 (2014)
84. E. Kuznetsova, A. Zenchuk, Phys. Lett. A **376**, 1029 (2012)

NMR Contributions to the Study of Quantum Correlations

Isabela A. Silva, Jefferson G. Filgueiras, Ruben Auccaise, Alexandre M. Souza, Raimund Marx, Steffen J. Glaser, Tito J. Bonagamba, Roberto S. Sarthour, Ivan S. Oliveira and Eduardo R. deAzevedo

Abstract In this chapter we review the contributions of Nuclear Magnetic Resonance to the study of quantum correlations, including its capabilities to prepare initial states, generate unitary transformations, and characterize the final state. These are the three main demands to implement quantum information processing in a physical system, which NMR offers, nearly to perfection, though for a small number of qubits. Our main discussion will concern liquid samples at room temperature.

1 NMR Fundamentals

1.1 Classical NMR

Nuclear Magnetic Resonance (NMR) was first reported in 1939 by Isidor Rabi and co-workers, as a method to measure nuclear magnetic moments, a work inspired by the earlier Stern-Gerlach experiment [1]. For that work Rabi received the Nobel Prize in Physics in 1944. In 1946 Felix Bloch and Edward Purcell [2, 3] demonstrated, independently, the NMR phenomenon in matter, solid and liquid, and reached an adequate parametric mathematical formulation known as the *Bloch Equations*. For that work they received the Nobel Prize in Physics in 1952. In 1950 a giant step

I.A. Silva · J.G. Filgueiras · T.J. Bonagamba · E.R. deAzevedo
Instituto de Física de São Carlos, Universidade de São Paulo, 369, São Carlos, SP 13560-970, Brazil

R. Marx · S.J. Glaser
Department of Chemistry, Technische Universität München, Lichtenbergstr. 4, 85747 Garching, Germany

A.M. Souza · R.S. Sarthour · I.S. Oliveira (✉)
Centro Brasileiro de Pesquisas Físicas, Rua Dr. Xavier Sigaud 150, Rio de Janeiro, RJ 22290-180, Brazil
e-mail: ivan@cbpf.br

R. Auccaise
Department of Fis, Universidade Estadual Ponta Grossa, Ponta Grossa, PR 84030900, Brazil

F.F. Fanchini et al. (eds.), *Lectures on General Quantum Correlations and their Applications*, Quantum Science and Technology,
DOI 10.1007/978-3-319-53412-1_24

was given by Erwin Hahn, a step that would turn NMR, in the years to come, into one of the main experimental techniques in Physics, Chemistry and Biology with a revolutionary application to Medicine. Hahn discovered the phenomenon of *spin echoes* [4], inaugurating *Pulsed NMR*. For further developments and applications of pulsed NMR, the Nobel Prize in Chemistry in 1991 was granted to Richard Ernst [5]. In 2002 another NMR Nobel Prize in Chemistry was awarded to Kurt Wuthrich [6] and, in 2003, the Nobel Prize in Medicine went to Paul Laterbur and Peter Mansfield for the discovery of the NMR imaging technique [7, 8]. Therefore, since its discovery, NMR has been awarded five Nobel Prizes, three of them due to contributions after the discovery of its pulsed version.

In a basic NMR experiment [9], an ensemble of nuclear magnetic moments are subject to a magnetic field given by:

$$\mathbf{B}(t) = B_0\mathbf{k} + B_1\{\cos(\omega t)\mathbf{i} + \sin(\omega t)\mathbf{j}\} \tag{1}$$

In this expression, B_0 is the magnitude of a homogeneous static magnetic field, typically of the order of 10 Tesla, whereas B_1 is the magnitude of a *radiofrequency field* (RF), typically four to five orders of magnitude below B_0. Therefore, B_1 can be considered a perturbation over B_0. It is worth mentioning that, although B_0 is considered homogeneous in the above equation, the description of the NMR phenomenon necessarily include a field *inhomogeneity* [9].

In the classical description of NMR the field described by Eq. (1) interacts with the nuclear magnetization producing a torque on it. The magnetization rotates with a characteristic frequency given by $\omega_0/2\pi = (\gamma_n/2\pi)B_0$, where γ_n is a nuclear parameter called *gyromagnetic ratio*, different for each isotopic specimen. This frequency, called the *Larmor frequency* of the system, can range from a few to many hundreds of MHz. The presence of time-dependent terms in the field complicates the description of the time evolution of the magnetization in the laboratory frame. Fortunately, due to the special geometrical arrangement, it is possible to make a transformation to a rotating frame in which the total field is static [9]. By adding the relaxation terms, we arrive at the Bloch Equations in the rotating frame, which can be conveniently written in the matrix form:

$$\frac{\partial \mathbf{M}}{\partial t} + \tilde{\mathbf{A}}\mathbf{M} = \mathbf{f} \tag{2}$$

where:

$$\tilde{\mathbf{A}} = \begin{pmatrix} 1/T_2 & -\Delta\omega & 0 \\ +\Delta\omega & 1/T_2 & -\omega_1 \\ 0 & \omega_1 & 1/T_1 \end{pmatrix}; \quad \mathbf{M} = \begin{pmatrix} M_x \\ M_y \\ M_z \end{pmatrix}; \quad \mathbf{f} = \begin{pmatrix} 0 \\ 0 \\ M_0/T_1 \end{pmatrix} \tag{3}$$

In this equation, M_0 is the equilibrium magnetization, T_1 and T_2 are, respectively, the longitudinal and transverse relaxation times, and $\Delta\omega = \omega - \omega_0$ is the *offset* between the RF and the resonance frequencies. The resonance condition is given by $\omega = \omega_0$. Finally, $\omega_1 = \gamma_n B_1$ is the rotating frequency of the magnetization about B_1 in the rotating frame. From the microscopic point of view, the longitudinal relaxation describes processes in which the nuclear spins interact with a bath and decay towards equilibrium by releasing energy (heat). The transverse relaxation is more subtle: it describes processes in which the nuclear magnetization loses coherence (quantum and classical) due to the random interactions between the nuclear magnetic moments; the energy is conserved and there is no heat transfer to the bath in the process.

The general solution of Eq. (2) is:

$$\mathbf{M}(t) = \mathbf{M}_\infty + \exp(-\tilde{\mathbf{A}}t) \times \{\mathbf{M}(0) - \mathbf{M}_\infty\} \tag{4}$$

where $\mathbf{M}_\infty$ is the stationary solution:

$$\mathbf{M}_\infty = \tilde{\mathbf{A}}^{-1}\mathbf{f} \tag{5}$$

We see that the dynamics of the magnetization is governed by the exponential operator on the transient (second) term of Eq. (4). In spite of its apparent simplicity, it is not an easy task to obtain an analytical expression for the magnetization for arbitrary T_1, T_2 and $\Delta\omega$ [52]. Figure 1 shows the trajectory of the magnetization, calculated numerically from Eq. (4), for three different regimes of relaxation: slow, intermediate and fast, and $\Delta\omega \neq 0$. Figure 2 shows the time evolution of the components of the magnetization with transient and stationary regimes. We see that at resonance, $M_x = 0$ and the magnetization rotates only about the field B_1. We also see that in the regime of fast relaxation and/or low RF power (small B_1^2), $M_z \approx M_0$, which means the system absorbs RF energy and releases it very fast as heat to the bath. On the other hand, in the regime of negligibly slow relaxation, the solutions of Bloch equations can be found easily:

$$\begin{aligned} M_x(t) &= -2M_0\frac{\omega_1\Delta\omega}{\Omega^2}\sin^2\left(\frac{\Omega t}{2}\right) \\ M_y(t) &= -M_0\frac{\omega_1}{\Omega}\sin\left(\frac{\Omega t}{2}\right) \\ M_z(t) &= M_0\left[1 - 2\frac{\omega_1^2}{\Omega^2}\sin^2\left(\frac{\Omega t}{2}\right)\right] \end{aligned} \tag{6}$$

where $\Omega = \sqrt{\Delta\omega^2 + \omega_1^2}$. As long as T_1 is much longer than the duration of a RF pulse, we can consider the system isolated. In this regime, using the above results we can calculate the work done by a RF pulse of duration τ to rotate the magnetization. This will be simply the difference between final and initial internal energy:

$$W = U_{final} - U_{initial} = -B_0M_z(\tau) + B_0M_0 - 2B_0M_0\frac{\omega_1^2}{\Omega^2}\sin^2\left(\frac{\Omega\tau}{2}\right) \tag{7}$$

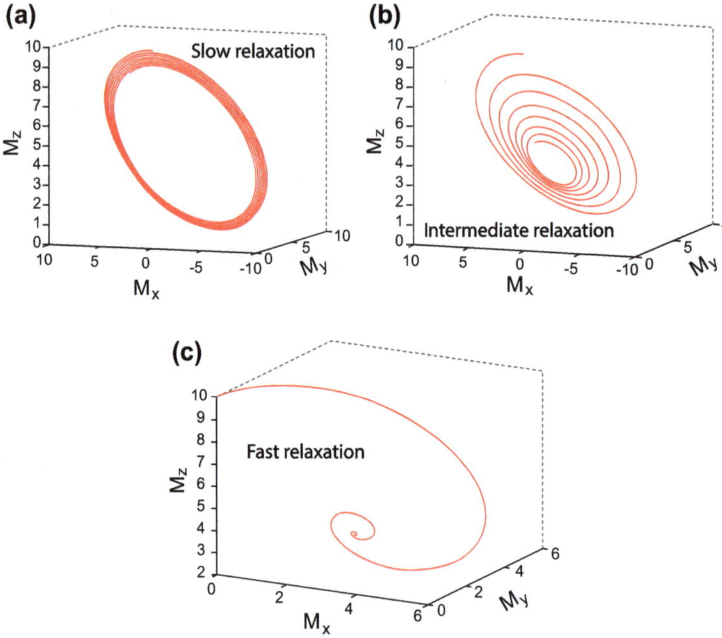

Fig. 1 Trajectories of the magnetization for different relaxation regimes

1.2 Quantum NMR

NMR has also a very cool quantum description. For an isolated spin-1/2, which encodes a quantum bit (qubit) in quantum information processing experiments, under a rotating framc hamiltonian [9]:

$$H = -\frac{1}{2}\hbar\Omega\sigma_u \tag{8}$$

where σ_u is the component of the spin along the direction of the effective field:

$$\sigma_u = \frac{\Delta\omega}{\Omega}\sigma_z - \frac{\omega_1}{\Omega}\sigma_x \tag{9}$$

Assuming that a spin is initially in the state $|\uparrow\rangle$, aligned with B_0, a RF pulse of duration τ takes the state to:

$$|\psi(\tau)\rangle = e^{-i(\Omega\tau/2)\sigma_u}|\uparrow\rangle = \left\{I\cos\left(\frac{\Omega\tau}{2}\right) - i\frac{\Delta\omega}{\Omega}\sin\left(\frac{\Omega\tau}{2}\right)\right\}|\uparrow\rangle + \frac{\omega_1}{\Omega}\sin\left(\frac{\Omega\tau}{2}\right)|\downarrow\rangle \tag{10}$$

The expectation value of the z component, and energy at τ are:

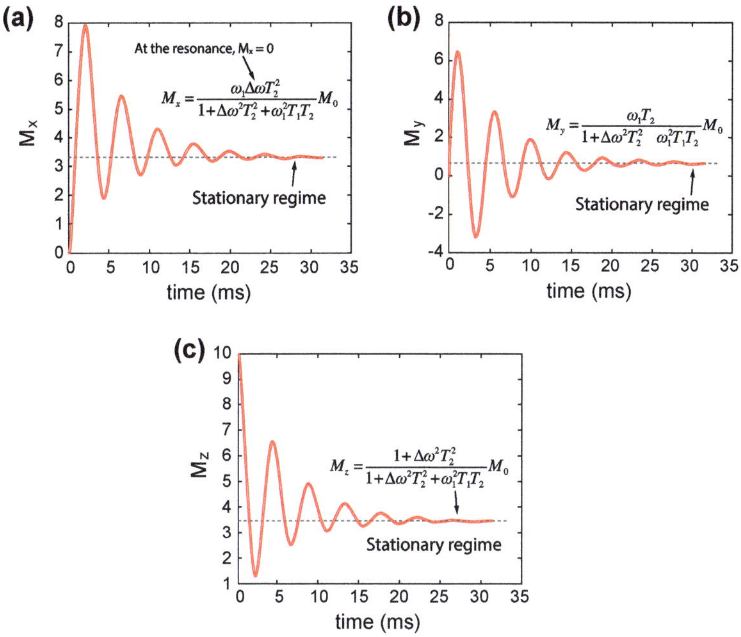

Fig. 2 Time evolution of the magnetization with transient and stationary regimes

$$\langle \sigma_z \rangle(\tau) = \langle \psi(\tau)|\sigma_z|\psi(\tau)\rangle = 1 - 2\frac{\omega_1^2}{\Omega^2}\sin^2\left(\frac{\Omega\tau}{2}\right) \tag{11}$$

and

$$E_f = -\frac{\hbar\omega_0}{2} + \hbar\omega_0\frac{\omega_1^2}{\Omega^2}\sin^2\left(\frac{\Omega\tau}{2}\right) \tag{12}$$

Thcrcforc, thc work produccd by thc pulsc is:

$$W = E_f - E_i = \hbar\omega_0\frac{\omega_1^2}{\Omega^2}\sin^2\left(\frac{\Omega\tau}{2}\right) \tag{13}$$

which is the same as the classical result, Eq. (7), if we remember that for a spin-1/2, $\hbar\omega_0 = 2\mu_N B_0$, where μ_N is the nuclear magneton.

The above calculation was made for an isolated spin. If we allow a thermal contact with a bath at equilibrium the initial magnetization will be $\overline{\langle\sigma_z\rangle}_0$. Besides, if there are fluctuations in the work performed by the pulse, the *average work* will be given by [12]:

$$\langle W \rangle = 2\overline{\langle\sigma_z\rangle}_0 B_0\frac{\omega_1^2}{\Omega^2}\sin^2\left(\frac{\Omega\tau}{2}\right) \tag{14}$$

That is, in the presence of fluctuations, the quantum average work equals the classical expression in the absence of relaxation. An equivalent quantum expression for the average work in the presence of relaxation, is still lacking in the literature.

NMR samples for quantum information processing applications are mainly liquids, typically 0.6 cc, in a glass tube at room temperature. The tube is positioned at the center of a RF coil, which is in turn placed in a static magnetic field. After a few seconds the nuclear spins reach thermal equilibrium in the field, and the initial state of the system is given by the thermal density matrix:

$$\rho_{eq} = \frac{e^{-H/k_B T}}{\mathcal{Z}} \tag{15}$$

where $\mathcal{Z}$ is the partition function. The energy scale of the hamiltonian is $\mu_N B_0 \approx 10^{-6}$ eV, much smaller than thermal energy, $k_B T \approx 0.01$ eV for room temperature. Therefore, the equilibrium density matrix can be expanded to first order:

$$\rho_{eq} \approx \frac{I}{2^N} - \frac{H}{2^N k_B T} \tag{16}$$

where I is the identity matrix, $\mathcal{Z} \approx 2^N$ and N is the number of quantum bits (qubits) in the system. The application of RF to the sample can generally be represented by an unitary transformation, U, of the density matrix:

$$\rho' = U \rho_{eq} U^\dagger = \frac{I}{2^N} - \frac{U H U^\dagger}{2^N k_B T} \tag{17}$$

Therefore, NMR state transformations turn out to be hamiltonian transformations. If after the transformation the mean value of the spin component σ_u is measured, we get the nuclear magnetization in the direction u:

$$M_u = \mathrm{Tr}\left\{\sigma_u \rho'\right\} = -\frac{1}{2^N k_B T} \mathrm{Tr}\left\{\sigma_u U H U^\dagger\right\} \tag{18}$$

This result is extremely important for NMR quantum information processing: the measured signal is not sensitive to the first term, proportional to the identity. This term represents a huge amount of noise, but which is simply invisible to NMR! By combining unitary transformations and space or time averages, the state (15) can be transformed to [13]:

$$\rho_{pp} = \frac{1-\epsilon}{2^N} I + \epsilon |\psi\rangle\langle\psi| \tag{19}$$

where $\epsilon \approx \hbar\omega_0 / k_B T \approx 10^{-6}$ for liquid samples at room temperature at magnetic field of the order of 10 Tesla, and $|\psi\rangle$ a pure quantum state. This is a non-equilibrium state known as *pseudopure state*. It is so-called because under unital operations it transforms as a pure state. Besides, if a measurement of $\langle\sigma_u\rangle$ is performed for this state, the detected signal will come only from the $\epsilon|\psi\rangle\langle\psi|$ part of the pseudopure state:

$$M_u = \epsilon \mathrm{Tr}\{\sigma_u|\psi\rangle\langle\psi|\} \tag{20}$$

The fact that ϵ is small is immaterial for both, transformation of states and detection of NMR signal. In a standard NMR quantum information processing task, the initial density matrix is the pseudopure state:

$$\rho_{pp}^0 = \frac{1-\epsilon}{2^N} I + \epsilon|00\cdots000\rangle\langle00\cdots00|, \tag{21}$$

a protocol represented by unitary transformations $U_1, U_2, \ldots U_N$ is applied:

$$\rho_{pp}' = \frac{1-\epsilon}{2^N} I + \epsilon\left\{U_N \cdots U_2 U_1|00\cdots0\rangle\langle00\cdots00|U_1^\dagger U_2^\dagger \cdots U_N^\dagger\right\}, \tag{22}$$

and the NMR signal is measured:

$$M_u = \epsilon \mathrm{Tr}\left[\sigma_u\left\{U_N \cdots U_2 U_1|00\cdots0\rangle\langle00\cdots00|U_1^\dagger U_2^\dagger \cdots U_N^\dagger\right\}\right] \tag{23}$$

By performing multiple measurements, the full density matrix can be reconstructed, the process called *Quantum State Tomography* [13]. Among the many examples in the literature on the application of this general procedure are NMR testing of Bell-inequality violation [15], violation of Leggert–Garg inequality [16], and an NMR quantum complementarity principle study [17, 18].

2 NMR Classical and Quantum Correlations

The NMR observable is the transverse nuclear magnetization, following a sequence of RF pulses. It is a classical quantity. Suppose a 90° pulse is applied along the x axis to an ensemble of nuclear magnetic moments initially at thermal equilibrium with a static and homogeneous magnetic field along the z axis. If the initial magnetization is M_0, after the pulse the magnetization will be $M_y = M_0$ and $M_z = 0$. Neglecting longitudinal relaxation, in the absence of field inhomogeneity and spin-spin interactions, the magnetization would rotate permanently about the z direction, maintaining its initial magnitude M_0. That is, all the magnetic moments in the sample, something like 10^{18} along the sensitive region of the sample holder, would rotate in perfect synchronism keeping their relative initial phase difference. This is a classical NMR coherent state. However, because field inhomogeneities and spin-spin interactions are always present, the initial coherent state $M_y = M_0$ will dephase, first due to field inhomogeneity and then due to spin-spin interaction. The first effect is reversible in a spin-echo experiment, but the second is not. It is worth mentioning that until quite recently the debate whether the spin-echo phenomenon violates the Second Law of Thermodynamics could be found in the literature [19, 20].

Those two independent relaxation phenomena are well described applying the operator-sum formalism [10], which is possible since every transformation that is given by a complete positive map admits a representation like

$$\sigma(t) = \sum_k E_k(t)\sigma(0)E_k^\dagger(t), \tag{24}$$

with $E_k(t)$ being the Kraus operators satisfying

$$\sum_k E_k^\dagger(t)E_k(t) = 1. \tag{25}$$

The transverse relaxation process in liquid state NMR is exactly analogous to the quantum information's phase damping (PD) channel, mathematically described as

$$E_1 = \sqrt{1-\frac{q(t)}{2}}\mathbb{I}, \quad E_2 = \sqrt{\frac{q(t)}{2}}\begin{bmatrix} 1 & 0 \\ 0 & -1 \end{bmatrix}, \tag{26}$$

where $q(t) = 1 - e^{-t/T_2}$, and T_2 being the NMR transversal relaxation characteristic time.

Moreover, the longitudinal relaxation is also known as Generalized Amplitude Damping (GAD) channel and has the mathematical description

$$\begin{aligned} E_1 = \sqrt{p}\begin{bmatrix} 1 & 0 \\ 0 & \sqrt{1-\gamma} \end{bmatrix}, \ E_2 = \sqrt{p}\begin{bmatrix} 0 & \sqrt{\gamma} \\ 0 & 0 \end{bmatrix} \\ E_3 = \sqrt{1-p}\begin{bmatrix} \sqrt{1-\gamma} & 0 \\ 0 & 1 \end{bmatrix}, \ E_4 = \sqrt{1-p}\begin{bmatrix} 0 & 0 \\ \sqrt{\gamma} & 0 \end{bmatrix}, \end{aligned} \tag{27}$$

where $\gamma = 1 - e^{-t/T_1}$, $p \approx (1-\alpha)/2$, $\alpha = \hbar\omega_L/k_B T$ and T_1 the NMR longitudinal relaxation characteristic time.

2.1 NMR Entanglement

NMR contributions to quantum information processing appeared right after the discovery of pseudopure states in 1997 [21, 22]. In the year after, the question of NMR entanglement was raised [23], and further elaborated [24]. From Eq. (20) we see that the signal measured from a peudopure state is proportional to spins in a single quantum state. Therefore if $|\psi\rangle$ is entangled, the NMR signal will bring the signature of spins in an entangled state, as demonstrated by quantum state tomography in various works [25]. However, if we consider the whole density matrix, Eq. (19), for ϵ below a threshold, the state will be separable. As an example consider two qubits in a pseudo-entangled state. From (20):

$$\rho_{pp} = \begin{pmatrix} (1+\epsilon)/4 & 0 & 0 & 0 \\ 0 & (1+\epsilon)/4 & -\epsilon/2 & 0 \\ 0 & -\epsilon/2 & (1+\epsilon)/4 & 0 \\ 0 & 0 & 0 & (1+\epsilon)/4 \end{pmatrix} \tag{28}$$

Calculating the eigenvalues of the partially transposed matrix, we obtain $\lambda_1 = \lambda_2 = \lambda_3 = (1+\epsilon)/4$ and $\lambda_4 = (1-3\epsilon)/4$. Therefore, according to Peres criterium [26], the state will be entangled for $\epsilon > 1/3$, much above that for room temperature liquid-state samples.

The question of NMR entanglement is quite subtle, and the whole thing has to do with the value of ϵ, which does not affect neither the unitary transformations, nor the measured signal besides its intensity. For instance, in Ref. [15] the NMR protocol for an experiment of Bell inequality violation produces a result which is indistinguishable from those obtained in a quantum optics experiment or quantum mechanical prediction. For that comparison the NMR data are normalized by the factor ϵ. Of course, if the normalization is not performed the curves cannot be compared with each other. The equivalent procedure in a quantum optics experiment is called *post-selection*, in which pairs of entangled photons are selected out from the total number of detected particles. If in a such experiment one pair of entangled photons is detected with probability 10^{-6} we would have the analog of a NMR experiment [27].

It has been shown that some of the aspects of entangled states cannot be tested by NMR, such as nonlocality [28]. However, the fact that NMR is capable of implementing all the basic steps for quantum information processing, state preparation, unitary evolution and quantum state tomography, makes the technique an unique laboratory system to test quantum information protocols in small systems. A non-exhaustive compilation of NMR earlier works involving entanglement can be found in [13]. For a recent compilation of NMR Quantum Information Processing from various research groups in the World, see [14].

One key aspect of NMR to study quantum correlations, in general, and entanglement, in particular, is the ability to prepare very specific initial states. In a two-qubit state scenario, an important class of states are the Bell diagonal states, mathematically written as

$$\rho = \frac{1}{4}\left[\mathbb{I} + \sum_{i=1}^{3} c_i(\sigma_i \otimes \sigma_i)\right], \tag{29}$$

where a full description only depends on the correlation triple $\vec{c} = \{c_1, c_2, c_3\}$, with $c_i = \mathrm{Tr}\{\rho(\sigma_i \otimes \sigma_i)\}$. In Ref. [11], this class of states was first prepared in NMR,

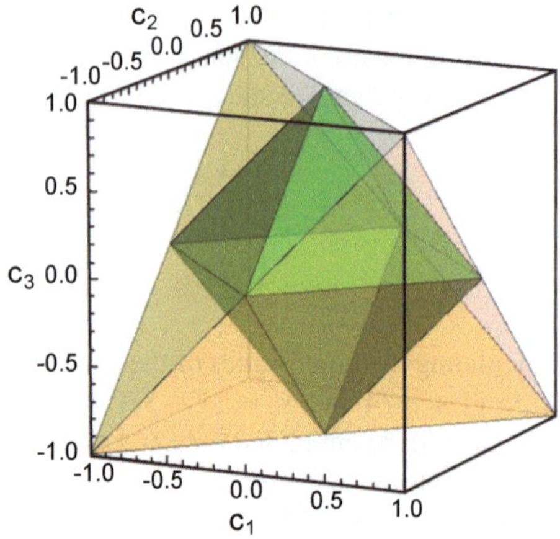

Fig. 3 Bell diagonal states geometry. The outside tetrahedron is the set of physical Bell diagonal states. The inside octahedron, characterized by $|c_1| + |c_2| + |c_3| \leq 1 (\lambda_l \leq 1/2)$ are the separable Bell diagonal states. The physical states outside the octahedron are entangled states. The maximally entangled Bell states are in the tetrahedron vertices. The classical states (diagonal in the product basis) belong to the cartesian axes

and the pulse sequence implemented, described in Fig. 6A, can be easily changed to produce a desired Bell diagonal state.

Figure 3 shows all physical Bell states. The states in the vertices are maximally entangled states. States inside the octahedron are separable, however presenting non-zero quantum correlations, measured by quantum discord, the subject of the next section.

In a more general scenario, this class of two-qubit states can be generalized to

$$\rho = \frac{1}{2^N}\left[\mathbb{I}^{\otimes N} + \sum_{i=1}^{3} c_i(\sigma_i^{\otimes N})\right], \tag{30}$$

which was named M_N^3 states [50], meaning N-qubit states with maximally mixed marginals. This general case will present similar properties as the Bell diagonal (M_2^3) one, as we shall discuss in the next sections.

2.2 NMR Discord

Correlation is a key concept in statistics. Two random variables are said to be correlated when knowledge about one of them can be gained by measuring the other. Since the seminal work of Shannon [36], information is quantified by *entropy*, a quantity which appears with different names in diverse contexts: thermodynamical entropy, statistical entropy, Shannon entropy, von Neumann entropy. The concept was proposed by Rudolf Clausius (1822–1888) as a "measure of the energy" in a thermodynamical system not available for the realization of work. The statistical entropy connects microscopic dynamics with macroscopic thermodynamical quantities. Shannon entropy is measured in bits: the entropy of one bit of information is equal to 1. But it is only the von Neumann entropy which captures the correlations present in quantum states.

It is possible to quantify quantum correlation, even in thermal systems in the presence of noise. Let us recall the definition of Shannon entropy, associated to two dichotomic random variables X and Y which can take the values $\{x, y\}$ with probabilities $\{p_x, p_y\}$:

$$S_X = -\sum_i p_x \log_2 p_x \quad \text{and} \quad S_Y = -\sum_i p_y \log_2 p_y \tag{31}$$

$S_{X(Y)}$ quantifies our uncertainty about $X(Y)$: the larger the entropy, the less we know about the system. If X and Y are correlated, a measurement of Y will yield information about X. Then, our knowledge about X must be "updated"; the new entropy of X after we got to know Y is represented by $S_{X|Y}$:

$$S_{X|Y} = S_{X,Y} - S_Y \tag{32}$$

where:

$$S_{X,Y} = -\sum_i p_{x,y} \log_2 p_{x,y} \tag{33}$$

In this expression, $p_{x,y}$ is the joint probability of obtaining x in a measurement of X, and y in a measurement of Y.

The content of information which belongs to both, X and Y, is called *mutual information*, $S_{X:Y}$, defined by:

$$S_{X:Y} = S_X + S_Y - S_{X,Y} = S_X - S_{X|Y} \tag{34}$$

For classical states the above equality is always valid, but for quantum correlated states it is not! For two-qubits an entangled state, for instance, $S_{X,Y} = 0$, whereas $S_X = S_Y = 1$. On the other hand, $S_{X|Y} = 0$. The difference between the classical and quantum mutual information is called *discord* [37]. A thorough review about discord and NMR is made in [38]. For a two-qubit system, ρ_{AB}, in a Bell diagonal state, Luo S. [39] found a simple analytical expression for this entropic-based quantum discord given by

$$D(\rho_{AB}) = 2 + \sum_{k=0}^{3} \lambda_k \log_2 \lambda_k - \frac{1-c}{2}\log_2(1-c) - \frac{1+c}{2}\log_2(1+c), \tag{35}$$

where λ_k is the k-th eigenvalue of ρ_{AB} and $c = \max\{|c_1|, |c_2|, |c_3|\}$.

However, as this entropic formulation depends on numerical extremizations, analytical expressions are known for only few classes of states. For that reason, and based on which were previously defined for entanglement measurements, discord quantifiers based on geometric arguments were proposed [31]. By definition, entanglement captures the non-separability degree of a global state ρ. An entanglement geometric quantifier is calculated through the distance between a state and its closest separable state, σ, which can be written as a convex combination of product states,

$$\sigma = \sum_{ij} p_{ij} \rho_i^A \otimes \rho_j^B \tag{36}$$

Analogously, a discord geometric quantifier is defined based on the distance between a state and its closest classical state. This classicality is associated with the application of a local projective measurement, then, in the two-qubit system example, this measurement can be applied on a subsystem only (asymmetric discord-type measures) or on both of them (symmetric discord-type measures). For an asymmetric discord-type geometric measure the distance is calculated to the closest classical-quantum state, defined as

$$\chi = \sum_{i} p_i \left|i\right\rangle \left\langle i\right| \otimes \sigma_i^B \tag{37}$$

The symmetric version, however, requires a classical-classical state

$$\chi = \sum_{i,j} p_{i,j} \left|i\right\rangle \left\langle i\right|^A \otimes \left|j\right\rangle \left\langle j\right|^B \tag{38}$$

To calculate this distance it is necessary to choose a metric, that is why a whole set of geometric-based discord quantifiers can be found in the literature nowadays. However, only some of them are proved to be *bona fide* [42].

2.2.1 NMR Discord and Relaxation Effects

Discord is an extremely useful quantity to detect quantum correlation. Figure 4, taken from Ref. [30] shows the time evolution of classical and quantum correlations of a two-qubit system under natural decoherence of a carefully prepared initial state. The picture shows the respective transverse relaxation times of the two-qubit system.

In Ref. [29] three general types of dynamics were identified for this two-qubit system under natural decoherence (phase flip, bit flip and bit-phase flip channels). These three categories depend on the relation between the correlation matrix elements. For the NMR natural phase flip decoherence, the three different types of dynamics are observed for (i) $|c_3| \geq |c_1|, |c_2|$, (ii) $|c_3| = 0$ and, the most interesting case, (iii) $|c_1| \geq |c_2|, |c_3|$ or $|c_2| \geq |c_1|, |c_3|$ and $|c_3| \neq 0$. Figure 5 (taken from Ref. [38]) shows these three dynamical possibilities for classical, quantum correlations and mutual information.

As pointed out in Ref. [29], the case (iii) reveals a peculiar sudden change in classical and quantum correlations. Afterwards, in Ref. [40] it was discovered that

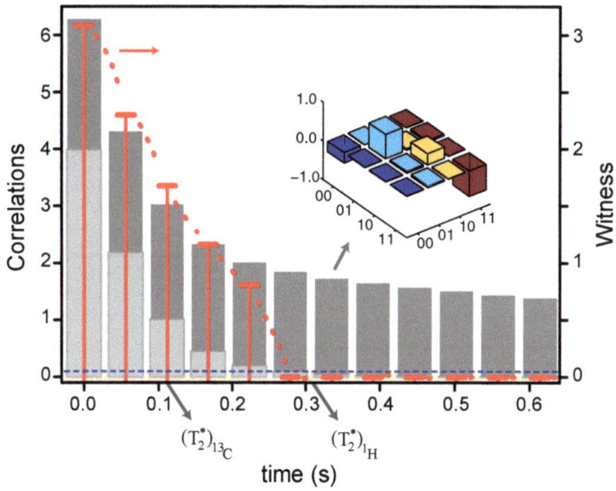

Fig. 4 The panel displays a witness and computed correlations of a quantum correlated state ($c_1 = 2\varepsilon$, $c_2 = 2\varepsilon$ and $c_3 = -2\varepsilon$) relaxed during a time interval, $t_n = n\delta t$ ($\delta t = 55.7\,\text{ms}$, $n = 0, 2, \ldots, 11$). The *red tick bars* represent the witness expectation value (proposed in Ref. [30]), the *grey bars* display the quantum mutual information (total correlation), the *dark grey* section represents the amount of classical correlation, and the *light grey* section represents the quantum discord. In each experimental run, the correlations quantifiers were computed from the tomography data while the witness was directly measured [30, 32]. The classicality bound is represented by the *blue doted line*. The *inset* image shows the real part of the deviation matrix elements reconstructed by QST for an intermediate classically correlated state. The effective transversal relaxation times, shown *below* the figure, are $T_2^* = 0.31\,\text{s}$ and $T_2^* = 0.12\,\text{s}$, for ^{1}H and ^{13}C nuclei, respectively. The correlations are displayed in units of $(\varepsilon^2/\ln 2)$bit. Taken from Ref. [30]

for a special class of initial states, quantum correlations are not destroyed by decoherence for times $t < \bar{t}$, while classical correlations decay. Then, for $t > \bar{t}$, classical correlations remain constant in time and quantum correlations are destroyed. This phenomenon was called Quantum Correlation Freezing. The classical correlation freezing is associated to the appearance of a pointer basis [41].

Ref. [32] shows the first experimental observation of quantum correlations freezing phenomenon, where a direct measurement procedure was applied. Until then, the NMR experiments were recorded applying full quantum state tomography. This work proved that, since discord quantifiers for Bell diagonal states depend only on the correlation function elements and these elements are proportional to the NMR signal (when a proper set of unitary rotations are applied), it is possible to avoid the expensive quantum state tomography and perform a direct measurement. Mathematically, it means

$$c_i = \text{Tr}\{(\sigma_i \otimes \sigma_i)\rho\} = \text{Tr}\{(\sigma_i \otimes I)\xi_i\}, \tag{39}$$

$$\xi_i = U_i \rho U_i^\dagger, \tag{40}$$

where U_i are the properly unitary rotations as described in Ref. [32] and $\text{Tr}\{(\sigma_i \otimes I)\xi_i\}$ is the detected NMR signal, as described in Eq. (20).

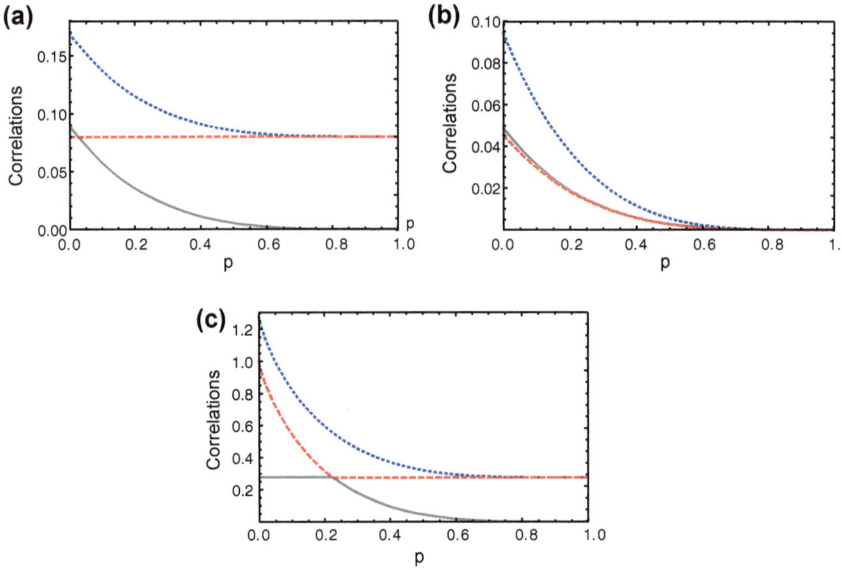

Fig. 5 Total (*blue dotted line*), classical (*red dashed line*), and quantum discord (*gray continuous line*) correlations for a Bell diagonal state evolving under local independent PD channels. In **a** the correlation triple is given by $\vec{c} = \{0.06, 0.30, 0.33\}$. In this case the classical correlation is not affected by the environment while the quantum correlation decays monotonically. In **b** $\vec{c} = \{0.25, 0.25, 0.00\}$ and all correlations decay monotonically. In **c** $\vec{c} = \{1.00, -0.60, 0.60\}$. For this state a sudden change occurs at $p_{SC} \approx 0.22$ and the quantum discord (classical correlation) remains constant (decays monotonically) for $p \leq 0.22$ with the opposite scenario taking place for $p \geq 0.22$. Taken from Ref. [38]

In the same line, Ref. [33] reported an NMR experiment of *double sudden change*, theoretically predicted in Ref. [34], where two different two-qubit NMR setups were applied. The first one was performed on a Varian 500 MHz spectrometer on a liquid state Carbon-13 enriched Chloroform sample ($^{13}CHCl_3$) at room temperature. In this case, the two-qubit are encoded in the ^1H and ^{13}C spin-1/2 nuclei. The measurement of characteristic relaxation times provided $T_1^C \approx 12.46$ s, $T_2^C \approx 0.15$ s, $T_1^H \approx 7.53$ s and $T_2^H \approx 0.27$ s. Since $T_1 >> T_2$ for both spins, and being the experiment evaluation time smaller than 0.5 s, the GAD channel effects can be neglected and the entire relaxation mechanism can be described effectively by a PD channel only. The prepared initial state corresponds to a Bell state with $\vec{c} = \{0.49, 0.20, 0.067\}$, which satisfies condition (iii). In Fig. 6 it is possible to observe clearly one sudden-change in classical correlations (C_G) and two points of sudden change in quantum correlation dynamics (Q_G).

The other system was a spin-3/2 NMR quadrupolar setup encoded in a liquid crystal sample with an NMR detectable sodium nuclei (more details about this setup is found in Ref. [44, 45]). The experiment was performed in a Varian 400 MHz spectrometer at room temperature. Considering that in this case a two-qubit system

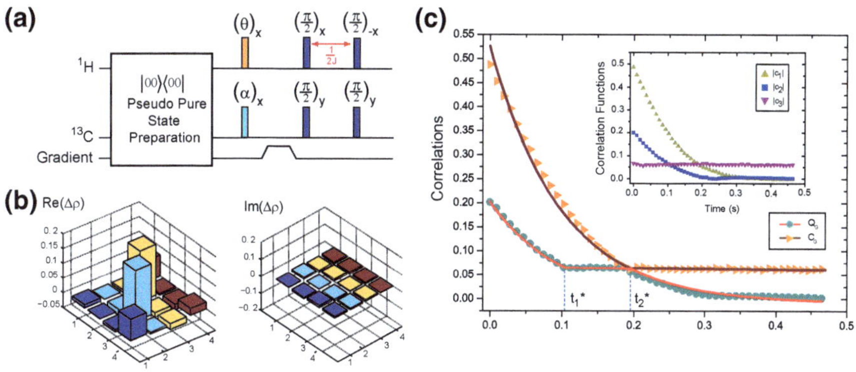

Fig. 6 **a** Schematic representation of the pulse sequence employed to obtain a deviation matrix in the form of a Bell diagonal state. **b** Experimentally reconstructed block diagrams for real and imaginary parts of the deviation matrix related to the Bell diagonal initial state with $c_1 = 0.49$, $c_2 = 0.20$ and $c_3 = 0.067$. The *curves* in **c** denote the time evolutions of quantum (Q_G, *bullet*) and classical (C_G, *triangle*) correlations, respectively. The *dots* represents the experimental results and the *solid lines* are the theoretical predictions. In the *inset* we detail the time evolutions of $|c_1|$ (*yellow upward triangles*), $|c_2|$ (*blue squares*), and $|c_3|$ (*purple downward triangles*) experimentally obtained for the PD decoherence process. Taken from Ref. [33]

is encoded in a single nuclei (which in the presence of a strong static magnetic field is described by four energy levels), the relaxation is described by the GAD channel and a modified PD channel called Global Phase Damping (GPD), which acts on both logical qubits simultaneously, and was proposed in Ref. [35]. However, GPD does not act on the cross-diagonal terms of the deviation matrix, meaning that Bell diagonal states are not affected by them and the decoherence is completely dictated by GAD channel. The initial Bell diagonal state prepared in this case corresponds to $\vec{c} = \{0.08, 0.14, 0.16\}$ and was implemented applying a strongly modulated pulse (SMP) method, where the radio frequency pulses are numerically optimized, as described in Ref. [46]. Although the relaxation is described by a non-conservative energy channel, it is also possible to observe a single and a double sudden change on classical and quantum correlations, respectively, as shown in Fig. 7. In this case classical correlations do not remain constant, so a pointer basis is not reached. And despite the quantum correlation changes its relaxation rate, it is destroyed during the entire process, while in the other system (under PD action) it is frozen between $t_1^\star$ and $t_2^\star$.

2.2.2 NMR Observation of Freezing Phenomenon

The frozen quantum correlation phenomenon was theoretically explained in detail in Ref. [42], where it was demonstrated that for an initial Bell diagonal state (two-qubit system) satisfying

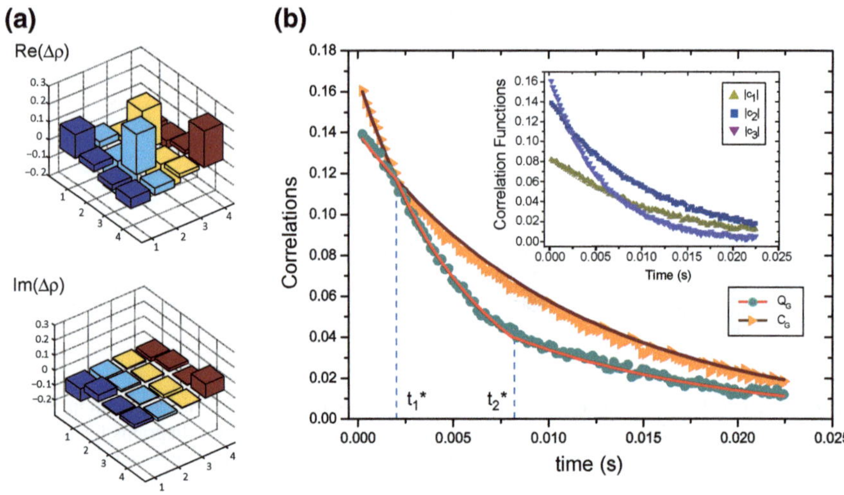

Fig. 7 **a** Experimentally reconstructed block diagrams for real and imaginary parts of the deviation matrix related to the Bell diagonal initial state with $c_1 = 0.08$, $c_2 = 0.14$ and $c_3 = 0.16$. The *curves* in **b** denote the time evolutions of quantum (Q_G, *bullet*) and classical (C_G, *triangle*) correlations, respectively. The *dots* represents the experimental results and the *solid lines* are the theoretical predictions. In the *inset* we detail the time evolutions of $|c_1|$ (*yellow upward triangles*), $|c_2|$ (*blue squares*), and $|c_3|$ (*purple downward triangles*) experimentally obtained for the GAD decoherence process. Taken from Ref. [33]

$$c_1(0) = \pm 1,\ c_2(0) = \mp c_3(0), \tag{41}$$

under a PD channel effect, for any reliable geometric quantum coherence quantifier, the freezing phenomenon will be observed from $t = 0$ (initial time) until $t = t^\star$,

$$t^* = -\frac{1}{2\gamma} \ln \frac{|c_3(0)|}{|c_1(0)|}, \tag{42}$$

after that the dynamical evolution occurs in an exponential way, and the relaxation rate is different for each adopted quantifier.

Figure 8 shows those experimental results, where the $^{13}CHCl_3$ sample was once more employed as a two-qubit system setup. A pulse sequence analogous to Fig. 6-a was applied, choosing θ and α appropriately to produce the initial state $c_1 = 1$, $c_2 = 0.7$ and $c_3 = -0.7$. Then, we observe that quantum discord calculated from entropic discord, trace, Bures and fidelity-based distances, is frozen until $t^\star = 0.04$ s.

For more general systems, $N > 2$ qubits, the quantifier called Global Quantum Discord, defined in Ref. [47, 48], predicts that frozen quantum correlations will be observed for systems with an even number of qubits, while it will never be observed for the odd case.

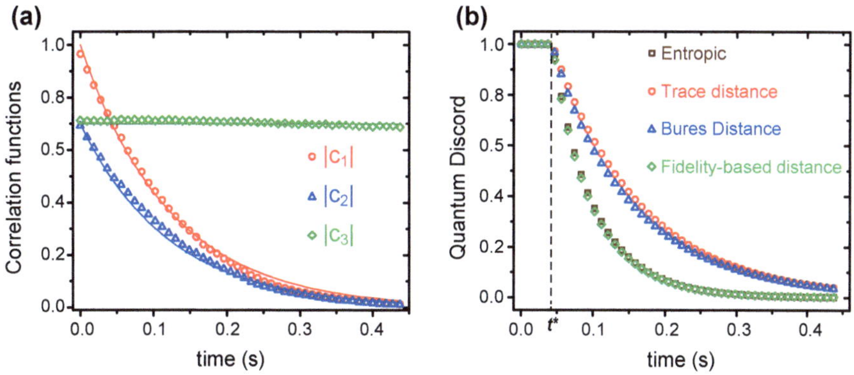

Fig. 8 **a** Time evolution of correlation function elements: $|c_1|$ (*circle*), $|c_2|$ (*triangle*), $|c_3|$ (*diamond*). **b** Time evolution of quantum discord quantified by entropic (*square*), trace distance (*circle*), Bures distance (*triangle*) and fidelity based distance (*diamond*) measures. The freezing phenomenon is observed for all quantifiers from $t = 0$ to $t = t^\star$, when an exponential evolution starts differently for each one [32]

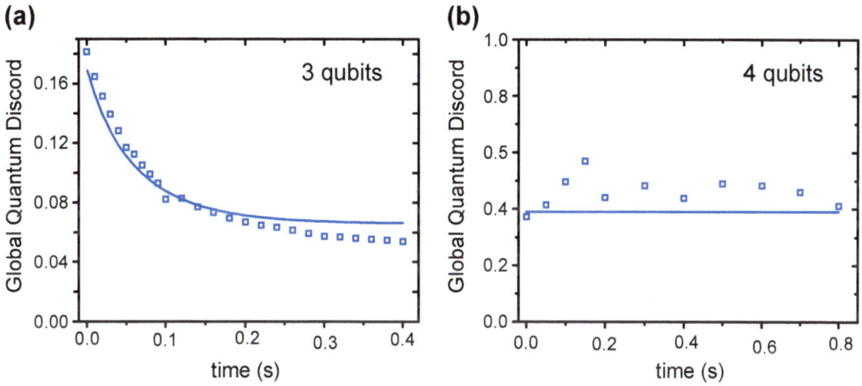

Fig. 9 Global Quantum Discord measured for NMR setups of **a** 3 and **b** 4 qubits. The *dots* correspond to experimental data and lines to theoretical predictions

In order to experimentally observe those predictions, see Fig. 9, a three-qubit system was encoded on a sample of Diethyl 2-Fluoromalonate-2-^{13}C dissolved in $CDCl_3$, for which the structural formula and coupling topology is shown in Fig. 10. The three-qubit were encoded in the ^{1}H, ^{19}F and ^{13}C nuclear spins. In this molecule each qubit is coupled to the two other ones. All scalar coupling constants were measured in E-COSY (exclusive correlation spectroscopy) type experiments: $J_{HF} \approx +48.2$, $J_{HC} \approx +159.7$ and $J_{FC} \approx -196.7$ Hz. This experiment was performed in a Bruker AVIII 600 MHz spectrometer, equipped with a QXI 600 MHz S3 five channels (^{2}H, ^{1}H, ^{13}C, ^{15}N, ^{19}F) probe with z-gradient, at room temperature.

The initial M_3^3 state,

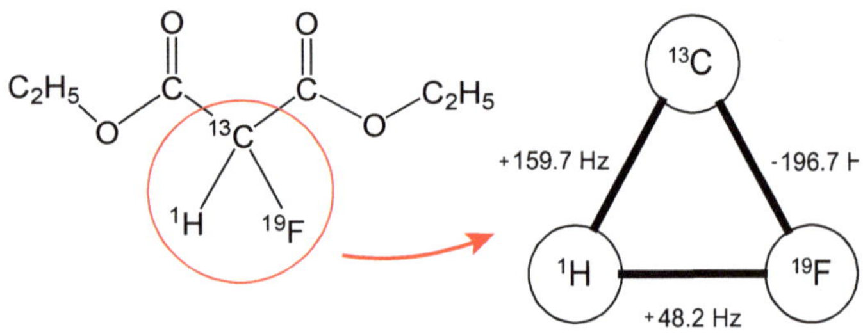

Fig. 10 Structural formula of Diethyl 2-Fluoromalonate-2-^{13}C. The three-qubit were encoded in ^{1}H, ^{19}F and ^{13}C. The coupling topology for this three-qubit system is shown

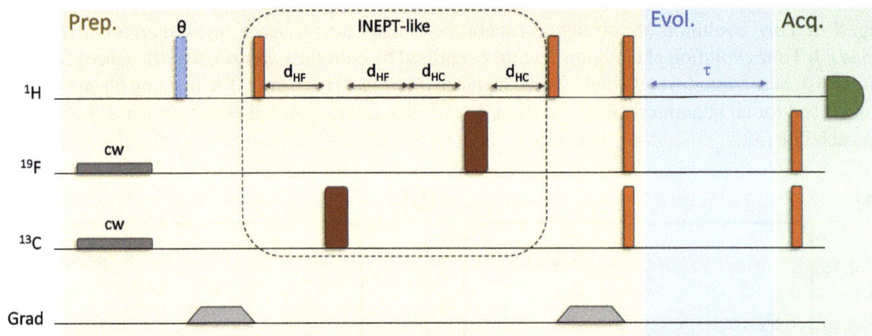

Fig. 11 Pulse sequence for the three-qubit experiment. First, a continuous wave (cw) pulse is applied to the ^{19}F and ^{13}C spins in order to avoid contributions from those nuclei. Then, a pulse with variable length (angle) is implemented to produce the correct scaling in the correlation function elements. Next, an INEPT-like sequence (where *thin bars* represent $\pi/2$ pulses and *large bars* π pulses) is applied to produce the desired multispin term (like $I_z^H I_z^F I_z^C$), where $d_{kl} = 1/(4J_{kl})$. After that, as at this point the multispin term is aligned along the longitudinal axis, a gradient pulse is applied to eliminate other terms and guarantee the state quality. Then, $\pi/2$ pulses are applied, with the appropriate phase, to produce the M_3^3 state term. The system is allowed to evolve freely and later $\pi/2$ pulses, with appropriate phases, are applied to produce an NMR detectable (single quantum) term in the ^{1}H channel

$$\rho = \frac{1}{8}\mathbb{I}^{\otimes 3} + c_1\sigma_x^{\otimes 3} + c_2\sigma_y^{\otimes 3} + c_3\sigma_z^{\otimes 3}, \tag{43}$$

was prepared applying the pulse sequence shown in Fig. 11. To distinguish the physical qubits, we will associate each $\sigma_i^{\otimes 3}$ term to the spin operators $I_i^H I_i^F I_i^C$. Each of these terms is independent of the others and also interacts independently with the environment (considering the same kind of decoherence process described before for the two-qubit system), therefore each term was prepared separately. First, a continuous wave (cw) pulse is applied to ^{19}F and ^{13}C to guarantee that all terms will be generated from ^{1}H magnetization. Then, a pulse with an appropriate angle, $\theta = \{0.8, 1.27, 1.27\}$ rad, is applied followed by a gradient pulse, which guaran-

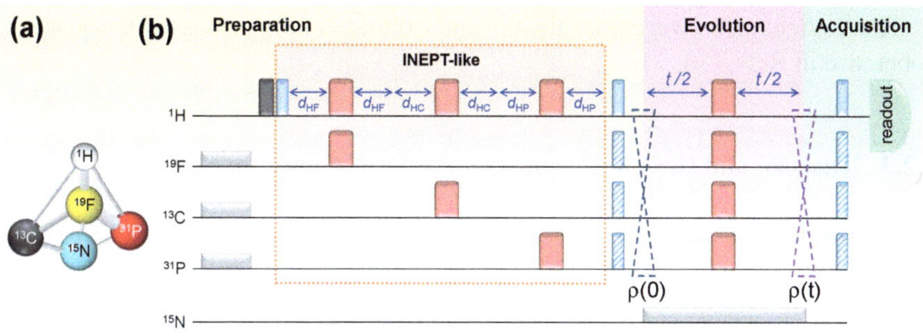

Fig. 12 Pulse sequence (with time flowing from *left* to *right*) to prepare four-qubit generalised Bell states encoded in the ^{1}H, ^{19}F, ^{13}C, and ^{31}P nuclear spins of the $^{13}C^O$-^{15}N-diethyl-(dimethylcarbamoyl)fluoromethyl-phosphonate molecule (whose coupling topology is illustrated in **a** by an INEPT-like procedure, where $d_{kl} = 1 = (4J_{kl})$ and J_{kl} is the scalar coupling between spins k and l. *Light-gray rectangles* denote continuous-wave pulses, used to decouple the ^{15}N nucleus. The *dark grey bar* denotes a variable pulse, applied to set the desired correlation triple $\{c_j(0)\}$. Thicker (*red*) and thinner (*blue*) bars denote π and $\pi/2$ pulses, respectively; the phases of the striped $\pi/2$ pulses were cycled to construct each density matrix element. After the preparation stage, the system was *left* to decohere in its environment; π pulses were applied in the middle of the evolution to avoid J_{kl} oscillations. The final $\pi/2$ pulses served to produce a detectable NMR signal in the ^{1}H spin channel. Taken from Ref. [51]

tees that only $\cos\theta$ of initial magnetization will survive, producing the correct scaling $\vec{c} = \{0.7, 0.3, 0.3\}$, after normalizing with full magnetization spectra. Then, an INEPT-like sequence produces the multispin term $(I_z^H I_z^F I_z^C)$ from ^1H, and a gradient pulse guarantees the initial state quality. The $\pi/2$ pulses applied on all nuclei, with appropriate phases, produce each density matrix term and are followed by a free evolution period. Finally, each element of the coherence triple is directly measured by producing single-quantum terms (like $I_x^H I_z^F I_z^C$), similarly to what was implemented in Ref. [32], where it was shown that a direct measure is as good as a full state tomography.

The even case was observed in a four-qubit system encoded in the $^{13}C^O$-^{15}N-diethyl-(dimethylcarbamoyl)fluoromethyl-phosphonate compound, whose coupling topology is shown in Fig. 12; a detailed description of its synthesis can be found in Ref. [49]. This molecule contains five NMR-active spins (^{1}H, ^{19}F, ^{13}C, ^{31}P and ^{15}N), so a continuous wave (cw) pulse was applied to decouple ^{15}N. The initial $\mathcal{M}_4^3$ (alias generalised BD) state,

$$\rho(0) = \frac{1}{16}\left(\mathbb{I}^{\otimes 4} + c_1(0)\sigma_1^{\otimes 4} + c_2(0)\sigma_2^{\otimes 4} + c_3(0)\sigma_3^{\otimes 4}\right), \tag{44}$$

was prepared according to the pulse sequence displayed in Fig. 12. Similarly to what was implemented for the three-qubit case, each term of these density matrix was prepared separately.

Those phenomena were theoretically and experimentally discussed for quantum coherence in Refs. [50, 51].

2.2.3 Discord and the Interferometric Power of Quantum States

Liquid state NMR was also employed to study quantum correlations in a quantum metrology scenario [43]. Unlike most common metrological applications, the object of investigation was the role of quantum discord in an interferometric configuration when only the spectrum of the generating Hamiltonian is known. In more details, we consider a bipartite probe state ρ_{AB} entering a two-arm channel, in which only the subsystem A is affected by a local unitary $U_A = e^{-i\varphi H_A} \otimes \mathbb{I}$, where φ is the parameter to be estimated and H_A is the generating Hamiltonian. The information about φ is obtained through the application of an estimator $\tilde{\varphi}$ over the output state $\rho_{AB}^{\varphi} = U_A \rho_{AB} U_A^{\dagger}$. If we have full knowledge of H_A, the maximum achievable precision is obtained when the input state has maximum coherence in the eigenbasis of H_A, and no quantum correlation between subsystems A and B is necessary [53]. A figure of merit to quantify the information about φ encoded in ρ_{AB}^{φ} is the quantum Fisher information [54]. The quantum Cramer–Rao bound sets a lower limit to the variance of $\tilde{\varphi}$, the estimated parameter, as $\mathrm{Var}_{\rho_{AB}^{\varphi}}(\tilde{\varphi}) \geq [\nu F(rho_{AB}, H_A)]^{-1}$, where ν is the number of repetitions of the experiment with identical copies and $F(\rho_{AB}, H_A)$ is the quantum Fisher information [55].

When a "black box" paradigm is considered, i.e., if only the spectrum of H_A is known a priori, some aspects of game are changed. One can imagine this setting as as a referee deciding the local transformation over A, after the probe state ρ_{AB} is prepared. After the application of U_A, the referee discloses his choice for H_A, allowing the experimenter to perform the respective optimal estimator. For the worst case scenario, the precision is minimal over all H_A and can be computed using the quantum Fisher information as

$$P^A(\rho_{AB}) = \frac{1}{4} \inf_{H_A} F(\rho_{AB}, H_A), \tag{45}$$

where the infimum is over all Hamiltonians with a given spectrum and the normalization factor $1/4$ is solely for convenience. $P^A(\rho_{AB})$, termed interferometric power of ρ_{AB}, is a well defined measure of quantum correlations of a bipartite state, as shown in [43] and discussed in details by Bogaert and Girolami [56]. On one hand, if ρ_{AB} is not discordant the interferometric power vanishes, since there is a H_A such that $[\rho_{AB}, H_A] = 0$, and no information about φ can be retrieved. On the other hand, the degree of quantum correlations of ρ_{AB} not only guarantees a minimal precision but also quantifies the usefulness of ρ_{AB} as a resource to estimate φ.

In the particular case of the subsystem A being a qubit, a computable and closed formula was obtained for the interferometric power [43]

$$P^A(\rho_{AB}) = \zeta_{\min}[M], \tag{46}$$

where $\zeta_{\min}[M]$ is the minimal eigenvalue of the 3 X 3 matrix M:

$$M_{mn} = \frac{1}{2} \sum_{i,l;q_i+q_l \neq 0} \frac{(q_i - q_l)^2}{q_i + q_l} \langle \psi_i | \sigma_{mA} \otimes \mathbb{I}_B | \psi_l \rangle \langle \psi_l | \mathbb{I}_A \otimes \sigma_{nB} | \psi_l \rangle, \quad (47)$$

and $\{q_i, |\psi_i\rangle\}$ is the set of eigenvalues and eigenvectors of ρ_{AB}, respectively.

The role of interferometric power was observed experimentally, in a proof-of-principle implementation, with the quantum state ρ_{AB} encoded in the nuclear spins of a ^{13}C-labelled chloroform sample. Two different classes of states were compared, a discordant and a classical-quantum ones. The chosen families are

$$\rho_{AB}^{Q} = \frac{1}{4} \begin{pmatrix} 1+p^2 & 0 & 0 & 2p \\ 0 & 1-p^2 & 0 & 0 \\ 0 & 0 & 1-p^2 & 0 \\ 2p & 0 & 0 & 1+p^2 \end{pmatrix}, \quad \rho_{AB}^{C} = \frac{1}{4} \begin{pmatrix} 1 & p^2 & p & p \\ p^2 & 1 & p & p \\ p & p & 1 & p^2 \\ p & p & p^2 & 1 \end{pmatrix}. (48)$$

The parameter p quantifies the purity of both classes, as $\mathrm{Tr}(\rho_{AB}^{(Q,C)})^2 = 1/4(1+p^2)^2$, and $0 \leq p \leq 1$. This allows a fair comparison between quantum and classical states, since for a given value of p we have the same degree of mixedness. The states ρ_{AB}^{C} are classically correlated, with $P^A(\rho_{AB}^{C}) = 0$ for any p, while ρ_{AB}^{Q} has discord monotonically increasing for $p > 0$.

The state preparation comprises three steps. First, the pseudopure state $|00\rangle\langle 00|$ is prepared using a spatial average [22]. Secondly, the purity is chosen with $(\theta)_{-y}$

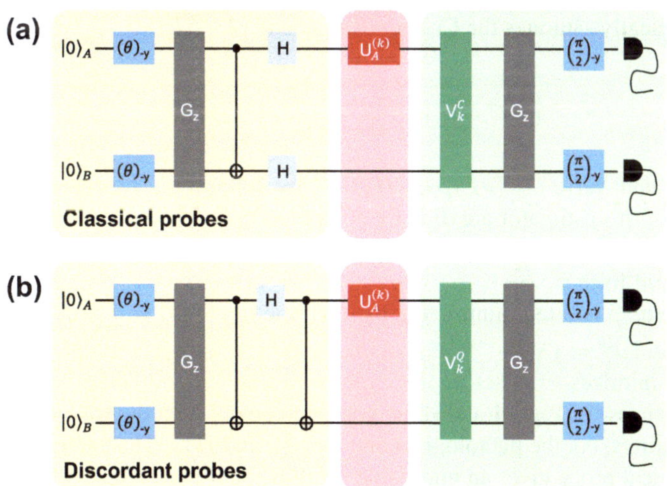

Fig. 13 Experimental scheme for *black box* parameter estimation with NMR. **a** Classical probes. **b** Discordant probes. The protocol is divided in three steps: probe state preparation (*yellow*); *black box* transformation (*red*); optimal measurement (*green*). Taken from Ref. [43]

pulses on both nuclei, followed by a pulsed gradient field G_z. The final step consists in the application of the combination of CNOT and Hadamard gates as depicted in Fig. 13a for classical probes and Fig. 13b for the quantum ones.

To calculate the interferometric power, a full quantum state tomography was performed to each prepared state, with a mean fidelity of (99.7 ± 0.2) with the theoretical density matrices of Eq. (48). The interferometric power was computed using the closed formula of Eq. (46) on the reconstructed states and is displayed in the first row of Fig. 14, with an excellent agreement with the theoretical expectation, $P^A(\rho_{AB}^Q) = p^2$.

For each fixed probe, three choices for H_A were performed. $H_A^{(1)} = \sigma_z^A \otimes \mathbb{I}_B$, $H_A^{(2)} = (\sigma_x^A + \sigma_y^A)/\sqrt{2} \otimes \mathbb{I}_B$ and $H_A^{(3)} = \sigma_x^A \otimes \mathbb{I}_B$. These three Hamiltonians encompass the worst and best settings for $H_A^{(1)}$ and $H_A^{(3)}$, respectively, while $H_A^{(2)}$ is an intermediate case [43]. In all experiments, the parameter φ was set to $\varphi_0 = \pi/4$.

Finally, the optimal measurement is carried out to estimate φ, for all six combinations of input states and black boxes. This set of estimators is given by the eigenvectors of the symmetric logarithmic derivative $L_\varphi = \sum l_j |\lambda_j\rangle\langle\lambda_j|$, which satisfies $\partial_\varphi \rho_{AB}^\varphi = \frac{1}{2}(\rho_{AB}^\varphi L_\varphi + L_\varphi \rho_{AB}^\varphi)$. The quantum Fisher information is given as $F(\rho_{AB}, H_a) = \mathrm{Tr}(\rho_{AB}^\varphi L_\varphi^2) = 4\sum_{i,l;q_i+q_l\neq 0} \frac{(q_i-q_l)^2}{q_i+q_l} |\langle\psi_i|H_A \otimes \mathbb{I}_\mathbb{B}|\psi_l\rangle|^2$, where $\{q_i, |\psi_i\rangle\}$ is the set of eigenvalues and eigenvectors of ρ_{AB} [57]. The readout procedure comprises a global rotation into the eigenbasis of L_φ, shown as $V_k^{(C,Q)}$[1] in Fig. 13, followed by a pulsed gradient field G_z, which performs an ensemble measurement of the expectation values $d_j = \langle\lambda_j|\rho_{AB}^\varphi|\lambda_j\rangle$. After the $(\pi/2)_{-y}$ rotations on both qubits, these d_j are obtained without the need of a full state reconstruction and resulting in the ensemble measured data d_j^{exp}.

The estimation is accomplished with an statistical estimator for φ, defined in such way that it asymptotically saturates the Cramer–Rao bound [43, 57]:

$$\tilde{\varphi} = \varphi_0 \mathbb{I} + \frac{L_\varphi}{\sqrt{\nu F(\rho_{AB}, H_A)}}, \tag{49}$$

such that $\langle\tilde{\varphi}\rangle = \varphi_0$ and $\mathrm{Var}(\tilde{\varphi}) = [\nu F(\rho_{AB}, H_A)]^{-1}$, by definition. The ensemble mean and variance of this estimator are directly computed from the measured d_j^{exp}, the initial probe states ρ_{AB} and the calculated eigenvalues l_φ of L_φ for each H_A, which are independent from φ.

The mean value for $\tilde{\varphi}$ is fitted minimizing the the function $\Theta(\varphi) = \sum_j [d_j^{exp} - d_j^{th}]^2$, where the model $d_j^{th} = \langle\lambda_j|(e^{-i\varphi H_A} \otimes \mathbb{I}_B)\rho_{AB}(e^{i\varphi H_A} \otimes \mathbb{I}_B)|\lambda_j\rangle)$ is employed. The value $\tilde{\varphi}$ that minimizes Θ is set as the expected value $\langle\tilde{\varphi}_{exp}\rangle$. These values are plotted in the last row of Fig. 14 and show good agreement with the true value $\varphi_0 = \pi/4$ for all settings, but the pathological one for ρ_{AB}^C and $H_A^{(3)}$. The latter case shows how the classical probe gives an unreliable result when it commutes with the generating Hamiltonian.

[1]All the $V_k^{(C,Q)}$ transformations, along their implementations, are shown in [43].

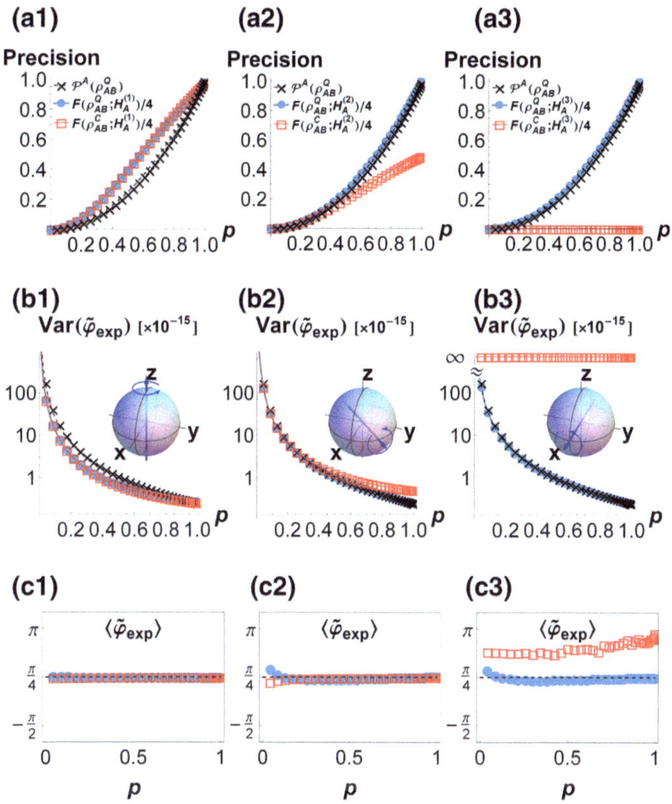

Fig. 14 Experimental results. Each column corresponds to a different *black box* setting $H_A^{(k)}$, $k = 1, 2, 3$, and the set directions are depicted in the *insets* of row (**b**). *Empty red squares* refer to data from classical probes ρ_{AB}^C and *filled blue circles* refer to data from discordant probes ρ_{AB}^Q. Both classes depend on the purity, quantified by the parameter p. The *first row* shows the measured quantum Fisher information normalized by a factor of 4 for each setting, along with the *lower bound* provided by $P^A(\rho_{AB}^Q)$. The *middle row* (**b**) presents the measured variances, together with a theoretical prediction for the saturation of the Cramer–Rao bound. *Last row* (**c**) depicts the inferred mean value for each setting. The *lines* refer to theoretical predictions. Taken from Ref. [43]

The experimental quantum Fisher information is obtained through the expansion of L_φ on its eigenbasis and using the measured data d_j^{exp}, as $F_{exp}(\rho_{AB}, H_A) = \sum_j (l_\varphi^j)^2 d_j^{exp}$. These values are shown in the first row of Fig. 14, along the lower bound given by $P^A(\rho_{AB}^Q)$. These quantum Fisher information are obtained from the output of our experiments, while the interferometric power is measured on the reconstructed input states. For $H_A^{(2)}$ and $H_A^{(3)}$, the quantum Fisher information saturates this lower bound defined by the interferometric power. It is also worth mentioning that $F(\rho_{AB}^C, H_A^{(3)}) = 0$, since $l_\varphi^j = 0$ for any j.

Finally, the variance of the optimal estimator is calculated replacing $\varphi_0\mathbb{I}$ with the experimental mean value $\tilde{\varphi}_{exp}$ in Eq. (49) and expanding it in terms of l_φ^j and the measured d_j^{exp} [43]. The resulting variances are shown in the row (b) of Fig. 14. This data is in excellent agreement with the relation $\mathrm{Var}(\tilde{\varphi}_{exp} = [\nu F(\rho_{AB}, H_A)]^{-1}$, allowing us to conclude that the optimal estimation strategy was performed in all settings.

Concluding, the experimental results show that quantum discord-type quantum correlations, via the interferometric power, offer a priori a minimal precision in the worst case scenario for any bipartite probe in a black box estimation.

2.2.4 Final Remarks

The recognition that nuclear spins in a magnetic field would be a clear representation of qubits made NMR a natural candidate for quantum information processing. Indeed, because of the capability of implementing all basic steps necessary for QIP using conventional spectrometers, NMR was used as a bench test for most of the pioneering experimental demonstrations of quantum gates and algorithms. Despite that, the lack of scalability of pseudopure states and the criticisms concerning the absence of entanglement in room temperature liquid state NMR systems led to the general feeling that the contribution of the technique to QIP would be limited to the first demonstrations. However, this also initiated a discussion about the quantumness of such systems, as discussed in Sect. 2.

Which aspects of NMR systems makes them useful in application to quantum information experiments? The discovery and study of quantum correlations in separable states, so-called general quantum correlations such as quantum discord, bring back the attention of the QIP community to NMR and led to a partial answer to the question of quantumness of NMR systems. Indeed, the facts that the system is highly mixed and subject to a very noisy environment played in favor of NMR because these are extreme conditions to test the quantum properties of a system. As shown in this chapter, such correlations can be observed and quantified even at room temperature in a highly mixed state. As examples, we presented bench tests on quantifiers of general quantum correlations, discussed phenomena like the freezing and sudden change of quantum discord in open systems and the role of discord in guaranteeing a minimum precision in black box estimation in interferometry.

In summary, NMR offers an excellent test bench for quantum information processing studies in few qubit systems, being able to perform with high precision experiments about quantum algorithms and foundations of quantum physics. However, the discovery of other types of quantum correlations, such as discord and the interest in the QIP community in understanding quantum phenomena associated to mixed states interacting with noisy environments brought liquid state NMR back as a method for demonstrating new concepts related to the quantum properties of these systems. As new perspectives, the new questions on the role of quantum coherences in quantum systems and in specific QIP procedures might benefit from NMR, since these concepts are explored in NMR techniques since its early days.

References

1. I.I. Rabi, S. Millman, P. Kusch, J.R. Zacharias, Phys. Rev. **55**, 526 (1939)
2. F. Bloch, Phys. Rev. **70**, 460 (1946)
3. E.M. Purcell, H.C. Torrey, R.V. Pound, Phys. Rev. **69**, 37 (1946)
4. E. Hahn, Phys. Rev. **80**, 580 (1950)
5. R.R. Ernst, Nobel Prize Lecture, http://www.nobelprize.org/nobel-prizes/chemistry/laureates/1991/ernst-lecture.html
6. K. Wuthrich, Nobel Prize Lecture, http://www.nobelprize.org/nobel-prizes/chemistry/laureates/2002/wuthrich-lecture.html
7. P.C. Lauterbur, Nobel Prize Lecture, http://www.nobelprize.org/nobel-prizes/medicine/laureates/2003/lauterbur-lecture.html
8. P. Mansfield, Nobel Prize Lecture, http://www.nobelprize.org/nobel-prizes/medicine/laureates/2003/lauterbur-lecture.html
9. C.P. Slichter, *Principles of Magnetic Resonance*, 3rd edn. (Springer, Berlin, 1996)
10. M.A. Nielsen, I.L. Chuang, *Quantum Computation and Quantum Information*, 1st edn. (Cambridge University Press, Cambridge, 2000)
11. I.L. Chuang, N. Gershenfeld, M.G. Kubinec, D.W. Leung, Proc. R. Soc. A **454**, 447 (1998)
12. G. Landi, F. Simeão, Quantum thermodynamics and work fluctuations with applications to magnetic resonance. Am. J. Phys. **84**, 948 (2016). doi:10.1119/1.4964111
13. I.S. Oliveira, T.J. Bonagamba, R.S. Sarthour, J.C.C. de Freitas, E.R. deAzevedo, *NMR Quantum Information Processing* (Elsevier, Amsterdam, 2011)
14. *Quantum Information Processing in NMR: Theory and Experiment*, ed. by I.S. Oliveira, R.M. Serra (Philosophical Transactions of the Royal Society, 2013)
15. A.M. Souza, A. Magalhes, J. Teles, T.J. Bonagamba, I.S. Oliveira, R.S. Sarthour, New J. Phys. **10**(3), 033020 (2008)
16. A.M. Souza, I.S. Oliveira, R.S. Sarthour, New J. Phys. **13**(5), 053023 (2011)
17. R. Auccaise, R.M. Serra, J.G. Filgueiras, R.S. Sarthour, I.S. Oliveira, L.C. Céleri, Phys. Rev A **85**(3), 032121 (2012)
18. R. Marx, A.F. Fahmy, J.M. Myers, W. Bermel, S.J. Glaser, Phys. Rev. A **62**, 012310 (2000)
19. J.S. Waugh, Chapter 6 in *Pulsed Magnetic Resonance: NMR, ESR and Optics*, ed. by D.M.S. Bagguley (Clarendon Press, Oxford, 1992)
20. C. Anastopoulus, N. Savvidou, Phys. Rev. E **83**, 021118 (2011)
21. N.A. Gershenfeld, I.L. Chuang, Science **275**, 350 (1997)
22. D.G. Cory, A.F. Fahmy, T.F. Havel, Proc. Nac. Acad. Sci. **94**(5), 1634 (1997)
23. S.L. Braunstein, C.M. Caves, R. Jozsa, N. Linden, S. Popescu, R. Schack, Phys. Rev. Lett. **83**, 1054 (1999)
24. N. Linden, S. Popescu, Phys. Rev. Lett. **87**, 047901–1 (2001)
25. F.A. Vind, A.M. Souza, R.S. Sarthour, I.S. Oliveira, Phys. Rev. A **90**(6), 062339 (2014). See, for instance
26. A. Peres, Phys. Rev. Lett. **77**, 1413 (1996)
27. A.F. Fahmy, R. Marx, W. Bermel, S.J. Glaser, Phys. Rev. A **78**, 022317 (2008)
28. N.C. Menicucci, C.M. Caves, Phys. Rev. Lett. **88**, 167901 (2002)
29. J. Maziero, L.C. Céleri, R.M. Serra, V. Vedral, Phys. Rev A **80**, 044102 (2009)
30. R. Auccaise, L.C. Céleri, D.O. Soares-Pinto, E.R. deAzevedo, J. Maziero, A.M. Souza, T.J. Bonagamba, R.S. Sarthour, I.S. Oliveira, R.M. Serra. Phys. Rev. Lett. **107**, 140403 (2011)

31. K. Modi, A. Brodutch, H. Cable, T. Paterek, V. Vedral, Rev. Mod. Phys. **84**, 1655 (2012)
32. I.A. Silva, D. Girolami, R. Auccaise, R.S. Sarthour, I.S. Oliveira, T.J. Bonagamba, E.R. deAzevedo, D.O. Soares-Pinto, G. Adesso. Phys. Rev. Lett. **110**, 140501 (2013)
33. F.M. Paula, I.A. Silva, J.D. Montealegre, A.M. Souza, E.R. deAzevedo, R.S. Sarthour, A. Saguia, I.S. Oliveira, D.O. Soares-Pinto, G. Adesso, M. Sarandy. Phys. Rev. Lett. **111**, 250401 (2013)
34. J.D. Montealegre, F.M. Paula, A. Saguia, M. Sarandy, Phys. Rev. A **87**, 042115 (2013)
35. A.M. Souza, A. Gavini-Viana, I.S. Oliveira, R.S. Sarthour, R. Auccaise, J. Teles, E.R. deAzevedo, T.J. Bonagamba, Quant. Info. Comput. **10**, 653 (2010)
36. C.E. Shannon, Bell Syst. Tech. J. **27**, 379 (1948)
37. H. Ollivier, W.H. Zurek, Phys. Rev. Lett. **88**, 017901 (2001)
38. J. Maziero, R. Auccaise, L.C. Céleri, D.O. Soares-Pinto, E.R. deAzevedo, T.J. Bonagamba, R.S. Sarthour, I.S. Oliveira, R.M. Serra. Braz. J. Phys. **43**, 86 (2013)
39. S. Luo, Phys. Rev. A **77**, 042303 (2008)
40. L. Mazzola, J. Piilo, S. Maniscalco, Phys. Rev. Lett. **104**, 200401 (2010)
41. M.F. Cornelio, O.J. Faras, F.F. Fanchini, I. Frerot, G.H. Aguilar, M.O. Hor-Meyll, M.C. de Oliveira, S.P. Walborn, A.O. Caldeira, P.H. Souto, Ribeiro. Phys. Rev. Lett. **109**, 190402 (2012)
42. M. Cianciaruso, T.R. Bromley, W. Roga, R. LoFranco, G. Adesso, Sci. Rep. **5**, 10177 (2015)
43. D. Girolami, A.M. Souza, V. Giovannetti, T. Tufarelli, J.G. Filgueiras, R.S. Sarthour, D.O. Soares-Pinto, I.S. Oliveira, G. Adesso, Phys. Rev. Lett. **112**, 210401 (2014)
44. K.D. Lawson, T.J. Flautt, J. Amer. Chem. Soc. **89**, 5489 (1967)
45. K. Radley, L.W. Reeves, A.S. Tracey, J. Chem. Phys. **80**, 176 (1976)
46. E.M. Fortunato, M.A. Pravia, N. Boulant, G. Teklemariam, T.F. Havel, D.G. Cory, J. Chem. Phys. **116**, 7599 (2002)
47. J. Xu, Phys. Lett. A **377**, 238 (2013)
48. C.C. Rulli, M.S. Sarandy, Phys. Rev. A **84**, 042109 (2011)
49. R. Marx, N. Pomplun, W. Bermel, H. Zeiger, F. Engelke, A.F. Fahmy, S.J. Glaser, Mag. Res. Chem. **53**, 442 (2015)
50. T.R. Bromley, M. Cianciaruso, G. Adesso, Phys. Rev. Lett. **114**, 210401 (2015)
51. I.A. Silva, A.M. Souza, T.R. Bromley, M. Cianciaruso, R. Marx, R.S. Sarthour, I.S. Oliveira, R. LoFranco, S.J. Glaser, E.R. deAzevedo, D.O. Soares-Pinto, G. Adesso. Phys. Rev. Lett. **117**, 160402 (2016)
52. A.D. Bain, C.K. Anand, Z. Nie, J. Magn. Reson. **206**, 227 (2010)
53. M. de Almeida, M. Gu, A. Fedrizzi, M.A. Broome, T.C. Ralph, A. White, Phys. Rev. A **89**, 042323 (2014)
54. S.L. Braunstein, C.M. Caves, Phys. Rev. Lett. **72**, 3439 (1994)
55. C.W. Helstrom, *Quantum Detection and Estimation Theory* (Academic Press, New York, 1976)
56. P. Bogaert, D. Girolami, arXiv:1609.02170 [quant-ph] (2016)
57. M.G.A. Paris, Int. J. Quant. Inf **07**, 125 (2009)

MIX
Papier aus verantwortungsvollen Quellen
Paper from responsible sources
FSC® C105338

If you have any concerns about our products,
you can contact us on
ProductSafety@springernature.com

In case Publisher is established outside the EU,
the EU authorized representative is:
Springer Nature Customer Service Center GmbH
Europaplatz 3, 69115 Heidelberg, Germany

Printed by Libri Plureos GmbH
in Hamburg, Germany